T0205261

Lecture Notes in Computer Science 12374

More information about this series at http://www.springer.com/series/7412

Andrea Vedaldi · Horst Bischof ·
Thomas Brox · Jan-Michael Frahm (Eds.)

Computer Vision – ECCV 2020

16th European Conference
Glasgow, UK, August 23–28, 2020
Proceedings, Part XXIX

 Springer

Editors
Andrea Vedaldi (iD)
University of Oxford
Oxford, UK

Horst Bischof (iD)
Graz University of Technology
Graz, Austria

Thomas Brox (iD)
University of Freiburg
Freiburg im Breisgau, Germany

Jan-Michael Frahm
University of North Carolina at Chapel Hill
Chapel Hill, NC, USA

ISSN 0302-9743   ISSN 1611-3349 (electronic)
Lecture Notes in Computer Science
ISBN 978-3-030-58525-9   ISBN 978-3-030-58526-6 (eBook)
https://doi.org/10.1007/978-3-030-58526-6

LNCS Sublibrary: SL6 – Image Processing, Computer Vision, Pattern Recognition, and Graphics

This Springer imprint is published by the registered company Springer Nature Switzerland AG
The registered company address is: Gewerbestrasse 11, 6330 Cham, Switzerland

Foreword

Hosting the European Conference on Computer Vision (ECCV 2020) was certainly an exciting journey. From the 2016 plan to hold it at the Edinburgh International Conference Centre (hosting 1,800 delegates) to the 2018 plan to hold it at Glasgow's Scottish Exhibition Centre (up to 6,000 delegates), we finally ended with moving online because of the COVID-19 outbreak. While possibly having fewer delegates than expected because of the online format, ECCV 2020 still had over 3,100 registered participants.

Although online, the conference delivered most of the activities expected at a face-to-face conference: peer-reviewed papers, industrial exhibitors, demonstrations, and messaging between delegates. In addition to the main technical sessions, the conference included a strong program of satellite events with 16 tutorials and 44 workshops.

Furthermore, the online conference format enabled new conference features. Every paper had an associated teaser video and a longer full presentation video. Along with the papers and slides from the videos, all these materials were available the week before the conference. This allowed delegates to become familiar with the paper content and be ready for the live interaction with the authors during the conference week. The live event consisted of brief presentations by the oral and spotlight authors and industrial sponsors. Question and answer sessions for all papers were timed to occur twice so delegates from around the world had convenient access to the authors.

As with ECCV 2018, authors' draft versions of the papers appeared online with open access, now on both the Computer Vision Foundation (CVF) and the European Computer Vision Association (ECVA) websites. An archival publication arrangement was put in place with the cooperation of Springer. SpringerLink hosts the final version of the papers with further improvements, such as activating reference links and supplementary materials. These two approaches benefit all potential readers: a version available freely for all researchers, and an authoritative and citable version with additional benefits for SpringerLink subscribers. We thank Alfred Hofmann and Aliaksandr Birukou from Springer for helping to negotiate this agreement, which we expect will continue for future versions of ECCV.

August 2020

Vittorio Ferrari
Bob Fisher
Cordelia Schmid
Emanuele Trucco

Preface

Welcome to the proceedings of the European Conference on Computer Vision (ECCV 2020). This is a unique edition of ECCV in many ways. Due to the COVID-19 pandemic, this is the first time the conference was held online, in a virtual format. This was also the first time the conference relied exclusively on the Open Review platform to manage the review process. Despite these challenges ECCV is thriving. The conference received 5,150 valid paper submissions, of which 1,360 were accepted for publication (27%) and, of those, 160 were presented as spotlights (3%) and 104 as orals (2%). This amounts to more than twice the number of submissions to ECCV 2018 (2,439). Furthermore, CVPR, the largest conference on computer vision, received 5,850 submissions this year, meaning that ECCV is now 87% the size of CVPR in terms of submissions. By comparison, in 2018 the size of ECCV was only 73% of CVPR.

The review model was similar to previous editions of ECCV; in particular, it was double blind in the sense that the authors did not know the name of the reviewers and vice versa. Furthermore, each conference submission was held confidentially, and was only publicly revealed if and once accepted for publication. Each paper received at least three reviews, totalling more than 15,000 reviews. Handling the review process at this scale was a significant challenge. In order to ensure that each submission received as fair and high-quality reviews as possible, we recruited 2,830 reviewers (a 130% increase with reference to 2018) and 207 area chairs (a 60% increase). The area chairs were selected based on their technical expertise and reputation, largely among people that served as area chair in previous top computer vision and machine learning conferences (ECCV, ICCV, CVPR, NeurIPS, etc.). Reviewers were similarly invited from previous conferences. We also encouraged experienced area chairs to suggest additional chairs and reviewers in the initial phase of recruiting.

Despite doubling the number of submissions, the reviewer load was slightly reduced from 2018, from a maximum of 8 papers down to 7 (with some reviewers offering to handle 6 papers plus an emergency review). The area chair load increased slightly, from 18 papers on average to 22 papers on average.

Conflicts of interest between authors, area chairs, and reviewers were handled largely automatically by the Open Review platform via their curated list of user profiles. Many authors submitting to ECCV already had a profile in Open Review. We set a paper registration deadline one week before the paper submission deadline in order to encourage all missing authors to register and create their Open Review profiles well on time (in practice, we allowed authors to create/change papers arbitrarily until the submission deadline). Except for minor issues with users creating duplicate profiles, this allowed us to easily and quickly identify institutional conflicts, and avoid them, while matching papers to area chairs and reviewers.

Papers were matched to area chairs based on: an affinity score computed by the Open Review platform, which is based on paper titles and abstracts, and an affinity score computed by the Toronto Paper Matching System (TPMS), which is based on the paper's full text, the area chair bids for individual papers, load balancing, and conflict avoidance. Open Review provides the program chairs a convenient web interface to experiment with different configurations of the matching algorithm. The chosen configuration resulted in about 50% of the assigned papers to be highly ranked by the area chair bids, and 50% to be ranked in the middle, with very few low bids assigned.

Assignments to reviewers were similar, with two differences. First, there was a maximum of 7 papers assigned to each reviewer. Second, area chairs recommended up to seven reviewers per paper, providing another highly-weighed term to the affinity scores used for matching.

The assignment of papers to area chairs was smooth. However, it was more difficult to find suitable reviewers for all papers. Having a ratio of 5.6 papers per reviewer with a maximum load of 7 (due to emergency reviewer commitment), which did not allow for much wiggle room in order to also satisfy conflict and expertise constraints. We received some complaints from reviewers who did not feel qualified to review specific papers and we reassigned them wherever possible. However, the large scale of the conference, the many constraints, and the fact that a large fraction of such complaints arrived very late in the review process made this process very difficult and not all complaints could be addressed.

Reviewers had six weeks to complete their assignments. Possibly due to COVID-19 or the fact that the NeurIPS deadline was moved closer to the review deadline, a record 30% of the reviews were still missing after the deadline. By comparison, ECCV 2018 experienced only 10% missing reviews at this stage of the process. In the subsequent week, area chairs chased the missing reviews intensely, found replacement reviewers in their own team, and managed to reach 10% missing reviews. Eventually, we could provide almost all reviews (more than 99.9%) with a delay of only a couple of days on the initial schedule by a significant use of emergency reviews. If this trend is confirmed, it might be a major challenge to run a smooth review process in future editions of ECCV. The community must reconsider prioritization of the time spent on paper writing (the number of submissions increased a lot despite COVID-19) and time spent on paper reviewing (the number of reviews delivered in time decreased a lot presumably due to COVID-19 or NeurIPS deadline). With this imbalance the peer-review system that ensures the quality of our top conferences may break soon.

Reviewers submitted their reviews independently. In the reviews, they had the opportunity to ask questions to the authors to be addressed in the rebuttal. However, reviewers were told not to request any significant new experiment. Using the Open Review interface, authors could provide an answer to each individual review, but were also allowed to cross-reference reviews and responses in their answers. Rather than PDF files, we allowed the use of formatted text for the rebuttal. The rebuttal and initial reviews were then made visible to all reviewers and the primary area chair for a given paper. The area chair encouraged and moderated the reviewer discussion. During the discussions, reviewers were invited to reach a consensus and possibly adjust their ratings as a result of the discussion and of the evidence in the rebuttal.

After the discussion period ended, most reviewers entered a final rating and recommendation, although in many cases this did not differ from their initial recommendation. Based on the updated reviews and discussion, the primary area chair then made a preliminary decision to accept or reject the paper and wrote a justification for it (meta-review). Except for cases where the outcome of this process was absolutely clear (as indicated by the three reviewers and primary area chairs all recommending clear rejection), the decision was then examined and potentially challenged by a secondary area chair. This led to further discussion and overturning a small number of preliminary decisions. Needless to say, there was no in-person area chair meeting, which would have been impossible due to COVID-19.

Area chairs were invited to observe the consensus of the reviewers whenever possible and use extreme caution in overturning a clear consensus to accept or reject a paper. If an area chair still decided to do so, she/he was asked to clearly justify it in the meta-review and to explicitly obtain the agreement of the secondary area chair. In practice, very few papers were rejected after being confidently accepted by the reviewers.

This was the first time Open Review was used as the main platform to run ECCV. In 2018, the program chairs used CMT3 for the user-facing interface and Open Review internally, for matching and conflict resolution. Since it is clearly preferable to only use a single platform, this year we switched to using Open Review in full. The experience was largely positive. The platform is highly-configurable, scalable, and open source. Being written in Python, it is easy to write scripts to extract data programmatically. The paper matching and conflict resolution algorithms and interfaces are top-notch, also due to the excellent author profiles in the platform. Naturally, there were a few kinks along the way due to the fact that the ECCV Open Review configuration was created from scratch for this event and it differs in substantial ways from many other Open Review conferences. However, the Open Review development and support team did a fantastic job in helping us to get the configuration right and to address issues in a timely manner as they unavoidably occurred. We cannot thank them enough for the tremendous effort they put into this project.

Finally, we would like to thank everyone involved in making ECCV 2020 possible in these very strange and difficult times. This starts with our authors, followed by the area chairs and reviewers, who ran the review process at an unprecedented scale. The whole Open Review team (and in particular Melisa Bok, Mohit Unyal, Carlos Mondragon Chapa, and Celeste Martinez Gomez) worked incredibly hard for the entire duration of the process. We would also like to thank René Vidal for contributing to the adoption of Open Review. Our thanks also go to Laurent Charling for TPMS and to the program chairs of ICML, ICLR, and NeurIPS for cross checking double submissions. We thank the website chair, Giovanni Farinella, and the CPI team (in particular Ashley Cook, Miriam Verdon, Nicola McGrane, and Sharon Kerr) for promptly adding material to the website as needed in the various phases of the process. Finally, we thank the publication chairs, Albert Ali Salah, Hamdi Dibeklioglu, Metehan Doyran, Henry Howard-Jenkins, Victor Prisacariu, Siyu Tang, and Gul Varol, who managed to compile these substantial proceedings in an exceedingly compressed schedule. We express our thanks to the ECVA team, in particular Kristina Scherbaum for allowing open access of the proceedings. We thank Alfred Hofmann from Springer who again serve as the publisher. Finally, we thank the other chairs of ECCV 2020, including in particular the general chairs for very useful feedback with the handling of the program.

August 2020   Andrea Vedaldi
                                                               Horst Bischof
                                                               Thomas Brox
                                                               Jan-Michael Frahm

Organization

General Chairs

Vittorio Ferrari   Google Research, Switzerland
Bob Fisher   University of Edinburgh, UK
Cordelia Schmid   Google and Inria, France
Emanuele Trucco   University of Dundee, UK

Program Chairs

Andrea Vedaldi   University of Oxford, UK
Horst Bischof   Graz University of Technology, Austria
Thomas Brox   University of Freiburg, Germany
Jan-Michael Frahm   University of North Carolina, USA

Industrial Liaison Chairs

Jim Ashe   University of Edinburgh, UK
Helmut Grabner   Zurich University of Applied Sciences, Switzerland
Diane Larlus   NAVER LABS Europe, France
Cristian Novotny   University of Edinburgh, UK

Local Arrangement Chairs

Yvan Petillot   Heriot-Watt University, UK
Paul Siebert   University of Glasgow, UK

Academic Demonstration Chair

Thomas Mensink   Google Research and University of Amsterdam, The Netherlands

Poster Chair

Stephen Mckenna   University of Dundee, UK

Technology Chair

Gerardo Aragon Camarasa   University of Glasgow, UK

Tutorial Chairs

Carlo Colombo   University of Florence, Italy
Sotirios Tsaftaris   University of Edinburgh, UK

Publication Chairs

Albert Ali Salah   Utrecht University, The Netherlands
Hamdi Dibeklioglu   Bilkent University, Turkey
Metehan Doyran   Utrecht University, The Netherlands
Henry Howard-Jenkins   University of Oxford, UK
Victor Adrian Prisacariu   University of Oxford, UK
Siyu Tang   ETH Zurich, Switzerland
Gul Varol   University of Oxford, UK

Website Chair

Giovanni Maria Farinella   University of Catania, Italy

Workshops Chairs

Adrien Bartoli   University of Clermont Auvergne, France
Andrea Fusiello   University of Udine, Italy

Area Chairs

Lourdes Agapito   University College London, UK
Zeynep Akata   University of Tübingen, Germany
Karteek Alahari   Inria, France
Antonis Argyros   University of Crete, Greece
Hossein Azizpour   KTH Royal Institute of Technology, Sweden
Joao P. Barreto   Universidade de Coimbra, Portugal
Alexander C. Berg   University of North Carolina at Chapel Hill, USA
Matthew B. Blaschko   KU Leuven, Belgium
Lubomir D. Bourdev   WaveOne, Inc., USA
Edmond Boyer   Inria, France
Yuri Boykov   University of Waterloo, Canada
Gabriel Brostow   University College London, UK
Michael S. Brown   National University of Singapore, Singapore
Jianfei Cai   Monash University, Australia
Barbara Caputo   Politecnico di Torino, Italy
Ayan Chakrabarti   Washington University, St. Louis, USA
Tat-Jen Cham   Nanyang Technological University, Singapore
Manmohan Chandraker   University of California, San Diego, USA
Rama Chellappa   Johns Hopkins University, USA
Liang-Chieh Chen   Google, USA

Yung-Yu Chuang   National Taiwan University, Taiwan
Ondrej Chum   Czech Technical University in Prague, Czech Republic
Brian Clipp   Kitware, USA
John Collomosse   University of Surrey and Adobe Research, UK
Jason J. Corso   University of Michigan, USA
David J. Crandall   Indiana University, USA
Daniel Cremers   University of California, Los Angeles, USA
Fabio Cuzzolin   Oxford Brookes University, UK
Jifeng Dai   SenseTime, SAR China
Kostas Daniilidis   University of Pennsylvania, USA
Andrew Davison   Imperial College London, UK
Alessio Del Bue   Fondazione Istituto Italiano di Tecnologia, Italy
Jia Deng   Princeton University, USA
Alexey Dosovitskiy   Google, Germany
Matthijs Douze   Facebook, France
Enrique Dunn   Stevens Institute of Technology, USA
Irfan Essa   Georgia Institute of Technology and Google, USA
Giovanni Maria Farinella University of Catania, Italy
Ryan Farrell   Brigham Young University, USA
Paolo Favaro   University of Bern, Switzerland
Rogerio Feris   International Business Machines, USA
Cornelia Fermuller   University of Maryland, College Park, USA
David J. Fleet   Vector Institute, Canada
Friedrich Fraundorfer   DLR, Austria
Mario Fritz   CISPA Helmholtz Center for Information Security,
                          Germany
Pascal Fua   EPFL (Swiss Federal Institute of Technology
                          Lausanne), Switzerland
Yasutaka Furukawa   Simon Fraser University, Canada
Li Fuxin   Oregon State University, USA
Efstratios Gavves   University of Amsterdam, The Netherlands
Peter Vincent Gehler   Amazon, USA
Theo Gevers   University of Amsterdam, The Netherlands
Ross Girshick   Facebook AI Research, USA
Boqing Gong   Google, USA
Stephen Gould   Australian National University, Australia
Jinwei Gu   SenseTime Research, USA
Abhinav Gupta   Facebook, USA
Bohyung Han   Seoul National University, South Korea
Bharath Hariharan   Cornell University, USA
Tal Hassner   Facebook AI Research, USA
Xuming He   Australian National University, Australia
Joao F. Henriques   University of Oxford, UK
Adrian Hilton   University of Surrey, UK
Minh Hoai   Stony Brooks, State University of New York, USA
Derek Hoiem   University of Illinois Urbana-Champaign, USA

Haibin Ling	Stony Brooks, State University of New York, USA
Jiaying Liu	Peking University, China
Ming-Yu Liu	NVIDIA, USA
Si Liu	Beihang University, China
Xiaoming Liu	Michigan State University, USA
Huchuan Lu	Dalian University of Technology, China
Simon Lucey	Carnegie Mellon University, USA
Jiebo Luo	University of Rochester, USA
Julien Mairal	Inria, France
Michael Maire	University of Chicago, USA
Subhransu Maji	University of Massachusetts, Amherst, USA
Yasushi Makihara	Osaka University, Japan
Jiri Matas	Czech Technical University in Prague, Czech Republic
Yasuyuki Matsushita	Osaka University, Japan
Philippos Mordohai	Stevens Institute of Technology, USA
Vittorio Murino	University of Verona, Italy
Naila Murray	NAVER LABS Europe, France
Hajime Nagahara	Osaka University, Japan
P. J. Narayanan	International Institute of Information Technology (IIIT), Hyderabad, India
Nassir Navab	Technical University of Munich, Germany
Natalia Neverova	Facebook AI Research, France
Matthias Niessner	Technical University of Munich, Germany
Jean-Marc Odobez	Idiap Research Institute and Swiss Federal Institute of Technology Lausanne, Switzerland
Francesca Odone	Universita di Genova, Italy
Takeshi Oishi	The University of Tokyo, Tokyo Institute of Technology, Japan
Vicente Ordonez	University of Virginia, USA
Manohar Paluri	Facebook AI Research, USA
Maja Pantic	Imperial College London, UK
In Kyu Park	Inha University, South Korea
Ioannis Patras	Queen Mary University of London, UK
Patrick Perez	Valeo, France
Bryan A. Plummer	Boston University, USA
Thomas Pock	Graz University of Technology, Austria
Marc Pollefeys	ETH Zurich and Microsoft MR & AI Zurich Lab, Switzerland
Jean Ponce	Inria, France
Gerard Pons-Moll	MPII, Saarland Informatics Campus, Germany
Jordi Pont-Tuset	Google, Switzerland
James Matthew Rehg	Georgia Institute of Technology, USA
Ian Reid	University of Adelaide, Australia
Olaf Ronneberger	DeepMind London, UK
Stefan Roth	TU Darmstadt, Germany
Bryan Russell	Adobe Research, USA

Mathieu Salzmann	EPFL, Switzerland
Dimitris Samaras	Stony Brook University, USA
Imari Sato	National Institute of Informatics (NII), Japan
Yoichi Sato	The University of Tokyo, Japan
Torsten Sattler	Czech Technical University in Prague, Czech Republic
Daniel Scharstein	Middlebury College, USA
Bernt Schiele	MPII, Saarland Informatics Campus, Germany
Julia A. Schnabel	King's College London, UK
Nicu Sebe	University of Trento, Italy
Greg Shakhnarovich	Toyota Technological Institute at Chicago, USA
Humphrey Shi	University of Oregon, USA
Jianbo Shi	University of Pennsylvania, USA
Jianping Shi	SenseTime, China
Leonid Sigal	University of British Columbia, Canada
Cees Snoek	University of Amsterdam, The Netherlands
Richard Souvenir	Temple University, USA
Hao Su	University of California, San Diego, USA
Akihiro Sugimoto	National Institute of Informatics (NII), Japan
Jian Sun	Megvii Technology, China
Jian Sun	Xi'an Jiaotong University, China
Chris Sweeney	Facebook Reality Labs, USA
Yu-wing Tai	Kuaishou Technology, China
Chi-Keung Tang	The Hong Kong University of Science and Technology, SAR China
Radu Timofte	ETH Zurich, Switzerland
Sinisa Todorovic	Oregon State University, USA
Giorgos Tolias	Czech Technical University in Prague, Czech Republic
Carlo Tomasi	Duke University, USA
Tatiana Tommasi	Politecnico di Torino, Italy
Lorenzo Torresani	Facebook AI Research and Dartmouth College, USA
Alexander Toshev	Google, USA
Zhuowen Tu	University of California, San Diego, USA
Tinne Tuytelaars	KU Leuven, Belgium
Jasper Uijlings	Google, Switzerland
Nuno Vasconcelos	University of California, San Diego, USA
Olga Veksler	University of Waterloo, Canada
Rene Vidal	Johns Hopkins University, USA
Gang Wang	Alibaba Group, China
Jingdong Wang	Microsoft Research Asia, China
Yizhou Wang	Peking University, China
Lior Wolf	Facebook AI Research and Tel Aviv University, Israel
Jianxin Wu	Nanjing University, China
Tao Xiang	University of Surrey, UK
Saining Xie	Facebook AI Research, USA
Ming-Hsuan Yang	University of California at Merced and Google, USA
Ruigang Yang	University of Kentucky, USA

Kwang Moo Yi	University of Victoria, Canada
Zhaozheng Yin	Stony Brook, State University of New York, USA
Chang D. Yoo	Korea Advanced Institute of Science and Technology, South Korea
Shaodi You	University of Amsterdam, The Netherlands
Jingyi Yu	ShanghaiTech University, China
Stella Yu	University of California, Berkeley, and ICSI, USA
Stefanos Zafeiriou	Imperial College London, UK
Hongbin Zha	Peking University, China
Tianzhu Zhang	University of Science and Technology of China, China
Liang Zheng	Australian National University, Australia
Todd E. Zickler	Harvard University, USA
Andrew Zisserman	University of Oxford, UK

Technical Program Committee

Sathyanarayanan
  N. Aakur
Wael Abd Almgaeed
Abdelrahman
  Abdelhamed
Abdullah Abuolaim
Supreeth Achar
Hanno Ackermann
Ehsan Adeli
Triantafyllos Afouras
Sameer Agarwal
Aishwarya Agrawal
Harsh Agrawal
Pulkit Agrawal
Antonio Agudo
Eirikur Agustsson
Karim Ahmed
Byeongjoo Ahn
Unaiza Ahsan
Thalaiyasingam Ajanthan
Kenan E. Ak
Emre Akbas
Naveed Akhtar
Derya Akkaynak
Yagiz Aksoy
Ziad Al-Halah
Xavier Alameda-Pineda
Jean-Baptiste Alayrac

Samuel Albanie
Shadi Albarqouni
Cenek Albl
Hassan Abu Alhaija
Daniel Aliaga
Mohammad
  S. Aliakbarian
Rahaf Aljundi
Thiemo Alldieck
Jon Almazan
Jose M. Alvarez
Senjian An
Saket Anand
Codruta Ancuti
Cosmin Ancuti
Peter Anderson
Juan Andrade-Cetto
Alexander Andreopoulos
Misha Andriluka
Dragomir Anguelov
Rushil Anirudh
Michel Antunes
Oisin Mac Aodha
Srikar Appalaraju
Relja Arandjelovic
Nikita Araslanov
Andre Araujo
Helder Araujo

Pablo Arbelaez
Shervin Ardeshir
Sercan O. Arik
Anil Armagan
Anurag Arnab
Chetan Arora
Federica Arrigoni
Mathieu Aubry
Shai Avidan
Angelica I. Aviles-Rivero
Yannis Avrithis
Ismail Ben Ayed
Shekoofeh Azizi
Ioan Andrei Bârsan
Artem Babenko
Deepak Babu Sam
Seung-Hwan Baek
Seungryul Baek
Andrew D. Bagdanov
Shai Bagon
Yuval Bahat
Junjie Bai
Song Bai
Xiang Bai
Yalong Bai
Yancheng Bai
Peter Bajcsy
Slawomir Bak

Mahsa Baktashmotlagh
Kavita Bala
Yogesh Balaji
Guha Balakrishnan
V. N. Balasubramanian
Federico Baldassarre
Vassileios Balntas
Shurjo Banerjee
Aayush Bansal
Ankan Bansal
Jianmin Bao
Linchao Bao
Wenbo Bao
Yingze Bao
Akash Bapat
Md Jawadul Hasan Bappy
Fabien Baradel
Lorenzo Baraldi
Daniel Barath
Adrian Barbu
Kobus Barnard
Nick Barnes
Francisco Barranco
Jonathan T. Barron
Arslan Basharat
Chaim Baskin
Anil S. Baslamisli
Jorge Batista
Kayhan Batmanghelich
Konstantinos Batsos
David Bau
Luis Baumela
Christoph Baur
Eduardo
   Bayro-Corrochano
Paul Beardsley
Jan Bednavr'ik
Oscar Beijbom
Philippe Bekaert
Esube Bekele
Vasileios Belagiannis
Ohad Ben-Shahar
Abhijit Bendale
Róger Bermúdez-Chacón
Maxim Berman
Jesus Bermudez-cameo

Florian Bernard
Stefano Berretti
Marcelo Bertalmio
Gedas Bertasius
Cigdem Beyan
Lucas Beyer
Vijayakumar Bhagavatula
Arjun Nitin Bhagoji
Apratim Bhattacharyya
Binod Bhattarai
Sai Bi
Jia-Wang Bian
Simone Bianco
Adel Bibi
Tolga Birdal
Tom Bishop
Soma Biswas
Mårten Björkman
Volker Blanz
Vishnu Boddeti
Navaneeth Bodla
Simion-Vlad Bogolin
Xavier Boix
Piotr Bojanowski
Timo Bolkart
Guido Borghi
Larbi Boubchir
Guillaume Bourmaud
Adrien Bousseau
Thierry Bouwmans
Richard Bowden
Hakan Boyraz
Mathieu Brédif
Samarth Brahmbhatt
Steve Branson
Nikolas Brasch
Biagio Brattoli
Ernesto Brau
Toby P. Breckon
Francois Bremond
Jesus Briales
Sofia Broomé
Marcus A. Brubaker
Luc Brun
Silvia Bucci
Shyamal Buch

Pradeep Buddharaju
Uta Buechler
Mai Bui
Tu Bui
Adrian Bulat
Giedrius T. Burachas
Elena Burceanu
Xavier P. Burgos-Artizzu
Kaylee Burns
Andrei Bursuc
Benjamin Busam
Wonmin Byeon
Zoya Bylinskii
Sergi Caelles
Jianrui Cai
Minjie Cai
Yujun Cai
Zhaowei Cai
Zhipeng Cai
Juan C. Caicedo
Simone Calderara
Necati Cihan Camgoz
Dylan Campbell
Octavia Camps
Jiale Cao
Kaidi Cao
Liangliang Cao
Xiangyong Cao
Xiaochun Cao
Yang Cao
Yu Cao
Yue Cao
Zhangjie Cao
Luca Carlone
Mathilde Caron
Dan Casas
Thomas J. Cashman
Umberto Castellani
Lluis Castrejon
Jacopo Cavazza
Fabio Cermelli
Hakan Cevikalp
Menglei Chai
Ishani Chakraborty
Rudrasis Chakraborty
Antoni B. Chan

Kwok-Ping Chan
Siddhartha Chandra
Sharat Chandran
Arjun Chandrasekaran
Angel X. Chang
Che-Han Chang
Hong Chang
Hyun Sung Chang
Hyung Jin Chang
Jianlong Chang
Ju Yong Chang
Ming-Ching Chang
Simyung Chang
Xiaojun Chang
Yu-Wei Chao
Devendra S. Chaplot
Arslan Chaudhry
Rizwan A. Chaudhry
Can Chen
Chang Chen
Chao Chen
Chen Chen
Chu-Song Chen
Dapeng Chen
Dong Chen
Dongdong Chen
Guanying Chen
Hongge Chen
Hsin-yi Chen
Huaijin Chen
Hwann-Tzong Chen
Jianbo Chen
Jianhui Chen
Jiansheng Chen
Jiaxin Chen
Jie Chen
Jun-Cheng Chen
Kan Chen
Kevin Chen
Lin Chen
Long Chen
Min-Hung Chen
Qifeng Chen
Shi Chen
Shixing Chen
Tianshui Chen

Weifeng Chen
Weikai Chen
Xi Chen
Xiaohan Chen
Xiaozhi Chen
Xilin Chen
Xingyu Chen
Xinlei Chen
Xinyun Chen
Yi-Ting Chen
Yilun Chen
Ying-Cong Chen
Yinpeng Chen
Yiran Chen
Yu Chen
Yu-Sheng Chen
Yuhua Chen
Yun-Chun Chen
Yunpeng Chen
Yuntao Chen
Zhuoyuan Chen
Zitian Chen
Anchieh Cheng
Bowen Cheng
Erkang Cheng
Gong Cheng
Guangliang Cheng
Jingchun Cheng
Jun Cheng
Li cheng
Ming-Ming Cheng
Yu Cheng
Ziang Cheng
Anoop Cherian
Dmitry Chetverikov
Ngai-man Cheung
William Cheung
Ajad Chhatkuli
Naoki Chiba
Benjamin Chidester
Han-pang Chiu
Mang Tik Chiu
Wei-Chen Chiu
Donghyeon Cho
Hojin Cho
Minsu Cho

Nam Ik Cho
Tim Cho
Tae Eun Choe
Chiho Choi
Edward Choi
Inchang Choi
Jinsoo Choi
Jonghyun Choi
Jongwon Choi
Yukyung Choi
Hisham Cholakkal
Eunji Chong
Jaegul Choo
Christopher Choy
Hang Chu
Peng Chu
Wen-Sheng Chu
Albert Chung
Joon Son Chung
Hai Ci
Safa Cicek
Ramazan G. Cinbis
Arridhana Ciptadi
Javier Civera
James J. Clark
Ronald Clark
Felipe Codevilla
Michael Cogswell
Andrea Cohen
Maxwell D. Collins
Carlo Colombo
Yang Cong
Adria R. Continente
Marcella Cornia
John Richard Corring
Darren Cosker
Dragos Costea
Garrison W. Cottrell
Florent Couzinie-Devy
Marco Cristani
Ioana Croitoru
James L. Crowley
Jiequan Cui
Zhaopeng Cui
Ross Cutler
Antonio D'Innocente

Rozenn Dahyot
Bo Dai
Dengxin Dai
Hang Dai
Longquan Dai
Shuyang Dai
Xiyang Dai
Yuchao Dai
Adrian V. Dalca
Dima Damen
Bharath B. Damodaran
Kristin Dana
Martin Danelljan
Zheng Dang
Zachary Alan Daniels
Donald G. Dansereau
Abhishek Das
Samyak Datta
Achal Dave
Titas De
Rodrigo de Bem
Teo de Campos
Raoul de Charette
Shalini De Mello
Joseph DeGol
Herve Delingette
Haowen Deng
Jiankang Deng
Weijian Deng
Zhiwei Deng
Joachim Denzler
Konstantinos G. Derpanis
Aditya Deshpande
Frederic Devernay
Somdip Dey
Arturo Deza
Abhinav Dhall
Helisa Dhamo
Vikas Dhiman
Fillipe Dias Moreira
  de Souza
Ali Diba
Ferran Diego
Guiguang Ding
Henghui Ding
Jian Ding

Mingyu Ding
Xinghao Ding
Zhengming Ding
Robert DiPietro
Cosimo Distante
Ajay Divakaran
Mandar Dixit
Abdelaziz Djelouah
Thanh-Toan Do
Jose Dolz
Bo Dong
Chao Dong
Jiangxin Dong
Weiming Dong
Weisheng Dong
Xingping Dong
Xuanyi Dong
Yinpeng Dong
Gianfranco Doretto
Hazel Doughty
Hassen Drira
Bertram Drost
Dawei Du
Ye Duan
Yueqi Duan
Abhimanyu Dubey
Anastasia Dubrovina
Stefan Duffner
Chi Nhan Duong
Thibaut Durand
Zoran Duric
Iulia Duta
Debidatta Dwibedi
Benjamin Eckart
Marc Eder
Marzieh Edraki
Alexei A. Efros
Kiana Ehsani
Hazm Kemal Ekenel
James H. Elder
Mohamed Elgharib
Shireen Elhabian
Ehsan Elhamifar
Mohamed Elhoseiny
Ian Endres
N. Benjamin Erichson

Jan Ernst
Sergio Escalera
Francisco Escolano
Victor Escorcia
Carlos Esteves
Francisco J. Estrada
Bin Fan
Chenyou Fan
Deng-Ping Fan
Haoqi Fan
Hehe Fan
Heng Fan
Kai Fan
Lijie Fan
Linxi Fan
Quanfu Fan
Shaojing Fan
Xiaochuan Fan
Xin Fan
Yuchen Fan
Sean Fanello
Hao-Shu Fang
Haoyang Fang
Kuan Fang
Yi Fang
Yuming Fang
Azade Farshad
Alireza Fathi
Raanan Fattal
Joao Fayad
Xiaohan Fei
Christoph Feichtenhofer
Michael Felsberg
Chen Feng
Jiashi Feng
Junyi Feng
Mengyang Feng
Qianli Feng
Zhenhua Feng
Michele Fenzi
Andras Ferencz
Martin Fergie
Basura Fernando
Ethan Fetaya
Michael Firman
John W. Fisher

Matthew Fisher
Boris Flach
Corneliu Florea
Wolfgang Foerstner
David Fofi
Gian Luca Foresti
Per-Erik Forssen
David Fouhey
Katerina Fragkiadaki
Victor Fragoso
Jean-Sébastien Franco
Ohad Fried
Iuri Frosio
Cheng-Yang Fu
Huazhu Fu
Jianlong Fu
Jingjing Fu
Xueyang Fu
Yanwei Fu
Ying Fu
Yun Fu
Olac Fuentes
Kent Fujiwara
Takuya Funatomi
Christopher Funk
Thomas Funkhouser
Antonino Furnari
Ryo Furukawa
Erik Gärtner
Raghudeep Gadde
Matheus Gadelha
Vandit Gajjar
Trevor Gale
Juergen Gall
Mathias Gallardo
Guillermo Gallego
Orazio Gallo
Chuang Gan
Zhe Gan
Madan Ravi Ganesh
Aditya Ganeshan
Siddha Ganju
Bin-Bin Gao
Changxin Gao
Feng Gao
Hongchang Gao

Jin Gao
Jiyang Gao
Junbin Gao
Katelyn Gao
Lin Gao
Mingfei Gao
Ruiqi Gao
Ruohan Gao
Shenghua Gao
Yuan Gao
Yue Gao
Noa Garcia
Alberto Garcia-Garcia
Guillermo
  Garcia-Hernando
Jacob R. Gardner
Animesh Garg
Kshitiz Garg
Rahul Garg
Ravi Garg
Philip N. Garner
Kirill Gavrilyuk
Paul Gay
Shiming Ge
Weifeng Ge
Baris Gecer
Xin Geng
Kyle Genova
Stamatios Georgoulis
Bernard Ghanem
Michael Gharbi
Kamran Ghasedi
Golnaz Ghiasi
Arnab Ghosh
Partha Ghosh
Silvio Giancola
Andrew Gilbert
Rohit Girdhar
Xavier Giro-i-Nieto
Thomas Gittings
Ioannis Gkioulekas
Clement Godard
Vaibhava Goel
Bastian Goldluecke
Lluis Gomez
Nuno Gonçalves

Dong Gong
Ke Gong
Mingming Gong
Abel Gonzalez-Garcia
Ariel Gordon
Daniel Gordon
Paulo Gotardo
Venu Madhav Govindu
Ankit Goyal
Priya Goyal
Raghav Goyal
Benjamin Graham
Douglas Gray
Brent A. Griffin
Etienne Grossmann
David Gu
Jiayuan Gu
Jiuxiang Gu
Lin Gu
Qiao Gu
Shuhang Gu
Jose J. Guerrero
Paul Guerrero
Jie Gui
Jean-Yves Guillemaut
Riza Alp Guler
Erhan Gundogdu
Fatma Guney
Guodong Guo
Kaiwen Guo
Qi Guo
Sheng Guo
Shi Guo
Tiantong Guo
Xiaojie Guo
Yijie Guo
Yiluan Guo
Yuanfang Guo
Yulan Guo
Agrim Gupta
Ankush Gupta
Mohit Gupta
Saurabh Gupta
Tanmay Gupta
Danna Gurari
Abner Guzman-Rivera

JunYoung Gwak
Michael Gygli
Jung-Woo Ha
Simon Hadfield
Isma Hadji
Bjoern Haefner
Taeyoung Hahn
Levente Hajder
Peter Hall
Emanuela Haller
Stefan Haller
Bumsub Ham
Abdullah Hamdi
Dongyoon Han
Hu Han
Jungong Han
Junwei Han
Kai Han
Tian Han
Xiaoguang Han
Xintong Han
Yahong Han
Ankur Handa
Zekun Hao
Albert Haque
Tatsuya Harada
Mehrtash Harandi
Adam W. Harley
Mahmudul Hasan
Atsushi Hashimoto
Ali Hatamizadeh
Munawar Hayat
Dongliang He
Jingrui He
Junfeng He
Kaiming He
Kun He
Lei He
Pan He
Ran He
Shengfeng He
Tong He
Weipeng He
Xuming He
Yang He
Yihui He

Zhihai He
Chinmay Hegde
Janne Heikkila
Mattias P. Heinrich
Stéphane Herbin
Alexander Hermans
Luis Herranz
John R. Hershey
Aaron Hertzmann
Roei Herzig
Anders Heyden
Steven Hickson
Otmar Hilliges
Tomas Hodan
Judy Hoffman
Michael Hofmann
Yannick Hold-Geoffroy
Namdar Homayounfar
Sina Honari
Richang Hong
Seunghoon Hong
Xiaopeng Hong
Yi Hong
Hidekata Hontani
Anthony Hoogs
Yedid Hoshen
Mir Rayat Imtiaz Hossain
Junhui Hou
Le Hou
Lu Hou
Tingbo Hou
Wei-Lin Hsiao
Cheng-Chun Hsu
Gee-Sern Jison Hsu
Kuang-jui Hsu
Changbo Hu
Di Hu
Guosheng Hu
Han Hu
Hao Hu
Hexiang Hu
Hou-Ning Hu
Jie Hu
Junlin Hu
Nan Hu
Ping Hu

Ronghang Hu
Xiaowei Hu
Yinlin Hu
Yuan-Ting Hu
Zhe Hu
Binh-Son Hua
Yang Hua
Bingyao Huang
Di Huang
Dong Huang
Fay Huang
Haibin Huang
Haozhi Huang
Heng Huang
Huaibo Huang
Jia-Bin Huang
Jing Huang
Jingwei Huang
Kaizhu Huang
Lei Huang
Qiangui Huang
Qiaoying Huang
Qingqiu Huang
Qixing Huang
Shaoli Huang
Sheng Huang
Siyuan Huang
Weilin Huang
Wenbing Huang
Xiangru Huang
Xun Huang
Yan Huang
Yifei Huang
Yue Huang
Zhiwu Huang
Zilong Huang
Minyoung Huh
Zhuo Hui
Matthias B. Hullin
Martin Humenberger
Wei-Chih Hung
Zhouyuan Huo
Junhwa Hur
Noureldien Hussein
Jyh-Jing Hwang
Seong Jae Hwang

Sung Ju Hwang
Ichiro Ide
Ivo Ihrke
Daiki Ikami
Satoshi Ikehata
Nazli Ikizler-Cinbis
Sunghoon Im
Yani Ioannou
Radu Tudor Ionescu
Umar Iqbal
Go Irie
Ahmet Iscen
Md Amirul Islam
Vamsi Ithapu
Nathan Jacobs
Arpit Jain
Himalaya Jain
Suyog Jain
Stuart James
Won-Dong Jang
Yunseok Jang
Ronnachai Jaroensri
Dinesh Jayaraman
Sadeep Jayasumana
Suren Jayasuriya
Herve Jegou
Simon Jenni
Hac-Gon Jeon
Yunho Jeon
Koteswar R. Jerripothula
Hueihan Jhuang
I-hong Jhuo
Dinghuang Ji
Hui Ji
Jingwei Ji
Pan Ji
Yanli Ji
Baoxiong Jia
Kui Jia
Xu Jia
Chiyu Max Jiang
Haiyong Jiang
Hao Jiang
Huaizu Jiang
Huajie Jiang
Ke Jiang

Lai Jiang
Li Jiang
Lu Jiang
Ming Jiang
Peng Jiang
Shuqiang Jiang
Wei Jiang
Xudong Jiang
Zhuolin Jiang
Jianbo Jiao
Zequn Jie
Dakai Jin
Kyong Hwan Jin
Lianwen Jin
SouYoung Jin
Xiaojie Jin
Xin Jin
Nebojsa Jojic
Alexis Joly
Michael Jeffrey Jones
Hanbyul Joo
Jungseock Joo
Kyungdon Joo
Ajjen Joshi
Shantanu H. Joshi
Da-Cheng Juan
Marco Körner
Kevin Köser
Asim Kadav
Christine Kaeser-Chen
Kushal Kafle
Dagmar Kainmueller
Ioannis A. Kakadiaris
Zdenek Kalal
Nima Kalantari
Yannis Kalantidis
Mahdi M. Kalayeh
Anmol Kalia
Sinan Kalkan
Vicky Kalogeiton
Ashwin Kalyan
Joni-kristian Kamarainen
Gerda Kamberova
Chandra Kambhamettu
Martin Kampel
Meina Kan

Christopher Kanan
Kenichi Kanatani
Angjoo Kanazawa
Atsushi Kanehira
Takuhiro Kaneko
Asako Kanezaki
Bingyi Kang
Di Kang
Sunghun Kang
Zhao Kang
Vadim Kantorov
Abhishek Kar
Amlan Kar
Theofanis Karaletsos
Leonid Karlinsky
Kevin Karsch
Angelos Katharopoulos
Isinsu Katircioglu
Hiroharu Kato
Zoltan Kato
Dotan Kaufman
Jan Kautz
Rei Kawakami
Qiuhong Ke
Wadim Kehl
Petr Kellnhofer
Aniruddha Kembhavi
Cem Keskin
Margret Keuper
Daniel Keysers
Ashkan Khakzar
Fahad Khan
Naeemullah Khan
Salman Khan
Siddhesh Khandelwal
Rawal Khirodkar
Anna Khoreva
Tejas Khot
Parmeshwar Khurd
Hadi Kiapour
Joe Kileel
Chanho Kim
Dahun Kim
Edward Kim
Eunwoo Kim
Han-ul Kim

Hansung Kim
Heewon Kim
Hyo Jin Kim
Hyunwoo J. Kim
Jinkyu Kim
Jiwon Kim
Jongmin Kim
Junsik Kim
Junyeong Kim
Min H. Kim
Namil Kim
Pyojin Kim
Seon Joo Kim
Seong Tae Kim
Seungryong Kim
Sungwoong Kim
Tae Hyun Kim
Vladimir Kim
Won Hwa Kim
Yonghyun Kim
Benjamin Kimia
Akisato Kimura
Pieter-Jan Kindermans
Zsolt Kira
Itaru Kitahara
Hedvig Kjellstrom
Jan Knopp
Takumi Kobayashi
Erich Kobler
Parker Koch
Reinhard Koch
Elyor Kodirov
Amir Kolaman
Nicholas Kolkin
Dimitrios Kollias
Stefanos Kollias
Soheil Kolouri
Adams Wai-Kin Kong
Naejin Kong
Shu Kong
Tao Kong
Yu Kong
Yoshinori Konishi
Daniil Kononenko
Theodora Kontogianni
Simon Korman

Adam Kortylewski
Jana Kosecka
Jean Kossaifi
Satwik Kottur
Rigas Kouskouridas
Adriana Kovashka
Rama Kovvuri
Adarsh Kowdle
Jedrzej Kozerawski
Mateusz Kozinski
Philipp Kraehenbuehl
Gregory Kramida
Josip Krapac
Dmitry Kravchenko
Ranjay Krishna
Pavel Krsek
Alexander Krull
Jakob Kruse
Hiroyuki Kubo
Hilde Kuehne
Jason Kuen
Andreas Kuhn
Arjan Kuijper
Zuzana Kukelova
Ajay Kumar
Amit Kumar
Avinash Kumar
Suryansh Kumar
Vijay Kumar
Kaustav Kundu
Weicheng Kuo
Nojun Kwak
Suha Kwak
Junseok Kwon
Nikolaos Kyriazis
Zorah Lähner
Ankit Laddha
Florent Lafarge
Jean Lahoud
Kevin Lai
Shang-Hong Lai
Wei-Sheng Lai
Yu-Kun Lai
Iro Laina
Antony Lam
John Wheatley Lambert

Xiangyuan lan
Xu Lan
Charis Lanaras
Georg Langs
Oswald Lanz
Dong Lao
Yizhen Lao
Agata Lapedriza
Gustav Larsson
Viktor Larsson
Katrin Lasinger
Christoph Lassner
Longin Jan Latecki
Stéphane Lathuilière
Rynson Lau
Hei Law
Justin Lazarow
Svetlana Lazebnik
Hieu Le
Huu Le
Ngan Hoang Le
Trung-Nghia Le
Vuong Le
Colin Lea
Erik Learned-Miller
Chen-Yu Lee
Gim Hee Lee
Hsin-Ying Lee
Hyungtae Lee
Jae-Han Lee
Jimmy Addison Lee
Joonseok Lee
Kibok Lee
Kuang-Huei Lee
Kwonjoon Lee
Minsik Lee
Sang-chul Lee
Seungkyu Lee
Soochan Lee
Stefan Lee
Taehee Lee
Andreas Lehrmann
Jie Lei
Peng Lei
Matthew Joseph Leotta
Wee Kheng Leow

Gil Levi
Evgeny Levinkov
Aviad Levis
Jose Lezama
Ang Li
Bin Li
Bing Li
Boyi Li
Changsheng Li
Chao Li
Chen Li
Cheng Li
Chenglong Li
Chi Li
Chun-Guang Li
Chun-Liang Li
Chunyuan Li
Dong Li
Guanbin Li
Hao Li
Haoxiang Li
Hongsheng Li
Hongyang Li
Houqiang Li
Huibin Li
Jia Li
Jianan Li
Jianguo Li
Junnan Li
Junxuan Li
Kai Li
Ke Li
Kejie Li
Kunpeng Li
Lerenhan Li
Li Erran Li
Mengtian Li
Mu Li
Peihua Li
Peiyi Li
Ping Li
Qi Li
Qing Li
Ruiyu Li
Ruoteng Li
Shaozi Li

Sheng Li
Shiwei Li
Shuang Li
Siyang Li
Stan Z. Li
Tianye Li
Wei Li
Weixin Li
Wen Li
Wenbo Li
Xiaomeng Li
Xin Li
Xiu Li
Xuelong Li
Xueting Li
Yan Li
Yandong Li
Yanghao Li
Yehao Li
Yi Li
Yijun Li
Yikang LI
Yining Li
Yongjie Li
Yu Li
Yu-Jhe Li
Yunpeng Li
Yunsheng Li
Yunzhu Li
Zhe Li
Zhen Li
Zhengqi Li
Zhenyang Li
Zhuwen Li
Dongze Lian
Xiaochen Lian
Zhouhui Lian
Chen Liang
Jie Liang
Ming Liang
Paul Pu Liang
Pengpeng Liang
Shu Liang
Wei Liang
Jing Liao
Minghui Liao

Renjie Liao
Shengcai Liao
Shuai Liao
Yiyi Liao
Ser-Nam Lim
Chen-Hsuan Lin
Chung-Ching Lin
Dahua Lin
Ji Lin
Kevin Lin
Tianwei Lin
Tsung-Yi Lin
Tsung-Yu Lin
Wei-An Lin
Weiyao Lin
Yen-Chen Lin
Yuewei Lin
David B. Lindell
Drew Linsley
Krzysztof Lis
Roee Litman
Jim Little
An-An Liu
Bo Liu
Buyu Liu
Chao Liu
Chen Liu
Cheng-lin Liu
Chenxi Liu
Dong Liu
Feng Liu
Guilin Liu
Haomiao Liu
Heshan Liu
Hong Liu
Ji Liu
Jingen Liu
Jun Liu
Lanlan Liu
Li Liu
Liu Liu
Mengyuan Liu
Miaomiao Liu
Nian Liu
Ping Liu
Risheng Liu

Sheng Liu
Shu Liu
Shuaicheng Liu
Sifei Liu
Siqi Liu
Siying Liu
Songtao Liu
Ting Liu
Tongliang Liu
Tyng-Luh Liu
Wanquan Liu
Wei Liu
Weiyang Liu
Weizhe Liu
Wenyu Liu
Wu Liu
Xialei Liu
Xianglong Liu
Xiaodong Liu
Xiaofeng Liu
Xihui Liu
Xingyu Liu
Xinwang Liu
Xuanqing Liu
Xuebo Liu
Yang Liu
Yaojie Liu
Yebin Liu
Yen-Cheng Liu
Yiming Liu
Yu Liu
Yu-Shen Liu
Yufan Liu
Yun Liu
Zheng Liu
Zhijian Liu
Zhuang Liu
Zichuan Liu
Ziwei Liu
Zongyi Liu
Stephan Liwicki
Liliana Lo Presti
Chengjiang Long
Fuchen Long
Mingsheng Long
Xiang Long

Yang Long
Charles T. Loop
Antonio Lopez
Roberto J. Lopez-Sastre
Javier Lorenzo-Navarro
Manolis Lourakis
Boyu Lu
Canyi Lu
Feng Lu
Guoyu Lu
Hongtao Lu
Jiajun Lu
Jiasen Lu
Jiwen Lu
Kaiyue Lu
Le Lu
Shao-Ping Lu
Shijian Lu
Xiankai Lu
Xin Lu
Yao Lu
Yiping Lu
Yongxi Lu
Yongyi Lu
Zhiwu Lu
Fujun Luan
Benjamin E. Lundell
Hao Luo
Jian-Hao Luo
Ruotian Luo
Weixin Luo
Wenhan Luo
Wenjie Luo
Yan Luo
Zelun Luo
Zixin Luo
Khoa Luu
Zhaoyang Lv
Pengyuan Lyu
Thomas Möllenhoff
Matthias Müller
Bingpeng Ma
Chih-Yao Ma
Chongyang Ma
Huimin Ma
Jiayi Ma

K. T. Ma
Ke Ma
Lin Ma
Liqian Ma
Shugao Ma
Wei-Chiu Ma
Xiaojian Ma
Xingjun Ma
Zhanyu Ma
Zheng Ma
Radek Jakob Mackowiak
Ludovic Magerand
Shweta Mahajan
Siddharth Mahendran
Long Mai
Ameesh Makadia
Oscar Mendez Maldonado
Mateusz Malinowski
Yury Malkov
Arun Mallya
Dipu Manandhar
Massimiliano Mancini
Fabian Manhardt
Kevis-kokitsi Maninis
Varun Manjunatha
Junhua Mao
Xudong Mao
Alina Marcu
Edgar Margffoy-Tuay
Dmitrii Marin
Manuel J. Marin-Jimenez
Kenneth Marino
Niki Martinel
Julieta Martinez
Jonathan Masci
Tomohiro Mashita
Iacopo Masi
David Masip
Daniela Massiceti
Stefan Mathe
Yusuke Matsui
Tetsu Matsukawa
Iain A. Matthews
Kevin James Matzen
Bruce Allen Maxwell
Stephen Maybank

Helmut Mayer
Amir Mazaheri
David McAllester
Steven McDonagh
Stephen J. Mckenna
Roey Mechrez
Prakhar Mehrotra
Christopher Mei
Xue Mei
Paulo R. S. Mendonca
Lili Meng
Zibo Meng
Thomas Mensink
Bjoern Menze
Michele Merler
Kourosh Meshgi
Pascal Mettes
Christopher Metzler
Liang Mi
Qiguang Miao
Xin Miao
Tomer Michaeli
Frank Michel
Antoine Miech
Krystian Mikolajczyk
Peyman Milanfar
Ben Mildenhall
Gregor Miller
Fausto Milletari
Dongbo Min
Kyle Min
Pedro Miraldo
Dmytro Mishkin
Anand Mishra
Ashish Mishra
Ishan Misra
Niluthpol C. Mithun
Kaushik Mitra
Niloy Mitra
Anton Mitrokhin
Ikuhisa Mitsugami
Anurag Mittal
Kaichun Mo
Zhipeng Mo
Davide Modolo
Michael Moeller

Pritish Mohapatra
Pavlo Molchanov
Davide Moltisanti
Pascal Monasse
Mathew Monfort
Aron Monszpart
Sean Moran
Vlad I. Morariu
Francesc Moreno-Noguer
Pietro Morerio
Stylianos Moschoglou
Yael Moses
Roozbeh Mottaghi
Pierre Moulon
Arsalan Mousavian
Yadong Mu
Yasuhiro Mukaigawa
Lopamudra Mukherjee
Yusuke Mukuta
Ravi Teja Mullapudi
Mario Enrique Munich
Zachary Murez
Ana C. Murillo
J. Krishna Murthy
Damien Muselet
Armin Mustafa
Siva Karthik Mustikovela
Carlo Dal Mutto
Moin Nabi
Varun K. Nagaraja
Tushar Nagarajan
Arsha Nagrani
Seungjun Nah
Nikhil Naik
Yoshikatsu Nakajima
Yuta Nakashima
Atsushi Nakazawa
Seonghyeon Nam
Vinay P. Namboodiri
Medhini Narasimhan
Srinivasa Narasimhan
Sanath Narayan
Erickson Rangel
    Nascimento
Jacinto Nascimento
Tayyab Naseer

Lakshmanan Nataraj
Neda Nategh
Nelson Isao Nauata
Fernando Navarro
Shah Nawaz
Lukas Neumann
Ram Nevatia
Alejandro Newell
Shawn Newsam
Joe Yue-Hei Ng
Trung Thanh Ngo
Duc Thanh Nguyen
Lam M. Nguyen
Phuc Xuan Nguyen
Thuong Nguyen Canh
Mihalis Nicolaou
Andrei Liviu Nicolicioiu
Xuecheng Nie
Michael Niemeyer
Simon Niklaus
Christophoros Nikou
David Nilsson
Jifeng Ning
Yuval Nirkin
Li Niu
Yuzhen Niu
Zhenxing Niu
Shohei Nobuhara
Nicoletta Noceti
Hyeonwoo Noh
Junhyug Noh
Mehdi Noroozi
Sotiris Nousias
Valsamis Ntouskos
Matthew O'Toole
Peter Ochs
Ferda Ofli
Seong Joon Oh
Seoung Wug Oh
Iason Oikonomidis
Utkarsh Ojha
Takahiro Okabe
Takayuki Okatani
Fumio Okura
Aude Oliva
Kyle Olszewski

Björn Ommer
Mohamed Omran
Elisabeta Oneata
Michael Opitz
Jose Oramas
Tribhuvanesh Orekondy
Shaul Oron
Sergio Orts-Escolano
Ivan Oseledets
Aljosa Osep
Magnus Oskarsson
Anton Osokin
Martin R. Oswald
Wanli Ouyang
Andrew Owens
Mete Ozay
Mustafa Ozuysal
Eduardo Pérez-Pellitero
Gautam Pai
Dipan Kumar Pal
P. H. Pamplona Savarese
Jinshan Pan
Junting Pan
Xingang Pan
Yingwei Pan
Yannis Panagakis
Rameswar Panda
Guan Pang
Jiahao Pang
Jiangmiao Pang
Tianyu Pang
Sharath Pankanti
Nicolas Papadakis
Dim Papadopoulos
George Papandreou
Toufiq Parag
Shaifali Parashar
Sarah Parisot
Eunhyeok Park
Hyun Soo Park
Jaesik Park
Min-Gyu Park
Taesung Park
Alvaro Parra
C. Alejandro Parraga
Despoina Paschalidou

Nikolaos Passalis
Vishal Patel
Viorica Patraucean
Badri Narayana Patro
Danda Pani Paudel
Sujoy Paul
Georgios Pavlakos
Ioannis Pavlidis
Vladimir Pavlovic
Nick Pears
Kim Steenstrup Pedersen
Selen Pehlivan
Shmuel Peleg
Chao Peng
Houwen Peng
Wen-Hsiao Peng
Xi Peng
Xiaojiang Peng
Xingchao Peng
Yuxin Peng
Federico Perazzi
Juan Camilo Perez
Vishwanath Peri
Federico Pernici
Luca Del Pero
Florent Perronnin
Stavros Petridis
Henning Petzka
Patrick Peursum
Michael Pfeiffer
Hanspeter Pfister
Roman Pflugfelder
Minh Tri Pham
Yongri Piao
David Picard
Tomasz Pieciak
A. J. Piergiovanni
Andrea Pilzer
Pedro O. Pinheiro
Silvia Laura Pintea
Lerrel Pinto
Axel Pinz
Robinson Piramuthu
Fiora Pirri
Leonid Pishchulin
Francesco Pittaluga

Daniel Pizarro
Tobias Plötz
Mirco Planamente
Matteo Poggi
Moacir A. Ponti
Parita Pooj
Fatih Porikli
Horst Possegger
Omid Poursaeed
Ameya Prabhu
Viraj Uday Prabhu
Dilip Prasad
Brian L. Price
True Price
Maria Priisalu
Veronique Prinet
Victor Adrian Prisacariu
Jan Prokaj
Sergey Prokudin
Nicolas Pugeault
Xavier Puig
Albert Pumarola
Pulak Purkait
Senthil Purushwalkam
Charles R. Qi
Hang Qi
Haozhi Qi
Lu Qi
Mengshi Qi
Siyuan Qi
Xiaojuan Qi
Yuankai Qi
Shengju Qian
Xuelin Qian
Siyuan Qiao
Yu Qiao
Jie Qin
Qiang Qiu
Weichao Qiu
Zhaofan Qiu
Kha Gia Quach
Yuhui Quan
Yvain Queau
Julian Quiroga
Faisal Qureshi
Mahdi Rad

Filip Radenovic
Petia Radeva
Venkatesh
  B. Radhakrishnan
Ilija Radosavovic
Noha Radwan
Rahul Raguram
Tanzila Rahman
Amit Raj
Ajit Rajwade
Kandan Ramakrishnan
Santhosh
  K. Ramakrishnan
Srikumar Ramalingam
Ravi Ramamoorthi
Vasili Ramanishka
Ramprasaath R. Selvaraju
Francois Rameau
Visvanathan Ramesh
Santu Rana
Rene Ranftl
Anand Rangarajan
Anurag Ranjan
Viresh Ranjan
Yongming Rao
Carolina Raposo
Vivek Rathod
Sathya N. Ravi
Avinash Ravichandran
Tammy Riklin Raviv
Daniel Rebain
Sylvestre-Alvise Rebuffi
N. Dinesh Reddy
Timo Rehfeld
Paolo Remagnino
Konstantinos Rematas
Edoardo Remelli
Dongwei Ren
Haibing Ren
Jian Ren
Jimmy Ren
Mengye Ren
Weihong Ren
Wenqi Ren
Zhile Ren
Zhongzheng Ren

Zhou Ren
Vijay Rengarajan
Md A. Reza
Farzaneh Rezaeianaran
Hamed R. Tavakoli
Nicholas Rhinehart
Helge Rhodin
Elisa Ricci
Alexander Richard
Eitan Richardson
Elad Richardson
Christian Richardt
Stephan Richter
Gernot Riegler
Daniel Ritchie
Tobias Ritschel
Samuel Rivera
Yong Man Ro
Richard Roberts
Joseph Robinson
Ignacio Rocco
Mrigank Rochan
Emanuele Rodolà
Mikel D. Rodriguez
Giorgio Roffo
Grégory Rogez
Gemma Roig
Javier Romero
Xuejian Rong
Yu Rong
Amir Rosenfeld
Bodo Rosenhahn
Guy Rosman
Arun Ross
Paolo Rota
Peter M. Roth
Anastasios Roussos
Anirban Roy
Sebastien Roy
Aruni RoyChowdhury
Artem Rozantsev
Ognjen Rudovic
Daniel Rueckert
Adria Ruiz
Javier Ruiz-del-solar
Christian Rupprecht

Chris Russell
Dan Ruta
Jongbin Ryu
Ömer Sümer
Alexandre Sablayrolles
Faraz Saeedan
Ryusuke Sagawa
Christos Sagonas
Tonmoy Saikia
Hideo Saito
Kuniaki Saito
Shunsuke Saito
Shunta Saito
Ken Sakurada
Joaquin Salas
Fatemeh Sadat Saleh
Mahdi Saleh
Pouya Samangouei
Leo Sampaio
  Ferraz Ribeiro
Artsiom Olegovich
  Sanakoyeu
Enrique Sanchez
Patsorn Sangkloy
Anush Sankaran
Aswin Sankaranarayanan
Swami Sankaranarayanan
Rodrigo Santa Cruz
Amartya Sanyal
Archana Sapkota
Nikolaos Sarafianos
Jun Sato
Shin'ichi Satoh
Hosnieh Sattar
Arman Savran
Manolis Savva
Alexander Sax
Hanno Scharr
Simone Schaub-Meyer
Konrad Schindler
Dmitrij Schlesinger
Uwe Schmidt
Dirk Schnieders
Björn Schuller
Samuel Schulter
Idan Schwartz

William Robson Schwartz	Hailin Shi	Roger
Alex Schwing	Miaojing Shi	D. Soberanis-Mukul
Sinisa Segvic	Yemin Shi	Kihyuk Sohn
Lorenzo Seidenari	Zhenmei Shi	Francesco Solera
Pradeep Sen	Zhiyuan Shi	Eric Sommerlade
Ozan Sener	Kevin Jonathan Shih	Sanghyun Son
Soumyadip Sengupta	Shiliang Shiliang	Byung Cheol Song
Arda Senocak	Hyunjung Shim	Chunfeng Song
Mojtaba Seyedhosseini	Atsushi Shimada	Dongjin Song
Shishir Shah	Nobutaka Shimada	Jiaming Song
Shital Shah	Daeyun Shin	Jie Song
Sohil Atul Shah	Young Min Shin	Jifei Song
Tamar Rott Shaham	Koichi Shinoda	Jingkuan Song
Huasong Shan	Konstantin Shmelkov	Mingli Song
Qi Shan	Michael Zheng Shou	Shiyu Song
Shiguang Shan	Abhinav Shrivastava	Shuran Song
Jing Shao	Tianmin Shu	Xiao Song
Roman Shapovalov	Zhixin Shu	Yafei Song
Gaurav Sharma	Hong-Han Shuai	Yale Song
Vivek Sharma	Pushkar Shukla	Yang Song
Viktoriia Sharmanska	Christian Siagian	Yi-Zhe Song
Dongyu She	Mennatullah M. Siam	Yibing Song
Sumit Shekhar	Kaleem Siddiqi	Humberto Sossa
Evan Shelhamer	Karan Sikka	Cesar de Souza
Chengyao Shen	Jae-Young Sim	Adrian Spurr
Chunhua Shen	Christian Simon	Srinath Sridhar
Falong Shen	Martin Simonovsky	Suraj Srinivas
Jie Shen	Dheeraj Singaraju	Pratul P. Srinivasan
Li Shen	Bharat Singh	Anuj Srivastava
Liyue Shen	Gurkirt Singh	Tania Stathaki
Shuhan Shen	Krishna Kumar Singh	Christopher Stauffer
Tianwei Shen	Maneesh Kumar Singh	Simon Stent
Wei Shen	Richa Singh	Rainer Stiefelhagen
William B. Shen	Saurabh Singh	Pierre Stock
Yantao Shen	Suriya Singh	Julian Straub
Ying Shen	Vikas Singh	Jonathan C. Stroud
Yiru Shen	Sudipta N. Sinha	Joerg Stueckler
Yujun Shen	Vincent Sitzmann	Jan Stuehmer
Yuming Shen	Josef Sivic	David Stutz
Zhiqiang Shen	Gregory Slabaugh	Chi Su
Ziyi Shen	Miroslava Slavcheva	Hang Su
Lu Sheng	Ron Slossberg	Jong-Chyi Su
Yu Sheng	Brandon Smith	Shuochen Su
Rakshith Shetty	Kevin Smith	Yu-Chuan Su
Baoguang Shi	Vladimir Smutny	Ramanathan Subramanian
Guangming Shi	Noah Snavely	Yusuke Sugano

Masanori Suganuma
Yumin Suh
Mohammed Suhail
Yao Sui
Heung-Il Suk
Josephine Sullivan
Baochen Sun
Chen Sun
Chong Sun
Deqing Sun
Jin Sun
Liang Sun
Lin Sun
Qianru Sun
Shao-Hua Sun
Shuyang Sun
Weiwei Sun
Wenxiu Sun
Xiaoshuai Sun
Xiaoxiao Sun
Xingyuan Sun
Yifan Sun
Zhun Sun
Sabine Susstrunk
David Suter
Supasorn Suwajanakorn
Tomas Svoboda
Eran Swears
Paul Swoboda
Attila Szabo
Richard Szeliski
Duy-Nguyen Ta
Andrea Tagliasacchi
Yuichi Taguchi
Ying Tai
Keita Takahashi
Kouske Takahashi
Jun Takamatsu
Hugues Talbot
Toru Tamaki
Chaowei Tan
Fuwen Tan
Mingkui Tan
Mingxing Tan
Qingyang Tan
Robby T. Tan

Xiaoyang Tan
Kenichiro Tanaka
Masayuki Tanaka
Chang Tang
Chengzhou Tang
Danhang Tang
Ming Tang
Peng Tang
Qingming Tang
Wei Tang
Xu Tang
Yansong Tang
Youbao Tang
Yuxing Tang
Zhiqiang Tang
Tatsunori Taniai
Junli Tao
Xin Tao
Makarand Tapaswi
Jean-Philippe Tarel
Lyne Tchapmi
Zachary Teed
Bugra Tekin
Damien Teney
Ayush Tewari
Christian Theobalt
Christopher Thomas
Diego Thomas
Jim Thomas
Rajat Mani Thomas
Xinmei Tian
Yapeng Tian
Yingli Tian
Yonglong Tian
Zhi Tian
Zhuotao Tian
Kinh Tieu
Joseph Tighe
Massimo Tistarelli
Matthew Toews
Carl Toft
Pavel Tokmakov
Federico Tombari
Chetan Tonde
Yan Tong
Alessio Tonioni

Andrea Torsello
Fabio Tosi
Du Tran
Luan Tran
Ngoc-Trung Tran
Quan Hung Tran
Truyen Tran
Rudolph Triebel
Martin Trimmel
Shashank Tripathi
Subarna Tripathi
Leonardo Trujillo
Eduard Trulls
Tomasz Trzcinski
Sam Tsai
Yi-Hsuan Tsai
Hung-Yu Tseng
Stavros Tsogkas
Aggeliki Tsoli
Devis Tuia
Shubham Tulsiani
Sergey Tulyakov
Frederick Tung
Tony Tung
Daniyar Turmukhambetov
Ambrish Tyagi
Radim Tylecek
Christos Tzelepis
Georgios Tzimiropoulos
Dimitrios Tzionas
Seiichi Uchida
Norimichi Ukita
Dmitry Ulyanov
Martin Urschler
Yoshitaka Ushiku
Ben Usman
Alexander Vakhitov
Julien P. C. Valentin
Jack Valmadre
Ernest Valveny
Joost van de Weijer
Jan van Gemert
Koen Van Leemput
Gul Varol
Sebastiano Vascon
M. Alex O. Vasilescu

Subeesh Vasu
Mayank Vatsa
David Vazquez
Javier Vazquez-Corral
Ashok Veeraraghavan
Erik Velasco-Salido
Raviteja Vemulapalli
Jonathan Ventura
Manisha Verma
Roberto Vezzani
Ruben Villegas
Minh Vo
MinhDuc Vo
Nam Vo
Michele Volpi
Riccardo Volpi
Carl Vondrick
Konstantinos Vougioukas
Tuan-Hung Vu
Sven Wachsmuth
Neal Wadhwa
Catherine Wah
Jacob C. Walker
Thomas S. A. Wallis
Chengde Wan
Jun Wan
Liang Wan
Renjie Wan
Baoyuan Wang
Boyu Wang
Cheng Wang
Chu Wang
Chuan Wang
Chunyu Wang
Dequan Wang
Di Wang
Dilin Wang
Dong Wang
Fang Wang
Guanzhi Wang
Guoyin Wang
Hanzi Wang
Hao Wang
He Wang
Heng Wang
Hongcheng Wang

Hongxing Wang
Hua Wang
Jian Wang
Jingbo Wang
Jinglu Wang
Jingya Wang
Jinjun Wang
Jinqiao Wang
Jue Wang
Ke Wang
Keze Wang
Le Wang
Lei Wang
Lezi Wang
Li Wang
Liang Wang
Lijun Wang
Limin Wang
Linwei Wang
Lizhi Wang
Mengjiao Wang
Mingzhe Wang
Minsi Wang
Naiyan Wang
Nannan Wang
Ning Wang
Oliver Wang
Pei Wang
Peng Wang
Pichao Wang
Qi Wang
Qian Wang
Qiaosong Wang
Qifei Wang
Qilong Wang
Qing Wang
Qingzhong Wang
Quan Wang
Rui Wang
Ruiping Wang
Ruixing Wang
Shangfei Wang
Shenlong Wang
Shiyao Wang
Shuhui Wang
Song Wang

Tao Wang
Tianlu Wang
Tiantian Wang
Ting-chun Wang
Tingwu Wang
Wei Wang
Weiyue Wang
Wenguan Wang
Wenlin Wang
Wenqi Wang
Xiang Wang
Xiaobo Wang
Xiaofang Wang
Xiaoling Wang
Xiaolong Wang
Xiaosong Wang
Xiaoyu Wang
Xin Eric Wang
Xinchao Wang
Xinggang Wang
Xintao Wang
Yali Wang
Yan Wang
Yang Wang
Yangang Wang
Yaxing Wang
Yi Wang
Yida Wang
Yilin Wang
Yiming Wang
Yisen Wang
Yongtao Wang
Yu-Xiong Wang
Yue Wang
Yujiang Wang
Yunbo Wang
Yunhe Wang
Zengmao Wang
Zhangyang Wang
Zhaowen Wang
Zhe Wang
Zhecan Wang
Zheng Wang
Zhixiang Wang
Zilei Wang
Jianqiao Wangni

Anne S. Wannenwetsch
Jan Dirk Wegner
Scott Wehrwein
Donglai Wei
Kaixuan Wei
Longhui Wei
Pengxu Wei
Ping Wei
Qi Wei
Shih-En Wei
Xing Wei
Yunchao Wei
Zijun Wei
Jerod Weinman
Michael Weinmann
Philippe Weinzaepfel
Yair Weiss
Bihan Wen
Longyin Wen
Wei Wen
Junwu Weng
Tsui-Wei Weng
Xinshuo Weng
Eric Wengrowski
Tomas Werner
Gordon Wetzstein
Tobias Weyand
Patrick Wieschollek
Maggie Wigness
Erik Wijmans
Richard Wildes
Olivia Wiles
Chris Williams
Williem Williem
Kyle Wilson
Calden Wloka
Nicolai Wojke
Christian Wolf
Yongkang Wong
Sanghyun Woo
Scott Workman
Baoyuan Wu
Bichen Wu
Chao-Yuan Wu
Huikai Wu
Jiajun Wu

Jialin Wu
Jiaxiang Wu
Jiqing Wu
Jonathan Wu
Lifang Wu
Qi Wu
Qiang Wu
Ruizheng Wu
Shangzhe Wu
Shun-Cheng Wu
Tianfu Wu
Wayne Wu
Wenxuan Wu
Xiao Wu
Xiaohe Wu
Xinxiao Wu
Yang Wu
Yi Wu
Yiming Wu
Ying Nian Wu
Yue Wu
Zheng Wu
Zhenyu Wu
Zhirong Wu
Zuxuan Wu
Stefanie Wuhrer
Jonas Wulff
Changqun Xia
Fangting Xia
Fei Xia
Gui-Song Xia
Lu Xia
Xide Xia
Yin Xia
Yingce Xia
Yongqin Xian
Lei Xiang
Shiming Xiang
Bin Xiao
Fanyi Xiao
Guobao Xiao
Huaxin Xiao
Taihong Xiao
Tete Xiao
Tong Xiao
Wang Xiao

Yang Xiao
Cihang Xie
Guosen Xie
Jianwen Xie
Lingxi Xie
Sirui Xie
Weidi Xie
Wenxuan Xie
Xiaohua Xie
Fuyong Xing
Jun Xing
Junliang Xing
Bo Xiong
Peixi Xiong
Yu Xiong
Yuanjun Xiong
Zhiwei Xiong
Chang Xu
Chenliang Xu
Dan Xu
Danfei Xu
Hang Xu
Hongteng Xu
Huijuan Xu
Jingwei Xu
Jun Xu
Kai Xu
Mengmeng Xu
Mingze Xu
Qianqian Xu
Ran Xu
Weijian Xu
Xiangyu Xu
Xiaogang Xu
Xing Xu
Xun Xu
Yanyu Xu
Yichao Xu
Yong Xu
Yongchao Xu
Yuanlu Xu
Zenglin Xu
Zheng Xu
Chuhui Xue
Jia Xue
Nan Xue

Tianfan Xue	Yanchao Yang	Ke Yu
Xiangyang Xue	Yee Hong Yang	Lequan Yu
Abhay Yadav	Yezhou Yang	Ning Yu
Yasushi Yagi	Zhenheng Yang	Qian Yu
I. Zeki Yalniz	Anbang Yao	Ronald Yu
Kota Yamaguchi	Angela Yao	Ruichi Yu
Toshihiko Yamasaki	Cong Yao	Shoou-I Yu
Takayoshi Yamashita	Jian Yao	Tao Yu
Junchi Yan	Li Yao	Tianshu Yu
Ke Yan	Ting Yao	Xiang Yu
Qingan Yan	Yao Yao	Xin Yu
Sijie Yan	Zhewei Yao	Xiyu Yu
Xinchen Yan	Chengxi Ye	Youngjae Yu
Yan Yan	Jianbo Ye	Yu Yu
Yichao Yan	Keren Ye	Zhiding Yu
Zhicheng Yan	Linwei Ye	Chunfeng Yuan
Keiji Yanai	Mang Ye	Ganzhao Yuan
Bin Yang	Mao Ye	Jinwei Yuan
Ceyuan Yang	Qi Ye	Lu Yuan
Dawei Yang	Qixiang Ye	Quan Yuan
Dong Yang	Mei-Chen Yeh	Shanxin Yuan
Fan Yang	Raymond Yeh	Tongtong Yuan
Guandao Yang	Yu-Ying Yeh	Wenjia Yuan
Guorun Yang	Sai-Kit Yeung	Ye Yuan
Haichuan Yang	Serena Yeung	Yuan Yuan
Hao Yang	Kwang Moo Yi	Yuhui Yuan
Jianwei Yang	Li Yi	Huanjing Yue
Jiaolong Yang	Renjiao Yi	Xiangyu Yue
Jie Yang	Alper Yilmaz	Ersin Yumer
Jing Yang	Junho Yim	Sergey Zagoruyko
Kaiyu Yang	Lijun Yin	Egor Zakharov
Linjie Yang	Weidong Yin	Amir Zamir
Meng Yang	Xi Yin	Andrei Zanfir
Michael Ying Yang	Zhichao Yin	Mihai Zanfir
Nan Yang	Tatsuya Yokota	Pablo Zegers
Shuai Yang	Ryo Yonetani	Bernhard Zeisl
Shuo Yang	Donggeun Yoo	John S. Zelek
Tianyu Yang	Jae Shin Yoon	Niclas Zeller
Tien-Ju Yang	Ju Hong Yoon	Huayi Zeng
Tsun-Yi Yang	Sung-eui Yoon	Jiabei Zeng
Wei Yang	Laurent Younes	Wenjun Zeng
Wenhan Yang	Changqian Yu	Yu Zeng
Xiao Yang	Fisher Yu	Xiaohua Zhai
Xiaodong Yang	Gang Yu	Fangneng Zhan
Xin Yang	Jiahui Yu	Huangying Zhan
Yan Yang	Kaicheng Yu	Kun Zhan

Xiaohang Zhan
Baochang Zhang
Bowen Zhang
Cecilia Zhang
Changqing Zhang
Chao Zhang
Chengquan Zhang
Chi Zhang
Chongyang Zhang
Dingwen Zhang
Dong Zhang
Feihu Zhang
Hang Zhang
Hanwang Zhang
Hao Zhang
He Zhang
Hongguang Zhang
Hua Zhang
Ji Zhang
Jianguo Zhang
Jianming Zhang
Jiawei Zhang
Jie Zhang
Jing Zhang
Juyong Zhang
Kai Zhang
Kaipeng Zhang
Ke Zhang
Le Zhang
Lei Zhang
Li Zhang
Lihe Zhang
Linguang Zhang
Lu Zhang
Mi Zhang
Mingda Zhang
Peng Zhang
Pingping Zhang
Qian Zhang
Qilin Zhang
Quanshi Zhang
Richard Zhang
Rui Zhang
Runze Zhang
Shengping Zhang
Shifeng Zhang

Shuai Zhang
Songyang Zhang
Tao Zhang
Ting Zhang
Tong Zhang
Wayne Zhang
Wei Zhang
Weizhong Zhang
Wenwei Zhang
Xiangyu Zhang
Xiaolin Zhang
Xiaopeng Zhang
Xiaoqin Zhang
Xiuming Zhang
Ya Zhang
Yang Zhang
Yimin Zhang
Yinda Zhang
Ying Zhang
Yongfei Zhang
Yu Zhang
Yulun Zhang
Yunhua Zhang
Yuting Zhang
Zhanpeng Zhang
Zhao Zhang
Zhaoxiang Zhang
Zhen Zhang
Zheng Zhang
Zhifei Zhang
Zhijin Zhang
Zhishuai Zhang
Ziming Zhang
Bo Zhao
Chen Zhao
Fang Zhao
Haiyu Zhao
Han Zhao
Hang Zhao
Hengshuang Zhao
Jian Zhao
Kai Zhao
Liang Zhao
Long Zhao
Qian Zhao
Qibin Zhao

Qijun Zhao
Rui Zhao
Shenglin Zhao
Sicheng Zhao
Tianyi Zhao
Wenda Zhao
Xiangyun Zhao
Xin Zhao
Yang Zhao
Yue Zhao
Zhichen Zhao
Zijing Zhao
Xiantong Zhen
Chuanxia Zheng
Feng Zheng
Haiyong Zheng
Jia Zheng
Kang Zheng
Shuai Kyle Zheng
Wei-Shi Zheng
Yinqiang Zheng
Zerong Zheng
Zhedong Zheng
Zilong Zheng
Bineng Zhong
Fangwei Zhong
Guangyu Zhong
Yiran Zhong
Yujie Zhong
Zhun Zhong
Chunluan Zhou
Huiyu Zhou
Jiahuan Zhou
Jun Zhou
Lei Zhou
Luowei Zhou
Luping Zhou
Mo Zhou
Ning Zhou
Pan Zhou
Peng Zhou
Qianyi Zhou
S. Kevin Zhou
Sanping Zhou
Wengang Zhou
Xingyi Zhou

Yanzhao Zhou
Yi Zhou
Yin Zhou
Yipin Zhou
Yuyin Zhou
Zihan Zhou
Alex Zihao Zhu
Chenchen Zhu
Feng Zhu
Guangming Zhu
Ji Zhu
Jun-Yan Zhu
Lei Zhu
Linchao Zhu
Rui Zhu
Shizhan Zhu
Tyler Lixuan Zhu

Wei Zhu
Xiangyu Zhu
Xinge Zhu
Xizhou Zhu
Yanjun Zhu
Yi Zhu
Yixin Zhu
Yizhe Zhu
Yousong Zhu
Zhe Zhu
Zhen Zhu
Zheng Zhu
Zhenyao Zhu
Zhihui Zhu
Zhuotun Zhu
Bingbing Zhuang
Wei Zhuo

Christian Zimmermann
Karel Zimmermann
Larry Zitnick
Mohammadreza
    Zolfaghari
Maria Zontak
Daniel Zoran
Changqing Zou
Chuhang Zou
Danping Zou
Qi Zou
Yang Zou
Yuliang Zou
Georgios Zoumpourlis
Wangmeng Zuo
Xinxin Zuo

Additional Reviewers

Victoria Fernandez
    Abrevaya
Maya Aghaei
Allam Allam
Christine
    Allen-Blanchette
Nicolas Aziere
Assia Benbihi
Neha Bhargava
Bharat Lal Bhatnagar
Joanna Bitton
Judy Borowski
Amine Bourki
Romain Brégier
Tali Brayer
Sebastian Bujwid
Andrea Burns
Yun-Hao Cao
Yuning Chai
Xiaojun Chang
Bo Chen
Shuo Chen
Zhixiang Chen
Junsuk Choe
Hung-Kuo Chu

Jonathan P. Crall
Kenan Dai
Lucas Deecke
Karan Desai
Prithviraj Dhar
Jing Dong
Wei Dong
Turan Kaan Elgin
Francis Engelmann
Erik Englesson
Fartash Faghri
Zicong Fan
Yang Fu
Risheek Garrepalli
Yifan Ge
Marco Godi
Helmut Grabner
Shuxuan Guo
Jianfeng He
Zhezhi He
Samitha Herath
Chih-Hui Ho
Yicong Hong
Vincent Tao Hu
Julio Hurtado

Jaedong Hwang
Andrey Ignatov
Muhammad
    Abdullah Jamal
Saumya Jetley
Meiguang Jin
Jeff Johnson
Minsoo Kang
Saeed Khorram
Mohammad Rami Koujan
Nilesh Kulkarni
Sudhakar Kumawat
Abdelhak Lemkhenter
Alexander Levine
Jiachen Li
Jing Li
Jun Li
Yi Li
Liang Liao
Ruochen Liao
Tzu-Heng Lin
Phillip Lippe
Bao-di Liu
Bo Liu
Fangchen Liu

Hanxiao Liu	Ketul Shah	Yunyang Xiong
Hongyu Liu	Rajvi Shah	An Xu
Huidong Liu	Hengcan Shi	Chi Xu
Miao Liu	Xiangxi Shi	Yinghao Xu
Xinxin Liu	Yujiao Shi	Fei Xue
Yongfei Liu	William A. P. Smith	Tingyun Yan
Yu-Lun Liu	Guoxian Song	Zike Yan
Amir Livne	Robin Strudel	Chao Yang
Tiange Luo	Abby Stylianou	Heran Yang
Wei Ma	Xinwei Sun	Ren Yang
Xiaoxuan Ma	Reuben Tan	Wenfei Yang
Ioannis Marras	Qingyi Tao	Xu Yang
Georg Martius	Kedar S. Tatwawadi	Rajeev Yasarla
Effrosyni Mavroudi	Anh Tuan Tran	Shaokai Ye
Tim Meinhardt	Son Dinh Tran	Yufei Ye
Givi Meishvili	Eleni Triantafillou	Kun Yi
Meng Meng	Aristeidis Tsitiridis	Haichao Yu
Zihang Meng	Md Zasim Uddin	Hanchao Yu
Zhongqi Miao	Andrea Vedaldi	Ruixuan Yu
Gyeongsik Moon	Evangelos Ververas	Liangzhe Yuan
Khoi Nguyen	Vidit Vidit	Chen-Lin Zhang
Yung-Kyun Noh	Paul Voigtlaender	Fandong Zhang
Antonio Norelli	Bo Wan	Tianyi Zhang
Jaeyoo Park	Huanyu Wang	Yang Zhang
Alexander Pashevich	Huiyu Wang	Yiyi Zhang
Mandela Patrick	Junqiu Wang	Yongshun Zhang
Mary Phuong	Pengxiao Wang	Yu Zhang
Bingqiao Qian	Tai Wang	Zhiwei Zhang
Yu Qiao	Xinyao Wang	Jiaojiao Zhao
Zhen Qiao	Tomoki Watanabe	Yipu Zhao
Sai Saketh Rambhatla	Mark Weber	Xingjian Zhen
Aniket Roy	Xi Wei	Haizhong Zheng
Amelie Royer	Botong Wu	Tiancheng Zhi
Parikshit Vishwas	James Wu	Chengju Zhou
Sakurikar	Jiamin Wu	Hao Zhou
Mark Sandler	Rujie Wu	Hao Zhu
Mert Bülent Sarıyıldız	Yu Wu	Alexander Zimin
Tanner Schmidt	Rongchang Xie	
Anshul B. Shah	Wei Xiong	

Contents – Part XXIX

Procrustean Regression Networks: Learning 3D Structure of Non-rigid Objects from 2D Annotations

Sungheon Park[1] , Minsik Lee[2] , and Nojun Kwak[3]([⊠])

[1] Samsung Advanced Institute of Technology (SAIT), Yongin-si, Korea
sungheonpark@snu.ac.kr
[2] Hanyang University, Seoul, Korea
mleepaper@hanyang.ac.kr
[3] Seoul National University, Seoul, Korea
nojunk@snu.ac.kr

Abstract. We propose a novel framework for training neural networks which is capable of learning 3D information of non-rigid objects when only 2D annotations are available as ground truths. Recently, there have been some approaches that incorporate the problem setting of non-rigid structure-from-motion (NRSfM) into deep learning to learn 3D structure reconstruction. The most important difficulty of NRSfM is to estimate both the rotation and deformation at the same time, and previous works handle this by regressing both of them. In this paper, we resolve this difficulty by proposing a loss function wherein the suitable rotation is automatically determined. Trained with the cost function consisting of the reprojection error and the low-rank term of aligned shapes, the network learns the 3D structures of such objects as human skeletons and faces during the training, whereas the testing is done in a single-frame basis. The proposed method can handle inputs with missing entries and experimental results validate that the proposed framework shows superior reconstruction performance to the state-of-the-art method on the Human 3.6M, 300-VW, and SURREAL datasets, even though the underlying network structure is very simple.

1 Introduction

Inferring 3D poses from several 2D observations is inherently an underconstrained problem. Especially, for non-rigid objects such as human faces or bodies, it is harder to retrieve the 3D shapes than for rigid objects due to their shape deformations.

S. Park, M. Lee—Authors contributed equally.

Electronic supplementary material The online version of this chapter (https://doi.org/10.1007/978-3-030-58526-6_1) contains supplementary material, which is available to authorized users.

© Springer Nature Switzerland AG 2020
A. Vedaldi et al. (Eds.): ECCV 2020, LNCS 12374, pp. 1–18, 2020.
https://doi.org/10.1007/978-3-030-58526-6_1

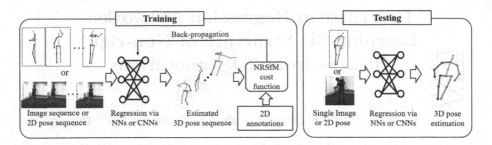

Fig. 1. Illustration of PRN. During the training, sequences of images or 2D poses are fed to the network, and their 3D shapes are estimated as the network outputs. The network is trained using the cost function which is based on an NRSfM algorithm. Testing is done by a simple feed-forward operation in a single-frame basis.

There are two distinct ways to retrieve 3D shapes of non-rigid objects from 2D observations. The first approach is to use a 3D reconstruction algorithm. Non-rigid structure from motion (NRSfM) algorithms [2,4,9,12,21] are designed to reconstruct 3D shapes of non-rigid objects from a sequence of 2D observations. Since NRSfM algorithms are not based on any learned models, the algorithms should be applied to each individual sequence, which makes the algorithm time-consuming when there are numerous number of sequences. The second approach is to learn the mappings from 2D to 3D with 3D ground truth training data. Prior knowledge can be obtained by dictionary learning [43,44], but neural networks or convolutional neural networks (CNNs) are the most-used methods to learn the 2D-to-3D or image-to-3D mappings [24,30], recently. However, 3D ground truth data are essential to learn those mappings, which requires large amounts of costs and efforts compared to the 2D data acquisition.

There is another possibility: With the framework which combines those two different frameworks, *i.e.*, NRSfM and neural networks, it is possible to overcome the limitations and to take advantages of both. There have been a couple of works that implement NRSfM using deep neural networks [6,19], but these methods mostly focus on the structure-from-category (SfC) problem, in which the 3D shapes of different rigid subjects in a category are reconstructed, and the deformation between subjects are not very diverse. Experiments on the CMU MoCap data in [19] show that, for data with diverse deformations, its generalization performance is not very good. Recently, Novotny *et al.* [27] proposed a neural network that reconstructs 3D shapes from monocular images by canonicalizing 3D shapes so that the 3D rigid motion is registered. This method has shown successful reconstruction results for data with more diverse deformations, which has been used in traditional NRSfM research. Wang *et al.* [38] also proposed knowledge distillation method that incorporate NRSfM algorithms as a teacher, which showed promising results on learning 3D human poses from 2D points.

The main difficulty of NRSfM is that one has to estimate both the rigid motion and the non-rigid shape deformation, which has been discussed extensively in the field of NRSfM throughout the past two decades. Especially, motion and deformation can get mixed up and some parts of rigid motions can be mistaken to be deformations. This has been first pointed out in [21], in which conditions derived from the generalized Procrustes analysis (GPA) has been adopted to resolve the problem. Meanwhile, all recent neural-network-based NRSfM approaches attempt to regress both the rigid motion and non-rigid deformation at the same time. Among these, only Novotny *et al.* [27] deals with the motion-deformation-separation problem in NRSfM, which is addressed as "shape transversality." Their solution is to register motions of different frames using an auxiliary neural network.

In this paper, we propose an alternative to this problem: First, we prove that a set of Procrustes-aligned shapes is transversal. Based on this fact, rather than explicitly estimating rigid motions, we propose a novel loss, in which suitable motions are determined automatically based on Procrustes alignment. This is achieved by modifying the cost function recently proposed in Procrustean regression (PR) [29], an NRSfM scheme that shares similar motivations with our work, which is used to train neural networks via back-propagation. Thanks to this new loss function, the network can concentrate only on the 3D shape estimation, and accordingly, the underlying structure of the proposed neural network is quite simple. The proposed framework, *Procrustean Regression Network* (PRN), learns to infer 3D structures of deformable objects using only 2D ground truths as training data.

Figure 1 illustrates the flow of the proposed framework. PRN accepts a set of image sequences or 2D point sequences as inputs at the training phase. The cost function of PRN is formulated to minimize the reprojection error and the nuclear norm of aligned shapes. The whole training procedure is done in an end-to-end manner, and the reconstruction result for an individual image is generated at the test phase via a simple forward propagation without requiring any post processing step for 3D reconstruction. Unlike the conventional NRSfM algorithms, PRN robustly estimates 3D structure of unseen test data with feed-forward operations in the test phase, taking the advantage of neural networks. The experimental results verify that PRN effectively reconstructs the 3D shapes of non-rigid objects such as human faces and bodies.

2   Related Works

The underlying assumption of NRSfM methods is that the 3D shape or the 3D trajectory of a point is interpreted as a weighted sum of several bases [2, 4]. 3D shapes are obtained by factorizing a shape matrix or a trajectory matrix so that the matrix has a pre-defined rank. Improvements have been made by several works which use probabilistic principal components analysis [34], metric constraints [28], course-to-fine reconstruction algorithm [3], complementary-space modeling [12], block sparse dictionary learning [18], or force-based models [1]. The major disadvantage of early NRSfM methods is that the number of basis should be determined explicitly while the optimal number of bases is usually unknown and is different from sequence to sequence. NRSfM methods using low-rank optimization have been proposed to overcome this problem [9,11].

It was proven that shape alignment also helps to increase the performance of NRSfM [7,20–22]. Procrustean normal distribution (PND) [21] is a powerful framework to separate rigid shape variations from the non-rigid ones. The expectation-maximization-based optimization algorithm applied to PND, EM-PND, showed superior performance to other NRSfM algorithms. Based on this idea, Procrustean Regression (PR) [29] has been proposed to optimize an NRSfM cost function via a simple gradient descent method. In [29], the cost function consists of a data term and a regularization term where low-rankness is imposed not directly on the reconstructed 3D shapes but on the aligned shapes with respect to the reference shape. Any type of differentiable function can be applied for both terms, which has allowed its applicability to perspective NRSfM.

On the other hand, along with recent rise of deep learning, there have been efforts to solve 3D reconstruction problems using CNNs. Object reconstruction from a single image with CNNs is an active field of research. The densely reconstructed shapes are often represented as 3D voxels or depth maps. While some works use ground truth 3D shapes [8,33,39], other works enable the networks to learn 3D reconstruction from multiple 2D observations [10,36,41,42]. The networks used in aforementioned works include a transformation layer that estimates the viewpoint of observations and/or a reprojection layer to minimize the error between input images and projected images. However, they mostly restrict the class of objects to ones that are rigid and have small amounts of deformations within each class, such as chairs and tables.

The 3D interpreter network [40] took a similar approach to NRSfM methods in that it formulates 3D shapes as the weighted sum of base shapes, but it used synthetic 3D models for network training. Warpnet [16] successfully reconstructs 3D shapes of non-rigid objects without supervision, but the results are only provided for birds datasets which have smaller deformations than human skeletons. Tulsiani et al. [35] provided a learning algorithm that automatically localize and reconstruct deformable 3D objects, and Kanazawa et al. [17] also infer 3D shapes as well as texture information from a single image. Although those methods output dense 3D meshes, the reconstruction is conducted on rigid objects or birds which do not contain large deformations. Our method provides a way to learn 3D structure of non-rigid objects that contain relatively large deformations and pose variations such as human skeletons or faces.

Training a neural network using the loss function based on NRSfM algorithms has been rarely studied. Kong and Lucey [19] proposed to interpret NRSfM as multi-layer sparse coding, and Cha et al. [6] proposed to estimate multiple basis shapes and rotations from 2D observations based on a deep neural network. However, they mostly focused on solving SfC problems which have rather small deformations, and the generalization performance of Kong and Lucey [19] is not very good for unseen data with large deformations. Recently, Novotny et al. [27] proposed a network structure which factors object deformation and viewpoint changes. Even though many existing ideas in NRSfM are nicely implemented in

[27], this in turn makes the network structure quite complicated. Unlike [27], the 3D shapes are aligned to the mean of aligned shapes in each minibatch in PRN, which enables the use of a simple network structure. Moreover, PRN does not need to set the number of basis shapes explicitly, because it is adjusted automatically in the low-rank loss.

3   Method

We briefly review PR [29] in Sect. 3.1, which is a regression problem based on Procrustes-aligned shapes and is the basis of PRN. Here, we also introduce the concept of "shape transversality" proposed by Novotny et al. [27] and prove that a set of Procrustes-aligned shapes is transversal, which means that Procrustes alignment can determine unique motions and eliminate the rigid motion components from reconstructed shapes. The cost function of PRN and its derivatives are explained in Sect. 3.2. The data term and the regularization term for PRN are proposed in Sect. 3.3. Lastly, network structures and training strategy is described in Sect. 3.4.

3.1   Procrustean Regression

NRSfM aims to recover 3D positions of the deformable objects from 2D correspondences. Concretely, given 2D observations of n_p points $\mathbf{U}_i (1 \leq i \leq n_f)$ in n_f frames, NRSfM reconstructs 3D shapes of each frame $\mathbf{X}_i$. PR [29] formulated NRSfM as a regression problem. The cost function of PR consists of data term that corresponds to the reprojection error and the regularization term that minimizes the rank of the aligned 3D shapes, which has the following form:

$$\mathcal{J} - \sum_{i=1}^{n_f} f(\mathbf{X}_i) + \lambda g(\widetilde{\mathbf{X}}, \overline{\mathbf{X}}). \tag{1}$$

Here, $\mathbf{X}_i$ is a $3 \times n_p$ matrix of the reconstructed 3D shapes on the ith frame, and $\overline{\mathbf{X}}$ is a reference shape for Procrustes alignment. $\widetilde{\mathbf{X}}$ is a $3n_p \times n_f$ matrix which is defined as $\widetilde{\mathbf{X}} \triangleq [\text{vec}(\widetilde{\mathbf{X}}_1) \, \text{vec}(\widetilde{\mathbf{X}}_2) \cdots \text{vec}(\widetilde{\mathbf{X}}_{n_f})]$, where $\text{vec}(\cdot)$ is a vectorization operator. $\widetilde{\mathbf{X}}_i$ is an aligned shape of the ith frame. The aligned shapes are retrieved via Procrustes analysis without scale alignment. In other words, the aligning rotation matrix for each frame is calculated as

$$\mathbf{R}_i = \underset{\mathbf{R}}{\text{argmin}} \, \|\mathbf{R}\mathbf{X}_i\mathbf{T} - \overline{\mathbf{X}}\| \quad \text{s.t.} \quad \mathbf{R}^T\mathbf{R} = \mathbf{I}. \tag{2}$$

Here, $\mathbf{T} \triangleq \mathbf{I}_{n_p} - \frac{1}{n_p}\mathbf{1}_{n_p}\mathbf{1}_{n_p}^T$ is the translation matrix that makes the shape centered at origin. $\mathbf{I}_n$ is an $n \times n$ identity matrix, and $\mathbf{1}_n$ is an all-one vector of size n. The aligned shape of the ith frame becomes $\widetilde{\mathbf{X}}_i = \mathbf{R}_i\mathbf{X}_i\mathbf{T}$.

In [29], (1) is optimized for variables $\mathbf{X}_i$ and $\overline{\mathbf{X}}$ and it is shown that their gradients for (1) can be analytically derived. Hence, any gradient-based optimization method can be applied for large choices of f and g. What the above formulation implies is that we can impose a regularization loss based on the alignment of reconstructed shapes, and therefore, we can enforce certain properties only to non-rigid deformations in which rigid motions are excluded.

To back up the above claim, we introduce the transversal property introduced in [27]:

Definition 1. The set $\mathcal{X}_0 \subset \mathbb{R}^{3 \times n_p}$ has the transversal property if, for any pair $\mathbf{X}, \mathbf{X}' \in \mathcal{X}_0$ related by a rotation $\mathbf{X}' = \mathbf{R}\mathbf{X}$, then $\mathbf{X} = \mathbf{X}'$.

The above definition basically defines a set of shapes that do not contain any non-trivial rigid transforms of its elements, and its elements can be interpreted as having canonical rigid poses. In other words, if two shapes in the set are distinctive, then they should not be identical up to a rigid transform. Here, we prove that the set of Procrustes-aligned shapes is indeed a transversal set. First, we need an assumption: Each shape should have a unique Procrustes alignment w.r.t. the reference shape. This condition might not be satisfied in some cases, e.g., degenerate shapes such as co-linear shapes.

Lemma 1. *A set $\mathcal{X}_P$ of Procrustes-aligned shapes w.r.t. a reference shape $\overline{\mathbf{X}}$ is transversal if the shapes are not degenerate.*

Proof. Suppose that there are $\mathbf{X}, \mathbf{X}' \in \mathcal{X}_P$ that satisfy $\mathbf{X}' = \mathbf{R}\mathbf{X}$. Based on the assumption, $\min_{\mathbf{R}'} \|\mathbf{R}'\mathbf{X}'\mathbf{T} - \overline{\mathbf{X}}\|^2$ will have a unique minimum at $\mathbf{R}' = \mathbf{I}$. Hence, $\min_{\mathbf{R}'} \|\mathbf{R}'\mathbf{R}\mathbf{X}\mathbf{T} - \overline{\mathbf{X}}\|^2$ will also have a unique minimum at the same point, which indicates that $\min_{\mathbf{R}''} \|\mathbf{R}''\mathbf{X}\mathbf{T} - \overline{\mathbf{X}}\|^2$ will have one at $\mathbf{R}'' = \mathbf{R}$. Based on the assumption, $\mathbf{R}''$ has to be $\mathbf{I}$, and hence $\mathbf{R} = \mathbf{I}$. $\qquad\square$

In [27], an arbitrary registration function f is introduced to ensure the transversality of a given set, which is implemented as an auxiliary neural network that has to be trained together with the main network component. We can interpret the Procrustes alignment in this work as a replacement of f that does not need training and has analytic gradients. Accordingly, the underlying network structure of PRN can become much simpler at the cost of a more complicated loss function.

3.2   PR Loss for Neural Networks

One may directly use the gradients of (1) to train neural networks by designing a neural network that estimates both the 3D shapes $\mathbf{X}_i$ and the reference shape $\overline{\mathbf{X}}$. However, the reference shape here incurs some problems when we are to handle it in a neural network. If the class of objects that we are interested in does not contain large deformations, then imposing this reference shape as a global parameter can be an option. On the contrary, if there can be a large deformation, then optimizing the cost function with minibatches of similar shapes or sequences of shapes can be vital for the success of training. In this case, a separate network module to estimate a good 3D reference shape is inevitable. However, designing a network module that estimates mean shapes may make the network structure more complex and training procedure harder. To keep it concise, we excluded the reference shape from (1) and defined the reference shape as the mean of the aligned output 3D shapes. The mean shape $\overline{\mathbf{X}}$ in (1) is simply replaced with $\sum_{j=1}^{n_f} \mathbf{R}_j \mathbf{X}_j \mathbf{T}$. Now, $\mathbf{X}_i$ is the only variable in the cost function, and the derivative of the cost function with respect to the estimated 3D shapes, $\frac{\partial \mathcal{J}}{\partial \mathbf{X}_i}$, is derived analytically.

The cost function of PRN can be written as follows:

$$\mathcal{J} = \sum_{i=1}^{n_f} f(\mathbf{X}_i) + \lambda g(\widetilde{\mathbf{X}}). \tag{3}$$

The alignment constraint is also changed to

$$\mathbf{R} = \underset{\mathbf{R}}{\operatorname{argmin}} \sum_{i=1}^{n_f} \|\mathbf{R}_i \mathbf{X}_i \mathbf{T} - \frac{1}{n_f} \sum_{j=1}^{n_f} \mathbf{R}_j \mathbf{X}_j \mathbf{T}\| \qquad \text{s.t.} \quad \mathbf{R}_i^T \mathbf{R}_i = \mathbf{I}. \tag{4}$$

where $\mathbf{R}$ is the concatenation of all rotation matrices, i.e., $\mathbf{R} = [\mathbf{R}_1, \mathbf{R}_2, \cdots, \mathbf{R}_{n_f}]$. Let us define $\mathbf{X}$ and $\widetilde{\mathbf{X}}$ as $\mathbf{X} \triangleq [\operatorname{vec}(\mathbf{X}_1), \operatorname{vec}(\mathbf{X}_2), \cdots, \operatorname{vec}(\mathbf{X}_{n_f})]$ and $\widetilde{\mathbf{X}} \triangleq [\operatorname{vec}(\widetilde{\mathbf{X}}_1), \operatorname{vec}(\widetilde{\mathbf{X}}_2), \cdots, \operatorname{vec}(\widetilde{\mathbf{X}}_{n_f})]$ respectively. The gradient of $\mathcal{J}$ with respect to $\mathbf{X}$ while satisfying the constraint (4) is

$$\frac{\partial \mathcal{J}}{\partial \mathbf{X}} = \frac{\partial f}{\partial \mathbf{X}} + \lambda \left\langle \frac{\partial g}{\partial \widetilde{\mathbf{X}}}, \frac{\partial \widetilde{\mathbf{X}}}{\partial \mathbf{X}} \right\rangle, \tag{5}$$

where $\langle \cdot, \cdot \rangle$ denotes the inner product. $\frac{\partial f}{\partial \mathbf{X}}$ and $\frac{\partial g}{\partial \mathbf{X}}$ are derived once f and g are determined. The derivation process of $\frac{\partial \widetilde{\mathbf{X}}}{\partial \mathbf{X}}$ is analogous to [29]. We explained detailed process in the supplementary material and provide only the results here, which has the form of

$$\frac{\partial \widetilde{\mathbf{X}}}{\partial \mathbf{X}} = (\mathbf{A}\mathbf{B}^{-1}\mathbf{C} + \mathbf{I}_{3n_p n_f})\mathbf{D}. \tag{6}$$

$\mathbf{A}$ is a $3n_p n_f \times 3n_f$ block diagonal matrix expressed as

$$\mathbf{A} = \operatorname{blkdiag}((\mathbf{X}_1'^T \otimes \mathbf{I}_3)\mathbf{L}, (\mathbf{X}_2'^T \otimes \mathbf{I}_3)\mathbf{L}, \cdots, (\mathbf{X}_{n_f}'^T \otimes \mathbf{I}_3)\mathbf{L}), \tag{7}$$

where $\operatorname{blkdiag}(\cdot)$ is the block-diagonal operator, $\otimes$ denotes the Kronecker product. $\mathbf{X}_i'^T = \hat{\mathbf{R}}_i \mathbf{X}_i \mathbf{T}$, where $\hat{\mathbf{R}}_i$ is the current rotation matrix before the gradient evaluation, and $\mathbf{L}$ is a 9×3 matrix that implies the orthogonality constraint of a rotation matrix [29], whose values are

$$\mathbf{L} = \begin{bmatrix} 0 & 0 & 0 & 0 & 0 & -1 & 0 & 1 & 0 \\ 0 & 0 & 1 & 0 & 0 & 0 & -1 & 0 & 0 \\ 0 & -1 & 0 & 1 & 0 & 0 & 0 & 0 & 0 \end{bmatrix}^T. \tag{8}$$

$\mathbf{B}$ is a $3n_f \times 3n_f$ matrix whose block elements are

$$\mathbf{b}_{ij} = \begin{cases} \mathbf{L}^T(\sum_{k \neq i} \mathbf{X}_k'^T \mathbf{X}_i'^T \otimes \mathbf{I}_3)\mathbf{L} & i = j \\ \mathbf{L}^T(\mathbf{I}_3 \otimes \mathbf{X}_i' \mathbf{X}_j'^T)\mathbf{E}\mathbf{L} & i \neq j \end{cases} \tag{9}$$

where $\mathbf{b}_{ij}$ means the (i,j)-th 3×3 submatrix of $\mathbf{B}$, i and j are integers ranging from 1 to n_f, and $\mathbf{E}$ is a permutation matrix that satisfies $\mathbf{E}\mathrm{vec}(\mathbf{H}) = \mathrm{vec}(\mathbf{H}^T)$. $\mathbf{C}$ is a $3n_f \times 3n_f n_p$ matrix whose block elements are

$$\mathbf{c}_{ij} = \begin{cases} -\mathbf{L}^T(\sum_{k \neq i} \mathbf{X}'_k \otimes \mathbf{I}_3) & i = j \\ -\mathbf{L}^T(\mathbf{I}_3 \otimes \mathbf{X}'_i)\mathbf{E} & i \neq j \end{cases} \tag{10}$$

where $\mathbf{c}_{ij}$ means the (i,j)-th 3×3 submatrix of $\mathbf{C}$. Finally, $\mathbf{D}$ is a $3n_f n_p \times 3n_f n_p$ block-diagonal matrix expressed as

$$\mathbf{D} = \mathrm{blkdiag}(\mathbf{T} \otimes \hat{\mathbf{R}}_1, \mathbf{T} \otimes \hat{\mathbf{R}}_2, \cdots, \mathbf{T} \otimes \hat{\mathbf{R}}_{n_f}). \tag{11}$$

Even though the size of $\partial \widetilde{\mathbf{X}}/\partial \mathbf{X}$ is quite large, i.e., $3n_f n_p \times 3n_f n_p$, we don't actually have to construct it explicitly since the only thing we need is the ability to backpropagate. Memory space and computations can be largely saved based on clever utilization of batch matrix multiplications and reshapes. In the next section, we will discuss about the design of the functions f and g and their derivatives.

3.3 Design of f and g

In PRN, the network produces the 3D position of each joint of a human body. The network output is fed into the cost function, and the gradients are calculated to update the network. For the data term f, we use the reprojection error between the estimated 3D shapes and the ground truth 2D points. We only consider the orthographic projection in this paper, but the framework can be easily extended to the perspective projection. The function f corresponding to the data term has the following form.

$$f(\mathbf{X}) = \sum_{i=1}^{n_f} \frac{1}{2} \|(\mathbf{U}_i - \mathbf{P}_o \mathbf{X}_i) \odot \mathbf{W}_i\|_F^2. \tag{12}$$

Here, $\mathbf{P}_o = \begin{bmatrix} 1 & 0 & 0 \\ 0 & 1 & 0 \end{bmatrix}$ is an 2×3 orthographic projection matrix, and $\mathbf{U}_i$ is a $2 \times n_p$ 2D observation matrix (ground truth). $\mathbf{W}_i$ is a $2 \times n_p$ weight matrix whose ith column represents the confidence of the position of ith point. $\mathbf{W}_i$ has values between 0 and 1, where 0 means the keypoint is not observable due to occlusion. Scores from 2D keypoint detectors can be used as values of $\mathbf{W}_i$. Lastly, $\|\cdot\|_F$ and $\odot$ denotes the Frobenius norm and element-wise multiplication respectively. The gradient of (12) is

$$\frac{\partial f}{\partial \mathbf{X}} = \sum_{i=1}^{n_f} \mathbf{P}_o^T((\mathbf{P}_o \mathbf{X}_i - \mathbf{U}_i) \odot \mathbf{W}_i \odot \mathbf{W}_i). \tag{13}$$

For the regularization term, we imposed a low-rank constraint to the aligned shapes. Log-determinant or the nuclear norm are two widely used functions and we choose the nuclear norm, i.e.,

$$g(\widetilde{\mathbf{X}}) = \|\widetilde{\mathbf{X}}\|_*, \tag{14}$$

where $\|\cdot\|_*$ stands for the nuclear norm of a matrix. The subgradient of a nuclear norm can be calculated as

$$\frac{\partial g}{\partial \widetilde{\mathbf{X}}} = \mathbf{U}\mathrm{sign}(\mathbf{\Sigma})\mathbf{V}^T, \tag{15}$$

where $\mathbf{U}\mathbf{\Sigma}\mathbf{V}^T$ is the singular value decomposition of $\widetilde{\mathbf{X}}$ and $\mathrm{sign}(\cdot)$ is the sign function. Note that the sign function is to deal with zero singular values. $\partial g/\partial \widetilde{\mathbf{X}}_i$ is easily obtained by reordering $\partial g/\partial \widetilde{\mathbf{X}}$.

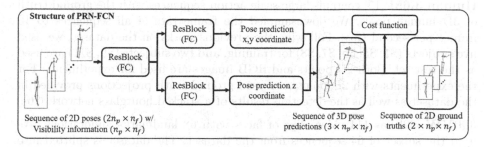

Fig. 2. Structure of FCNs used in this paper. Structure of CNNs are the same except that ResNet-50 is used as a backbone network instead of a fully connected ResBlock.

3.4 Network Structure

By susbstituting (6), (13), and (15) into (5), the gradient of the cost function of PRN with respect to the 3D shape $\mathbf{X}_i$ can be calculated. Then, the gradient for the entire parameters in the network can also be calculated by back-propagation. We experimented two different structures of PRN in Sect. 4: fully connected networks (FCNs) and convolutional neural networks (CNNs). For the FCN structure, inputs are the 2D point sequences. Each minibatch has a size of $2n_p \times n_f$, and the network produces the 3D positions of the input sequences. We use two stacks of residual modules [13] as the network structure. The prediction parts of x, y coordinates and z coordinates in the network are separated as illustrated in Fig. 2, which achieved better performance in our empirical experience.

For the CNNs, sequences of RGB images are fed into the networks. ResNet-50 [13] is used as a backbone network. The features of the final convolutional layers consisting of 2,048 feature maps of 7×7 size are connected to a network with the same structure as in the previous FCN to produce the final 3D output. We initialize the weights in the convolutional layers to those of the ImageNet [31] pre-trained network. More detailed hyperparameter settings are described in the supplementary material.

4   Experiments

The proposed framework is applied to reconstruct 3D human poses, 3D human faces, and dense 3D human meshes, all of which are the representative types of non-rigid objects. Additional qualitative results and experiments of PRN including comparison with the other methods on the datasets can be found in the supplementary materials.

4.1   Datasets

Human 3.6M [15] contains large-scale action sequences with the ground truth of 3D human poses. We downsampled the frame rate of all sequences to 10 frames per second (fps). Following the previous works on the dataset, we used five subjects (S1, S5, S6, S7, S8) for training, and two subjects (S9, S11) are used as the test set. Both 2D points and RGB images are used for experiments. For the experiments with 2D points, we used ground truth projections provided in the dataset as well as the detection results of a stacked hourglass network [26].

300VW [32] has 114 video clips of faces with 68 landmarks annotations. We used the subset of 64 sequences from the dataset. The dataset is splitted into train and test sets, each of which consists of 32 sequences. 63,205 training images and 60,216 test images are used for the experiment. Since 300-VW dataset only provides 2D annotations and no 3D ground truth data exists, we used the data provided in [5] as 3D ground truths.

SURREAL [37] dataset is used to validate our framework on dense 3D human shapes. It contains 3D human meshes which are created by fitting SMPL body model [23] on CMU Mocap sequences. Each mesh is comprised of 6,890 vertices. We selected 25 sequences from the dataset and split into training and test sets which consist of 5,000 and 2,401 samples respectively. The meshes are randomly rotated around y-axis, and orthographic projection is applied to generate 2D points.

4.2   Implementation Details

The parameter λ is set to $\lambda = 0.05$ for all experiments. The datasets used for our experiment consist of video sequences from fixed monocular cameras. However, most NRSfM algorithms including PRN requires moderate rotation variations in a sequence. To this end, for the Human 3.6M dataset where the sequences are taken by 4 different cameras, we alternately sample the frames or 2D poses from different cameras for consecutive frames. We set the time interval of the samples from different cameras to 0.5 s. Meanwhile, 300-VW dataset does not have multi-view sequences, and each sequence does not have enough rotations. Hence, we randomly sample the inputs in a minibatch from different sequences. For SURREAL datasets, we used 2D poses from consecutive frames.

On the other hand, the rotation alignment is applied to the samples within the same mini-batch. Therefore, if we select the samples in a mini-batch from a single sequence, the samples within the mini-batch does not have enough variations which affects training speed and performance. To alleviate this problem, we divided a mini-batch into 4 groups and calculated the gradients of the cost function for each group during the training of Human 3.6 M and SURREAL datasets. In addition, since only a small number of different sequences are used in each mini-batch during the training and frames in the same mini-batch are highly correlated as a result, batch normalization [14] may make the training unstable. Hence, we train the networks using batch normalization with moving average for the 70% of the training, and the rest of the iterations are trained with fixed average values as in the test phase.

4.3   Results

The performance of PRN on Human3.6M is evaluated in terms of mean per joint position error (MPJPE) which is the widely used metric in the literature. Meanwhile, we used normalized error as the error metric of 300-VW and SURREAL datasets since the dataset does not provide absolute scales of 3D points. MPJPE and normalized error(NE) are defined as

$$\text{MPJPE}(\hat{\mathbf{X}}_i, \mathbf{X}_i^*) = \frac{1}{n_p} \sum_{j=1}^{n_p} \|\hat{\mathbf{X}}_{ij} - \mathbf{X}_{ij}^*\|, \quad \text{NE}(\hat{\mathbf{X}}_i, \mathbf{X}_i^*) = \frac{\|\hat{\mathbf{X}}_i - \mathbf{X}_i^*\|_F}{\|\mathbf{X}_i^*\|_F}, \quad (16)$$

where $\hat{\mathbf{X}}_i$ and $\mathbf{X}_i^*$ denote the reconstructed 3D shape and the ground truth 3D shape on the ith frame, respectively, and $\hat{\mathbf{X}}_{ij}$ and $\mathbf{X}_{ij}^*$ are the jth keypoint of $\hat{\mathbf{X}}_i$ and $\mathbf{X}_i^*$, respectively. Since orthographic projection has reflection ambiguity, we measure the error also for the reflected shapes and choose the shape that has a smaller error.

To verify the effectiveness of PRN in the fully-connected network architecture (PRN-FCN), we first applied PRN to the task of 3D reconstruction given an input of 2D points. First, we trained PRN-FCN on the Human 3.6M dataset using either ground truth 2D generated by orthographic projection or perspective projection (GT-ortho, GT-persp) or keypoints detected using Stacked hourglass networks (SH). The detailed results for different actions are illustrated in Table 1. For comparison, we also show the results of C3DPO from [27] under the same training setting. As a baseline, we also provide the performance of FCN trained only on the reprojection error (PRN w/o reg). We also trained the neural nets using the 3D shapes reconstructed from existing NRSfM methods, CSF2 [12] and SPM [9] to compare our framework with NRSfM methods. We applied the NRSfM methods to each sequence with the same strides and camera settings as done in training PRN. The trained networks also have the same structure as the one used for PRN.

Here, we can confirm that the regularization term helps estimating depth information more accurately and drops the error significantly. Moreover, PRN-FCN significantly outperforms the NRSfM methods and is also superior to the

Table 1. MPJPE with 2D inputs on Human 3.6M dataset with different 2D inputs: (GT-ortho) Orthographic projection of 3D GT points. (GT-persp) Perspective projection of 3D GT. (SH) 2D keypoint detection results of a stacked hourglass network either from [27] (SH [27]) or from the network fine-tuned on Human 3.6M (SH-FT). PRN-FCN-W used weighted reprojection error based on the keypoint detection score.

Method (GT-ortho)	Direct	Discuss	Eating	Greet	Phone	Pose	Purch	Sitting	SitingD	Smoke	Photo	Wait	Walk	WalkD	WalkT	Avg
PRN w/o reg	138.1	139.7	146.5	145.2	140	127.6	149.4	170.4	188.4	138.3	150.9	133.1	125.6	143.9	139.3	144.8
CSF2 [12] + NN	87.2	90.1	96.1	95.9	102.9	92.1	99.3	129.8	136.7	99.5	120.1	95.2	90.8	102.4	89.2	101.6
SPM [9] + NN	65.3	68.7	82.0	70.1	95.3	65.1	71.9	117.0	136.0	84.3	88.9	71.2	59.5	73.3	68.3	82.3
C3DPO [27]	**56.1**	**55.6**	62.2	**66.4**	63.2	**62.0**	**62.9**	**76.3**	**85.8**	**59.9**	88.7	**63.3**	71.1	70.7	72.3	67.8
PRN-FCN	65.3	58.2	**60.5**	73.8	**60.7**	71.5	64.6	79.8	90.2	60.3	**81.2**	67.1	**54.4**	**61.2**	**65.6**	**66.7**
Method (GT-persp)	**Direct**	**Discuss**	**Eating**	**Greet**	**Phone**	**Pose**	**Purch**	**Sitting**	**SitingD**	**Smoke**	**Photo**	**Wait**	**Walk**	**WalkD**	**WalkT**	**Avg**
C3DPO [27]	96.8	85.7	85.8	107.1	86.0	96.8	93.9	94.9	96.7	86.0	124.3	90.7	95.2	93.4	101.3	95.6
PRN-FCN	**93.1**	**83.3**	**76.2**	**98.6**	**78.8**	**91.7**	**81.4**	**87.4**	**91.6**	**78.2**	**104.3**	**89.6**	**83.0**	**80.5**	**95.3**	**86.4**
Method (SH)	**Direct**	**Discuss**	**Eating**	**Greet**	**Phone**	**Pose**	**Purch**	**Sitting**	**SitingD**	**Smoke**	**Photo**	**Wait**	**Walk**	**WalkD**	**WalkT**	**Avg**
C3DPO [27]	131.1	137.4	125.2	146.4	143.2	141.4	137.3	141.4	163.8	136.2	161	143.4	145.9	153.2	168.6	145.0
PRN-FCN (SH [27])	127.2	115.1	109.2	130.0	126.9	122.3	116.4	128.4	149.3	117.3	140.7	124.0	123.9	115.3	140.4	124.5
PRN-FCN (SH FT)	**100.2**	89.4	83.8	105.5	93.0	97.2	89.2	114.0	141.2	89.1	**114.8**	97.3	**91.0**	88.3	107.2	99.1
PRN-FCN-W (SH FT)	100.3	**88.8**	**82.8**	**105.2**	**91.4**	**96.7**	**88.1**	**102.1**	**113.2**	**87.4**	115.1	**96.5**	91.7	**87.6**	**106.4**	**95.9**

2D inputs   PRN-FCN   GT   2D inputs   PRN-FCN   GT

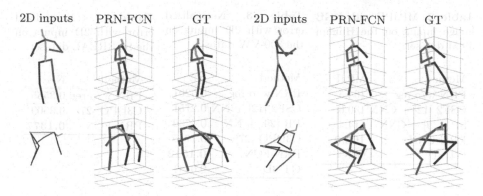

Fig. 3. Qualitative results of PRN-FCN on Human 3.6M dataset. PRN successfully reconstructs 3D shapes from 2D points under various rotations and poses. Left arms and legs are shown in blue, and right arms and legs are shown in red. (Color figure online)

recently proposed work [27] for both ground truth inputs and inputs from keypoint detectors, which proves the effectiveness of the alignment and the low-rank assumption for similar shapes. While PRN-FCN is silghtly better than [27] under orthographic projections, it largely outperforms [27] when trained using 2D points with perspective projections, which indicates that PRN is also robust to the noisy data. The results from the neural networks trained with NRSfM tend to have large variations depending on the types of sequences. This is mainly because the label data comes from NRSfM methods does not show prominent reconstruction results, and this erroneous signal limits the performance of the network in difficult sequences. On the other hand, PRN-FCN robustly reconstruct 3D shapes across all sequences. More interestingly, when the scores of keypoint detectors are used as a weight(PRN-FCN-W), PRN showed improved performance. This result implies that PRN is also robust to inputs with structured missing points since occluded keypoints have lower scores. Although we did not provide the confidence information as input signals, lower weight in the cost function makes the keypoints with lower confidence rely more on the regularization term. As a consequence, PRN-FCN-W performs especially better on the sequences that have complex pose variations such as *Sitting* or *SittingDown*.

Qualitative results for PRN and comparison with the ground truth shapes are illustrated in Fig. 3. It is shown that PRN accurately reconstructs 3D shapes of human bodies from various challenging 2D poses.

Next, we apply PRN to the CNNs to learn the 3D shapes directly from RGB images. MPJPE on the Human 3.6M test set are provided in Table 2. For comparison, we also trained the networks using only the reprojection error and excluding the regularization term in the cost function of PRN (PRN w/o reg). Moreover, we also trained the networks using the 3D shapes reconstructed from existing NRSfM methods, CSF2 [12] and PR [29] since SPM [9] diverged for many sequences in this dataset. Estimating 3D poses from RGB images directly

Table 2. MPJPE with RGB image inputs on the Human 3.6M dataset.

Method	MPJPE
PRN w/o reg	164.5
CSF2 [12] + CNN	130.6
SPM [9] + CNN	114.4
PRN-CNN	**108.9**
GT 3D	98.8

Table 3. Normalized error with 2D inputs on the 300-VW dataset.

Method	NE
PRN w/o reg	0.5201
CSF2 [12] + NN	0.2751
PR [29] + NN	0.2730
C3DPO [27]	0.1715
PRN-FCN	**0.1512**
GT 3D	0.0441

Table 4. Normalized error with 2D inputs on the SURREAL dataset.

Method	NE
PRN w/o reg	0.3565
C3DPO [27]	0.3509
PRN-FCN	**0.1377**

is more challenging than using 2D points as inputs because 2D information as well as depth information should also be learned, and images also contain photometric variations or self-occlusions. PRN largely outperforms the model without regularization term and shows better results than the CNNs trained using NRSfM reconstruction results. It can be observed that the CNN trained with ground truth 3D still has large errors. The performance may be improved if recently-proposed networks for 3D human pose estimation [25, 30] is applied here. However, a large network structure reduces the batch size, which can ruin the entire training process of PRN. Therefore, we instead used the largest network we can afford with maintaining the batch size to at least 16. Even though this limits the performance gain due to network structure, we can still compare the results from other similar-sized networks to verify that the proposed training strategy is effective. Qualitative results of PRN-CNN are provided in the supplementary materials.

Next, for the task of 3D face reconstruction, we used the 300-VW dataset [32] which has a video sequence of human faces. We used the reconstruction results from [5] as 3D ground truths. The reconstruction performance is evaluated in terms of normalized error, and the results are illustrated in Table 3. PRN-FCN is also superior to the other methods, including C3DPO [27], in 300-VW datasets. Qualitative results are shown in the two leftmost columns of Fig. 4. Both PRN and C3DPO output plausible results, but C3DPO tends to have larger depth ranges than ground truth depths, which led to increase the normalized errors.

Lastly, we validated the effectiveness of PRN on dense 3D models. Human meshes in SURREAL datasets consist of 6890 3D points for each shape. Since calculating the cost function on dense 3D data imposes heavy computational burden, we subdivided the 3D points into a few groups and compute the cost function for a small set of points. The groups are randomly organized in every iteration. Normalized errors on the SURREAL dataset is shown in Table 4. As it can be seen in Table 4 and the two rightmost columns of Fig. 4, PRN-FCN effectively reconstruct 3D human mesh models from 2D inputs while C3DPO [27] fails to recover depth information.

300-VW dataset   SURREAL datset

2D inputs

C3DPO [27]

PRN-FCN

GT

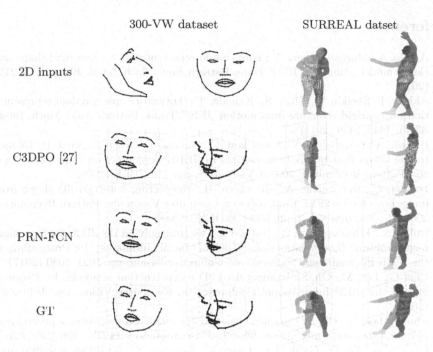

Fig. 4. Qualitative results of PRN-FCN on 300-VW datasets (two leftmost columns) and SURREAL datasets (two rightmost columns).

5   Conclusion

In this paper, a novel framework for training neural networks to estimate 3D shapes of non-rigid objects based on only 2D annotations is proposed. 3D shapes of an image can be rapidly estimated using the trained networks unlike existing NRSfM algorithms. The performance of PRN can be improved by adopting different network architectures. For example, CNNs based on heatmap representations may provide accurate 2D poses and improve reconstruction performance. Moreover, the flexibility for designing the data term and the regularization term in PRN makes it easier to extend the framework to handle perspective projection. Nonetheless, the proposed PRN with simple network structures outperforms the existing state-of-the-art. Although solving NRSfM with deep learning still has some challenges, we believe that the proposed framework establishes the connection between NRSfM algorithms and deep learning which will be useful for future research.

Acknowledgement. This work was supported by grants from IITP (No.2019-0-01367, Babymind) and NRF Korea (2017M3C4A7077582, 2020R1C1C1012479), all of which are funded by the Korea government (MSIT).

References

1. Agudo, A., Moreno-Noguer, F.: Force-based representation for non-rigid shape and elastic model estimation. IEEE Trans. Pattern Anal. Mach. Intell. **40**(9), 2137–2150 (2018)
2. Akhter, I., Sheikh, Y., Khan, S., Kanade, T.: Trajectory space: a dual representation for nonrigid structure from motion. IEEE Trans. Pattern Anal. Mach. Intell. **33**(7), 1442–1456 (2011)
3. Bartoli, A., Gay-Bellile, V., Castellani, U., Peyras, J., Olsen, S., Sayd, P.: Coarse-to-fine low-rank structure-from-motion. In: IEEE Conference on Computer Vision and Pattern Recognition 2008, CVPR 2008, pp. 1–8. IEEE (2008)
4. Bregler, C., Hertzmann, A., Biermann, H.: Recovering non-rigid 3D shape from image streams. In: IEEE Conference on Computer Vision and Pattern Recognition 2000, Proceedings, vol. 2, pp. 690–696. IEEE (2000)
5. Bulat, A., Tzimiropoulos, G.: How far are we from solving the 2D & 3D face alignment problem? (and a dataset of 230,000 3D facial landmarks). In: Proceedings of the IEEE International Conference on Computer Vision, pp. 1021–1030 (2017)
6. Cha, G., Lee, M., Oh, S.: Unsupervised 3D reconstruction networks. In: Proceedings of the IEEE International Conference on Computer Vision, pp. 3849–3858 (2019)
7. Cho, J., Lee, M., Oh, S.: Complex non-rigid 3D shape recovery using a procrustean normal distribution mixture model. Int. J. Comput. Vis. **117**(3), 226–246 (2016)
8. Choy, C.B., Xu, D., Gwak, J.Y., Chen, K., Savarese, S.: 3D-R2N2: a unified approach for single and multi-view 3D object reconstruction. In: Leibe, B., Matas, J., Sebe, N., Welling, M. (eds.) ECCV 2016. LNCS, vol. 9912, pp. 628–644. Springer, Cham (2016). https://doi.org/10.1007/978-3-319-46484-8_38
9. Dai, Y., Li, H., He, M.: A simple prior-free method for non-rigid structure-from-motion factorization. Int. J. Comput. Vis. **107**(2), 101–122 (2014)
10. Gadelha, M., Maji, S., Wang, R.: 3D shape induction from 2D views of multiple objects. arXiv preprint arXiv:1612.05872 (2016)
11. Garg, R., Roussos, A., Agapito, L.: Dense variational reconstruction of non-rigid surfaces from monocular video. In: Proceedings of the IEEE Conference on Computer Vision and Pattern Recognition, pp. 1272–1279 (2013)
12. Gotardo, P.F., Martinez, A.M.: Non-rigid structure from motion with complementary rank-3 spaces. In: 2011 IEEE Conference on Computer Vision and Pattern Recognition (CVPR), pp. 3065–3072. IEEE (2011)
13. He, K., Zhang, X., Ren, S., Sun, J.: Deep residual learning for image recognition. In: Proceedings of the IEEE Conference on Computer Vision and Pattern Recognition, pp. 770–778 (2016)
14. Ioffe, S., Szegedy, C.: Batch normalization: accelerating deep network training by reducing internal covariate shift. In: Proceedings of the 32nd International Conference on International Conference on Machine Learning-Volume 37, pp. 448–456. JMLR.org (2015)
15. Ionescu, C., Papava, D., Olaru, V., Sminchisescu, C.: Human3.6M: large scale datasets and predictive methods for 3D human sensing in natural environments. IEEE Trans. Pattern Anal. Mach. Intell. **36**(7), 1325–1339 (2014)
16. Kanazawa, A., Jacobs, D.W., Chandraker, M.: WarpNet: weakly supervised matching for single-view reconstruction. In: The IEEE Conference on Computer Vision and Pattern Recognition (CVPR), June 2016

17. Kanazawa, A., Tulsiani, S., Efros, A.A., Malik, J.: Learning category-specific mesh reconstruction from image collections. In: Ferrari, V., Hebert, M., Sminchisescu, C., Weiss, Y. (eds.) ECCV 2018. LNCS, vol. 11219, pp. 386–402. Springer, Cham (2018). https://doi.org/10.1007/978-3-030-01267-0_23

18. Kong, C., Lucey, S.: Prior-less compressible structure from motion. In: Proceedings of the IEEE Conference on Computer Vision and Pattern Recognition, pp. 4123–4131 (2016)

19. Kong, C., Lucey, S.: Deep non-rigid structure from motion. In: Proceedings of the IEEE International Conference on Computer Vision, pp. 1558–1567 (2019)

20. Lee, M., Cho, J., Oh, S.: Procrustean normal distribution for non-rigid structure from motion. IEEE Trans. Pattern Anal. Mach. Intell. **39**(7), 1388–1400 (2017). https://doi.org/10.1109/TPAMI.2016.2596720

21. Lee, M., Cho, J., Choi, C.H., Oh, S.: Procrustean normal distribution for non-rigid structure from motion. In: 2013 IEEE Conference on Computer Vision and Pattern Recognition (CVPR), pp. 1280–1287. IEEE (2013)

22. Lee, M., Choi, C.H., Oh, S.: A procrustean Markov process for non-rigid structure recovery. In: Proceedings of the IEEE Conference on Computer Vision and Pattern Recognition, pp. 1550–1557 (2014)

23. Loper, M., Mahmood, N., Romero, J., Pons-Moll, G., Black, M.J.: SMPL: a skinned multi-person linear model. ACM Trans. Graph. **34**(6), 248:1–248:16 (2015). (Proc. SIGGRAPH Asia)

24. Martinez, J., Hossain, R., Romero, J., Little, J.J.: A simple yet effective baseline for 3D human pose estimation. In: Proceedings of the IEEE International Conference on Computer Vision, pp. 2640–2649 (2017)

25. Mehta, D., et al.: VNect: real-time 3D human pose estimation with a single RGB camera. ACM Trans. Graph. (TOG) **36**(4), 44 (2017)

26. Newell, A., Yang, K., Deng, J.: Stacked hourglass networks for human pose estimation. In: Leibe, B., Matas, J., Sebe, N., Welling, M. (eds.) ECCV 2016. LNCS, vol. 9912, pp. 483–499. Springer, Cham (2016). https://doi.org/10.1007/978-3-319-46484-8_29

27. Novotny, D., Ravi, N., Graham, B., Neverova, N., Vedaldi, A.: C3DPO: canonical 3D pose networks for non-rigid structure from motion. In: Proceedings of the IEEE International Conference on Computer Vision, pp. 7688–7697 (2019)

28. Paladini, M., Del Bue, A., Stosic, M., Dodig, M., Xavier, J., Agapito, L.: Factorization for non-rigid and articulated structure using metric projections. In: IEEE Conference on Computer Vision and Pattern Recognition 2009, CVPR 2009, pp. 2898–2905. IEEE (2009)

29. Park, S., Lee, M., Kwak, N.: Procrustean regression: a flexible alignment-based framework for nonrigid structure estimation. IEEE Trans. Image Process. **27**(1), 249–264 (2018)

30. Pavlakos, G., Zhou, X., Derpanis, K.G., Daniilidis, K.: Coarse-to-fine volumetric prediction for single-image 3D human pose. In: Proceedings of the IEEE Conference on Computer Vision and Pattern Recognition, pp. 7025–7034 (2017)

31. Russakovsky, O., et al.: ImageNet large scale visual recognition challenge. Int. J. Comput. Vis. **115**(3), 211–252 (2015). https://doi.org/10.1007/s11263-015-0816-y

32. Shen, J., Zafeiriou, S., Chrysos, G.G., Kossaifi, J., Tzimiropoulos, G., Pantic, M.: The first facial landmark tracking in-the-wild challenge: benchmark and results. In: Proceedings of the IEEE International Conference on Computer Vision Workshops, pp. 50–58 (2015)

18   S. Park et al.

33. Tatarchenko, M., Dosovitskiy, A., Brox, T.: Multi-view 3D models from single images with a convolutional network. In: Leibe, B., Matas, J., Sebe, N., Welling, M. (eds.) ECCV 2016. LNCS, vol. 9911, pp. 322–337. Springer, Cham (2016). https://doi.org/10.1007/978-3-319-46478-7_20
34. Torresani, L., Hertzmann, A., Bregler, C.: Nonrigid structure-from-motion: estimating shape and motion with hierarchical priors. IEEE Trans. Pattern Anal. Mach. Intell. **30**(5), 878–892 (2008)
35. Tulsiani, S., Kar, A., Carreira, J., Malik, J.: Learning category-specific deformable 3D models for object reconstruction. IEEE Trans. Pattern Anal. Mach. Intell. **39**(4), 719–731 (2017)
36. Tulsiani, S., Zhou, T., Efros, A.A., Malik, J.: Multi-view supervision for single-view reconstruction via differentiable ray consistency. In: CVPR, vol. 1, p. 3 (2017)
37. Varol, G., et al.: Learning from synthetic humans. In: CVPR (2017)
38. Wang, C., Kong, C., Lucey, S.: Distill knowledge from NRSfM for weakly supervised 3D pose learning. In: Proceedings of the IEEE International Conference on Computer Vision, pp. 743–752 (2019)
39. Wu, J., Wang, Y., Xue, T., Sun, X., Freeman, B., Tenenbaum, J.: MarrNet: 3D shape reconstruction via 2.5 D sketches. In: Advances in Neural Information Processing Systems, pp. 540–550 (2017)
40. Wu, J., et al.: Single image 3D interpreter network. In: Leibe, B., Matas, J., Sebe, N., Welling, M. (eds.) ECCV 2016. LNCS, vol. 9910, pp. 365–382. Springer, Cham (2016). https://doi.org/10.1007/978-3-319-46466-4_22
41. Yan, X., Yang, J., Yumer, E., Guo, Y., Lee, H.: Perspective transformer nets: Learning single-view 3D object reconstruction without 3D supervision. In: Advances in Neural Information Processing Systems, pp. 1696–1704 (2016)
42. Zhang, D., Han, J., Yang, Y., Huang, D.: Learning category-specific 3D shape models from weakly labeled 2D images. In: Proceedings of the CVPR, pp. 4573–4581 (2017)
43. Zhou, X., Leonardos, S., Hu, X., Daniilidis, K.: 3D shape estimation from 2D landmarks: a convex relaxation approach. In: Proceedings of the IEEE Conference on Computer Vision and Pattern Recognition, pp. 4447–4455 (2015)
44. Zhou, X., Zhu, M., Leonardos, S., Derpanis, K.G., Daniilidis, K.: Sparseness meets deepness: 3D human pose estimation from monocular video. In: The IEEE Conference on Computer Vision and Pattern Recognition (CVPR), June 2016

Learning to Learn Parameterized Classification Networks for Scalable Input Images

Duo Li[1,2], Anbang Yao[2(✉)], and Qifeng Chen[1(✉)]

[1] The Hong Kong University of Science and Technology, Kowloon, Hong Kong
duo.li@connect.ust.hk, cqf@ust.hk
[2] Intel Labs, Beijing, China
anbang.yao@intel.com

Abstract. Convolutional Neural Networks (CNNs) do not have a predictable recognition behavior with respect to the input resolution change. This prevents the feasibility of deployment on different input image resolutions for a specific model. To achieve efficient and flexible image classification at runtime, we employ meta learners to generate convolutional weights of main networks for various input scales and maintain privatized Batch Normalization layers per scale. For improved training performance, we further utilize knowledge distillation on the fly over model predictions based on different input resolutions. The learned meta network could dynamically parameterize main networks to act on input images of arbitrary size with consistently better accuracy compared to individually trained models. Extensive experiments on the ImageNet demonstrate that our method achieves an improved accuracy-efficiency trade-off during the adaptive inference process. By switching executable input resolutions, our method could satisfy the requirement of fast adaption in different resource-constrained environments. Code and models are available at https://github.com/d-li14/SAN.

Keywords: Efficient neural networks · Visual classification · Scale deviation · Meta learning · Knowledge Distillation

1 Introduction

Although CNNs have demonstrated their dominant power in a wide array of computer vision tasks, their accuracy does not scale up and down with respect to the corresponding input resolution change. Typically, modern CNNs are constructed by stacking convolutional modules in the body, a Global Average Pooling (GAP)

D. Li–Intern at Intel Labs China.

Electronic supplementary material The online version of this chapter (https://doi.org/10.1007/978-3-030-58526-6_2) contains supplementary material, which is available to authorized users.

© Springer Nature Switzerland AG 2020
A. Vedaldi et al. (Eds.): ECCV 2020, LNCS 12374, pp. 19–35, 2020.
https://doi.org/10.1007/978-3-030-58526-6_2

layer and a Fully-Connected (FC) layer in the head. When input images with different sizes are fed to a CNN model, the convolutional feature maps also vary in their size accordingly, but the subsequent GAP operation could reduce all the incoming features into a tensor with 1×1 spatial size and equal amount of channels. Thanks to the GAP layer, even trained on specific-sized input images, modern CNNs are also amenable to processing images of other sizes during the inference phase. However, the primary concern lies in that their performance is vulnerable to *scale deviation*[1], exhibiting severe deterioration when evaluating images of varying sizes at the inference time, as illustrated in Fig. 1. Therefore, as done in a series of efficient network designs [12,13,16,20,27,36], in order to adapt to real-time computational demand from the aspect of input resolution, it is necessary to train a spectrum of models from scratch using input images of different resolutions. These pre-trained models are then put into a storage pool and individually reserved for future usage. Client-side devices have to retrieve pertinent models based on requirements and available resources at runtime, which will largely impede the flexibility of deployment due to inevitable downloading and offloading overheads. To flexibly handle the real-time demand on various resource-constrained application platforms, a question arises: under the premise of not sacrificing or even improving accuracy, is it possible to learn a controller to dynamically parameterize different visual classification models having a shared CNN architecture conditioned on the input resolutions at runtime?

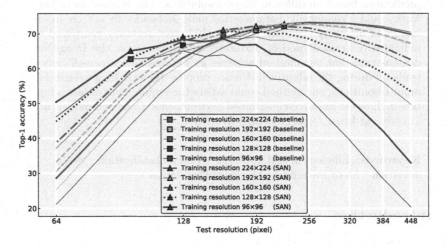

Fig. 1. Validation accuracy envelope of our proposed SAN with MobileNetV2 on ImageNet. Curves with the same color/style represent the results of models trained with the same input resolutions. The x-axis is plotted in the logarithmic scale. ▲ and ■ indicate the spots when test resolution meets the training one.

To echo the question above, a scale-adaptive behavior is anticipated for the controller to acquire. That implies, given a CNN structure, to instantiate

[1] The concept of *scale deviation* will be discussed detailedly in Sect. 2.

different main networks for each certain image resolution, the controller should have prior knowledge about switching between scaled data distributions and tactfully bridging the scaling coefficients with network parameters. We propose that appropriate data scheduling and network layer manipulations could lend this attribute to the controller. Specifically, we synthesize image patches with a set of training resolutions and employ meta networks as the controller to integrate diverse knowledge from these varying resolutions. The meta learners redistribute the gathered scale knowledge by generating convolutional kernels for multiple main networks conditioned on their assigned input resolutions respectively. Due to the tight relationship between Batch Normalization (BN) [14] layers and scaled activation distributions, all parameters and statistics of BN layers in each main network are maintained in a privatized manner. The main networks are collaboratively optimized following a mixed-scale training paradigm. The meta networks are optimized via collecting gradients passed through different main networks, such that information from multi-scale training flows towards the controller. Furthermore, aiming at leveraging the knowledge extracted with large resolutions to advance the performance on smaller ones, a scale distillation technique is utilized on the fly via taking the probabilistic prediction based on large resolutions as smoothed supervision signals. These aforementioned ingredients are coherently aggregated, leading to our proposed Scale Adaptive Network (SAN) framework, which is scalable by design and also generally applicable to prevailing CNN architectures.

During inference, given an input image, the learned controller could parameterize the visual classification model according to its resolution, showing consistently improved accuracy compared with the model individually trained on the corresponding resolution, as demonstrated in Fig. 1. Therefore, client devices almost merely need to reserve a single meta network and the computation graph of the backbone architecture whose parameters could be dynamically generated based on the evaluation resolution. Depending on the ability to flexibly switch target resolutions, the inference process is controllable to meet the on-device latency requirements and the accuracy expectation. Furthermore, the generated models could adapt smoothly to a wide range of input image sizes, even under the circumstances with severe problems of scale deviation. Benefiting from in-place parameterization and performance resilience on a spectrum of evaluated resolutions for each individual model, the one-model-fits-all style addresses the major obstacle of application to various scenarios.

Our main contributions can be summarized as follows:

❑ We employ meta networks to decide the main network parameters conditioned on its input resolution at runtime. Little research attention has been paid to this kind of meta-level information before us. We also extend the scope of knowledge distillation, based on the same image instance with different resolutions, which is also a rarely explored data-driven application scenario.

❑ We develop a new perspective of efficient network design by combining the aforementioned two components to permit adaptive inference on universally scaled input images, make a step forward in pushing the boundary of the resolution-accuracy spectrum and facilitating flexible deployment of visual classification models among switchable input resolutions.

2 Related Work

We briefly summarize related methodologies in previous literature and analyze their relations and differences with our approach as follows.

Scale Deviation. FixRes [30] sheds light on the distribution shift between the train and test data, and quantitatively analyzes its effect on apparent object sizes and activation statistics, which arises from inconsistent data pre-processing protocols during training and testing time. The discrepancy between train and test resolution is defined as *scale deviation* in this work. We would like to clarify that the issue of scale deviation exists across universally scaled images in the visual classification task, which shows a clear distinction compared to another popular phenomenon named *scale variance*. The problem of *scale variance* is more commonly identified among instances of different sizes within a single image, especially in the images of the MS COCO [19] benchmark for object detection. Typically hierarchical or parallel multi-scale feature fusion approaches [6,17,18,29,38,39] are utilized to address this problem, which has a loose connection with our research focus nevertheless. To handle the initial problem of *scale deviation*, FixRes proposes a simple yet effective strategy that prefers increased test crop size and fine-tunes the pre-trained model on the training data with the test resolution as a post-facto compensation. Notably, FixRes lays emphasis on calibrating BN statistics over the training data modulated by the test-size distribution, which is of vital importance to remedy the activation distribution shift. In comparison, we could use a proxy or data-free inference method to avoid the calibration of BN statistics, thus no post-processing steps are involved after end-to-end network training. We further provide empirical comparison in Sect. 4. Specializing BN layers for network adaption between a few different tasks has been adopted in domain adaption [25], transfer learning [21], adversarial training [31] and optimization of sub-models in a super net [33,34]. Inspired by them, we also overcome the shortcoming of statistic discrepancy through privatizing statistics and trainable parameters in the BN layers of each main network.

Meta Learning. Meta learning, or learning to learn, has come into prominence progressively in the field of artificial intelligence, which is supposed to be the learning mechanism in a more human-like manner. The target of meta learning is to advance the learning procedure at two coupled levels, enabling a *meta learning* algorithm to adapt to unseen tasks efficiently via utilizing another *base learning* algorithm to extract transferable prior knowledge within a set of auxiliary tasks. The hypernetwork [9] is proposed to efficiently generate the weights of the large main network using a small auxiliary network, in a relaxed form of weight-sharing across layers. Bertinetto *et al.* [4] concentrate on the one-shot learning problem and use a *learnet* to map a single training exemplar to weights of the main network in one go. Andrychowicz *et al.* [1] replace the standard optimization algorithm with an LSTM optimizer to generate an update for the main network in a self-adaptive manner. Meta-LSTM [24] further develops this idea by revealing the resemblance between cell states in an LSTM and network weights in a CNN model with respect to their update process. SMASH [5] applies the HyperNet to Neural Architecture Search (NAS) by transforming a binary encoding of the candidate network architecture to its corresponding weights. In our framework, meta networks, or to say hypernetworks, are responsible for generating weights of convolutional layers in the main network according to scale-related meta knowledge of the input data.

Knowledge Distillation. Knowledge Distillation (KD) is based on a teacher-student knowledge transfer framework in which the optimization of a lower-capacity student neural network is guided by imitating the soft targets [11,37] or intermediate representation [26,32,35] from a large and powerful pre-trained teacher model. Inspired by curriculum learning [3], RCO [15] promotes the student network to mimic the entire route sequence that the teacher network passed by. RKD [22] and CCKD [23] both exploit structural relation knowledge hidden in the embedding space to achieve correlation congruence. Several recent works also demonstrate the effectiveness of KD in improving the teacher model itself by self-distillation [2,8]. We introduce the scale distillation into our framework to transfer the scale and structural information of the same object instances from large input images to smaller ones at each step throughout the whole training procedure. The knowledge distillation process emerges among different main networks with the same network structure that travel along the same route sequence of optimization, without the assistance of external teacher models.

3 Approach

A schematic overview of our method is presented in Fig. 2. The key innovations lie in the employment of meta networks in the body and scale distillation at the end. In this section, we elaborate on the insights and formulations of them.

3.1 Network Architecture

Motivated by the target of achieving the scale-adaptive peculiarity of CNNs, we speculate that scale-invariant knowledge should be exploited regarding each input image in its heterogeneous modes of scales. Although the same object instances pose significant scale variation in these modes, they could share common structural relations and geometric characteristics. A meta learner built upon the top of CNNs is expected to extract and analyze this complementary knowledge across scales. Then it injects this prior knowledge into underlying CNNs

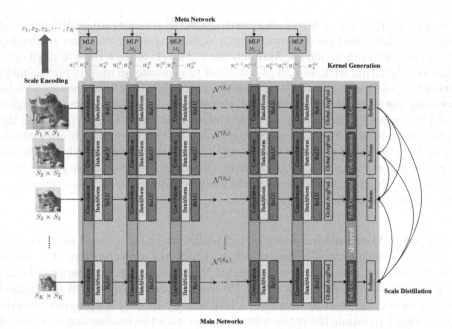

Fig. 2. Schematic illustration of our proposed Scale Adaptive Network framework. The input image scales $S_1, S_2, \cdots, S_K$ are linearly transformed into a set of encoding scalars $\varepsilon_1, \varepsilon_2, \cdots, \varepsilon_K$, which are fed into the MLP meta network $\mathcal{M}_l, l = 1, 2, \cdots, L$, generating weights $\{W_1^{(l)}, W_2^{(l)}, \cdots, W_K^{(l)}\}$ of the lth convolutional layer for each main network $\mathcal{N}^{(S_i)}$ associated with one certain input scale $S_i, i = 1, 2, \cdots, K$. BN layers are privatized and FC layers are shared among different main networks. The Scale Distillation process is performed in a top-down fashion. Best viewed in electronic version.

through parameterization, making them quickly adapt to a wide range of new tasks of visual classification which may include unseen scale transformations.

In this spirit, we denote the model optimized on training images with the resolution $S \times S$ as $\mathcal{N}^{(S)}$ and consider a group of main networks optimized on a set of input resolutions $\mathbb{S} = \{S_1, S_2, \cdots, S_k\}$ respectively. We also construct a cohort of meta networks $\mathcal{M} = \{\mathcal{M}_1, \mathcal{M}_2, \cdots, \mathcal{M}_L\}$ to generate convolutional kernels conditioned on the scale S of input images for an L-layer convolutional neural network $\mathcal{N}^{(S)}$. As illustrated in Fig. 2, bidirectional information flow is established between the shared meta network and each individual main network by the means of **Scale Encoding** and **Kernel Generation** respectively. All meta networks are instantiated by the same network structure of a Multi-Layer Perceptron (MLP).

Scale Encoding. The input to each MLP is an encoding scalar that contains information about the relative scale of training examples. Since the downsampling rate of prevalent CNNs is 1/32, we heuristically apply a normalization policy to linearly transform the input scale S to an encoding $\varepsilon = 0.1 \times S/32$.

Kernel Generation. The output dimension of an MLP equals the dimension of its corresponding convolutional kernel but in the flattened form. For example, a 1-in, $(C_{out}C_{in}K^2)$-out MLP is built to adaptively generate C_{out} groups of convolutional kernels, each group containing C_{in} kernels with the same size of $K \times K$. Compared with hypernetworks [9] that map each learnable embedding vector to the weights of its corresponding layer with one auxiliary network, our mechanism assigns each meta learner to generate weights for the corresponding layer with one common input encoding scalar regarding a specific main network.

With these well-defined meta learners, parameters of convolutional layers in the main network $\mathcal{N}^{(S)}$ can be generated and associated with a specific input scale S. As emphasized by previous works [30], BN layers should be delicately handled for the sake of inconsistency between data distributions of varying scales. Note that parameters in the BN layers usually occupy a negligible portion of the parameters within the whole model, e.g., less than 1% of the total amount in most cases, we opt for a straightforward yet effective strategy by maintaining individual BN layers, denoted as $\text{BN}^{(S)}$, for each scale-specified main network $\mathcal{N}^{(S)}$. Injecting these conditional normalization layers can lend more inherent adaptability and scalability to intermediate feature representations to accommodate scale discriminative knowledge. By contrast, with regard to the last FC layer that occupies a considerable amount of parameters, a shared one is designated to fit any potential main networks.

To take full advantage of various sources of scale information, we propose a mixed-scale optimization scheme accordingly. With the label-preserved transformation, each image in the training set $\mathcal{D} = \{(\boldsymbol{x}_1, y_1), (\boldsymbol{x}_2, y_2), \cdots, (\boldsymbol{x}_N, y_N)\}$ is resized to a series of scales, e.g., $S_1, S_2, \cdots, S_k$, where $y_j, j = 1, 2, \cdots, N$ is the ground truth category label of the sampled image x_j. For a certain scale S_i, we encode it as $\varepsilon^{(S_i)}$ and map the encoding scalar to a fully parameterized main network $\mathcal{N}^{(S_i)}$ through the meta learners. Transformed samples with the same size S_i are assembled to form a resized version of the original training set denoted as $\mathcal{D}^{(S_i)}$ and fed into the corresponding main network $\mathcal{N}^{(S_i)}$. For each pair of $\mathcal{D}^{(S_i)}$ and $\mathcal{N}^{(S_i)}$, it follows the standard optimization procedure of convolutional neural networks via minimizing the cross-entropy loss. Since our objective is to optimize the overall accuracy under different settings of scales, no scale is privileged and the total classification loss is an un-weighted sum of the individual losses, represented as

$$\mathcal{L}_{CE} = \sum_{i=1}^{k} \sum_{j=1}^{N} \mathcal{L}_{CE}(\boldsymbol{\theta}^{(S_i)}; \boldsymbol{x}_j^{(S_i)}, y_j), \tag{1}$$

where k is the number of resize transformations and $\boldsymbol{\theta}^{(S_i)} = \{\mathcal{M}_l(\varepsilon^{(S_i)}) \mid l = 1, 2, \cdots, L\} \cup \{\text{BN}^{(S_i)}, \text{FC}\}$ are weights of the network $\mathcal{N}^{(S_i)}$ where convolutional layers are directly generated by meta learners. The parameters in meta networks $\mathcal{M}$ are optimized simultaneously following the chain rule as the aforementioned weight generation operations are completely differentiable.

The number of hidden layers and units in the MLP could be tuned. In our main experiments, the meta network is chosen as a single layer MLP for the purpose of effectiveness and efficiency (validated by ablation experiments later). It could be represented as an FC layer with the weight $\mathbf{W}_l$ and bias $\mathbf{b}_l$

$$\mathcal{M}_l(\varepsilon^{(S_i)}) = \varepsilon^{(S_i)}\mathbf{W}_l + \mathbf{b}_l. \tag{2}$$

Due to the existence of the bias term, the output convolutional kernels are *not* simply distinct up to a scaling factor $\varepsilon^{(S_i)}$ across different main networks. To be cautious, we also examine the value of $\mathbf{W}_l$ and $\mathbf{b}_l$ regarding each layer. They are of the same size and the same order of magnitude in almost all cases, indicating that the weights and biases have an equivalently important influence on the learning dynamics. The ratio of $\mathbf{W}_l$ to $\mathbf{b}_l$ in each layer is reported in the supplementary materials.

3.2 Scale Distillation

High-resolution representations [28] can contain finer local feature descriptions and more discernable semantic meaning than the lower-resolution ones, hence it is appropriate to utilize the probabilistic prediction over larger-scale inputs to offer auxiliary supervision signals for smaller ones, which will be referred to as *Scale Distillation* in this context. Specifically, a Kullback-Leibler divergence $D_{\text{KL}}(\cdot\|\cdot)$ is calculated between each pair of output probability distributions among all main networks, which leads to an additional representation mimicking loss shaped in a top-down manner as follows

$$\mathcal{L}_{\text{SD}} = \sum_{i=1}^{k} \sum_{\substack{j=1 \\ S_j < S_i}}^{k} D_{\text{KL}}(\boldsymbol{p}^{(S_i)}\|\boldsymbol{p}^{(S_j)}), \tag{3}$$

where $\boldsymbol{p}^{(S_i)}$ denotes the probabilistic distribution prediction with respect to object categories outputted by the main network $\mathcal{N}^{(S_i)}$.

Our mixed-scale training mechanism naturally supports scale distillation as an in-place operation. During each training step, we take the predicted labels of one model and fix them as the soft training labels for other models processing smaller input samples, which can be implemented on the fly without introducing further memory overheads in practice. Compared with conventional KD methodology [11], a main network in our framework may be both the teacher model and the student model, depending on its matched counterpart model. Furthermore, the cohort of main networks are of the same model size and optimized in a single-shot rather than two-stage manner.

Then, the overall optimization objective of our framework is to minimize the combined loss function

$$\mathcal{L} = \alpha\mathcal{L}_{\text{CE}} + \beta\mathcal{L}_{\text{SD}}, \tag{4}$$

where α and β are positive weight coefficients to balance the cross-entropy loss and the scale distillation loss. In our main experiments, we do not make much investment in tuning these hyperparameters and find that simply setting $\alpha =$

$\beta = 1$ leads to satisfactory performance, which demonstrates the robustness of our proposed optimization scheme in some sense.

Intuitively condensing networks for different purposes into a shared framework tends to bring about performance degradation compared to individually trained ones, since they might exhibit inconsistent learning dynamics. However, we surmise that our performance improvements could originate from a relaxed form of knowledge transfer across different scales. According to Eq. 2, weights and biases of the meta networks respectively enforce convolution parameter scaling and sharing across different main networks. The generated weights for one scale also depend on the information from any other different scales due to the joint training. By feat of the shared meta networks and a collaborative training regime, the knowledge interaction process between models may be interpreted as an *implicit distillation*. In addition to the implicit information sharing mechanism above, we further develop an *explicit distillation* technique to aggressively advance this knowledge transfer process, presented as Scale Distillation.

3.3 Inference

For inference, let the selected training resolution range be $\mathbb{S} = \{S_1, S_2, \cdots, S_k\}$ and the test resolution be T, we first search the nearest resolution $S(T) \in \mathbb{S}$ for T, then feed the scale encoding $\varepsilon^{(T)}$ to the pre-trained meta network to parameterize convolutional layers for the main network and the BN layers reserved for $S(T)$ during training could be applied directly. Finally the *ideal inference* is realized by sending the test image to the network parameterized by

$$\theta_{\text{ideal}}^{(T)} = \{\mathcal{M}_l(\varepsilon^{(T)})|l = 1, 2, \cdots, L\} \cup \{\text{BN}^{(S(T))}, \text{FC}\}. \tag{5}$$

If $S(T)$ is *not* close enough to T regarding its value, the trainable parameters of BN (scale and shift parameter) could still be applied directly but calibration of BN statistics (running mean and variance) is necessary for retaining a decent performance (will be shown later in the ablation studies). Intriguingly, we find that simply substituting the scale encoding $\varepsilon^{(T)}$ by $\varepsilon^{(S(T))}$ to match its corresponding BN layers is ready to make compensation. It says that the runtime network could also be dynamically parameterized by

$$\theta_{\text{proxy}}^{(T)} = \{\mathcal{M}_l(\varepsilon^{(S(T))})|l = 1, 2, \cdots, L\} \cup \{\text{BN}^{(S(T))}, \text{FC}\}, \tag{6}$$

to achieve *approximate* performance as that above (recognition accuracy by using $\theta_{\text{ideal}}^{(T)}$ and $\theta_{\text{proxy}}^{(T)}$ for inference is comprehensively compared in the ablation experiments). Since using $\theta_{\text{proxy}}^{(T)}$ for inference would be relatively convenient once the pre-trained main network weights $\mathcal{M}(\varepsilon^{(S(T))})$ are already on hand, we opt for this *proxy inference* method as an alternative for performance benchmark throughout Sect. 4, if no further specification. In practice, clients could choose either option to achieve similar performance, depending on whether the desired convolutional weights are handily accessible or even on hand.

Summarily, the **ideal inference** in Eq. 5 uses test-specified encoding and calibration for BN statistics (if necessary) while the **proxy inference** in Eq. 6 uses training-specified encoding and no calibration. The concrete algorithms of optimization and inference are provided in the supplementary materials.

4 Experiments

We present extensive experimental results on ImageNet using various prevailing CNN architectures and conduct controlled experiments for introspection.

4.1 Main Results

Our method is evaluated on the large-scale ImageNet [7] dataset, which is a very challenging image recognition benchmark including over 1.2 million training images and 50,000 validation images belonging to 1,000 object categories. We follow the standard practice for data augmentation [10], utilizing random resizing and cropping operations together with horizontal flipping to generate image patches with the desired resolutions. During evaluation, we crop the center region from each transformed image of which the shorter side is resized to satisfy the crop ratio of 0.875. We report top-1 validation error in all the following tables.

Table 1. Comparison of ResNet-18, MobileNetV2 and ResNet-50 (*from top to bottom*) baseline models (*left panel*) and SAN (*right panel*) on a spectrum of test resolutions.

Left panel (baseline):

Train \ Test	256	224	208	192	176	160	144	128	112	96	64
224	71.90	70.97	69.14	68.72	66.20	64.68	60.36	56.85	50.17	42.55	20.40
192	71.71	71.25	69.64	69.75	67.70	67.04	63.23	60.47	54.21	47.73	25.10
160	70.72	70.70	69.50	70.14	68.47	68.48	65.61	63.80	58.54	53.78	31.03
128	67.41	68.51	67.83	69.06	67.81	68.56	66.51	66.36	61.88	58.29	38.58
96	54.68	58.87	58.46	62.39	61.62	64.84	63.42	65.51	62.39	62.56	47.48
Train \ Test	**256**	**224**	**208**	**192**	**176**	**160**	**144**	**128**	**112**	**96**	**64**
224	73.04	72.19	71.13	70.09	68.48	66.10	63.13	58.63	52.50	43.91	21.23
192	72.67	72.26	71.47	71.08	69.68	68.18	65.84	61.61	56.80	48.81	25.37
160	70.83	71.39	70.17	71.08	69.31	69.50	66.63	65.05	59.60	54.08	30.56
128	66.70	68.18	68.19	69.14	68.65	68.98	67.78	66.74	63.73	58.69	37.14
96	51.48	56.73	57.06	60.85	61.13	63.99	63.55	65.13	63.40	62.70	46.15
Train \ Test	**256**	**224**	**208**	**192**	**176**	**160**	**144**	**128**	**112**	**96**	**64**
224	77.95	77.15	76.43	75.47	74.29	72.44	70.32	66.85	62.38	54.84	32.10
192	77.56	77.26	76.81	76.41	75.27	74.39	72.32	69.77	65.70	59.22	37.23
160	76.69	76.93	76.60	76.61	75.81	75.31	73.91	72.12	69.11	63.87	43.74
128	74.14	75.13	75.16	75.66	75.46	75.53	74.63	73.53	71.58	67.67	51.40
96	68.81	70.97	71.29	72.68	72.66	73.64	73.43	73.53	72.35	70.61	59.09

Right panel (SAN):

Train \ Test	256	224	208	192	176	160	144	128	112	96	64
224	72.96	72.65	71.89	71.89	70.39	69.62	67.15	65.36	59.96	55.57	34.98
192	72.60	72.79	72.10	72.32	71.09	70.77	68.44	67.25	62.61	58.77	38.76
160	71.51	72.15	71.61	72.12	71.04	71.31	69.36	68.60	64.69	61.53	43.02
128	69.04	70.30	69.78	70.90	70.19	71.01	69.35	69.28	65.96	63.87	47.33
96	62.96	65.52	65.37	67.65	66.89	68.89	67.60	68.73	65.78	65.30	51.32
Train \ Test	**256**	**224**	**208**	**192**	**176**	**160**	**144**	**128**	**112**	**96**	**64**
224	73.26	72.80	72.24	71.47	70.35	68.64	66.50	63.08	58.43	51.01	28.64
192	72.91	72.86	72.42	72.22	71.33	70.26	68.43	65.83	61.81	55.44	33.24
160	71.68	72.16	71.79	72.31	71.63	71.16	69.67	67.96	64.61	59.62	38.75
128	68.61	70.13	69.94	71.16	70.51	70.94	69.80	69.14	66.55	63.05	44.84
96	60.55	64.11	64.21	66.91	66.80	68.58	67.82	68.41	66.59	65.07	50.41
Train \ Test	**256**	**224**	**208**	**192**	**176**	**160**	**144**	**128**	**112**	**96**	**64**
224	78.93	78.57	78.32	77.81	77.02	76.11	74.67	72.47	69.44	64.30	44.91
192	78.63	78.67	78.43	78.44	77.64	77.10	75.86	74.00	71.29	67.05	48.91
160	77.66	78.07	77.99	78.26	77.73	77.60	76.60	75.35	73.15	69.49	53.62
128	75.18	76.22	76.39	77.17	77.01	77.38	76.50	76.04	74.24	71.56	57.84
96	69.91	72.24	72.69	74.18	74.33	75.39	75.12	75.30	74.28	72.83	62.14

Several different choices of main networks are explored to demonstrate the effectiveness and scalability of our approach. Specifically, we select three network architectures including ResNet [10] with 18/50 layers and MobileNetV2 [27] in view of their strong track record. It is noted that we consider both large-scale networks and a very lightweight one, which also feature regular and inverted residual blocks respectively. Furthermore, we prove that our meta learners could smoothly learn to generate the kernels of both standard convolutions and depthwise separable convolutions. The ResNet family is trained using the default SGD optimizer with the momentum of 0.9 and the weight decay of 1e-4 for 120 epochs.

Table 2. Comparison of ResNet-18-based SAN, evaluated on interpolated resolutions with different inference configurations, including (i) ideal inference: before (*left*) and after (*middle*) BN calibration; (ii) proxy inference: no need for BN calibration (*right*).

Train \ Test	208	176	144	112	Train \ Test	208	176	144	112	Train \ Test	208	176	144	112
224	71.71	67.44	57.47	39.46	224	72.03	70.32	66.90	59.18	224	71.89	70.39	67.15	59.96
192	71.51	70.72	65.22	52.15	192	72.13	71.13	68.48	62.14	192	72.10	71.09	68.44	62.61
160	66.21	70.52	69.07	61.10	160	71.41	71.07	69.40	64.59	160	71.61	71.04	69.36	64.69
128	49.45	64.49	68.78	65.57	128	68.95	69.89	69.52	66.16	128	69.78	70.19	69.35	65.96
96	19.71	45.75	61.74	65.26	96	62.49	65.86	67.35	66.26	96	65.37	66.89	67.60	65.78

The learning rate initiates from 0.1 and decays to zero following a half cosine annealing schedule. The batch size is set to 256. The lightweight MobileNetV2 is trained using almost the same optimization hyperparameters and learning rate decay strategy but with a smaller initial learning rate of 0.05 and a smaller weight decay of 4e-5 to suppress underfitting. The optimization procedure lasts for 150 epochs for full convergence. All baselines and our SAN-based models are trained using the above scheme for fair comparisons. The training resolutions of SAN models are set to $\mathbb{S} = \{S_1, S_2, S_3, S_4, S_5\} = \{224, 192, 160, 128, 96\}$.

We take independently trained models as the baselines and evaluate them among a wide range of test resolutions on the ImageNet validation set. The baseline results of three networks are shown in the left panel of Table 1. We report corresponding results using our SAN in the right panel of Table 1, where the results in the jth row is evaluated with the scale encoding $\varepsilon^{(S_j)}$ and BN layers $\text{BN}^{(S_j)}$ without the calibration of statistics. Therefore, the shaded numeric value (in the jth row, s.t. $S(T) = S_j$) points out the inference result using the *proxy inference* method for the test resolution T in each column, as stated in Sect. 3.3. It is evident that our dynamically parameterized models achieve consistent accuracy improvement over the individually trained baselines. Such performance enhancement emerges not only on the training resolutions but also on the interpolated and extrapolated ones of the training range, demonstrating the universal applicability of our meta learner. Furthermore, it should be noticed that the generated classifier for one specific resolution also generalizes well on other resolutions (indicated by those numeric values without being shaded in each row) compared to the individually trained model, even obtaining over 10% compensation for those baseline models in several cases with severe scale deviation (around the corner of the table).

The recognition accuracy curves of our SAN with MobileNetV2 and corresponding baseline models are depicted in Fig. 1. It showcases a clear trend that the curves of SAN models envelop those of their baseline counterparts. Our SAN also guarantees much milder performance drop when input samples pose a resolution discrepancy between optimization and inference.

4.2 Ablation Studies

We conduct comprehensive ablation experiments to analyze the influence of different configurations and provide empirical validation for our design.

BN Calibration for Ideal Inference. The encoding scalars used for training are bound to their privatized BN layers. If these scalars are altered according to input resolutions at test time following the ideal inference method, evaluation performance might suffer from incompatibility, as demonstrated in the left panel of Table 2. For compensation, we apply the post-hoc calibration for BN statistics. Specifically, we recalculate the running mean and variance of BN layers over training samples of the test resolution with exact averages rather than moving averages. The evaluation performance after BN calibration is reported in the middle panel of Table 2, showing amelioration for those before calibration. To deserve to be mentioned, when the training resolution approaches the test one, the performance gap would be relatively minor, crediting to the smoothness of the linear meta modeling space and the BN parameter space. Anyway, the above dissection reiterates the critical value of Batch Normalization.

Proxy Inference. We empirically justify that using θ_{proxy} for inference is a credible alternative for model evaluation. For test resolutions included in the training resolution range, $i.e.$, $224, 192, 160, 128$ and 96, $S(T) = T$, thus $\theta_{\text{proxy}} = \theta_{\text{ideal}}$ and the inference process will be identical for the two proposed methods. Thus, we lay analytic emphasis on the test resolutions sandwiched between two training resolutions, $i.e.$, $208, 176, 144$ and 112, and report their performance in the right panel of Table 2. Since the scale encoding reverts to the ones during training in the proxy inference method, calibration is not needed here. We notice that using proxy inference for each test resolution (as shaded in the right panel of Table 2) leads to nearly the same results as those with the ideal inference method (as shaded in the middle panel of Table 2). These comparisons provide empirical support for our claim in Sect. 3.3.

Table 3. Accuracy of MobileNetV2-based SAN (w/o SD) (*left*) and MobileNetV2-based SAN (w/o SD) with shared Batch Normalization layers (*right*). Please refer to Table 1 (*middle-right*) for accuracy of intact MobileNetV2-based SAN.

Train \ Test	256	224	208	192	176	160	144	128	112	96	64
224	73.14	72.53	71.50	70.94	69.26	67.91	64.50	61.84	55.18	48.60	24.94
192	72.85	72.58	71.45	71.56	69.93	69.36	66.32	64.15	58.23	52.52	28.62
160	71.30	71.78	70.66	71.54	69.92	70.00	67.35	66.09	60.85	56.20	32.94
128	67.42	69.11	67.64	69.95	68.37	69.65	67.43	67.19	62.64	59.54	38.26
96	56.25	60.34	58.50	63.81	61.86	66.25	63.87	66.23	62.25	61.86	44.05

Train \ Test	256	224	208	192	176	160	144	128	112	96	64
224	2.82	2.95	2.88	3.02	2.93	3.04	2.90	2.77	2.59	2.29	1.13
192	27.27	29.07	29.03	29.91	29.30	29.24	28.09	25.70	23.24	18.64	8.45
160	66.41	66.89	66.46	66.52	65.62	65.02	63.08	60.61	56.812	50.66	29.64
128	49.17	50.86	50.26	51.43	50.10	50.70	47.92	46.73	41.52	37.71	19.82
96	0.76	0.90	0.85	1.03	0.94	1.23	1.13	1.28	1.07	1.38	1.07

Scale Distillation. For the purpose of ablating the influence of our proposed Scale Distillation, we further train a MobileNetV2-based SAN without this technique. Accuracy records of the original SAN are presented in Table 1 (middle-right) and the accuracy of SAN (w/o SD) is provided in Table 3 for comparison. We observe that the accuracy drop increases in the test scenarios with smaller input scales, which empirically validates the importance of transferring knowledge from high resolutions to lower ones by introducing the extra supervision

Table 4. Accuracy of parallel ResNet-18 models with privatized convolution for each resolution and scale distillation across resolutions. Please refer to Table 1 (*top-left* and *top-right*) for results of the baseline and SAN.

Train\Test	256	224	208	192	176	160	144	128	112	96	64
224		72.06	71.10	69.54	68.87	66.23	64.54	60.47	56.89	50.11	42.60	20.32
192		72.53	71.94	70.41	70.79	68.28	67.93	64.17	61.97	55 33	49.39	25.96
160		70.93	71.37	70.34	71.21	69.55	69.82	66.94	65.62	60.57	55.92	33.51
128		67.57	69.11	68.56	69.89	68.78	69.89	67.81	67.87	63.84	61.46	41.96
96		56.80	60.81	60.75	64.31	63.53	66.67	65.22	67.32	64.35	64.53	50.62

signal. As a side benefit, the performance of models over interpolated resolutions is ameliorated to a large extent after the application of Scale Distillation, which may be attributed to the improved interaction of multi-scale information among the same image instances with different resolutions.

Privatized BN. Sharing similar philosophy with [30] and [34], we privatize BN layers by design to eliminate incompatibility between various scaled distributions. For the purpose of ablation, we take the above MobileNetV2-based SAN (w/o SD) as an exemplar and substitute shared BN layers for privatized ones. The consequent accuracy is provided in Table 3, where the benefit of BN is skewed to resolutions in the middle of the training resolution range while the performance in each case still deteriorates compared to its counterpart with privatized BNs. We observe that the training process is stable but the validation accuracy is extremely low because during training the mean and variance of the current mini-batch with a specific resolution are applied while the validation process depends on the moving average statistics from all resolutions. The experimental results further prove the point that a unified BN is insufficient to strike a balance among a broad range of resolutions concurrently.

Parameterized Convolution. A quick question is that why not simply use privatized convolutional layers for each main network rather than employing an auxiliary meta learner? We take the ResNet18 model as an example and show the accuracy with privatized convolutional layers in Table 4 (scale distillation is still applicable in this context). In comparison, the performance are mostly superior to baseline but inferior to our proposed SAN. The meta learning algorithm is adept at integrating prior experience from multiple existing tasks and achieving fast adaption to unseen tasks. In accordance to this rationale, the comparison results also speak to the advantage of utilizing meta networks to collect multi-scale information and dynamically parameterize the convolutional layers for arbitrary image resolutions at runtime.

Architecture of Meta Learner. We explore different MLP structures of the meta network design. Besides the single FC layer adopted in our main exper-

Table 5. Absolute accuracy drop of ResNet-18-based SAN with non-linear meta learners using 8/16 (*left*/*right*) hidden units in the FC layer. Please refer to Table 1 (*top-right*) for results of the original SAN equipped with linear meta learners. Note that negative numbers indicate accuracy increment.

Train \ Test	256	224	208	192	176	160	144	128	112	96	64	Train \ Test	256	224	208	192	176	160	144	128	112	96	64
224	-0.22	-0.33	0.11	0.12	0.15	0.01	0.26	0.19	0.29	0.76	1.50	224	-0.48	-0.31	-0.24	0.01	-0.17	-0.11	0.21	0.67	0.55	1.35	2.20
192	-0.21	-0.07	0.19	0.25	0.27	0.20	0.36	0.56	0.92	1.30	2.27	192	-0.39	-0.13	-0.03	0.09	-0.13	0.08	0.34	1.03	1.04	1.80	2.74
160	-0.08	-0.03	0.18	0.02	0.19	0.14	0.27	0.29	0.62	0.88	2.16	160	-0.33	-0.18	-0.05	-0.11	-0.23	0.00	0.02	0.42	0.57	1.25	2.27
128	0.22	0.19	0.16	-0.02	0.43	0.09	0.22	-0.00	0.25	0.46	1.24	128	0.11	0.11	-0.11	-0.06	-0.12	0.08	-0.26	0.03	0.03	0.42	1.32
96	0.93	0.80	1.00	0.54	0.76	0.36	0.66	0.26	0.09	0.11	-0.07	96	1.26	1.04	0.87	0.74	0.27	0.38	0.02	0.23	-0.28	0.12	0.05

iments, we draw into the non-linear activation to construct a two-layer MLP, where consecutive linear FC layers are interleaved with a non-linear ReLU activation operation and the number of hidden units is set to 8 or 16 in order to avoid introducing heavy computational overheads. The ResNet-18 model is selected for this experiment in view of its relatively low memory consumption. The absolute accuracy drop after substituting the non-linear meta leaner with 8 hidden units for the default linear one is summarized in Table 5 (left). We could find that the final performance of these two designs is in close proximity to each other on the medium resolutions, which implies that though conceptually simple, a linear modeling space of meta networks is sufficient for giving rise to desired scale-adaptive characteristics and satisfactory performance. Nevertheless, the decline is not negligible on quite large and small resolutions, which might be attributed to over-fitting problems caused by expansion of the non-linear modeling space. It is probably difficult for the SAN (w/ non-linearity) to generalize well to these situations with large scale deviation. We also explore larger hidden-unit space via doubling its unit number and observe that the tendency of overfitting to marginal resolutions becomes more obvious, as shown in Table 5 (right).

Generalization to Large Resolutions. We shift the range of training resolutions to a larger magnitude, using ResNet-18 as the test case. As shown in Table 6, the accuracy improvement of SAN presents a similar trend within a range of large resolutions $\mathbb{S} = \{384, 320, 256, 224, 192\}$ compared to the smaller ones discussed in Sect. 4.1. The results empirically show the effectiveness and robustness of our method in handling images with different ranges of resolutions.

Comparison to FixRes. FixRes [30] is merely akin to us in the motivation of alleviating *scale deviation*, but it focuses on further pushing the performance upper boundary of large-scale models with resolution adaption. In stark contrast, we pay attention to enabling efficient and flexible inference conditioned on the input image resolution for evaluation at runtime. A performance comparison is applicable for these two methods using the ResNet-50 model trained on the image resolution of 224×224, as summarized in Table 7. We are able to beat FixRes when test resolutions are relatively small, which is especially beneficial for resource-limited environments. Furthermore, due to the limitation of available computational resources now, our training resolution range could be at most set to $\{224, 192, 160, 128, 96\}$. If we could enlarge this range, *e.g.*, including training images with larger resolutions such as 384, 448 and 480, our SAN may achieve better performance on these large scales.

Table 6. Comparison of ResNet-18 baseline models (*top*) and ResNet-18-based SAN with a range of larger training resolutions (*bottom*). The shaded numeric values have the same meaning as Table 1.

Train \ Test	384	320	256	224	208	192	176	160	144	128	112	96	64
384	73.20	71.23	67.33	63.83	60.65	58.62	54.71	50.58	45.09	38.86	32.05	23.96	8.89
320	73.42	72.56	70.03	67.44	64.92	63.23	59.47	56.07	51.17	45.89	38.64	30.27	11.74
256	72.75	72.68	71.71	70.24	68.00	67.27	64.05	62.12	57.20	53.27	45.45	38.00	17.00
224	71.87	72.36	71.90	70.97	69.14	68.72	66.20	64.68	60.36	56.85	50.17	42.55	20.40
192	69.95	71.07	71.71	71.25	69.64	69.75	67.69	67.04	63.23	60.47	54.21	47.73	25.10
Train \ Test	384	320	256	224	208	192	176	160	144	128	112	96	64
384	74.69	74.12	72.14	70.28	68.65	67.06	64.77	61.97	58.25	53.41	47.33	39.51	18.66
320	74.61	74.50	73.28	71.72	70.39	69.07	67.10	64.64	62.28	56.94	51.22	43.63	21.85
256	73.41	74.05	73.71	72.87	71.85	70.86	69.43	67.53	64.42	60.86	55.56	48.26	25.80
224	72.27	73.39	73.59	73.00	72.29	71.52	70.16	68.52	65.64	62.32	57.36	50.54	28.13
192	70.44	72.15	73.11	72.73	72.21	71.74	70.72	69.36	66.74	63.80	59.20	52.72	30.52

Table 7. Comparison of FixRes and our SAN using ResNet-50 with training image size of 224×224 on ImageNet. The results of FixRes are extracted from the original publication. The better result for each test resolution is highlighted in **bold**.

Method\Test resolution	64	128	224	288	384	448	480
FixRes [30]	41.7	67.7	77.1	**78.6**	**79.0**	**78.4**	**78.1**
SAN (ours)	**44.9**	**72.5**	**78.4**	**78.6**	77.0	75.3	74.3

More Ablation and Analysis. We provide a great many additional results in the supplementary materials, including analysis of complementary information across resolutions, the superiority of switching input resolutions over network widths, border and round-off effects, results of inference on more dense sampled resolutions, visualization of privatized BN layers, a data-free form of ideal inference, comparison to stronger baselines, and so on.

5 Conclusion

In this paper, we have proposed a Scale Adaptive Network (SAN) framework. For each main network encompassed into the meta learning algorithm, convolutional kernels are dynamically generated by meta networks and BN layers are specially treated. The meta learner could parameterize neural networks for visual classification conditioned on the input resolution at runtime, achieving considerably better performance compared with individually trained models. It is compatible with any CNN backbone architectures, providing an adaptive resolution-accuracy trade-off for fast adaption to environments with different real-time demands. The generality of the proposed framework makes it promising to be translated well to other application domains, such as object detection and semantic segmentation, which is expected to appear in the future work.

Acknowledgement. We sincerely thank Ming Lu and Aojun Zhou from Intel Labs China for their insightful discussions, warm encouragement and valuable feedback. We would like to thank SenseTime Group Limited for their support of computational resource.

References

1. Andrychowicz, M., et al.: Learning to learn by gradient descent by gradient descent. In: NIPS (2016)
2. Bagherinezhad, H., Horton, M., Rastegari, M., Farhadi, A.: Label refinery: improving imagenet classification through label progression. arXiv e-prints arXiv:1805.02641, May 2018
3. Bengio, Y., Louradour, J., Collobert, R., Weston, J.: Curriculum learning. In: ICML (2009)
4. Bertinetto, L., Henriques, J.F., Valmadre, J., Torr, P., Vedaldi, A.: Learning feed-forward one-shot learners. In: NIPS (2016)
5. Brock, A., Lim, T., Ritchie, J., Weston, N.: SMASH: one-shot model architecture search through hypernetworks. In: ICLR (2018)
6. Chen, L.-C., Zhu, Y., Papandreou, G., Schroff, F., Adam, H.: Encoder-decoder with atrous separable convolution for semantic image segmentation. In: Ferrari, V., Hebert, M., Sminchisescu, C., Weiss, Y. (eds.) ECCV 2018. LNCS, vol. 11211, pp. 833–851. Springer, Cham (2018). https://doi.org/10.1007/978-3-030-01234-2_49
7. Deng, J., Dong, W., Socher, R., Li, L.J., Li, K., Fei-Fei, L.: ImageNet: a large-scale hierarchical image database. In: CVPR (2009)
8. Furlanello, T., Lipton, Z., Tschannen, M., Itti, L., Anandkumar, A.: Born again neural networks. In: ICML (2018)
9. Ha, D., Dai, A.M., Le, Q.V.: Hypernetworks. In: ICLR (2017)
10. He, K., Zhang, X., Ren, S., Sun, J.: Deep residual learning for image recognition. In: CVPR (2016)
11. Hinton, G., Vinyals, O., Dean, J.: Distilling the knowledge in a neural network. arXiv e-prints arXiv:1503.02531, March 2015
12. Howard, A., et al.: Searching for MobileNetV3. In: ICCV (2019)
13. Howard, A.G., et al.: MobileNets: efficient convolutional neural networks for mobile vision applications. arXiv e-prints arXiv:1704.04861, April 2017
14. Ioffe, S., Szegedy, C.: Batch normalization: accelerating deep network training by reducing internal covariate shift. In: ICML (2015)
15. Jin, X., et al.: Knowledge distillation via route constrained optimization. In: ICCV (2019)
16. Li, D., Zhou, A., Yao, A.: HBONet: harmonious bottleneck on two orthogonal dimensions. In: ICCV (2019)

17. Li, Y., Chen, Y., Wang, N., Zhang, Z.: Scale-aware trident networks for object detection. In: ICCV (2019)
18. Lin, T.Y., Dollar, P., Girshick, R., He, K., Hariharan, B., Belongie, S.: Feature pyramid networks for object detection. In: CVPR (2017)
19. Lin, T.-Y., et al.: Microsoft COCO: common objects in context. In: Fleet, D., Pajdla, T., Schiele, B., Tuytelaars, T. (eds.) ECCV 2014. LNCS, vol. 8693, pp. 740–755. Springer, Cham (2014). https://doi.org/10.1007/978-3-319-10602-1_48
20. Ma, N., Zhang, X., Zheng, H.-T., Sun, J.: ShuffleNet V2: practical guidelines for efficient CNN architecture design. In: Ferrari, V., Hebert, M., Sminchisescu, C., Weiss, Y. (eds.) Computer Vision – ECCV 2018. LNCS, vol. 11218, pp. 122–138. Springer, Cham (2018). https://doi.org/10.1007/978-3-030-01264-9_8
21. Mudrakarta, P.K., Sandler, M., Zhmoginov, A., Howard, A.: K for the price of 1: parameter efficient multi-task and transfer learning. In: ICLR (2019)
22. Park, W., Kim, D., Lu, Y., Cho, M.: Relational knowledge distillation. In: CVPR (2019)
23. Peng, B., et al.: Correlation congruence for knowledge distillation. In: ICCV (2019)
24. Ravi, S., Larochelle, H.: Optimization as a model for few-shot learning. In: ICLR (2017)
25. Rebuffi, S.A., Bilen, H., Vedaldi, A.: Learning multiple visual domains with residual adapters. In: NIPS (2017)
26. Romero, A., Ballas, N., Kahou, S.E., Chassang, A., Gatta, C., Bengio, Y.: FitNets: hints for thin deep nets. In: ICLR (2015)
27. Sandler, M., Howard, A., Zhu, M., Zhmoginov, A., Chen, L.C.: MobileNetV2: inverted residuals and linear bottlenecks. In: CVPR (2018)
28. Sun, K., Xiao, B., Liu, D., Wang, J.: Deep high-resolution representation learning for human pose estimation. In: CVPR (2019)
29. Szegedy, C., Vanhoucke, V., Ioffe, S., Shlens, J., Wojna, Z.: Rethinking the inception architecture for computer vision. In: CVPR (2016)
30. Touvron, H., Vedaldi, A., Douze, M., Jégou, H.: Fixing the train-test resolution discrepancy. In: NeurIPS (2019)
31. Xie, C., Tan, M., Gong, B., Wang, J., Yuille, A.L., Le, Q.V.: Adversarial examples improve image recognition. In: CVPR (2020)
32. Yim, J., Joo, D., Bae, J., Kim, J.: A gift from knowledge distillation: fast optimization, network minimization and transfer learning. In: CVPR (2017)
33. Yu, J., Huang, T.S.: Universally slimmable networks and improved training techniques. In: ICCV (2019)
34. Yu, J., Yang, L., Xu, N., Yang, J., Huang, T.: Slimmable neural networks. In: ICLR (2019)
35. Zagoruyko, S., Komodakis, N.: Paying more attention to attention: improving the performance of convolutional neural networks via attention transfer. In: ICLR (2017)
36. Zhang, X., Zhou, X., Lin, M., Sun, J.: ShuffleNet: an extremely efficient convolutional neural network for mobile devices. In: CVPR (2018)
37. Zhang, Y., Xiang, T., Hospedales, T.M., Lu, H.: Deep mutual learning. In: CVPR (2018)
38. Zhao, H., Shi, J., Qi, X., Wang, X., Jia, J.: Pyramid scene parsing network. In: CVPR (2017)
39. Zhou, P., Ni, B., Geng, C., Hu, J., Xu, Y.: Scale-transferrable object detection. In: CVPR (2018)

Stereo Event-Based Particle Tracking Velocimetry for 3D Fluid Flow Reconstruction

Yuanhao Wang(✉), Ramzi Idoughi, and Wolfgang Heidrich

King Abdullah University of Science and Technology, Thuwal, Saudi Arabia
{yuanhao.wang,ramzi.idoughi,wolfgang.heidrich}@kaust.edu.sa

Abstract. Existing Particle Imaging Velocimetry techniques require the use of high-speed cameras to reconstruct time-resolved fluid flows. These cameras provide high-resolution images at high frame rates, which generates bandwidth and memory issues. By capturing only changes in the brightness with a very low latency and at low data rate, event-based cameras have the ability to tackle such issues. In this paper, we present a new framework that retrieves dense 3D measurements of the fluid velocity field using a pair of event-based cameras. First, we track particles inside the two event sequences in order to estimate their 2D velocity in the two sequences of images. A stereo-matching step is then performed to retrieve their 3D positions. These intermediate outputs are incorporated into an optimization framework that also includes physically plausible regularizers, in order to retrieve the 3D velocity field. Extensive experiments on both simulated and real data demonstrate the efficacy of our approach.

Keywords: Fluid imaging · Event-based camera · Particle Imaging Velocimetry · Stereo-PTV · Optimization

1 Introduction

Fluid imaging is a topic of interest for several scientific and engineering areas like fluid dynamic, combustion, biology, computer vision and graphics. The capture of the 3D fluid flow is a common requirement to characterize the fluid and its motion regardless the application domain. Despite the large number of contributions to this field, retrieving a 3D dense measurement of the velocity vector over the fluid remains a challenging task. Different techniques have been proposed to capture and measure the fluid motion. The most commonly used approach involves introducing tracers such as dye, smoke, or particles into the studied fluid. Then by tracking the advected motion of the tracers, the fluid flows are retrieved.

Electronic supplementary material The online version of this chapter (https://doi.org/10.1007/978-3-030-58526-6_3) contains supplementary material, which is available to authorized users.

A. Vedaldi et al. (Eds.): ECCV 2020, LNCS 12374, pp. 36–53, 2020.
https://doi.org/10.1007/978-3-030-58526-6_3

Particle Tracking Velocimetry (PTV) and Particle Imaging Velocimetry (PIV) are the two most popular techniques among the tracer-based approaches [1, 13,52]. The PTV methods follow a Lagrangian formalism, where each particle is tracked individually. On the other hand, the PIV techniques retrieve the velocity in an Eulerian fashion by tracking the particles as a group, like a texture patch.

In basic PIV approaches [46], a thin slice of the fluid is illuminated by a laser sheet. From an in-plane tracking of the illuminated particles, a dense measurement of two components of the motion field can be retrieved inside the illuminated slice of the volume. However, turbulent or unsteady flows cannot be fully characterized with such planar measurements. To solve this issue, several variants of PIV have been proposed to extend the velocity retrieval to the third spatial dimension (a detailed discussion about such techniques is provided in the next section). Overall, these techniques can be regrouped into two families: multiple-cameras based approaches like tomographic PIV (tomo-PIV) [18,61] and single-camera based techniques such as light-field (or plenoptic) PIV [19,42,66] or structured-light PIV [2,69,70]. Among these, tomo-PIV is considered the established reference technology for the velocity 3D measurement, since it provides high spatial and temporal resolutions. However, it requires to have a precise calibration and synchronization of the used cameras. Moreover, for the tomo-PIV setups, the depth-of-field is a real limitation on the size of the volume of interest. On top of that, setup needed to reach a high temporal resolution can be very costly. Finally, due to space limitation on the hardware setup, only a few cameras (4–6) can be used, which limits the reconstruction quality. Although the proposed single-camera based techniques overcame the main shortcomings of the tomo-PIV (calibration, synchronization and the space limitation), they still have some limitations. The plenoptic PIV systems have significantly lower spatial resolution. Moreover, the current light-field cameras have low frame rates, which reduces the temporal resolution of the retrieved flow field. Furthermore, the main limitation of the structured-light PIV methods, like the RainbowPIV [70], is also the limited spatial resolution along the axial dimension.

To deal with many common flows, the cameras used in PIV systems should have a sufficiently high frame rate. However, the high speed cameras are relatively expensive, which particularly impacts the cost of multi-camera setups like tomo-PIV. The large bandwidth and storage requirements of such solutions pose additional difficulties. In order to address these issues, this paper introduces a new stereo-PIV technique based on two event cameras. Thanks to some outstanding properties like a very low latency ($1\,\mu s$) and a low power consumption [8,30,49,64], event cameras are well suited for fast motions detection and tracking. After capturing two sequences of event images using two different cameras, we propose a new framework that retrieves the velocity field of the fluid with a high time resolution. Indeed, the events based cameras can react very fast to motions given their very low latency. Thus, no motion blur can be observed with such cameras. In addition, we have a complete control over the time interval over which the events are aggregated.

The main technical contributions of this work are:

1. To the best of our knowledge, we propose the first event-camera based stereo-PIV setup for measuring time-resolved fluid flows.
2. We formulate a pertinent data term that links the event images to the 3D fluid velocity vector field.
3. We propose an optimization framework to retrieve the fluid velocity field from the event images. This framework include physically-based priors to solve the ill-posed inverse problem.
4. We demonstrate the accuracy of our approach on both simulated and real fluid flows.

2   Related Work

Fluid imaging is an active and challenging research topic for several domains such as fluid dynamics, combustion, biology, computer vision and graphics. To capture a fluid and its flow, several techniques have been applied in order to retrieve some of the fluid characteristics like the temperature, the density, the species concentration (scalar fields), the velocity or the vorticity (vector fields) [67]. In computer vision and graphics, an initial effort was focused on retrieving some physical properties of the fluid, from which a good visualization can be obtained. Thus, the captured properties vary from one fluid to another. As examples, light emission, refractive index, scattering density, and dye concentration have been reconstructed to visualize flames [26,33], air plumes [4,34], liquid surfaces [32,44,68], smoke [25,27] and fluid mixtures [23,24]. More recently, the interest has shifted to the estimation of velocity field, in order to improve the scalar density reconstruction [16,17,68,72], or as the final output [4,23,37,69,70].

Fluid velocity estimation techniques are mostly tracer-based approaches. Tracers like particles or dye are introduced into the fluid. Then, the velocity field of the fluid is recovered by tracking those tracers. However, tracer-free methods have been also investigated. Examples of such methods, Background Oriented Schlieren (BOS) [4,51,56] and Schlieren PIV [10,35], use the variations of the refractive index of the fluid as a "tracer" of the fluid motion.

Among the tracer-based methods, Particle Imaging Velocimetry (PIV) is the most commonly used [1,52]. During the last three decades, several techniques have been proposed to extend the standard PIV [46], where only two components of the velocity are measured in a thin slice of the volume (2D-2C). Stereoscopic PIV (Stereo-PIV) [50] records the region of interest using two synchronized cameras, which allows to retrieve the out-of-plane velocity component (2D-3C). The 3D scanning PIV (SPIV) technique [12,31], performs a standard PIV reconstruction on a large set of parallel light-sheet planes that samples the volume. A fast scanning laser is used to illuminate those planes at a high scanning rate. This approach however still retrieves only two-dimensions of the velocity (3D-2C). Moreover, at each time only one depth layer is scanned. More sophisticated techniques that can resolve 3D volumes include holographic PIV [29,43], defocusing digital PIV [47,48,71], synthetic aperture PIV [6], tomographic PIV (tomo-PIV) [18,61], structured-light PIV [2,69,70], and plenoptic (light-field) PIV [19,42,66]. These approaches can be multi-view based, like the widely used tomo-PIV. In this case, the hardware setup induces some difficulties like the calibration and synchronization of the cameras, as well as a limited space. Moreover, the reconstructed volume is usually small, since it should be included in the field of view of all cameras. On the other hand, mono-camera PIV approaches encode the depth information using color/intensity or using light-field camera. In the case of Rainbow PIV [70], the lower sensitivity of the camera to the wavelength change and the light scattering limit the depth resolution of the retrieved velocity field. Similar drawback can be observed with the intensity-coded PIV setup [2]. Otherwise, the plenoptic PIV systems have a more limited spatial resolution, since they capture on the same image several copies of the volume from different angles. They also suffer from a reduced temporal resolution, given the frame rate of current light-field cameras.

On the other hand, Particle Tracking Velocimetry (PTV) techniques [13,39] work by tracking individual particles to retrieve their velocity. The obtained flow field is then sampled according to these particles. Some variational approaches [21,28,63] were proposed to optimize the 3D flow field from particle tracks in order to estimate the flow in all the volume. Another advantage of using variational approaches is to incorporate physical constraints as prior in the optimization framework. This has been done in several PTV and PIV techniques [3,28,38,58–60,70].

The major limitation of PTV approaches has been the low number of particles that can be tracked. However, the recent shake-the-box method [62] succeeded into tracking particles with high densities, similar to those in PIV techniques.

All of these techniques require the use of high speed cameras, in order to reconstruct many real-world flow phenomena. Combined with the high resolution requirement, the large data generated at each capture (typically a rate of 2 GD/s) induces high bandwidth and large memory specifications for the used camera(s). To address this issue, [15] presented a proof-of-concept study about the use of dynamic vision sensor (DVS) in the capture and the tracking of particles. The proposed algorithm is suitable only for the 2D tracking of a sparse set of fully resolved particles (10 pixels diameter) in the volume. Recently, [11] proposed another approach based on Kalman filters to track neutrally buoyant soap bubbles from three cameras. This technique can handle the tracking and 3D reconstruction of under-resolved particles, which allows the larger studied volumes. However, this approach is not well-suited for high particle seeding densities, and it is not able to provide a dense measurement of the velocity field. In this paper, we propose a new framework that reconstructs a dense time-resolved 3D fluid flows from two event-based cameras.

Event cameras (a.k.a dynamic vision sensors) were first developed by [41] to mimic the retina of eyes, which is more sensitive to motions. These cameras respond only to the brightness change in the scene asynchronously and independently for each pixel. Event cameras have a very low latency (up to 1 μs), a low power consumption, and a high dynamic range [8,30,49,64]. These properties are major assets to fulfill several computer vision tasks. For instance, event cameras were introduced for object and feature tracking [14,22], depth estimation [53,57], optical flow estimation [5,7,74], high dynamic range imaging [55] and many other applications. An exhaustive survey about event camera applications can be found in [20]. In our framework, the event cameras are used for the particle tracking and optical flow estimation.

To track micro-particles using an event-camera, [45] estimate the particles' positions using an event-based Hough circle transform, combined with a centroid (a centre-of-mass algorithm). [9] apply an event-based visual flow algorithm [7] in order to track particles imaged by a full-field Optical Coherence Tomography (OCT) setup. This visual flow algorithm estimates the normal flow by fitting the events to a plane in the x-y-t space. The optical flow is then estimated as the slope of this plane. However, these two approaches reconstruct only 2D velocities, and are limited to the tracking of sparse particle densities.

To improve the optical flow estimation from event sequences, [73,75] propose to regroup events into features in a probabilistic way. This assignment is governed by the length of the optical flow. The latter is computed as a maximization of the expectation of all these assignments. An affine fit is used to model the features deformation. In our approach we use this technique to compute the optical flow over the event sequences.

3   Proposed Method

3.1   System Overview

We propose a new particle tracking velocimetry technique for the reconstruction of dense 3D fluid flows captured by two event-cameras. Our framework, illustrated in Fig. 1, is mainly composed of four modules: (1) a **Cameras calibration step**, which entails estimation of the camera calibration matrices for both event cameras. (2) **Event feature tracking** for a 2D particle velocity reconstruction. In this step, we apply feature tracking algorithms proposed by [73,75] in order to track the particles in the two captured sequences of event images. Then the 2D particle velocity in the two cameras image planes is recovered. (3) A **Stereo matching step** is performed using a double triangulation method to find the position of the particles in the 3D volume. (4) **3D velocity field reconstruction**. This last module is an optimization framework that includes our derived data fitting term, and a physically constrained 3D optical flow model. In the following section, we provide a detailed presentation of the three main modules of our framework.

3.2   Image Formation Model

In this section, we derive a model that links the 3D fluid velocity field to the captured event sequences.

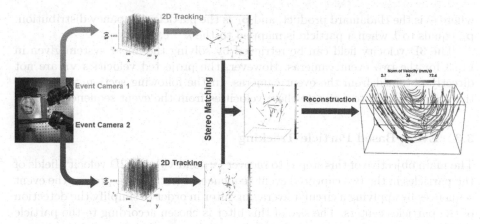

Fig. 1. Overview of the architecture of our stereo-event PTV framework. The two event cameras capture the motion of the particles inside the fluid. They generate two sequence of events, represented here in the x-y-t space. A 2D tracking step provide the 2D velocity of the captured particles for each sequence. Then, using a stereo matching step we build a sparse 3D velocity field that we use in order to estimate the dense 3D fluid flow.

Event Camera Model. Each pixel (x, y) of an event camera independently generates an event $e_i = (x, y, t_i, \rho_i)$, when detecting a brightness change higher than a pre-defined threshold τ:

$$|\mathrm{L}(x, y, t_i) - \mathrm{L}(x, y, t_{previous})| \geq \tau \tag{1}$$

where $\mathrm{L}(x, y, t_i)$ is the brightness (log intensity) of the pixel (x, y) at time t_i, $t_{previous}$ is the time of occurrence of the previous event and $\rho_i = \pm 1$ is the event polarity corresponding to the sign of the brightness change.

Camera Calibration. Each point (X, Y, Z) in the fluid is projected onto a pixel (x^k, y^k) of the kth camera image plane:

$$\alpha \cdot \left[x^k, y^k, 1\right]^T = \mathbf{M}_k \left[X, Y, Z, 1\right]^T \tag{2}$$

where α is a scale factor, and $\mathbf{M}_k$ is a 4×3 matrix describing the kth camera calibration matrix (intrinsic + extrinsic parameters). This matrix is obtained during the calibration step. Note that we use lowercase letters for 2D pixel coordinates and the Uppercase for the 3D voxel positions.

These camera calibration matrices are also used to project the 3D fluid velocity field $\mathbf{u} = [\mathbf{u}_X, \mathbf{u}_Y, \mathbf{u}_Z]^T$ onto the image planes of the cameras. The obtained 2D velocity fields are denoted $\mathbf{v}_k$. Given that the velocity can only be measured for the positions where particles are present, we can write:

$$\mathbf{v}_k = \mathbf{p}_k \odot (\mathbf{M}_k \mathbf{u}) \tag{3}$$

where $\odot$ is the Hadamard product, and $\mathbf{p}_k$ is the particle occupancy distribution. $\mathbf{p}_k$ equals to 1 when a particle is mapped to the kth camera.

The 3D velocity field can be retrieved by solving the linear system given in Eq. 3 for the two event cameras. However, the projected velocities $\mathbf{v}_k$ are not directly obtained from the event cameras. In the following section, we explain the approach used to estimate these velocities from the event sequences.

3.3   Event-Based Particle Tracking

The main objective of this step is to recover ($\mathbf{v}_1$ and $\mathbf{v}_2$), the 2D velocity fields of the particles in the two captured event sequences. First, we pre-process the event sequences by applying a circular averaging filter in order to simplify the detection of the particles centers. The size of this filter is chosen according to the particle size in the images. At this stage the center coordinates of each particle in the two event sequences can be easily determined. After this prepossessing step, we use the event-based optical flow method introduced by [73,75], in order to track the particles in the image planes, and then retrieve the velocities $\mathbf{v}_1$ and $\mathbf{v}_2$. In this approach, the events $\{e_i\}$ are associated with a set of features representing the particles in our case. All events associated with a given particle P are within a spatio-temporal window, where the average flow $v(P,t)$ is assumed to be constant if the temporal dimension $[t, t + \Delta t]$ of the window is small enough. This window can be written as:

$$W(R,t) := \{e_i \mid t_i \leq t, \ \|(x_i, y_i) - t_i v(P,t) - (x_c, y_c)\| \leq R\}, \qquad (4)$$

where R and (x_c, y_c) are respectively the spatial extension (radius) and the coordinate of the particle center in the image plane.

The association of events to particles is defined in terms of the flow $v(P,t)$ that we would like to estimate: events corresponding to the same 3D point should propagate backward onto the same image position. [73] propose an Expectation Maximization algorithm to solve this flow constraint. In the first step (E step), they update the association between events and the particles, given a fixed flow $v(P,t)$. Then, this flow is updated in the second step (M step), using the new matches between events and particles. More details about the implementation of this algorithm can be found in [73,75]. By applying this algorithm for tracking all particles and for all time stamps, we can recover the 2D velocity fields $\mathbf{v}_1$ and $\mathbf{v}_2$.

3.4   Stereo Matching

The aim of this step is on one hand to find the particle positions in 3D space. On the other hand, the retrieved 2D velocity field can be backprojected to get an estimation of the 3D velocity field $\mathbf{u}$ for the positions corresponding to the particles. We perform this stereo matching step using a triangulation procedure [36,40]. The main idea is to build for each particle a pixel-to-line transformation and then find the 3D positions that minimizes the total distances to all the lines.

For each identified particle P_i in an image captured by camera 1, its center $(x^1_{i,c}, y^1_{i,c})$ is used to backproject a line of sight to the different planes of the volume of interest. Then, the intersection points of this line with the different planes of the volume will be reprojected on the corresponding image frame captured by camera 2. Candidate particles P_j are selected in camera 2 only if the distance d^{1-2}_{ij} between the particle's center on the camera 2 and the reprojected points corresponding to the center of particles P_i on the camera 1 are under a given threshold distance (2 pixels, for example). Similarly, we perform the inverse mapping, a particle P_i on the camera 2 is backprojected to the fluid volume then reprojected to the image plane of the camera 1 to find candidates particles P_i, under the constraint that the distance d^{2-1}_{ij} is less than the fixed threshold. The correspondence between the two cameras are obtained by minimize the summation of all the distances. This is formulated as a simple linear assignment problem, and solved by using the Hungarian algorithm:

$$\min C_{ij} d^{1-2}_{ij} + C_{ij} d^{2-1}_{ij} \tag{5}$$

$$\text{subject to} \begin{cases} \sum_j C_{ij} \le 1 \\ \sum_i C_{ij} \le 1 \\ C_{ij} \in \{0, 1\} \end{cases}$$

From this stereo matching step, we estimate the particles 3D position as well as the velocity of those particles. However in practice, because of occlusions and noise, some of the particles may not be matched or, worse, they might be mismatched. This is what motivate us to use a variational approach to improve the particles' velocity estimation and also to extend the velocity estimation to whole volume of interest.

3.5   3D Velocimetry Reconstruction

We propose to reconstruct the 3D fluid velocity field $\mathbf{u} = [\mathbf{u}_X, \mathbf{u}_Y, \mathbf{u}_Z]^T$ for each voxel of the volume of interest, by solving Eq. 3 using the two event cameras, and by combining all time frames. In order to handle this ill-posed inverse problem, we introduce several regularizer terms, directly derived from the physical properties of the fluid.

Data Fitting-Term. As mentioned previously, we define the data-fitting term from Eq. 3. This term translates that the projection of the 3D velocity field to each camera image plane, should be consistent with the 2D velocity field observed with that camera. The data-fitting term can then be written as follows:

$$E_{data}(\mathbf{u}) = \tfrac{1}{2} \| \mathbf{p} \odot (\mathbf{M}_1 \mathbf{u}) - \mathbf{v}_1 \|^2_2 + \tfrac{1}{2} \| \mathbf{p} \odot (\mathbf{M}_2 \mathbf{u}) - \mathbf{v}_2 \|^2_2, \tag{6}$$

where $\mathbf{p} = \mathbf{p}_1 \odot \mathbf{p}_2$ is the particle occupancy distribution, that take into account only the matched particles between the 2 cameras.

Spatial Smoothness. The second term of our optimization is a spatial smoothness on the 3D velocity field. The advantage of this term is to help in giving a better interpolation for the voxels where no particles are detected.

$$E_{smooth}(\mathbf{u}) = \|\nabla_S \mathbf{u}\|_2^2 \tag{7}$$

Incompressibility. In the case of incompressible fluid simulation or capture, it is common to constrain the flow field to be divergence-free [16,23,70]. This constraint is derived directly from the mass-conservation equation for the fluid. Usually the divergence-free regularization is applied by projecting the velocity field onto the space of divergence-free velocity field. However, we notice that in the case of lower spatial resolution, the discretization of the divergence operator may introduce some divergence to the flow. Therefore, we prefer to include the incompressibility prior as a soft (L-2) constraint instead of a hard projection.

$$E_{div}(\mathbf{u}) = \|div(\mathbf{u})\|_2^2 \tag{8}$$

Temporal Coherence. In the absence of external forces, the Navier-Stokes equation for a non-viscous fluid can be simplified as follows:

$$\frac{\partial \mathbf{u}}{\partial t} + (\mathbf{u} \cdot \nabla)\mathbf{u} = 0 \tag{9}$$

This equation can be used as an approximation of the temporal evolution of the fluid flow. We can then, advect the velocity field at a given time stamp by itself to deduce an estimate of the field at the next time stamp. This advection is applied in a forward and backward manners. This yields the following term:

$$\begin{aligned} E_{TC}(\mathbf{u}_t) = &\|\mathbf{p} \odot (\mathbf{u}_t - \text{advect}(\mathbf{u}_{t-1}, \Delta t))\|_2^2 \\ &+ \|\mathbf{p} \odot (\mathbf{u}_t - \text{advect}(\mathbf{u}_{t+1}, -\Delta t))\|_2^2 \end{aligned} \tag{10}$$

where $\mathbf{u}_t$ is the velocity field at the time t. The particle occupancy $\mathbf{p}$ is used here as a mask, in order to take into account only the regions of the volume where particles have been observed, since these are the only regions with reliable velocity estimates.

Optimization Framework. The general optimization framework is then expressed as:

$$(\mathbf{u}^*) = \underset{\mathbf{u}}{\arg\min} \; E_{data}(\mathbf{u}) + \lambda_1 E_{smooth}(\mathbf{u}) + \lambda_2 E_{TC}(\mathbf{u}) + \lambda_3 E_{div}(\mathbf{u}) \tag{11}$$

Equation 11 is composed only of L-2 terms. We solve it then using the conjugate gradient method. To handle the large velocities in the fluid, we build a multi-scale coarse-to-fine scheme [70]. More details about our implementation are given in the supplement material.

4   Experiments

Experiments on both simulated and captured fluid flows were conducted to evaluate our approach. We implemented our framework in matlab[1]. All the experiments were conducted on a computer with an Intel Xeon E5-2680 CPU processor and 128 GB RAM. The reconstruction time for the simulated dataset which contains 8 frames (with $80 \times 80 \times 80$ voxels) was around 28 min. Furthermore, the parameter settings for the optimization (see Eq. 11) were kept the same for all the datasets ($\lambda_1 = 2.5 \times 10^{-5}$, $\lambda_2 = 0.025$, $\lambda_3 = 2.5 \times 10^{-5}$).

4.1   Synthetic Data

To quantitatively assess our method, we simulated a fluid undergoing a rigid-body-like vortex, with a fixed angular speed. A volume with a size of $20\,\text{mm} \times 20\,\text{mm} \times 20\,\text{mm}$ was seeded randomly with particles of an averaged size of $0.1\,\text{mm}$ (1% variance). This fluid is captured by two different simulated event cameras having a spatial resolution of 800×800, which is similar to the real experiment setup in Fig. 5. Different vortex speeds and particle densities were simulated. The approach introduced by [65] was applied to advect the particles over time using the vortex velocity field. Moreover, we simulated different frame rate images. Finally, we used E-sim code [54] to generate the event sequences observed by the simulated sensors.

Ablation Study. In order to illustrate the impact of each of our priors, we conducted an ablation study. We compare our method without the use of the temporal coherence and the divergence terms (**w/o E_{TC} & E_{div}**), our method without the divergence term (**w/o E_{div}**), and our proposed method (**Ours**). For the quantitative comparison, we use two metrics: the average angular error (AAE), i.e. the average discrepancy in the flow direction, and the average endpoint error (AEE), i.e. the average Euclidean norm of the difference between the real and estimated flow vectors.

In Fig. 2, we illustrate the velocity field reconstruction using our method versus the ground truth. Except for the borders, the reconstruction is very accurate. The numerical results of the ablation study are shown in the Fig. 3. As expected, both the AAE and the EPE errors are improved when adding the different priors. Moreover, these errors are almost constant from one time frame to another. We need to point out that the temporal coherence term might not improve too much the reconstruction for all frames. However, in the general case, it smoothes the result in the temporal domain, which is important for visual quality in frame-based or time-based data processing.

In Fig. 4 we illustrate for a 2D slice the end point error as well as the divergence of the velocity field for different methods. The mean error for the different methods is 0.182, 0.178, 0.171 respectively. The error will generally become

---

[1] The code is available on: https://github.com/vccimaging/StereoEventPTV.

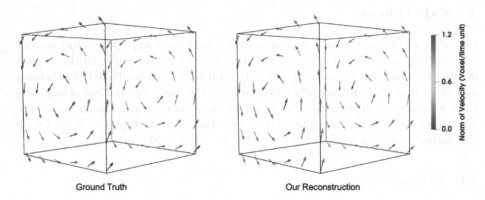

Fig. 2. Ground truth (left) and reconstruction result using our method (right) for a simulated rigid-body-like vortex.

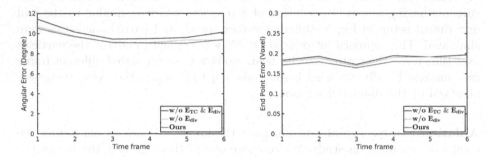

Fig. 3. Quantitative comparisons with ground truth velocity field for different reconstruction using different priors. Left: Average angular error (in degree). Right: Average end-point error (in voxel).

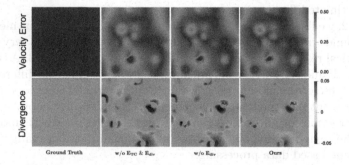

Fig. 4. End point error (first row) and the divergence of the flow (second row) computed on a 2D slice. From left to right: Ground truth, Our method without the incompresibility and the temporal coherence terms, Our method without the incompresibility term, and Our proposed method

smaller gradually as expected. The mean absolute divergence for the three different method is 0.0096, 0.0100, 0.0071. We notice that the temporal coherence term introduces some divergence to the flow. It can be explained by the fact that the temporal smoothness might propagate wrong stereo matching to adjacent time frame. However, the incompressibility constraint will reduce the divergence and bring it closer to zero.

Table 1. Quantitative evaluation (AAE in degree/EPE in voxel) of different particle densities and at different rotation speeds.

Angluar Speed (rad/s)	Particle density (ppp)				
	0.006	0.012	0.018	0.024	0.03
5	9.73/0.046	10.22/0.042	9.56/0.043	15.38/0.052	12.83/0.048
10	10.20/0.097	11.12/0.090	10.72/0.092	13.34/0.097	12.12/0.087
15	10.58/0.146	10.38/0.131	9.94/0.123	13.77/0.150	11.77/0.133
20	9.73/0.187	10.37/0.164	9.76/0.168	12.13/0.178	11.49/0.180

Particle Densities and Vortex Speed Impact. We also evaluated our method for different particle densities and different angular speeds of the vortex. The results are shown in Table 1. These experiments have been conducted for the same duration. As expected the larger the speed, the larger the EPE. However, the angular error is in the same range independent of the vortex speed. On the other hand, from these experiments we can deduce that our method can handle a wide range of particle densities. These experiments show that our method can be used in very different situations with a wide range of particle densities and different fluid velocities.

4.2   Captured Data

Experimental Setup. The experimental setup used for the event-based fluid imaging is shown in Fig. 5. To capture the stereo events at the same time, we utilized two synchronized Prophesee cameras (Model: PEK3SHEM, Sensor: CSD3SVCD [7.2 mm × 5.4 mm], 480 × 360 pixels), with an angle of 60° between the two optical axes. Two lenses with a focal length of 85 mm and a 3D printed extension tubes were attached to the cameras. The aperture was set to $f/16$ to have a depth-of-field of 10 mm. The tank was seeded with white particles (White Polyethylene Microspheres) having a diameter in the range [90, 106 μm]. The size of the particles on the image plane is approximately 6.7 pixels. By applying downsampling with a downsample factor of 6 and stereo matching, we reconstruct a volume with: 78 × 48 × 42 voxels. For the calibration step, a 17 × 16 checkerboard where each square has an edge length of 0.5 mm was attached on a glass slide. We used a controllable translation stage to modify the distance of the checkerboard to the cameras. More details about the calibration can be found in the supplement material.

Fig. 5. Left: Illustration of our experimental setup. A collimated white light source illuminate the hexagonal tank. A vortex generator is used to control the speed of the vortex during the experiments. Right: illustration of the calibration step, where images of a small check board are captured for several positions. A controlled translation stage is used to change the positions.

Controlled Vortex Flow. The first experiment we performed was a controlled vortex flow. We used a magnetic stirring rod (Model: Stuart CB162) to generate different vortices by controlling the rotation speed of the stirring rod. We evaluate our reconstruction method over the different vortices. The reconstructed streamlines for two examples are shown in Fig. 6. We can see that our reconstruction offers a good representation of the vortex structure, and the velocity norm seems to be reliable given the speed of the stirring rod. Please refer to the supplement for more results.

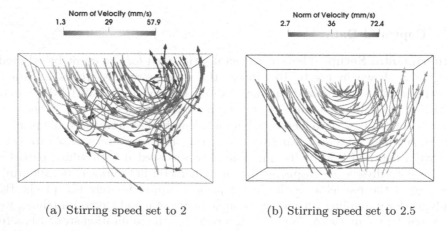

(a) Stirring speed set to 2   (b) Stirring speed set to 2.5

Fig. 6. Streamline visualization for controlled vortex flows. Left: The stirrer speed was set to 2. Right: The stirrer speed was set to 2.5.

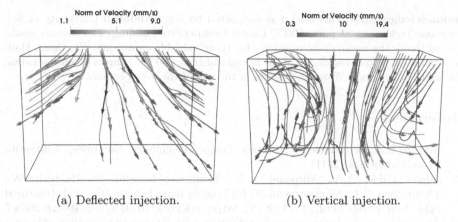

(a) Deflected injection.   (b) Vertical injection.

Fig. 7. Streamline visualization for a fast fluid injections using a syringe. By controlling the syringe orientation we have captured: (a) a deflected injection and (b) a vertical injection.

Fluid Injection. Finally, we conducted another set of experiments, consisting of a relatively fast fluid injections into the tank using a syringe. As shown in Fig. 7, different speeds of the flow and different injection directions can be easily distinguished from our reconstructed results. Additional results and illustrations are presented in the supplemental material.

5   Conclusions

We have introduced a stereo event-based camera system coupled with 3D fluid flow reconstruction strategies in this paper. Instead of using image based optical flow reconstruction in the traditional tomographic PTV, our approach is based on generating the two dimensional flow from the event information, and then matching the resulting trajectories in 3D to obtain full 3D-3C flow fields.

Both the numerical and experimental assessment confirm the effectiveness of our approach. By simulating different particle numbers in the tank that usually used in the PTV, we found that our method works on a wide range of particle densities. Furthermore, by controlling the stirring speed of the vortex, we found that our approach can deal with fast fluid flow.

There are some drawbacks to our approach. First of all, the spatial resolution of currently available event cameras is quite low, which also adversely impacts the spatial resolution of the reconstruction. Second, due to the high dynamic range of the event camera, the light intensity and the camera sensitivity should be carefully selected to have a good measurements in real experiments. Last but not least, the bandwidth of the event-camera is limited, the method fails when the speed of the controlled vortex exceeds a certain threshold, in which case the bus of the camera was saturated. However, with future improvements of event camera hardware, we believe these shortcomings can be overcome, making our method an attractive option for 3D-3C fluid imaging.

Acknowledgments. This work was supported by King Abdullah University of Science and Technology as part of VCC Center Competitive Funding. The authors would like to thank the anonymous reviewers for their valuable comments. We thank Hadi Amata for his help in the design of the hexagonal tank and the camera extension tubes. We also thank Congli Wang for helping in the use of the event cameras.

References

1. Adrian, R.J., Westerweel, J.: Particle Image Velocimetry. Cambridge University Press, Cambridge (2011)
2. Aguirre-Pablo, A.A., Aljedaani, A.B., Xiong, J., Idoughi, R., Heidrich, W., Thoroddsen, S.T.: Single-camera 3D PTV using particle intensities and structured light. Exp. Fluids **60**(2), 1–13 (2019). https://doi.org/10.1007/s00348-018-2660-7
3. Álvarez, L., et al.: A new energy-based method for 3D motion estimation of incompressible PIV flows. Comput. Vis. Image Underst. **113**(7), 802–810 (2009)
4. Atcheson, B., et al.: Time-resolved 3D capture of non-stationary gas flows. ACM Trans. Graph. **27**(5), 132 (2008)
5. Bardow, P., Davison, A.J., Leutenegger, S.: Simultaneous optical flow and intensity estimation from an event camera. In: Proceedings of the CVPR, pp. 884–892 (2016)
6. Belden, J., Truscott, T.T., Axiak, M.C., Techet, A.H.: Three-dimensional synthetic aperture particle image velocimetry. Meas. Sci. Technol. **21**(12), 125403 (2010)
7. Benosman, R., Clercq, C., Lagorce, X., Ieng, S.H., Bartolozzi, C.: Event-based visual flow. IEEE Trans. Neural Netw. Learn. Syst. **25**(2), 407–417 (2013)
8. Berner, R., Brandli, C., Yang, M., Liu, S.C., Delbruck, T.: A 240 × 180 10mW 12μs latency sparse-output vision sensor for mobile applications. In: 2013 Symposium on VLSI Circuits, pp. C186–C187. IEEE (2013)
9. Berthelon, X., Chenegros, G., Libert, N., Sahel, J.A., Grieve, K., Benosman, R.: Full-field OCT technique for high speed event-based optical flow and particle tracking. Opt. Express **25**(11), 12611–12621 (2017)
10. Biswas, S.: Schlieren image velocimetry (SIV). Physics of Turbulent Jet Ignition. ST, pp. 35–64. Springer, Cham (2018). https://doi.org/10.1007/978-3-319-76243-2_3
11. Borer, D., Delbruck, T., Rösgen, T.: Three-dimensional particle tracking velocimetry using dynamic vision sensors. Exp. Fluids **58**(12), 1–7 (2017). https://doi.org/10.1007/s00348-017-2452-5
12. Brücker, C.: 3D scanning PIV applied to an air flow in a motored engine using digital high-speed video. Meas. Sci. Technol. **8**(12), 1480 (1997)
13. Dabiri, D., Pecora, C.: Particle Tracking Velocimetry. IOP Publishing, Bristol (2020)
14. Delbruck, T., Lang, M.: Robotic goalie with 3 ms reaction time at 4% CPU load using event-based dynamic vision sensor. Front. Neurosci. **7**, 223 (2013)
15. Drazen, D., Lichtsteiner, P., Häfliger, P., Delbrück, T., Jensen, A.: Toward real-time particle tracking using an event-based dynamic vision sensor. Exp. Fluids **51**(5), 1465 (2011)
16. Eckert, M.L., Heidrich, W., Thürey, N.: Coupled fluid density and motion from single views. In: CGF, vol. 37, pp. 47–58. Wiley (2018)
17. Eckert, M.L., Um, K., Thuerey, N.: ScalarFlow: a large-scale volumetric data set of real-world scalar transport flows for computer animation and machine learning. ACM Trans. Graph. **38**(6), 1–16 (2019)

18. Elsinga, G.E., Scarano, F., Wieneke, B., van Oudheusden, B.W.: Tomographic particle image velocimetry. Exp. Fluids **41**(6), 933–947 (2006)
19. Fahringer, T.W., Lynch, K.P., Thurow, B.S.: Volumetric particle image velocimetry with a single plenoptic camera. Meas. Sci. Technol. **26**(11), 115201 (2015)
20. Gallego, G., et al.: Event-based vision: a survey. arXiv preprint arXiv:1904.08405 (2019)
21. Gesemann, S., Huhn, F., Schanz, D., Schröder, A.: From noisy particle tracks to velocity, acceleration and pressure fields using B-splines and penalties. In: 18th International Symposium on Applications of Laser and Imaging Techniques to Fluid Mechanics, Lisbon, Portugal, pp. 4–7 (2016)
22. Glover, A., Bartolozzi, C.: Event-driven ball detection and gaze fixation in clutter. In: Proceedings of the IROS, pp. 2203–2208. IEEE (2016)
23. Gregson, J., Ihrke, I., Thuerey, N., Heidrich, W.: From capture to simulation: connecting forward and inverse problems in fluids. ACM Trans. Graph. **33**(4), 139 (2014)
24. Gregson, J., Krimerman, M., Hullin, M.B., Heidrich, W.: Stochastic tomography and its applications in 3D imaging of mixing fluids. ACM Trans. Graph. **31**(4), 52:1–52:10 (2012)
25. Gu, J., Nayar, S.K., Grinspun, E., Belhumeur, P.N., Ramamoorthi, R.: Compressive structured light for recovering inhomogeneous participating media. IEEE Trans. PAMI **35**(3), 1 (2012)
26. Hasinoff, S.W., Kutulakos, K.N.: Photo-consistent reconstruction of semitransparent scenes by density-sheet decomposition. IEEE Trans. PAMI **29**(5), 870–885 (2007)
27. Hawkins, T., Einarsson, P., Debevec, P.: Acquisition of time-varying participating media. Technical report, University of Southern California Marina del Rey CA Institute for Creative (2005)
28. Heitz, D., Mémin, E., Schnörr, C.: Variational fluid flow measurements from image sequences: synopsis and perspectives. Exp. Fluids **48**(3), 369–393 (2010). https://doi.org/10.1007/s00348-009-0778-3
29. Hinsch, K.D.: Holographic particle image velocimetry. Meas. Sci. Technol. **13**(7), R61 (2002)
30. Hofstatter, M., Schön, P., Posch, C.: A SPARC-compatible general purpose address-event processor with 20-bit 10ns-resolution asynchronous sensor data interface in 0.18 μm CMOS. In: Proceedings of 2010 IEEE International Symposium on Circuits and Systems, pp. 4229–4232. IEEE (2010)
31. Hori, T., Sakakibara, J.: High-speed scanning stereoscopic PIV for 3D vorticity measurement in liquids. Meas. Sci. Technol. **15**(6), 1067 (2004)
32. Ihrke, I., Goidluecke, B., Magnor, M.: Reconstructing the geometry of flowing water. In: Proceedings of the ICCV, vol. 2, pp. 1055–1060. IEEE (2005)
33. Ihrke, I., Magnor, M.: Image-based tomographic reconstruction of flames. In: Proceedings of the SCA, pp. 365–373 (2004)
34. Ji, Y., Ye, J., Yu, J.: Reconstructing gas flows using light-path approximation. In: Proceedings of the CVPR, pp. 2507–2514 (2013)
35. Jonassen, D.R., Settles, G.S., Tronosky, M.D.: Schlieren "PIV" for turbulent flows. Opt. Lasers Eng. **44**(3–4), 190–207 (2006)
36. Knutsen, A.N., Lawson, J.M., Dawson, J.R., Worth, N.A.: A laser sheet self-calibration method for scanning PIV. Exp. Fluids **58**(10), 1–13 (2017). https://doi.org/10.1007/s00348-017-2428-5

37. Lasinger, K., Vogel, C., Pock, T., Schindler, K.: 3D fluid flow estimation with integrated particle reconstruction. Int. J. Comput. Vis. **128**(4), 1012–1027 (2020). https://doi.org/10.1007/s11263-019-01261-6
38. Lasinger, K., Vogel, C., Schindler, K.: Volumetric flow estimation for incompressible fluids using the stationary stokes equations. In: Proceedings of the ICCV, pp. 2565–2573 (2017)
39. Maas, H., Gruen, A., Papantoniou, D.: Particle tracking velocimetry in three-dimensional flows. Exp. Fluids **15**(2), 133–146 (1993). https://doi.org/10.1007/BF00223406
40. Machicoane, N., Aliseda, A., Volk, R., Bourgoin, M.: A simplified and versatile calibration method for multi-camera optical systems in 3D particle imaging. Rev. Sci. Instrum. **90**(3), 035112 (2019)
41. Mahowald, M.: VLSI analogs of neuronal visual processing: a synthesis of form and function. Ph.D. thesis, California Institute of Technology Pasadena (1992)
42. Mei, D., Ding, J., Shi, S., New, T.H., Soria, J.: High resolution volumetric dual-camera light-field PIV. Exp. Fluids **60**(8), 1–21 (2019). https://doi.org/10.1007/s00348-019-2781-7
43. Meng, H., Hussain, F.: Holographic particle velocimetry: a 3D measurement technique for vortex interactions, coherent structures and turbulence. Fluid Dyn. Res. **8**(1–4), 33 (1991)
44. Morris, N.J., Kutulakos, K.N.: Dynamic refraction stereo. IEEE Trans. Pattern Anal. Mach. Intell. **33**(8), 1518–1531 (2011)
45. Ni, Z., Pacoret, C., Benosman, R., Ieng, S., Régnier, S.: Asynchronous event-based high speed vision for microparticle tracking. J. Microsc. **245**(3), 236–244 (2012)
46. Okamoto, K., Nishio, S., Saga, T., Kobayashi, T.: Standard images for particle-image velocimetry. Meas. Sci. Technol. **11**(6), 685 (2000)
47. Pereira, F., Gharib, M., Dabiri, D., Modarress, D.: Defocusing digital particle image velocimetry: a 3-component 3-dimensional DPIV measurement technique. Application to bubbly flows. Exp. Fluids **29**(1), S078–S084 (2000). https://doi.org/10.1007/s003480070010
48. Pereira, F., Gharib, M.: Defocusing digital particle image velocimetry and the three-dimensional characterization of two-phase flows. Meas. Sci. Technol. **13**(5), 683 (2002)
49. Posch, C., Matolin, D., Wohlgenannt, R.: A QVGA 143 dB dynamic range frame-free PWM image sensor with lossless pixel-level video compression and time-domain CDS. IEEE J. Solid-State Circuits **46**(1), 259–275 (2010)
50. Prasad, A.K.: Stereoscopic particle image velocimetry. Exp. Fluids **29**(2), 103–116 (2000)
51. Raffel, M.: Background-oriented schlieren (BOS) techniques. Exp. Fluids **56**(3), 1–17 (2015). https://doi.org/10.1007/s00348-015-1927-5
52. Raffel, M., Willert, C.E., Scarano, F., Kähler, C.J., Wereley, S.T., Kompenhans, J.: Particle Image Velocimetry. Springer, Cham (2018). https://doi.org/10.1007/978-3-319-68852-7
53. Rebecq, H., Gallego, G., Mueggler, E., Scaramuzza, D.: EMVS: event-based multi-view stereo-3D reconstruction with an event camera in real-time. Int. J. Comput. Vis. **126**(12), 1394–1414 (2018)
54. Rebecq, H., Gehrig, D., Scaramuzza, D.: ESIM: an open event camera simulator. In: Conference on Robot Learning, pp. 969–982 (2018)
55. Rebecq, H., Ranftl, R., Koltun, V., Scaramuzza, D.: High speed and high dynamic range video with an event camera. IEEE Trans. Pattern Anal. Mach. Intell. (2019)

56. Richard, H., Raffel, M.: Principle and applications of the background oriented schlieren (BOS) method. Meas. Sci. Technol. **12**(9), 1576 (2001)
57. Rogister, P., Benosman, R., Ieng, S.H., Lichtsteiner, P., Delbruck, T.: Asynchronous event-based binocular stereo matching. IEEE Trans. Neural Netw. Learn. Syst. **23**(2), 347–353 (2011)
58. Ruhnau, P., Guetter, C., Putze, T., Schnörr, C.: A variational approach for particle tracking velocimetry. Meas. Sci. Technol. **16**(7), 1449 (2005)
59. Ruhnau, P., Schnörr, C.: Optical stokes flow estimation: an imaging-based control approach. Exp. Fluids **42**(1), 61–78 (2007). https://doi.org/10.1007/s00348-006-0220-z
60. Ruhnau, P., Stahl, A., Schnörr, C.: On-line variational estimation of dynamical fluid flows with physics-based spatio-temporal regularization. In: Franke, K., Müller, K.-R., Nickolay, B., Schäfer, R. (eds.) DAGM 2006. LNCS, vol. 4174, pp. 444–454. Springer, Heidelberg (2006). https://doi.org/10.1007/11861898_45
61. Scarano, F.: Tomographic PIV: principles and practice. Meas. Sci. Technol. **24**(1), 012001 (2012)
62. Schanz, D., Gesemann, S., Schröder, A.: Shake-The-Box: Lagrangian particle tracking at high particle image densities. Exp. Fluids **57**(5), 1–27 (2016). https://doi.org/10.1007/s00348-016-2157-1
63. Schneiders, J.F.G., Scarano, F.: Dense velocity reconstruction from tomographic PTV with material derivatives. Exp. Fluids **57**(9), 1–22 (2016). https://doi.org/10.1007/s00348-016-2225-6
64. Serrano-Gotarredona, T., Linares-Barranco, B.: A 128 × 128 1.5% contrast sensitivity 0.9% FPN 3 µs latency 4 mW asynchronous frame-free dynamic vision sensor using transimpedance preamplifiers. IEEE J. Solid-State Circuits **48**(3), 827–838 (2013)
65. Stam, J.: Stable fluids. In: Proceedings of the 26th Annual Conference on Computer Graphics and Interactive Techniques, pp. 121–128 (1999)
66. Tan, Z.P., Thurow, B.S.: Time-resolved 3D flow-measurement with a single plenoptic-camera. In: AIAA Scitech 2019 Forum, p. 0267 (2019)
67. Tropea, C., Yarin, A.L.: Springer Handbook of Experimental Fluid Mechanics. SHB. Springer, Heidelberg (2007). https://doi.org/10.1007/978-3-540-30299-5
68. Wang, H., Liao, M., Zhang, Q., Yang, R., Turk, G.: Physically guided liquid surface modeling from videos. ACM Trans. Graph. (TOG) **28**(3), 1–11 (2009)
69. Xiong, J., Fu, Q., Idoughi, R., Heidrich, W.: Reconfigurable rainbow PIV for 3D flow measurement. In: Proceedings of the ICCP, pp. 1–9. IEEE (2018)
70. Xiong, J., et al.: Rainbow particle imaging velocimetry for dense 3D fluid velocity imaging. ACM Trans. Graph. **36**(4), 36 (2017)
71. Yoon, S.Y., Kim, K.C.: 3D particle position and 3D velocity field measurement in a microvolume via the defocusing concept. Meas. Sci. Technol. **17**(11), 2897 (2006)
72. Zang, G., et al.: TomoFluid: reconstructing dynamic fluid from sparse view videos. In: Proceedings of the CVPR, pp. 1870–1879 (2020)
73. Zhu, A.Z., Atanasov, N., Daniilidis, K.: Event-based feature tracking with probabilistic data association. In: 2017 IEEE International Conference on Robotics and Automation (ICRA) pp. 4465–4470. IEEE (2017)
74. Zhu, A.Z., Yuan, L., Chaney, K., Daniilidis, K.: EV-FlowNet: self-supervised optical flow estimation for event-based cameras. arXiv preprint arXiv:1802.06898 (2018)
75. Zihao Zhu, A., Atanasov, N., Daniilidis, K.: Event-based visual inertial odometry. In: Proceedings of the CVPR, pp. 5391–5399 (2017)

Simplicial Complex Based Point Correspondence Between Images Warped onto Manifolds

Charu Sharma[(✉)] and Manohar Kaul[(✉)]

Indian Institute of Technology Hyderabad, Hyderabad, India
{cs16resch11007,mkaul}@iith.ac.in

Abstract. Recent increase in the availability of warped images projected onto a curved manifold, especially omnidirectional spherical ones, coupled with the success of higher-order assignment methods, has sparked an interest in the search for improved higher-order matching algorithms on warped images due to projection. Although, currently, several existing methods "flatten" such 3D images to use planar graph/hypergraph matching methods, they still suffer from severe distortions and other undesired artifacts, which result in inaccurate matching. Alternatively, current planar methods cannot be trivially extended to effectively match points on images warped on curved manifold. Hence, matching on these warped images persists as a formidable challenge. In this paper, we pose the assignment problem as finding a bijective map between two graph induced simplicial complexes, which are higher-order analogues of graphs. We propose a constrained quadratic assignment problem (QAP) that matches each p-skeleton of the simplicial complexes, iterating from the highest to the lowest dimension. The accuracy and robustness of our approach are illustrated on both synthetic and real-world spherical/warped (projected) images with known ground-truth correspondences. We significantly outperform existing state-of-the-art spherical matching methods on a diverse set of datasets.

1 Introduction

There exists a longstanding line of research on finding bijective correspondences (i.e., assignments/matchings[1]) between two sets of visual features. Notable applications include stereo matching [14], structure from motion (SfM) [34], and image registration [31], to name a few. Traditionally, when matching points between multiple images of a fixed environment from various viewpoints, most approaches recover matchings and relative camera geometry (e.g. fundamental matrix) using

---

[1] *assignment* and *matching* are used interchangeably in this paper.

Electronic supplementary material The online version of this chapter (https://doi.org/10.1007/978-3-030-58526-6_4) contains supplementary material, which is available to authorized users.

A. Vedaldi et al. (Eds.): ECCV 2020, LNCS 12374, pp. 54–70, 2020.
https://doi.org/10.1007/978-3-030-58526-6_4

a robust technique such as RANSAC [16]. On the other hand, when matching between different instances of the same category, graph matching methods [38] using *unary* and *pairwise* constraints have been successfully utilized. More recently, graph matching has been subsumed by *hypergraph matching* using *higher-order* constraints [12,19]. An important appeal of higher-order matching methods is their ability to coherently match compact local geometric features from the source space to similar compact regions in the target space, despite the presence of noise, outliers, and incomplete data, thus achieving accurate matches that are also *local structure-preserving* in nature.

The recent proliferation of spherical images (e.g., omnidirectional and panoramic images captured from cameras mounted on drones and autonomous vehicles) and more generally, images warped onto *curved manifolds*, has sparked a heightened interest in assignment algorithms on such datasets due to the challenges they present in terms of curvature, both uniform and non-uniform [18,32, 35,36]. Although assignment problems have been well studied for decades in computer vision, a majority of the work has only focused on matching points between *planar (flat) images*. Therefore, matching points on images with warping transformations which fall into the category of projective parametric models remains a challenging task, mainly due to the introduction of undesirable artifacts like severe distortions in pairwise distances between landmark points, non-linear distortions in local geometries, noise, illumination, blur, and occlusions [3,8], when flattening.

When dealing with matchings on curved geometries, primarily two types of methods are employed. Some putative matchings are computed to estimate a *fundamental matrix* [10,16] that captures the *epipolar geometry* of the 3D image. *Stereo rectification* [5] uses this fundamental matrix to re-project the two images on the same flat plane with row images aligned in parallel, followed by a re-matching to improve matching accuracy. Alternatively, *geometric alignment* on the fundamental matrix is used to *verify* and distinguish *inliers* from *outliers*, so that outliers can be pruned post matching to further boost accuracy [34]. Elements warped on the curved manifold cannot be metrically sampled in such methods and hence severe distortions are introduced [7], which is also consistent with the findings in our empirical studies.

Applications. An interesting and noteworthy application of higher-order matching on spherical images arises in the area of biomedical imaging, especially in *retinal imaging* using *optical coherence tomography* (OCT). To investigate a wider *field of view*, 3D *fundus* images of the eye are captured, matched, and "stitched" together to form an *OCT montage* [22,25]. This matching operation must additionally preserve *regions of interest* such as the *optic cup/disc, fovea, macula, vessels, and microaneurysms*, to name a few [29]. In addition to the standard noise, occlusion, and artifacts in these OCT fundus images, the data also suffers from data shifts due to axial eye motions and unpredictability between *eye positions* and *instrument alignment* across various scans [22]. Therefore, OCT datasets cannot easily be matched using rigid 3D transformations. Such images are not limited to merely spherical ones, but also arise in more general warped images due to projection. For instance, 3D sonograms depict the cervix as a *conic frustum (truncated cone)* [1] and clustered nanofluid microflow patterns in elastic micro-tubes are tracked via matching between *cylindrical* images in a time-lapse [33].

Our Method. In this paper, we focus on exploiting the intrinsic higher-order geometric relationships between landmark points on images warped onto curved manifolds. We capture these higher-order connections by constructing a combinatorial topological structure (simplicial complex) which is induced by a graph, whose *vertices* are the landmark points embedded on the warped image and whose *edges* are *geodesic curves* between selected vertex pairs. Next, we pose the assignment problem as a multi-dimensional *quadratic assignment problem (QAP)* between two graph-induced simplicial complexes.

Our Contributions. (i) To the best of our knowledge, we are the first to propose matching landmark points on warped images projected onto curved manifolds. (ii) In an attempt to break away from other works which solely focus on flat or spherical images, we propose a novel graph induced simplicial complex that efficiently captures higher order structures in a succinct manner, considering the inherent properties of the underlying curved manifold on which the landmark points are embedded. (iii) We uniquely formulate the assignment problem as a multi-dimensional combinatorial matching between two graph induced simplicial complexes, propose a novel algorithm to solve it, and analyze the time-complexity of our algorithm. (iv) Finally, to illustrate the robustness of our proposed method, we perform extensive experiments by comparing to planar matching methods, both *existing* and *extended by us* as *naive baselines* for matching on manifolds. We compare our method against existing *graph matching* and *spherical matching* (both boosted using *rectification* and *verification* techniques) [12,21,27,30,37–39] on warped images and interestingly observe that not only does our method significantly outperform these matching methods on warped images onto curved manifolds (with up to 49.7% matching error reduction), but it also outperforms existing planar matching algorithms on "flat" planar images too (with up to 42.2% matching error reduction), due to the ability to naturally capture higher-order relationships by the simplicial complex.

2   Preliminaries

In this section, we introduce our notation and provide the necessary background for our higher-order assignment algorithm on curved manifolds. We begin by introducing certain standard definitions followed by our problem definition.

Let $\mathcal{M}$ denote a *curved manifold*, i.e., with zero genus and no boundaries. On a plane, the shortest distance between any two points is a straight line, i.e., a curve whose derivative to its tangent vectors is zero. We extend this notion of a "straight line" to *curved manifolds* by defining the shortest path (on $\mathcal{M}$) between its endpoints u and v placed on $\mathcal{M}$, as a *geodesic curve* $\gamma(u, v)$.

Simplicial Complex. We begin by providing some general definitions before we can formally define a simplicial complex. More background can be found in [24]. Given a set $V = \{v_0, \ldots, v_n\}$ of $(n+1)$ affinely independent points in $\mathbb{R}^{n+1}$, a *n-dimensional simplex* (also called *n*-simplex) $\sigma^{(n)}$ with *vertices* V is the *convex hull* of V, i.e., more formally

$$\sigma^{(n)} = \left\{ (k_0, k_1, \ldots, k_n) \in \mathbb{R}^{n+1} \mid \sum_{i=0}^{n} k_i = 1, \; k_i \geq 0 \; \forall i \right\}$$

The dimension of *n*-simplex $\sigma^{(n)}$ is denoted by $dim(\sigma^{(n)})$. For example, a *point/vertex* (0-simplex), an *edge* (1-simplex), and a *triangle* (2-simplex) are represented as $\sigma^{(0)}$, $\sigma^{(1)}$, and $\sigma^{(2)}$, respectively. For $0 \leq i \leq n$, the *i*-th *facet* f_i of the *n*-simplex $\sigma^{(n)}$ is the $(n-1)$-simplex $\sigma^{(n-1)}$, whose vertices are those underlying $\sigma^{(n)}$, except the *i*-th vertex. For example, a 2-simplex (triangle) has three 1-simplices (edges) as *facets*. The *boundary* $\partial\sigma^{(n)}$ of the *n*-simplex $\sigma^{(n)}$ is $\bigcup_{i=0}^{n} f_i$. Finally, a *simplicial complex* $\mathcal{K}$ is a set of simplices that satisfy the following conditions: (i). Any face of a simplex in $\mathcal{K}$ is a simplex in $\mathcal{K}$ and (ii). Intersection of distinct simplices σ_i and σ_j in $\mathcal{K}$, is a *common face* of both σ_i and σ_j[2]. The *p-skeleton* $\mathcal{K}^{(p)} \subset \mathcal{K}$ is formed by the set of *k*-simplices $\sigma^{(k)}$, where $k \leq p$. Additionally, we denote by $\mathcal{K}_k$ the set of *k*-simplices in $\mathcal{K}$. The dimension $dim(\mathcal{K})$ of a simplicial complex $\mathcal{K}$ is the maximum of the dimensions of its constituent simplices.

Problem Definition. Our problem consists of first constructing *geometric simplicial complexes* between landmark points given on curved manifolds, followed by finding an optimal (i.e., least cost) assignment between a pair of such geometric simplicial complexes by matching simplices of the same dimension, one dimension at a time. More formally, Let P and P' denote two sets of *landmark points* on curved manifolds $\mathcal{M}$ and $\mathcal{M}'$, respectively. We construct *geometric simplicial complexes* $\mathcal{K}$ and $\mathcal{K}'$ whose set of vertices (0-simplices) are P and P'. The edges/arcs (1-simplices) in $\mathcal{K}$ and $\mathcal{K}'$ are given by geodesics between select few pairs of vertices, from their corresponding vertex sets.

Given two simplicial complexes $\mathcal{K}$ and $\mathcal{K}'$, we assume without loss of generality, that the number of simplices of each corresponding dimension are equal in both complexes. Then, our goal is to find a set of h bijective *matching functions* $\{m_k\}_{k=0}^{h} : \mathcal{K} \longrightarrow \mathcal{K}'$ that match the set of *k*-simplices in $\mathcal{K}$ (i.e., $\mathcal{K}_k$) to *k*-simplices in $\mathcal{K}'$ (i.e., $\mathcal{K}'_k$), for dimensions $k = 0 \ldots h$, to minimize the overall objective function

$$\arg \min_{m_0, \ldots, m_h} \sum_{k=0}^{h} \sum_{i=1}^{|\mathcal{K}_k|} c(\sigma_i^{(k)}, m_k(\sigma_i^{(k)})) \tag{1}$$

---

[2] For ease of notation, we drop the dimension superscript and index subscript for a simplex when it is understood from context.

where $c(\cdot, \cdot)$ is the *geometric matching cost* between a k-simplex $\sigma^{(k)}$ in $\mathcal{K}$ to a k-simplex $m(\sigma^{(k)})$ in $\mathcal{K}'$ and simplicial complex dimension $h = \min(dim(\mathcal{K}), dim(\mathcal{K}'))$. Unlike formulations proposed in graph matching methods [38], where only node and pairwise geometric relations are considered, our *combinatorial optimization* formulation takes into consideration higher-order geometric constraints, which better excludes ambiguous matchings. In subsequent sections, we show how we construct such geometric simplicial complexes from the landmark points on curved manifolds (Sect. 3), followed by a detailed explanation of our assignment algorithm (Sect. 4).

3   Building a Simplicial Complex on a Curved Manifold

In this section, inspired by the work of Dey et al. [11], we similarly construct a *graph-induced* simplicial complex, which is built upon a graph connecting the landmark points. We begin by describing the process of constructing the *underlying graph*.

Graph Construction. Let (P, g) denote the set of landmark points P with a metric g that denotes the geodesic distance between a pair of points on $\mathcal{M}$. Additionally, let the k-neighborhood $\mathcal{N}_k(u)$ denote the set of k nearest neighbors of landmark point $u \in P$ (inclusive of u) on manifold $\mathcal{M}$ according to the geodesic metric g.

Considering all ordered pairs (u, v), where $u, v \in P$, an undirected *edge/arc* is introduced between points u and v, when their corresponding k-neighborhoods $\mathcal{N}_k(u)$ and $\mathcal{N}_k(v)$ have a non-empty intersection, i.e., $\mathcal{N}_k(u) \cap \mathcal{N}_k(v) \neq \emptyset$. All such edges are collected into a set denoted by E. This completes the construction of our underlying graph $G = (P, E)$. Observe that the vertex set (landmarks) P form the 0-skeleton $\mathcal{K}^{(0)}(G)$ and the sets E and P together form the 1-skeleton $\mathcal{K}^{(1)}(G)$, of our graph-induced simplicial complex that we will denote by $\mathcal{K}(G)$.

Recall that a n-clique in a graph is a complete subgraph between n vertices, i.e., it consists of n vertices and $\binom{n}{2}$ edges.

Graph-Induced Complex. $\mathcal{K}(G)$ is defined as the simplicial complex where a n-simplex $\sigma^{(n)} = \{p_1, p_2, \ldots, p_{n+1}\}$ is in $\mathcal{K}(G)$, if and only if there exists a $(n+1)$-clique $\{p_1, p_2, \ldots, p_{n+1}\} \subseteq P$ in the underlying graph $G = (P, E)$. In words, the *cliques* of the underlying graph $G = (P, E)$ form the *simplices* in $\mathcal{K}(G)$ because cliques satisfy both conditions of being a simplicial complex (which can be trivially verified). In order to be used in our assignment algorithm, we must represent the graph-induced simplicial complex $\mathcal{K}(G)$ as a set of *boundary* matrices, which we present next.

Algorithm 1. Matching graph induced simplicial complexes

Input: $\mathcal{K}(G) = \{M_p\}_{p=1}^h$ and $\mathcal{K}(G') = \{M'_p\}_{p=1}^h$

1: **for** $p = h \ldots 1$ **do**
2:     Build cost matrix $\mathcal{L}^{(p)}$ for M_p and M'_p (*account for $\mathcal{L}^{(p-1)}$)
3:     $X_p^* :=$ **Solve QAP** $(M_p, M'_p, \mathcal{L}^{(p)})$
4:     $\mathcal{L}^{(p-1)} :=$ Build cost matrix of $(p-1)$-faces
5: from successful p-simplex matches.
6: **end for**

Return: $\{X_1^*, \ldots, X_h^*\}$ # set of permutation matrices

---

Matrix Representation of $\mathcal{K}(G)$: Given $\mathcal{K}(G)$ and its p-skeleton $\mathcal{K}^{(p)}(G)$ that contains cliques upto size $p+1$, we represent it as a *boundary matrix* $M_p \in \mathbb{Z}^{n \times m}$ defined as

$$M_p = \begin{array}{c} \\ \tau_1^{(p-1)} \\ \vdots \\ \tau_n^{(p-1)} \end{array} \overset{\begin{array}{ccc} \sigma_1^{(p)} & \ldots & \sigma_m^{(p)} \end{array}}{\begin{pmatrix} a_{11} & \ldots & a_{1m} \\ \vdots & \ddots & \vdots \\ a_{n1} & \ldots & a_{nm} \end{pmatrix}}$$

where $a_{ij} = 1$ if and only if the i-th $(p-1)$-simplex $\tau_i^{(p-1)}$ is a *facet* of the j-th p-simplex $\sigma_j^{(p)}$, otherwise $a_{ij} = 0$. Then, the boundary of a j-th p-simplex is given by $\partial_p \sigma_j^{(p)} = \sum_{i=1}^n a_{ij} \tau_i^{(p-1)}$.

Observe that the p-th boundary matrix M_p captures all possible relationships between p-simplices and their $(p-1)$-simplex boundaries (or facets). Boundary matrix M_p is made for each p-skeleton and therefore $\mathcal{K}(G)$ is expressed as a set of boundary matrices $\{M_p\}_{p=1}^h$, where $h = dim(\mathcal{K}(G))$.

Remark 1. Our underlying graph G already contains as a *subgraph* a simple k-nearest neighbor graph which is constructed by introducing edges between a vertex in question and its k nearest neighbors. Therefore, our underlying graph G has more edges and thus has a higher likelihood to form higher-order relations between vertices. On the other hand, while the Delaunay triangulation is simple to compute and is a good vehicle for extracting topology of sampled spaces, its size becomes prohibitively large for reasonable computations and thus adversely affects the QAP matching algorithm.

In summary, our underlying graph G which is inspired by the *Vietoris-Rips* complex construction provides a good *proximity structure*, which is neither *too sparse* (like simple k-NN graphs) or *too dense* (like Delaunay triangulated graphs) and encodes useful higher-order information about local relations of points in P.

4   Assignment Algorithm

Recall our problem definition (Sect. 2) of trying to find a set of assignments/matching functions between two graph-induced simplicial complexes $\mathcal{K}(G)$ and $\mathcal{K}(G')$. Here, we outline the details of our assignment algorithm.

Given a boundary matrix $M_p \in \mathbb{Z}^{n \times m}$ that represents a p-skeleton $\mathcal{K}^{(p)}(G)$, we first capture the *geodesic neighborhood geometry* of simplices in M_p. We begin by defining an *adjacency operator* $\sim$ between two simplices followed by a definition of a *neighborhood of a simplex*. This neighborhood of a simplex is then elegantly captured by *affine weight vectors*, which are later used in the matching algorithm.

Definition 1 (adjacency relation). *Given two simplices $\sigma^{(d)}$ and $\sigma'^{(d')}$, each of arbitrary dimension d and d', we consider them to be* adjacent *to one another if and only if they share a* common simplex. *We denote this adjacency relation by $\sigma^{(d)} \sim \sigma'^{(d')}$. The dimension of the common simplex can take values from 0 to $\min(d, d')$.*

For example, two 2-simplices/triangles $\sigma^{(2)}$ and $\sigma'^{(2)}$ could either be connected at a common 0-simplex/vertex or share a common 1-simplex/edge; both cases would result in the simplices being *adjacent*, i.e., $\sigma^{(2)} \sim \sigma'^{(2)}$.

Simplex Neighborhood. The boundary matrix M_p's columns encode p-simplices $\sigma_1^{(p)}, \ldots, \sigma_m^{(p)}$ and its rows encode $(p-1)$-simplices $\tau_1^{(p-1)}, \ldots, \tau_n^{(p-1)}$. The computation of the neighborhood $\mathfrak{N}(\cdot)$ for p-simplices and $(p-1)$-simplices differ slightly. The neighborhood of a p-simplex consists of p-simplices (same dimension) and $(p-1)$-simplices (one dimension lower) that are adjacent to it. While, the neighborhood of a $(p-1)$-simplex consists of $(p-1)$-simplices (same dimension) and p-simplices (one dimension higher) that are adjacent to it. More formally, the neighborhood of the i-th p-simplex $\sigma_i^{(p)}$ is $\mathfrak{N}(\sigma_i^{(p)}) = \{\sigma_j^{(p)} \mid \sigma_j^{(p)} \sim \sigma_i^{(p)}\} \cup \{\tau_j^{(p-1)} \mid \tau_j^{(p-1)} \sim \sigma_i^{(p)}\}$ and the neighborhood of the i-th $(p-1)$-simplex $\tau_i^{(p-1)}$ is $\mathfrak{N}(\tau_i^{(p-1)})\{\tau_j^{(p-1)} \mid \tau_j^{(p-1)} \sim \tau_i^{(p-1)}\} \cup \{\sigma_j^{(p)} \mid \sigma_j^{(p)} \sim \tau_i^{(p-1)}\}$ Such neighborhoods are computed for all the p- and $(p-1)$-simplices in M_p, where $i \neq j$.

Affine Weight Vectors. For a p-simplex $\sigma^{(p)}$, let $\mathcal{B}(\sigma^{(p)})$ denote the set of all the *barycenters* $\{b_1, \ldots, b_{|\mathfrak{N}(\sigma^{(p)})|}\}$ of the simplices in the neighborhood $\mathfrak{N}(\sigma^{(p)})$. Then, $\sigma^{(p)}$ is represented as an *affine combination* of the barycenters in $\mathcal{B}(\sigma^{(p)})$, i.e., $\sum_{i=1}^{|\mathcal{B}(\sigma^{(p)})|} \alpha_i b_i$, where $\sum_{i=1}^{|\mathcal{B}(\sigma^{(p)})|} \alpha_i = 1$ (i.e., weights α_i's must sum to 1). Therefore, $\sigma^{(p)}$ is expressed as an *affine weight vector* $\alpha(\sigma^{(p)})$ of dimension $(n+m)$, with $|\mathcal{B}(\sigma^{(p)})|$ positions corresponding to $\mathfrak{N}(\sigma^{(p)})$ filled with non-empty affine weights and the rest set to zero. Such an affine weight vector is computed for every simplex of dimension p and $(p-1)$ contained in M_p. Among all possible affine representations of a simplex, we chose to use *least squares* to guarantee minimal error under L2-norm, and furthermore it assigns non-zero weights to each of its adjacent simplex barycenters, thereby better capturing the local geometric properties in its neighborhood.

Remark 2. The affine weight vectors act as *locally affine invariant descriptors* that can handle complex and natural transformations of the underlying manifold $\mathcal{M}$. Additionally, it allows for much fewer variables and can be much more easily linearized in the subsequent QAP formulation. Furthermore, the inclusion of barycenters from neighborhoods of each simplex act as higher-order geometric constraints that easily excludes ambiguous matchings. In comparison, simple matching models that rely on just a distance matrix with pairwise geodesic distances on the manifold are not invariant to local and global affine transformations and completely disregard higher-order relationships.

Cost Matrix Construction. Next, we describe the construction of a *cost matrix* that is needed to compute assignments between $M_p \in \mathbb{Z}^{n \times m}$ and $M'_p \in \mathbb{Z}^{n' \times m'}$. We begin by constructing two cost matrices $\mathcal{C}^{(p-1)} \in \mathbb{R}^{n \times n'}$ and $\mathcal{C}^{(p)} \in \mathbb{R}^{m \times m'}$ to measure the Euclidean distance between the affine weight vectors of $(p-1)$-simplices and the Euclidean distance between the affine weight vectors of p-simplices, respectively.

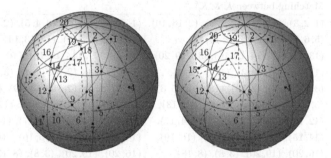

Fig. 1. Pair of spheres with simplicial complexes constructed between the landmark points on the spheres along with assignments between cliques. .

More specifically, $c_{ii'}^{(p-1)} = \|\alpha(\tau_i^{(p-1)}) - \alpha(\tau_{i'}^{(p-1)})\|_2$, measures the Euclidean distance between the affine weight vectors of the i-th $(p-1)$-simplex of M_p and the i'-th $(p-1)$-simplex of M'_p, while $c_{kk'}^{(p)} = \|\alpha(\sigma_k^{(p)}) - \alpha(\sigma_{k'}^{(p)})\|_2$, measures the Euclidean distance between the affine weight vectors of the k-th p-simplex of M_p and the k'-th p-simplex of M'_p.

Similar to the *affinity matrix* construction in [38], we combine both the cost matrices in a single geodesic-cost matrix $\mathcal{L}^{(p)} = (l_{ii',jj'}) \in \mathbb{R}^{nn' \times mm'}$ as

$$
l_{ii',jj'}^{(p)} = \begin{cases} c_{ii'}^{(p-1)} & i = j \,,\, i' = j' \\ c_{kk'}^{(p)} & i \neq j \,,\, i' \neq j' \,,\, a_{ik} a_{jk} a'_{i'k'} a'_{j'k'} = 1 \\ 0 & \text{otherwise} \end{cases}
$$

The *diagonal* and *off-diagonal* entries of matrix $\mathcal{L}^{(p)}$ capture the *Euclidean distances between the affine weight vectors of* $(p-1)$*-simplices* and the *Euclidean distances between the affine weight vectors of p-simplices*, respectively. Therefore, our QAP can now be formulated as

$$\arg\min_{X_1,\ldots,X_h} \sum_{p=1}^{h} vec(X_p)^T \mathcal{L}^{(p)} vec(X_p)$$

$$\text{subject to} \quad \forall p \leq h, \mathbb{1}^T X_p = \mathbb{1}, X_p^T \mathbb{1} = \mathbb{1} \tag{2}$$

where X_p is a permutation matrix and $vec(X_p)$ is it's vector representation. Our solution to Eq. 2 is concisely outlined in Algorithm 1. As we solve a QAP from highest to lowest dimension p-skeleton, we track the $(p-1)$-simplices whose matchings are *induced* by higher order simplex matches. On finding $(p-1)$-simplices that have the lowest cost and cannot be improved by solving a lower level QAP, we eliminate such simplices, causing the size of the matrix to shrink in subsequent iterations, leading to substantial speedups. Also, we use a *spectral relaxation* proposed by Lordeneu et al. [20] to solve our QAP efficiently.

Table 1. Matchings of 3, 2-cliques of simplicial complexes $\mathcal{K}$ and $\mathcal{K}'$ shown in Fig. 1.

k-Clique	Matching between $\mathcal{K}$ & $\mathcal{K}'$	
3-Cliques	$(1,2,3)$, $(1,3,4)$, $(3,4,5)$, $(2,18,19)$, $(5,6,7)$, $(6,8,9)$, $(12,13,14)$, $(13,14,17)$, $(14,15,16)$, $(16,19,20)$	$(1,2,3)$, $(1,3,4)$, $(3,4,5)$, $(2,3,19)$, $(5,6,8)$, $(6,8,9)$, $(12,13,14)$, $(13,14,17)$, $(14,15,16)$, $(16,19,20)$.
2-Cliques	$(1,2)$, $(1,3)$, $(2,3)$, $(1,4)$, $(3,4)$, $(3,5)$, $(4,5)$, $(2,19)$, $(18,19)$, $(5,6)$, $(5,7)$, $(6,8)$, $(6,9)$, $(8,9)$, $(12,13)$, $(12,14)$, $(13,14)$, $(13,17)$, $(14,17)$, $(14,15)$, $(14,16)$, $(15,16)$, $(16,19)$, $(16,20)$, $(19,20)$, $(3,8)$, $(8,18)$, $(9,10)$, $(9,13)$, $(17,18)$	$(1,2)$, $(1,3)$, $(2,3)$, $(1,4)$, $(3,4)$, $(3,5)$, $(4,5)$, $(2,19)$, $(3,19)$, $(5,6)$, $(5,8)$, $(6,8)$, $(6,9)$, $(8,9)$, $(12,13)$, $(12,14)$, $(13,14)$, $(13,17)$, $(14,17)$, $(14,15)$, $(14,16)$, $(15,16)$, $(16,19)$, $(16,20)$, $(19,20)$, $(3,8)$, $(8,13)$, $(9,12)$, $(9,13)$, $(17,8)$

Example. We illustrate with an example the bijective assignment produced by our algorithm between cliques/simplices of a pair of graph-induced *spherical* simplicial complexes, as shown in Fig. 1. We consider two simplicial complexes $\mathcal{K}$ and $\mathcal{K}'$ each embedded on $\mathcal{S}^2$, with 20 and 16 vertices, respectively. Matching of corresponding 3-cliques and 2-cliques are mentioned in the Table 1. Matching between vertices (1-cliques) is shown by marking them with the same label on both spheres.

Time Complexity Analysis. The major cost incurred by our algorithm arises from matching cliques between two simplicial complexes. Therefore, we first derive an upper bound on the number of cliques that need to be matched as follows (proof in supplementary notes).

Lemma 1. *Let $\mathcal{K}(G)$ represent the simplicial complex induced from graph G with n and m number of vertices and edges, respectively. Let h denote the maximum order of cliques in G and δ be the maximum degree of a vertex in G. Then, the total number of k-cliques in $\mathcal{K}(G)$ for $k = (1, \ldots, h)$, are at most*

$$n + \frac{2m}{\delta(\delta+1)} \left[\min \left\{ (\delta+1)^h + 1, \left(\frac{e(\delta+1)}{h} \right)^h \right\} - \delta - 2 \right]$$

Neglecting lower order terms, the number of cliques are of order $O(n + m(\delta^{h-2} - \delta))$. We know that the *spectral relaxation* proposed by Lordeneu et al. [20] has a complexity of $O(n^{3/2})$, where n is the number of points to match on each side. Our higher order matching of cliques then has a time complexity of $O(\{n + m(\delta^{h-2} - \delta)\}^{3/2})$. In practice, for maximum order of cliques, $h = 3$ (triangles) and $h = 4$ (tetrahedrons), observe that the complexity drops to $O(n^{3/2})$ and $O(\{n + m\delta^2\}^{3/2})$, respectively, which is very efficient.

5 Experiments

For our experiments, we considered *synthetic* and *real-world* datasets that cover both *spherical* and *planar* images. Spherical images can broadly be categorized as: *parabolic omnidirectional* (360°), *fish-eye*, and *panoramic* images. Note that our matching algorithm does not require any calibration parameters of cameras.

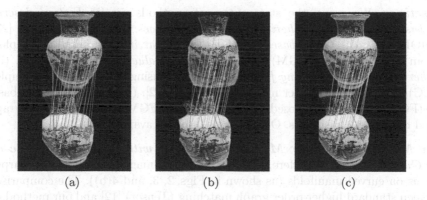

(a)	(b)	(c)

Fig. 2. Instances of matchings between (a) *Chinese vase* images for Tensor based method, (b) flat version of *Chinese vase* images for Tensor based method, and (c) *Chinese vase* images for our method. Green/red lines show correct/incorrect matches respectively. Isolated points show no matches. (Color figure online)

To evaluate our matching algorithm, we compared against three main categories. (i) *Planar matching methods extended with geodesic metric on 3D manifolds*: Here, we *extended* the factorized graph matching (FGM) [38] algorithm by feeding it a k-NN graph based on geodesic distances between points to serve as our naive baseline method (called "FGM+geodesic"). The rest of the methods were feature-descriptor based. (ii) *Planar matching methods on 2D projected (unwrapped[3]*. (iii) *Planar matching methods on 2D planar images*: Here, we proposed a *flat* version of our algorithm with Euclidean distance as the underlying metric (called "OurPlanar") to work on flat 2D images.

Furthermore, we also perform experiments using RANSAC [16] for geometric verification and rectification. In our ablative studies, we analyze the robustness of our algorithm under affine transformations (rotation, reflection, scaling, and shear).

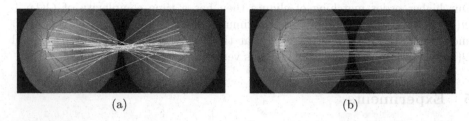

(a)   (b)

Fig. 3. Instances of matchings between (a) *Fundus* images for Tensor based method, (b) *Fundus* images for our method. Green/yellow lines show correct/incorrect matches respectively. Isolated points show no matches. (Color figure online)

Baselines. We group the state-of-the-art methods as: (i) *Feature descriptor based matching for spherical and planar images:* BRISK [21], ORB [27], SPHORB [37]. (ii) *Graph based matching for planar images:* based on employing an *affinity matrix* (FGM) [38,39] and *eigenvalues* (EigenAlign) [13]. (iii) *Higher-order based matching for planar images:* using random clique complex (RCC) [30] and higher-order matching (Tensor) [12]. (iv) Finally, a naive baseline (FGM+geodesic) proposed by us that extends FGM by constructing a graph based on geodesic distances. Our code[4] is publicly available.

Our Method vs. Planar Matching Methods with Geodesic Metric on 3D Curved Manifolds. Here, we match pairwise images directly on the warped images on curved manifolds (as shown in Figs. 2, 3, and 4(b)). The comparison between standard higher-order graph matching (Tensor) [12] and our method on manifold is shown in Figs. 2 and 3 using Chinese vases[5] and Fundus images [17], respectively. We observe from Figs. 2(a) and 3(a) that the Tensor based method does not perform well on warped images. Although, the matching does improve when images are flattened to reduce the effect of curvature in Fig. 2(b). Our

---

[3] *unwrapped:* planar projection of a spherical image with minimal distortion [9,15] manifolds.

[4] Our Method.

[5] From Google images.

Table 2. Error (%) of pairwise matching between spherical images (omnidirectional, fish-eye and panorama) of five datasets for different methods.

Algorithms	Kamaishi	Chessboard	Desktop	Parking	Table
OurWarped	$0.79 \pm 0.0\%$	$3.89 \pm 0.0\%$	$0.32 \pm 0.0\%$	$0.0 \pm 0.0\%$	$0.74 \pm 0.0\%$
FGM+geo	$55.6 \pm 0.10\%$	$79.2 \pm 1.21\%$	$23.3 \pm 0.03\%$	$37.5 \pm 0.0\%$	$64.3 \pm 6.58\%$
SPHORB	$90.0 \pm 0.0\%$	$58.5 \pm 0.0\%$	$91.1 \pm 0.0\%$	$95.0 \pm 0.0\%$	$78.5 \pm 0.0\%$
BRISK	$85.6 \pm 0.0\%$	$53.6 \pm 0.0\%$	$78.9 \pm 0.0\%$	$81.6 \pm 0.0\%$	$69.2 \pm 0.0\%$
ORB	$90.2 \pm 0.0\%$	$53.8 \pm 0.0\%$	$51.7 \pm 0.0\%$	$71.1 \pm 0.0\%$	$64.4 \pm 0.0\%$
Tensor	$37.7 \pm 0.69\%$	$60.5 \pm 0.41\%$	$23.9 \pm 1.7\%$	$23.7 \pm 7.5\%$	$85.1 \pm 1.05\%$
FGM	$53.3 \pm 0.21\%$	$80.0 \pm 0.11\%$	$31.9 \pm 0.12\%$	$36.0 \pm 1.5\%$	$65.5 \pm 0.01\%$

method outperforms the baseline and has a maximum number of correct matches in Figs. 2(c) and 3(b).

The error percentages of our warped image matching algorithm (OurWarped) are shown in the first row of Table 2. We observe that our method outperforms all other matching methods, including spherical feature descriptor based ones as well. Additional multimodal warped-planar matching experiments can be found in our supplementary notes.

For matches between spherical and planar images, we find two variants which match between a spherical and a planar image (Fig. 4(a)) and matching between different types of spherical images (Fig. 4(b)). In Table 2, there is a slight increase in error percentages when matching across different types of spherical images, i.e., 3.89% for Chessboard, as compared to matching similar types, i.e., 0.32% for Desktop, due to differences in distortion levels. In spite of this, we find that our method significantly outperforms naive baseline and other matching methods on spherical images.

Table 3. Error (%) of pairwise matching between unwrapped equirectangular version of spherical (omnidirectional and fish-eye) images of four datasets for different methods including graph matching methods on flat surfaces.

Algorithms	Chessboard	Desktop	Parking	Table
OurWarped	$3.64 \pm 0.0\%$	$1.06 \pm 0.0\%$	$0.0 \pm 0.0\%$	$0.57 \pm 0.0\%$
RCC	$28.6 \pm 0.94\%$	$11.6 \pm 0.74\%$	$13.2 \pm 11.8\%$	$11.6 \pm 0.57\%$
EigenAlign	$98.47 \pm 0.0\%$	$95.24 \pm 0.0\%$	$97.5 \pm 0.0\%$	$97.9 \pm 0.0\%$
Tensor	$68.9 \pm 0.16\%$	$26.1 \pm 0.58\%$	$19.0 \pm 3.75\%$	$72.4 \pm 0.67\%$
FGM	$84.0 \pm 0.0\%$	$31.0 \pm 0.0\%$	$38.0 \pm 0.0\%$	$52.0 \pm 0.0\%$
SPHORB	$58.6 \pm 0.0\%$	$90.3 \pm 0.0\%$	$97.5 \pm 0.0\%$	$79.2 \pm 0.0\%$
BRISK	$54.9 \pm 0.0\%$	$84.9 \pm 0.0\%$	$100.0 \pm 0.0\%$	$74.2 \pm 0.0\%$
ORB	$49.5 \pm 0.0\%$	$78.2 \pm 0.0\%$	$82.5 \pm 0.0\%$	$70.3 \pm 0.0\%$

(a)   (b)

Fig. 4. Instances of matchings between (a) *Desktop* omnidirectional and planar images and (b) *Chessboard* omnidirectional and fish-eye images. Green/red lines show correct/incorrect matches, respectively. Isolated points show no matches. (Color figure online)

Our Method vs. Planar Matching Methods on 2D-Projected Curved Manifolds. Matching between spherical images can also be performed by applying planar graph matching methods on unwrapped equirectangular versions of spherical images. This makes the image flat and standard planar matching algorithms can then be employed. However, any kind of projection (on a flat surface in this case) introduces distortions in the resulting image. We flattened spherical images for four datasets mentioned in Table 3. We used two different methods to flatten omnidirectional and fish-eye images. The 360° image is unwrapped by dividing it into four parts (quadrants) and concatenated into a single flat image. On the other hand, fish-eye images do not cover the complete view of the scene and add distortion to the image due to curved mirrors and lenses of the cameras. We try to reduce the distortion by removing curves and flattening the image using calibration techniques outlined in [28]. Since any projection will lead to distortion, we can compare the results from Table 2 with Table 3. Both the experimental outcomes are based on the same set of spherical images. Our matching algorithm significantly outperforms its competitors on both the spherical images and on curved manifolds.

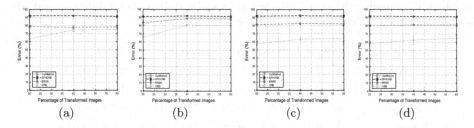

(a)   (b)   (c)   (d)

Fig. 5. Error (%) in matching when varying the percentage (20% to 60%) of transformed images in the set of spherical images of *Desktop* (a)–(d). (a) 40° rotation, (b) reflection, (c) scaling and (d) shear.

Our Method vs. Planar Matching Methods on 2D Planar Images. Our proposed method can also be reduced to a higher-order planar graph matching method. To show the importance of higher-order combinatorial matching not only with geodesic neighborhood, but also with euclidean neighborhood, we run our planar variant (OurPlanar) on popular 2D image datasets, competing with standard matching algorithms. We pick four well-known difficult matching datasets (Books, Building, Magazine, and Butterfly) that suffer from heavy occlusions and non-affine transformations [30]. Results for such an experiment are shown in Table 4. From the results, we observe that our method also serves as a powerful planar matching method and is still competitive using an euclidean neighborhood for our affine weight vectors. It significantly outperforms both the popular planar matching methods.

Table 4. Error (%) of pairwise matching between planar images of four datasets for different methods.

Algorithms	Magazine	Building	Books	Butterfly
OurPlanar	**0.0 ± 0.0%**	**1.03 ± 0.01%**	**19.72 ± 0.20%**	**0.0 ± 0.0%**
FGM	**0.0 ± 0.0%**	74.87 ± 0.07%	97.54 ± 0.01%	16.12 ± 0.0%
Tensor	**0.0 ± 0.0%**	43.24 ± 2.98%	32.35 ± 0.15%	1.07 ± 0.17%

RANSAC: Geometric Verification and Rectification. We also performed fundamental matrix based geometric verification using RANSAC algorithm [16] after descriptor based matching on two datasets for spherical and their planar versions for SPHORB and ORB in Table 5. We observed that the results are improved (but still not better than our proposed method) in some cases but prune a lot of matches. Nearly 40–50% of matches are considered as outliers which makes it difficult to handle the noise. On the other hand, our method performs much better in any case while considering outliers.

Table 5. Error (%) of pairwise matching between spherical images of Desktop and Parking datasets and on their unwrapped versions for verification.

Algorithms	Desktop	Desktop_flat	Parking	Parking_flat
OurWarped	**0.32 ± 0.0%**	1.06 ± 0.0%	0.0 ± 0.0%	0.0 ± 0.0%
SPHORB+RANSAC	96.1 ± 0.0%	93.9 ± 0.0%	95.0 ± 0.0%	100.0 ± 0.0%
ORB+RANSAC	29.3 ± 0.0%	70.6 ± 0.0%	55.0 ± 0.0%	100.0 ± 0.0%

We performed rectification [5] on spherical images of Desktop dataset followed by BRISK descriptor for matching. The results improved from 78.9% (in Table 2) to 52.11% error. However, we observed that despite these improvements, our method still outperforms them. Also, in most of the cases, the rectification algorithm does not perform well and outputs noisy or distorted images. So, there is no guarantee to find the best solution.

Ablative Studies (Effect of Affine Transformation). We *remove* completely at random 40–80% of landmark points on the Desktop dataset, and introduce affine transformations on these points. Figure 5 shows the results of affine transformation like *rotation, reflection, scaling,* and *shear.* We rotated images (clockwise) by 40° and performed matching for four algorithms. Then, we generated mirror images along the x-axis from the same dataset to introduce reflection. We also conducted transformation by scaling and shear of 360° images. We resized images in both the directions with scales 0.5 and 1.5 randomly. For shearing, we stretched images with 0.5 factor along y-axis. For all types of transformations, we observe that the results shown in Fig. 5 clearly indicates that our method is robust to all kinds of affine transformations and easily outperforms other state-of-the-art methods.

6   Conclusion

We presented a bijective assignment between sets of landmark points embedded on a pair of images warped onto curved manifolds by the following steps. First, we built a *graph induced simplicial complex* on the warped images. Second, we proposed a constrained QAP that matches corresponding co-dimensional simplices between two simplicial complexes along with an efficient algorithm to solve the constrained QAP. Finally, we conducted extensive experiments, broadly grouped as *comparative matching* and *ablative studies*, in order to gain insight into the accuracy and robustness of our method. We are currently exploring the possibility of integrating such high-dimensional combinatorial structures into *Spherical CNNs* [6] to capture higher-order and latent structure.

References

1. Ahmed, A.I., et al.: Sonographic measurement of cervical volume in pregnant women at high risk of preterm birth using a geometric formula for a frustum versus 3-dimensional automated virtual organ computer-aided analysis. J. Ultrasound Med. **36**(11), 2209–2217 (2017)
2. Alahi, A., Ortiz, R., Vandergheynst, P.: FREAK: fast retina keypoint. In: CVPR, pp. 510–517. IEEE (2012)
3. Azevedo, R.G.d.A., Birkbeck, N., De Simone, F., Janatra, I., Adsumilli, B., Frossard, P.: Visual distortions in 360-degree videos. arXiv preprint arXiv:1901.01848 (2019)
4. Bay, H., Tuytelaars, T., Van Gool, L.: SURF: speeded up robust features. In: Leonardis, A., Bischof, H., Pinz, A. (eds.) ECCV 2006. LNCS, vol. 3951, pp. 404–417. Springer, Heidelberg (2006). https://doi.org/10.1007/11744023_32
5. Bradski, G., Kaehler, A.: Learning OpenCV: Computer Vision with the OpenCV Library. O'Reilly Media Inc., Sebastopol (2008)
6. Cohen, T.S., Geiger, M., Köhler, J., Welling, M.: Spherical CNNs. In: ICLR (2018)
7. Colombo, C., Bimbo, A.D., Pernici, F.: Image mosaicing from uncalibrated views of a surface of revolution. In: Proceedings of the BMVC, pp. 43.1–43.10 (2004)

8. Coors, B., Condurache, A.P., Geiger, A.: SphereNet: learning spherical representations for detection and classification in omnidirectional images. In: Ferrari, V.,
   Hebert, M., Sminchisescu, C., Weiss, Y. (eds.) ECCV 2018. LNCS, vol. 11213, pp.
   525–541. Springer, Cham (2018). https://doi.org/10.1007/978-3-030-01240-3_32
9. Cruz-Mota, J., Bogdanova, I., Paquier, B., Bierlaire, M., Thiran, J.P.: Scale invariant feature transform on the sphere: theory and applications. Int. J. Comput.
   Vision 98(2), 217–241 (2012). https://doi.org/10.1007/s11263-011-0505-4
10. Cyganek, B.: An Introduction to 3D Computer Vision Techniques and Algorithms.
    Wiley, Hoboken (2007)
11. Dey, T.K., Fan, F., Wang, Y.: Graph induced complex on point data. Comput.
    Geom. Theory Appl. 48(8), 575–588 (2015)
12. Duchenne, O., Bach, F., Kweon, I.S., Ponce, J.: A tensor-based algorithm for high-
    order graph matching. IEEE Trans. Pattern Anal. Mach. Intell. (PAMI) 33(12),
    2383–2395 (2011)
13. Feizi, S., Quon, G., Recamonde-Mendoza, M., Médard, M., Kellis, M., Jadbabaie,
    A.: Spectral alignment of networks. arXiv preprint arXiv:1602.04181 (2016)
14. Goesele, M., Snavely, N., Curless, B., Hoppe, H., Seitz, S.M.: Multi-view stereo for
    community photo collections. In: ICCV, pp. 1–8. IEEE (2007)
15. Guan, H., Smith, W.A.: BRISKS: binary features for spherical images on a geodesic
    grid (2017)
16. Hartley, R., Zisserman, A.: Multiple View Geometry in Computer Vision. Cam-
    bridge University Press, Cambridge (2003)
17. Kalesnykiene, V., et al.: DIARETDB0: evaluation database and methodology for
    diabetic retinopathy algorithms (2006)
18. Kaminsky, R.S., Snavely, N., Seitz, S.M., Szeliski, R.: Alignment of 3D point clouds
    to overhead images. In: 2009 IEEE Computer Society Conference on Computer
    Vision and Pattern Recognition Workshops, pp. 63–70. IEEE (2009)
19. Leordeanu, M., Zanfir, A., Sminchisescu, C.: Semi-supervised Learning and Opti-
    mization for hypergraph matching. In: IEEE International Conference on Com-
    puter Vision (ICCV), November 2011
20. Leordeanu, M., Hebert, M.: A spectral technique for correspondence problems using
    pairwise constraints. In: ICCV, vol. 2, pp. 1482–1489. IEEE (2005)
21. Leutenegger, S., Chli, M., Siegwart, R.Y.: BRISK: binary robust invariant scalable
    keypoints. In: ICCV, pp. 2548–2555. IEEE (2011)
22. Li, Y., Gregori, G., Lam, B.L., Rosenfeld, P.J.: Automatic montage of SD-OCT
    data sets. Opt. Express 19(27), 26239–26248 (2011)
23. Lowe, D.G.: Distinctive image features from scale-invariant keypoints. Int. J. Com-
    put. Vision 60(2), 91–110 (2004)
24. Munkres, J.R.: Elements of Algebraic Topology. Addison-Wesley, Boston (1984)
25. Pauly, O., Unal, G., Slabaugh, G., Carlier, S., Fang, T.: Semi-automatic matching
    of OCT and IVUS images for image fusion (2008). https://doi.org/10.1117/12.
    773805
26. Rosten, E., Drummond, T.: Machine learning for high-speed corner detection. In:
    Leonardis, A., Bischof, H., Pinz, A. (eds.) ECCV 2006. LNCS, vol. 3951, pp. 430–
    443. Springer, Heidelberg (2006). https://doi.org/10.1007/11744023_34
27. Rublee, E., Rabaud, V., Konolige, K., Bradski, G.: ORB: an efficient alternative
    to SIFT or SURF. In: ICCV, pp. 2564–2571. IEEE (2011)
28. Scaramuzza, D., Martinelli, A., Siegwart, R.: A toolbox for easily calibrating omni-
    directional cameras. In: 2006 IEEE/RSJ International Conference on Intelligent
    Robots and Systems, pp. 5695–5701. IEEE (2006)

29. Sengupta, S., Singh, A., Leopold, H.A., Lakshminarayanan, V.: Ophthalmic diagnosis and deep learning - a survey (2018)
30. Sharma, C., Nathani, D., Kaul, M.: Solving partial assignment problems using random clique complexes. In: ICML, pp. 4593–4602 (2018)
31. Shen, D., Davatzikos, C.: HAMMER: hierarchical attribute matching mechanism for elastic registration. IEEE Trans. Med. Imaging **21**(11), 1421–1439 (2002)
32. Starck, J., Hilton, A.: Spherical matching for temporal correspondence of non-rigid surfaces. In: Tenth IEEE International Conference on Computer Vision (ICCV 2005) Volume 1, vol. 2, pp. 1387–1394. IEEE (2005)
33. Sung, B., Kim, S.H., Lee, S., Lim, J., Lee, J.K., Soh, K.S.: Nanofluid transport in a living soft microtube. J. Phys. D Appl. Phys. **48** (2015)
34. Szeliski, R.: Computer Vision: Algorithms and Applications. TCS. Springer, London (2011). https://doi.org/10.1007/978-1-84882-935-0
35. Yang, J., Li, H., Jia, Y.: Go-ICP: solving 3D registration efficiently and globally optimally. In: Proceedings of the IEEE International Conference on Computer Vision, pp. 1457–1464 (2013)
36. Zeng, A., Song, S., Nießner, M., Fisher, M., Xiao, J., Funkhouser, T.: 3DMatch: Learning local geometric descriptors from RGB-D reconstructions. In: Proceedings of the IEEE Conference on Computer Vision and Pattern Recognition, pp. 1802–1811 (2017)
37. Zhao, Q., Feng, W., Wan, L., Zhang, J.: SPHORB: a fast and robust binary feature on the sphere. Int. J. Comput. Vision **113**(2), 143–159 (2015)
38. Zhou, F., De la Torre, F.: Factorized graph matching. IEEE Trans. Pattern Anal. Mach. Intell. (PAMI) **38**(9), 1774–1789 (2016)
39. Zhou, F., De la Torre, F.: Deformable graph matching. In: CVPR, pp. 2922–2929 (2013)

Representation Learning on Visual-Symbolic Graphs for Video Understanding

Effrosyni Mavroudi[✉][ID], Benjamín Béjar Haro[ID], and René Vidal[ID]

Mathematical Institute for Data Science, Johns Hopkins University,
Baltimore, MD, USA
{emavrou1,bbejar,rvidal}@jhu.edu

Abstract. Events in natural videos typically arise from spatio-temporal interactions between actors and objects and involve multiple co-occurring activities and object classes. To capture this rich visual and semantic context, we propose using two graphs: (1) an attributed spatio-temporal visual graph whose nodes correspond to actors and objects and whose edges encode different types of interactions, and (2) a symbolic graph that models semantic relationships. We further propose a graph neural network for refining the representations of actors, objects and their interactions on the resulting hybrid graph. Our model goes beyond current approaches that assume nodes and edges are of the same type, operate on graphs with fixed edge weights and do not use a symbolic graph. In particular, our framework: a) has specialized attention-based message functions for different node and edge types; b) uses visual edge features; c) integrates visual evidence with label relationships; and d) performs global reasoning in the semantic space. Experiments on challenging video understanding tasks, such as temporal action localization on the Charades dataset, show that the proposed method leads to state-of-the-art performance.

1 Introduction

The field of video understanding has been moving towards increasing levels of complexity, from classifying a single action in short videos to detecting multiple complex activities performed by multiple actors interacting with objects in untrimmed videos. Therefore, there is a need to develop algorithms that can effectively model spatio-temporal visual and semantic context. One way of capturing such context is to use graph-based modeling, which has a rich history in computer vision. Traditional graph-based approaches to video understanding, e.g., using probabilistic graphical models [26,27,58,68], focused mainly on modeling context at the level of symbols rather than visual signals or representations. However,

Electronic supplementary material The online version of this chapter (https://doi.org/10.1007/978-3-030-58526-6_5) contains supplementary material, which is available to authorized users.

recent advances have enabled *representation learning on graph-structured data* using deep architectures called Graph Neural Networks (GNNs), which learn how to refine node representations by aggregating messages from their neighbors [25].

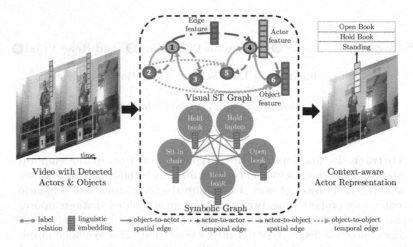

Fig. 1. Cues for video understanding: (1) *visual spatio-temporal interactions* between actors and objects and (2) commonsense *relationships between labels*, such as co-occurrences. These cues can be encoded in a hybrid spatio-temporal visual and symbolic attributed graph. In this work, we perform representation learning on this hybrid graph to obtain context-aware representations of detected semantic entities, such as actors and objects, that can be used to solve downstream video understanding tasks, such as multi-label action recognition.

Videos can be represented as visual spatio-temporal attributed graphs (visual st-graphs) whose nodes correspond to regions obtained by an object detector and whose edges capture interactions between such regions. GNNs have recently been designed for refining the local node/edge features, typically extracted by a convolutional neural network, based on the spatio-temporal context captured by the graph. Although representation learning on visual st-graphs has lead to significant advances in video understanding [3,14,19,52,57,62,63], there are four key limitations of state-of-the-art approaches that prevent them from fully exploiting the rich structure of these graphs. First, the visual st-graph is a *heterogeneous* graph that has distinct node types (*actor, object,* etc.) and distinct edge types (*object-to-actor spatial, actor-to-actor temporal,* etc.), with each type being associated with a feature of potentially different dimensionality and semantics, as shown in the example of Fig. 1. However, most GNNs assume nodes/edges of the same type. Therefore, recent attempts at explicitly modeling actors and objects have resorted to applying separate GNNs for each node/edge type [12,63]. Second, most methods operate on a graph of fixed edge weights with dense connectivity between detected regions. In practice, only a few of the edges capture meaningful interactions. Third, current approaches do not incorporate edge features, such as *geometric relations between regions*, for updating the node representations. Finally, despite modeling local visual context, existing approaches do not reason at a global video level or exploit semantic label relationships, which have been shown to be beneficial in the image recognition domain [5,33].

In this work, in an effort to address these limitations, we propose a novel Graph Neural Network (GNN) model, called Visual Symbolic - Spatio Temporal - Message Passing Neural Network (VS-ST-MPNN), that performs representation learning on visual st-graphs to obtain context-aware representations of detected actors and objects (Fig. 1). Our model handles heterogeneous graphs by employing *learnable message functions that are specialized for each edge type*. We also adapt the visual edge weights with an *attention mechanism* specialized for each type of interaction. For example, an actor node will separately attend to actor nodes at the previous frame and object nodes at the current frame. Furthermore, we use *edge features* to refine the actor and object representations and to compute the attention coefficients that determine the connection strength between regions. Intuitively, nodes which are close to each other or are interacting should be strongly connected. Finally, one of our key contributions is incorporating an attributed *symbolic graph* whose nodes correspond to semantic labels, such as actions, described by word embeddings and whose edges capture label relationships, such as co-occurrence. We fuse the information of the two graphs with learnable association weights between their nodes and learn global semantic interaction-aware features. Importantly, we do not require ground truth annotations of objects, tracks or semantic labels for each visual node.

In summary, the contributions of this work are three-fold. First, we model contextual cues for video understanding by combining a symbolic graph, capturing semantic label relationships, with a visual st-graph, encoding interactions between detected actors and objects. Second, we introduce a novel GNN that can perform joint representation learning on the heterogeneous visual-symbolic graph, in order to obtain visual and semantic context-aware representations of actors, objects and their interactions in a video, which can then be used to solve downstream recognition tasks. Finally, to demonstrate the effectiveness and generality of our method, we evaluate it on tasks such as multi-label temporal activity localization, object affordance detection and grounded video description on three challenging datasets and show that it achieves state-of-the-art performance.

2   Related Work

Visual Context for Video Understanding. Context and its role in vision has been studied for a long time [42]. There are two major, complementary ways of utilizing context in video understanding tasks: (a) extracting global *representations from whole frames* by applying convolutional neural networks to short video segments [4,51,54,56,65] followed by long-term temporal models [29, 43,65] and (b) extracting *mid-level representations* based on semantic parts, such as body parts [7,39], latent attributes [35], secondary regions [15], human-object interactions [45,67] and object-object interactions [3,36,65]. Our proposed method falls into the latter category, using GNNs to obtain representations of detected semantic entities based on interactions captured by visual and symbolic graphs.

Graph Neural Networks for Video Understanding. The first approach applying a deep network on a visual graph for video understanding was the Structured Inference Machine [11], which introduced actor feature refinement with message passing, and trainable gating functions for filtering out spurious interactions, but only captured spatial relationships between actors. Another early approach was the S-RNN [21], which introduced the concept of weight-sharing between nodes or edges of the same type, but did not iteratively refine node representations. With the advent of GNNs, many researchers have explored modeling whole frames [64], tracklets [63], feature map columns [14,41,52] or object proposals [19,57,62] as graph nodes and using off-the-shelf GNNs, such as MPNNs [13], GCNs [25] and Relation Networks [3,20,52,64] to refine the node or edge representations, obtaining significant performance gains. However, most of these GNNs are unable to handle edge features, directed edges and distinct node and edge types. Therefore, applying existing GNNs to visual st-graphs requires treating every node and edge in the same way [3,14], or focusing only on one edge type [19,20,36,52,64], or using separate GNNs for each type of interaction [12,57,63], hence completely ignoring or sub-optimally handling their rich graph structure. In contrast, our proposed method can be directly applied to any st-graph and supports message passing in heterogeneous graphs. The benefit of such fine-grained modeling has already been established in fields such as computational pharmacology and relational databases [16,48,69], but remains relatively unexplored in computer vision. Furthermore, similar to [14,46], our method iteratively adapts the visual edge weights, but employs an attention mechanism that is specialized for different edge types and takes edge features into account.

Symbolic Graphs. There is a long line of work on exploiting external knowledge encoded in label relation graphs for visual recognition tasks. Semantic label hierarchies, such as co-occurrence, have been leveraged for improving object recognition [8,10,37,38], multi-label zero-shot learning [30] and other image-based visual tasks [31,47]. Much fewer papers utilize knowledge graphs for video understanding [1,23,24], possibly due to the limited number of semantic classes in traditional video datasets. However, most of these methods directly perform inference on the symbolic graph. For example, the SINN [24] performs graph-based inference in a hierarchical label space for action recognition. Rather, we aim to use the semantics of labels to integrate prior knowledge about the inter-class relationships and facilitate the computation of semantic context-aware region features. In a similar vein, Liang et al. [33] enhance feature maps extracted from images by using a symbolic graph, while [6,32] use a latent interaction graph. In contrast, we seek to improve the representation of visual st-graph nodes rather than enhance convolutional features on a regular grid. Fusing information from multiple graphs using GNNs is an exciting new line of research [2,53,59]. Similar to our approach, Chen et al. [5] combine a visual graph instantiated on objects with a symbolic graph and perform graph representation learning, while [22] enforce the scalar edge weights between visual regions to be consistent with the edges of the symbolic graph. However, they operate on simple spatial graphs and assume access to semantic labels of regions during training.

3 Method

In this section, we describe the overall architecture of our proposed VS-ST-MPNN model, shown in Fig. 2. Our goal is to refine the features of detected actors, objects and their interactions based on the contextual information captured in two graphs: a visual st-graph and a symbolic graph. The refinement is performed by a novel GNN, which is designed to exploit the rich structure of the visual st-graph by utilizing edge features and learning specialized attention-based neighborhood aggregation functions for different node and edge types. In addition, our model enables the fusion with the symbolic graph, by

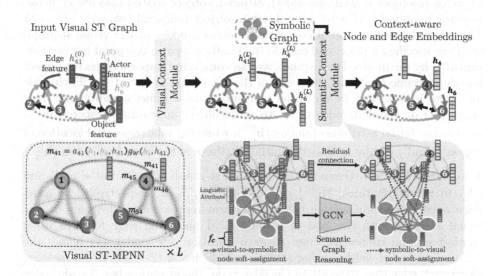

Fig. 2. Overview of our VS-ST-MPNN model that performs representation learning on a hybrid visual-symbolic graph. Given an input video that is represented as a visual st-graph, with nodes corresponding to detected actors and objects and edges capturing latent interactions, our framework has two modules that integrate context in the local representations of its nodes and edges: (a) a Visual Context Module (Sect. 3.1) that performs L rounds of node and edge updates on the visual graph, with specialized neighborhood aggregation functions that depend on the type of an edge and (b) a Semantic Context Module (Sect. 3.2) that integrates visual evidence with semantic knowledge encoded in an external symbolic graph and learns global semantic interaction-aware features.

incorporating graph convolutions, to learn semantic relation-aware features, and soft-assignment weights, to connect visual and symbolic graph nodes without requiring access to ground-truth semantic labels of regions during training. Our model can be trained jointly with recognition networks to solve downstream video understanding tasks.

3.1 Visual Context Module

Visual st-graph. Our input is a sequence of T frames with detected actor and object regions. Let $G^v = (V^v, E^v)$ be a spatio-temporal attributed directed graph, called the *visual st-graph*, where V^v is a finite set of vertices and $E^v \subseteq V^v \times V^v$ is a set of edges. Nodes correspond to actor and object detections, while edges model latent interactions. There are M actors and N objects per frame. Figure 2 illustrates a toy example with $M = 1$, $N = 2$ and $T = 2$. The graph is both node- and edge-typed with $\mathcal{N}$ node types and $\mathcal{E}$ edge types, i.e., each node (edge) is associated with a single node (edge) type. Specifically, the node types are *actor* and *object* ($\mathcal{N} = 2$) and the edge types are object-to-actor spatial (*obj-act-s*), actor-to-object spatial (*act-obj-s*), object-to-object spatial (*obj-obj-s*), actor-to-actor temporal (*act-act-t*) and object-to-object temporal (*obj-obj-t*) ($\mathcal{E} = 5$). The allowed spatio-temporal connections between nodes of the visual st-graph (E^v) are specified a priori and encode the family of spatio-temporal interactions captured by the model. For instance, we can constrain temporal edges to connect a node at frame t with another node of the same type at time $t - 1$. Each node and edge is described by an initial attribute, whose dimensionality may vary depending on the node/edge type. An actor/object appearance feature can be used as the initial attribute of node i ($\mathbf{h}_i^{(0)}$), while the relative spatial location of regions i and j can be used as the initial attribute of the edge from j to i ($\mathbf{h}_{ij}^{(0)}$).

Visual ST-MPNN. Given the input visual st-graph G^v with initial node and edge attributes/features, $\{\mathbf{h}_i^{(0)}\}_{i \in V^v}$ and $\{\mathbf{h}_{ij}^{(0)}\}_{(i,j) \in E^v}$, respectively, we introduce novel GNN propagation rules to perform representation learning on the visual st-graph with the goal of refining local node and edge features using spatio-temporal contextual cues. At each iteration our model: (1) refines the scalar visual edge weights using attention coefficients; (2) computes a message along each edge that depends on the edge type, the attention-based scalar edge weight, the features of the connected nodes and the edge feature; (3) updates the feature of every node by aggregating messages from incoming edges; and (4) updates the feature of every edge by using the message that was computed alongside it. Next, we describe each one of these steps in more detail.

– *Attention Mechanism:* At each iteration l of the MPNN, we first refine the strength of region connections by computing *attention coefficients*, a_{ij}, that capture the relevance of node j (message sender) for the update of node i (message receiver). In contrast to GAT [55], our model learns an attention mechanism specialized for each type of interaction and it utilizes edge features for its computation. The attention coefficients for the l-th iteration are computed as follows:

$$a_{ij}^{(l)} = \exp\left(\gamma_{ij}^{(l)}\right) / \left(\sum_{k \in N_{\epsilon_{ij}}^{v}(i)} \exp\left(\gamma_{ik}^{(l)}\right)\right),$$ (1)

$$\gamma_{ij}^{(l)} = \rho\left((\mathbf{v}_a^{\epsilon_{ij}})^T \left[W_r^{\nu_i}\mathbf{h}_i^{(l-1)}; W_s^{\nu_j}\mathbf{h}_j^{(l-1)}; \beta W_e^{\epsilon_{ij}}\mathbf{h}_{ij}^{(l-1)}\right]\right).$$ (2)

Here, ϵ_{ij} is the type of the edge from node j to node i, $N_{\epsilon_{ij}}^{v}(i)$ is the set of visual nodes connected with node i via an incoming edge of type ϵ_{ij}, $\mathbf{h}_i^{(l-1)}$ is the feature of the i-th node at the previous iteration, $\mathbf{h}_{ij}^{(l-1)}$ is the feature of the edge from j to i at the previous iteration, ν_i is the type of node i and ρ is a non-linearity, such as Leaky-ReLU [18]. The parameter $\beta \in \{0,1\}$ denotes whether edge features will be used for computing the attention coefficients ($\beta = 1$) or not ($\beta = 0$). $W_r^{\nu_i}$, $W_s^{\nu_j}$ and $W_e^{\epsilon_{ij}}$ are learnable projection weights and are shared between nodes (edges) of the same type. All projection matrices linearly transform the current node (edge) feature to a refined feature of fixed dimensionality d_l. $\mathbf{v}_a^{\epsilon_{ij}}$ is a learnable attention vector. For improved readability we have dropped the layer index (l) from the attention and projection weights.

– *Message Computation:* After computing the attention coefficients, we compute a message along each edge. The message from node j to node i is:

$$\mathbf{m}_{ij}^{(l)} = a_{ij}^{(l)}\left(\lambda_v W_s^{\nu_j}\mathbf{h}_j^{(l-1)} + \lambda_e W_e^{\epsilon_{ij}}\mathbf{h}_{ij}^{(l-1)}\right),$$ (3)

where the parameters $\lambda_e, \lambda_v \in \{0,1\}$ denote whether the edge feature and the sender node feature will be used for the message computation, respectively.

– *Node and Edge Update:* Following the message computation, the node feature is updated using an aggregation of incoming messages from different edge types and a residual connection, while the edge feature is set to be equal to the message:

$$\mathbf{h}_i^{(l)} = \mathbf{h}_i^{(l-1)} + \sigma\left(\sum_{j \in N^v(i)} \mathbf{m}_{ij}^{(l)}\right), \mathbf{h}_{ij}^{(l)} = \mathbf{m}_{ij}^{(l)},$$ (4)

where $N^v(i)$ is the set of visual nodes that are connected with node i, $\sigma(\cdot)$ is a non-linearity, such as ReLU. After L layers of the spatio-temporal MPNN (or equivalently L rounds of node and edge updates), we obtain refined, visual context-aware node and edge features: $\mathbf{h}_i^{(L)} \in \mathbb{R}^{d_L}$ and $\mathbf{h}_{ij}^{(L)} \in \mathbb{R}^{d_L}$.

3.2 Semantic Context Module

Symbolic Graph. Let $G^s = (V^s, E^s)$, be the input *symbolic* graph, where V^s and E^s denote the symbol set and edge set, respectively. The nodes of this graph correspond to semantic labels, such as action labels or object labels. Each symbolic node c is associated with a semantic attribute, such as the linguistic embedding of the label ($\mathbf{s}_c \in \mathbb{R}^K$). Edges in the symbolic graph are associated with scalar weights, which encode label relationships, such as co-occurrence. These edge weights are summarized in the fixed adjacency matrix $L^s \in \mathbb{R}^{|V^s| \times |V^s|}$.

– *Integration of Visual Evidence with the Symbolic Graph:* As a first step, we update the attributes of the symbolic graph using visual evidence, i.e., the visual context-aware representations of the nodes of the visual st-graph. To achieve this, without requiring access to the ground-truth semantic labels of regions, we learn associations between the nodes of the visual st-graph and those of the symbolic graph. These associations are the edges of the bipartite graph $G^{vs} = (V^{vs}, E^{vs})$, with $V^{vs} = V^v \cup V^s$ and $E^{vs} \subseteq V^v \times V^s$. Although latent, we can specify a priori the allowed visual-to-symbolic node connections (edges). For example, when symbolic nodes correspond to action classes, we can remove edges between object and symbolic nodes. The learnable association weight $\phi_{c,i}^{vs}$ represents the confidence of assigning the feature from visual node i to the symbolic node c:

$$\phi_{c,i}^{vs} = \frac{\exp\left((\mathbf{w}_c^{vs})^T \mathbf{h}_i^{(L)}\right)}{\sum_{c' \in N^{vs}(i)} \exp\left((\mathbf{w}_{c'}^{vs})^T \mathbf{h}_i^{(L)}\right)}, \tag{5}$$

where $\mathbf{w}_c^{vs} \in \mathbb{R}^{d_L}$ is a trainable weight vector and $N^{vs}(i)$ is the neighborhood of visual node i on the bipartite graph G^{vs}. After computing the voting weights, each symbolic node is associated with a weighted sum of projected visual node features: $\mathbf{f}_c = \sigma(\sum_i \phi_{c,i}^{vs} W_p^{vs} \mathbf{h}_i^{(L)})$, where $W_p^{vs} \in \mathbb{R}^{D_s \times d_L}$ is a learnable projection weight matrix. The new representation of each node c is computed as the concatenation of the linguistic embedding and the visual feature: $\mathbf{s}_c^{(0)} = [\mathbf{s}_c; \mathbf{f}_c] \in \mathbb{R}^{K+D_s}$.

– *Semantic Graph Convolutions:* We obtain semantic relation-aware features by applying a vanilla GCN [25] to the nodes of the symbolic graph. More specifically, by iteratively applying the propagation rule $S^{(r+1)} = \text{GCN}(S^{(r)}, L^s)$, where $S^{(r)}$ denotes the matrix of symbolic node embeddings at iteration $r = 1, \ldots, R$, the GCN yields evolved symbolic node features $S^{(R)} \in \mathbb{R}^{|V^s| \times D_s}$.

– *Update of Visual st-graph:* The evolved symbolic node representations obtained after R iterations of graph convolutions on the symbolic graph can be mapped back to the visual st-graph, so that the representation of the visual nodes can be enriched by global semantic context. To achieve this we compute mapping weights (attention coefficients) from symbolic nodes to visual nodes:

$$\phi_{i,c}^{sv} = \frac{\exp\left((\mathbf{v}_a^{sv})^T \left[\mathbf{s}_c^{(R)}; \mathbf{h}_i^{(L)}\right]\right)}{\sum_{c' \in N^{vs}(i)} \exp\left((\mathbf{v}_a^{sv})^T \left[\mathbf{s}_{c'}^{(R)}; \mathbf{h}_i^{(L)}\right]\right)}, \tag{6}$$

where $\mathbf{v}_a^{sv} \in \mathbb{R}^{d_L+D_s}$ is a learnable attention vector. The final visual node feature representation is then obtained using a residual connection: $\mathbf{h}_i = \mathbf{h}_i^{(L)} + \sigma\left(\sum_{c' \in V^s} \phi_{i,c'}^{sv} W_p^{sv} \mathbf{s}_{c'}^{(R)}\right)$. These context-aware representations can be fed to recognition networks to solve downstream video understanding tasks.

4   Experiments

To demonstrate the effectiveness and generality of our method, we conduct experiments on three challenging video understanding tasks that require reasoning about interactions between semantic entities and relationships between classes: a) sub-activity and object affordance classification (Sect. 4.1), b) multi-label temporal action localization (Sect. 4.2) and c) grounded video description (Sect. 4.3).

Table 1. Results on CAD-120 [27] for sub-activity and object affordance detection, measured via F1-score. Our results are averaged over five random runs, with the standard deviation reported in parentheses.

Method	Detection F1-score (%)	
	Sub-activity	Object affordance
ATCRF [27]	80.4	81.5
S-RNN [21]	83.2	88.7
S-RNN [21] (multitask)	82.4	**91.1**
GPNN [46]	88.9	88.8
STGCN [12]	88.5	–
VS-ST-MPNN (ours)	**90.4** (±0.8)	*89.2* (±0.3)
Only visual graph (ours)	89.6 (±1.1)	88.6 (±0.6)

4.1   Experiments on CAD-120

CAD-120. This dataset provides 120 RGB-D videos, with each video showing a daily activity comprised of a sequence of sub-activities (e.g., *moving, drinking*) and object affordances (e.g., *reachable, drinkable*) [27]. Given temporal segments, the task is to classify each actor in each segment into one of 10 sub-activity classes and each object into one of 12 affordance classes. Evaluation is performed with 4-fold, leave-one-subject-out, cross-validation using F1-scores averaged over all classes as an evaluation metric. With a visual st-graph provided by the dataset [27] (including hand-crafted features of actors and objects and geometric relations), it is a particularly good test-bed for comparing different GNNs.

Implementation Details. The visual st-graph provided with the dataset is instantiated on the actors and objects of each temporal segment of the input video and contains 5 edge types: *obj-obj-s*, *obj-act-s*, *act-obj-s*, *act-act-t* and *obj-obj-t*. We construct a symbolic graph that has nodes corresponding to the 10 sub-activity and 12 affordance classes, with edge weights capturing per-frame class co-occurrences in training data. The attribute of each symbolic node is obtained by using off-the-shelf word2vec [40] class embeddings of size $K = 300$. Actor (object) nodes are connected to sub-activity (affordance) symbolic nodes. The following hyperparameters are used in the our model: $L = 4$, $R = 1$, $\lambda_v = 1, \lambda_e = 1$ and $\beta = 1$. All messages are of size 256. We use the sum of cross-entropy losses per node to jointly train our model and the sub-activity and affordance classifiers applied at each node of the st-graph. We train for 100 epochs with a batch size of 5 sequences and use the Adam learning rate scheduler with an initial learning rate of 0.001. Dropout with a rate of 0.5 is applied to all fully connected layers.

Comparison with the State of the Art. Table 1 compares the sub-activity and affordance detection performance of our method with prior work. Our method obtains state-of-the-art results for sub-activity detection, with an average performance of **90.4%** and a best of **91.3%**, and the second best result on affordance detection (89.2%) - being only second to the S-RNN (multi-task) [21]. The S-RNN was trained on the joint task of detection and anticipation and we outperform it by 8% in the sub-activity classification task. Even without using the symbolic graph, our method improves upon recent GNNs, which were applied on the same attributed visual st-graph, validating our novel layer propagation rules.

Ablation Analysis. In Fig. 3, we show the effect of attention, edge features and number of visual node updates on the recognition performance. First, we compare the performance of a model using a fixed binary adjacency matrix with that of a model using our attention mechanism. Clearly, attention benefits performance in both tasks. Second, we conclude that using the attributes of both the neighboring nodes and adjacent edges is better than using only those of the neighboring nodes, validating the usefulness of edge features. Finally, increasing the number of ST-MPNN layers improves performance, which saturates after 4–5 layers.

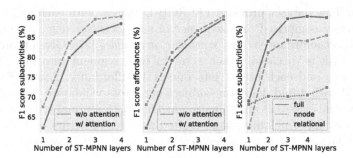

Fig. 3. Effect of attention mechanism and node update type on CAD-120 detection performance. Using an attention mechanism outperforms using a fixed visual adjacency matrix. Updating nodes based on both neighboring node and incoming edge attributes (*full*) is superior to updating them using just the nodes (*nnode*) or edges (*relational*).

4.2 Experiments on Charades

Charades. Charades [50] is a large dataset with 9848 RGB videos and temporal annotations for 157 action classes, many of which involve human-object interactions. Each video contains an average of 6.8 activity instances, many of which are co-occurring. Following [50], multi-label action temporal localization performance is measured in terms of mean Average Precision (mAP), evaluating per-frame predictions for 25 equidistant frames in each one of the 1.8k validation videos.

Implementation Details. To tackle the challenging problem of multi-label temporal action localization, we perform a late fusion of a global model, operating on whole frames, and a local model, operating on actors and objects. The global model is an I3D RGB model [4] fine-tuned on Charades [43], combined with a two-layer biGRU of size 256, similar to existing baselines on this dataset. The proposed VS-ST-MPNN is used as the local model. To build the visual st-graph we detect actors and objects using a Faster-RCNN [17] trained on the MS-COCO [34] dataset. We rank detections based on their score and we keep the top-2 person detections and top-10 object detections per frame. We pool features from the Mixed_4f 3D feature map of the I3D for each detected region using RoIAlign [17] and max-pooling in space. This yields an attribute of size 832 for the actor/object regions for the frames of the original video sampled at 1.5 FPS. We use 3 types of visual edges: *obj-act-s*, *act-obj-s* and *act-act-t* and

Table 2. Multi-label temporal action localization results on Charades [50]. Performance is measured via per-frame mAP. R: RGB, F: optical flow. Our method yields a relative improvement of 6% over the state-of-the-art method by using **only raw RGB frames**.

Method	Feat	Input	mAP (%)
Predictive-corrective [9]	VGG	R	8.9
Two-stream [49]	VGG	R+F	8.94
Two-stream + LSTM [49]	VGG	R+F	9.6
R-C3D [60]	VGG	R+F	12.7
ATF [49]	VGG	R+F	12.8
RGB I3D [43]	I3D	R	15.63
I3D [43]	I3D	R+F	17.22
I3D + LSTM [43]	I3D	R+F	18.12
RGB I3D + super-events [43]	I3D	R	18.64
I3D + super-events [43]	I3D	R+F	19.41
STGCN [12]	I3D	R+F	19.09
I3D + biGRU	I3D	R	21.7
I3D + 3TGMs + super-events [44]	I3D	R+F	22.3
I3D + biGRU + VS-ST-MPNN (Ours)	I3D	**R**	**23.7**(±0.2)

describe each edge with the relative position of the connected regions. Our symbolic graph has nodes corresponding to the 157 action classes and edge weights corresponding to per-frame label co-occurrences in training data. Only actor nodes are connected to symbolic nodes. Obtaining a linguistic attribute for each symbolic node is not trivial, since action names often contain multiple words. To circumvent that, each action class is separated into a verb and an object and the average of their word2vec [40] embeddings ($K = 300$) is used as the initial node attribute. The hyperparameters are: $L = 3$, $d_L = 512$, $R = 1$, $D_s = 256$, $\lambda_v = 1, \lambda_e = 1$ and $\beta = 1$. For performing per-frame multi-label action classification, we average the learned actor node and edge representations at each frame, we input them to a two-layer biGRU of size 512, and we feed the resulting hidden states to binary action classifiers for per-frame multi-label action classification. We jointly train the VS-ST-MPNN and biGRU for 40 epochs with a binary cross-entropy loss applied per frame, using a batch size of 16 sequences. We also apply dropout with a rate of 0.5 on all fully connected layers and use the Adam scheduler, with an initial learning rate of $1e^{-4}$.

Comparison with Prior Work. As shown in Table 2, our framework outperforms all other methods on temporal action localization, with a mAP of **23.7%** (averaged across 3 random runs) by using only raw RGB frames. It yields a relative improvement of 24% over the alternative graph-based approach [12], which uses both RGB and optical flow inputs, as well as additional actor embeddings trained at the ImSitu dataset [61].

Table 3. Ablation analysis on Charades [50]. *Visual*: Visual Context Module. *Semantic*: Semantic Context Module. *Long Term*: long-term temporal modeling.

ID	Visual	Semantic	Long Term	mAP (%)	mAP (%)
					+ Global model
1	✓	✓	✓	**18.6**	**23.4**
2	–	–	✓	15.2	22.2
3	✓	✓	–	15.3	22.0
4	✓	–	–	13.7	21.8
5	–	✓	–	11.7	21.8
6	–	–	–	10.7	20.9

Impact of Each Graph. In Table 3, we report the baseline result (10.7%) obtained by classifying activities based on local actor features (ID: 6). Refining these features by using our Visual Context Module improves performance by 3%. As shown quantitatively in the supplementary material, both our specialized attention mechanism and the usage of edge features improve the performance, outperforming a vanilla GNN. Representation learning on the hybrid graph yields a significant absolute improvement of **5% over the baseline**. Additionally, modeling long-term temporal context and global context leads to the final state-of-the-art performance, indicating that the representations learned by our model are complementary to holistic scene cues and temporal dynamics.

Per-class Improvement Analysis. To gain a better understanding of the benefits of representation learning on the visual graph, we highlight in Fig. 4 the activity classes with the highest positive and negative difference in performance when adding *obj-to-act-s* messages. By harnessing visual human-object interaction cues, our model is able to better recognize actions such as *Watching television.*

Impact of Semantic Context Module. Comparing the models with IDs 3 and 4 in Table 3, we observe that adding the semantic context module improves mAP by 2%. Notably, updating the visual nodes by attending over the initial symbolic node features (linguistic) instead of the evolved features did not improve performance in our experiments, showing the importance of semantic graph convolutions. The semantic module seems to particularly help with rare classes, such as *Holding a vacuum*, which has only 213 training examples (3% of available annotated segments), and classes with strong co-occurrences (Fig. 5). The t-SNE visualization shows that, although the visual context-aware actor embeddings are already capturing meaningful label relationships (e.g., *open* and *hold book*), the integration of semantic relationships via the symbolic graph results in more tightly clustered embeddings and well-defined groups, facilitating action recognition.

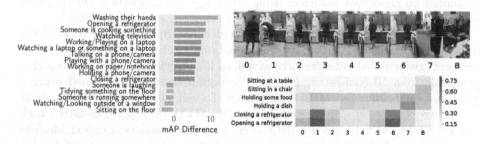

Fig. 4. Qualitative results on Charades. (*left*) The classes with the highest positive and negative performance difference after adding *object-to-actor spatial* messages. Incorporating spatial structure benefits actions that involve interactions with distant objects, such as *watching television* or *cooking*. (*right*) Action predictions of our model (ID: 3) for 9 frames of a sample Charades video.

Model Complexity. Since our visual st-graph is designed to capture only local spatio-temporal interactions, we can compute messages in parallel and process the entire Charades validation set (around 2K videos at 1.5FPS) in 2 min on a single Titan XP GPU, given initial features pooled from actor/object regions.

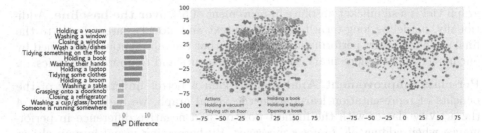

Fig. 5. Qualitative evaluation of the Semantic Context Module (SCM). (*left*) Classes with the highest positive and negative performance difference when adding the semantic module. (*right*) t-SNE visualization of actor node embeddings from Charades validation set obtained before and after adding the SCM. We show 1121 random samples per class for 5 selected action classes. (*Best viewed zoomed in and in color.*) (Color figure online)

4.3 Experiments on ActivityNet Entities

ActivityNet Entities. The task in the recently released ActivityNet Entities [65] dataset, containing 15k videos and more than 158k annotated bounding boxes, is to generate a sentence describing the event in a ground-truth video segment, and to spatially localize all the generated nouns that belong to a vocabulary of 432 object classes. Following Zhou et al. [65], the quality of generated captions is measured using standard metrics, such as Bleu (B@1, B@4), METEOR (M), CIDEr (C), and SPICE (S), whereas the quality of object localization is evaluated on generated sentences using the $F1_{all}$, $F1_{loc}$ metrics. Object localization results on the test set were obtained using the evaluation server[1].

The current state-of-the-art grounded video description model (GVD) [65] uses a hierarchical LSTM decoder that generates a sentence describing a video segment, given global video features as well as local region features of 100 region proposals from 10 equidistant frames. The region features are refined using a *multi-head self-attention* (MHA) mechanism. To validate the effectiveness of our model, we experiment with three variants of the GVD: (a) replace the MHA with our VS-ST-MPNN; (b) use the MHA along with our Semantic Context Module; (c) the same as before but with visual-to-symbolic node assignment weights initialized based on knowledge transfer from the Visual Genome dataset [28]. We use a symbolic graph whose nodes correspond to object classes. As shown in Table 4, replacing MHA with our visual module does not improve captioning, but it improves localization accuracy with a relative improvement of 4% (24.1 → 25.2). Adding our Semantic Context Module to GVD leads to an improvement across all captioning and localization metrics, which is even more pronounced in the test set (improving CIDEr from 45.5 to 47.7). Note that the initial region features already captured semantic information by including object class probabilities. Therefore, the improvement in captioning cannot be attributed solely to

---

[1] https://competitions.codalab.org/competitions/20537.

Table 4. Grounded video description results on ActivityNet Entities [65]. MHA: multi-head self-attention. SCM-VG: our semantic context module with visual-to-symbolic node correspondences pre-trained on Visual Genome.

	$B@1$	$B@4$	M	C	S	$F1_{all}$	$F1_{loc}$
Validation set							
GVD (MHA) [65]	**23.9**	2.59	11.2	47.5	15.1	7.11	24.1
GVD (VCM + SCM) (ours)	23.4	2.41	11.1	47.3	14.8	7.28	25.2
GVD (MHA + SCM) (ours)	23.8	2.67	**11.3**	48.6	**15.2**	**7.35**	**25.3**
GVD (MHA + SCM-VG) (ours)	**23.9**	**2.78**	11.3	49.1	15.1	7.15	24.0
Test set							
Masked Transformer [66]	22.9	2.41	10.6	46.1	13.7	–	–
Bi-LSTM+TempoAttn [66]	22.8	2.17	10.2	42.2	11.8	–	–
GVD (MHA) [65]	23.6	2.35	11.0	45.5	14.9	7.59	25.0
GVD (VCM + SCM) (ours)	23.1	2.34	10.9	46.1	14.5	–	–
GVD (MHA + SCM) (ours)	23.6	2.54	11.2	47.7	15.0	7.30	24.4
GVD (MHA + SCM-VG) (ours)	**24.1**	**2.63**	**11.4**	**49.0**	**15.1**	**7.81**	**27.1**

the inclusion of semantic context, but rather to our semantic reasoning framework. Finally, from the superior captioning performance of our third variant, we conclude that prior knowledge about correspondences between visual and symbolic nodes, if available, can possibly facilitate representation learning on the hybrid graph.

5   Conclusions

In this paper, we proposed a novel deep learning framework for video understanding that performs joint representation learning on a hybrid graph, composed of a symbolic graph and a visual st-graph, for obtaining context-aware visual node and edge features. We obtained state-of-the-art performance on three challenging datasets, demonstrating the effectiveness and generality of our framework.

Acknowledgements. The authors thank Carolina Pacheco Oñate, Paris Giampouras and the anonymous reviewers for their valuable comments. This research was supported by the IARPA DIVA program via contract number D17PC00345.

References

1. Assari, S.M., Zamir, A.R., Shah, M.: Video classification using semantic concept co-occurrences. In: IEEE Conference on Computer Vision and Pattern Recognition (2014)
2. Bajaj, M., Wang, L., Sigal, L.: G3raphGround: graph-based language grounding. In: IEEE International Conference on Computer Vision, pp. 4281–4290 (2019)

3. Baradel, F., Neverova, N., Wolf, C., Mille, J., Mori, G.: Object level visual reasoning in videos. In: Ferrari, V., Hebert, M., Sminchisescu, C., Weiss, Y. (eds.) ECCV 2018. LNCS, vol. 11217, pp. 106–122. Springer, Cham (2018). https://doi.org/10.1007/978-3-030-01261-8_7

4. Carreira, J., Zisserman, A.: Quo vadis, action recognition? A new model and the kinetics dataset. In: IEEE Conference on Computer Vision and Pattern Recognition, pp. 4724–4733 (2017). https://doi.org/10.1109/CVPR.2017.502

5. Chen, X., Li, L., Fei-Fei, L., Gupta, A.: Iterative visual reasoning beyond convolutions. In: IEEE Conference on Computer Vision and Pattern Recognition, pp. 7239–7248 (2018). https://doi.org/10.1109/CVPR.2018.00756

6. Chen, Y., Rohrbach, M., Yan, Z., Shuicheng, Y., Feng, J., Kalantidis, Y.: Graph-based global reasoning networks. In: IEEE Conference on Computer Vision and Pattern Recognition (2019)

7. Chéron, G., Laptev, I., Schmid, C.: P-CNN: pose-based CNN features for action recognition. In: IEEE International Conference on Computer Vision, pp. 3218–3226 (2015). https://doi.org/10.1109/ICCV.2015.368

8. Choi, M.J., Lim, J.J., Torralba, A., Willsky, A.S.: Exploiting hierarchical context on a large database of object categories. In: IEEE Conference on Computer Vision and Pattern Recognition, pp. 129–136 (2010). https://doi.org/10.1109/CVPR.2010.5540221

9. Dave, A., Russakovsky, O., Ramanan, D.: Predictive-corrective networks for action detection. In: IEEE Conference on Computer Vision and Pattern Recognition, pp. 2067–2076 (2017). https://doi.org/10.1109/CVPR.2017.223

10. Deng, J., et al.: Large-scale object classification using label relation graphs. In: Fleet, D., Pajdla, T., Schiele, B., Tuytelaars, T. (eds.) ECCV 2014. LNCS, vol. 8689, pp. 48–64. Springer, Cham (2014). https://doi.org/10.1007/978-3-319-10590-1_4

11. Deng, Z., Vahdat, A., Hu, H., Mori, G.: Structure inference machines: recurrent neural networks for analyzing relations in group activity recognition. In: IEEE Conference on Computer Vision and Pattern Recognition, pp. 4772–4781 (2016). https://doi.org/10.1109/CVPR.2016.516

12. Ghosh, P., Yao, Y., Davis, L., Divakaran, A.: Stacked spatio-temporal graph convolutional networks for action segmentation. In: IEEE Winter Applications of Computer Vision Conference (2020)

13. Gilmer, J., Schoenholz, S.S., Riley, P.F., Vinyals, O., Dahl, G.E.: Neural message passing for quantum chemistry. In: International Conference on Machine Learning, pp. 1263–1272 (2017)

14. Girdhar, R., Carreira, J., Doersch, C., Zisserman, A.: Video action transformer network. In: IEEE Conference on Computer Vision and Pattern Recognition (2019)

15. Gkioxari, G., Girshick, R., Malik, J.: Contextual action recognition with R*CNN. In: IEEE International Conference on Computer Vision, pp. 1080–1088 (2015). https://doi.org/10.1109/ICCV.2015.129

16. Gong, L., Cheng, Q.: Exploiting edge features for graph neural networks. In: IEEE Conference on Computer Vision and Pattern Recognition (2019)

17. He, K., Gkioxari, G., Dollar, P., Girshick, R.: Mask R-CNN. IEEE Trans. Pattern Anal. Mach. Intell., 1 (2018). https://doi.org/10.1109/TPAMI.2018.2844175

18. He, K., Zhang, X., Ren, S., Sun, J.: Delving deep into rectifiers: surpassing human-level performance on ImageNet classification. In: IEEE International Conference on Computer Vision (2015)

19. Huang, H., Zhou, L., Zhang, W., Xu, C.: Dynamic graph modules for modeling higher-order interactions in activity recognition. In: British Machine Vision Conference (2019)
20. Ibrahim, M.S., Mori, G.: Hierarchical relational networks for group activity recognition and retrieval. In: Ferrari, V., Hebert, M., Sminchisescu, C., Weiss, Y. (eds.) ECCV 2018. LNCS, vol. 11207, pp. 742–758. Springer, Cham (2018). https://doi.org/10.1007/978-3-030-01219-9_44
21. Jain, A., Zamir, A.R., Savarese, S., Saxena, A.: Structural-RNN: deep learning on spatio-temporal graphs. In: IEEE Conference on Computer Vision and Pattern Recognition, pp. 5308–5317 (2016)
22. Jiang, C., Xu, H., Liang, X., Lin, L.: Hybrid knowledge routed modules for large-scale object detection. In: Neural Information Processing Systems, pp. 1552–1563 (2018)
23. Jiang, Y.G., Wu, Z., Wang, J., Xue, X., Chang, S.F.: Exploiting feature and class relationships in video categorization with regularized deep neural networks. IEEE Trans. Pattern Anal. Mach. Intell. 40(2), 352–364 (2018). https://doi.org/10.1109/TPAMI.2017.2670560
24. Junior, N.I.N., Hu, H., Zhou, G., Deng, Z., Liao, Z., Mori, G.: Structured label inference for visual understanding. IEEE Trans. Pattern Anal. Mach. Intell., 1 (2019). https://doi.org/10.1109/TPAMI.2019.2893215
25. Kipf, T.N., Welling, M.: Semi-supervised classification with graph convolutional networks. In: International Conference on Learning Representations (2017)
26. Koller, D., et al.: Towards robust automatic traffic scene analysis in real-time. In: IEEE Conference on Computer Vision and Pattern Recognition (1994)
27. Koppula, H.S., Gupta, R., Saxena, A.: Learning human activities and object affordances from RGB-D videos. Int. J. Rob. Res. 32(8), 951–970 (2013). https://doi.org/10.1177/0278364913478446
28. Krishna, R., et al.: Visual genome: connecting language and vision using crowdsourced dense image annotations. Int. J. Comput. Vis. 123(1), 32–73 (2017). https://doi.org/10.1007/s11263-016-0981-7
29. Lea, C., Flynn, M.D., Vidal, R., Reiter, A., Hager, G.D.: Temporal convolutional networks for action segmentation and detection. In: IEEE Conference on Computer Vision and Pattern Recognition, pp. 1003–1012 (2017). https://doi.org/10.1109/CVPR.2017.113
30. Lee, C., Fang, W., Yeh, C., Wang, Y.F.: Multi-label zero-shot learning with structured knowledge graphs. In: IEEE Conference on Computer Vision and Pattern Recognition, pp. 1576–1585 (2018). https://doi.org/10.1109/CVPR.2018.00170
31. Li, R., Tapaswi, M., Liao, R., Jia, J., Urtasun, R., Fidler, S.: Situation recognition with graph neural networks. In: IEEE International Conference on Computer Vision (2017)
32. Li, Y., Gupta, A.: Beyond grids: learning graph representations for visual recognition. In: Bengio, S., Wallach, H., Larochelle, H., Grauman, K., Cesa-Bianchi, N., Garnett, R. (eds.) Neural Information Processing Systems, pp. 9225–9235 (2018)
33. Liang, X., Hu, Z., Zhang, H., Lin, L., Xing, E.P.: Symbolic graph reasoning meets convolutions. In: Neural Information Processing Systems, pp. 1853–1863. Curran Associates, Inc. (2018)
34. Lin, T.-Y., et al.: Microsoft COCO: common objects in context. In: Fleet, D., Pajdla, T., Schiele, B., Tuytelaars, T. (eds.) ECCV 2014. LNCS, vol. 8693, pp. 740–755. Springer, Cham (2014). https://doi.org/10.1007/978-3-319-10602-1_48

35. Liu, J., Kuipers, B., Savarese, S.: Recognizing human actions by attributes. In: IEEE Conference on Computer Vision and Pattern Recognition, pp. 3337–3344 (2011). https://doi.org/10.1109/CVPR.2011.5995353
36. Ma, C., Kadav, A., Melvin, I., Kira, Z., AlRegib, G., Graf, H.P.: Attend and interact: higher-order object interactions for video understanding. In: IEEE Conference on Computer Vision and Pattern Recognition, pp. 6790–6800 (2018). https://doi.org/10.1109/CVPR.2018.00710
37. Marszalek, M., Schmid, C.: Semantic hierarchies for visual object recognition. In: IEEE Conference on Computer Vision and Pattern Recognition, pp. 1–7 (2007). https://doi.org/10.1109/CVPR.2007.383272
38. Marszałek, M., Schmid, C.: Constructing category hierarchies for visual recognition. In: Forsyth, D., Torr, P., Zisserman, A. (eds.) ECCV 2008. LNCS, vol. 5305, pp. 479–491. Springer, Heidelberg (2008). https://doi.org/10.1007/978-3-540-88693-8_35
39. Mavroudi, E., Tao, L., Vidal, R.: Deep moving poselets for video based action recognition. In: IEEE Winter Applications of Computer Vision Conference, pp. 111–120 (2017). https://doi.org/10.1109/WACV.2017.20
40. Mikolov, T., Sutskever, I., Chen, K., Corrado, G.S., Dean, J.: Distributed representations of words and phrases and their compositionality. In: Neural Information Processing Systems, pp. 3111–3119 (2013)
41. Nicolicioiu, A., Duta, I., Leordeanu, M.: Recurrent space-time graph neural networks. In: Neural Information Processing Systems (2019)
42. Oliva, A., Torralba, A.: The role of context in object recognition. Trends Cogn. Sci. 11(12), 520–527 (2007). https://doi.org/10.1016/j.tics.2007.09.009
43. Piergiovanni, A., Ryoo, M.S.: Learning latent super-events to detect multiple activities in videos. In: IEEE Conference on Computer Vision and Pattern Recognition, pp. 5304–5313 (2018). https://doi.org/10.1109/CVPR.2018.00556
44. Piergiovanni, A.J., Ryoo, M.S.: Temporal gaussian mixture layer for videos. In: International Conference on Machine learning (2019)
45. Prest, A., Ferrari, V., Schmid, C.: Explicit modeling of human-object interactions in realistic videos. IEEE Trans. Pattern Anal. Mach. Intell. (2013)
46. Qi, S., Wang, W., Jia, B., Shen, J., Zhu, S.-C.: Learning human-object interactions by graph parsing neural networks. In: Ferrari, V., Hebert, M., Sminchisescu, C., Weiss, Y. (eds.) ECCV 2018. LNCS, vol. 11213, pp. 407–423. Springer, Cham (2018). https://doi.org/10.1007/978-3-030-01240-3_25
47. Ramanathan, V., et al.: Learning semantic relationships for better action retrieval in images. In: IEEE Conference on Computer Vision and Pattern Recognition, pp. 1100–1109 (2015). https://doi.org/10.1109/CVPR.2015.7298713
48. Schlichtkrull, M., Kipf, T.N., Bloem, P., van den Berg, R., Titov, I., Welling, M.: Modeling relational data with graph convolutional networks. In: Gangemi, A., et al. (eds.) ESWC 2018. LNCS, vol. 10843, pp. 593–607. Springer, Cham (2018). https://doi.org/10.1007/978-3-319-93417-4_38
49. Sigurdsson, G.A., Divvala, S., Farhadi, A., Gupta, A.: Asynchronous temporal fields for action recognition. In: IEEE Conference on Computer Vision and Pattern Recognition, pp. 5650–5659 (2017). https://doi.org/10.1109/CVPR.2017.599
50. Sigurdsson, G.A., Varol, G., Wang, X., Farhadi, A., Laptev, I., Gupta, A.: Hollywood in homes: crowdsourcing data collection for activity understanding. In: Leibe, B., Matas, J., Sebe, N., Welling, M. (eds.) ECCV 2016. LNCS, vol. 9905, pp. 510–526. Springer, Cham (2016). https://doi.org/10.1007/978-3-319-46448-0_31

51. Simonyan, K., Zisserman, A.: Two-stream convolutional networks for action recognition in videos. In: Ghahramani, Z., Welling, M., Cortes, C., Lawrence, N.D., Weinberger, K.Q. (eds.) Neural Information Processing Systems, pp. 568–576. Curran Associates, Inc. (2014)
52. Sun, C., et al.: Actor-centric relation network. In: Ferrari, V., Hebert, M., Sminchisescu, C., Weiss, Y. (eds.) ECCV 2018. LNCS, vol. 11215, pp. 335–351. Springer, Cham (2018). https://doi.org/10.1007/978-3-030-01252-6_20
53. Teney, D., Liu, L., van den Hengel, A.: Graph-structured representations for visual question answering. In: IEEE Conference on Computer Vision and Pattern Recognition, pp. 1–9 (2017)
54. Tran, D., Bourdev, L., Fergus, R., Torresani, L., Paluri, M.: Learning spatiotemporal features with 3D convolutional networks. In: IEEE International Conference on Computer Vision (2015)
55. Veličković, P., Cucurull, G., Casanova, A., Romero, A., Liò, P., Bengio, Y.: Graph attention networks. In: International Conference on Learning Representations (2018)
56. Wang, L., et al.: Temporal segment networks: towards good practices for deep action recognition. In: Leibe, B., Matas, J., Sebe, N., Welling, M. (eds.) ECCV 2016. LNCS, vol. 9912, pp. 20–36. Springer, Cham (2016). https://doi.org/10.1007/978-3-319-46484-8_2
57. Wang, X., Gupta, A.: Videos as space-time region graphs. In: Ferrari, V., Hebert, M., Sminchisescu, C., Weiss, Y. (eds.) ECCV 2018. LNCS, vol. 11209, pp. 413–431. Springer, Cham (2018). https://doi.org/10.1007/978-3-030-01228-1_25
58. Wang, X., Ji, Q.: Video event recognition with deep hierarchical context model. In: IEEE Conference on Computer Vision and Pattern Recognition (2015)
59. Xiong, Y., Huang, Q., Guo, L., Zhou, H., Zhou, B., Lin, D.: A graph-based framework to bridge movies and synopses. In: IEEE International Conference on Computer Vision (2019)
60. Xu, H., Das, A., Saenko, K.: R-C3D: region convolutional 3D network for temporal activity detection. In: IEEE International Conference on Computer Vision, pp. 5794–5803 (2017). https://doi.org/10.1109/ICCV.2017.617
61. Yatskar, M., Zettlemoyer, L., Farhadi, A.: Situation recognition: visual semantic role labeling for image understanding. In: IEEE Conference on Computer Vision and Pattern Recognition (2016)
62. Yuan, Y., Liang, X., Wang, X., Yeung, D., Gupta, A.: Temporal dynamic graph LSTM for action-driven video object detection. In: IEEE International Conference on Computer Vision, pp. 1819–1828 (2017). https://doi.org/10.1109/ICCV.2017.200
63. Zhang, Y., Tokmakov, P., Hebert, M., Schmid, C.: A structured model for action detection. In: IEEE Conference on Computer Vision and Pattern Recognition (2019)
64. Zhou, B., Andonian, A., Oliva, A., Torralba, A.: Temporal relational reasoning in videos. In: Ferrari, V., Hebert, M., Sminchisescu, C., Weiss, Y. (eds.) ECCV 2018. LNCS, vol. 11205, pp. 831–846. Springer, Cham (2018). https://doi.org/10.1007/978-3-030-01246-5_49
65. Zhou, L., Kalantidis, Y., Chen, X., Corso, J.J., Rohrbach, M.: Grounded video description. In: IEEE Conference on Computer Vision and Pattern Recognition (2019)
66. Zhou, L., Zhou, Y., Corso, J.J., Socher, R., Xiong, C.: End-to-end dense video captioning with masked transformer. In: IEEE Conference on Computer Vision and Pattern Recognition, pp. 8739–8748 (2018)

67. Zhou, Y., Ni, B., Tian, Q.: Interaction part mining: a mid-level approach for fine-grained action recognition. In: IEEE Conference on Computer Vision and Pattern Recognition, pp. 3323–3331 (2015). https://doi.org/10.1109/CVPR.2015.7298953
68. Zhu, Y., Nayak, N.M., Roy-Chowdhury, A.K.: Context-aware modeling and recognition of activities in video. In: IEEE Conference on Computer Vision and Pattern Recognition (2013)
69. Zitnik, M., Agrawal, M., Leskovec, J.: Modeling polypharmacy side effects with graph convolutional networks. Bioinformatics, 457–466 (2018)

Distance-Normalized Unified Representation for Monocular 3D Object Detection

Xuepeng Shi[1], Zhixiang Chen[1(✉)], and Tae-Kyun Kim[1,2]

[1] Imperial College London, London, England
{x.shi19,zhixiang.chen,tk.kim}@imperial.ac.uk
[2] Korea Advanced Institute of Science and Technology, Daejeon, South Korea

Abstract. Monocular 3D object detection plays an important role in autonomous driving and still remains challenging. To achieve fast and accurate monocular 3D object detection, we introduce a single-stage and multi-scale framework to learn a unified representation for objects within different distance ranges, termed as UR3D. UR3D formulates different tasks of detection by exploiting the scale information, to reduce model capacity requirement and achieve accurate monocular 3D object detection. Besides, distance estimation is enhanced by a distance-guided NMS, which automatically selects candidate boxes with better distance estimates. In addition, an efficient fully convolutional cascaded point regression method is proposed to infer accurate locations of the projected 2D corners and centers of 3D boxes, which can be used to recover object physical size and orientation by a projection-consistency loss. Experimental results on the challenging KITTI autonomous driving dataset show that UR3D achieves accurate monocular 3D object detection with a compact architecture.

Keywords: Monocular 3D object detection · Unified representation across different distance ranges · Distance-guided NMS · Fully convolutional cascaded point regression

1 Introduction

Object detection is a fundamental and challenging problem in computer vision [25]. In the past years, with the emergence of deep learning [11,18] and the availability of large-scale annotated datasets [6,24], the state of the art in 2D object detection has improved significantly [4,10,23,27,34,40]. Object detection in the 2D image plane, however, is not sufficient for autonomous driving, which often requires accurate 3D localization of targets in the scene. Currently, the foremost methods [17,36,45,49] on 3D object detection heavily rely on expensive LiDAR sensors to provide accurate depth information as input. Monocular 3D object detection [2,3,16,19,26,30,31,44] is a promising low-cost solution, but it is much harder due to the ill-posed nature, i.e., lack of depth cues.

© Springer Nature Switzerland AG 2020
A. Vedaldi et al. (Eds.): ECCV 2020, LNCS 12374, pp. 91–107, 2020.
https://doi.org/10.1007/978-3-030-58526-6_6

The performance gap between LiDAR-based approaches and monocular methods is still substantial.

One key challenge for monocular 3D object detection is in handling large distance variations so that the detector can estimate 3D locations accurately. Learning the distance-specific feature requires specific sophisticated designs [1, 29, 33, 42], while simply learning the feature covering all possible locations is difficult and costs much capacity of the model, resulting in a heavy and slow model for good accuracy. In this work, we solve the learning efficiency problem by introducing a single-stage and multi-scale framework that learns a unified representation of objects in different scales and distance ranges, termed as UR3D. The deep model is relieved from learning different representations for objects within different scale and distance ranges, which significantly reduces the cost of network capacity. Besides, the unified object representation reduces the number of learnable parameters and thus prevents overfitting. Consequently, we achieve accurate monocular 3D object detection with a lightweight network.

An important step for monocular 3D object detection is Non-Maximum Suppression (NMS), which is usually based on the confidence from the classification branch [1, 33]. This may cause omissions of candidate boxes with high-quality 3D information prediction, because the higher classification confidence doesn't always interpret as the better 3D information prediction. To solve the mismatch, we propose a distance-guided NMS, which automatically selects candidate boxes with better distance estimations. With the distance-guided NMS, UR3D achieves better distance estimation and 3D detection accuracy.

Another challenge for monocular 3D object detection is recovering object physical sizes. Such physical parameters are abstract 3D quantities not directly linked to how objects appear in images [14]. It is thus hard to directly predict the physical sizes of 3D bounding boxes by CNNs. Besides, estimating orientations of 3D boxes is shown imprecise by direct regression [1, 14, 31]. To tackle this problem, we propose a fully convolutional cascaded point regression to estimate the projected 2D center points and corner points of 3D boxes accurately and efficiently. Then the predicted keypoints are used to post-optimize the physical sizes and orientations by minimizing a projection-consistency loss [14], which improves the estimates. The contributions of the proposed UR3D are summarised below:

1. UR3D is a single stage and multi-scale framework that can learn a unified representation of objects within different distance ranges for monocular 3D object detection, which leads to a compact and robust network.
2. A distance-guided NMS is proposed, which selects the candidate boxes with better distance estimations.
3. A fully convolutional cascaded point regression is proposed to estimate the projected 2D center points and corner points precisely and efficiently. The predicted keypoints are used to post-optimize the estimated physical sizes and orientations by minimizing a projection-consistency loss.
4. Experimental results on the KITTI [9] autonomous driving dataset show that our method achieves accurate monocular 3D object detection with a compact architecture.

2 Related Work

2.1 2D Object Detection

Scale-Aware Designs. Large scale variation is one of the key challenges for 2D object detection. Image pyramid [20,37,39,41,48] is a classical solution, but not efficient enough. Faster RCNN [34] utilizes multi-scale anchor boxes to achieve multi-scale object detection. SSD [27] further uses multi-scale features to approximate the image pyramid. Recent works [21–23,40] not only adopt multi-scale features, but also share the convolutional weights of detection heads on different layers to get better object representation. However, learning the unified object representation across different scales and distance ranges for monocular 3D object detection is not a trivial problem. The reason is that the quantities for 3D boxes are much more complicated, especially the distance is highly nonlinear. Our UR3D learns robust and compact distance-normalized unified object representation via proposed designs.

Score Mismatch in NMS. [12,13] find that probabilities for class labels naturally reflect classification confidence instead of localization confidence, thus they predict the score or uncertainty of bounding box regression, which can be used to guide the NMS procedure to preserve accurately localized bounding boxes. We reveal the severe score mismatch problem in the NMS of monocular 3D object detection and propose distance-guided NMS to tackle it.

2.2 Monocular 3D Object Detection

Distance-Aware Designs. Handling large distance variations in monocular 3D object detection is challenging, which requires distance-specific representation. MonoDIS [38] uses a two-stage architecture for monocular 3D object detection, in which the 2D module first detects objects then all the detected objects are fed into a 3D detection head to predict 3D parameters. MonoDIS further disentangles dependencies of different parameters by introducing a loss enabling to handle groups of parameters separately. MonoGRNet [33] is a multi-stage method consisting of four specialized modules for different tasks: 2D detection, instance depth estimation, 3D location estimation and local corner regression. MonoGRNet first predicts objects' 3D locations progressively and then estimates the corner coordinates locally.

MonoPSR [16] uses a network to jointly compute 3D bounding boxes from 2D ones and estimate instance point clouds to help recover shape and scale information. Pseudo-Lidar [42] and AM3D [29] convert the estimated depth image into 3D point clouds to utilize the geometry information, then LiDAR-based 3D object detection methods are employed.

To help the spatial feature learning, OFTNet [35] proposes an orthographic feature transform to map image-level feature into a 3D voxel map, which is then reduced to 2D bird's eye view representation. M3D-RPN [1] is a single-stage framework that exploits 3D anchor boxes to utilize 3D location priors and proposes depth-aware convolution to generate distance-specific feature, which eases the difficulty of learning the distance-information in the full possible range.

To learn the spatial location information, previous works utilize careful multi-stage designs [33,38], point cloud feature [16,29,42], or feature transformation [1,35]. Prior methods directly learn object representation covering all possible distance locations, without considering the feature reuse between different distance ranges. UR3D solves the learning efficiency problem by learning a unified representation for objects within different distance ranges.

3D Box Fitting via Projection-Consistency. Deep3DBox [31] and M3D-RPN [1] fit better 3D boxes by constraining the consistency between the projected 2D boxes from camera coordinate to image coordinate and the network-predicted 2D boxes. SS3D [14] improves the accuracy of 3D box estimation in the similar way. SS3D further optimizes the 3D location, physical size and orientation together. As a comparison, our UR3D solves the projection-consistency loss of corner points and center points as a post-optimization, but only optimizes physical size and orientation prediction.

2.3   Cascaded Point Regression

Cascaded point regression is a classical mechanism for keypoint regression [5, 28,47]. [28,47] predict facial keypoints by a multi-stage cascaded structure, i.e., a global stage to predict coarse shapes and local stages using shape-indexed feature as input to predict fine shapes. Previous works mainly focus on cascaded point regression with a single object input, which are inefficient when predicting keypoints for thousands of candidates simultaneously. In contrast, our proposed fully convolutional cascaded point regression makes dense prediction efficient.

3   Proposed UR3D

We first detail the overall framework, then present the three key components, i.e., distance-normalized unified representation, followed by the distance-guided NMS, and finally the fully convolutional cascaded point regression and projection-consistency based post-optimization. We term our method as UR3D and the main architecture is illustrated in Fig. 1.

3.1   Basic Framework

We address the problem of monocular 3D object detection, which predicts the 3D bounding boxes of targets in camera coordinate from a RGB image. As commonly assumed [9], we only consider yaw angles, and set roll and pitch angles as zero. We also assume that per-image calibration parameters are available both at training and testing phase [9]. For a given RGB image $\mathbf{x} \in \mathbb{R}^{H \times W \times 3}$, UR3D reports all objects of concerned categories, and the output for each object is the

1. class label *cls* and confidence *score*,

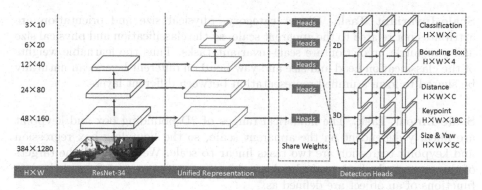

Fig. 1. Framework of our UR3D. UR3D learns a compact and robust unified representation for objects within different distance ranges, which relieves the model from learning the complicated distance-specific representation covering all possible locations.

2. 2D bounding box represented by its top-left and bottom-right corners $\mathbf{b} = (a_1, b_1, a_2, b_2)$,
3. 2D projected center point and eight corner points in image coordinate of 3D box in camera coordinate, encoded as $\mathbf{p} = (x_0, x_1, .., x_8, y_0, y_1, .., y_8)$,
4. distance of center point of the 3D bounding box, in image coordinate, encoded as z_0,
5. 3D bounding box parameters encoded as $\mathbf{m} = (w, h, l, \sin(\theta), \cos(\theta))$, where w, h, l are the physical dimensions, and θ is the allocentric pose of the 3D box. UR3D predicts $\sin(\theta)$ and $\cos(\theta)$, then converts them to θ.

UR3D predicts the center point (x_0, y_0, z_0) in image coordinate and converts it to camera coordinate using the calibration parameters during the testing phase.

UR3D is a single-stage and multi-scale architecture (Fig. 1). During the training stage, we assign targets onto five different layers based on their scales. With the rules, we make the scale range of objects assigned on a layer is larger than that of objects assigned on the previous layer. Since the distance is related to scale, objects within different distance ranges are also assigned to different layers. Detailed assignment rules can be found in Sect. 3.5.

3.2 Distance-Normalized Unified Representation

At this part we detail the distance-normalized unified representation. As shown in Fig. 1, there are five different detection heads on each detection layer, corresponding to five tasks, i.e., classification, bounding box regression, distance estimation, keypoint regression and physical size and yaw angle prediction. To learn a unified representation for objects assigned on different detection layers, we first share the learnable weights of the detection heads on different layers, then we normalize each task's training targets on different layers to a same range according to their relationships with scale, details as follows:

Scale-Invariant Task. Object category, physical size and orientation are attributes not related to the apparent scale, so the classification and physical size and yaw angle prediction are scale-invariant tasks. Thus the learnable weights of the classification head and size and yaw head on different layers can naturally be shared to form a unified representation between different layers.

Scale-Linear Task. The numerical ranges of 2D bounding box and keypoint are linearly dependent on the apparent scale, so the bounding box regression and keypoint regression are two tasks linear to scale. We normalize the targets of these two tasks by introducing learnable parameters α_i and β_i, and the loss functions of an object are defined as:

$$L_{bbox} = loss(\hat{\mathbf{b}}_i, \mathbf{b}_i) = loss(\hat{\mathbf{b}}_i, \alpha_i \mathbf{b}_i'), \tag{1}$$

$$L_{point} = loss(\hat{\mathbf{p}}_i, \mathbf{p}_i) = loss(\hat{\mathbf{p}}_i, \beta_i \mathbf{p}_i'), \tag{2}$$

where $i = 0, 1, 2, 3, 4$ denotes the index of the object-assigned detection layer, $\hat{\mathbf{b}}_i$ and $\hat{\mathbf{p}}_i$ are groundtruths of bounding box regression and keypoint regression respectively, $\mathbf{b}_i'$ and $\mathbf{p}_i'$ are network-predicted bounding box regression result and keypoint regression result respectively, $0 < \alpha_0 < \alpha_1 < \alpha_2 < \alpha_3 < \alpha_4$ and $0 < \beta_0 < \beta_1 < \beta_2 < \beta_3 < \beta_4$. During the training phase, the network learns the best normalization parameters α_i and β_i automatically. During the testing phase, we use $\mathbf{b}_i = \alpha_i \mathbf{b}_i'$ and $\mathbf{p}_i = \beta_i \mathbf{p}_i'$ as outputs for the bounding box regression and keypoint regression respectively.

Scale-Nonlinear Task. To investigate the relationship between distance values and apparent scales, we show some statistics of the car category in KITTI training set [9] in Fig. 2(a). The left figure shows the relationship of distance vs. height, the middle figure shows the curve of depth value of the center point vs. height, and the right figure shows their difference vs. height. The depth images are generated by a monocular depth estimation model [8] as in [29,42]. Apparently the relationships of distance vs. height and depth vs. height are highly nonlinear but in the similar trends (left figure and middle figure), i.e., subtracting the depth can reduce the degree of nonlinearity of distance (right figure).

To get accurate distance estimation, we first introduce learnable parameters γ_i multiplied with the output of i_{th} distance head to use a piece-wise linear curve to fit the nonlinear distance curve. However, the capacity of our piece-wise linear distance estimation model consisting of only five parts is limited, and we still cannot fit the highly nonlinear distance precisely. We further subtract the depth value of a low resolution depth image with the same size of the distance head (Fig. 2(b)), to reduce the degree of nonlinearity of distance, which significantly eases the distance learning. The distance loss of an object is defined as:

$$L_{dist} = loss(\hat{z}_{0i}, z_{0i}) = loss(\hat{z}_{0i}, \gamma_i z_{0i}' + depth), \tag{3}$$

where $i = 0, 1, 2, 3, 4$ denotes the index of the object-assigned detection layer, $\hat{z}_{0i}$ is the groundtruth distance, z_{0i}' is the network-predicted distance result,

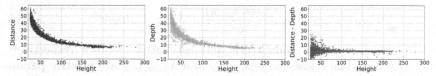

(a) Statistics of car category in KITTI [9] training set. The curves of distance vs. height and depth vs. height are highly nonlinear but in the similar trends. Distance and depth are in image coordinate, height is in pixels. The curves are general conclusions not only limited to KITTI [9], under the assumption of autonomous driving application.

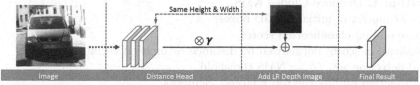

(b) Illustration of the distance head. We first introduce the learnable parameter γ multiplied with the output to use a piece-wise linear curve to fit the nonlinear distance curve, then we further add the depth value from a estimated depth image to reduce the nonlinearity degree of distance. Such designs ease the distance learning significantly.

Fig. 2. Illustration of distance estimation method.

$\gamma_0 > \gamma_1 > \gamma_2 > \gamma_3 > \gamma_4 > 0$, and *depth* is the depth value from the corresponding position of the low resolution depth image. During the training phase, the network can learn the best slope parameters γ_i automatically. During the testing phase, we use $z_{0i} = \gamma_i z_{0i}' + depth$ as output for distance estimation. For both train and test, we run the depth estimation model [8] once and downsample the depth map five times to feed into each distance head, and the maximum size of depth maps we need is only one eighth of the size of depth maps required by [29, 42].

3.3 Distance-Guided NMS

In this part, we detail the distance-guided NMS. Firstly, to get the score of distance estimation, we extend an uncertainty-aware regression loss [15] for distance estimation, as follows:

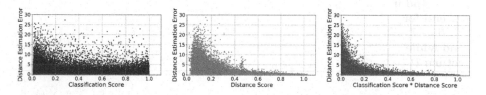

Fig. 3. Illustration for distance estimation error vs. different scores of candidate boxes. Classification score × distance score can best push boxes with inaccurate estimates to the left side. Statistics are based on a UR3D model trained with KITTI [9] car class.

$$L_{dist}(\hat{z}_0, z_0) = \lambda_{dist}\frac{loss(\hat{z}_0, z_0)}{\sigma^2} + \lambda_{uncertain}log(\sigma^2), \qquad (4)$$

where $\hat{z}_0$ and z_0 are the groundtruth and estimated distance respectively, $loss(\hat{z}_0, z_0)$ is a normal regression loss, λ_{dist} and $\lambda_{uncertain}$ are positive parameters to balance the two parts. σ^2 is a positive learnable parameter and $\frac{1}{\sigma^2}$ can be regarded as the score of distance estimation.

Algorithm 1. Distance-Guided NMS.

$\mathcal{B}$: $N \times 27$ matrix of initial 2D/3D boxes,

$\mathcal{S}$: corresponding classification scores,

$\mathcal{C}$: classification scores normalized by distance estimation variances,

$\mathcal{D}$: final detection set, Ω_{nms}: NMS threshold,

$top(k)$: function finding the top k largest elements,

$Ave(z_1, .., z_K, w_1, .., w_K)$: function returning the average of $z_1, .., z_K$ weighted by $w_1, .., w_K$,

K: number of boxes participating the average.

The lines in blue and in red are traditional NMS and Distance-Guided NMS respectively.

Input: $\mathcal{B} = \{o_1, o_2, .., o_N\}$,

    $\mathcal{S} = \{score_1, score_2, .., score_N\}$,

    $\mathcal{C} = \{\frac{score_1}{\sigma_1^2}, \frac{score_2}{\sigma_2^2}, .., \frac{score_N}{\sigma_N^2}\}$, K, Ω_{nms},

Output: $\mathcal{D} \leftarrow \{\}$

  while $\mathcal{B} \neq$ empty **do**

    $m \leftarrow \arg\max \mathcal{S}$

    $\mathcal{D} \leftarrow \mathcal{D} \bigcup o_m$

    $\mathcal{B} \leftarrow \mathcal{B} - o_m$

    $m_1, m_2, .., m_K \leftarrow \arg top(K)\mathcal{C}$

    $temp \leftarrow o_{m_1}$

    $temp.z_0 =$

    $Ave(o_{m_1}.z_0, .., o_{m_K}.z_0, \frac{score_{m_1}}{\sigma_{m_1}^2}, .., \frac{score_{m_K}}{\sigma_{m_K}^2})$

    $\mathcal{D} \leftarrow \mathcal{D} \bigcup temp$

    $\mathcal{B} \leftarrow \mathcal{B} - o_{m_1}$

    $\mathcal{T} \leftarrow \mathcal{B}$

    for $o_i \in \mathcal{B}$ **do**

      if $IoU(o_m, o_i) > \Omega_{nms}$ **then**

        $\mathcal{T} \leftarrow \mathcal{T} - o_i$

      end if

    end for

    $\mathcal{B} \leftarrow \mathcal{T}$

  end while

  return $\mathcal{D}$

In Fig. 3, we show the correlations between the distance estimation error of predicted 3D bounding boxes and corresponding $score$, $\frac{1}{\sigma^2}$, $\frac{score}{\sigma^2}$. As can be seen, $\frac{score}{\sigma^2}$ best pushes candidates with inaccurate distance estimates to the left side. Traditional NMS does not select the candidate boxes with better distance estimates, we propose Distance-Guided NMS (Algorithm 1) to solve the problem.

3.4 Fully Convolutional Cascaded Point Regression

The proposed efficient fully convolutional cascaded point regression (Fig. 4) is adapted from [4] and consists of two stages. In the first stage, we directly regress the positions of center point and eight corner points, and the results of position q are encoded as:

$$\mathbf{p}_0 = \{p_0, p_1, \ldots, p_8\} = \{(x_0, y_0), (x_1, y_1), \ldots, (x_8, y_8)\},$$

In the second stage, we extract the shape-indexed feature guided by $\mathbf{p}_0$, and predict the residual values of keypoints. The extraction of shape-indexed feature can be formulated as an efficient convolutional layer as in [4], instead of traditional time-consuming multi-patch extraction [28, 47]. Let the nine positions of a 3×3 convolutional kernels correspond to the nine keypoints. The convolutional layer for the extraction consists of two steps: 1) sampling using $\mathbf{p}_0$ as the kernel point positions over the input feature map $\mathbf{f}_{in}$; 2) summation of sampled values weighted by kernel weights $\mathbf{w}$ to get the output feature map $\mathbf{f}_{out}$, i.e.,

$$\mathbf{f}_{out}(q) = \sum_{i=0}^{8} \mathbf{w}(i) \cdot \mathbf{f}_{in}(p_i). \tag{5}$$

The sampling is on the irregular locations. As the location p_i is typically fractional, Eq. (5) $\mathbf{f}_{in}(p_i)$ is obtained by bilinear interpolation. The detailed implementation is similar to [4]. Note during the training, the gradients will not be backpropagated to p_i through Eq. (5), because p_i has its own supervised loss. The keypoint losses for two stages are:

$$L_{point_0} = loss(\hat{\mathbf{p}}, \mathbf{p}_0), \tag{6}$$

$$L_{point_1} = loss(\hat{\mathbf{p}}, \mathbf{p}) = loss(\hat{\mathbf{p}}, \mathbf{p}_0 + \mathbf{p}_1), \tag{7}$$

where $\hat{\mathbf{p}}$ is the groundtruth of keypoint regression, $\mathbf{p}_0$ and $\mathbf{p}_1$ are the outputs of the first and second stage respectively, and $\mathbf{p} = \mathbf{p}_0 + \mathbf{p}_1$ is the final output of keypoint regression.

Fully convolutional cascaded point regression achieves accurate prediction of thousands of candidates simultaneously. Then we use the estimated keypoints to post-optimize the physical size and yaw angle prediction. Given a set of center point (x_0, y_0, z_0), physical size w, h, l, and yaw angle θ, we calculate the center and corner points of corresponding 3D bounding box in camera coordinate with calibration parameters. Denote the calculation function as $\mathbf{F}(x_0, y_0, z_0, w, h, l, \theta)$. We try to find a set of w', h', l', θ' to minimize the objective function:

$$\arg\min_{w', h', l', \theta'} \lambda_{post} \|\mathbf{F}(x_0, y_0, z_0, w', h', l', \theta') - \mathbf{p}\|_2^2$$
$$+ \left[(w' - w)^2 + (h' - h)^2 + (l' - l)\right]^2, \tag{8}$$

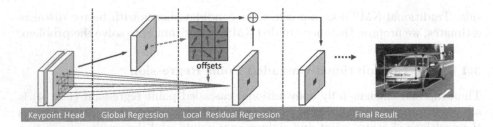

Fig. 4. Illustration of fully convolutional cascaded point regression, which formulates dense cascaded point regression as an efficient convolutional layer.

where $x_0, y_0, z_0, w, h, l, \theta$ are the network-predicted results, w', h', l', θ' are the post-optimized results. This is a standard nonlinear optimization problem, which can be solved by an optimization toolbox.

3.5   Implementation Details

Object Assignment Rule. During the training stage, we assign a position q on a detection layer $\mathbf{f}_i$ ($i = 0, 1, 2, 3, 4$) to an object, if 1) q falls in the object, 2) the maximum distance from q to the boundaries of the object is within a given range $\mathbf{r}_i$, and 3) the distance from q to the center of the object is less than a given value $\mathbf{d}_i$. $\mathbf{r}_i$ denotes the scale range of objects assigned on each detection layer [40], and $\mathbf{d}_i$ defines the radius of positive samples on each detection layer. $\mathbf{r}_i$ is $[0, 64], [64, 128], [128, 256], [256, 512], [512, 1024]$ for the five layers, and $\mathbf{d}_i$ is $12, 24, 48, 96, 192$ respectively, all in pixels. Positions without assigning to any object will be regarded as negative samples, except that the positions adjacent with the positive samples are treated as ignored samples.

Network Architecture. The backbone of UR3D is ResNet-34 [11]. All the head depth of the detection heads is two. Images are scaled to a fixed height of 384 pixels for both training and testing.

Loss. We use the focal loss [23] for classification task, IoU loss [46] for bounding box regression, smooth L_1 loss [10] for keypoint regression, and Wing loss [7] for distance, size and orientation estimation. The loss weights are $1, 1, 0.003, 0.1, 0.05, 0.1, 0.001$ for the classification, bounding box regression, keypoint regression, distance estimation, distance variance estimation, size and orientation estimation, and post-optimization, respectively.

Optimization. We adopt the step strategy to adjust a learning rate. At first the learning rate is fixed to 0.01 and reduced by 50 times every 3×10^4 iterations. The total iteration number is 9×10^4 with batch size 5. The only augmentation we perform is random mirroring. We implement our framework using Python and PyTorch [32]. All the experiments run on a server with 2.6 GHz CPU and GTX Titan X.

4 Experiments

We evaluate our method on KITTI [9] dataset with the car class under the two 3D localization tasks: Bird's Eye View (BEV) and 3D Object Detection. The method is comprehensively tested on two validation splits [3,43] and the official test dataset. We further present analyses on the impacts of individual components of the proposed UR3D. Finally we visualize qualitative examples of UR3D on KITTI (Fig. 5).

4.1 KITTI

The KITTI [9] dataset provides multiple widely used benchmarks for computer vision problems in autonomous driving. The BirdEye View (BEV) and 3D Object Detection tasks are used to evaluate 3D localization performance. These two tasks are characterized by 7481 training and 7518 test images with 2D and 3D annotations for cars, pedestrians, cyclists, etc. Each object is assigned with a difficulty level, i.e., easy, moderate or hard, based on its occlusion level and truncation degree.

We conduct experiments on three common data splits including val1 [3], val2 [43], and the official test split [9]. Each split contains images from non-overlapping sequences such that no data from an evaluated frame, or its neighbors, are used for training. We report the $AP|_{R_{11}}$ and $AP|_{R_{40}}$ on val1 and val2, and $AP|_{R_{40}}$ on test subset. We use the car class, the most representative, and the official IoU criteria for cars, i.e., 0.7.

Val Set Results. We evaluate UR3D on val1 and val2 as detailed in Table 1 and Table 2. Using the same monocular depth estimator [8] as in AM3D [29] and Pseudo-LiDAR [42], UR3D can compete with them on the two splits. The time cost of depth map generation of our UR3D can be much smaller than that of [29,42], since the size of depth maps we need is only one eighth of the size of depth maps required by them. We use depth priors to normalize the learning targets of distance instead of converting to point clouds as in [29,42], leading to a more compact and efficient architecture.

Test Set Results. We evaluate the results on test set in Table 3. Compared with FQNet [26], ROI-10D [30], GS3D [19], and MonoGRNet [33], UR3D outperforms them significantly in all indicators. Compared with MonoDIS [38], UR3D outperforms it by a large margin in three indicators, i.e., AP_{3D} of easy subset, AP_{3D} of moderate subset and AP_{BEV} of easy subset. Note MonoDIS [38] is a two-stage method while ours is a more compact single-stage method. Compared with another single-stage method, M3D-RPN [1], UR3D outperforms it on two indicators, i.e., AP_{3D} and AP_{BEV} of easy subset, with a more lightweight backbone. Compared with AM3D [29], UR3D runs with a much faster speed.

Table 1. Bird's Eye View. Comparisons on the Bird's Eye View task (AP_{BEV}) on val1 [3] and val2 [43] of KITTI [9].

Method	Time (ms)	$AP|_{R_{11}}$ [val1 / val2]			$AP|_{R_{40}}$ [val1 / val2]		
		Easy	Mod	Hard	Easy	Mod	Hard
ROI-10D [30]	200	14.76/−	9.55/−	7.57/−	−/−	−/−	−/−
MonoPSR [16]	200	20.63/21.52	18.67/18.90	14.45/14.94	−/−	−/−	−/−
MonoGRNet [33]	60	24.97/−	19.44/−	16.30/−	19.72/−	12.81/−	10.15/−
M3D-RPN [1]	160	25.94/26.86	21.18/21.15	17.90/17.14	20.85/21.36	15.62/15.22	11.88/11.28
Pseudo-LiDAR [42]	−	40.60/−	26.30/−	22.90/−	−/−	−/−	−/−
AM3D [29]	400	43.75/−	28.39/−	23.87/−	−/−	−/−	−/−
UR3D (Ours)	120	37.35/36.15	26.01/25.25	20.84/20.12	33.07/32.35	20.84/20.05	15.25/14.4

Table 2. 3D Detection. Comparisons on the 3D Detection task (AP_{3D}) on val1 [3] and val2 [43] of KITTI [9].

Method	Time (ms)	$AP|_{R_{11}}$ [val1 / val2]			$AP|_{R_{40}}$ [val1 / val2]		
		Easy	Mod	Hard	Easy	Mod	Hard
ROI-10D [30]	200	10.25/−	6.39/−	6.18/−	−/−	−/−	−/−
MonoPSR [16]	200	12.75/13.94	11.48/12.24	8.59/10.77	−/−	−/−	−/−
MonoGRNet [33]	60	13.88/−	10.19/−	7.62/−	11.90/−	7.56/−	5.76/−
M3D-RPN [1]	160	20.27/20.40	17.06/16.48	15.21/13.34	14.53/14.57	11.07/10.07	8.65/7.51
Pseudo-LiDAR [42]	−	28.20/−	18.50/−	16.40/−	−/−	−/−	−/−
AM3D [29]	400	32.23/−	21.09/−	17.26/−	−/−	−/−	−/−
UR3D (Ours)	120	28.05/26.30	18.76/16.75	16.55/13.60	23.24/22.15	13.35/11.10	10.15/9.15

Table 3. Test Set Results. Comparisons of our UR3D to SOTA methods of monocular 3D object detection on the test set of KITTI [9].

Method	Reference	Time (ms)	$AP|_{R_{40}}$ [Easy/Mod/Hard]	
			AP_{3D}	AP_{BEV}
FQNet [26]	CVPR 2019	500	2.77/1.51/1.01	5.40/3.23/2.46
ROI-10D [30]	CVPR 2019	200	4.32/2.02/1.46	9.78/4.91/3.74
GS3D [19]	CVPR 2019	2000	4.47/2.90/2.47	8.41/6.08/4.94
MonoGRNet [33]	AAAI 2019	60	9.61/5.74/4.25	18.19/11.17/8.73
MonoDIS [38]	ICCV 2019	−	10.37/7.94/6.40	17.23/13.19/11.12
M3D-RPN [1]	ICCV 2019	160	14.76/9.71/7.42	21.02/13.67/10.23
AM3D [29]	ICCV 2019	400	16.50/10.74/9.52	25.03/17.32/14.91
UR3D (Ours)		120	15.58/8.61/6.00	21.85/12.51/9.20

Learned Parameters. We initialize α_i and β_i with $32, 64, 128, 256, 512$, and $16, 8, 4, 2, 1$ for γ_i. The learned results on val1 split are $5.7, 10.6, 20.7, 41.0, 82.3$ for α_i, $5.3, 10.4, 20.6, 41.4, 82.2$ for β_i, and $2.3, 1.4, 0.8, 0.3, 0.2$ for γ_i.

Table 4. Ablations. We ablate the effects of key components of UR3D with respect to accuracy and inference time.

Setting	Time (ms)	$AP\|_{R_{40}}$ [Easy/Mod/Hard]	
		AP_{3D}	AP_{BEV}
Baseline	90	11.26/6.52/4.25	18.26/10.25/8.55
+ LR Depth Image	90	18.57/10.65/7.90	27.65/15.25/12.52
+ Distance Guided NMS ($K = 1$)	90	19.75/11.35/8.50	30.55/18.47/14.18
+ Distance Guided NMS ($K = 2$)	90	20.20/11.59/8.85	31.69/19.68/14.81
+ Post-Optimization	110	22.50/12.95/9.90	32.58/20.54/15.05
+ Cascaded Regression (UR3D)	120	23.24/13.35/10.15	33.07/20.84/15.25

4.2 Ablation Study

We conduct ablation experiments to examine how each proposed component affects the final performance of UR3D. We evaluate the performance by first setting a simple baseline which doesn't adopt proposed components, then adding the proposed designs one-by-one, as shown in Table 4. For all ablations we use the KITTI val1 dataset split and evaluate based on the car class. From the results listed in Table 4, some promising conclusions can be summed up as follows:

Distance-Normalized Unified Representation Is Crucial. The results of "+ LR Depth Image" show that adding the low resolution depth image to help normalize the distance improves the AP_{3D} and AP_{BEV} of baseline a lot, which indicates that reducing the nonlinear degree of distance estimation eases the unified object representation learning dramatically.

Distance-Guided NMS Is Promising. The AP_{3D} and AP_{BEV} of "+ Distance-Guided NMS ($K = 1$)" are much better than the results of "+ LR Depth Image". It supports that our distance-guided NMS can select the candidate boxes with better distance estimates automatically and effectively. Increasing the number of candidates participating the average (from $K = 1$ to $K = 2$) also helps, suggesting that the candidate with the best distance estimate may not be the top one but among the top K due to the noise of distance score.

Fully Convolutional Cascaded Point Regression Is Effective. The results of "+ Post-Optimization" illustrate that introducing the projection-consistency based post-optimization improves AP_{3D} and AP_{BEV}. The results of "+ Cascaded Regression" show that adding the fully convolutional cascaded point regression further improves AP_{3D} and AP_{BEV}. The fully convolutional cascaded point regression only costs $10\,ms$ with a non-optimized Python implementation.

Fig. 5. Qualitative Examples. We visualize qualitative examples of UR3D. All illustrated images are from the val1 [3] split and not used for training. Bird's eye view results (right) are also provided and the red lines indicate the yaw angles of cars. (Color figure online)

5   Conclusions

In this work, we present a monocular 3D object detector, i.e., UR3D, which learns a distance-normalized unified object representation, in contrast to prior works which learns to represent objects in full possible range. UR3D is uniquely designed to learn the shared representation across different distance ranges, which is robust and compact. We further propose a distance-guided NMS to select candidate boxes with better distance estimates and a fully convolutional cascaded point regression predicting accurate keypoints to post-optimize the 3D boxes parameters, both of which improve the accuracy. Collectively, our method achieves accurate monocular 3D object detection with a compact architecture.

Acknowledgment. The authors are partly funded by Huawei.

References

1. Brazil, G., Liu, X.: M3D-RPN: monocular 3D region proposal network for object detection. In: ICCV, pp. 9287–9296 (2019)
2. Chen, X., Kundu, K., Zhang, Z., Ma, H., Fidler, S., Urtasun, R.: Monocular 3D object detection for autonomous driving. In: CVPR, pp. 2147–2156 (2016)
3. Chen, X., et al.: 3D object proposals for accurate object class detection. In: NeurIPS, pp. 424–432 (2015)

4. Dai, J., et al.: Deformable convolutional networks. In: ICCV, pp. 764–773 (2017)
5. Dollár, P., Welinder, P., Perona, P.: Cascaded pose regression. In: CVPR, pp. 1078–1085 (2010)
6. Everingham, M., et al.: The pascal visual object classes challenge: a retrospective. Int. J. Comput. Vision **111**(1), 98–136 (2015)
7. Feng, Z., Kittler, J., Awais, M., Huber, P., Wu, X.: Wing loss for robust facial landmark localisation with convolutional neural networks. In: CVPR, pp. 2235–2245 (2018)
8. Fu, H., Gong, M., Wang, C., Batmanghelich, K., Tao, D.: Deep ordinal regression network for monocular depth estimation. In: CVPR, pp. 2002–2011 (2018)
9. Geiger, A., Lenz, P., Urtasun, R.: Are we ready for autonomous driving? The KITTI vision benchmark suite. In: CVPR, pp. 3354–3361 (2012)
10. Girshick, R.B.: Fast R-CNN. In: ICCV, pp. 1440–1448 (2015)
11. He, K., Zhang, X., Ren, S., Sun, J.: Deep residual learning for image recognition. In: CVPR, pp. 770–778 (2016)
12. He, Y., Zhu, C., Wang, J., Savvides, M., Zhang, X.: Bounding box regression with uncertainty for accurate object detection. In: CVPR, pp. 2888–2897 (2019)
13. Jiang, B., Luo, R., Mao, J., Xiao, T., Jiang, Y.: Acquisition of localization confidence for accurate object detection. In: Ferrari, V., Hebert, M., Sminchisescu, C., Weiss, Y. (eds.) Computer Vision – ECCV 2018. LNCS, vol. 11218, pp. 816–832. Springer, Cham (2018). https://doi.org/10.1007/978-3-030-01264-9_48
14. Jörgensen, E., Zach, C., Kahl, F.: Monocular 3d object detection and box fitting trained end-to-end using intersection-over-union loss. CoRR abs/1906.08070 (2019)
15. Kendall, A., Gal, Y., Cipolla, R.: Multi-task learning using uncertainty to weigh losses for scene geometry and semantics. In: CVPR, pp. 7482–7491 (2018)
16. Ku, J., Pon, A.D., Waslander, S.L.: Monocular 3D object detection leveraging accurate proposals and shape reconstruction. In: CVPR, pp. 11867–11876 (2019)
17. Lang, A.H., Vora, S., Caesar, H., Zhou, L., Yang, J., Beijbom, O.: PointPillars: fast encoders for object detection from point clouds. In: CVPR, pp. 12697–12705 (2019)
18. LeCun, Y., Bengio, Y., Hinton, G.E.: Deep learning. Nature **521**(7553), 436–444 (2015)
19. Li, B., Ouyang, W., Sheng, L., Zeng, X., Wang, X.: GS3D: an efficient 3D object detection framework for autonomous driving. In: CVPR, pp. 1019–1028 (2019)
20. Li, H., Lin, Z., Shen, X., Brandt, J., Hua, G.: A convolutional neural network cascade for face detection. In: CVPR, pp. 5325–5334 (2015)
21. Li, Y., Chen, Y., Wang, N., Zhang, Z.: Scale-aware trident networks for object detection. In: CVPR, pp. 6054–6063 (2019)
22. Lin, T., Dollár, P., Girshick, R.B., He, K., Hariharan, B., Belongie, S.J.: Feature pyramid networks for object detection. In: CVPR, pp. 936–944 (2017)
23. Lin, T., Goyal, P., Girshick, R.B., He, K., Dollár, P.: Focal loss for dense object detection. In: ICCV, pp. 2999–3007 (2017)
24. Lin, T.-Y., et al.: Microsoft COCO: common objects in context. In: Fleet, D., Pajdla, T., Schiele, B., Tuytelaars, T. (eds.) ECCV 2014. LNCS, vol. 8693, pp. 740–755. Springer, Cham (2014). https://doi.org/10.1007/978-3-319-10602-1_48
25. Liu, L., et al.: Deep learning for generic object detection: a survey. Int. J. Comput. Vision **128**(2), 261–318 (2020)

26. Liu, L., Lu, J., Xu, C., Tian, Q., Zhou, J.: Deep fitting degree scoring network for monocular 3D object detection. In: CVPR, pp. 1057–1066 (2019)
27. Liu, W., et al.: SSD: single shot MultiBox detector. In: Leibe, B., Matas, J., Sebe, N., Welling, M. (eds.) ECCV 2016. LNCS, vol. 9905, pp. 21–37. Springer, Cham (2016). https://doi.org/10.1007/978-3-319-46448-0_2
28. Lv, J., Shao, X., Xing, J., Cheng, C., Zhou, X.: A deep regression architecture with two-stage re-initialization for high performance facial landmark detection. In: CVPR, pp. 3691–3700 (2017)
29. Ma, X., Wang, Z., Li, H., Ouyang, W., Zhang, P.: Accurate monocular 3D object detection via color-embedded 3D reconstruction for autonomous driving. In: ICCV, pp. 6851–6860 (2019)
30. Manhardt, F., Kehl, W., Gaidon, A.: ROI-10D: monocular lifting of 2D detection to 6d pose and metric shape. In: CVPR, pp. 2069–2078 (2019)
31. Mousavian, A., Anguelov, D., Flynn, J., Kosecka, J.: 3D bounding box estimation using deep learning and geometry. In: CVPR, pp. 5632–5640 (2017)
32. Paszke, A., et al.: Pytorch: an imperative style, high-performance deep learning library. In: NeurIPS, pp. 8024–8035 (2019)
33. Qin, Z., Wang, J., Lu, Y.: MonoGRNet: a geometric reasoning network for monocular 3D object localization. In: AAAI, pp. 8851–8858 (2019)
34. Ren, S., He, K., Girshick, R.B., Sun, J.: Faster R-CNN: towards real-time object detection with region proposal networks. In: NeurIPS, pp. 91–99 (2015)
35. Roddick, T., Kendall, A., Cipolla, R.: Orthographic feature transform for monocular 3D object detection. In: British Machine Vision Conference (2019)
36. Shi, S., Wang, X., Li, H.: PointRCNN: 3D object proposal generation and detection from point cloud. In: CVPR, pp. 770–779 (2019)
37. Shi, X., Shan, S., Kan, M., Wu, S., Chen, X.: Real-time rotation-invariant face detection with progressive calibration networks. In: CVPR, pp. 2295–2303 (2018)
38. Simonelli, A., Bulò, S.R., Porzi, L., López-Antequera, M., Kontschieder, P.: Disentangling monocular 3D object detection. In: ICCV, pp. 1991–1999 (2019)
39. Singh, B., Davis, L.S.: An analysis of scale invariance in object detection SNIP. In: CVPR, pp. 3578–3587 (2018)
40. Tian, Z., Shen, C., Chen, H., He, T.: FCOS: fully convolutional one-stage object detection. In: CVPR, pp. 9627–9636 (2019)
41. Viola, P.A., Jones, M.J.: Robust real-time face detection. Int. J. Comput. Vision 57(2), 137–154 (2004)
42. Wang, Y., Chao, W., Garg, D., Hariharan, B., Campbell, M.E., Weinberger, K.Q.: Pseudo-lidar from visual depth estimation: bridging the gap in 3D object detection for autonomous driving. In: CVPR, pp. 8445–8453 (2019)
43. Xiang, Y., Choi, W., Lin, Y., Savarese, S.: Subcategory-aware convolutional neural networks for object proposals and detection. In: WACV, pp. 924–933 (2017)
44. Xu, B., Chen, Z.: Multi-level fusion based 3D object detection from monocular images. In: CVPR, pp. 2345–2353 (2018)
45. Yang, B., Luo, W., Urtasun, R.: PIXOR: real-time 3d object detection from point clouds. In: CVPR, pp. 7652–7660 (2018)
46. Yu, J., Jiang, Y., Wang, Z., Cao, Z., Huang, T.S.: Unitbox: an advanced object detection network. In: ACM MM, pp. 516–520. ACM (2016)

47. Zhang, J., Shan, S., Kan, M., Chen, X.: Coarse-to-fine auto-encoder networks (CFAN) for real-time face alignment. In: Fleet, D., Pajdla, T., Schiele, B., Tuytelaars, T. (eds.) ECCV 2014. LNCS, vol. 8690, pp. 1–16. Springer, Cham (2014). https://doi.org/10.1007/978-3-319-10605-2_1

48. Zhang, K., Zhang, Z., Li, Z., Qiao, Y.: Joint face detection and alignment using multitask cascaded convolutional networks. IEEE Signal Process. Lett. **23**(10), 1499–1503 (2016)

49. Zhou, Y., Tuzel, O.: VoxelNet: end-to-end learning for point cloud based 3D object detection. In: CVPR, pp. 4490–4499 (2018)

Sequential Deformation for Accurate Scene Text Detection

Shanyu Xiao[1], Liangrui Peng[1]([✉]), Ruijie Yan[1], Keyu An[1],
Gang Yao[1], and Jaesik Min[2]

[1] Department of Electronic Engineering, Beijing National Research Center for
Information Science and Technology, Tsinghua University, Beijing, China
{xiaosy19,yrj17,aky19,yg19}@mails.tsinghua.edu.cn, penglr@tsinghua.edu.cn
[2] Hyundai Motor Group AIRS Company, Seoul, Korea
jaesik.min@hyundai.com

Abstract. Scene text detection has been significantly advanced over
recent years, especially after the emergence of deep neural network. However, due to high diversity of scene texts in scale, orientation, shape and
aspect ratio, as well as the inherent limitation of convolutional neural network for geometric transformations, to achieve accurate scene text detection is still an open problem. In this paper, we propose a novel sequential
deformation method to effectively model the line-shape of scene text. An
auxiliary character counting supervision is further introduced to guide
the sequential offset prediction. The whole network can be easily optimized through an end-to-end multi-task manner. Extensive experiments
are conducted on public scene text detection datasets including ICDAR
2017 MLT, ICDAR 2015, Total-text and SCUT-CTW1500. The experimental results demonstrate that the proposed method has outperformed
previous state-of-the-art methods.

Keywords: Scene text detection · Deep neural network · Sequential
deformation

1 Introduction

Scene text detection has attracted growing research attention in the computer
vision field due to its wide range of real-world applications including automatic
driving navigation, instant translation and image retrieval. Scene text's unique-
ness of high diversity in geometric transformations including scale, orientation,
shape and aspect ratio also makes it distinct from generic objects. It is obvious
that scene text detection is a challenging research topic.

Recently, the community has witnessed substantial advancements in scene
text detection [1,18,20,26,37,38,49], especially after the emergence of deep neural network. For scene text detection, a straightforward approach is to model
text instance (word or text line) as a special kind of object and adopt frameworks of generic object detection, such as SSD [19] and Faster R-CNN [6]. These

© Springer Nature Switzerland AG 2020
A. Vedaldi et al. (Eds.): ECCV 2020, LNCS 12374, pp. 108–124, 2020.
https://doi.org/10.1007/978-3-030-58526-6_7

methods yield great performance on standard benchmarks. However, their performance and generalization ability are undermined by standard convolution's fixed receptive field and limited capability for large geometric transformations.

Text instance is composed of similar components (e.g. text segments, characters and strokes), where component is spatially smaller and has less geometrical transformations. Consequently, some methods [1,31] predict text components rather than the whole text instance. As pixel can be regarded as the finest-grained component, many methods [15,38] localize text based on instance segmentation. These methods are more flexible in modeling, and have a lower requirement for the receptive field, achieving remarkable results on localizing texts with arbitrary shape. Nevertheless, an additional component grouping operation like pixel clustering or segment connecting is always indispensable, where error propagation from wrong component prediction and lack of end-to-end optimization make barriers for the optimal performance.

Component detecting and grouping are also exploited by human visual system [42]. Our eyes first localize one endpoint of a text instance, then sequentially sweeps through the text center line and gazes only a part of the text at one time. Finally, we group different parts into text instance along the sweeping path.

(a)   (b)   (c)   (d)

Fig. 1. Demonstration of the sampling locations for standard convolution and the proposed SDM. For clearer visualization, all sampling locations are mapped on input images. (a) (b) regular sampling locations in standard convolution. The yellow point indicates the center location of convolution. (c) (d) Sequential sampling procedure of our SDM. The yellow point indicates the start location, and each blue arrow indicates the predicted offset of one iteration. Two deformation branches are used in our work. (Color figure online)

Inspired by above observations, we propose an end-to-end trainable Sequential Deformation Module (SDM) for accurate scene text detection, which sequentially groups feature-level components to effectively extend the modeling capability for text's geometrical configuration and learn informative instance-level semantic representations along text line. The SDM first samples features iteratively from a start location. SDM runs densely, regarding each integral location on the input feature map as the start location to fit for unique geometric configurations for different instances. As depicted in Fig. 1, by performing sampling in a sequential manner, a much larger effective receptive field than the standard convolutional layer could be achieved. After that, SDM performs weighted summation on all sampled features to aggregates features and capture the adaptive instance-level representations without complicated grouping post-processing. Besides, we introduce an auxiliary character counting supervision to guide SDM's sequential sampling and learn richer representations. The character counting task is modeled as a sequence-to-sequence problem [33], and the counting network receives all the SDM's sampled features and predict a valid sequence, whose length is expected to equal the character number of corresponding text instance.

The main contributions of this work are three-fold: (1) We propose a novel end-to-end sequential deformation module for accurate detection of arbitrary-shaped scene text, which adaptively enhances the modeling capability for text's geometric configuration and learns informative instance-level semantic representations; (2) We introduce an auxiliary character counting supervision which facilitates the sequential offset predicting and learning of generic features; (3) Integrating the sequential deformation module and auxiliary character counting into Mask R-CNN, the whole network is optimized through an end-to-end multi-task manner without any complicated grouping post-processing. Experiments on benchmarks for multi-lingual, oriented, and curved text detection demonstrate that our method achieves the state-of-the-art performance.

2   Related Work

Scene text detection has been widely studied in the last few years, especially with the popularity of deep learning. In this section, we review related works of two different categories of deep learning based methods according to their modeling granularity, then we look back on relevant works for learning spatial deformation in the convolutional neural network.

Instance-level detection methods [14,44,49] follow the routine of generic object detection, viewing the text instance as a specific kind of object. TextBoxes [14] modifies SSD [19] by adding default boxes and filters with larger aspect ratios to handle the text's significant variation of aspect ratio. EAST [49] and Deep Regression [9] directly regress the rotated rectangles or quadrangles of text without the priori of anchors. SPCNET [44] augments Mask R-CNN [7] with the guidance of semantic information and sharing FPN, suppressing false positive detections. These methods achieve excellent performances on standard benchmarks but face problems such as CNN's limited receptive field and incapability for geometric transformation.

Component-level methods [5,15,26,31,34,38] decompose instance into components such as characters, text segments or the finest-grained pixels, addressing the problems faced by instance-level modeling. SegLink [31] decomposes text into locally detectable segments and links and combines them into the final oriented detection. TextDragon [5] describes text's shape with a series of local quad-rangles to adaptively spot arbitrary-shaped texts. PAN [39] adopts a learnable post-processing implemented by semantic segmentation and embedding to precisely aggregate text pixels. Tian et al. [35] propose to learn shape-aware pixel embedding to ease separating adjacent instances and detecting large instances. For these methods, a grouping post-processing is required, where the error propagation of wrong component prediction and the lack of end-to-end training could harm the robustness.

Spatial deformation methods [3, 12, 37] enable the network to adaptively capture the geometric transformations. STN [12] rectifies the image or feature maps via global parametric transformations. Deformable ConvNet [3] augments the spatial sampling locations in convolutional layers with additional predicted offsets. ITN [37] also augments convolution but constrains it as affine transformation to learn the geometry-aware representation for scene text.

Different from existing methods, a sequential deformation module to group feature components is proposed in our paper, adaptively enhancing the instance-level Mask R-CNN without complicated grouping post-processing.

3   Methodology

In this section, we first elaborate the sequential deformation module and auxiliary character counting task. Then we describe the Mask R-CNN equipped with sequential deformation module.

3.1   Sequential Deformation Module

For a standard convolution with weight w and the input feature map x, it first samples features on x using a fixed rectangular sampling grid $\mathcal{R} = \{\mathbf{p}_1, \ldots, \mathbf{p}_N\}$ (e.g. 3×3 grid $\{(-1, -1), (-1, 0), \ldots, (0, 1), (1, 1)\}$ with $N = 9$). Then the weighted summation of sampled features is calculated using weight w. For every location $\mathbf{p}$ on the output feature map y, we have:

$$y(\mathbf{p}) = \sum_{n=1}^{N} \mathrm{w}(n) \cdot \mathrm{x}(\mathbf{p} + \mathbf{p}_n), \tag{1}$$

As depicted in Fig. 1, the standard convolution is insufficient in scene text detection because of the mismatch of shape and size between the fixed receptive field and text instance. On one hand, to capture the whole instance, the fixed rectangular receptive field is required to completely cover the text's circumscribed rectangle rather than only the text region, while much undesired background information is included. On the other hand, the fixed-sized receptive field is incapable of well-extracting representations for instances with different scale and aspect ratio. This results in imprecise classification and regression, especially for instance-level models that detect text in one or a few stages.

Detecting text component, which is spatially smaller and less geometrically transformed, relieves the above-mentioned problems. Inspired by the insight of detecting text component and spatial deformation learning in CNN, we propose an end-to-end trainable Sequential Deformation Module (SDM). SDM first

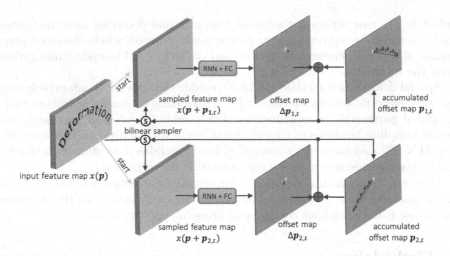

Fig. 2. Illustration of sequential sampling in SDM. Since the SDM runs densely, the notations of all 3D tensor are represented by the corresponding 1D vector at location **p**. From each integral start location **p**, SDM samples features along two separate sampling paths.

performs sampling in a sequential manner, and then, like the standard convolution, SDM performs weighted summation on all sampled features to aggregates features and capture the adaptive instance-level representations. The procedure of sequential sampling is illustrated in Fig. 2. Regarding each integral location $\mathbf{p} \in \{(0,0), \ldots, (H-1, W-1)\}$ on input feature map x (height H and width W) as the start location, the relative sampling locations $\mathcal{S} = \{\mathbf{p}_t | t = 1, \ldots, T\}$ are sequentially generated by offsets accumulation:

$$\mathbf{p}_{t+1} = \mathbf{p}_t + \triangle\mathbf{p}_t, \ t = 0, ..., T-1 \tag{2}$$

where $\mathbf{p}_0 = (0, 0)$, $\triangle\mathbf{p}_t$ denotes the current 2D offset and T denotes the predefined iteration number. The relative sampling locations $\mathcal{S}$ form a sampling path.

At each step, we get a new sampling location $\mathbf{p} + \mathbf{p}_t$ from current accumulated offset $\mathbf{p}_t$. The sampled feature x($\mathbf{p} + \mathbf{p}_t$) represents a feature-level component of text observed at current step, and the whole text instance is gradually grouped naturally through the step-by-step samplings. It's notable that, from any start location within the text instance, we should "sweep" along two opposite directions to capture the whole text instance, wherefore adopting two separate sampling paths $\mathcal{S}_d = \{\mathbf{p}_{d,t} | t = 1, ..., T\}$ ($d = 1, 2$) are more suitable. For two directions $d = 1, 2$, Eq. 2 becomes:

$$\mathbf{p}_{d,t+1} = \mathbf{p}_{d,t} + \triangle\mathbf{p}_{d,t}, \ t = 0, ..., T-1. \tag{3}$$

In this work, the sequential sampling network is realized through a recurrent neural network (RNN) followed by a linear layer, and two separate sampling paths are generated by two separate sequential sampling networks, so $\triangle \mathbf{p}_{d,t}$ is conditioned on previous sampled features $\{\mathrm{x}(\mathbf{p} + \mathbf{p}_{d,0}), ..., \mathrm{x}(\mathbf{p} + \mathbf{p}_{d,t})\}$:

$$h_{d,t} = RNN_d(\mathrm{x}(\mathbf{p} + \mathbf{p}_{d,t}), \ h_{d,t-1}) \tag{4}$$

$$\triangle \mathbf{p}_{d,t} = Linear_d(h_{d,t}). \tag{5}$$

The RNN stores the historical shape information, and stabilizes the training process and moderately boosts the performance. Based on previous observations, the network will adaptively calibrate magnitude and orientation of $\triangle \mathbf{p}_t$ to ensure the largest possible covering of whole text instance. Sequential sampling network for one direction d is shared across all start locations.

After sequential sampling, the SDM calculates the weighted summation of feature at the start location and all sampled features to generate the output feature map y:

$$y(\mathbf{p}) = \mathrm{w}_0 \cdot \mathrm{x}(\mathbf{p}) + \sum_{d=1}^{2} \sum_{t=1}^{T} \mathrm{w}(d, t) \cdot \mathrm{x}(\mathbf{p} + \mathbf{p}_{d,t}). \tag{6}$$

In practical implementation, Eq. 6 is equivalently implemented as following:

$$\mathrm{m}(\mathbf{p}) = Concat(\{\mathrm{x}(\mathbf{p} + \mathbf{p}_{d,t}) \mid d = 1, 2, \ t = 1, ..., T\} \cup \{\mathrm{x}(\mathbf{p})\}), \tag{7}$$

$$y(\mathbf{p}) = Conv_{1 \times 1}(\mathrm{m}(\mathbf{p})), \tag{8}$$

where the intermediate feature map m is the concatenation of all sequentially sampled feature maps and the input feature. C is the channel of input feature map and the intermediate feature map m has a channel of $(2T + 1) \cdot C$, corresponding to $2T$ times sampling and the original input. Bilinear interpolation is used to compute $\mathrm{x}(\mathbf{p} + \mathbf{p}_{d,t})$ owing to fractional sampling locations. As bilinear interpolation is differentiable, the gradients can be back-propagated to the sampling locations as well as the predicted offsets, and the training of SDM is conducted via a weakly-supervised end-to-end optimization.

3.2   Auxiliary Character Counting Supervision

The SDM paves a way to adaptively capture the whole text instance, but its sequential sampling is inevitably undermined by error accumulation in offsets. Besides, without any explicit supervision, the model's training stability is unsatisfactory.

In many common datasets, the text transcription labels and naturally the character number labels are provided. Therefore, we further introduce an auxiliary character counting supervision to guide the SDM's precise sequential sampling. This simultaneously enables the model to learn character-level semantic information. Instead of text recognition, we adopt the language-agnostic character counting as the extra supervision, because text recognition task has a large search space and there is a large gap between the convergence rates of scene text detection and recognition. On real datasets where samples for recognition is

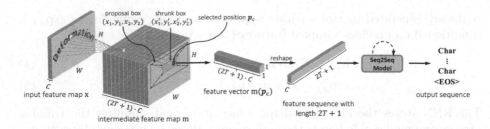

Fig. 3. Sequence-to-sequence based character counting.

very insufficient, jointly training recognition and detection is hard to maximize the performance. By reducing the text recognition to the character counting, the search space is greatly reduced and the optimization is much easier. As a result, we can use character counting to boost our detector on real datasets without additional large synthetic dataset or pre-trained recognition model.

The character counting task is modeled as a sequence-to-sequence (seq2seq) problem [33]. In essence, provided a selected start location, the feature at start location and all the sequentially sampled features could form a input feature sequence, and a seq2seq-based model predicts a valid sequence, whose length is expected to equal the character number of corresponding text instance. The detailed counting process is depicted in Fig. 3.

In SDM, an intermediate feature map m that is the concatenation of all sampled feature maps and the input feature map is obtained first, as described in Sect. 3.1. The map m contains instance-level modeling information for different text instances, and it is capitalized by the consequent character counting.

Then, we select some training samples to optimize the seq2seq-based counting network. In this work, we integrate our SDM into the anchor-based Mask R-CNN [7], where SDM is inserted before the region proposal network (RPN). The effective scheme is adopting the feature vectors around the center area of positive proposal boxes from RPN as training samples. More specifically, we first randomly select K proposals from all positive proposals from RPN. Next, for a selected positive proposal box (x_1, y_1, x_2, y_2), we employ centered shrinkage upon it with shrunk ratio σ. The shrunk box (x_1', y_1', x_2', y_2') (orange area in Fig. 3) lies in the center field of the proposal box, so the features within the shrunk box have higher probability to identify the global existence of text instance. We designate the shrunk box as the distribution region of training sample. We randomly select a position $\mathbf{p}_c = (x_c, y_c)$ from the shrunk box:

$$x_c \sim U(x_1', x_2'), \ y_c \sim U(y_1', y_2'), \tag{9}$$

where U denotes the uniform distribution. Given the select position $\mathbf{p}_c$, we get the feature vector $\mathrm{m}(\mathbf{p}_c)$ with channel $(2T + 1) \cdot C$ and reshape it to a sequence with length $2T + 1$ and channel C, which composes one training sample for character counting. Bilinear interpolation is also used to compute $\mathrm{m}(\mathbf{p}_c)$.

Finally, a one-layer transformer [36] is utilized as our seq2seq model to predict the number of character, and it makes classification for four symbols at each time step, including start-of-sequence symbol "<SOS>", end-of-sequence symbol "<EOS>", padding symbol "<PAD>" and a "Char" symbol. The "Char" symbol represents the existence of one character. Receiving a "<SOS>" symbol and the feature sequence reshaped from feature vector $m(\mathbf{p}_c)$, the model is expected to maximizing the log probability of the target sequence s. The target sequence s contains consecutive "Char" symbols, whose amount equals the character number of corresponding text instance, and ends with a "< EOS>" symbol. Hence the counting loss is:

$$\mathcal{L}_{cnt} = -\log p(s \mid reshape(m(\mathbf{p}_c))). \tag{10}$$

The auxiliary counting task can be simply extended to the scene text recognition task by forcing the network to discriminate different characters at each step. The comparison between character counting and recognition is described in Sect. 4.3.

3.3   Mask R-CNN with SDM

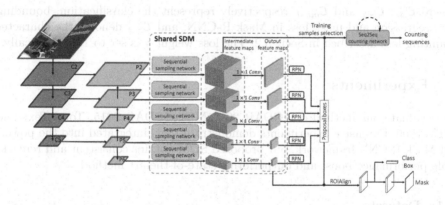

Fig. 4. The architecture of Mask R-CNN equipped with the proposed SDM. C_n ($n = 2, ..., 5$) and P_n ($n = 2, ..., 6$) respectively denote the feature maps (stride 2^n) from the backbone network and feature pyramid network (FPN). SDM is inserted before the region proposal network (RPN) and shared between different levels. The seq2seq counting network is only used in the training phase and is also shared between different levels.

In this work, modeling scene text detection as an instance segmentation task, we leverage the powerful Mask R-CNN [7] with feature pyramid network (FPN) [17] as our baseline detector and equip it with the proposed SDM, as shown in Fig. 4. We re-implement the Mask R-CNN tailored for scene text detection in [18]. The main modifications to standard Mask R-CNN in [18] include: (1) Flipping, resizing and cropping training augmentations; (2) Fine-tuned RPN anchor aspect ratios {0.17, 0.44, 1.13, 2.90, 7.46}; (3) convolution layers with dilation=2 and bilinear upsampling layer in the mask branch; (4) Online hard example mining (OHEM) [32] in bounding box branch. Moreover, we carry out additional color and geometrical augmentations to further enhance the generalization ability. Color augmentations include hue, saturation, brightness and contrast [48] and geometrical augmentation is random rotation in range $[-10°, +10°]$. All these augmentations are performed with a probability of 0.5 independently. Our implemented baseline detector (ResNet-50) achieves a F1-score of 77.07% on MLT2017 dataset.

The SDMs are inserted before RPNs for different feature levels, and, following the practice in FPN, the SDMs are shared between different levels. With respect to the auxiliary character counting, a proposal box at level i generates a counting training sample from the corresponding i-th intermediate feature map m_i, and the seq2seq character counting network is also shared across different feature levels. Meanwhile, the RoIAlign layer extracts region features from the output feature map of SDM (i.e. y in Eq. 6). The network is trained in an end-to-end manner using the following objective:

$$\mathcal{L} = \mathcal{L}_{cls} + \mathcal{L}_{box} + \mathcal{L}_{mask} + \gamma \mathcal{L}_{cnt}, \qquad (11)$$

where $\mathcal{L}_{cls}$, $\mathcal{L}_{box}$ and $\mathcal{L}_{mask}$ respectively represent the classification, bounding box regression and mask loss in Mask R-CNN, and $\mathcal{L}_{cnt}$ denotes the character counting loss described in Sect. 3.2. The loss weight γ is set to 1.0 empirically.

4   Experiments

We evaluate our method on ICDAR 2017 MLT, ICDAR 2015, Total-Text and CTW1500. Extensive experiments demonstrate that, integrated into the powerful Mask R-CNN framework, our proposed SDM obtains consistent and remarkable performance boost and outperforms state-of-the-art methods.

4.1   Datasets

ICDAR 2017 MLT [30] is a multi-oriented, multi-scripting, and multi-lingual scene text dataset. It consists of 7200 training images, 1800 validation images, and 9000 test images, respectively. The text regions are annotated as quadrangles in word-level or line-level for different languages.

ICDAR 2015 [13] is an incidental multi-oriented text detection dataset for English. It consists of 1000 training images, 500 validation images, and 500 test images, respectively. The text regions are labeled as word-level quadrangles.

Total-Text [2] is a dataset not only contains horizontal and multi-oriented text but also specially features curved-oriented text for English. The dataset is split into training and testing sets with 1255 and 300 images, respectively, and all the text regions are labeled as a polygon in word-level.

CTW1500 [21] is a dataset mainly consisting of curved text with both English and Chinese instances. Each image has at least one curved text when horizontal and multi-oriented texts are also contained in this dataset. The dataset contains 1000 training images and 500 test images. Each text is labeled as a polygon in line-level with 14 vertexes.

4.2 Implementation Details

The main configurations for the re-implementation of Mask R-CNN baseline are described in Sect. 3.3. In the SDM, the iteration number T is set to 5, and the hidden size of RNN in the sequential sampling network is 64. For auxiliary character counting, the proposal box's shrunk ratio σ is empirically set to 0.1 and 0.3 for ResNet-18 and ResNet-50, respectively. The one-layer transformer has one attention head and a model dimension of 256. An additional polygon NMS with threshold 0.2 is applied to suppress redundant polygons. We adopt the SGD optimizer with batch size 32, momentum 0.9 and weight decay 1×10^{-4}. During training stage, on all datasets except ICDAR 2015, image's two sides are resized independently in ranges of $[640, 2560]$, $[640, 1600]$ and $[512, 1024]$, respectively. And for ICDAR 2015, image's long side is resized in range $[640, 2560]$, preserving its aspect ration, and then the height is rescaled from 0.8 to 1.2 while the width keeps unchanged. Horizontal flipping with a probability of 0.5 is applied, and a 640×640 patch is cropped for training. For single scale testing, image's long side is resized to 1600, 1920, 1024 and 768 on four datasets, respectively. For multi-scale testing, the long side is resized to $\{960, 1600, 2560\}$, $\{1280, 1920, 2560\}$, $\{640, 1024, 1600\}$ and $\{512, 768, 1024\}$ on four datasets, respectively.

4.3 Ablation Study

To demonstrate the effectiveness of our approach, extensive ablation studies are conducted on ICDAR 2017 MLT dataset considering its high variety in text and multi-lingual challenge. We evaluate two essential components in our model: Sequential Deformation Module (SDM) and Auxiliary Character Counting (ACC). Results are shown in Table 1.

Baseline. The baseline model is built on Mask R-CNN, which is described in Sect. 3.3. It achieves an F-measure of 77.07%.

Sequential Deformation Module. The experimental results show that the Sequential Deformation Module brings a gain of 0.68% and 0.55% for ResNet-18 and ResNet-50 backbones. This shows the effectiveness of SDM to handle multi-oriented and multi-lingual text.

Auxiliary Character Counting. To verify the effectiveness of auxiliary character counting, we introduce character counting and recognition upon SDM. For recognition, we extended the character counting to recognition by simply replacing the "Char" symbol with symbols of full characters. The counting supervision achieves an improvement of 0.5% on F-measure, while recognition brings

Table 1. Effectiveness of Sequential Deformation Module (SDM) and Auxiliary Character Counting (ACC) on ICDAR 2017 MLT dataset. "P", "R", and "F" refer to precision, recall and F-measure, respectively.

Method	Backbone	P(%)	R(%)	F(%)
Baseline	ResNet-18	80.36	70.04	74.84
Baseline + SDM (w/o ACC)	ResNet-18	81.80	70.31	75.62
Baseline + SDM (w/ ACC)	ResNet-18	82.14	70.72	76.00
Baseline	ResNet-50	82.10	72.62	77.07
Baseline + SDM (w/o ACC)	ResNet-50	83.34	72.64	77.62
Baseline + SDM (w/ ACC)	ResNet-50	84.16	72.82	78.08
Baseline + SDM (w/ recognition)	ResNet-50	82.98	72.95	77.64

no gains. Besides, counting network converges as the training of detector, but the recognition network hardly converges. Figure 5 indicates the guidance of auxiliary character counting for sequential sampling.

Fig. 5. Top: Sequential sampling without and with auxiliary character counting. Bottom: Ablations for iteration number T in sequential sampling on MLT 2017.

Iteration Number for Sequential Sampling. As mentioned in Sect. 3.1, the sequential sampling's iteration number T is pre-set and is critical for expanding SDM's sampling range and associated receptive field. We adopt the ResNet-18 backbone and Fig. 5 shows the performances as T changes. The F-measure firstly increases and then saturates for $T \geq 5$. Thus, we use 5 in the remaining experiments. Meanwhile, irrespective of T, all the sampling paths are able to adaptively cover the text regions and avoid going out of the instance. Surprisingly, even when $T = 0$, i.e. using just one feature vector rather sequence to predict character number, a promising improvement of 0.4% is observed, suggesting that the auxiliary character counting is essentially beneficial for detection.

4.4   Comparative Results on Public Benchmarks

Detecting Multi-lingual Text. On ICDAR 2017 MLT, we train the network on 9000 training and validation images for 140 epochs with the weight pre-trained on ImageNet [4]. The learning rate is initialized as 4×10^{-2} and reduced by a factor of 10 at epoch 80 and 125. For single-scale testing, our models with ResNet-18 and ResNet-50 achieve F-measures of 76.00% and 78.08%. For multi-scale testing, the model with ResNet-50 achieves 80.61% F-measure, outperforming all the state-of-the-art methods. Even though a weak backbone (ResNet-18) is adopted, our model is also very competitive compared with the best PMTD [18] (79.13% vs 80.13%). The results are listed in Table 2. Some qualitative results are shown in Fig. 6(a), showcasing the SDM's great robustness for multi-lingual text, long text and complicated background.

Table 2. Comparative Results on ICDAR 2017 MLT and ICDAR 2015 datasets. * denotes the results based on multi-scale testing. "P", "R", and "F" refer to precision, recall and F-measure, respectively.

Datasets	ICDAR 2017 MLT			ICDAR 2015		
Method	P(%)	R(%)	F(%)	P(%)	R(%)	F(%)
EAST [49]	–	–	–	83.27	78.33	80.72
TextSnake [26]	–	–	–	84.90	80.40	82.60
RRD* [16]	–	–	–	88.00	80.00	83.80
Lyu et al.* [28]	74.30	70.60	72.40	89.50	79.70	84.30
LOMO* [47]	79.10	60.20	68.40	87.80	87.60	87.70
PSENet [38]	77.01	68.40	72.45	89.30	85.22	87.21
SPCNET* [44]	80.60	68.60	74.10	–	–	–
FOTS* [20]	81.86	62.30	70.75	91.85	87.92	89.84
PMTD [18]	85.15	72.77	78.48	91.30	87.43	89.33
PMTD* [18]	84.42	**76.25**	80.13	–	–	–
Ours (ResNet-18)	82.14	70.72	76.00	91.14	84.69	87.80
Ours* (ResNet-18)	85.44	73.68	79.13	90.15	88.16	89.14
Ours (ResNet-50)	84.16	72.82	78.08	88.70	88.44	88.57
Ours* (ResNet-50)	**86.79**	75.26	**80.61**	**91.96**	**89.22**	**90.57**

Detecting Oriented English Text. On ICDAR2015, the weights trained on ICDAR 2017 MLT are used to initialize the models. We fine-tune the network for 80 epochs with learning rate 4×10^{-3} in the first 40 epoch and 4×10^{-4} in the remaining 40 epoch. As shown in Table 2, our model with ResNet-50 even surpasses FOTS [20], which is trained with both detection and recognition supervision. The visualization in Fig. 6(b) shows our SDM can effectively tackle challenging situations including skewed viewpoint and low resolution. Notably, our model could accurately locate different text instances under the crowded scene.

Table 3. Comparative Results on Total-Text and CTW1500 datasets. * denotes the results based on multi-scale testing. "P", "R", and "F" refer to precision, recall and F-measure, respectively.

Datasets	Total-Text			CTW1500		
Method	P(%)	R(%)	F(%)	P(%)	R(%)	F(%)
CTD + TLOC [21]	74.30	69.80	73.40	–	–	–
TextSnake [26]	82.70	74.50	78.40	85.30	67.90	75.60
Mask TextSpotter [27]	69.00	55.00	61.30	–	–	–
PSENet [38]	84.02	77.96	80.87	84.84	79.73	82.20
CRAFT [1]	87.60	79.90	83.60	86.00	81.10	83.50
DB-ResNet-50 [15]	87.10	82.50	84.70	86.90	80.20	83.4
PAN [39]	89.30	81.00	85.00	86.40	81.20	83.70
PAN Mask R-CNN [11]	–	–	–	86.80	83.20	85.00
CharNet* [45]	88.00	85.00	86.50	–	–	–
Baseline (ResNet-50)	87.44	84.93	86.16	84.16	81.99	83.06
Ours (ResNet-50)	89.24	84.70	86.91	85.82	82.27	84.01
Ours* (ResNet-50)	**90.85**	**86.03**	**88.37**	**88.40**	**84.42**	**86.36**

Detecting Curved Text. We evaluate our method on the Total-Text and CTW1500 to validate SDM's ability to detect curved text. For Total-Text, the network is also initialized with weights pre-trained on ICDAR 2017 MLT. Considering there's no text transcription in CTW1500, we initialize the network with weights trained on Total-Text and disable the character counting task. All models are fine-tuned for 140 epochs with learning rate 4×10^{-3} in the first 80 epochs and 4×10^{-4} in the remaining epochs.

As visualized in Fig. 7, our SDM is capable of capturing various shapes, which leverages the model's weakness on geometric transformation. The quantitative results for the curved datasets are shown in Table 3. Our model respectively brings an absolute improvement of 0.75% and 0.95% on Total-Text and CTW1500 Datasets, and also surpasses all previous methods. Especially, our model outperforms the state-of-the-art on Total-Text by a large margin of 1.87%. This verifies the SDM's generalization ability on arbitrary-shaped scene text detection.

4.5   Discussion for SDM's Adaptability

Owing to the iteration number T in SDM is pre-set, we expect the SDM has sufficient adaptability for geometrical variance and start location. From the visualization in Fig. 7, it shows that: 1) The sampling range will automatically expand and shrink to fit for different instances and different start locations within the same instance, and avoid going out of the instance; 2) For the more

(a) MLT2017   (b) ICDAR 2015   (c) Total-Text   (d) CTW1500

Fig. 6. Qualitative results of the proposed method on four public datasets.

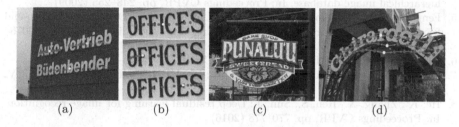

(a)   (b)   (c)   (d)

Fig. 7. Examples of sequential sampling locations (red points) for different start locations (yellow points). Sampling locations for (a) different instances, (b) different start locations within the same instance, (c) different curved text instances at multi-level feature maps (larger point indicates higher feature map) and (d) start locations inside and outside the text region are visualized. In (d), the sequential relations are also visualized through blue arrows to distinguish different sampling paths. (Color figure online)

challenging curved texts, the SDM still performs competently to capture the curved shape; 3) For a start location outside the text, the SDM will try to follow the text center line. These imply that the SDM has high adaptability and dynamically calibrates itself to capture the text instance as far as possible, enriching CNN's capability to model geometrical transformations.

5   Conclusion

In this paper, we propose a novel end-to-end sequential deformation module for accurate scene text detection, which adaptively enhances the modeling capability for text's geometric configuration without any post-processing. We also introduce an auxiliary character counting supervision to facilitate the sequential sampling and features learning. The effectiveness of our method has been demonstrated on several public benchmarks for multi-lingual, multi-oriented and curved text.

Acknowledgement. The authors would like to thank the reviewers for their valuable comments to improve the quality of the paper. This research is supported by a joint research project between Hyundai Motor Group AIRS Company and Tsinghua University. The second author is partially supported by National Key R&D Program of China and a grant from the Institute for Guo Qiang, Tsinghua University.

References

1. Baek, Y., Lee, B., Han, D., Yun, S., Lee, H.: Character region awareness for text detection. In: Proceedings CVPR, pp. 9365–9374 (2019)
2. Ch'ng, C.K., Chan, C.S.: Total-text: a comprehensive dataset for scene text detection and recognition. In: Proceedings ICDAR, pp. 935–942 (2017)
3. Dai, J., et al.: Deformable convolutional networks. In: Proceedings ICCV, pp. 764–773 (2017)
4. Deng, J., Dong, W., Socher, R., Li, L.J., Li, K., Li, F.F.: ImageNet: a large-scale hierarchical image database. In: Proceedings CVPR, pp. 248–255 (2009)
5. Feng, W., He, W., Yin, F., Zhang, X.Y., Liu, C.L.: TextDragon: an end-to-end framework for Arbitrary shaped text spotting. In: Proceedings ICCV, pp. 9076–9085 (2019)
6. Girshick, R.: FastR-CNN. In: Proceedings CVPR, pp. 1440–1448 (2015)
7. He, K., Gkioxari, G., Dollár, P., Girshick, R.: Mask R-CNN. In: Proceedings ICCV, pp. 2961–2969 (2017)
8. He, K., Zhang, X., Ren, S., Sun, J.: Deep residual learning for image recognition. In: Proceedings CVPR, pp. 770–778 (2016)
9. He, W., Zhang, X.Y., Yin, F., Liu, C.L.: Deep direct regression for multi-oriented scene text detection. In: Proceedings ICCV, pp. 745–753 (2017)
10. Huang, L., Yang, Y., Deng, Y., Yu, Y.: DenseBox: unifying landmark localization with end to end object detection. arXiv preprint arXiv:1509.04874 (2015)
11. Huang, Z., Zhong, Z., Sun, L., Huo, Q.: Mask R-CNN with pyramid attention network for scene text detection. In: Proceedings of the IEEE Winter Conference on Applications of Computer Vision (WACV), pp. 764–772 (2019)
12. Jaderberg, M., Simonyan, K., Zisserman, A., et al.: Spatial transformer networks. In: Proceedings NIPS, pp. 2017–2025 (2015)
13. Karatzas, D., et al.: ICDAR 2015 competition on robust reading. In: Proceedings ICDAR, pp. 1156–1160 (2015)
14. Liao, M., Shi, B., Bai, X., Wang, X., Liu, W.: TextBoxes: a fast text detector with a single deep neural network. In: Proceedings AAAI, pp. 4161–4167 (2017)
15. Liao, M., Wan, Z., Yao, C., Chen, K., Bai, X.: Real-time Scene Text Detection with Differentiable Binarization. arXiv preprint arXiv:1911.08947 (2019)
16. Liao, M., Zhu, Z., Shi, B., Xia, G.S., Bai, X.: Rotation-sensitive regression for oriented scene text detection. In: Proceedings CVPR, pp. 5909–5918 (2018)
17. Lin, T.Y., Dollar, P., Girshick, R., He, K., Hariharan, B., Belongie, S.: Feature pyramid networks for object detection. In: Proceedings CVPR, pp. 2117–2125 (2017)
18. Liu, J., Liu, X., Sheng, J., Liang, D., Li, X., Liu, Q.: Pyramid mask text detector. arXiv preprint arXiv:1903.11800 (2019)
19. Liu, W., et al.: SSD: single shot MultiBox detector. In: Leibe, B., Matas, J., Sebe, N., Welling, M. (eds.) ECCV 2016. LNCS, vol. 9905, pp. 21–37. Springer, Cham (2016). https://doi.org/10.1007/978-3-319-46448-0_2
20. Liu, X., Liang, D., Yan, S., Chen, D., Qiao, Y., Yan, J.: FOTS: fast oriented text spotting with a unified network. In: Proceedings CVPR, pp. 5676–5685 (2018)

21. Liu, Y., Jin, L., Zhang, S., Zhang, S.: Detecting curve text in the wild: new dataset and new solution. arXiv preprint arXiv:1712.02170 (2017)
22. Liu, Y., Zhang, S., Jin, L., Xie, L., Wu, Y., Wang, Z.: Omnidirectional scene text detection with sequential-free box discretization. arXiv preprint arXiv:1906.02371 (2019)
23. Liu, Z., Lin, G., Yang, S., Liu, F., Lin, W., Goh, W.L.: Towards robust curve text detection with conditional spatial expansion. In: Proceedings CVPR, pp. 7269–7278 (2019)
24. Long, J., Shelhamer, E., Darrell, T.: Fully convolutional networks for semantic segmentation. In: Proceedings CVPR, pp. 3431–3440 (2015)
25. Long, S., He, X., Yao, C.: Scene text detection and recognition: the deep learning era. arXiv preprint arXiv:1811.04256 (2018)
26. Long, S., et al.: TextSnake: a flexible representation for detecting text of arbitrary shapes. In: Ferrari, V., Hebert, M., Sminchisescu, C., Weiss, Y. (eds.) ECCV 2018. LNCS, vol. 11206, pp. 19–35. Springer, Cham (2018). https://doi.org/10.1007/978-3-030-01216-8_2
27. Lyu, P., Liao, M., Yao, C., Wu, W., Bai, X.: Mask TextSpotter: an end-to-end trainable neural network for spotting text with arbitrary shapes. In: Ferrari, V., Hebert, M., Sminchisescu, C., Weiss, Y. (eds.) Computer Vision – ECCV 2018. LNCS, vol. 11218, pp. 71–88. Springer, Cham (2018). https://doi.org/10.1007/978-3-030-01264-9_5
28. Lyu, P., Yao, C., Wu, W., Yan, S., Bai, X.: Multi-oriented scene text detection via corner localization and region segmentation. In: Proceedings CVPR, pp. 7553–7563 (2018)
29. Ma, J., et al.: Arbitrary-oriented scene text detection via rotation proposals. IEEE Trans. Multimedia, 3111–3122 (2018)
30. Nayef, N., et al.: ICDAR2017 robust reading challenge on multi-lingual scene text detection and script identification-RRC-MLT. In: Proceedings ICDAR, pp. 1454–1459 (2017)
31. Shi, B., Bai, X., Belongie, S.: Detecting oriented text in natural images by linking segments. In: Proceedings CVPR, pp. 2550–2558 (2017)
32. Shrivastava, A., Gupta, A., Girshick, R.: Training region-based object detectors with online hard example mining. In: Proceedings CVPR, pp. 761–769 (2010)
33. Sutskever, I., Vinyals, O., Le, Q.V.: Sequence to sequence learning with neural networks. In: Proceedings NIPS, pp. 3104–3112 (2014)
34. Tian, Z., Huang, W., He, T., He, P., Qiao, Yu.: Detecting text in natural image with connectionist text proposal network. In: Leibe, B., Matas, J., Sebe, N., Welling, M. (eds.) ECCV 2016. LNCS, vol. 9912, pp. 56–72. Springer, Cham (2016). https://doi.org/10.1007/978-3-319-46484-8_4
35. Tian, Z., et al.: Learning shape-aware embedding for scene text detection. In: Proceedings CVPR, pp. 4234–4243 (2019)
36. Vaswani, A., et al.: Attention is all you need. In: Proceedings NIPS, pp. 5998–6008 (2017)
37. Wang, F., Zhao, L., Li, X., Wang, X., Tao, D.: Geometry-aware scene text detection with instance transformation network. In: Proceedings CVPR, pp. 1381–1389 (2018)
38. Wang, W., et al.: Shape robust text detection with progressive scale expansion network. In: Proceedings CVPR, pp. 9336–9345 (2019)
39. Wang, W., et al.: Efficient and accurate arbitrary-shaped text detection with pixel aggregation network. In: Proceedings CVPR, pp. 8440–8449 (2019)

40. Wang, X., Jiang, Y., Luo, Z., Liu, C.L., Choi, H., Kim, S.: Arbitrary shape scene text detection with adaptive text region representation. In: Proceedings CVPR, pp. 6449–6458 (2019)
41. Wigington, C., Tensmeyer, C., Davis, B., Barrett, W., Price, B., Cohen, S.: Start, follow, read: end-to-end full-page handwriting recognition. In: Ferrari, V., Hebert, M., Sminchisescu, C., Weiss, Y. (eds.) ECCV 2018. LNCS, vol. 11210, pp. 372–388. Springer, Cham (2018). https://doi.org/10.1007/978-3-030-01231-1_23
42. Wikipedia: Eye movement in reading. https://en.wikipedia.org/wiki/Eye_movement_in_reading
43. Wu, W., Xing, J., Zhou, H.: TextCohesion: detecting text for arbitrary shapes. arXiv preprint arXiv:1904.12640 (2019)
44. Xie, E., Zang, Y., Shao, S., Yu, G., Yao, C., Li, G.: Scene text detection with supervised pyramid context network. In: Proceedings AAAI, pp. 9038–9045 (2019)
45. Xing, L., Tian, Z., Huang, W., Scott, M.R.: Convolutional character networks. In: Proceedings ICCV, pp. 9126–9136 (2019)
46. Xue, C., Lu, S., Zhan, F.: Accurate scene text detection through border semantics awareness and bootstrapping. In: Ferrari, V., Hebert, M., Sminchisescu, C., Weiss, Y. (eds.) ECCV 2018. LNCS, vol. 11220, pp. 370–387. Springer, Cham (2018). https://doi.org/10.1007/978-3-030-01270-0_22
47. Zhang, C., et al.: Look more than once: an accurate detector for text of arbitrary shapes. In: Proceedings CVPR, pp. 10552–10561 (2019)
48. Zhang, Z., He, T., Zhang, H., Zhang, Z., Xie, J., Li, M.: Bag of freebies for training object detection neural networks. arXiv preprint arXiv:1902.04103 (2019)
49. Zhou, X., et al.: EAST: an efficient and accurate scene text detector. In: Proceedings CVPR, pp. 5551–5560 (2017)
50. Zhu, X., Hu, H., Lin, S., Dai, J.: Deformable convnets v2: more deformable, better results. In: Proceedings CVPR, pp. 9308–9316 (2019)

Where to Explore Next? ExHistCNN for History-Aware Autonomous 3D Exploration

Yiming Wang[1]([⊠]) [iD] and Alessio Del Bue[1,2] [iD]

[1] Visual Geometry and Modelling (VGM), Fondazione Istituto Italiano
di Tecnologia (IIT), Genoa, Italy
{yiming.wang,alessio.delbue}@iit.it
[2] Pattern Analysis and Computer Vision (PAVIS),
Fondazione Istituto Italiano di Tecnologia (IIT), Genova, Italy
https://github.com/IIT-PAVIS/ExHistCNN

Abstract. In this work we address the problem of autonomous 3D exploration of an unknown indoor environment using a depth camera. We cast the problem as the estimation of the Next Best View (NBV) that maximises the coverage of the unknown area. We do this by reformulating NBV estimation as a classification problem and we propose a novel learning-based metric that encodes both, the current 3D observation (a depth frame) and the history of the ongoing reconstruction. One of the major contributions of this work is about introducing a new representation for the 3D reconstruction history as an auxiliary utility map which is efficiently coupled with the current depth observation. With both pieces of information, we train a light-weight CNN, named ExHistCNN, that estimates the NBV as a set of directions towards which the depth sensor finds most unexplored areas. We perform extensive evaluation on both synthetic and real room scans demonstrating that the proposed ExHistCNN is able to approach the exploration performance of an oracle using the complete knowledge of the 3D environment.

Keywords: Next best view · CNN · 3D exploration · 3D reconstruction

1 Introduction

Being able to perceive the surrounding 3D world is essential for autonomous systems in order to navigate and operate safely in any environment. Often, areas where agents move and interact are unknown, i.e. no previous 3D information is available. There is the need to develop autonomous systems with the ability to

Electronic supplementary material The online version of this chapter (https://doi.org/10.1007/978-3-030-58526-6_8) contains supplementary material, which is available to authorized users.

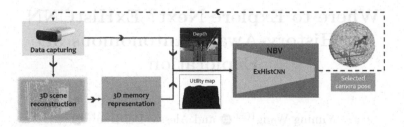

Fig. 1. The overall procedure for autonomous 3D exploration following a Next Best View (NBV) paradigm. Our contributed modules are highlighted in blue. (Color figure online)

explore and cover entirely an environment without human intervention. Even 3D reconstruction from RGBD data is a mature technology [3,7,12,14,27], it relies mainly on a user manually moving the camera to cover completely an area. Less attention has been posed to the problem of obtaining a full coverage of the 3D structure of an unknown space without human intervention. This task has strong relations with the longstanding *Next Best View* (NBV) problem [4,16,20,23] but with the additional issue of having no *a priori* knowledge of the environment where the autonomous system is located.

In this paper, we address this 3D exploration task using a single depth camera as shown in Fig. 1. At each time step, the system captures a new depth image which is then passed into a general purpose online 3D reconstruction module [29]. The previously reconstructed 3D scene together with the current depth observation provide the NBV module with hints which are often represented as a utility function, whose objective is to select the next view among a set of candidate views that explores most unseen areas of the environment [5,15,17,18,24]. Often the set of candidate views ensure not only the physical reachability but also sufficient overlap with the current view in order to guarantee a feasible 3D reconstruction. How to model the utility is essential in all NBV related literature either in a hand-crafted [5,15,17,18,24] or learning-based manner [8,25].

In such autonomous 3D exploration framework, we propose a novel learning-based NBV method to encode both current observation and reconstruction history with a new CNN model named **ExHistCNN**. We avoid formulating the task as a regression problem using a 3D CNN [8], since it requires a large number of parameters to optimise along with extensive training data. Instead, our key advantage is that we formulate NBV as a classification problem using a lightweight 2D CNN architecture which needs less training data and computation. Our ExHistCNN takes as input the current depth image and the neighbourhood reconstruction status, and outputs the direction that suggests the largest unexplored surface. We exploit ray tracing to produce binary utility maps that encode the neighbourhood reconstruction status and we further propose various data formats to combine the depth and the utility maps to facilitate the history encoding. We train and evaluate our proposed CNN using a novel dataset built on top of the publicly available dataset which are SUNCG [22] for synthetic rooms and Matterport3D [2] for real rooms. With experiments, we prove that the proposed CNN and data representation show great potential to encode reconstruction history during 3D exploration and can approach the exploration performance of an oracle strategy with the complete 3D knowledge of the tested environments. The performance is comparable to the state-of-the-art methods [18,25], with a consistent boost of the scene coverage at the early exploration.

To summarise, our three major contributions are: 1) We study and evaluate new data embedding to encode history of previously explored areas in the context of NBV estimation; 2) We propose the light-weight ExHistCNN with a careful design of input data, supervision and network, and prove its effectiveness for addressing the 3D exploration problem in unknown environments, and 3) we build a novel dataset based on SUNCG [22] and Matterport3D [2] to train and evaluate NBV methods in both synthetic and real environments.

2   Related Work

In this section, we will cover related works on NBV (following the observe-decide-move cycle at each step) for 3D exploration and mapping with the focus on the modelling of information utility.

The information modelling greatly depends on how the 3D environment is represented, which can be categorised as surface-based [1] and volume-based representations [21]. The volumetric representation is often employed for online motion planning for its compactness and efficiency in visibility operations [5]. Multiple volumetric information metrics have been proposed for selecting the NBV, often through ray tracing. A common idea is to provide statistics on the voxels [18,24,26,28], where one can either count the unknown voxels [18], or count only the frontier voxels, which are the voxels on the boundary between the known free space and the unexplored space [26,28]. Occlusion is further taken into account by counting the *occuplane* (a contraction for occlusion plane) voxels that are defined as bordering free and occluded space [24].

In addition to counting-based metrics, there are also metrics based on probabilistic occupancy estimation that accounts for the measurement uncertainty [9,13,15]. The main method for computing probabilistic information metrics is based on information entropy [5,11,15,17]. As a ray traverses the map, the information gain of each ray is the accumulated gain of all visible voxels in the form of either a sum [5] or an average [11]. The sum favours views with rays traversing deeper into the map, while the average favours more on the uncertainty of voxels regardless the ray depth. Moreover, inaccurate prediction of the new measurement probability during ray tracing can be an issue for computing the information gain if occlusion is not considered. To address this issue, Potthast and Sukhatme [17] utilise a Hidden Markov Model to estimate the likelihood of an unknown voxel being visible at any viewpoints. The work in [5] accounts for the probability of a voxel being occluded via weighting the information gain by the product of the emptiness probability of all voxels before reaching that voxel.

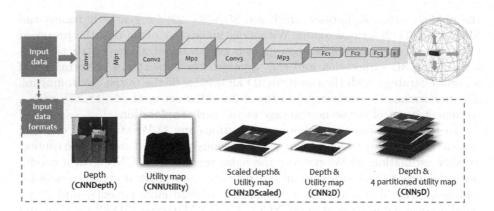

Fig. 2. ExHistCNN architecture and the proposed formats of input data with the aim of encoding local reconstruction status in the context of 3D exploration.

Although the computation can be different, the heuristic behind both [5,17] is similar, i.e. a voxel with a large unobserved volume between its position and the candidate view position is more likely to be occluded and therefore contributes less information gain.

Recent works have shifted their focus towards learning-based methods [8,10,25]. Hepp *et al.* train a 3D CNN with the known 3D models of the scene, and the utility is defined as the decrease in uncertainty of surface voxels with a new measurement. The learnt metric shows better mapping performance compared to the hand-crafted information gain metrics. However, the method can be demanding for data preparation and heavy for training. Wang *et al.* [25] instead propose a 2D CNN to learn the information gain function directly from a single-shot depth image and combine the learnt metric with the hand-crafted metric that encodes the reconstruction history. However due to the fact that a single depth cannot encode reconstruction status, the heuristic-based combination strategies struggle to outperform the strategy using the hand-crafted metric. Jayaraman *et al.* [10] address a related but different task, visual observation completion instead of 3D exploration and mapping, where they exploit reinforcement learning trained with RGB images instead of 3D data.

3   Proposed Method

We represent the reconstructed scene using octomap [9], which is an efficient volumetric representation of 3D environments. The space is quantised into voxels that are associated with an occupancy probability $o^i \in [0, 1]$. A higher o^i value indicates that the voxel is more likely to be occupied, while the lower indicates a higher likelihood to be empty. Based on the occupancy probability, each voxel is thus expressed as having one of the three states: Unknown, Free and Occupied.

At each time step k, o_k^i is updated with the new range measurement coming from the depth image $\mathbf{D}_k$. Let $\mathbf{P}_k = [\mathbf{t}_k, \mathbf{R}_k]$ be the camera pose at k with $\mathbf{t}_k$ for

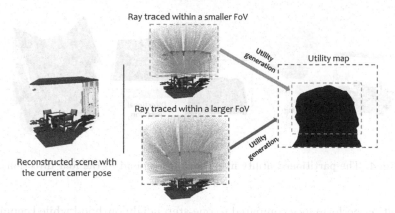

Fig. 3. The generation of the binary utility map through ray tracing with different FoV settings. A smaller FoV (highlighted in red) may lead to local solutions when multiple directions expose similar unexplored areas. (Color figure online)

translation and $\mathbf{R}_k$ for rotation. Let Ω_k be the set of candidate poses that are accessible and also satisfy the view overlapping constraints for 3D reconstruction at $k + 1$. The proposed ExHistCNN predicts m_k, the direction of the NBV which leads to the largest reconstructed surface voxels (see Fig. 2). With the CNN-predicted direction m_k, the next best pose $\mathbf{P}_k^* \in \Omega_k$ is selected as the furthest position at direction m_k among the candidate poses. More specifically, let $\mathbf{e}_k^m = \mathbf{R}_k \mathbf{e}^m$ be the unit vector of the m_k at the current pose $\mathbf{P}_k$, where $\mathbf{e}^m$ is the unit vector of selected direction m_k in world coordinate. For each $\mathbf{P}_j \in \Omega_k$, we define $\Delta \mathbf{t}_{j,k} = \mathbf{t}_j - \mathbf{t}_k$ as the vector originating from the position of current pose $\mathbf{t}_k$ and the position of candidate pose $\mathbf{t}_j$. The projection of $\mathbf{e}_k^m$ and $\Delta \mathbf{t}_{j,k}$ can be computed as the dot product $s_{j,k} = \Delta \mathbf{t}_{j,k} \mathbf{e}_k^m$. Finally, the pose with the largest projection $s_{j,k}$ is selected as the NBV pose $\mathbf{P}_k^*$.

3.1 Representation for 3D Reconstruction History

We encode the reconstruction history into a utility map through ray-tracing (see Fig. 3). Given a camera pose $\mathbf{P}_k$, we trace a set of rays in a discretised manner within a defined FoV originating from the camera pose towards the 3D space. Each ray corresponds to a utility value. The utility should encourage the NBV towards most unexplored area, we therefore set the utility value to be 1 if the ray does not encounter any Free or Occupied voxels, otherwise 0. The resulting utility map is in practice a binary image and each pixel corresponds to a ray with the value of zero for visited areas, and the value of one for unexplored areas.

The FoV of ray tracing defines the extension of the reconstruction area that can be used for making the NBV decision. As shown in Fig. 3, the camera can see the area within its one-step neighbourhood (in red box), two-step neighbourhood (in blue box) or even larger neighbourhoods. We experimentally prove that by seeing the reconstruction status with two-step neighbourhood can improve the

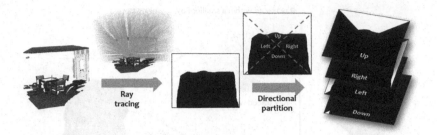

Fig. 4. The partitioned utility maps that correspond to the four directions.

exploration performance compared to one-step neighbourhood while keeping the method cost-effective. Note that the utility map in the following sections refers to the FoV that reflects the reconstruction status in two-step neighbourhood.

3.2   ExHistCNN

The proposed ExHistCNN makes use of a light-weight CNN architecture with convolutional layers, max pooling layers and fully connected layers (see Fig. 2) that functions as a next direction classifier. We consider four main movement directions, i.e. *up, down, left and right*, as the output of the classifier, which is then used to estimate the next camera pose distributed on a sphere surface. Note that the current movement setting is chosen with simplicity for efficient and repeatable datatset preparation and method evaluation.

In order to obtain the ground-truth direction label y, we introduce an *oracle* classifier with the access to ground-truth depth frames at each possible camera movement. In this way, the oracle can always decide the best move that maximises the coverage to unexplored areas at any given pose. ExHistCNN will then learn to imitate the oracle classifier without the access to ground truth depth frames. The network is trained by minimising the cross-entropy loss over the training set $\{\mathbf{X}_1, ..., \mathbf{X}_N\}$ where N is the number of training samples.

Regarding the input $\mathbf{X}$, we explored extensively its potential formats and its impact to facilitate the reconstruction history encoding. We firstly investigate the necessity of combining depth and utility map as input $\mathbf{X}$ by training basic models using only depth, **CNNDepth** (Fig. 2 (a)), and only utility map, **CNNUtility** (Fig. 2 (b)). Secondly, we further investigate the impact of various strategies in combining the depth image and utility map. As a straightforward option, we stack the depth image and utility map into a two-channel data. Moreover, considering the property of convolution that exploits the spatial information in the data, we train **CNN2DScaled** (Fig. 2 (c)) using the depth that is scaled (in this case, shrunk) based on the ratio of its FoV and the FoV for capturing the utility map with zero padding for the rest of the image, and **CNN2D** (Fig. 2 (d)) using the depth without the scaling.

Both **CNN2D** and **CNN2DScaled** use the utility map that contains the complete neighbourhood reconstruction history without an explicit indication of where the network should look at. In order to validate if an explicit division of the utility map based on the direction can facilitate the network learning, we also train **CNN4D** (Fig. 2 (e)) and **CNN5D** (Fig. 2 (f)). **CNN4D** only takes 4 partitioned utility maps with each corresponds to one direction. The partition is performed by dividing the utility map into four non-overlapping triangular areas (see Fig. 4). For each partitioned utility map, the image areas that correspond to other directions will be zero padded in order to not introduce additional information. Such partition choice is experimentally determined because of its better exploration performance compared to another overlapping partition choice. A detailed experiment session for the design choices is provided in Sect. 4.3. Finally, **CNN5D** stacks the depth with the four direction-specific utility maps.

3.3 NBV for 3D Exploration

In this section, we describe in detail the complete NBV pipeline for 3D exploration of one time step as shown in Fig. 1. The system starts with sensor capturing at the time step k, where an incoming depth $\mathbf{D}_k$ (or color-depth pair) frame is then passed into a general purpose online 3D reconstruction module [29] to form a volumetric representation of the scene [9]. Note that we are not bounded to any specific 3D reconstruction algorithm and implementation details are described in Sect. 4.3. With the availability of camera poses, which can be obtained by any SLAM algorithm [27], each depth (or color-depth pair) is registered and integrated to the reconstructed scene volume. We can then obtain the binary utility map by tracing rays into the scene volume with an enlarged FoV in order to have the exploration status for a two-step neighbourhood. We then pass through our proposed ExHistCNN the combined binary utility map and depth frame to predict the best movement direction m_k for the next view. Given m_k predicted by ExHistCNN, the system moves to the selected NBV pose $\mathbf{P}_k^*$ at the time step $k+1$ and repeats the pipeline, until certain termination criteria are met. In this work we terminate the system once a fixed number of steps is reached to allow most baseline methods to saturate their exploration performance.

4   Experiments

Section 4.1 first describes the dataset generation procedure for the training of ExHistCNN. Then, in Sect. 4.2 we perform the ablation study on the proposed ExHistCNN with various input data formats and network architectures. We finally report the 3D exploration performance using both synthetic rooms and real rooms in Sect. 4.3 with a detailed description on the evaluation dataset and the comparison between our proposed learnt strategies and baseline methods.

4.1 Dataset Generation

We produce a new dataset for training/testing the proposed ExHistCNN using in total 33 rooms from the synthetic SUNCG dataset [22]. Our dataset covers various room types including kitchens, living rooms, bedrooms, corridors and toilets. For each room, we rendered a set of viewpoints that are uniformly and isometrically distributed on a sphere with a radius of 20 cm at the height of 1.5m to simulate potential settings of any robotic platforms (e.g. a table-up robotic arm). The camera view is looking out from the sphere centre towards the environment. In particular, we consider a total of 642 viewpoints as the set used for selecting the viewpoints for the tested NBV strategies.

For each viewpoint, we compute a sub-set of neighbouring viewpoints which are within its circular neighbourhood of a radius r. In our experimental setup we set r to 5 cm because the overlapping view constraint is necessary for 3D reconstruction algorithms to work. In order to represent the neighbourhood reconstruction status of each view point in a tractable manner, we discretise status into six levels, i.e. 0%, 20%, 40%, 60%, 80% and 100% of the neighbourhood reconstruction. Each reconstruction level is approximated by selecting the corresponding percentage of neighbouring viewpoints. With the selected view points, we then reconstruct the scene using their corresponding depth frames. For instance, if viewpoint A has 10 neighbouring viewpoints, then 20% reconstruction status will be achieved by selecting 2 neighbouring viewpoints out of the overall 10 viewpoints for the reconstruction. Since the combinations of the selected viewpoints can be large due to many neighbouring viewpoints, thus to limit the amount of data produced, we constrain only up to 10 combinations for each reconstruction status.

For each viewpoint, we first perform the reconstruction using the neighbouring viewpoints under each neighbouring reconstruction status. We then generate the binary utility map that reflects a two-step neighbourhood reconstruction status through ray tracing. The ground-truth motion label is finally produced by integrating the depth frame that corresponds to the 4 directions in two steps, and selecting the direction that results in most surface voxels. This generation procedure produces 17,960 samples per room, where each sample is composed of a depth image for the current observation, a binary utility map for the neighbourhood reconstruction status and the direction label. We further organise the dataset in a balanced manner with 100K samples per direction class and the train-validation-test follows a 75-15-15 split.

4.2 ExHistCNN Ablation Study

We train ExHistCNN with various input data as described in Sect. 3.2, namely, CNNDepth, CNNUtility, CNN2DScaled, CNN2D, CNN4D, and CNN5D. As comparison, we also train a set of classifiers using ResNet101 [6] pretrained on ImageNet, that serves as a feature extractor. Each channel of the input (see Fig. 2) is repeated to 3 channels and fed to ResNet101. The extracted feature

Table 1. Direction classification result of multiple classifiers at test

	Recall				Avg precision	Avg recall	Avg F1
	Up	Down	Left	Right			
CNNDepth	0.469	0.58	0.32	0.41	0.446	0.445	0.445
CNNUtility	0.719	0.830	0.450	0.528	0.651	0.632	0.624
CNN4D	0.668	0.826	0.458	0.608	0.663	0.64	0.635
CNN2DScaled	0.765	0.851	0.379	0.446	0.649	0.61	0.595
CNN2D	0.707	0.861	0.536	0.449	0.666	0.638	0.632
CNN5D	0.617	0.871	0.511	0.576	**0.677**	0.644	0.642
MLP4D	0.664	0.709	0.58	0.554	0.626	0.627	0.625
MLP2DScaled	0.639	0.691	0.575	0.553	0.616	0.614	0.614
MLP2D	0.622	0.707	0.558	0.553	0.614	0.61	0.610
MLP5D	0.683	0.723	0.618	0.595	0.655	**0.654**	**0.654**

vectors are then concatenated and used as input to train a four-layered Multiple Layer Perceptron (MLP) classifier. According to the data input formats, we therefor train four MLP-based classifiers: **MLP2DScaled**, **MLP2D**, **MLP4D** and **MLP5D**. For all networks, we resize the input to 64×64. We apply techniques including batch norm and drop out during training and the batch size is set to maximise the usage of GPU. Stochastic gradient descent is used with learning rate $1e^{-3}$, 200 epochs and momentum 0.9. For testing, we use the model at the epoch where each network starts to saturate.

Table 1 shows the testing classification performance of multiple ExHistCNN models and MLP-based models. Regarding the average classification performance, CNNDepth performs the worst among all. In general, models that combine both depth and utility maps (CNN2D and CNN5D) are better than the models using only utility maps (CNNUtility and CNN4D). Moreover, we notice that partitioning the utility map into four directions, CNN4D is able to perform better than CNNUtility which uses only a single-channel utility map. Similarly, CNN5D with depth and partitioned utility maps is also marginally better than CNN2D. Interestingly, we observe that CNN2DScaled is not performing better than CNN2D. The reason might be that the depth after rescaling and resizing to 64×64, becomes a rather small patch which could be not very informative for the network to learn from. Moreover, we do notice that all the models perform better in particular directions, i.e. up and down is better than left and right. This can be due to the standard camera setting with wider horizontal FoV than the vertical FoV. Finally, the MLP-based models have a similar pattern as our CNN models, however are achieving a worse classification performances apart from MLP5D. One possible reason can be that the pretrained network extracts feature vectors with semantics bias from other datasets, while our CNN models is trained from scratch without the impact of external bias. In the following section

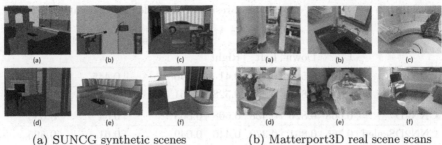

(a) SUNCG synthetic scenes   (b) Matterport3D real scene scans

Fig. 5. Rooms used for the exploration experiment. (a) six synthetic rooms from SUNCG dataset rendered with SUNCG Toolbox. (b) six real room scans from Matterport3D dataset rendered with HabitatSim.

we will use CNN2D and CNN5D as our ExHistCNN models for their best performance, to evaluate the 3D exploration performance. Other models including CNNUtility, CNN4D, MLP4D, MLP2D and MLP5D are also evaluated.

4.3 Autonomous 3D Exploration Performance

We apply different NBV strategies to indoor dataset for 3D exploration and report the surface coverage ratio, i.e. the number of the surface voxels generated using autonomous methods against the number of surface voxels of a complete reconstructed room. For a fair comparison, we evaluate all methods until a fixed number of steps (150 steps throughout the experiments). The step number is set to allow most strategies to saturate in their exploration performance, i.e. when the camera starts looping within a small area. The metric that achieves a larger coverage ratio within the fixed number of steps is considered better.

Evaluation Dataset. We perform experiments using dataset rendered from synthetic rooms in SUNCG dataset [22] (Fig. 5 (a)) and real room scans (Fig. 5 (b)) from Matterport3D [2] using HabitatSim [19]. The set of synthetic rooms for evaluating the exploration performance is different than the set of rooms used for training our ExHistCNN models. Moreover, to validate the generalisation of the model from synthetic to more realistic data, we use a publicly available tool, HabitatSim, to render depth (and color) data with real room scans from Matterposrt3D, following the same dome-shaped path as described in Sect. 4.1. In the experiments, scenes are reconstructed using a truncated signed distance function (TSDF) volume integration method with implementation tools provided in Open3D [29]. In particular, we consider that the dome-shaped path explores the complete room, i.e. the coverage ratio is 100%.

Justification of Design Choices. We first performed a set of experiments to justify two choices in the method and dataset design: 1) the FoV selection during ray-tracing to produce the utility map that reflects multiple-step neighbourhood reconstruction status and 2) the partitioning of the utility map to

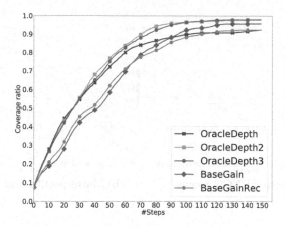

Fig. 6. The coverage ratio with time achieved with strategies that are used for justifying design choices.

enforce the directions in the input data. To justify the FoV selection, we perform 3D exploration using the oracle NBV strategies by integrating the depth frames in different time steps. **OracleDepth** integrates the depth frame corresponding to each candidate direction for the next step into the current volume, and the NBV is selected with the largest resulted surface voxels. Similarly, **OracleDepth2 (OracleDepth3)** integrates the depth frames corresponding to each candidate direction for the next two (three) steps into the current volume. To justify the partitioning of the utility map, we perform 3D exploration using the NBV selected based on the sum of each partitioned utility map. **BaseGain** divides the utility map into four non-overlapping triangular areas that correspond to the four candidate directions (see Fig. 4), while **BaseGainRec** divides the utility map by half for each direction, resulting in rectangular overlapping areas.

Figure 6 shows the coverage ratio with time using the above motioned strategies. Results are averaged by five independent runs on the six synthetic rooms. We observe that OracleDepth2 and OracleDepth3 outperform OracleDepth because being aware only one-step ahead the reconstruction status can be prone to early saturation due to local solutions. OracleDepth2 achieves almost the same exploration speed and coverage performance as OracleDepth3, but with reduced computational/storage cost for both offline dataset preparation and NBV estimation at runtime. We therefore perform ray-tracing to produce the utility map that reflects the two-step neighbourhood reconstruction status. Moreover, BaseGain achieves a higher coverage ratio compared to BaseGainRec, which makes the non-overlapping triangular partition a better choice.

Methods and Baselines Comparison. We performed the exploration experiments using our ExHistCNN models with various input: **CNNdepth, CNNUtility, CNN4D, CNN2D** and **CNN5D**, as well as the MLP-based models: **MLP4D, MLP2D** and **MLP5D** as described in Sect. 4.2. We compared the

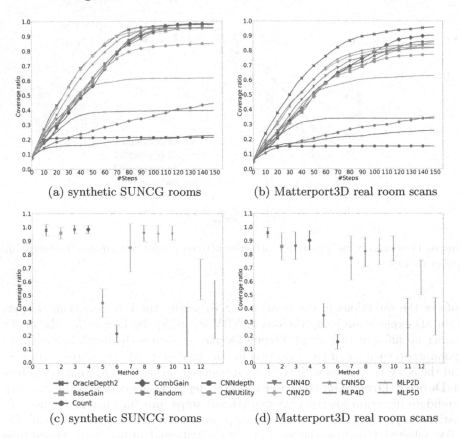

(a) synthetic SUNCG rooms (b) Matterport3D real room scans

(c) synthetic SUNCG rooms (d) Matterport3D real room scans

Fig. 7. The coverage ratio with time (upper row) and the averaged final coverage ratio and its standard deviation (lower row).

above mentioned learning-based strategies against: **Random** strategy that randomly selects the NBV, **BaseGain** that selects the view based on the sum of the partitioned utility maps, **Count** [18] that selects the NBV by counting unknown voxels for each candidate pose, and **CombGain** [25] that selects the NBV using both the output of CNNDepth for direction and the entropy-based utility maps computed using view-dependent descriptors for each candidate pose[1]. Finally, **OracleDepth2** serves as a reference for the best reachable result for our learning-based strategies.

Figure 7(a) shows the average coverage ratio over time for synthetic rooms. NBV with CNNDepth achieves the worst coverage because the camera moves without any knowledge of the reconstructions status, leading to repeated back and forth movement at a very early stage. We observe that CNNUtility with only the binary utility map as its input is worse than CNN4D which uses the partitioned utility maps. This result indicates that the partition of input data

---

[1] We are not able to compare with [8] as their dataset and code are not available.

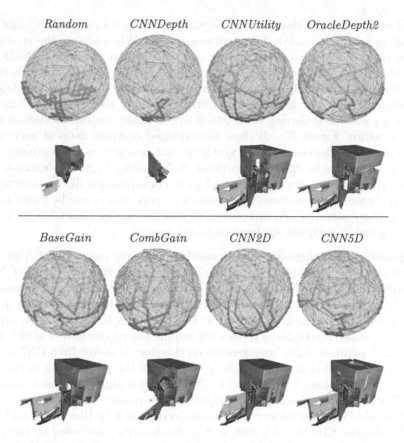

Fig. 8. Selected poses on the dome surface and their corresponding 3D reconstruction for a real room in Matterport3D (best viewed in color). (Color figure online)

Table 2. Processing time for different NBV strategies averaged over all steps tested with the two datasets from SUNCG and Matterport3D (Unit is in second).

	Random	BaseGain	Count	CombGain	CNNdepth	CNNUtility	CNN2D	CNN5D	MLP2D	MLP5D
SUNCG	0	0.028	0.016	0.040	0.018	0.023	0.024	0.058	0.098	
Matterport3D	0	0.022	0.016	0.047	0.026	0.030	0.031	0.064	0.105	
Average	0	0.025	0.016	0.044	0.007	0.022	0.027	0.028	0.061	0.102

can facilitate the network learning of the NBV directions and boost the exploration performance. Moreover, our variants CNN2D and CNN5D leverage the depth information to explore faster compared to CNN4D at the earlier stage, and saturate at a similar coverage ratio which approaches to the performance of OracleDepth2. In the early phase, CNN2D explores faster than CNN5D at a similar exploration speed as OracleDepth2. MLP-based strategies in general are worse than the ExHistCNN-based strategies. Even the best performed MLP2D model is almost 30% less than CNN2D. BaseGain, Count and CombGain use hand-crafted utility, and are slower in the beginning, however Count and CombGain are able to achieve a slightly better coverage ratio. When testing the strategies on real room scans, from Fig. 7(b), we can see that our variants CNN2D and CNN5D are still the fastest in the early phase compared to other strategies, although at saturation they are surpassed by BaseGain, Count and CombGain. This suggests that our learning-based strategies may require further domain adaptation when transferring from purely synthetically trained models to real-world scenarios. Figure 7(c, d) show the averaged coverage ratio at saturation and its standard deviation under synthetic and real scenes, respectively. The variance occurs due to different viewpoint initialisation. Figure 8 showcases the resulted paths on the dome-surface and their corresponding 3D reconstruction of one real room using various NBV strategies (more results can be found in the supplementary material). In this particular run, we can see that CNN2D is able perform a rather complete exploration of the room.

Computational Analysis. We performed experiments using a Dell Alienware Aurora with core i7. Table 2 shows the computational time for different NBV methods, where the processing time includes the time for utility map generation through ray-tracing and the time for NBV estimation. Random uses 0 s as the strategy is a random number generator. CNNDepth is fastest (0.007 s) among all learning-based strategies as it does not require the production of utility map. CNNUtility requires utility map generation therefore is slower than CNNDepth, however is the second fastest as it does not require the preparation of stacking depth and utility map. CNN2D and CNN5D requires almost the same time for the utility map generation and data preparation for passing through the network. CNN2D and CNN5D are slower than Count, comparable to BaseGain and faster than CombGain. CombGain is slow as it performs extra processing on the view-dependent descriptor update for each voxel before generating the utility maps. MLP-based strategies are the slowest as each channel of the input is processed through the pretrained network.

5 Conclusion

In this paper we proposed ExHistCNN, a light-weight learning-based solution to address autonomous 3D exploration of any unknown environment. With experiments using dataset from both synthetic and real rooms, we showed that our ExHistCNN, both CNN2D and CNN5D, are able to effectively encode depth observation and reconstruction history for the exploration task. ExHistCNN-based NBV strategies are computationally efficient and able to explore the space faster in the early phase while approaching the oracle NBV performance in the synthetic dataset. When testing with real room scans, our ExHistCNN even if purely trained using synthetic data maintains its property of fast exploration at the early stage, while achieving less final coverage compared to the soa method. As future work, we will further investigate domain adaptation techniques to boost the exploration performance in real-world scenarios.

Acknowledgements. We would like to thank Andrea Zunino, Maya Aghaei, Pietro Morerio, Stuart James and Waqar Ahmed for their valuable suggestions and support for this work.

References

1. Border, R., Gammell, J.D., Newman, P.: Surface edge explorer (see): planning next best views directly from 3d observations. In: Proceedings of IEEE International Conference on Robotics and Automation (ICRA), pp. 1–8, May 2018
2. Chang, A., et al.: Matterport3D: learning from RGB-D data in indoor environments. In: Proceedings of International Conference on 3D Vision (3DV), pp. 667–676, October 2017
3. Choi, S., Zhou, Q.Y., Koltun, V.: Robust reconstruction of indoor scenes. In: Proceedings of IEEE Conference on Computer Vision and Pattern Recognition (CVPR), pp. 5556–5565, June 2015
4. Connolly, C.: The determination of next best views. In: Proceedings of IEEE International Conference on Robotics and Automation (ICRA), vol. 2, pp. 432–435 (1985)
5. Delmerico, J., Isler, S., Sabzevari, R., Scaramuzza, D.: A comparison of volumetric information gain metrics for active 3D object reconstruction. Auton. Rob. **42**(2), 197–208, February 2018
6. He, K., Zhang, X., Ren, S., Sun, J.: Deep residual learning for image recognition. In: Proceedings of IEEE Conference on Computer Vision and Pattern Recognition (CVPR), pp. 770–778, June 2016
7. Henry, P., Krainin, M., Herbst, E., Ren, X., Fox, D.: RGB-D mapping: using kinect-style depth cameras for dense 3D modeling of indoor environments. Int. J. Rob. Res. **31**(5), 647–663 (2012)
8. Hepp, B., et al.: Learn-to-score: efficient 3D scene exploration by predicting view utility. In: Ferrari, V., Hebert, M., Sminchisescu, C., Weiss, Y. (eds.) ECCV 2018. LNCS, vol. 11219, pp. 455–472. Springer, Cham (2018). https://doi.org/10.1007/978-3-030-01267-0_27
9. Hornung, A., Wurm, K.M., Bennewitz, M., Stachniss, C., Burgard, W.: OctoMap: an efficient probabilistic 3D mapping framework based on octrees. Auton. Rob., 189–206 (2013)
10. Jayaraman, D., Grauman, K.: Learning to look around: Intelligently exploring unseen environments for unknown tasks. In: Proceedings of IEEE Conference on Computer Vision and Pattern Recognition (CVPR), pp. 1238–1247, June 2018
11. Kriegel, S., Rink, C., Bodenmüller, T., Suppa, M.: Efficient next-best-scan planning for autonomous 3D surface reconstruction of unknown objects. J. Real-Time Image Process. **10**(4), 611–631 (2015)
12. Meilland, M., Comport, A.I.: On unifying key-frame and voxel-based dense visual slam at large scales. In: Proceedings of IEEE/RSJ International Conference on Intelligent Robots and Systems (IROS), pp. 3677–3683 (2013)
13. Meng, Z., et al.: A two-stage optimized next-view planning framework for 3-D unknown environment exploration, and structural reconstruction. IEEE Rob. Autom. Lett. **2**(3), 1680–1687 (2017)
14. Newcombe, R.A., et al.: KinectFusion: Real-time dense surface mapping and tracking. In: Proceedings of IEEE International Symposium on Mixed and Augmented Reality (ISMAR), pp. 127–136, October 2011

15. Palazzolo, E., Stachniss, C.: Effective exploration for MAVs based on the expected information gain. Drones **2**(1) (2018)
16. Pito, R.: A solution to the next best view problem for automated surface acquisition. IEEE Trans. Pattern Anal. Mach. Intell. **21**(10), 1016–1030 (1999)
17. Potthast, C., Sukhatme, G.S.: A probabilistic framework for next best view estimation in a cluttered environment. J. Vis. Commun. Image Represent. **25**(1), 148–164 (2014)
18. Quin, P., Paul, G., Alempijevic, A., Liu, D., Dissanayake, G.: Efficient neighbourhood-based information gain approach for exploration of complex 3d environments. In: Proceedings of IEEE International Conference on Robotics and Automation (ICRA), pp. 1343–1348, May 2013
19. Savva, M., et al.: Habitat: a platform for embodied AI research. In: Proceedings of the IEEE/CVF International Conference on Computer Vision (CVPR), June 2019
20. Schmid, K., Hirschmüller, H., Dömel, A., Grixa, I., Suppa, M., Hirzinger, G.: View planning for multi-view stereo 3D reconstruction using an autonomous multicopter. J. Intell. Rob. Syst. **65**(1), 309–323, January 2012
21. Scott, W.R., Roth, G., Rivest, J.F.: View planning for automated three-dimensional object reconstruction and inspection. ACM Comput. Surve. **35**(1), 64–96 (2003)
22. Song, S., Yu, F., Zeng, A., Chang, A.X., Savva, M., Funkhouser, T.: Semantic scene completion from a single depth image. In: Proceedings of IEEE Conference on Computer Vision and Pattern Recognition (CVPR), June 2017
23. Surmann, H., Nüchter, A., Hertzberg, J.: An autonomous mobile robot with a 3D laser range finder for 3D exploration and digitalization of indoor environments. Rob. Autono. Syst. **45**(3–4), 181–198 (2003)
24. Vasquez-Gomez, J.I., Sucar, L.E., Murrieta-Cid, R., Lopez-Damian, E.: Volumetric next-best-view planning for 3D object reconstruction with positioning error. Int. J. Adv. Rob. Syst. **11**(10), 159 (2014)
25. Wang, Y., James, S., Stathopoulou, E.K., Beltrán-González, C., Konishi, Y., Del Bue, A.: Autonomous 3D reconstruction, mapping and exploration of indoor environments with a robotic arm. IEEE Rob. Autom. Lett. **6**, 3340–3347 (2019)
26. Wettach, J., Berns, K.: Dynamic frontier based exploration with a mobile indoor robot. In: Proceedings of International Symposium on Robotics (ISR) and German Conference on Robotics (ROBOTIK), pp. 1–8, June 2010
27. Whelan, T., Salas-Moreno, R.F., Glocker, B., Davison, A.J., Leutenegger, S.: ElasticFusion: real-time dense SLAM and light source estimation. Int. J. Rob. Res. **35**(14), 1697–1716 (2016)
28. Yamauchi, B.: A frontier-based approach for autonomous exploration. In: Proceedings of IEEE International Symposium on Computational Intelligence in Robotics and Automation (CIRA), pp. 146–151, July 1997
29. Zhou, Q.Y., Park, J., Koltun, V.: Open3D: a modern library for 3D data processing. arXiv:1801.09847 (2018)

Semi-supervised Segmentation Based on Error-Correcting Supervision

Robert Mendel[1]([✉]), Luis Antonio de Souza Jr.[2], David Rauber[1],
João Paulo Papa[3], and Christoph Palm[1]

[1] Ostbayerische Technische Hochschule Regensburg, Regensburg, Germany
{robert1.mendel,david.rauber,christoph.palm}@oth-regensburg.de
[2] Federal University of São Carlos, São Carlos, Brazil
luis.souza@dc.ufscar.br
[3] São Paulo State University, Bauru, Brazil
papa@fc.unesp.br

Abstract. Pixel-level classification is an essential part of computer vision. For learning from labeled data, many powerful deep learning models have been developed recently. In this work, we augment such supervised segmentation models by allowing them to learn from unlabeled data. Our semi-supervised approach, termed Error-Correcting Supervision, leverages a collaborative strategy. Apart from the supervised training on the labeled data, the segmentation network is judged by an additional network. The secondary correction network learns on the labeled data to optimally spot correct predictions, as well as to amend incorrect ones. As auxiliary regularization term, the corrector directly influences the supervised training of the segmentation network. On unlabeled data, the output of the correction network is essential to create a proxy for the unknown truth. The corrector's output is combined with the segmentation network's prediction to form the new target. We propose a loss function that incorporates both the pseudo-labels as well as the predictive certainty of the correction network. Our approach can easily be added to supervised segmentation models. We show consistent improvements over a supervised baseline on experiments on both the Pascal VOC 2012 and the Cityscapes datasets with varying amounts of labeled data.

1 Introduction

One factor that led to the reemergence of neural networks as an active topic of research is the availability of large datasets to researchers today. Starting with Krizhevsky et al. [20] significantly improving the classification accuracy on the ImageNet dataset [6], and the many impressive results in the domains of vision, natural language processing, and control that followed, neural networks have proven to be an incredibly effective tool, when enough labeled data is available. Large amounts of labeled data is already accessible for generic object detection tasks or can be gathered if enough resources are on-hand to cope with such a process. However, in some computer vision domains, the availability still poses a problem.

© Springer Nature Switzerland AG 2020
A. Vedaldi et al. (Eds.): ECCV 2020, LNCS 12374, pp. 141–157, 2020.
https://doi.org/10.1007/978-3-030-58526-6_9

In medical imaging, data is commonly sparse, and labeling it is costly. Additionally, many problems are semantic segmentation problems, a task where each pixel in the image needs to be classified. Annotating image data for a segmentation task is more time consuming, and in some domains like medical imaging, has to be done by experts.

In this work, we propose to learn from unlabeled data with Error-Correcting Supervision (ECS). ECS takes the form of an extension of the supervised segmentation task, where an additional model is used to assess and correct the agreement between an image and its segmentation. The insights of this second model are then used as a proxy for the truth on unlabeled data. At first glance ECS borrows concepts from Generative Adversarial Networks (GANs) [8]. But contrary to GANs, our framework profits from the primary and secondary models collaborating, instead of competing. By using both labeled and unlabeled data, our framework allows for efficient utilization of all data available, which is especially important in domains where data gathering is nontrivial.

In summary, our contributions are as follows:

- A collaborative approach for semi-supervised segmentation leveraging two networks without the need for weakly labeled data.
- Stating the secondary model's task as fine-grained error correction, to fit the semi-supervised objective.
- An augmented loss function, which utilizes both the secondary model's prediction and the certainty in it, to adaptively adjust the contributions when training on unlabeled data.
- An end-to-end approach which can augment the training of existing segmentation networks and is not reliant on post-processing during validation.

2   Related Work

2.1   Supervised Semantic Segmentation

The most widespread approach for designing a deep neural network for semantic segmentation as fully convolutional was proposed by Long et al. [29]. Today most models build upon this concept and employ either an encoder-decoder [1,28] structure or some form of spatial pyramid pooling [2,10,34]. PSPNets [34] use several pooling kernels to capture representations at various resolutions. Instead of reducing the resolution, the DeepLab Family [2–4] employs an Atrous Spatial Pyramid Pooling (ASPP) module, with dilated convolutions [33] to capture multi-scale relationships in the input data. With DeepLabv3+ [4], they have transitioned from just the ASPP module and extend their design with a decoder. Their architecture combines low and high-level features to detect sharper object boundaries.

2.2   Weakly-Supervised Segmentation

Weakly-supervised segmentation models generate dense classification maps despite only image level [32,35] or bounding box annotations [17] being present.

Some methods can use both weak and strong signals [24]. Decoupled Neural Networks [13] split the network into classification and segmentation models, resembling the encoder-decoder structure. The segmentation branch then performs binary pixel-wise classification, to separates foreground from background, for each of the identified classes. Additional bounding box annotations are leveraged by [15]. A self-correcting network learns to combine the prediction of two individual segmentation networks. One trained on densely labeled data and the other with the bounding box annotations.

2.3 Semi-supervised Segmentation

Generative Adversarial Networks (GAN) [8] are generative models that try to capture high dimensional implicit distributions. They consist of a generator and discriminator network, realizing the idea of an adversarial two-player game, where each player tries to outperform the other. This adversarial structure has been applied to semi-supervised learning in various ways. Concerning the classification setting, generated samples are either grouped as *fake* or as one of the classes contained in the dataset. For unlabeled images, the sum of the probabilities of the true classes should surpass the probability of it being *fake*.

Souly et al. [30] transferred this approach to semi-supervised segmentation, by keeping a generator that produces artificial samples but choosing a segmentation architecture for the discriminator. With their approach, each pixel is classified as either generated or as one of the true classes. Qi et al. [26] extend this approach to include a more advanced architecture, as well as the addition of Knowledge Graph Embeddings to enforce semantic consistency.

Luc et al. [22] proposed adversarial regularization on the supervised loss. Their discriminator network predicts if an image and label map pair are *real* or *fake*. These labels are chosen depending on the label map showing the ground-truth or being the output of the segmentation network. However, this approach does not learn from unlabeled data.

The approach introduced by Hung et al. [14] is overall similar in execution to our proposed method, but varies on a conceptional level. Unlike [30], the segmentation network assumes the role of the generator, and a new discriminator is added. This Fully Convolutional Discriminator is designed to approximate the space of possible label distributions without directly including the base image or any information whether the segmentation is correct on a pixel level. The label maps used to optimize the discriminator just describe whether the given label map is *real* or *fake* so originating from the ground-truth or the output of the segmentation network. The discriminator is used for an adversarial loss function, in the form of a regularizing term during supervised training, similar to Luc et al. [22]. For unlabeled data, the prediction of the discriminator is compared with a threshold value. If a region is predicted as *real* with a probability above a given threshold, it is accepted as true and used to optimize the segmentation network.

The work of Mittal et al. [23] extends [22] for semi-supervised learning. The classification result of the discriminator is used to flag unlabeled images and their segmentation for self-training. If the prediction of an image–segmentation pair being *real* surpasses a chosen threshold, this segmentation is used for supervised training. Additionally, they apply a Mean Teacher model [31] during validation, deactivating classes which are found to be absent in the image. This post-processing step only is applied, when the dataset features a background class.

Zhou et al. [36] explore a collaborative approach to semi-supervised segmentation in the medical domain, with influences from adversarial learning. A model pretrained for diabetic retinopathy lesion segmentation produces segmentations for a large set of weakly labeled data of the same domain. One component of their approach discriminates image and segmentation pairs between data that has pixel-level and image-level annotations. In addition, a lesion attention model produces segmentation maps for the weakly labeled data and can be utilized to further fine-tune the primary model.

But semi-supervised segmentation can not only be modeled with adversarial approaches. Kalluri et al. [16] extend the segmentation network with an entropy module. Minimizing the entropy of the similarity of the outputs of the traditional decoder and entropy model, within and across domains, allows their universal approach to learning from labeled and unlabeled data, beyond just one domain.

3   Error-Correcting Supervision

Error-Correcting Supervision is, at first glance, inspired by the GAN-Framework. In addition to a base segmentation network, a secondary model is optimized with the available labeled data. However, instead of classifying a given segmentation as either *real* or *fake*, the additional network in ECS, termed corrector or correction network, judges how well the given image–segmentation pair match on a pixel level, as well as offering corrections for areas where the outputs do not seem to agree. Then, the corrector's predictions are used as a proxy for the truth on unlabeled data and incorporated in the semi-supervised update. The interaction between correction and segmentation network on unlabeled data is controlled by a specific loss function, which individually weights the contribution of the pseudo-labels proportional to the corrector's certainty.

A single training iteration with ECS consists of three parts: The error-correcting, supervised and semi-supervised training steps, which are indicated by the backdrop color in Fig. 1. In each step, the weights of the affected model are updated. During these stages, the labeled training data and correction maps, shown in Fig. 2, as well as pseudo-labels for unlabeled images are utilized.

In contrast to competing methods, the relationship between the segmentation network and the corrector is collaborative instead of adversarial. This allows us to use arbitrarily powerful network architectures for the corrector since the common case of the generator being overpowered is not possible here.

Notice that ECS does not require any weakly-labeled data such as image-level or bounding box annotations. The only requirement is that the additional unlabeled and labeled training data belong to the same domain.

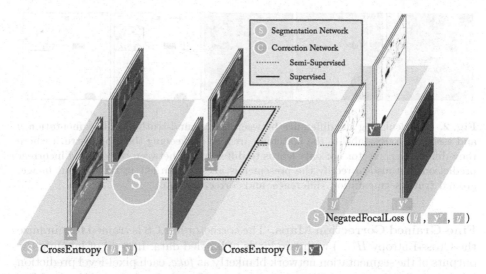

Fig. 1. Overview of Error-Correcting Supervision. Apart from the supervised training, the method is comprised of two additional steps (represented by the backdrop color). The Error-Correcting network C is trained with the image–ground-truth $(\mathbf{x}, \mathbf{y})$ and image–segmentation $(\mathbf{x}, \hat{y})$ pairs (only the latter is shown). For the semi-supervised step, the segmentation network learns to minimize the Negated Focal Loss between the segmentation of an unlabeled image and a pseudo-label $\mathbf{y}^p$. This generated label is a combination of the corrector's output and the original segmentation.

Notation. $\mathbf{x}$ represents an image, and $\mathbf{y}$ its corresponding discrete label map. In cases where the label map is used as input for the corrector it is transformed to a one-hot representation. The continuous output of the segmentation network $S(\cdot)$ given $\mathbf{x}$ is denoted as $\hat{y}$. An added subscript i is used to identify an individual value. $\mathbf{x}, \mathbf{y} \sim \mathcal{D}$ implies sampling from the labeled training data, $\mathcal{D}_u$ denotes the unlabeled data.

3.1   Error-Correcting Network

The correction network $C(\cdot, \cdot)$ transforms a given image–segmentation pair into a segmentation map of depth $N + 1$, where N is the number of classes in the dataset. The $(N + 1)^{th}$ class indicates whether the input segmentation matches the content shown in the input image.

An important distinction to the previous works that apply adversarial learning for semi-supervised segmentation, is related to how the labels to train the secondary model are chosen. In Hung et al.[14] all outputs of the segmentation network are always tagged as *fake*. Although their discriminator is designed as fully convolutional and produces pixel-level confidence for a given input, this information is not incorporated to distinguish whether parts of the output segmentation are correct or not.

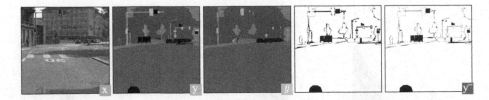

Fig. 2. By calculating the difference between the ground-truth y and segmentation ŷ and assigning a new label to the matching areas while keeping the ground-truth where they differ, the corrector not only learns to differentiate between correct and incorrect predictions but also to rectify the present mistakes. (From left to right: input image, ground truth, segmentation, difference, and correction map)

Fine-Grained Correction Maps. The corrector in ECS is trained to minimize the Cross-Entropy $H(\cdot, \cdot)$ with two kinds of labeled data. Instead of labeling the outputs of the segmentation network blanketly as *fake*, each pixel-level prediction is compared with the ground-truth to produce a fine-grained correction map $\mathbf{y}^{cor}$. For all matching pixels the corresponding values in the correction-map are set to the added class $N + 1$, whereas all differing pixels adopt the class given by the ground-truth:

$$\mathbf{y}^{cor} = \begin{cases} N + 1 & \text{if } \hat{y}_i = y_i \\ y_i & \text{otherwise.} \end{cases} \tag{1}$$

The involved components as well as the resulting $\mathbf{y}^{cor}$ are shown in Fig. 2. The second labeled samples are the image–ground-truth pairs with the corresponding label map $\mathbf{y}^t$. As by definition the ground-truth matches the image, $\mathbf{y}^t$ is filled exclusively with the added $(N + 1)^{th}$ class. Generating a fine-grained correction map drives the correction network to spot actual mismatches between image and segmentation instead of just recognizing indicators that reveal the origin. The full correction loss is given by:

$$\mathcal{L}_{cor} := \mathbb{E}_{(\mathbf{x},\mathbf{y}) \sim \mathcal{D}}[H(C(\mathbf{x}, \hat{\mathbf{y}}), \mathbf{y}^{cor}) + H(C(\mathbf{x}, \mathbf{y}), \mathbf{y}^t)]. \tag{2}$$

Setting the corrector's labels to be identical to the ground-truth and omitting the correction maps could theoretically lead to the same results with no classes added. In practice, however, this proved unsuccessful. We suspect that without the transfer required by identifying matching regions, the corrector would simply try to copy the input segmentation. This would yield accuracies as high as the preceding segmentation network with no learned understanding of the data. Designing the corrector to pick if a given prediction matches the content avoids this unstable case.

Weighted Cross-Entropy to Counter Class Imbalances. Assigning the $(N + 1)^{th}$ class to all accurately detected regions leads to extremely imbalanced label distributions, as can be seen in Fig. 2. This could result in the correction network not learning anything since always predicting the $(N + 1)^{th}$ class would lead to high accuracies on average. An imbalanced class distribution is not an uncommon setting and can be alleviated by individually weighing the contribution each class has on the overall loss. Instead of recalculating the class frequencies at runtime to proportionally weight each class, a fixed weighting scheme $\alpha \in \{1, 2\}$ is adopted. Balancing all regular dataset classes with a weight of $\alpha = 2$, and the additional $(N + 1)^{th}$ class with weight $\alpha = 1$ penalizes the misclassifications and forces the correction network to acknowledge these low-frequency regions.

3.2 Supervised Training with an Auxiliary Objective

Following a standard supervised approach, the Cross-Entropy between the output segmentation and the ground-truth labeled data is minimized. Additionally, the network is constrained to produce segmentations, which the error-correcting network interprets as *correct* with high probability:

$$\mathcal{L}_{sup} := \mathbb{E}_{(\mathbf{x},\mathbf{y}) \sim \mathcal{D}}[H(S(\mathbf{x}), \mathbf{y}) + \lambda_{cor} H(C(\mathbf{x}, S(\mathbf{x})), \mathbf{y}^t)]. \tag{3}$$

The contribution of the second term on the overall loss is controlled by the parameter λ_{cor}. This auxiliary objective regularizes the segmentation network and at a first glance resembles the concept of an adversarial loss. However, the relationship between the correction and the segmentation network is collaborative in nature. With GANs, an improving generator will increase the discriminator loss. On the contrary, as the segmentation in ECS improves and $\mathcal{L}_{sup}$ approaches the minimum, $\mathbf{y}^{cor}$ approaches $\mathbf{y}^t$. This collapses both terms in Eq. 2 and shows that ultimately the goals of both networks align.

3.3 Semi-supervised Step

The concept behind ECS involves the corrector judging the agreement between a given image and its segmentation, and offering a proposal for a correction if areas do not seem to match. For semi-supervised training, these predicted corrections receive an additional processing step to form the pseudo-labels $\mathbf{y}^p$. The continuous outputs $\hat{\mathbf{y}} = S(\mathbf{x}^u)$ and $\mathbf{y}^c = C(\mathbf{x}^u, \hat{\mathbf{y}})$ are both transformed to their discrete label representations $\mathbf{y}^u$ and $\mathbf{y}^c$. All areas predicted as $N + 1$ in $\mathbf{y}^c$ are replaced with the corresponding values of $\mathbf{y}^u$ while the remaining corrections are kept:

$$\mathbf{y}^p = \begin{cases} y_i^u & \text{if } y_i^c = N + 1 \\ y_i^c & \text{otherwise.} \end{cases} \tag{4}$$

Negated Focal Loss. The idea behind the Focal Loss [21] is to reduce the contribution of an easily classified example. This is achieved by weighting the Cross-Entropy with the negated probability of the true class. Likewise, our proposed loss for learning from the proxy labels $\mathbf{y}^p$ takes the form of a weighted Cross-Entropy but does *not* use negated probabilities:

$$NFL(\cdot, \cdot, \boldsymbol{j}) = \max(\boldsymbol{j})^\gamma H(\cdot, \cdot), \tag{5}$$

where j is a probability distributions, whose influence on the loss is smoothed by the focusing parameter γ. Instead of weighting the loss with the probabilistic output of the segmentation network, j is set to the correction network's predictions. This regularization measure ensures that the influence of the loss calculated with the pseudo-labels is proportional to the *certainty* of the corrector in its decision. Thus, high entropy predictions will be down-weighted and have a reduced effect. The complete semi-supervised objective is given by:

$$\mathcal{L}_{ecs} := \mathbb{E}_{\mathbf{x}^u \sim \mathcal{D}_u}[NFL(S(\mathbf{x}^u), \mathbf{y}^p, C(\mathbf{x}^u, S(\mathbf{x}^u)))]. \tag{6}$$

4   Experiments and Analysis

The following section gives brief insight into both datasets, parameter settings and evaluation metrics that were used for the analysis of our approach.

4.1   Cityscapes

Cityscapes [5] is a large scale dataset depicting urban scenes and environments which can be used for pixel-level and instance-level labeling tasks. Of the video sequences recorded in 50 cities, 5000 images have high quality annotations. 2975 of the 2048×1024 pixel images are contained in the training set, and 500 compose the validation set. The remaining 1525 images compose the test set, for which the label maps are not publicly available. The dataset contains 30 classes, of which 19 are used for training and evaluation purposes. All reported results are on the Cityscapes validation set.

4.2   Pascal VOC 2012

The second dataset is Pascal VOC 2012 [7], which consists of 1464 training, 1449 validation, and 1456 test images showing objects from 20 foreground classes and a single background class in varying resolutions. The SBD dataset [9] extends the original dataset, adding 9118 densely labeled training images, showing the same object categories. For our experiments, all models are trained on the combined Pascal VOC 2012 and SBD training sets. As with Cityscapes, all reported results are on the validation set.

4.3   Model Architecture

The segmentation network used in most experiments is a DeepLabv3+ [4] with a ResNet50 backbone [11,12]. Dilated convolutions with a dilation rate of two are applied to the last three residual blocks, such that the output features of the ResNet are 16 times smaller than the resolution of the input. The correction network acts as a secondary segmentation on the data. Thus, instead of utilizing architectures described in prior literature or producing a unique design, we decided to use DeepLabv3+ as well. As a result, all recent advances in network design for semantic segmentation are present in the corrector. Both correction and segmentation network employ the Atrous Spatial Pyramid Pooling (ASPP) intrinsic to DeepLabv3+, they only differ in the depth of the ResNet backbone. Here the corrector utilizes a smaller ResNet34. Apart from the model depth, the input dimensions of the first convolutional layer are extended. The first layer of the corrector takes RGB image data and concatenates it with the one-hot encoded labels or segmentations. Both networks feature a softmax layer at the end. Although the main experiments are run with DeepLabv3+, the proposed method is completely agnostic to the network design, as long as it is suitable for segmentation.

4.4 Setup

Both models contain a ResNet backbone with the ASPP and decoder as described in [4], implemented in Pytorch [25]. For all experiments, the segmentation network is trained with Stochastic Gradient Descent [18,27] with a learning rate of 0.01, momentum 0.9 and $1e - 4$ weight decay. The correction network is optimized with Adam [19], with $\beta_1 = 0.9$ and $\beta_2 = 0.99$, and the same weight decay as with SGD. The initial learning rate is set to $1e-4$. For both optimizers polynomial learning rate decay $lr = lr_{initial} \cdot (1 - \frac{iter}{maxiter})^{0.9}$ is applied. For all experiments λ_{cor} is set to 0.1 and γ in to 2. Both ResNets are initialized to the publicly available pre-trained ImageNet model contained in the PyTorch repository. The extended initial layer of the correction network is initialized according to [11]. Correction and segmentation networks are trained in tandem and the pseudo-labels are incorporated from the beginning with no warm-up phase.

Every reported result is the mean value of 10 individual trials, initialized with a random seed ascending from 0. This seed controls the training data distribution and ensures that the supervised and semi-supervised experiments are run with the same labeled data. For the experiments, the labeled data is limited to ratios ranging between 1/8 and 1/2 of the available dataset. The remaining images are used for semi-supervised training.

The same training data augmentation scheme as in [4] is used. On the Cityscapes dataset, the images are flipped horizontally at random. For training square, 768 pixel crops are randomly extracted. The validation images are kept at full resolution with only normalization being applied. The models are trained for 15000 iterations with a batch-size of 6. The Pascal VOC 2012 images are randomly cropped with a size of 512 and zero-padded if necessary. As with Cityscapes, the images are horizontally flipped and the validation images remain unaltered. The models are trained for 22000 iterations with a batch-size of 14.

These specific iterations numbers were chosen to result in comparable training duration to [14]. Differently from [4], no form of multi-resolution or mirroring steps are employed for validation.

The models were trained on Nvidia Titan RTX and Quadro RTX 6000 GPUs, and with the given batch sizes and input resolutions consumed 22 GB and 24 GB of memory on Cityscapes and Pascal respectively.

Table 1. Overview of the per-class Intersection over Union on Cityscapes (top) and Pascal VOC 2012 (bottom). The highlighted results indicate that IoU values with ECS are larger than the supervised counterpart, in the row above.

Cityscapes	Road	Sidewalk	Building	Wall	Fence	Pole	T-Light	T-Sign	Vegetation	Terrain	Sky	Person	Rider	Car	Truck	Bus	Train	Motorcycle	Bicycle	mIoU
100% Base	97.88	83.33	91.59	45.62	57.35	60.89	66.23	75.32	91.96	62.70	94.40	79.79	59.54	94.10	68.01	84.12	70.88	61.56	75.31	74.76 (±0.08)
50% ECS	97.74	82.25	91.05	43.84	55.04	58.50	63.26	72.95	91.63	61.51	93.90	78.29	56.47	93.64	65.69	81.02	65.53	59.45	73.20	72.89 (±0.54)
50% Base	97.56	81.10	90.77	41.83	53.77	57.33	62.12	72.07	91.40	59.69	93.74	77.80	55.54	93.23	61.32	76.20	58.49	56.01	73.29	71.22 (±0.74)
25% ECS	97.59	81.22	90.67	42.50	51.33	57.78	62.05	71.70	91.36	59.29	93.67	77.41	54.73	93.17	60.02	74.06	55.80	56.41	72.45	70.70 (±0.68)
25% Base	97.31	79.28	90.16	36.81	49.04	56.24	60.92	70.47	90.95	56.14	93.17	76.75	53.01	92.04	49.01	61.52	43.64	50.20	72.20	67.31 (±0.61)
12.5% ECS	97.37	79.62	90.21	41.26	48.00	56.11	59.95	70.11	90.98	58.41	93.03	76.31	52.39	92.34	48.57	67.07	33.50	53.37	71.58	67.38 (±0.96)
12.5% Base	96.94	76.84	89.20	33.11	40.83	54.06	57.71	67.57	90.38	53.22	92.36	75.20	48.71	90.78	34.64	51.62	30.26	45.06	70.87	63.12 (±0.79)

Pascal	Background	Aeroplane	Bicycle	Bird	Boat	Bottle	Bus	Car	Cat	Chair	Cow	Diningtable	Dog	Horse	Motorbike	Person	Pottedplant	Sheep	Sofa	Train	TVMonitor	mIoU
100%	93.87	86.99	42.48	87.34	65.58	77.64	94.21	84.99	92.09	36.36	87.69	55.54	87.31	87.14	83.11	84.88	60.64	87.50	48.38	83.88	74.58	76.29 (±0.12)
50% ECS	93.62	87.70	41.68	86.39	68.62	76.79	92.62	84.67	90.26	32.81	81.81	57.69	85.04	81.48	81.73	83.86	57.70	83.53	45.67	82.85	70.64	74.63 (±0.19)
50% Base	93.32	85.95	41.45	85.99	66.40	74.19	90.50	83.75	90.26	33.07	80.33	51.25	84.27	80.66	81.46	83.15	57.17	81.36	45.05	81.61	70.72	73.42 (±0.49)
25% ECS	93.20	86.85	40.70	84.96	63.92	72.12	91.13	83.00	89.11	31.29	76.49	57.98	82.71	79.15	80.17	82.67	56.07	78.73	44.84	81.81	67.68	72.60 (±0.44)
25% Base	92.62	84.78	40.52	83.54	62.96	66.38	86.93	80.11	87.20	30.14	74.28	46.37	80.86	75.24	77.08	81.73	53.73	76.50	40.17	78.00	66.41	69.78 (±0.59)
12.5% ECS	92.54	85.74	39.66	83.34	65.23	67.67	88.92	80.51	86.29	30.03	70.58	55.01	78.85	75.57	76.49	81.01	52.88	76.84	43.03	80.24	64.26	70.22 (±0.75)
12.5% Base	91.85	82.09	39.56	80.24	59.74	61.37	82.78	77.19	83.64	26.07	63.32	38.85	77.09	67.23	73.49	79.70	47.02	69.60	35.14	73.86	59.49	65.20 (±0.86)

Evaluation Metric. The quality of the models is assessed with the commonly used mean Intersection-over-Union (mIoU).

4.5 Results

In the following sections, we establish a supervised baseline and compare ECS with competing methods. Methods that operate on weakly labeled data are not part of our evaluation, as weakly supervised models cover a fundamentally different use-case.

Baseline. To give context to our results, we compare our implementation of DeepLabv3+ with the results stated in [4]. Especially on the Pascal VOC 2012 dataset, our results of 76.29 are close to the mIoU of 78.85 reported in the original paper, considering a smaller ResNet50 backbone was used for replication instead of a larger ResNet101. On Cityscapes the discrepancy is larger, which is likely due to the very different backbone architectures. Here our baseline of 74.76 can not as closely match the 78.79 mIoU that was achieved with an Xception style network backbone. Pushing the state-of-the-art in the Cityscapes or Pascal VOC 2012 benchmark is not the intention of this work, but to develop novel methods for training on unlabeled data. Therefore, with DeepLabv3+-ResNet50, an architecture was chosen to provide high quality and competitive segmentations, while still having moderate hardware demands.

Error-Correcting Supervision. Table 1 shows the per class Intersection over Union for both datasets with 1/8, 1/4, 1/2 of the labeled data as well as fully supervised. Here ECS consistently improves over the supervised baseline. In the case of training with 1/8 of the labeled data, ECS performs as well as a purely supervised model with 1/4th. Therefore, especially when only small amounts of labeled data are available ECS provides a significant improvement in mIoU. As expected, the more labeled data and consequently less unlabeled data is present, the less pronounced the benefits of ECS become.

4.6 Ablation Study

To highlight the individual contributions to the overall performance, we provide an ablation study for a set selection of model configurations. Table 2 compares the effectiveness of the proposed Negated Focal Loss, with directly using the Cross-Entropy loss H in $\mathcal{L}_{ecs}$ (Eq. 6). Further, we present the effect decreasing values for λ_{cor} have on the quality of the output, with and without the semi-supervised objective. Just the auxiliary regularization in Eq. 3 without any semi-supervised learning improves the mIoU but only slightly. This result implies that the ECS is not simply a regularization scheme on the supervised objective, but that the main contribution is from the corrector's pseudo-labels. Setting λ_{cor} to 0, i.e. using just the pseudo-labels leads to the second-best performance, and reinforces this observation. Admittedly the pseudo-labels alone are not sufficient. Minimizing the Cross-Entropy instead of the proposed Negated Focal Loss does not lead to optimal results. In the case of Pascal VOC 2012, it even falls below the supervised baseline. The pseudo-labels y^p would be accepted as fact, and contribute an equal amount to the gradient update as the supervised objective. The Negated Focal Loss's weighting scheme, which incorporates the certainty of the correction network, is essential and leads to the overall best results.

Table 2. Hyperparameter study for ECS trained with 1/4 the data on both datasets. The second best result on both datasets is with just semi-supervised learning, without auxiliary regularization.

λ_{corr}	H	NFL	Pascal	Cityscapes
0.0			69.79	67.31
0.0		✓	71.97	70.39
0.1			69.93	68.22
0.1	✓		68.39	69.27
0.1		✓	**72.60**	**70.70**

Corrector Evaluation. To evaluate the weighting scheme discussed in Sect. 3.1, we compared the $N + 1$ class mIoU values the correction network can achieve on the Cityscapes and Pascal VOC 2012 validation sets. Figure 3 shows that independently of the loss functions, penalizing the original N classes offers large improvements in mIoU. While the Negated Focal Loss is essential in the semi-supervised step, the choice of the standard Focal Loss to train the corrector in Eq. 2 is less conclusive. In the full ECS model, we found no statistically significant improvements of one loss function over the other. They effectively perform the same.

4.7 Comparison with Existing Methods

We considered two approaches to compare ECS with competing methods.

DeepLabv3+. The results of training the publicly available code for [14] with the same segmentation network that was used with ECS can be seen in Table 3. Apart from Cityscapes with 1/8 the labeled data, the method improves over the supervised baseline, but is consistently outdone by ECS.

But the two approaches still differ in the discriminator's architecture. Using a DeepLabv3+ discriminator with [14]'s method does not lead to further improvements. Experiments clearly showed signs of the discriminator overpowering the segmentation network. The discriminator loss approaches zero after 10% of the trained iterations, whereas with their architecture the loss hovers consistently above zero. Similarily less than 1% of the predictions on the unlabeled data are classified above the set acceptance threshold by the DeepLabv3+ discriminator. This effectively leads to most unlabeled data being ignored. Additionally, such experiments stay behind the supervised baseline with mIoU values 65.83 and 65.79 on Pascal VOC 2012 and Cityscapes with 1/4 the labeled data.

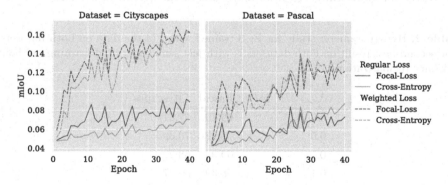

Fig. 3. Training just the corrector to evaluate the predictions of a pretrained segmentation network. The $N+1$ class mIoU is plotted for each epoch. Assigning an increased weighting to the original N classes of the dataset has a substantial positive effect on the mIoU.

DeepLabv2. Here, ECS is trained with DeepLabv2 [2] used in [14,23] and a DeepLabv3+ corrector. Comparing with the published results in Table 4, we achieve an improved supervised baseline on Pascal VOC 2012 but a very similar result for Cityscapes. Again, ECS consistently outperforms the competition.

4.8   Relation Between Correction and Truth

The comparison between [14] and [23] in combination with the ablation study implies that the correction network is potent in providing quality approximations for the truth. This hypothesis is reinforced when the correlation between prediction and truth is studied on the unlabeled data. Figure 4 depicts the Spearman's rank correlation coefficient between the squared probability of the correction network accepting the output as correct and the negative Cross-Entropy loss between the segmentation and ground-truth labels. The Cross-Entropy loss, in this case, is used as a measure of proximity between truth and segmentation. The correlation is computed for results on both the labeled and unlabeled data, for each dataset ratio on both Pascal VOC 2012 and Cityscapes. Especially for Cityscapes, there is an evident correlation. As expected, it is stronger on the training set, as the corrector is optimized with this data. The fact that there still is a positive correlation on the unlabeled dataset, illustrates why this semi-supervised learning approach is effective. On Pascal VOC 2012, while still positive, the correlation is decreased for both labeled and unlabeled data.

Table 3. Training the publicly available code from [14] with the same segmentation network and data distribution as our model. Although the method provides a statistically significant improvement over the supervised baseline in most cases, ECS outperforms it.

Dataset	#Data	[14]	Ours
Pascal	1/8	66.22 (±1.27)	**70.22** (±0.75)
	1/4	70.48 (±0.56)	**72.60** (±0.44)
Cityscapes	1/8	63.21 (±0.81)	**67.38** (±0.96)
	1/4	68.43 (±0.52)	**70.70** (±0.68)

Comparing the correlation with the individual IoU values for each class in Table 1 gives additional insight. There is a negative correlation in the unlabeled data for the *Chair* class in the Pascal VOC 2012 dataset. For 1/2 the labeled data this leads to a decrease in IoU with ECS compared to the supervised baseline. Again with half the data, there is a small positive correlation for *Pottedplant* on the unlabeled images, compared to the training set. The IoU for this class is only improved by 0.53 when ECS is applied. However, analyzing this correlation coefficient does not fully explain the benefits of our model. In some cases, the performance decreases with ECS, although a positive correlation on labeled and unlabeled images is present.

Table 4. Comparison between [14], [23] and our method trained with a DeepLabv2. For each method, the first row presents the mIoU of the supervised baseline and the second row the results when the respective approach is applied. The results for [23] on Pascal VOC 2012 include the Mean Teacher model.

Dataset	#Data		[14]	[23]	Ours
Pascal	1/8	sup	66.0	65.2	67.36 (±1.16)
		semi	69.5	71.4	**72.95** (±0.72)
	1/4	sup	68.1	-	71.61 (±0.48)
		semi	72.1	-	**74.68** (±0.37)
Cityscapes	1/8	sup	55.5	56.2	55.96 (±0.86)
		semi	58.8	59.3	**60.26** (±0.84)
	1/4	sup	59.9	60.2	60.54 (±0.85)
		semi	62.3	61.9	**63.77** (±0.65)

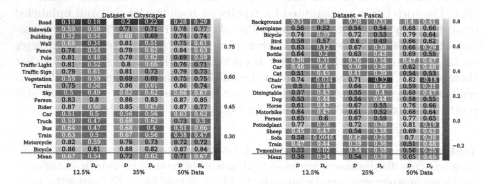

Fig. 4. Spearman's rank correlation coefficient describing the monotonic relationship between the squared prediction of correction network whether the segmentation is correct and the negative cross entropy loss between segmentation and truth. The positive correlation especially on the unlabeled data explains the effectiveness of our approach.

5 Conclusion

Error-Correcting Supervision offers a novel approach for semi-supervised segmentation, and can easily be added to existing supervised models if additional unlabeled data is present. Being model agnostic and reusing the segmentation architecture for the correction network, eliminates architecture search time. The utilization of the same architecture for both tasks ensures that the corrector is expressive enough for the underlying problem.

We have shown that our approach consistently outperforms a supervised baseline, as well as competing methods. It is the most effective when the ratio between labeled and unlabeled data is heavily in favor of the latter. Cityscapes and Pascal VOC 2012 are very different datasets. Pascal VOC 2012 mainly features foreground objects, most of the image being labeled as background.

Cityscapes, on the other hand, is densely labeled, and only some regions of the image are ignored. ECS is effective on both tasks which indicates the generality of the approach. Framing the interaction between the two involved networks as collaboration instead of competition, allows us to profit from complex corrector architectures without the danger of the segmentation network being overpowered and degenerating the results.

References

1. Badrinarayanan, V., Kendall, A., Cipolla, R.: SegNet: a deep convolutional encoder-decoder architecture for image segmentation. IEEE Trans. Pattern Anal. Mach. Intell. **39**(12), 2481–2495 (2017)
2. Chen, L., Papandreou, G., Kokkinos, I., Murphy, K., Yuille, A.L.: DeepLab: semantic image segmentation with deep convolutional nets, atrous convolution, and fully connected CRFs. IEEE Trans. Pattern Anal. Mach. Intell. **40**(4), 834–848 (2018)
3. Chen, L.C., Papandreou, G., Schroff, F., Adam, H.: Rethinking atrous convolution for semantic image segmentation. arXiv:1706.05587 (2017)
4. Chen, L.-C., Zhu, Y., Papandreou, G., Schroff, F., Adam, H.: Encoder-decoder with atrous separable convolution for semantic image segmentation. In: Ferrari, V., Hebert, M., Sminchisescu, C., Weiss, Y. (eds.) ECCV 2018. LNCS, vol. 11211, pp. 833–851. Springer, Cham (2018). https://doi.org/10.1007/978-3-030-01234-2_49
5. Cordts, M., et al.: The cityscapes dataset for semantic urban scene understanding. In: 2016 IEEE Conference on Computer Vision and Pattern Recognition (CVPR), pp. 3213–3223, June 2016
6. Deng, J., Dong, W., Socher, R., Li, L.J., Li, K., Fei-Fei, L.: ImageNet: a large-scale hierarchical image database. In: CVPR09 (2009)
7. Everingham, M., Eslami, S.M.A., Van Gool, L., Williams, C.K.I., Winn, J., Zisserman, A.: The PASCAL visual object classes challenge: a retrospective. Int. J. Comput. Vis. **111**(1), 98–136 (2015). https://doi.org/10.1007/s11263-014-0733-5
8. Goodfellow, I., et al.: Generative adversarial nets. In: Ghahramani, Z., Welling, M., Cortes, C., Lawrence, N.D., Weinberger, K.Q. (eds.) Advances in Neural Information Processing Systems 27, pp. 2672–2680. Curran Associates, Inc. (2014)
9. Hariharan, B., Arbeláez, P., Bourdev, L., Maji, S., Malik, J.: Semantic contours from inverse detectors. In: 2011 International Conference on Computer Vision, pp. 991–998 (2011)
10. He, K., Zhang, X., Ren, S., Sun, J.: Spatial pyramid pooling in deep convolutional networks for visual recognition. IEEE Trans. Pattern Anal. Mach. Intell. **37**(9), 1904–1916 (2015)
11. He, K., Zhang, X., Ren, S., Sun, J.: Delving deep into rectifiers: Surpassing human-level performance on ImageNet classification. In: Proceedings of the IEEE International Conference on Computer Vision (ICCV), December 2015
12. He, K., Zhang, X., Ren, S., Sun, J.: Deep residual learning for image recognition. In: Proceedings of the IEEE Conference on Computer Vision and Pattern Recognition, pp. 770–778 (2016)
13. Hong, S., Noh, H., Han, B.: Decoupled deep neural network for semi-supervised semantic segmentation. In: Cortes, C., Lawrence, N.D., Lee, D.D., Sugiyama, M., Garnett, R. (eds.) Advances in Neural Information Processing Systems 28, pp. 1495–1503. Curran Associates, Inc. (2015)

14. Hung, W.C., Tsai, Y.H., Liou, Y.T., Lin, Y.Y., Yang, M.H.: Adversarial learning for semi-supervised semantic segmentation. In: Proceedings of the British Machine Vision Conference (BMVC) (2018)
15. Ibrahim, M.S., Vahdat, A., Macready, W.G.: Weakly supervised semantic image segmentation with self-correcting networks. arXiv:1811.07073 (2018)
16. Kalluri, T., Varma, G., Chandraker, M., Jawahar, C.: Universal semi-supervised semantic segmentation. In: The IEEE International Conference on Computer Vision (ICCV), October 2019
17. Khoreva, A., Benenson, R., Hosang, J., Hein, M., Schiele, B.: Simple does it: weakly supervised instance and semantic segmentation. In: Proceedings of the IEEE Conference on Computer Vision and Pattern Recognition, pp. 876–885 (2017)
18. Kiefer, J., Wolfowitz, J.: Stochastic estimation of the maximum of a regression function. Ann. Math. Stat. **23**(3), 462–466 (1952)
19. Kingma, D.P., Ba, J.: Adam: a method for stochastic optimization. arXiv:1412.6980 (2014)
20. Krizhevsky, A., Sutskever, I., Hinton, G.E.: ImageNet classification with deep convolutional neural networks. In: Pereira, F., Burges, C.J.C., Bottou, L., Weinberger, K.Q. (eds.) Advances in Neural Information Processing Systems 25, pp. 1097–1105. Curran Associates, Inc. (2012)
21. Lin, T.Y., Goyal, P., Girshick, R., He, K., Dollar, P.: Focal loss for dense object detection. In: Proceedings of the IEEE International Conference on Computer Vision (ICCV), October 2017
22. Luc, P., Couprie, C., Chintala, S., Verbeek, J.: Semantic segmentation using adversarial networks. arXiv:1611.08408 (2016)
23. Mittal, S., Tatarchenko, M., Brox, T.: Semi-supervised semantic segmentation with high-and low-level consistency. arXiv:1908.05724 (2019)
24. Papandreou, G., Chen, L.C., Murphy, K.P., Yuille, A.L.: Weakly- and semi-supervised learning of a deep convolutional network for semantic image segmentation. In: The IEEE International Conference on Computer Vision (ICCV), December 2015
25. Paszke, A., et al.: Pytorch: an imperative style, high-performance deep learning library. In: Advances in Neural Information Processing Systems 32, pp. 8024–8035. Curran Associates, Inc. (2019)
26. Qi, M., Wang, Y., Qin, J., Li, A.: KE-GAN: knowledge embedded generative adversarial networks for semi-supervised scene parsing. In: The IEEE Conference on Computer Vision and Pattern Recognition (CVPR), June 2019
27. Robbins, H., Monro, S.: A stochastic approximation method. Ann. Math. Stat. **22**(3), 400–407 (1951)
28. Ronneberger, O., Fischer, P., Brox, T.: U-Net: convolutional networks for biomedical image segmentation. In: Navab, N., Hornegger, J., Wells, W.M., Frangi, A.F. (eds.) MICCAI 2015. LNCS, vol. 9351, pp. 234–241. Springer, Cham (2015). https://doi.org/10.1007/978-3-319-24574-4_28
29. Shelhamer, E., Long, J., Darrell, T.: Fully convolutional networks for semantic segmentation. IEEE Trans. Pattern Anal. Mach. Intell. **39**(4), 640–651 (2017)
30. Souly, N., Spampinato, C., Shah, M.: Semi supervised semantic segmentation using generative adversarial network. In: The IEEE International Conference on Computer Vision (ICCV), October 2017
31. Tarvainen, A., Valpola, H.: Mean teachers are better role models: weight-averaged consistency targets improve semi-supervised deep learning results. In: Guyon, I., et al. (eds.) Advances in Neural Information Processing Systems 30, pp. 1195–1204. Curran Associates, Inc. (2017)

32. Wei, Y., Xiao, H., Shi, H., Jie, Z., Feng, J., Huang, T.S.: Revisiting dilated convolution: a simple approach for weakly- and semi-supervised semantic segmentation. In: Proceedings of the IEEE Conference on Computer Vision and Pattern Recognition (CVPR), June 2018
33. Yu, F., Koltun, V.: Multi-scale context aggregation by dilated convolutions. arXiv:1511.07122 (2015)
34. Zhao, H., Shi, J., Qi, X., Wang, X., Jia, J.: Pyramid scene parsing network. In: Proceedings of the IEEE Conference on Computer Vision and Pattern Recognition (CVPR), July 2017
35. Zhou, B., Khosla, A., Lapedriza, A., Oliva, A., Torralba, A.: Learning deep features for discriminative localization. In: Proceedings of the IEEE Conference on Computer Vision and Pattern Recognition (CVPR), June 2016
36. Zhou, Y., et al.: Collaborative learning of semi-supervised segmentation and classification for medical images. In: The IEEE Conference on Computer Vision and Pattern Recognition (CVPR), June 2019

Quantum-Soft QUBO Suppression
for Accurate Object Detection

Junde Li[(✉)] and Swaroop Ghosh

The Pennsylvania State University, University Park, PA 16802, USA
{jul1512,szg212}@psu.edu

Abstract. Non-maximum suppression (NMS) has been adopted by default for removing redundant object detections for decades. It eliminates false positives by only keeping the image $\mathcal{M}$ with highest detection score and images whose overlap ratio with $\mathcal{M}$ is less than a predefined threshold. However, this greedy algorithm may not work well for object detection under occlusion scenario where true positives with lower detection scores are possibly suppressed. In this paper, we first map the task of removing redundant detections into Quadratic Unconstrained Binary Optimization (QUBO) framework that consists of detection score from each bounding box and overlap ratio between pair of bounding boxes. Next, we solve the QUBO problem using the proposed Quantum-soft QUBO Suppression (QSQS) algorithm for fast and accurate detection by exploiting quantum computing advantages. Experiments indicate that QSQS improves mean average precision from 74.20% to 75.11% for PASCAL VOC 2007. It consistently outperforms NMS and soft-NMS for *Reasonable* subset of benchmark pedestrian detection CityPersons.

Keywords: Object detection · Quantum computing · Pedestrian detection · Occlusion

1 Introduction

Object detection helps image semantic understanding by locating and classifying objects in many applications such as, image classification [16], face recognition [6,30], autonomous driving [7,8], and surveillance [20]. It has been developed from handcrafted features to Convolutional Neural Network (CNN) features, and from sliding windows to region proposals. Redundant bounding boxes are eliminated by default using Non-Maximum Suppression (NMS) for decades. However, greedy NMS can lead to detection misses for detection of partially occluded objects by completely suppressing neighbouring bounding boxes when overlap threshold is reached. While improvements have been made in [5,14,17,26], a better algorithm is still a necessity for accurately retaining true positive object locations, especially for pedestrian detection where occluded people are common. This is particularly important for autonomous systems that need to make accurate decisions even under occluded situations to guarantee safe operations.

© Springer Nature Switzerland AG 2020
A. Vedaldi et al. (Eds.): ECCV 2020, LNCS 12374, pp. 158–173, 2020.
https://doi.org/10.1007/978-3-030-58526-6_10

Quadratic Unconstrained Binary Optimization (QUBO) framework has been used for removing redundant object detections in [26], but neither its detection accuracy in mAP metric nor detection delay are as good as traditional non-maximum suppression [5,14]. While no further research on QUBO for object detection has been performed since [26], its potential for accurate object detection by incorporating multiple detection scores and overlap ratios is worth being explored. Importantly, QUBO problem can be efficiently solved by quantum computing than any classical algorithms. Recent advances in quantum computing can be exploited to address this issue.

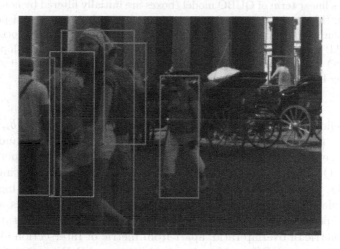

Fig. 1. Pedestrian detection using Non-Maximum Suppression (NMS) for removing false positive detections. Detection indicated by green box is eliminated under harsh suppression threshold due to significant overlap with front detection. (Color figure online)

Quantum advantage (supremacy) has been specifically demonstrated in [2,12], and practically in applications of flight gate scheduling [28], machine learning [25], and stereo matching [9]. In this study, we focus on another real-world application of removing redundant object detections using quantum annealer (QA). Quantum annealing system first embeds binary variables in QUBO framework to physical qubits, and outputs optimized binary string indicating whether corresponding bounding boxes to be retained or removed. For example, D-Wave 2000Q quantum annealer supports embedding of up to 2048 qubits which is sufficient for object detection task of this work.

NMS recognizes objects less well for crowded and partially occluded scenes as occluded objects are likely to be completely suppressed due to significant overlap with front objects that have higher detection scores (Fig. 1). QUBO framework was developed for addressing this challenge [26], with its linear term modeling detection score of each bounding box b_i and quadratic term (aka pairwise

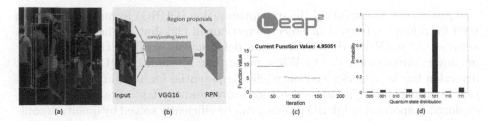

Fig. 2. Formulation and solution steps of QUBO problem for removing redundant detections using our proposed QSQS algorithm: (a) generate detection score from detector to serve as linear term of QUBO model (boxes are initially filtered by non-maximum suppression); (b) obtain regions of interest through a series of conv and pooling layers and Region Proposal Network (RPN); (c) use quantum annealing process through D-Wave LEAP service for filtering out false detections; (d) retrieve (near-)optimal solution of QUBO problem through quantum state readout (for example bit string '101' indicates only first and third bounding boxes are kept).

term) modeling overlap ratio of each pair of bounding boxes $or(b_i, b_j)$. However, QUBO framework indicated degraded accuracy than the standard NMS [14] from a generic detection perspective. In order to improve its performance, we propose Quantmum-soft QUBO Suppression (QSQS) by adapting the idea of Soft-NMS [5] for decaying classification score of b_i which has high overlap with target detection $\mathcal{M}$ (i.e., detection with highest score). Besides, we further improve its performance by incorporating spatial correlation features [17] as additional metric of overlap ratio, apart from metric of Intersection Over Union (IOU). Finally, the QUBO problem is solved using real D-Wave 2000Q quantum system with unmatched computing efficiency compared to classical solvers. The overall detection framework using soft-QUBO and quantum annealing is illustrated in Fig. 2. The contributions of this paper are three-fold: (1) We propose an novel hybrid quantum-classical algorithm, QSQS, for removing redundant object detections for occluded situations; (2) We implement the hybrid algorithm on both GPU of classical computer for running convolutional neural network, and QPU (quantum processing unit) of quantum annealer for harnessing quantum computing advantage; (3) The proposed QSQS improves mean average precision (mAP) from 74.20 to 75.11% for generic object detection, and outperforms NMS and soft-NMS consistently for pedestrian detection on CityPersons *Reasonable* subset. However, our method takes 1.44X longer average inference time than NMS on PASCAL VOC 2007.

2   Related Work

Non-Maximum Suppression: NMS is an integral part of modern object detectors (one-stage or two-stage) for removing false positive raw detections. NMS greedily selects bounding box $\mathcal{M}$ with highest detection score, and all close-by boxes whose overlap with $\mathcal{M}$ reach hand-crafted threshold are suppressed.

This simple and fast algorithm is widely used in many computer vision applications, however, it hurts either precision or recall depending on predefined parameter, especially for crowded scenes where close-by high scoring true positives are very likely to be removed. Soft-NMS [5] improves the detection accuracy by decaying detection scores of close-by boxes by a linear or Gaussian rescoring function. Learning NMS [15] eliminates non-maximum suppression post-processing stage so as to achieve true end-to-end training. However, the accuracy of these false positive suppression schemes are not high enough.

Quadratic Unconstrained Binary Optimization (QUBO): Quadratic binary optimization is the problem of finding a binary string vector that maximizes the objective function $C(x)$ which is composed of linear and paired terms. QUBO framework [26] has been proposed in past for suppressing false positive detections, and achieved better accuracy than NMS for pedestrian detection. However, it has been concluded that QUBO may not be optimal for other detection situations. QUBO framework is inherently suitable for crowded and partially occluded object detection as classification scores of all bounding boxes and overlaps among all pairs of them are equally considered. In this study, we aim to improve QUBO performance on detection accuracy by incorporating spatial features into pairwise terms of the cost function, and generalize its detection accuracy robustness for both crowded and non-crowded dataset.

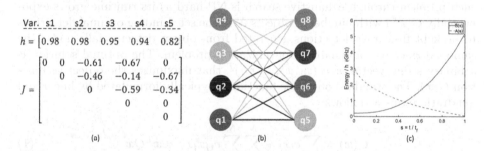

Fig. 3. (a) Flux biases h on each qubit s_i and couplings J_{ij} between qubits of Ising Hamiltonian of an example object detection instance with 5 bounding boxes; (b) one optimal minor embedding of these 5 binary variables of Ising model to 6 physical qubits in one unit cell of D-Wave 2X quantum annealer (the cluster of 2 physical qubits q_3 and q_5 represents binary variable s_3, and variables s_1, s_2, s_4, and s_5 are represented by physical qubits q_1, q_2, q_6, and q_7, respectively); (c) typical profile of annealing functions of tunneling energy $A(s)$ and Ising Hamiltonian energy $B(s)$ (t_f is the total annealing time).

Quantum Computing: Quantum supremacy has been recently demonstrated by Google [2] using a quantum processor with 53 superconducting qubits. It has been shown that quantum computer is more than billion times faster in sampling one instance of a quantum circuit than a classical supercomputer. Quantum computing is first proposed for object detection using Quantum Approximation Optimization Algorithm (QAOA) implemented on a universal gate based quantum simulator [3,18]. However, it is only an exploratory study rather than a real-world object detection application due to gate error and limited number of qubits on gate based quantum computer. Compared to QAOA, we propose a novel hybrid quantum-classical algorithm, QSQS, specifically for object detection task using quantum annealing. In this study, the NP-hard QUBO problem is solved by quantum annealing with pronounced computing advantage, in terms of speed and accuracy, over classical algorithms such as, Tabu and Greedy Search.

3   Approach

This section briefly explains the redundant detection suppression task of object detection addressed by QUBO framework and its suitability as standard NMS alternative. The rationale behind the quantum advantage of quantum annealing over classical algorithms is also explained.

3.1   Quadratic Unconstrained Binary Optimization

QUBO is an unifying framework for solving combinatorial optimization problem which aims to find an optimal subset of objects from a given finite set. Solving such problem through exhaustive search is NP-hard as its runtime grows exponentially $\mathcal{O}(2^N)$ with number of objects N in the set. Finding optimal detections to be kept from raw detections generated from object detector falls under the same category of combinatorial optimization problem. The optimal solution is a binary string vector $\boldsymbol{x} = (x_1, x_2, ..., x_N)^T$ that maximizes the objective function $C(\boldsymbol{x})$. The standard objective function is typically formulated by linear and quadratic terms as follows:

$$C(\boldsymbol{x}) = \sum_{i=1}^{N} c_i x_i + \sum_{i=1}^{N} \sum_{j=i}^{N} c_{ij} x_i x_j = \boldsymbol{x}^T Q \boldsymbol{x} \tag{1}$$

where $\forall i, x_i \in \{0, 1\}$ are binary variables, c_i are linear coefficients and c_{ij} are quadratic coefficients. Each object detection instance is represented by a specific upper triangular matrix Q, where c_i terms make up the diagonal elements and c_{ij} terms constitute off-diagonal elements of the matrix. For object detection problem of this work, N denotes the number of detection bounding boxes, c_i denotes the detection score of each b_i, and c_{ij} denotes negative value of $or(b_i, b_j)$ for penalizing high overlap between pair of bounding boxes.

Theoretically, QUBO is a suitable alternative to greedy NMS for redundancy suppression as overlap $or(b_i, b_j)$ between each pair of detections is fully considered in QUBO while NMS only considers overlap $or(\mathcal{M}, b_i)$ between highest-scoring detection and others. Besides, QUBO is also able to uniformly combine multiple linear and quadratic metrics from different strategies and even detectors. Put differently, matrix Q can be formed by the sum of multiple weighted linear and pair matrices for more accurate detection. We adopt a single linear matrix $\mathcal{L}$ and two pair matrices $\mathcal{P}_1$ and $\mathcal{P}_2$ as follows:

$$Q = w_1\mathcal{L} - w_2\mathcal{P}_1 - w_3\mathcal{P}_2 \tag{2}$$

where $\mathcal{L}$ is diagonal matrix of objectness scores, $\mathcal{P}_1$ is one pair matrix for measuring IOU overlap, and $\mathcal{P}_2$ is another pair matrix for measuring spatial overlap [17]. The additional spatial overlap feature expects to enhance the detection accuracy by comparing positional similarity between each pair of the proposed regions.

3.2 Quantum Annealing

Quantum Annealing has been shown to be advantageous over classical algorithms in terms of computational time for QUBO problems [1]. Classical algorithm such as Tabu Search [23] reduces time complexity from exponential to polynomial at the expense of reduced accuracy. Few problem instances with varying number of variables are run to compare the accuracy between Tabu Search and quantum annealing. Tabu Search shows 87.87% accuracy for 9-variable problem relative to ground truth solution using brute-force, and its accuracy decreases to 66.63% for 15-variable problem. However, quantum annealing with 1000 shots achieves the optimal strings with ground truth solutions for these problem instances. Therefore, only quantum algorithm is considered as QUBO solver in this study.

Quantum Annealing is a metaheuristic optimization algorithm running on quantum computational devices for specially solving QUBO problems. Every QUBO instance is first mapped to its Ising Hamiltonian through equation $s_i = 2x_i - 1$ for making use of flow of the currents in the superconductor loops [4]. The Ising Hamiltonian of an example object detection case with 5 bounding boxes is shown in Fig. 3(a). Minor embedding is performed to map binary variables to clusters of physical qubits due to limited connectivity of D-Wave QPU architecture. Figure 3(b) displays one optimal minor embedding of the 5 binary variables of the Ising Hamiltonian in Fig. 3(a). The cluster of 2 physical qubits q_3 and q_5 represents a single binary variable s_3, while q_1, q_2, q_6, and q_7 represent binary variables s_1, s_2, s_4, and s_5, respectively. This embedding only takes one unit cell of Chimera graph of D-Wave QPU. In Fig. 3(b), physical qubits q_4 and q_8 are unused, and couplings between logical qubits are denoted with black edges and yellow edge denotes coupling within a chain of qubits.

Quantum annealing is evolved under the control of two annealing functions $A(s)$ and $B(s)$. The time-dependent Hamiltonian of annealing follows the equation below:

$$H(t) = A(s)H_0 + B(s)H_1 \tag{3}$$

where H_1 is the Ising Hamiltonian of QUBO problem and H_0 is the initial Hamiltonian fixed in D-Wave system. As shown in Fig. 3(c), Ising Hamiltonian energy increasingly dominates system energy in accordance with profile configuration of $A(s)$ and $B(s)$. Note, t_f is the total annealing time.

3.3 Quantum-Soft QUBO Suppression

We develop a novel hybrid quantum-classical algorithm i.e., Quantum-soft QUBO Supression (QSQS) for accurately suppressing redundant raw detection boxes. QSQS provides a means to uniformly combine multiple linear and pair terms as a QUBO problem, and harnesses quantum computational advantage for solving such problem. As shown in Algorithm 1, it contains three tunable weights w_1, w_2 and w_3 for detection confidence, intersection over union, and spatial overlap feature (refer to [17]), respectively.

Algorithm 1. Quantum-soft QUBO Suppression (QSQS)

Input: $\mathcal{B}$ - N x 4 matrix of detection boxes; $\mathcal{S}$ - N x 1 vector of corresponding detection scores; O_t - detection confidence threshold.
Output: $\mathcal{D}$ - final set of detections.

$\mathcal{B} = \{b_1, b_2, ..., b_N\}, \mathcal{S} = \{s_1, s_2, ..., s_N\}$
$\mathcal{D} = \{\}, Q = \{\}$
for $i \leftarrow 1$ **to** N **do**
  $Q[i, i] \leftarrow w_1 S_i$
  for $j \leftarrow i$ **to** N **do**
    $Q[i, j] \leftarrow -w_2 IoU(b_i, b_j)$
    $SpatFeat \leftarrow getSpatialOverlapFeature(b_i, b_j)$
    $Q[i, j] \leftarrow Q[i, j] - w_3 SpatFeat$
  end
end
$(\mathcal{B}_{kept}, \mathcal{B}_{soft}) \leftarrow quantumAnnealing(-Q)$
$\mathcal{D} \leftarrow \mathcal{D} \cup \mathcal{B}_{kept}$
for b_i **in** $\mathcal{B}_{soft}$ **do**
  $b_m \leftarrow argmax(IoU(\mathcal{B}_{kept}, b_i))$
  $s_i \leftarrow s_i f(IoU(b_m, b_i))$
  if $s_i \geq O_t$ **then**
    $\mathcal{D} \leftarrow \mathcal{D} \cup b_i$
  end
end
return $\mathcal{D}, \mathcal{S}$

QUBO problems are formulated from initial detection boxes $\mathcal{B}$ and corresponding scores $\mathcal{S}$. We first negate all elements in square matrix Q of the QUBO instance to convert original maximization to a minimization problem to match with the default setting of the D-Wave system. Then a binary quadratic model is formed using negated diagonal and off-diagonal entries in matrix Q through D-Wave cloud API dimod. Optimal or near-optimal binary string vector $x = (x_1, x_2, ..., x_N)^T$ is the string vector with highest sampling frequency after 1000 quantum state readouts. Value of 1 or 0 for each x_i indicates keep or removal of corresponding detection b_i.

Instead of removing all boxes $\mathcal{B}_{soft}$ returned from the annealer, a rescoring function $f(IoU(b_m, b_i))$ is applied for mitigating probability of detection misses. The Gaussian weighting function [5] is as follows:

$$s_i = s_i exp(-\frac{IoU(b_m, b_i)^2}{\sigma}), \forall b_i \in \mathcal{B}_{soft} \tag{4}$$

where σ is fixed at 0.5 in this study. Average precision drop caused by picking only highest scoring detection may be avoided through this soft rescoring step.

4   Experiments

In this section, we first introduce evaluation datasets and metrics. Next, we describe the implementation details followed by ablation study of our approach. Finally, the comparison with the state-of-the-art is presented. All experiments and analyses are conducted on two application scenarios, i.e., generic object detection and pedestrian detection, to validate the proposed quantum-soft QUBO suppression method.

4.1   Datasets and Evaluation Metrics

Datasets. We choose two datasets for generic object detection evaluation. The PASCAL VOC 2007 [11] dataset has 20 object categories and 4952 images from test set, while the large-scale MS-COCO [19] dataset contains 80 object categories and 5000 images from minival set.

The CityPersons dataset [29] is a popular pedestrian detection benchmark, which is a challenging dataset for its large diversity in terms of countries and cities in Europe, seasons, person poses, and occlusion levels. The dataset has 2795, 1575, and 500 images for Train, Test, and Val subsets, respectively. It contains six object categories, namely, ignored region, person, group of people, sitter, rider, and other. All objects are classified into bare, reasonable, partially occluded and heavily occluded categories based on visibility of body part. The Reasonable (**R**) occlusion level subset contains pedestrian examples with visibility greater than 65%, and height no less than 50 pixels for evaluation. The Heavy occlusion (**HO**) subset has visibility range from 20 to 65%.

Metrics. Following the widely used evaluation protocol, generic object detection is measured by the metric of mean average precision (mAP) over all object categories, together with average recall (AR) for MS-COCO.

Only pedestrian examples taller than 50 pixels and within *Reasonable* occlusion range are used for evaluating our approach. We report detection performance using log-average miss rate (MR) metric which is computed by averaging miss rates at nine False Positives Per Image (FPPI) rates evenly spaced between 10^{-2} to 10^0 in log space [10].

4.2   Implementation Details

D-Wave quantum annealer currently supports over 2000 physical qubits however the minor embedding process takes nontrivial time when more than 45 qubits are fed for quantum processing. In order to reduce detection latency, raw detections for each object category is initially suppressed using NMS with predefined threshold and filtered by top detection scores, to ensure number of bounding boxes are within qubit upper bound of quantum processing we preset. Fortunately, such setting doesn't hurt real application much as it is rare to encounter the image case where there are over 45 objects of the same category.

Training. We base on the two-stage Faster R-CNN [24] baseline detector to implement quantum-soft QUBO suppression. For generic object detection, the detector is trained with backbone ResNet101 CNN architecture [13] starting from a publicly available pretrained model for generic object detection datasets. We train the detector with mini-batch of 1 image using stochastic gradient descent with 0.9 momentum and 0.0005 weight decay on a single RTX 2080 Ti GPU. The learning rate is initially set to 0.001, and decays at a factor of 0.1 for every 5 epochs with 20 epochs in total for training. We set 7 epochs for large-scale MS-COCO training.

Inference. Our QSQS method only applies at suppression stage of testing phase. The greedy-NMS is replaced with Algorithm 1 where the weight parameters $w_1 + w_2 + w_3 = 1$ defined in Eq. 2 are tuned using pattern search algorithm [22]. After a group of searches, we fix them with $w_1 = 0.4$, $w_2 = 0.3$, and $w_3 = 0.3$ for the entire experiment. The original QUBO model is enhanced by introducing a score s_i penalty term and an overlap $or(b_i, b_j)$ reward term if objectiveness score is smaller than confidence threshold. Sensitivity analysis is conducted for determining optimal pre-suppression NMS threshold before quantum processing. Bit string solution is retrieved from D-Wave annealer, then detection score above 0.01 after Gaussian rescoring function [5] is applied to output final true locations by the detector.

4.3   Ablation Study

In this section, we conduct an ablative analysis to evaluate the performance of our proposed method on generic object detection datasets.

Why Quantum QUBO Is Enhanced? Quantum-soft QUBO suppression is enhanced with three contributing factors, namely quantum QUBO (QQS), soft-NMS (QSQS), and adjustment terms (QSQS+enh), each of which either reduces inference latency or increases detection accuracy. Classical solver execution time is exponentially proportional to the number of binary variables to solve the NP-hard QUBO problem, while quantum annealing solves the larger problem in a divide-and-conquer paradigm by breaking it down to multiple sub-problems and finally produces a (near-)optimal solution [27].

Two potential drawbacks of QUBO suppression were pointed out in [17], i.e. blindness of equally selecting all detections and high penalty for occluded objects. However, these two issues can be efficiently solved by the enhanced model by introducing two adjustment factors for locations with detection cores lower than predefined objectiveness threshold. The penalty factor of 0.1 penalizes those lower-scored bounding boxes as they are likely to be false positives, while the reward factor of 0.7 reduces overlap between such lower-scored box and neighboring boxes accordingly. Impact of the enhanced model (QSQS+enh) with adjustment factors is shown in Table 1 on MS-COCO.

Table 1. Average precision (AP) results for MS-COCO test-dev set for various recall rates. The baseline Faster R-CNN uses standard NMS, while following three lines denote basic QUBO, soft-QUBO and enhanced QUBO, respectively. Best results are shown in bold. QSQS-enh shows no significant improvement over QSQS in MS-COCO.

Method	AP 0.5:0.95	AP @0.5	AP small	AP medium	AP large	AR @10	AR @100
F-RCNN* [24]	25.6	43.9	9.6	29.1	40.0	37.6	38.4
F-RCNN + QQS	21.8	35.4	6.6	24.3	35.0	27.0	27.0
F-RCNN + QSQS	25.7	44.2	**9.8**	29.2	40.3	**38.5**	**39.4**
F-RCNN + QSQS-enh	**25.8**	**44.3**	9.7	**29.3**	**40.4**	**38.5**	**39.4**

F-RCNN* refers to F-RCNN [24] trained in the present study.

Soft-NMS has been adopted in few study and competition projects as it consistently generates better performance than greedy-NMS. The soft-QUBO idea in this paper is not the same as [5]. However, the same rescoring function is applied to discard detections from D-Wave annealer. Impact of the soft model (QSQS) is also shown in Table 1 compared to basic QUBO model (QQS).

Did Quantum Era Arrive for Object Detection? We may provide clues to this question from three perspectives, i.e. cost, accuracy, and inference latency of our QSQS method. The D-Wave Systems provides developers real-time quantum cloud service at commercial price of $2000 per hour access to quantum processing unit. However, only problems with less than 45 qubits are mapped to quantum annealer for optimization in this study, and these small problems can be solved instantly such that the annealing process is undetectable and without time count. Since there is no extra QPU access cost, our method shares the same level cost with other existing methods.

Accuracy result shown in Table 1 indicates that QSQS surpasses the standard greedy-NMS embedded in most state-of-the-art detectors. More accuracy results are shown in Sect. 4.4.

The frame rate of QSQS is satisfying provided that Algorithm 1 is sort of more complicated than standard NMS. Even powered by quantum supremacy, extra inference latency caused by D-Wave cloud service connection, problem mapping, and annealing process is non-negligible. In order to better depict frame rate, we conduct the sensitivity analysis on inference latency vs. accuracy with varying initial overlap threshold N_t (not lower than 0.3) and number of qubits (not higher than 45) on PASCAL VOC 2007. We choose initial overlap threshold from 0.3 to 0.7 with step size 0.1 and qubit upper bound from 15 to 45 with step size 5. Greedy-NMS achieves mAP of 74.20 (percent) over 20 object categories, and average inference latency is 65.87 ms. QSQS achieves best performance with overlap threshold N_t of 0.5 and 35 qubits upper bound.

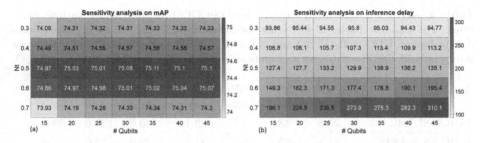

Fig. 4. Sensitivity analysis on detection accuracy mAP and average inference latency. QSQS achieves best accuracy with configuration of 0.5 initial overlap ratio and 35 qubit upper bound, at which the corresponding average inference latency is 138.9 ms.

The study results are shown in Fig. 4. Inference delay significantly relates to the initial overlap threshold, and the limit on number of qubits at higher N_t. Interestingly, QSQS underperforms at both high and low overlap threshold ends. Poor performance at low end 0.3 indicates that QSQS works better than greedy-NMS by introducing individual and correlation features. Performance at high end 0.7 indicates QSQS cannot handle well with large number of false positives. QSQS has performance gains with 28 ms extra latency, and best performance with 73.02 ms extra latency. It is also worth noting that we have submitted 223 problem instances with more than 45 qubits during this work, they are solved in 12 ms on average by the D-Wave annealer. This indicates that the main latency overhead originates from initial false positive suppression and quantum cloud service connection.

The hybrid quantum-classical false positive suppression algorithm incurs no extra quantum computing cost, has better detection accuracy, and does not hurt much real-time detection application. In this sense, quantum era for object detection has already arrived.

4.4 Results

Parameters in our QSQS method are well tuned after sensitivity analysis. Its performance on generic object detection and pedestrian detection are compared with the state-of-the-art detectors on PASCAL VOC 2007 and CityPersons.

Generic Object Detection. Faster R-CNN with QSQS is sufficiently trained on PASCAL VOC 2007 for fair comparison with greedy-NMS. The detection comparison over 20 object categories is summarized in Fig. 5. It can be seen that performance increases a lot by adapting soft-NMS mechanism to basic QUBO model, and QSQS-enh performs slightly better only for few object categories. Interestingly, all our three schemes predict `table` and `sofa` significantly worse than standard NMS, but it becomes more accurate from QQS to QSQS-enh. Specifically, average precision for categories such as, `bike`, `motorbike`, and `person` are fairly improved by our method.

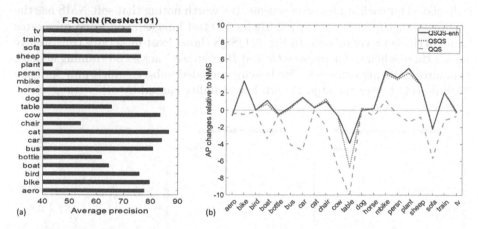

Fig. 5. (a) Baseline detector results for 20 object categories; (b) average precision changes for our three layer methods relative to the baseline. Curves above 0 show precision gains while that below 0 line show precision losses.

Figure 7 shows the qualitative results of generic object detections. Overlap threshold of 0.3 is set for greedy-NMS. Images 1–5 show examples where QSQS performs better than NMS, and images 8–9 show cases where NMS does better. Both do not detect well for images like 6–7. QSQS helps for images where false positives are partially overlapped with true detections, and it detects relatively better for image 3. False positive detection scores are fairly reduced for QSQS for images like 8 and 9. Bad detections in images 6 and 7 are essentially caused by proposed regions of interests, instead of post-processing schemes.

Pedestrian Detection. We embed QSQS on Center and Scale Prediction (CSP) detector [21] as the enhanced post-processing scheme. CSP achieves best

Table 2. Best log-average Miss Rate for three schemes across 35 training epochs. Total evaluation time for 500 test images are shown for each scheme. Best values are shown in bold for each metric column.

Method	Backbone	Time (s)	Reasonable	Bare	Partial	Heavy
CSP* [21] + NMS	ResNet-50	**83.63**	27.48	20.58	**27.34**	**65.78**
CSP + Soft-NMS	ResNet-50	92.07	28.13	21.44	28.57	67.12
CSP + QSQS	ResNet-50	90.91	**27.33**	**20.52**	27.48	66.38

CSP* refers to CSP [21] trained in our own schedule.

detection result in terms of log-average Miss Rate metric for two challenging pedestrian benchmark datasets. Table 2 and Fig. 6 show MR^{-2} comparison among NMS, Soft-NMS and QSQS on the more challenging CityPersons across 35 training epochs. Results are evaluated on standard *Reasonable* subset, together with three other subsets with various occlusion levels. The best MR^{-2} is displayed for each suppression scheme. It's worth noting that soft-NMS has the longest average inference delay (92.07s for 500 test images), and worst miss rates for all occlusion level subsets. In Fig. 6, QSQS shows least miss rates consistently among three schemes for *Reasonable* and *Bare* subsets across 35 training epochs. Standard NMS performs best for heavily occluded subset, which may indicate NMS detects better the objects with low visibility caused by other categories.

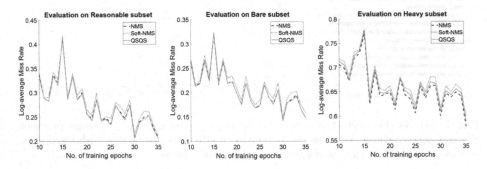

Fig. 6. Log-average Miss Rate comparison for three occlusion subsets across 35 training epochs. All three schemes show decreasing MR^{-2} trends along increasing training epochs. QSQS outperforms NMS and soft-NMS for *Reasonable* and *Bare* subsets.

It's worth mentioning that QSQS is a false positive suppression scheme that performs well on both generic object detection and pedestrian detection application scenarios. The flexibility on composing score term of individual bounding box and correlation term between neighboring boxes entails its superiority of handling distinct tasks.

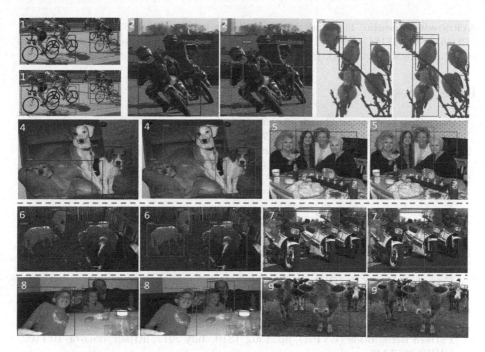

Fig. 7. Qualitative results. Image pairs in the top row shows cases where QSQS (in red boxes) performs better than NMS (in blue boxes). Images in bottom row show cases where NMS detects better. Pairs in the middle row are cases that cannot be properly detected by both methods due to poor raw detections. (Color figure online)

5 Conclusion

In this paper, we propose a novel hybrid quantum-classical QSQS algorithm for removing redundant object detections. The inspiration originates from basic QUBO suppression method, however it is impractical for real-world detection due to low accuracy and long inference latency. Detection accuracy in our work is improved by adapting soft-NMS method and introducing two adjustment factors for basic linear and quadratic terms. Furthermore, we leverage quantum advantage by solving QUBO problems (which are NP-hard) in D-Wave annealing system instead of classical solver that cannot solve such problems for real-time applications. The proposed algorithm incurs no extra QPU access cost.

We report the experimental results on generic object detection datasets, namely, MS-COCO and PASCAL VOC 2007, and pedestrian detection CityPersons datasets. The results show the proposed QSQS method improves mAP from 74.20% to 75.11% for PASCAL VOC 2007 through two level enhancements. For benchmark pedestrian detection CityPersons, it consistently outperforms NMS and soft-NMS for bothe *Reasonable* and *Bare* subsets.

Acknowledgments. This work was supported in part by SRC (2847.001), NSF (CNS-1722557, CCF-1718474, DGE-1723687 and DGE-1821766), Institute for Computational and Data Sciences and Huck Institutes of the Life Sciences at Penn State.

References

1. Ajagekar, A., You, F.: Quantum computing for energy systems optimization: challenges and opportunities. Energy **179**, 76–89 (2019)
2. Arute, F., et al.: Quantum supremacy using a programmable superconducting processor. Nature **574**(7779), 505–510 (2019)
3. Ash-Saki, A., Alam, M., Ghosh, S.: Qure: qubit re-allocation in noisy intermediate-scale quantum computers. In: 2019 Proceedings of the 56th Annual Design Automation Conference, pp. 1–6 (2019)
4. Biswas, R., et al.: A NASA perspective on quantum computing: opportunities and challenges. Parallel Comput. **64**, 81–98 (2017)
5. Bodla, N., Singh, B., Chellappa, R., Davis, L.S.: Soft-NMS - improving object detection with one line of code. In: 2017 IEEE International Conference on Computer Vision (ICCV), pp. 5562–5570, October 2017. https://doi.org/10.1109/ICCV.2017.593
6. Cao, Z., Simon, T., Wei, S., Sheikh, Y.: Realtime multi-person 2D pose estimation using part affinity fields. In: 2017 IEEE Conference on Computer Vision and Pattern Recognition (CVPR), pp. 1302–1310, July 2017. https://doi.org/10.1109/CVPR.2017.143
7. Chen, C., Seff, A., Kornhauser, A., Xiao, J.: DeepDriving: learning affordance for direct perception in autonomous driving. In: 2015 IEEE International Conference on Computer Vision (ICCV), pp. 2722–2730, December 2015. https://doi.org/10.1109/ICCV.2015.312
8. Chen, X., Ma, H., Wan, J., Li, B., Xia, T.: Multi-view 3D object detection network for autonomous driving. In: 2017 IEEE Conference on Computer Vision and Pattern Recognition (CVPR), pp. 6526–6534, July 2017. https://doi.org/10.1109/CVPR.2017.691
9. Cruz-Santos, W., Venegas-Andraca, S., Lanzagorta, M.: A GUBO formulation of the stereo matching problem for D-wave quantum annealers. Entropy **20**(10), 786 (2018)
10. Dollar, P., Wojek, C., Schiele, B., Perona, P.: Pedestrian detection: an evaluation of the state of the art. IEEE Trans. Pattern Anal. Mach. Intell. **34**(4), 743–761 (2011)
11. Everingham, M., Van Gool, L., Williams, C.K.I., Winn, J., Zisserman, A.: The PASCAL Visual Object Classes Challenge 2007 (VOC2007) Results. http://www.pascal-network.org/challenges/VOC/voc2007/workshop/index.html
12. Harrow, A.W., Montanaro, A.: Quantum computational supremacy. Nature **549**(7671), 203–209 (2017)
13. He, K., Zhang, X., Ren, S., Sun, J.: Deep residual learning for image recognition. In: Proceedings of the IEEE Conference on Computer Vision and Pattern Recognition, pp. 770–778 (2016)
14. He, Y., Zhu, C., Wang, J., Savvides, M., Zhang, X.: Bounding box regression with uncertainty for accurate object detection. In: The IEEE Conference on Computer Vision and Pattern Recognition (CVPR), June 2019
15. Hosang, J., Benenson, R., Schiele, B.: Learning non-maximum suppression. In: The IEEE Conference on Computer Vision and Pattern Recognition (CVPR), July 2017

16. Krizhevsky, A., Sutskever, I., Hinton, G.E.: ImageNet classification with deep convolutional neural networks. Commun. ACM **60**(6), 84–90 (2017). https://doi.org/10.1145/3065386
17. Lee, D., Cha, G., Yang, M.-H., Oh, S.: Individualness and determinantal point processes for pedestrian detection. In: Leibe, B., Matas, J., Sebe, N., Welling, M. (eds.) ECCV 2016. LNCS, vol. 9910, pp. 330–346. Springer, Cham (2016). https://doi.org/10.1007/978-3-319-46466-4_20
18. Li, J., Alam, M., Ash-Saki, A., Ghosh, S.: Hierarchical improvement of quantum approximate optimization algorithm for object detection. In: 2020 21th International Symposium on Quality Electronic Design (ISQED), pp. 335–340. IEEE (2020)
19. Lin, T.-Y., et al.: Microsoft COCO: common objects in context. In: Fleet, D., Pajdla, T., Schiele, B., Tuytelaars, T. (eds.) ECCV 2014. LNCS, vol. 8693, pp. 740–755. Springer, Cham (2014). https://doi.org/10.1007/978-3-319-10602-1_48
20. Liu, S., Wang, C., Qian, R., Yu, H., Bao, R., Sun, Y.: Surveillance video parsing with single frame supervision. In: 2017 IEEE Conference on Computer Vision and Pattern Recognition (CVPR), pp. 1013–1021, July 2017. https://doi.org/10.1109/CVPR.2017.114
21. Liu, W., Liao, S., Ren, W., Hu, W., Yu, Y.: High-level semantic feature detection: a new perspective for pedestrian detection. In: Proceedings of the IEEE Conference on Computer Vision and Pattern Recognition, pp. 5187–5196 (2019)
22. Meinshausen, N., Bickel, P., Rice, J., et al.: Efficient blind search: optimal power of detection under computational cost constraints. Ann. Appl. Stat. **3**(1), 38–60 (2009)
23. Palubeckis, G.: Multistart tabu search strategies for the unconstrained binary quadratic optimization problem. Ann. Oper. Res. **131**(1–4), 259–282 (2004). https://doi.org/10.1023/B:ANOR.0000039522.58036.68
24. Ren, S., He, K., Girshick, R., Sun, J.: Faster R-CNN: towards real-time object detection with region proposal networks. In: Advances in Neural Information Processing Systems, pp. 91–99 (2015)
25. Ristè, D., et al.: Demonstration of quantum advantage in machine learning. NPJ Quantum Inf. **3**(1), 16 (2017)
26. Rujikietgumjorn, S., Collins, R.T.: Optimized pedestrian detection for multiple and occluded people. In: 2013 IEEE Conference on Computer Vision and Pattern Recognition (CVPR), pp. 3690–3697, June 2013. https://doi.org/10.1109/CVPR.2013.473
27. Shaydulin, R., Ushijima-Mwesigwa, H., Negre, C.F., Safro, I., Mniszewski, S.M., Alexeev, Y.: A hybrid approach for solving optimization problems on small quantum computers. Computer **52**(6), 18–26 (2019)
28. Stollenwerk, T., Lobe, E., Jung, M.: Flight gate assignment with a quantum annealer. In: Feld, S., Linnhoff-Popien, C. (eds.) QTOP 2019. LNCS, vol. 11413, pp. 99–110. Springer, Cham (2019). https://doi.org/10.1007/978-3-030-14082-3_9
29. Zhang, S., Benenson, R., Schiele, B.: CityPersons: a diverse dataset for pedestrian detection. In: Proceedings of the IEEE Conference on Computer Vision and Pattern Recognition, pp. 3213–3221 (2017)
30. Zhu, X., Ramanan, D.: Face detection, pose estimation, and landmark localization in the wild. In: 2012 IEEE Conference on Computer Vision and Pattern Recognition (CVPR), pp. 2879–2886, June 2012

Label-Similarity Curriculum Learning

Ürün Dogan[1]($\boxtimes$), Aniket Anand Deshmukh[1], Marcin Bronislaw Machura[1],
and Christian Igel[2]

[1] Microsoft Ads, Microsoft, Sunnyvale, CA, USA
{udogan,andeshm,mmachura}@microsoft.com
[2] Department of Computer Science, University of Copenhagen,
Copenhagen, Denmark
igel@di.ku.dk

Abstract. Curriculum learning can improve neural network training by guiding the optimization to desirable optima. We propose a novel curriculum learning approach for image classification that adapts the loss function by changing the label representation.

The idea is to use a probability distribution over classes as target label, where the class probabilities reflect the similarity to the true class. Gradually, this label representation is shifted towards the standard one-hot-encoding. That is, in the beginning minor mistakes are corrected less than large mistakes, resembling a teaching process in which broad concepts are explained first before subtle differences are taught.

The class similarity can be based on prior knowledge. For the special case of the labels being natural words, we propose a generic way to automatically compute the similarities. The natural words are embedded into Euclidean space using a standard word embedding. The probability of each class is then a function of the cosine similarity between the vector representations of the class and the true label.

The proposed label-similarity curriculum learning (LCL) approach was empirically evaluated using several popular deep learning architectures for image classification tasks applied to five datasets including ImageNet, CIFAR100, and AWA2. In all scenarios, LCL was able to improve the classification accuracy on the test data compared to standard training. Code to reproduce results is available at https://github.com/speedystream/LCL.

Keywords: Curriculum learning · Deep learning · Multi-modal learning · Classification

1 Introduction

When educating humans, the teaching material is typically presented with increasing difficulty. *Curriculum learning* adopts this principle for

Electronic supplementary material The online version of this chapter (https://doi.org/10.1007/978-3-030-58526-6_11) contains supplementary material, which is available to authorized users.

machine learning to guide an iterative optimization method to a desirable optimum. In curriculum learning for neural networks as proposed by Bengio et al. [1], the training examples are weighted. In the beginning of the training, more weight is put on "easier" examples. The weighting is gradually changed to uniform weights corresponding to the canonical objective function.

Inspired by Bengio et al. [1], we propose *label-similarity curriculum learning* (LCL) as another way to "learn easier aspects of the task or easier sub-tasks, and then gradually increase the difficulty level." If a toddler who is just learning to speak points at a car and utters "cow", a parent will typically react with some teaching signal. However, a young infant is not expected to discriminate between a cheetah and a leopard, and mixing up the two would only lead to a very mild correction signal – if at all. With increasing age, smaller errors will also be communicated.

We transfer this approach to neural network training for classification tasks. Instead of a one-hot-encoding, the target represents a probability distribution over all possible classes. The probability of each class depends on the similarity between the class and the true label. That is, instead of solely belonging to its true class, each input can also belong to similar classes to a lesser extent. Gradually, this label representation is shifted towards the standard one-hot-encoding, where targets representing different classes are orthogonal. In the beginning of training, the targets of inputs with labels `cheetah` and `leopard` should almost be the same, but always be very different from `car`. During the training process, the label representation is gradually morphed into the one-hot encoding, decreasing the entropy of the distribution encoded by the target over time. That is, in the beginning small mistakes – in the sense that similar classes are mixed up – are corrected less than big mistakes, resembling a teaching process in which broad concepts are explained first before subtle differences are taught.

The question arises how to define a proper similarity between classes. One can get a label-similarity matrix based on prior knowledge or some known structure. For the case where the similarity is not explicitly given and the labels correspond to natural language words, we propose a way to automatically infer a representation that reflects semantic similarity. We map the labels into a Euclidean space using a word embedding. Concretely, this is done by applying a generic document embedding to a document explaining the label (its Wikipedia entry). Then the cosine similarities between the vector representations of the label and all possible classes are computed. Based on these values, a distribution over the possible classes is defined which serves as the learning target.

Our way to define the target representation resembles the idea of *hierarchical loss functions* [27,32] ("We define a metric that, inter alia, can penalize failure to distinguish between a sheepdog and a skyscraper more than failure to distinguish between a sheepdog and a poodle." [32]). However, there are two decisive differences. First, we propose to gradually shift from a "hierarchical loss" to a "flat loss". Second, unlike in [32], our approach does not necessarily presume a given hierarchy. When dealing with natural language labels, we propose a way to automatically infer the similarity from a generic word embedding under the assumption that exploiting semantic similarity can be helpful in guiding the learning process.

For evaluating label-similarity curriculum learning (LCL), we need data with some structure in the label space that curriculum learning can exploit. Furthermore, there should be sufficiently many classes and the task should not be easy to learn. To get label similarity based on word embeddings, we need a dataset with natural language labels. In this study, we focus on three popular benchmark datasets, ImageNet [4], CIFAR100 [17], and Animals with Attributes (AwA) [33]. To show the generality of our approach, we consider different deep learning architectures, and also different preprocessing and learning processes. The time schedule for increasing the "difficulty" of the learning task is an obvious hyperparameter, which we carefully study and show to have little importance.

The next section points to related literature and Sect. 3 introduces the new *label-similarity curriculum learning*. Section 4 describes the experiments and Sect. 5 the results before we conclude.

2   Related Work

Starting from the work by Bengio et al. [1], a variety of curriculum learning approaches has been studied. However, they all define a curriculum at the level of training examples. For instance, *self-paced learning* by Kumar et al. [18] introduces latent variables for modelling "easiness" of an examples. Graves et al. [10] consider example-based improvement measures as reward signals for multi-armed bandits, which then build stochastic syllabi for neural networks. Florensa et al. [5] study curriculum learning in the context of reinforcement learning in robotics. They propose to train a robot by gradually increasing the complexity of the task at hand (e.g., the robot learns to reach a goal by setting starting points increasingly far from the goal). In recent work, Weinshall et al. [31] consider learning tasks with convex linear regression loss and prove that the convergence rate of a perfect curriculum learning method increases with the difficulty of the examples. In addition, they propose a method which infers the curriculum using transfer learning from another network (e.g., ResNet-50) pretrained on a different task. They train a linear classifier using features extracted from the pretrained model and score each training example using the linear classifier's confidence (e.g., the margin of an SVM). Finally, they train a smaller deep neural network for the transfer learning task following a curriculum based on these scores.

Buciluă et al. [2] have proposed compressing a large model into a simple model which reduces space requirements and increases inference speed at the cost of a small performance loss. This idea has been revisited in [13] under the name *knowledge distillation* (KD) and received a significant amount of attention (e.g., [22,23,25,35,36]). KD methods typically require a pretrained model to start with or train a series of models on the same training data. Standard KD considers a teacher network and a student network. The powerful teacher network is used to support the training of the student network which may be less complex or may have access to less data for training. KD is related to curriculum learning methods because the teacher network guides the learning of student networks [13]. A variant of KD, *born again neural network*, trains a series of models, not only one [7].

Deep mutual learning (DML) is also loosely related to our proposed approach [13,38]. In DML, two models solve the same classification problem collaboratively and are jointly optimised [38]. Each model acts as a teacher for the other model, and each network is trained with two losses. The first loss is the standard cross-entropy between the model's predictions and target labels. The second is a *mimicry loss* that aligns both model's class posteriors with the class probabilities of the respective other model.

Another related approach is *CurriculumNet* [11], a clustering based curriculum strategy for learning from noisy data. CurriculumNet consists of three steps. First, a deep neural network is trained on the noisy label data. Second, features are extracted by using the model trained in the first step. Using clustering algorithms, these features are then grouped into different sets and sorted into easy and difficult examples. Finally, a new deep neural network is trained using example-weighted curriculum learning. Sorting of examples from easy to hard and clustering algorithms add many hyper-parameters (e.g., number of clusters), and one has to train two neural network models of almost the same size.

Our algorithm can be considered as a multi-modal deep learning method, where text data is used for estimating the class similarity matrix to improve image classification. However, it is different from standard multimodal methods as it does not use text data as an input to the deep neural network. The *DeVise* algorithm is a popular multi-modal method which utilizes the text modality in order to learn a mapping from an image classifier's feature space to a semantic space [6]. DeVise requires a pretrained deep neural network. Furthermore, as stated in [6], it does not improve the accuracy on the original task but aims at training a model for zero-shot learning.

There is an obvious relation between LCL and *label smoothing* (LS) [21], which we will discuss in Sect. 4.

The computational requirements of KD, DML, and CurriculumNet are significantly higher compared to our method, which is rather simple. Furthermore, our method does not require training more than one model and adds only a single hyper-parameter.

3   Method

We assume a discrete set of training examples $(\boldsymbol{x}_1, c_1), \ldots, (\boldsymbol{x}_\ell, c_\ell) \in \mathcal{X} \times \mathcal{C}$, with input space $\mathcal{X}$ and finite label space $\mathcal{C}$ with cardinality $|\mathcal{C}| = C$. Let $n : \mathcal{C} \to \{1, \ldots, C\}$ be a bijective mapping assigning each label to a unique integer. This allows a straight-forward definition of the one-hot encoding $\boldsymbol{y}_i \in \mathbb{R}^C$ for each training example $(\boldsymbol{x}_i, c_i)$. The j-th component of $\boldsymbol{y}_i$, which is denoted by $[\boldsymbol{y}_i]_j$, equals 1 if $n(c_i) = j$ and 0 otherwise.

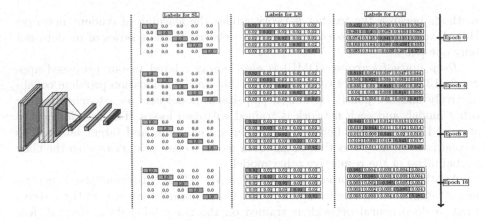

Fig. 1. A deep network (left) trained with three different encodings on a five-class dataset with labels Lion, Tiger, Aircraft Carrier, Alaskan Wolf and Mushroom. The SL (standard learning, see Sect. 4 for details) column shows the label matrix for one-hot-encoding. When using LS (label smoothing, see Sect. 4), the loss between the network output and a smoothed version of the label, which does not change over time, is minimized. We propose to use a probability distribution over classes as target label, where the class probabilities reflect the similarity to the true class. This is shown in the LCL column. Unlike LS the proposed label encoding changes during training and converges to the original optimization problem solved when using SL.

3.1 Document Embedding for Defining Label Similarity

Our learning curriculum is based on the pairwise similarities between the C classes, which are defined based on the semantic similarity of the class labels. Now assume that the labels are natural language words, for example $C = \{\ldots,$ flute$, \ldots,$ strawberry$, \ldots,$ backpack$, \ldots \}$. To quantify semantic similarity, we embed the natural language labels into Euclidean space using a word embedding [20] such that similar words are nearby in the new representation.

ImageNet labels are given by WordNet identifiers representing synsets, and we redefine the labels for other datasets in a similar way. First, we convert synsest to words, for example, n02119789 to "fox". Then, we find the Wikipedia article describing each word, for instance, "Orange (fruit)" was selected for orange. Then we apply doc2vec [19] for mapping the article into Euclidean space. We used a generic doc2vec embedding trained on the English Wikipedia corpus. This gives us the encoding $f_{\text{enc}} : C \rightarrow \mathbb{R}^d$, mapping each class label to the corresponding Wikipedia article and then computing the corresponding vector representation using doc2vec (with $d = 100$, see below). Now we can compute the similarity between two classes c_i and c_j by the cosine similarity

$$s(c_i, c_j) = \frac{\langle f_{\text{enc}}(c_i), f_{\text{enc}}(c_j) \rangle}{\|f_{\text{enc}}(c_i)\| \|f_{\text{enc}}(c_j)\|} \ , \tag{1}$$

which in our setting is always non-negative. The resulting label dissimilarity matrix for the ImageNet labels is visualized in the supplementary material.

3.2   Label Encoding

We adopt the formal definition of a curriculum from the seminal paper by Bengio et al. [1]. In [1], a weighting of the training data is adapted, so that in the beginning a larger weight is put on easy examples. To distinguish this work from our approach, we refer to it as *example-weighting curriculum*.

Let $t \geq 0$ denote some notion of training time (e.g., a counter of training epochs). In [1], there is a sequence of weights associated with each example $i = 1, \ldots, \ell$, which we denote by $w_i^{(t)} \in [0, 1]$. These weights are normalized so that $\sum_{i=1}^{\ell} w_i^{(t)} = 1$ to describe a proper probability distribution over the training examples.

For the weight sequence to be a proper (example-weighting) curriculum, Bengio et al. [1] demand that the entropy of the weights

$$H(\boldsymbol{w}^{(t)}) = - \sum_{i=1}^{\ell} w_i^{(t)} \ln w_i^{(t)} \tag{2}$$

is monotonically increasing with t (the weights should converge to the uniform distribution).

We define our *label-weighting curriculum* in a similar axiomatic way. Instead of a sequence of weights for the training examples varying with t, we have a sequence of label vectors for each training example. Let $\boldsymbol{v}_i^{(t)}$ denote the C-dimensional label vector for training pattern i at time t. For the sequence to be a label-weighting curriculum, the entropy of the label vector components

$$\forall i = 1, \ldots, \ell : H(\boldsymbol{v}_i^{(t)}) = - \sum_{c=1}^{C} [\boldsymbol{v}_i]_c^{(t)} \ln[\boldsymbol{v}_i]_c^{(t)} \tag{3}$$

should be monotonically *decreasing* under the constraints that for each label vector $\boldsymbol{v}_i^{(t)}$ we have $[\boldsymbol{v}]_j \geq 0$ for all j, $\|\boldsymbol{v}_i^{(t)}\|_1 = 1$, and $\operatorname{argmax}_j[\boldsymbol{v}]_j^{(t)} = n(c_i)$ for all t. The conditions imply that $\boldsymbol{v}_i$ is always an element of the probability simplex, the class label given in the training set always gets the highest probability, and $\boldsymbol{v}_i^{(t)}$ converges to $\boldsymbol{y}_i$.

We now give an example of how to adapt the label vectors. Similar as in [1], we define for each training example i the simple update rule:

$$[\boldsymbol{v}_i]_j^{(t+1)} = \begin{cases} \frac{1}{1+\epsilon \sum_{k \neq n(c_i)}[\boldsymbol{v}_i]_k^{(t)}} & \text{if } j = n(c_i) \\ \frac{\epsilon[\boldsymbol{v}_i]_j^{(t)}}{1+\epsilon \sum_{k \neq n(c_i)}[\boldsymbol{v}_i]_k^{(t)}} & \text{otherwise} \end{cases} \tag{4}$$

The constant parameter $0 < \epsilon < 1$ controls how quickly the label vectors converge to the one-hot-encoded labels. This update rule leads to a proper label-weighting curriculum. During learning, the entries for all components except $n(c_i)$ drop with $\mathcal{O}(\epsilon^t)$. Note that $[\boldsymbol{v}_i]_{n(c_i)}^{(t+1)} \geq [\boldsymbol{v}_i]_{n(c_i)}^{(t)}$. The vectors are initialized using the label similarity defined in (1):

$$[\boldsymbol{v}_i]_j^{(0)} = \frac{s\big(c_i, n^{-1}(j)\big)}{\sum_{k=1}^{C} s\big(c_i, n^{-1}(k)\big)} \tag{5}$$

Recall that $n^{-1}(j)$ denotes the "j-th" natural language class label.

3.3 Loss Function

Let $\mathcal{L}$ be a loss function between two probability distributions and $f_\theta(x)$ be the predicted distribution for example $\boldsymbol{x}$ for some model parameters θ. At time step t we optimize $J^{(t)}(\theta) = \sum_{i=1}^{n} \mathcal{L}(f_\theta(\boldsymbol{x}_i), \boldsymbol{v}_i^{(t)}) + \lambda r(\theta)$, where λ is a positive constant and $r(\theta)$ is a regularization function. In this paper, the networks are trained using the standard cross-entropy loss function with normalized targets $\boldsymbol{v}_i$ for the inputs $\boldsymbol{x}_i$, $i = 1, \ldots, \ell$. Hence, in the beginning, predicting the correct one-hot encoded label $\boldsymbol{y}_i$ causes an error signal. That is, initially it is less penalized if an object is not correctly classified with maximum confidence. Later in the training process, $\boldsymbol{v}_i$ converges to $\boldsymbol{y}_i$ and the classifier is then pushed to build up confidence.

4 Experiments

We evaluated our curriculum learning strategy by running extensive experiments on ImageNet [4], CIFAR100 [17], and AWA2 [33] plus additional experiments on CUB-200-2011 [30] and NABirds [29] (see supplementary material). On CUB-200-2011 and AwA2, we evaluated our approach using both the proposed semantic similarity of the labels as well as visual similarity. On NABirds, we evaluated our approach also using similarity based on the given (biological) hierarchy, where we used simrank [16] for calculating the similarity matrix.

Table 1. ℓ_{train} denotes the number of training images, ℓ_{test} denotes the number of test images and C the number of classes in a given dataset; DR indicates the data set sizes (percentage of ℓ_{train}) and ϵ the cooling parameters. The column Sim. indicates which similarity measures were used, where l stands for the semantic similarity using the word embedding of the labels, v for a measure based on the similarity of the images, and h for similarity based on a given label hierarchy. For each experimental setup, #Rep = 4 repetitions with different initializations/seeds were conducted.

	ℓ_{train}	ℓ_{test}	C	DR	ϵ	Sim.
AWA2	29865	7457	50	5%, 10%, 20%, 100%	0.9, 0.99, 0.999	l, v
CIFAR100	50000	10000	100	5%, 10%, 20%, 100%	0.9, 0.99, 0.999	l
ImageNet	1281167	50000	1000	5%, 10%, 20%, 100%	0.9, 0.99, 0.999	l
NABirds	23912	24615	555	100%	0.9, 0.99, .999	l, h
CUB-200-2011	5994	5794	201	100%	0.9, 0.99, .999	l, v

For each dataset we considered at least two different models and two different baselines. Descriptive statistics of the datasets and a summary of the experimental setup are given in Table 1 and Table 2. We considered different training set sizes, where DR $\in \{5\%, 10\%, 20\%, 100\%\}$ refers to the fraction of training data used. The remaining training data was discarded (i.e., not used in the training process at all); the test data were always the same.

We empirically compared the following algorithms:

1. Label-similarity curriculum learning (LCL): Proposed method with label update rule (4). The time step t is the epoch number.
2. Standard Learning (SL): This is a standard setup with fixed one-hot encoding.
3. Label Smoothing (LS): Label smoothing uses soft targets instead of one-hot encoding. It has been argued that LS prevents the network from becoming over-confident and improves the empirical performance of the algorithm [21]. For $0 \leq \alpha \leq 1$ label smoothing uses following label vector

$$[\boldsymbol{v}_i]_j^{(t)} = \begin{cases} (1 - \alpha) + \frac{\alpha}{C} & \text{if } j = n(c_i) \\ \frac{\alpha}{C} & \text{otherwise} \end{cases} \quad \text{for all } t. \tag{6}$$

We set $\alpha = 0.1$ for the evaluations in this study.
4. Deep Mutual Learning (DML): In DML, two models, referred to as DML_1 and DML_2, solve the same classification problem collaboratively and are optimised jointly [38]. It uses one hot-encoding along with cross-entropy loss as in SL but adds additional terms $\text{KL}(\hat{\boldsymbol{v}}_{\text{DML}_1}^{(t)} \| \hat{\boldsymbol{v}}_{\text{DML}_2}^{(t)}) + \text{KL}(\hat{\boldsymbol{v}}_{\text{DML}_2}^{(t)} \| \hat{\boldsymbol{v}}_{\text{DML}_1}^{(t)})$, where KL denotes the Kullback–Leibler divergence and $\hat{\boldsymbol{v}}_{\text{DML}_1}^{(t)}$ and $\hat{\boldsymbol{v}}_{\text{DML}_2}^{(t)}$ are the predicted label probability vectors for both models. We report the classification performance of both DML_1 and DML_2.
5. Knowledge Distillation (KD): In KD, one model is trained first using one-hot encoded targets, and then the class probabilities produced by the first model are used as "soft targets" for training the second model [13].

Table 2. ResNeXt-101 denotes ResNeXt-101 $(32 \times 8d)$, WRN denotes WRN (28-10-dropout), and DenseNet-BC denotes DenseNet-BC $(k = 40, \text{depth} = 190)$.

Model	Dataset	Baselines
ResNet-18 [34]	CUB-200-2011	LS, DML, SL, KD, CN
ResNet-34 [34]	CUB-200-2011, NABirds	LS, DML, SL, KD, CN
ResNet-50 [12]	ImageNet, NABirds	SL, LS, KD, CN
ResNeXt-101 [34]	ImageNet	SL, LS, KD, CN
SENet-154 [14]	ImageNet	SL, LS, KD, CN
ResNet-101 [12]	AWA2	LS, DML, SL, KD, CN
InceptionResNetV2 [28]	AWA2	LS, DML, SL, KD, CN
WRN [37]	CIFAR100	LS, DML, SL, KD, CN
DenseNet-BC [15]	CIFAR100	LS, DML, SL, KD, CN

6. Curriculum Net (CN): In CN, example-weighted curriculum is built by sorting examples from easy to hard [11].

For all architectures, we have followed the experimental protocols described in the original publications [12,14,15,28,34,37]. All experiments were conducted using the PyTorch deep learning library [24].[1] For all experiments, except ImageNet DR $= 100\%$, we used stochastic gradient descent (SGD) for optimization. For ImageNet DR $= 100\%$ we used the distributed SGD algorithm [9] with Horovod[2] [26] support because of the computational demands. The distributed SGD algorithm [9] is one of the state-of-the-art methods for large scale training. It is expected to lead to a slight loss in performance when a large batch size is used (see [9] for details).

Our approach introduces the hyperparameter ϵ, see (4). In order to assess the stability of the proposed method, we present results for $\epsilon \in \{0.9, 0.99, 0.999\}$.[3] We repeated all experiments four times. We report the top-1 and top-5 classification accuracy on the test datasets (standard deviations are reported in the supplementary material).

For estimating the label similarity matrix, we used *pretrained* doc2vec emebeddings with dimensions $d \in \{100, 300, 500\}$ with ResNet-50 and ResNet-101 . We did not observe any significant differences in the classification accuracies. The maximum difference between compatible settings were less than 0.06%. Hence, we only report results for the $d = 100$ dimensional doc2vec embeddings.

For each experiment, we used workstations having 4 Tesla P100 GPUs (with 16 GB GPU RAM) each. For network communication we used InfiniBand, which is a computer-networking communications standard designed for high throughput and low-latency scenarios.

We tuned the hyperparameters for the baseline method (SL) only. For 100% data, we took the hyperparameters from the original publications. For all other settings, we optimized learning rate, batch size and weight-decay for the standard baseline (SL). Then we use the very same parameters for our approach (we just varied the new parameter epsilon). Thus, hyperparameter tuning would rather increase the gain from using our method. Thus, one might argue that the new algorithm is using sub-optimal hyperparameters compared to the baselines. However, our goal was to show that the proposed algorithm can improve any model on different datasets without tuning hyperparameters.

---

[1] The code to reproduce our results is available in the supplementary material.

[2] Horovod is a method which uses large batches over multiple GPU nodes and some accuracy loss is expected for the baseline method and this is well established. For more details please see Table 1 and Table 2.c in [24].

[3] We have tried $\epsilon \in \{0.8, 0.9, 0.91, \ldots, 0.98, 0.99, 0.992, \ldots, 0.998, 0.999\}$ for ResNet-50 and ResNet-101 . The results showed that the search space for ϵ can be less granular and we have limited the search space accordingly.

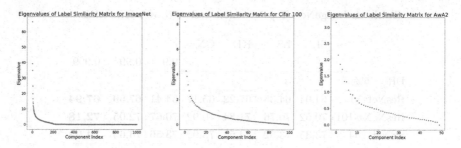

Fig. 2. Eigenvalue distributions of the class similarity matrices for ImageNet, CIFAR100, and AwA2.

5   Results and Discussion

We will focus on the results for ImageNet, CIFAR100, and AWA2 and the similarity measure introduced in Sect. 3, result tables for the other data sets and other similarity measures can be found in the supplementary material. Before we present the learning results, we will discuss the structure of the label similarity matrices for the data sets ImageNet, CIFAR100, and AWA2.

Table 3. ImageNet. Top-1 results, averaged over four trials.

	SL	LS	KD	CN	ϵ		
					0.9	0.99	0.999
DR = 5%							
ResNet-50	38.21	38.43	39.81	36.12	41.24	41.6	**42.21**
ResNeXt-101	45.46	45.71	46.21	44.14	46.1	46.92	**47.12**
SENet-154	48.29	48.57	48.44	46.21	49.8	50.04	**50.19**
DR = 10%							
ResNet-50	51.95	52.25	53.64	52.17	55.39	55.62	**55.64**
ResNeXt-101	58.63	58.92	58.94	57.64	59.78	**60.07**	59.92
SENet-154	60.61	60.74	60.82	60.14	60.99	61.18	**62.28**
DR = 20%							
ResNet-50	61.87	62.11	63.17	62.41	64.41	64.42	**64.44**
ResNeXt-101	67.96	68.13	68.29	68.14	68.48	68.47	**68.57**
SENet-154	67.77	67.71	67.64	67.43	68.14	**68.4**	68.33
DR = 100%							
ResNet-50	76.25	76.4	76.38	76.1	76.71	76.75	**76.89**
ResNeXt-101	78.05	78.17	78.21	77.94	78.31	78.5	**78.64**
SENet-154	79.33	79.65	79.44	79.44	80.11	80.03	**80.21**

Table 4. ImageNet. Top-5 results, averaged over four trials.

	SL	LS	KD	CN	ϵ		
					0.9	0.99	0.999
DR = 5%							
ResNet-50	64.04	64.35	67.22	65.12	64.41	67.59	**67.94**
ResNeXt-101	70.52	70.76	71.84	70.92	70.67	72.05	**72.18**
SENet-154	73.35	73.52	74.54	73.92	73.56	74.65	**74.92**
DR = 10%							
ResNet-50	76.86	77.04	79.73	78.14	77.1	**79.8**	79.69
ResNeXt-101	81.52	81.87	82.56	81.92	81.67	82.66	**82.74**
SENet-154	82.53	**83.71**	83.38	82.76	82.94	83.6	83.5
DR = 20%							
ResNet-50	84.41	84.57	86.1	85.36	84.91	**86.15**	86.14
ResNeXt-101	87.96	88.11	88.2	88.04	87.84	**88.36**	**88.36**
SENet-154	88.16	88.23	88.17	88.17	88.11	**88.37**	88.31
DR = 100%							
ResNet-50	92.87	92.91	92.94	92.41	92.84	**92.95**	92.93
ResNeXt-101	93.95	93.92	93.96	93.85	93.87	94.07	**96.15**
SENet-154	94.33	94.44	94.84	94.02	94.35	**94.93**	94.79

Label Similarities. For a better understanding of the label similarity matrices, we visualized their eigenspectra in Fig. 2. Consider two extreme scenarios: If a label similarity matrix has rank 1, all classes are exactly the same and there cannot be any discriminatory learning. In contrast, the full rank case with equal eigenvalues is the standard learning case where all classes are orthogonal to each other (one-hot-encoding). Figure 2 shows exponential eigenvalues decays, which means there are clusters of similar classes. Distinguishing between these clusters of classes is an easier task than distinguishing between classes within one cluster.

Classification Performance. We measured the top-1 and top-5 classification accuracy after the last epoch. The results are summarized in Table 3 and Table 4 for ImageNet, in Table 5 and Table 6 for CIFAR100, and in Table 7 and Table 8 for AWA2 (for standard deviations see the supplementary material). All results are averaged over four trials. It is important to note that we compare against baseline results achieved with architectures and hyperparameters tuned for excellent performance. Furthermore, we compare to baseline results from our own experiments, not to results taken from the literature. We ran each experiment 4 times with same seeds for all algorithms. This allows for a fair comparison. Our averaged results also provide a more reliable estimate of the performance of the systems compared to single trials reported in the original works.

Table 5. CIFAR100. Top-1 results, averaged over four trials.

	SL	LS	DML$_1$	DML$_2$	KD	CN	ϵ		
							0.9	0.99	0.999
DR = 5%									
WRN	40.2	40.31	40.16	40.47	40.94	38.14	41.36	41.86	**41.92**
DenseNet-BC	43.34	43.5	43.89	44.14	43.76	43.42	**44.66**	45.37	44.53
DR = 10%									
WRN	60.2	60.1	60.38	60.34	60.45	60.14	60.49	60.86	**61.19**
DenseNet-BC	60.85	61.1	61.22	61.34	60.81	59.83	61.5	61.4	**61.65**
DR = 20%									
WRN	71.05	71.25	71.61	71.65	71.53	71.37	71.64	71.67	**71.83**
DenseNet-BC	72.38	72.39	71.5	71.34	72.24	71.54	72.71	72.65	**72.87**
DR = 100%									
WRN	79.52	79.84	80.32	80.20	80.14	78.64	81.17	81.15	**81.25**
DenseNet-BC	82.85	83.01	82.91	82.57	82.67	81.14	82.96	83.11	**83.2**

Table 6. CIFAR100. Top-5 results, averaged over four trials.

	SL	LS	DML$_1$	DML$_2$	KD	CN	ϵ		
							0.9	0.99	0.999
DR = 5%									
WRN	68.82	68.95	68.74	68.89	68.94	69.12	69.47	69.39	**69.63**
DenseNet-BC	70.61	70.85	70.7	70.92	71.14	71.27	72.1	71.13	**72.52**
DR = 10%									
WRN	83.64	83.82	84.05	84.11	83.94	83.77	83.99	84.15	**84.3**
DenseNet-BC	84.07	84.21	84.27	84.45	84.07	84.31	84.34	**84.63**	84.52
DR = 20%									
WRN	90.52	90.38	90.3	90.27	90.71	90.45	91.02	**91.19**	90.95
DenseNet-BC	91.24	91.37	91.33	91.30	91.39	91.38	91.4	91.31	**91.47**
DR = 100%									
WRN	94.04	94.23	94.44	94.42	94.52	94.52	95.29	95.17	**95.49**
DenseNet-BC	95.22	95.28	95.34	95.27	95.37	95.63	95.72	95.74	**95.88**

The results show that for all datasets and in all experimental cases using LCL outperformed all baselines, with SeNet with DR = 10% and top-5 metric being the only exception. The improvement was more pronounced when DR < 100%. It is quite intuitive that a curriculum is much more important when the training data is limited (i.e., the learning problem is more difficult). Loosely speaking, the importance of a teacher decreases when a student has access to unlimited information without any computational and/or time budget. For example, for

Table 7. AWA2. Top-1 results, averaged over four trials.

	SL	LS	DML$_1$	DML$_2$	KD	CN	ϵ		
							0.9	0.99	0.999
DR = 5%									
ResNet-101	23.09	27.82	39.22	37.19	41.58	24.1	45.51	45.55	**45.78**
InceptionResNetV2	57.42	57.95	58.69	58.14	59.3	56.9	60.85	**61.07**	60.71
DR = 10%									
ResNet-101	41.86	44.98	48.92	50.5	44.02	43.12	47.21	51.67	**53.39**
InceptionResNetV2	71.47	71.86	71.82	72.37	71.49	72.01	72.61	72.97	**73.01**
DR = 20%									
ResNet-101	77.11	78.23	78.34	78.32	78.28	77.64	80.03	**80.07**	79.86
InceptionResNetV2	83.64	83.92	83.87	83.76	83.83	84.12	**84.27**	84.05	**84.27**
DR = 100%									
ResNet-101	88.73	89.25	89.01	89.11	89.17	88.92	89.44	**89.64**	89.63
InceptionResNetV2	89.69	89.94	90.05	90.22	89.94	89.29	**90.49**	90.34	90.47

Table 8. AWA2. Top-5 results, averaged over four trials.

	SL	LS	DML$_1$	DML$_2$	KD	CN	ϵ		
							0.9	0.99	0.999
DR = 5%									
ResNet-101	53.31	54.19	65.14	63.12	56.19	54.17	76.02	76.14	**76.47**
InceptionResNetV2	83.06	83.14	84.07	84.18	84.12	83.77	84.94	**85.24**	84.84
DR = 10%									
ResNet-101	72.59	72.43	75.07	76.14	76.61	76.34	77.04	**80.46**	80.11
InceptionResNetV2	91.37	91.42	91.35	91.43	91.48	91.35	**91.9**	91.89	91.71
DR = 20%									
ResNet-101	94.21	94.56	94.79	95.01	94.61	94.45	**95.2**	95.07	95.12
InceptionResNetV2	96.03	96.23	96.28	96.13	96.21	95.19	96.18	96.49	**96.57**
DR = 100%									
ResNet-101	97.85	97.92	98.1	97.95	97.43	97.32	98.11	**98.14**	98.1
InceptionResNetV2	98.01	98.07	98.25	98.17	97.67	97.56	98.25	**98.41**	98.2

ResNet-50 on ImageNet LCL improved the top-1 accuracy on average by 4 percentage points (p.p.) over the baseline when DR = 5%, and 2 p.p. in top-5 accuracy were gained with DR = 100% on ImageNet for the ResNeXt architecture. The biggest improvements were achieved on the AWA2 dataset. For ResNet-101 and DR = 5%, average improvements of more than 22 p.p. and 23 p.p. could be achieved in the top-1 and top-5 accuracy, respectively. As could be expected, the performance gains in the top-5 setting were typically smaller than for top-1. Still, nothing changed with respect to the ranking of the network architectures.

Larger values of ϵ mean slower convergence to the one-hot-encoding and therefore more emphasis on the curriculum learning. In most experiments, $\epsilon = 0.999$ performed best. The observation that larger ϵ values gave better results than small ones provides additional evidence that the curriculum really supports the learning process (and that we are not discussing random artifacts).

Under the assumption that the experimental scenarios are statistically independent, we performed a statistical comparisons of the classifiers over multiple data sets for all pairwise comparisons following [3, 8]. Using the Iman and Daveport test, all but one result were statistically significant. If we consider all ImageNet top-1 accuracy results, our method with $\epsilon = 0.999$ ranked best, followed by $\epsilon = 0.99$, $\epsilon = 0.9$, LS and then SL. This ranking was highly significant (Iman and Daveport test, $p < 0.001$). Similarly, our method with $\epsilon = 0.999$ was best for both CIFAR-100 and AWA2 ($p < 0.001$).

6 Conclusions

We proposed a novel curriculum learning approach referred to as label-similarity curriculum learning. In contrast to previous methods, which change the weighting of training examples, it is based on adapting the label representation during training. This adaptation considers the semantic similarity of labels. It implements the basic idea that at an early stage of learning it is less important to distinguish between similar classes compared to separating very different classes. The class similarity can be based on arbitrary *a priori* knowledge, in particular on additional information not directly encoded in the training data. For the case where the class labels are natural language words, we proposed a way to automatically define class similarity via a word embedding. We also considered other similarity measures for datasets where these similarity measures were available.

We extensively evaluated the approach on five datasets. For each dataset, two to three deep learning architectures proposed in the literature were considered. We looked at simple label smoothing and, for the two smaller datasets, also at deep mutual learning (DML) as additional baselines. In each case, we considered four different training data set sizes. Each experiment was repeated four times. The empirical results strongly support our approach. *Label-similarity curriculum learning was able to improve the average classification accuracy on the test data compared to standard training in all scenarios.* The improvements achieved by our method were more pronounced for smaller training data sets. When considering only 10% of the AWA2 training data, label-similarity curriculum learning increased the Resnet101 top-1 test accuracy by more than 22% points on average compared to the standard baseline. Our curriculum learning also outperformed simple label smoothing and DML in all but a single case. Our method turned out to be robust with respect to the choice of the single hyperparameter controlling how quickly the learning process converges to minimizing the standard cross-entropy loss. In contrast to related approaches such as knowledge distillation and DML, the additional computational and memory requirements can be neglected.

The proposed label-similarity curriculum learning is a general approach, which also works for settings where the class similarity is not based on the semantic similarity of natural language words (see supplementary material).

References

1. Bengio, Y., Louradour, J., Collobert, R., Weston, J.: Curriculum learning. In: International Conference on Machine Learning (ICML), pp. 41–48. ACM (2009)
2. Buciluă, C., Caruana, R., Niculescu-Mizil, A.: Model compression. In: Proceedings of the 12th ACM SIGKDD International Conference on Knowledge Discovery and Data Mining (KDD), pp. 535–541. ACM (2006)
3. Demšar, J.: Statistical comparisons of classifiers over multiple data sets. J. Mach. Learn. Res. **7**, 1–30 (2006)
4. Deng, J., Dong, W., Socher, R., Li, L.J., Li, K., Fei-Fei, L.: ImageNet: a large-scale hierarchical image database. In: Proceedings of the IEEE Conference on Computer Vision and Pattern Recognition (CVPR), pp. 248–255. IEEE (2009)
5. Florensa, C., Held, D., Wulfmeier, M., Zhang, M., Abbeel, P.: Reverse curriculum generation for reinforcement learning. In: Conference on Robot Learning (CoRL), pp. 482–495 (2017)
6. Frome, A., et al.: Devise: a deep visual-semantic embedding model. In: Advances in Neural Information Processing Systems (NeurIPS), pp. 2121–2129 (2013)
7. Furlanello, T., Lipton, Z.C., Tschannen, M., Itti, L., Anandkumar, A.: Born again neural networks (2018). arXiv:1805.04770 [stat.ML]
8. García, S., Herrera, F.: An extension on statistical "comparisons of classifiers over multiple data sets" for all pairwise comparisons. J. Mach. Learn. Res. **9**, 2677–2694 (2008)
9. Goyal, P., et al.: Accurate, large minibatch SGD: Training ImageNet in 1 hour (2017). arXiv:1706.02677v2 [cs.CV]
10. Graves, A., Bellemare, M.G., Menick, J., Munos, R., Kavukcuoglu, K.: Automated curriculum learning for neural networks. In: International Conference on Machine Learning (ICML), pp. 1311–1320. PMLR (2017)
11. Guo, S., et al.: CurriculumNet: weakly supervised learning from large-scale web images. In: Ferrari, V., Hebert, M., Sminchisescu, C., Weiss, Y. (eds.) ECCV 2018. LNCS, vol. 11214, pp. 139–154. Springer, Cham (2018). https://doi.org/10.1007/978-3-030-01249-6_9
12. He, K., Zhang, X., Ren, S., Sun, J.: Deep residual learning for image recognition. In: Proceedings of the IEEE Conference on Computer Vision and Pattern Recognition (CVPR), pp. 770–778. IEEE (2016)
13. Hinton, G., Vinyals, O., Dean, J.: Distilling the knowledge in a neural network. In: Deep Learning and Representation Learning Workshop: NIPS 2014 (2014)
14. Hu, J., Shen, L., Sun, G.: Squeeze-and-excitation networks. In: Proceedings of the IEEE Conference on Computer Vision and Pattern Recognition (CVPR), pp. 7132–7141. IEEE (2018)
15. Huang, G., Liu, Z., Van Der Maaten, L., Weinberger, K.Q.: Densely connected convolutional networks. In: Proceedings of the IEEE Conference on Computer Vision and Pattern Recognition (CVPR), pp. 4700–4708. IEEE (2017)

16. Jeh, G., Widom, J.: SimRank: a measure of structural-context similarity. In: Proceedings of the Eighth ACM SIGKDD International Conference on Knowledge Discovery and Data Mining, pp. 538–543 (2002)
17. Krizhevsky, A.: Learning multiple layers of features from tiny images. Technical report, University of Toronto (2009)
18. Kumar, M.P., Packer, B., Koller, D.: Self-paced learning for latent variable models. In: Advances in Neural Information Processing Systems (NeurIPS), pp. 1189–1197 (2010)
19. Le, Q., Mikolov, T.: Distributed representations of sentences and documents. In: International Conference on Machine Learning (ICML), pp. 1188–1196 (2014)
20. Mikolov, T., Chen, K., Corrado, G., Dean, J.: Efficient estimation of word representations in vector space (2013). arXiv:1301.3781v3 [cs.CL]
21. Müller, R., Kornblith, S., Hinton, G.: When does label smoothing help? (2019). arXiv:1906.02629 [cs.LG]
22. Orbes-Arteainst, M., et al.: Knowledge distillation for semi-supervised domain adaptation. In: Zhou, L., et al. (eds.) OR 2.0/MLCN -2019. LNCS, vol. 11796, pp. 68–76. Springer, Cham (2019). https://doi.org/10.1007/978-3-030-32695-1_8
23. Papernot, N., McDaniel, P., Wu, X., Jha, S., Swami, A.: Distillation as a defense to adversarial perturbations against deep neural networks. In: 2016 IEEE Symposium on Security and Privacy (SP), pp. 582–597. IEEE (2016)
24. Paszke, A., et al.: Automatic differentiation in PyTorch. In: NeurIPS 2017 Workshop Autodiff (2017)
25. Romero, A., Ballas, N., Kahou, S.E., Chassang, A., Gatta, C., Bengio, Y.: FitNets: hints for thin deep nets (2014). arXiv:1412.6550 [cs.LG]
26. Sergeev, A., Del Balso, M.: Horovod: fast and easy distributed deep learning in TensorFlow (2018). arXiv:1802.05799v3 [cs.LG]
27. Silla, C.N., Freitas, A.A.: A survey of hierarchical classification across different application domains. Data Min. Knowl. Disc. **22**(1), 31–72 (2011). https://doi.org/10.1007/s10618-010-0175-9
28. Szegedy, C., Ioffe, S., Vanhoucke, V., Alemi, A.A.: Inception-v4, Inception-ResNet and the impact of residual connections on learning. In: Thirty-First AAAI Conference on Artificial Intelligence (AAAI), pp. 4278–4284 (2017)
20. Van Horn, G., et al.: Building a bird recognition app and large scale dataset with citizen scientists: the fine print in fine-grained dataset collection. In: Proceedings of the IEEE Conference on Computer Vision and Pattern Recognition (CVPR) (2015)
30. Wah, C., Branson, S., Welinder, P., Perona, P., Belongie, S.: The Caltech-UCSD Birds-200-2011 dataset. Technical report. CNS-TR-2011-001, California Institute of Technology (2011)
31. Weinshall, D., Cohen, G., Amir, D.: Curriculum learning by transfer learning: theory and experiments with deep networks. In: International Conference on Machine Learning (ICML), pp. 5235–5243. PMLR (2018)
32. Wu, C., Tygert, M., LeCun, Y.: Hierarchical loss for classification (2017). arXiv:1709.01062v1 [cs.LG]
33. Xian, Y., Lampert, C.H., Schiele, B., Akata, Z.: Zero-shot learning-a comprehensive evaluation of the good, the bad and the ugly. IEEE Trans. Pattern Anal. Mach. Intell. **41**, 2251–2265 (2018)
34. Xie, S., Girshick, R., Dollár, P., Tu, Z., He, K.: Aggregated residual transformations for deep neural networks. In: Proceedings of the IEEE Conference on Computer Vision and Pattern Recognition (CVPR), pp. 1492–1500. IEEE (2017)

35. Yim, J., Joo, D., Bae, J., Kim, J.: A gift from knowledge distillation: fast optimization, network minimization and transfer learning. In: Proceedings of the IEEE Conference on Computer Vision and Pattern Recognition (CVPR), pp. 4133–4141 (2017)
36. Zagoruyko, S., Komodakis, N.: Paying more attention to attention: improving the performance of convolutional neural networks via attention transfer (2016). arXiv:1612.03928 [cs.CV]
37. Zagoruyko, S., Komodakis, N.: Wide residual networks. In: British Machine Vision Conference (BMVC). BMVA Press (2016)
38. Zhang, Y., Xiang, T., Hospedales, T.M., Lu, H.: Deep mutual learning. In: Proceedings of the IEEE Conference on Computer Vision and Pattern Recognition (CVPR), pp. 4320–4328 (2018)

Recurrent Image Annotation with Explicit Inter-label Dependencies

Ayushi Dutta[1]([✉]) [iD], Yashaswi Verma[2][iD], and C. V. Jawahar[3][iD]

[1] Target Corporation India Private Limited, Bangalore, India
ayushi.dutta@target.com
[2] Indian Institute of Technology, Jodhpur, India
yashaswi@iitj.ac.in
[3] IIIT Hyderabad, Hyderabad, India
jawahar@iiit.ac.in

Abstract. Inspired by the success of the CNN-RNN framework in the image captioning task, several works have explored this in multi-label image annotation with the hope that the RNN followed by a CNN would encode inter-label dependencies better than using a CNN alone. To do so, for each training sample, the earlier methods converted the ground-truth label-set into a sequence of labels based on their frequencies (e.g., rare-to-frequent) for training the RNN. However, since the ground-truth is an unordered *set* of labels, imposing a fixed and predefined sequence on them does not naturally align with this task. To address this, some of the recent papers have proposed techniques that are capable to train the RNN without feeding the ground-truth labels in a particular sequence/order. However, most of these techniques leave it to the RNN to implicitly choose one sequence for the ground-truth labels corresponding to each sample at the time of training, thus making it inherently biased. In this paper, we address this limitation and propose a novel approach in which the RNN is explicitly forced to learn multiple relevant inter-label dependencies, without the need of feeding the ground-truth in any particular order. Using thorough empirical comparisons, we demonstrate that our approach outperforms several state-of-the-art techniques on two popular datasets (MS-COCO and NUS-WIDE). Additionally, it provides a new perspecitve of looking at an unordered set of labels as equivalent to a collection of different permutations (sequences) of those labels, thus naturally aligning with the image annotation task. Our code is available at: https://github.com/ayushidutta/multi-order-rnn.

Keywords: Image annotation · Multi-label learning · CNN-RNN framework · Inter-label dependencies · Order-free training

---

A. Dutta—The author did most of this work while she was a student at IIIT Hyderabad, India.

© Springer Nature Switzerland AG 2020
A. Vedaldi et al. (Eds.): ECCV 2020, LNCS 12374, pp. 191–207, 2020.
https://doi.org/10.1007/978-3-030-58526-6_12

1 Introduction

Multi-label image annotation is a fundamental problem in computer vision and machine learning, with applications in image retrieval [7,37,53], scene recognition [1], object recognition [47], image captioning [8], etc. In the last few years, deep Convolution Neural Networks (CNNs) such as [15,22,39,40] have been shown to achieve great success in the single-label image classification task [38], which aims at assigning *one* label (or category) to an image from a fixed vocabulary. However, in the multi-label image annotation task, each image is associated with an unordered *subset* of labels from a vocabulary that corresponds to different visual concepts present in that image, such as objects (e.g., *shirt*), attributes (e.g., *green*), scene (e.g., *outdoor*), and other visual entities (e.g., *pavement, sky*, etc.). Further, these labels share rich semantic relationships among them (e.g., *forest* is related to *green*, *ferrari* is related to *car*, etc.), thus making it much more challenging than single-label classification.

To model inter-label dependencies, existing works have used a variety of techniques, such as nearest-neighbours based models [12,33,45], ranking-based models [2,14] probabilistic graphical models [27,28], structured inference models [17,23,47,48], and models comprising of a Recurrent Neural Network (RNN) following a CNN [4,18,31,48,48] (also referred to as CNN-RNN framework). Among these, CNN-RNN based models have received increasing attention in the recent years [4,5,17,18,26,48,50], particularly due to the capability of an RNN to capture higher-order inter-label relationships while keeping the computational complexity tractable. The earlier models in this direction were motivated by the success of the CNN-RNN framework in the image captioning task [21,46]. Analogous to the sequence/order of words in a caption, these models proposed to train the RNN for the image annotation task by imposing a fixed and predefined order on the labels based on their frequencies in the training data (e.g., frequent-to-rare or rare-to-frequent). In [18], Jin and Nakayama showed that the order of labels in the training phase had an impact on the annotation performance, and found that rare-to-frequent order worked the best, which was further validated in the subsequent papers such as [31,48]. However, such an ordering introduces a hard constraint on the RNN model. E.g., if we impose rare-to-frequent label order, the model would be forced to learn to identify the rare labels first, which is difficult since these labels have very few training examples. Further, in RNN, since the future labels are predicted based on the previously predicted ones, any error in the initial predictions would increase the likelihood of errors in the subsequent predictions. Similarly, if we impose frequent-to-rare label order, the model would get biased towards frequent labels and would have to make several correct predictions before predicting the correct rare label(s). In general, any frequency-based predefined label order does not reflect the true inter-label dependencies since when an image has multiple labels, each label is related to many other labels with respect to the global context of that image, though spatially a label may relate more strongly to only a few of them. Additionally, defining such an order makes the model biased towards the dataset-specific statistics.

To address these limitations, some recent papers [4,5,26,50] have proposed techniques that do not require to feed the ground-truth labels to the RNN in any particular sequence. However, these techniques allow the RNN to implicitly choose one out of many possible sequences, which in turn makes it inherently biased. In this paper, we address this limitation using a novel approach in which the RNN is explicitly forced to learn multiple relevant inter-label dependencies in the form of multiple label orders instead of a fixed and predefined one. Specifically, at any given time-step, we train the model to predict all the correct labels except the one it has selected as the most probable one in the previous time-step. During testing, we max-pool the prediction scores for each label across all the time-steps, and then pick the labels with scores above a threshold. In this way, the best prediction of a label is obtained from its individual prediction path. Additionally, allowing the model to learn and predict from multiple label paths also provides the advantage that in reality there may be more than one sequences that reflect appropriate inter-label dependencies. As one could observe, the proposed idea is closely related to the well-known Viterbi algorithm, and provides a new perspecitve of looking at an unordered set of labels as equivalent to a collection of different permutations of those labels, thus naturally aligning the inherent capability of an RNN (i.e., sequence prediction) with the objective of the image annotation task (i.e., unordered subset prediction). In our experiments on two large-scale multi-label image annotation datasets, we demonstrate that the proposed approach outperforms competing baselines and several state-of-the-art image annotation techniques.

2   Related Work

Multi-label image annotation has been an active area of research from the last two decades. The initial works such as [2,3,9,12,24,33,43,44,47] relied on hand-crafted local [3,9,24] and global [2,12,33,43,44] features, and explored a variety of techniques such as joint [9,24] and conditional [3] probabilistic models, nearest-neighbours based models [12,33,43], structured inference models [23] and ranking-based models [2,44].

With the advent of the deep learning era, most of the initial attempts were based on integrating the existing approaches with the powerful features made available by pre-trained deep CNN models. In [11], the authors used a deep CNN model pre-trained on the ImageNet dataset, and fine-tuned it for multi-label image annotation datasets using different loss functions such as softmax, pairwise ranking [19] and WARP [49]. Similarly, other works such as [35,42,45] revisited some of the state-of-the-art methods from the pre-deep-learning era, and re-evaluated them using the features extracted from the last fully-connected hidden layer of a pre-trained deep CNN model. Moving further on the similar ideas, Li et al. [29] introduced a smooth variant of the hinge-loss (called log-sum-exp pairwise loss, or LSEP loss) especially useful for the multi-label prediction task, and showed it to perform better than the previously known loss functions.

In parallel, there have also been attempts to explicitly model inter-label dependencies prominent in this task using end-to-end deep learning based techniques. One of the early attempts was by Andrea *et al.* [10] who proposed to learn a joint embedding space for images and labels using a deep neural network, thus allowing direct matching between visual (images) and textual (labels) samples in the learned common space, similar to [49]. To capture the underlying relationships between labels in a deep CNN framework, Feng *et al.* [52] proposed a spatial regularization network with learnable convolution filters and attention maps for individual labels, by making use of spatial features from the last convolution layer. On similar lines, several other works such as [13,32] have also explored the utility of local features and spatial attention. Apart from these, some of the works have also explored techniques such as deep metric learning for multi-label prediction [25], multi-modal learning [36] similar to [10], and Generative Adversarial Networks [41]. It is worth noting that while these approaches remained confined to the available training data, some of the works have demonstrated the advantage of using contextual knowledge coming from external sources. In [20], Johnson *et al.* used a non-parametric approach to find nearest neighbours of an image based on textual meta-data, and then aggregated visual information of an image and with its neighbours using a neural network to improve label prediction. In [17], Hu *et al.* proposed a deep structured neural network that consisted of multiple concept-layers based on the WordNet [34] hierarchy, and trained it to capture inter-label relationships across those layers.

Another class of algorithms that has become popular in the recent past is based on the CNN-RNN framework, that is motivated by the ability of an RNN to model complex inter-label relationships, and at the same time it offers a simple and scalable solution. The earlier attempts were simple adaptations of the CNN-RNN based encoder-decoder models proposed for the image captioning task [46]. These approaches treated multi-label prediction as a sequence prediction problem, where the RNN was trained to predict the labels for a given image in a sequential manner, analogous to predicting a caption [18,31,48]. As discussed above, such approaches required a predefined order among the labels at the time of training, and thus constrained the model to predict the labels in that order. Since this does not naturally align with the objective of the image annotation task, some of the recent approaches have proposed order-free techniques that do not require to feed the ground-truth labels to the RNN in any particular order at the time of training. The Order-free RNN model proposed by Shang *et al.* [4] was the first such model that used the concept of "candidate label pool". This pool initially contains all the true labels, and then at each time-step, the most confident label from this pool is used for feedback to the RNN and at the same time removed from this pool. Similar ideas have been proposed in the subsequent works such as [5,26,50]. However, these approaches are prone to internally choosing one particular order of labels at the initial time-step, and then iterating over the same sequence in the subsequent time-steps. To address this limitation, we propose a novel approach that forces the RNN model to predict all the correct labels at every time-step, except the one predicted in the previous time-step. As

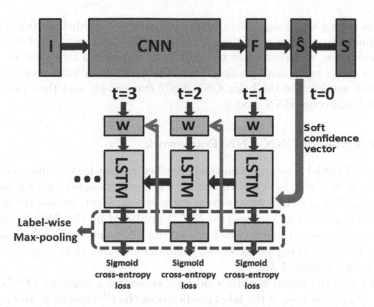

Fig. 1. Overview of the proposed approach. The first component of our model is a deep CNN that that is fine-tuned using the ground-truth (S) from a given dataset. The second component is an LSTM model that uses the soft confidence vector (Ŝ) from the CNN as its initial state. Given a sample, at every time-step, a cross-entropy loss is computed considering all the true labels except the one from the previous time-step as possible candidates for prediction at that time-step. Final predictions are obtained by max-pooling individual label scores across all the time-steps

we will show later in Sect. 3.3, this drives the RNN to learn complex inter-label dependencies in the form of multiple sequences among the labels arising from a given label-set, and thus we call it *Multi-order RNN*. The utility of our approach is also demonstrated in the empirical analysis in which it is shown to outperform all the existing CNN-RNN-based image annotation techniques.

3   Approach

Our CNN-RNN framework consists of two components: (i) a deep CNN model that provides a real-valued vectorial representation of an input image, and (ii) an RNN that models image-label and label-label relationships. Let there be N training images $\{I_1, \ldots, I_N\}$, such that each image I is associated with a ground-truth label vector $y = (y^1, y^2, \ldots, y^C)$, where y^c ($\forall c \in \{1, 2, \ldots, C\}$) is 1 if the image I has the cth label in its ground-truth and 0 otherwise, with C being the total number of labels. Also, let $\hat{y}$ denote the vector containing the scores predicted by the RNN corresponding to all the labels at some time-step.

During the training phase, for a given image I, the representation obtained by the CNN is initially fed to the RNN. Based on this, the RNN predicts a label score vector $\hat{y}_t$ at each time-step t in a sequential manner, and generates a prediction path $\pi = (a_1, a_2, ..., a_T)$, where a_t denotes the label corresponding to the maximum score from $\hat{y}_t$ at time t, and T is the total number of time-steps. During inference, we accumulate the scores for individual labels across all the prediction paths using max-pooling, and obtain the final label scores (Fig. 1). Below, first we describe the basic CNN-RNN framework, and then present the proposed Multi-order RNN model.

3.1   Background: CNN-RNN Framework

Given a CNN model pre-trained on the ImageNet dataset for the image classification task, the first step is to fine-tune it for a given multi-label image annotation dataset using the standard binary cross-entropy loss. Next, we use the soft confidence (label probability) scores ($\hat{s} \in \mathbb{R}_+^{C \times 1}$) predicted by the CNN for an input image, and use this as the interface between the CNN and the RNN. This allows to decouple the learning of these two components, and thus helps in a more efficient joint training [4,31].

Taking $\hat{s}$ as the input, the RNN decoder generates a sequence of labels $\pi = (a_1, a_2, ..., a_{n_I})$, where a_t is the label predicted at the tth time-step, and n_I is the total number of labels predicted. Analogous to the contemporary approaches, we use the Long Short-Term Memory (LSTM) [16] network as the RNN decoder, which controls message passing between time-steps with specialized gates. At time step t, the model uses its last prediction a_{t-1} as the input, and computes a distribution over the possible outputs:

$$x_t = E \cdot a_{t-1} \tag{1}$$

$$h_t = LSTM(x_t, h_{t-1}, c_{t-1}) \tag{2}$$

$$\hat{y}_t = W \cdot h_t + b \tag{3}$$

where E is the label embedding matrix, W and b are the weight and bias of the output layer, a_{t-1} denotes the *one-hot encoding* of the last prediction, c_t and h_t are the model's cell and hidden states respectively at time t, and $LSTM(\cdot)$ is a forward step of the unit. The output vector $\hat{y}_t$ defines the output scores at t, from which the next label a_t is sampled.

3.2   Multi-order RNN

As introduced earlier, let the ground-truth (binary) label vector of an image I be denoted by $y = (y^1, y^2, \ldots, y^C)$. Also, let y_t denote the ground-truth label vector at time t, and $\hat{y}_t$ be the corresponding predicted label score vector. In practice, since the ground-truth y_t at time-step t is unknown, one could assume that y_t at each time-step is the original label vector y. This would force the model to assign high scores to all the ground-truth labels instead of one particular ground-truth label at each time-step. E.g., let us assume that an image has the labels {*sky, clouds, person*} in its ground-truth, then the model would be forced to predict (i.e., assign high prediction scores to) all the three labels {*sky, clouds, person*} at each time-step based on the most confident label predicted in previous time-step. However, this poses the problem that if the most confident label predicted at time-step t is l, the model may end-up learning a dependency from l to l in the next time-step along with the dependencies from l to other labels. In other words, there is a high chance that a label which is easiest to predict would be the most confident prediction by RNN at every time-step, and thus the same label would then be repeatedly chosen for feedback to the RNN. To address this, we use a greedy approach that forces the model to explicitly learn to predict a different label. Specifically, if l is the most confident prediction at time-step t, we mask out l in the next time-step; i.e., in the next time-step, we treat l as a negative label rather than positive and learn a dependency from l to all other labels except itself. We explain this mathematically below.

Let l_t be the most confident label with the highest prediction score for an image I at time-step t:

$$l_t = \underset{c \in \{1,2,...,C\}}{\arg\max} \ \hat{y}_t^c \tag{4}$$

Let a_{t-1} be the one-hot encoding corresponding to l_{t-1}. Then we define a label mask $\tilde{a}_t$ at time-step t as:

$$\tilde{a}_t = \neg a_{t-1} \tag{5}$$

In other words, this label mask is a negation of the one-hot encoding of the most confident label from the previous time-step. The mask contains a 0 corresponding to the previously selected label index, and 1 for the rest. Using this, we define a modified ground-truth label vector at time-step t as:

$$y_t = \tilde{a}_t \odot y \tag{6}$$

where $\odot$ represents element-wise multiplication. At time-step $t = 0$, $\tilde{a}_0$ will be a vector with all ones. Using this modified ground-label vector and the predicted label scores at a particular time-step t, we compute the sigmoid cross-entropy loss at that time-step as:

$$\mathcal{L}_t = y_t \cdot log(\sigma(\hat{y}_t)) + (1 - y_t) \cdot log(1 - \sigma(\hat{y}_t)) \tag{7}$$

The above loss is aggregated over all the time-steps and summed over all the training samples to obtain the total loss. This loss is then used to train our model using a gradient descent approach.

Label Prediction. Once the model is trained, for a given test image, first we obtain the soft confidence (probability) scores from the CNN and initiate the LSTM using them. Then, the LSTM network is iterated for T time-steps, resulting in T prediction score vectors $(\hat{y}_1, \hat{y}_2, ..., \hat{y}_T)$, where each $\hat{y}_t = (\hat{y}_t^1, \hat{y}_t^2, ..., \hat{y}_t^C)$ denotes the scores for all the C labels at times-step t. We employ label-wise max-pooling to integrate the scores across all the time-steps into the final result $\hat{y} = (\hat{y}^1, \hat{y}^2, ..., \hat{y}^C)$, where: $\hat{y}^c = max(\hat{y}_1^c, \hat{y}_2^c, ..., \hat{y}_T^c)$, $\forall c = 1, ..., C$. The final predicted label probability distribution $\hat{p}$ is obtained as $\hat{p}_i = \sigma(\hat{y})$. Since we use the sigmoid function, finally we assign all those labels whose probability scores

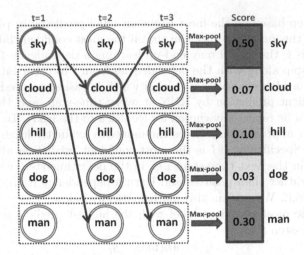

Fig. 2. An example of multiple inter-label dependencies that can be learned using the proposed approach. Please see Sect. 3.3 for details.

are greater than 0.5. We carry out the LSTM iterations for a fixed number of 'T' time-steps which is determined experimentally. Interestingly, unlike methods that do a predefined label order based training and sequential label prediction, our model does not require to predict an "end-of-sequence" ($< EOS >$) token.

3.3 Discussion

Now we will discuss how the proposed approach learns multiple relevant inter-label dependencies (and not all possible permutations, since many of them will not be meaningful). Algorithmically, in the first time-step, the LSTM is trained to predict all the true labels by the loss $\mathcal{L}_1$, analogous to a CNN that can be trained to predict all the true labels by a single look at an image. As the LSTM strives to predict all the labels, the most confident label predicted by it is given as the feedback for the next time-step. In the second time-step, the model is trained to predict all the true labels except the most confident label predicted by it in the previous time-step, and this continues for a fixed number of time-steps.

Let us try to understand this through an example illustrated in Fig. 2. Let {*sky, cloud, hill, dog, man*} be the complete label-set (vocabulary), and let {*sky, cloud, man*} be the ground-truth set of labels for a given image. **During training**, at $t = 1$, the model is forced to predict all the positive labels correctly. Suppose it selects *sky* as the most confident label. Then at $t = 2$, the model is trained to learn dependencies from *sky* to both *cloud* and *man*, while keeping *sky* as a negative label. Suppose the model now selects *cloud* as the label with the highest confidence. Then at $t = 3$, the model will be trained to predict both *sky* and *man*. In this way, the model explicitly learns multiple dependency paths *sky* $\rightarrow$ *cloud* $\rightarrow$ *sky* and *sky* $\rightarrow$ *cloud* $\rightarrow$ *man*. In other words, it learns not only to go back to *sky* from *cloud*, but also learns to predict *man* based on the confidence that the image already has {*sky, cloud*}. In this way, the label correlations that are hard to learn initially get learned at later time-steps based on their dependencies on other labels in the ground-truth. **During testing**, given an input image, the most confident label at each time-step is fed as input to the LSTM, and this is repeated for a fixed number of time-steps. At the end, the prediction scores across all the time-steps are label-wise max-pooled, and the labels above a threshold are assigned.

The proposed training approach is a greedy approach that is self-guided by the training procedure. Let $\{l_1, l_2, l_3, l_4, l_5\}$ be the complete set of labels, and let $\{l_1, l_3, l_5\}$ be the true labels for a given image. If the model is most confident at predicting l_1 at the first time-step, we train it to predict $\{l_3, l_5\}$ in the next time-step. If the model had predicted l_3 first, the training at the next time-step would have been for $\{l_1, l_5\}$. In case the model predicts an incorrect label, say l_2, as the most confident prediction, it will be given as the feedback to the LSTM for the next time-step. However, since we penalize all the true negative labels (not present in the ground-truth) at every time-step, the model learns not to predict them. By self-learning multiple inter-label dependencies, the model inherently tries to learn label correlations. E.g., if l_3 is chosen immediately after l_1 by the model, there is a strong likelihood that $\{l_1, l_3\}$ co-occur often in practice, and it makes more sense to predict l_3 when l_1 is present. In contrast, in a predefined order based training, the label dependencies are learned in some specified order, which may not be an appropriate order in practice. Also, there may not be a single specific order that reflects the label dependencies in an image. Since the proposed approach can learn multiple inter-label dependencies paths, it can mitigate both these issues.

As we can observe, there is a possibility that the sequence of the most confident labels that the LSTM model predicts as it iterates over time-steps is $\{l_1, l_5, l_1, l_5, \ldots\}$. Algorithmically, this implies that l_1 was the most confident label at $t = 1$, l_5 at $t = 2$, l_1 at $t = 3$, l_5 at $t = 4$, and so on. Intuitively, this would mean that at each time-step, while the previous most confident label guides the LSTM model, the model is still forced to learn that label's dependencies with all the remaining correct labels including the (current) most confident one. While this particular behavior is probably not the best one, this would still facilitate the model to encode multiple (even bidirectional) inter-label dependencies unlike any existing CNN-RNN-based method. At this point, if we had maintained a pool of all the true labels predicted until a certain time-step and forced the model to predict a label only from the remaining ones, then this would have resulted in forcing some improbable dependencies among the labels, which is not desirable. We would like to highlight that this is exactly what was proposed in [4], and our algorithm elegantly relaxes this hard constraint imposed on LSTM. As evident from the empirical analyses, this results in achieving significantly better performance than [4].

4 Experiments

4.1 Datasets

We experiment using two popular and large-scale image annotation datasets: MS-COCO [30] and NUS-WIDE [6]. The MS-COCO dataset has been used for various object recognition tasks in the context of natural scene understanding. It contains $82,783$ images in the training set, $40,504$ images in the validation set, and a vocabulary of 80 labels with around 2.9 labels per image. Since the ground-truth labels for the test set are not publicly available, we use the validation set in our comparisons following the earlier papers. The NUS-WIDE dataset contains $269,648$ images downloaded from Flickr. Its vocabulary contains 81 labels, with around 2.4 labels per image. Following earlier papers, we discard the images without any label, that leaves us with $209,347$ images. For empirical comparisons, we split into around $125,449$ images for training and $83,898$ for testing by adopting the split originally provided along with this dataset.

Table 1. Comparison of the proposed approach with a vanilla CNN using binary cross-entropy loss, and CNN-RNN models trained with different label-ordering methods on the MS-COCO dataset

Metric→	Per-label			Per-image		
Method↓	P_L	R_L	$F1_L$	P_I	R_I	$F1_I$
CNN (Binary Cross-Entropy)	59.30	58.60	58.90	61.70	65.00	63.30
CNN-RNN (frequent-first)	70.27	56.49	62.63	72.15	64.53	68.13
CNN-RNN (rare-first)	65.68	61.32	63.43	70.82	64.73	67.64
CNN-RNN (lexicographic order)	70.98	55.86	62.52	74.14	62.35	67.74
Multi-order RNN (Proposed)	**77.09**	**64.32**	**70.13**	**84.90**	**75.83**	**80.11**

4.2 Evaluation Metrics

For empirical analyses, we consider both per-label as well as per-image evaluation metrics. In case of per-label metrics, for a given label, let a_1 be the number of images that contain that label in the ground-truth, a_2 be the number of images that are assigned that label during prediction, and a_3 be the number of images with correct predictions ($a_3 \leq a_2$ and $a_3 \leq a_1$). Then, for that label, precision is given by $\frac{a_3}{a_2}$, and recall by $\frac{a_3}{a_1}$. These scores are computed for all the labels and averaged to get mean per-label precision P_L and mean per-label recall R_L. Finally, the mean per-label F1 score is computed as the harmonic mean of P_L and R_L; i.e., $F1_L = \frac{2 \times P_L \times R_L}{P_L + R_L}$.

In case of per-image metrics, for a given (test) image, let b_1 be the number of labels present in its ground-truth, b_2 be the number of labels assigned during prediction, and b_3 be the number of correctly predicted labels ($b_3 \leq b_2$ and

$b_3 \leq b_1$). Then, for that image, precision is given by $\frac{b_3}{b_2}$, and recall by $\frac{b_3}{b_1}$. These scores are computed for all the test images and averaged to get mean per-image precision P_I and mean per-image recall R_I. Finally, the mean per-image F1 score is computed as the harmonic mean of P_I and R_I; i.e., $F1_I = \frac{2 \times P_I \times R_I}{P_I + R_I}$.

4.3   Implementation Details

We use ResNet-101 [15] pre-trained on the ILSVRC12 1000-class classification dataset [38] as our CNN model, and it is fine-tuned for both NUS-WIDE and MS-COCO datasets separately. For our LSTM (RNN) model, we use 512 cells with the *tanh* activation function and 256-dimensional label-embedding. We train using a batch size of 32 with RMSProp Optimiser and learning-rate of $1e-4$ for 50 epochs, and recurse the LSTM for $T = 5$ time-steps.

Table 2. Comparison of the proposed approach with the state-of-the-art methods on the MS-COCO dataset

Metric→	Per-label			Per-image		
Method↓	P_L	R_L	$F1_L$	P_I	R_I	$F1_I$
CNN based — TagProp [12]	63.11	58.29	60.61	58.17	71.07	63.98
2PKNN [45]	63.77	55.70	59.46	54.13	66.95	59.86
MS-CNN+LQP [36]	67.48	60.93	64.04	70.22	67.93	69.06
WARP [11,49]	57.09	55.31	56.19	57.54	70.03	63.18
LSEP [29]	73.50	56.40	63.82	76.30	61.80	68.29
SRN [52]	85.20	58.80	67.40	87.40	62.50	72.90
S-Cls [32]	–	–	69.20	–	–	74.00
ACfs [13]	77.40	68.30	72.20	79.80	**73.10**	**76.30**
WGAN-gp [41]	70.50	58.70	64.00	72.30	64.60	68.20
RETDM [25]	**79.90**	55.50	65.50	81.90	61.10	70.00
CNN-RNN based — SR-CNN-RNN [31]	67.40	59.83	63.39	76.63	68.73	72.47
Order-free RNN [4]	71.60	54.80	62.10	74.20	62.20	67.70
Recurrent-Attention RL [5]	78.80	57.20	66.20	84.00	61.60	71.10
Attentive RNN [26]	71.90	59.60	65.20	74.30	69.70	71.80
MLA [50]	68.37	60.39	64.13	72.16	66.71	69.33
PLA [50]	70.18	61.96	65.81	73.75	67.74	70.62
Multi-order RNN (Proposed)	77.09	**64.32**	**70.13**	**84.90**	75.83	80.11

4.4   Results and Discussion

Comparison with Baselines. In Table 1, we compare the results of various baselines with our approach, including a CNN model trained with the standard binary-cross entropy loss, and three CNN-RNN style models trained using different schemes for ordering labels at the time of training the RNN module. Here, we observe that the CNN-RNN style models outperform the CNN model, indicating the advantage of using an RNN. We also notice that among the three baseline CNN-RNN models, CNN-RNN (frequent-first) outperforms others with the per-image metrics, and CNN-RNN (rare-first) outperforms others with the per-label metrics. This is expected since the rare-first order assigns more importance to less frequent labels, thus forcing the model to learn to predict them first.

In general, the proposed Multi-order RNN technique consistently outperforms all the baselines by a large margin, providing an improvement of 10.6% in terms of $F1_L$ and 17.6% in terms of $F1_I$ compared to the best performing baseline, thus validating the need for automatically identifying and learning with multiple label orders as being done in the proposed approach, rather than using a fixed and predefined one as in the baselines.

Table 3. Comparison of the proposed approach with the state-of-the-art methods on the NUS-WIDE dataset

	Metric→	Per-label			Per-image		
	Method↓	P_L	R_L	$F1_L$	P_I	R_I	$F1_I$
CNN based	TagProp [12]	49.45	59.13	53.86	52.37	**74.21**	61.41
	2PKNN [45]	47.94	**55.76**	51.55	51.90	73.00	60.67
	LSEP [29]	66.70	45.90	54.38	76.80	65.70	70.82
	WARP [11,49]	44.74	52.44	48.28	53.81	75.48	62.83
	CMA [51]	–	–	55.50	–	–	70.00
	MS-CMA [51]	–	–	55.70	–	–	69.50
	WGAN-gp [41]	**62.40**	50.50	55.80	71.40	70.90	71.20
CNN-RNN based	SR-CNN-RNN [31]	55.65	50.17	52.77	70.57	71.35	70.96
	Order-free RNN [4]	59.40	50.70	54.70	69.00	71.40	70.20
	Attentive RNN [26]	44.20	49.30	46.60	53.90	68.70	60.40
	PLA [50]	60.67	52.40	**56.23**	71.96	72.79	**72.37**
	Multi-order RNN (Proposed)	60.85	54.43	57.46	**76.50**	73.06	74.74

Comparison with the State-of-the-Art. In Table 2 and 3, we compare the performance of the proposed Multi-order RNN with both CNN based as well CNN-RNN based methods. In each column, we highlight the best result in red and the second best result in blue. Among the CNN-RNN based methods, SR-CNN-RNN [31] is the state-of-the-art method that uses rare-to-frequent order of labels for training the RNN, and Order-free RNN [4], Recurrent-Attention RL [5], Attentive RNN [26], MLA [50] and PLA [50] are the methods that do not require to feed the ground-truth labels in any particular order. Among the CNN based methods, TagProp [12] and 2PKNN [45] are state-of-the-art nearest-neighbour

Table 4. Comparison between SR-CNN-RNN [31] and the proposed Multi-order RNN approach based on top-1 accuracy

NUS-WIDE		MS-COCO	
SR-CNN-RNN	Multi-order RNN	SR-CNN-RNN	Multi-order RNN
68.06	84.05	81.44	93.49

based methods that are evaluated using the ResNet-101 features (extracted from the last fully-connected layer), and others are end-to-end trainable models, with WARP [11,49] and LSEP [29] being trained using a pairwise ranking loss. From the results, we can make the following observations: (a) CNN based methods achieve the maximum average precision in all the cases (SRN [52] on the MS-COCO dataset and LSEP [29] on the NUS-WIDE dataset). However, except ACfs [13], all other methods generally fail to manage the trade-off between precision and recall, thus resulting in low F1 scores. (b) On the MS-COCO dataset, the CNN based approaches perform better than the existing CNN-RNN based approaches. The proposed Multi-order RNN approach brings a big jump in the performance of this class of methods, making it either comparable to or better than the former one. (c) Compared to the existing CNN-RNN based approaches, Multi-order RNN not only achieves higher average precision and recall, but also manages the trade-off between the two better than others by achieving an increase in both average precision as well recall, thus achieving the best F1 score in all the cases. (d) In terms of $F1_L$, Multi-order RNN is inferior only to ACfs [13] by 2.07% on the MS-COCO dataset, and outperforms all the methods on the NUS-WIDE dataset with its score being 1.23% more than the second best method. In terms of $F1_I$, Multi-order RNN outperforms all the methods on both the datasets, with its score being 3.81% (on MS-COCO) and 2.37% (on NUS-WIDE) better than the second best methods.

GT: {person, window}
SR-CNN-RNN: <start>, temple, <end>
Multi-order RNN: {temple, person}

t=1	temple(0.69)*	person(0.43)	sky(0.21)	animal(0.02)	buildings(0.02)
t=2	**person***(0.61)	temple(0.21)	sky(0.14)	buildings(0.03)	animal(0.03)
t=3	temple(0.58)	sky(0.09)	statue(0.03)	buildings(0.02)	water(0.01)
t=4	person(0.54)	temple(0.13)	sky(0.05)	buildings(0.05)	animal(0.05)
t=5	temple(0.47)	sky(0.05)	statue(0.03)	buildings(0.03)	plants(0.01)

Fig. 3. Comparison between the prediction of SR-CNN-RNN and Multi-order RNN for an example image. For Multi-order RNN, we show the top five labels along with their probabilities obtained at each time-step. * indicates the max pooled values

Additional Analysis. As discussed before, earlier CNN-RNN based image annotation approaches [18,31,48] had advocated the use of rare-to-frequent label order at the time of training, and SR-CNN-RNN [31] is the state-of-the-art method from this class of methods. Here, we further analyze the performance of our model against SR-CNN-RNN and consider "top-1 accuracy" as the evaluation metric that denotes the percentage of images with the correct top-1 predicted label. This label is obtained by doing one iteration of LSTM at the time of inference for both the methods (note that the training process is unchanged for both the methods, and both of them use ResNet-101 as their CNN model). As we can see in Table 4, the top-1 accuracy of Multi-order RNN is much higher compared to SR-CNN-RNN. This is so because in practice, many of the rare labels correspond to specific concepts that are difficult to learn as well as predict. In such a scenario, if the first predicted label is wrong, it increases the likelihood of the subsequent ones also being wrong. However, our model explicitly learns to choose a salient label based on which it can predict other labels, and thus achieves a higher top-1 accuracy. We further illustrate this in Fig. 3 where for a given image, the SR-CNN-RNN approach predicts only one (incorrect) label *temple*. Interestingly, while Multi-order RNN also predicts *temple* as the most confident label at $t = 1$, it predicts a correct label *person* at $t = 2$ with probability > 0.5, showing its correlation with *temple* learned by the model.

Ground-truth	clouds, sky	clouds, sky, water, beach, buildings	clouds, sky, window	ocean, water, waves	animal, dog
Multi-order RNN	clouds, house, sky	clouds, sky, water	clouds, sky, vehicle, window	animal, ocean, water, waves	animal, dog, sky

Fig. 4. Annotations for example images from the NUS-WIDE dataset. The labels in blue are the ones that match with the ground-truth, and the labels in red are the ones that are depicted in the corresponding images but missing in their ground-truth. (Color figure online)

Finally, we present some qualitative results in Fig. 4. From these results, we can observe that our model correctly predicts most of the ground-truth labels. Moreover, the additional labels that are predicted but missing in the ground-truth are actually depicted in their corresponding images. These results further validate the capability of our model to learn complex inter-label relationships.

5   Summary and Conclusion

While recent CNN-RNN based multi-label image annotation techniques have been successful in training RNN without the need of feeding the ground-truth labels in any particular order, they implicitly leave it to RNN to choose one order for those labels and then force it to learn to predict them in that sequence. To overcome this constraint, we have presented a new approach called Multi-order RNN that provides RNN the flexibility to explore and learn multiple relevant inter-label dependencies on its own. Experiments demonstrate that Multi-order RNN consistently outperforms the existing CNN-RNN based approaches, and also provides an intuitive way of adapting a sequence prediction framework for the image annotation (subset prediction) task.

Acknowledgement. YV would like to thank the Department of Science and and Technology (India) for the INSPIRE Faculty Award 2017.

References

1. Boutell, M.R., Luo, J., Shen, X., Brown, C.M.: Learning multi-label scene classification. Pattern Recogn. **39**(9), 1757–1771 (2004)
2. Bucak, S.S., Jin, R., Jain, A.K.: Multi-label learning with incomplete class assignments. In: CVPR (2011)
3. Carneiro, G., Chan, A.B., Moreno, P.J., Vasconcelos, N.: Supervised learning of semantic classes for image annotation and retrieval. IEEE Trans. Pattern Anal. Mach. Intell. **29**(3), 394–410 (2007)
4. Chen, S.F., Chen, Y.C., Yeh, C.K., Wang, Y.C.F.: Order-free RNN with visual attention for multi-label classification. In: AAAI (2018)
5. Chen, T., Wang, Z., Li, G., Lin, L.: Recurrent attentional reinforcement learning for multi-label image recognition. In: AAAI. pp. 6730–6737 (2018)
6. Chua, T.S, Tang, J., Hong, R., Li, H., Luo, Z., Zheng, Y.: Nus-wide: a real-world web image database from national university of Singapore. In: In CIVR (2009)
7. Escalante, H.J., Hérnadez, C.A., Sucar, L.E., Montes, M.: Late fusion of heterogeneous methods for multimedia image retrieval. In: MIR (2008)
8. Fang, H., et al.: From captions to visual concepts and back. In: Proceedings of the IEEE Conference on Computer Vision and Pattern Recognition, pp. 1473–1482 (2015)
9. Feng, S.L., Manmatha, R., Lavrenko, V.: Multiple Bernoulli relevance models for image and video annotation. In: CVPR (2004)
10. Frome, A., et al.: Devise: a deep visual-semantic embedding model. In: Neural Information Processing Systems (NIPS) (2013)
11. Gong, Y., Jia, Y., Leung, T.K., Toshev, A., Ioffe, S.: Deep convolutional ranking for multilabel image annotation. In: ICLR (2014)
12. Guillaumin, M., Mensink, T., Verbeek, J., Schmid, C.: TagProp: discriminative metric learning in nearest neighbour models for image auto-annotation. In: ICCV (2009)
13. Guo, H., Zheng, K., Fan, X., Yu, H., Wang, S.: Visual attention consistency under image transforms for multi-label image classification. In: CVPR. pp. 729–739 (2019)
14. Hariharan, B., Zelnik-Manor, L., Vishwanathan, S.V.N., Varma, M.: Large scale max-margin multi-label classification with priors. In: ICML (2010)
15. He, K., Zhang, X., Ren, S., Sun, J.: Deep residual learning for image recognition. In: CVPR (2016)

16. Hochreiter, S., Schmidhuber, J.: Long short-term memory. Neural Comput. **9**(9), 1735–1780 (1997). https://doi.org/10.1162/neco.1997.9.8.1735
17. Hu, H., Zhou, G.T., Deng, Z., Liao, Z., Mori, G.: Learning structured inference neural networks with label relations. In: CVPR (2016)
18. Jin, J., Nakayama, H.: Annotation order matters: Recurrent image annotator for arbitrary length image tagging. In: ICPR (2016)
19. Joachims, T.: Optimizing search engines using clickthrough data. In: KDD (2002)
20. Johnson, J., Ballan, L., Fei-Fei, L.: Love thy neighbors: image annotation by exploiting image metadata. In: ICCV (2015)
21. Johnson, J., Karpathy, A., Fei-Fei, L.: Densecap: fully convolutional localization networks for dense captioning. In: CVPR, pp. 4565–4574 (2015)
22. Krizhevsky, A., Sutskever, I., Hinton, G.E.: Imagenet classification with deep convolutional neural networks. Adv. Neural Inf. Process. Syst. **25**, 1097–1105 (2012)
23. Lan, T., Mori, G.: A max-margin riffled independence model for image tag ranking. In: Computer Vision and Pattern Recognition (CVPR) (2013)
24. Lavrenko, V., Manmatha, R., Jeon, J.: A model for learning the semantics of pictures. In: NIPS (2003)
25. Li, C., Liu, C., Duan, L., Gao, P., Zheng, K.: Reconstruction regularized deep metric learning for multi-label image classification. IEEE Trans. Neural Netw. Learn. Syst. **31**(4), 2294–2303 (2019)
26. Li, L., Wang, S., Jiang, S., Huang, Q.: Attentive recurrent neural network for weak-supervised multi-label image classification. In: ACM Multimedia, pp. 1092–1100 (2018)
27. Li, Q., Qiao, M., Bian, W., Tao, D.: Conditional graphical lasso for multi-label image classification. In: CVPR (2016)
28. Li, X., Zhao, F., Guo, Y.: Multi-label image classification with a probabilistic label enhancement model. In: Proceedings Uncertainty in Artificial Intelligence (2014)
29. Li, Y., Song, Y., Luo, J.: Improving pairwise ranking for multi-label image classification. In: CVPR (2017)
30. Lin, T.Y., et al.: Microsoft COCO: common objects in context. In: Fleet, D., Pajdla, T., Schiele, B., Tuytelaars, T. (eds.) ECCV 2014. LNCS, vol. 8693, pp. 740–755. Springer, Cham (2014). https://doi.org/10.1007/978-3-319-10602-1_48
31. Liu, F., Xiang, T., Hospedales, T.M., Yang, W., Sun, C.: Semantic regularisation for recurrent image annotation. In: CVPR (2017)
32. Liu, Y., Sheng, L., Shao, J., Yan, J., Xiang, S., Pan, C.: Multi-label image classification via knowledge distillation from weakly-supervised detection. In: ACM Multimedia, pp. 700–708 (2018)
33. Makadia, A., Pavlovic, V., Kumar, S.: A new baseline for image annotation. In: Forsyth, D., Torr, P., Zisserman, A. (eds.) ECCV 2008. LNCS, vol. 5304, pp. 316–329. Springer, Heidelberg (2008). https://doi.org/10.1007/978-3-540-88690-7_24
34. Miller, G.A.: Wordnet: a lexical database for English. Commun. ACM (CACM) **38**(11), 39–41 (1995)
35. Murthy, V.N., Maji, S., Manmatha, R.: Automatic image annotation using deep learning representations. In: ICMR (2015)
36. Niu, Y., Lu, Z., Wen, J.R., Xiang, T., Chang, S.F.: Multi-modal multi-scale deep learning for large-scale image annotation. IEEE Trans. Image Process. **28**, 1720–1731 (2017)
37. Rasiwasia, N., et al.: A new approach to cross-modal multimedia retrieval. In: ACM MM (2010)
38. Russakovsky, O., et al.: Imagenet large scale visual recognition challenge. Int. J. Comput. Vision **115**(3), 211–252 (2015)

39. Simonyan, K., Zisserman, A.: Very deep convolutional networks for large-scale image recognition. In: ICLR (2015)
40. Szegedy, C., et al.: Going deeper with convolutions. In: CVPR (2015)
41. Tsai, C.P., Lee, Y.H.: Adversarial learning of label dependency: a novel framework for multi-class classification. ICASSP pp. 3847–3851 (2019)
42. Uricchio, T., Ballan, L., Seidenari, L., Bimbo, A.D.: Automatic image annotation via label transfer in the semantic space (2016). CoRR abs/1605.04770
43. Verma, Y., Jawahar, C.V.: Image annotation using metric learning in semantic neighbourhoods. In: Fitzgibbon, A., Lazebnik, S., Perona, P., Sato, Y., Schmid, C. (eds.) ECCV 2012. LNCS, vol. 7574, pp. 836–849. Springer, Heidelberg (2012). https://doi.org/10.1007/978-3-642-33712-3_60
44. Verma, Y., Jawahar, C.V.: Exploring SVM for image annotation in presence of confusing labels. In: BMVC (2013)
45. Verma, Y., Jawahar, C.V.: Image annotation by propagating labels from semantic neighbourhoods. Int. J. Comput. Vision **121**(1), 126–148 (2017)
46. Vinyals, O., Toshev, A., Bengio, S., Erhan, D.: Show and tell: a neural image caption generator. In: CVPR (2015)
47. Wang, C., Blei, D., Fei-Fei, L.: Simultaneous image classification and annotation. In: Proceedings CVPR (2009)
48. Wang, J., Yang, Y., Mao, J., Huang, Z., Huang, C., Xu, W.: CNN-RNN: a unified framework for multi-label image classification. In: CVPR (2016)
49. Weston, J., Bengio, S., Usunier, N.: WSABIE: scaling up to large vocabulary image annotation. In: IJCAI (2011)
50. Yazici, V.O., Gonzalez-Garcia, A., Ramisa, A., Twardowski, B., van de Weijer, J.: Orderless recurrent models for multi-label classification (2019). CoRR abs/1911.09996
51. You, R., Guo, Z., Cui, L., Long, X., Bao, Y., Wen, S.: Cross-modality attention with semantic graph embedding for multi-label classification (2019). CoRR abs/1912.07872
52. Zhu, F., Li, H., Ouyang, W., Yu, N., Wang, X.: Learning spatial regularization with image-level supervisions for multi-label image classification. In: CVPR, pp. 2027–2036 (2017)
53. Zhuang, Y., Yang, Y., Wu, F.: Mining semantic correlation of heterogeneous multimedia data for cross-media retrieval. IEEE Trans. Multimedia **10**(2), 221–229 (2008)

Cross-Attention in Coupled Unmixing Nets for Unsupervised Hyperspectral Super-Resolution

Jing Yao[1,2,4] [iD], Danfeng Hong[2(✉)] [iD], Jocelyn Chanussot[3] [iD], Deyu Meng[1,5] [iD], Xiaoxiang Zhu[2,4] [iD], and Zongben Xu[1]

[1] School of Mathematics and Statistics, Xi'an Jiaotong University, Xi'an, China
jasonyao@stu.xjtu.edu.cn, {dymeng,zbxu}@mail.xjtu.edu.cn
[2] Remote Sensing Technology Institute, German Aerospace Center,
Weßling, Germany
{danfeng.hong,xiaoxiang.zhu}@dlr.de
[3] University Grenoble Alpes, INRIA, CNRS, Grenoble INP, LJK, Grenoble, France
jocelyn.chanussot@grenoble-inp.fr
[4] Technical University of Munich, Munich, Germany
[5] Macau University of Science and Technology, Macao, China

Abstract. The recent advancement of deep learning techniques has made great progress on hyperspectral image super-resolution (HSI-SR). Yet the development of unsupervised deep networks remains challenging for this task. To this end, we propose a novel coupled unmixing network with a cross-attention mechanism, CUCaNet for short, to enhance the spatial resolution of HSI by means of higher-spatial-resolution multispectral image (MSI). Inspired by coupled spectral unmixing, a two-stream convolutional autoencoder framework is taken as backbone to jointly decompose MS and HS data into a spectrally meaningful basis and corresponding coefficients. CUCaNet is capable of adaptively learning spectral and spatial response functions from HS-MS correspondences by enforcing reasonable consistency assumptions on the networks. Moreover, a cross-attention module is devised to yield more effective spatial-spectral information transfer in networks. Extensive experiments are conducted on three widely-used HS-MS datasets in comparison with state-of-the-art HSI-SR models, demonstrating the superiority of the CUCaNet in the HSI-SR application. Furthermore, the codes and datasets are made available at: https://github.com/danfenghong/ECCV2020_CUCaNet.

Keywords: Coupled unmixing · Cross-attention · Deep learning · Hyperspectral super-resolution · Multispectral · Unsupervised

1 Introduction

Recent advances in hyperspectral (HS) imaging technology have enabled the availability of enormous HS images (HSIs) with a densely sampled spectrum

Danfeng Hong — Corresponding author.

[26]. Benefited from the abundant spectral information contained in those hundreds of bands measurement, HSI features great promise in delivering faithful representation of real-world materials and objects. Thus the pursuit of effective and efficient processing of HS data has long been recognized as a prominent topic in the field of computer vision [9,11].

Though physically, the insufficient spatial resolution of HS instruments, combined with an inherently intimate mixing effect, severely hampers the abilities of HSI in various real applications [2,35]. Fortunately, the multispectral (MS) imaging systems (e.g., RGB cameras, spaceborne MS sensors) are capable of providing complementary products, which preserve much finer spatial information at the cost of reduced spectral resolution [13]. Accordingly, the research on enhancing the spatial resolution (henceforth, resolution refers to the spatial resolution) of an observable low-resolution HSI (LrHSI) by merging a high-resolution MSI (HrMSI) under the same scene, which is referred to hyperspectral image super-resolution (HSI-SR), has been gaining considerable attention [15,16].

The last decade has witnessed a dominant development of optimization-based methods, from either deterministic or stochastic perspectives, to tackle the HSI-SR issue [37]. To mitigate the severe ill-posedness of such an inverse problem, the majority of prevailing methods put their focus on exploiting various handcrafted priors to characterize spatial and spectral information underlying the desired solution. Moreover, the dependency on the knowledge of relevant sensor characteristics, such as spectral response function (SRF) and point spread function (PSF), inevitably compromises their transparency and practicability.

More recently, a growing interest has been paid to leverage the tool of deep learning (DL) by exploiting its merit on low-level vision applications. Among them, the best result is achieved by investigators who resort to performing HSI-SR progressively in a supervised fashion [34]. However, the demand for sufficient training image pairs acquired with different sensors inevitably makes their practicability limited. On the other hand, though being rarely studied, the existing unsupervised works rely on either complicated multi-stage alternating optimization [25], or an external camera spectral response (CSR) dataset in the context of RGB image guidance [10], the latter of which also losses generality in confronting other kinds of data with higher spectral resolution than RGB one.

To address the aforementioned challenges, we propose a novel coupled unmixing network with cross-attention (CUCaNet) for unsupervised HSI-SR. The contributions of this paper are briefly summarized as follows:

1. We propose a novel unsupervised HSI-SR model, called CUCaNet, which is built on a coupled convolutional autoencoder network. CUCaNet models the physically mixing properties in HS imaging into the networks to transfer the spatial information of MSI to HSI and preserve the high spectral resolution itself simultaneously in a coupled fashion.
2. We devise an effective cross-attention module to extract and transfer significant spectral (or spatial) information from HSI (or MSI) to another branch, yielding more sufficient spatial-spectral information blending.

3. Beyond previous coupled HSI-SR models, the proposed CUCaNet is capable of adaptively learning PSFs and SRFs across MS-HS sensors with a high ability to generalize. To find the local optimum of the network more effectively, we shrink the solution space by designing a closed-loop consistency regularization in networks, acting on both spatial and spectral domains.

2　Related Work

Pioneer researches have emerged naturally by adapting the similar but extensively studied pansharpening techniques to HSI-SR [22,28], which usually fail to well capture the global continuity in the spectral profiles and thus brings unignorable performance degradation, leaving much room to be desired.

2.1　Conventional Methods

Apace with the advances in statistically modeling and machine learning, recent optimization-based methods has lifted the HSI-SR ratio evidently. According to a subspace assumption, Bayesian approach was first introduced by Eismann *et al.* utilizing a stochastic mixing model [8], and developed through subsequent researches by exploiting more inherent characteristics [27,33]. Another class of methods that have been actively investigated stems from the idea of spectral unmixing [14], which takes the intimate mixing effect into consideration. Yokoya *et al.* brought up coupled non-negative matrix factorization (CNMF) [38] to estimate the spectral signature of the underlying materials and corresponding coefficients alternately. On basis of CNMF, Kawakami *et al.* [4] employed sparse regularization and an effective projected gradient solver was devised by Lanaras *et al.* [20]. Besides, [2,7] adopted dictionary learning and sparse coding techniques in this context. Various kinds of tensor factorization strategies are also studied, such as Tucker decomposition adopted by Dian *et al.* [5] and Li *et al.* [21] to model non-local and coupled structure information, respectively.

2.2　DL-Based Methods

To avoid tedious hand-crafted priors modeling in conventional methods, DL-based methods have attracted increasing interest these years. In the class of supervised methods, Dian *et al.* [6] employed CNN with prior training to finely tune the result acquired by solving a conventional optimization problem, while Xie *et al.* [34] introduced a deep unfolding network based on a novel HSI degradation model. Unsupervised methods are more rarely studied. Qu *et al.* [25] developed an unsupervised HSI-SR net with Dirichlet distribution-induced layer embedded, which results in a multi-stage alternating optimization. Under the guidance of RGB image and an external CSR database, Fu *et al.* [10] designed an unified CNN framework with a particular CSR optimization layer. Albeit demonstrated to be comparatively effective, these methods require either large training data for supervision or the knowledge of PSFs or SRFs, which are both

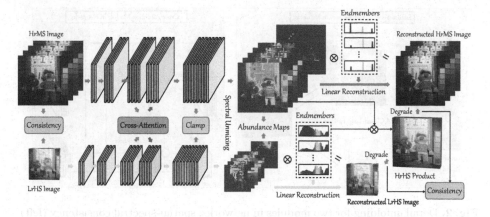

Fig. 1. An illustration of the proposed end-to-end CUCaNet inspired by spectral unmixing techniques, which mainly consists of two important modules: cross-attention and spatial-spectral consistency.

unrealistic in real HSI-SR scenario. Very recently, Zheng *et al.* [40] proposed a coupled CNN by adaptively learning the two functions of PSFs and SRFs for unsupervised HSI-SR. However, due to the lack of effective regularizations or constraints, the two to-be-estimated functions inevitably introduce more freedoms, limiting the performance to be further improved.

3 Coupled Unmixing Nets with Cross-Attention

In this section, we present the proposed coupled unmixing networks with a cross-attention module implanted, which is called CUCaNet for short. For mathematical brevity, we resort to a 2D representation of the 3D image cube, that is, the spectrum of each pixel is stacked row-by-row.

3.1 Method Overview

CUCaNet builds on a two-stream convolutional autoencoder backbone, which aims at decomposing MS and HS data into a spectrally meaningful basis and corresponding coefficients jointly. Inspired by CNMF, the fused HrHSI is obtained by feeding the decoder of the HSI branch with the encoded maps of the MSI branch. Two additional convolution layers are incorporated to simulate the spatial and spectral downsampling processes across MS-HS sensors. To guarantee that CUCaNet can converge to a faithful product through an unsupervised training, reasonable consistency, and necessary unmixing constraints, are integrated smoothly without imposing evident redundancy. Moreover, we introduced the cross-attention attention mechanism into the HSI-SR for the first time.

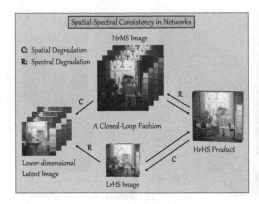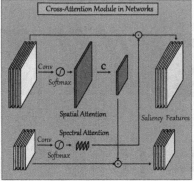

Fig. 2. Detail unfolding for two modules in networks: spatial-spectral consistency (left) and cross-attention (right).

3.2   Problem Formulation

Given the LrHSI $\mathbf{X} \in \mathbb{R}^{hw \times L}$, and the HrMSI $\mathbf{Y} \in \mathbb{R}^{HW \times l}$, the goal of HSI-SR is to recover the latent HrHSI $\mathbf{Z} \in \mathbb{R}^{HW \times L}$, where (h, w, s) are the reduced height, width, and number of spectral bands, respectively, and (H, W, S) are corresponding upsampled version. Based on the linear mixing model that well explains the phenomenon of *mixed pixels* involved in $\mathbf{Z}$, we then have the following NMF-based representation,

$$\mathbf{Z} = \mathbf{SA}, \tag{1}$$

where $\mathbf{A} \in \mathbb{R}^{K \times L}$ and $\mathbf{S} \in \mathbb{R}^{HW \times K}$ are a collection of spectral signatures of pure materials (or say, endmembers) and their fractional coefficients (or say, abundances), respectively.

On the other hand, the degradation processes in the spatial ($\mathbf{X}$) and the spectral ($\mathbf{Y}$) observations can be modeled as

$$\mathbf{X} \approx \mathbf{CZ} = \mathbf{CSA} = \tilde{\mathbf{S}}\mathbf{A}, \tag{2}$$

$$\mathbf{Y} \approx \mathbf{ZR} = \mathbf{SAR} = \mathbf{S}\tilde{\mathbf{A}}, \tag{3}$$

where $\mathbf{C} \in \mathbb{R}^{hw \times HW}$ and $\mathbf{R} \in \mathbb{R}^{L \times l}$ represent the PSF and SRF from the HrHSI to the HrMSI and the LrHSI, respectively. Since $\mathbf{C}$ and $\mathbf{R}$ are non-negative and normalized, $\tilde{\mathbf{S}}$ and $\tilde{\mathbf{A}}$ can be regarded as spatially downsampled abundances and spectrally downsampled endmembers, respectively. Therefore, an intuitive solution is to unmix $\mathbf{X}$ and $\mathbf{Y}$ based on Eq. (2) and Eq. (3) alternately, which is coupled with the prior knowledge of $\mathbf{C}$ and $\mathbf{R}$. Such a principle has been exploited in various optimization formulations, obtaining state-of-the-art fusion performance by linear approximation with converged $\mathbf{S}$ and $\mathbf{A}$.

Constraints. Still, the issued HSI-SR problem involves the inversions from $\mathbf{X}$ and $\mathbf{Y}$ to $\mathbf{S}$ and $\mathbf{A}$, which are highly ill-posed. To narrow the solution space, several physically meaningful constraints are commonly adopted, they are the abundance sum-to-one constraint (ASC), the abundance non-negative constraint (ANC), and non-negative constraint on endmembers, i.e.,

$$\mathbf{S1}_K = \mathbf{1}_{HW}, \; \mathbf{S} \succeq 0, \; \mathbf{A} \succeq 0, \tag{4}$$

where $\succeq$ marks element-wise inequality, and $\mathbf{1}_p$ represents p-length all-one vector. It is worth mentioning that the combination of ASC and ANC would promote the sparsity of abundances, which well characterizes the rule that the endmembers are sparsely contributing to the spectrum in each pixel.

Yet in practice, the prior knowledge of PSFs and SRFs for numerous kinds of imaging systems is hardly available. This restriction motivates us to extend the current coupled unmixing model to a fully end-to-end framework, which is only in need of LrHSI and HrMSI. To estimate $\mathbf{C}$ and $\mathbf{R}$ in an unsupervised manner, we introduce the following consistency constraint,

$$\mathbf{U} = \mathbf{XR} = \mathbf{CY}, \tag{5}$$

where $\mathbf{U} \in \mathbb{R}^{hw \times l}$ denotes the latent LrMSI.

3.3 Network Architecture

Inspired by the recent success of deep networks on visual processing tasks, we would like to first perform coupled spectral unmixing by the established two-stream convolutional autoencoder for the two-modal inputs, i.e., we consider two deep subnetworks, with $f(\mathbf{X}) = f_{de}(f_{en}(\mathbf{X}; \mathbf{W}_{f,en}); \mathbf{W}_{f,de})$ to self-express the LrHSI, $g(\mathbf{Y}) = g_{de}(g_{en}(\mathbf{Y}; \mathbf{W}_{g,en}); \mathbf{W}_{g,de})$ for the HrMSI, and the fused result can be obtained by $\hat{\mathbf{Z}} = f_{de}(g_{en}(\mathbf{Y}; \mathbf{W}_{g,en}); \mathbf{W}_{f,de})$, herein $\mathbf{W}$ collects the weights of corresponding subpart.

As shown in Fig. 1, both encoders f_{en} and g_{en} are constructed by cascading "Convolution+LReLU" blocks f_l with an additional 1×1 convolution layer. We set the sizes of convolutional kernels in f_{en} all as 1×1 while those in g_{en} are with larger but descending scales of the receptive field. The idea behind this setting is to consider the low fidelity of spatial information in LrHSI and simultaneously map the cross-channel and spatial correlations underlying HrMSI. Furthermore, to ensure that the encoded maps are able to possess the properties of abundances, an additional activation layer using the clamp function in the range of $[0, 1]$ is concatenated after each encoder. As for the structure of decoders f_{de} and g_{de}, we simply adopt a 1×1 convolution layer without any nonlinear activation, making the weights $\mathbf{W}_{f,de}$ and $\mathbf{W}_{g,de}$ interpretable as the endmembers $\mathbf{A}$ and $\tilde{\mathbf{A}}$ according to Eq. (2) and Eq. (3). By backward gradient descent-based optimization, our backbone network can not only avoid the need for good initialization for conventional unmixing algorithms but also enjoy the amelioration brought by its capability of local perception and nonlinear processing.

Cross-Attention. To further exploit the advantageous information from the two modalities, we devise an effective cross-attention module to enrich the features across modalities. As shown in Fig. 2, the cross-attention module is employed on high-level features within the encoder part, with three steps to follow. First, we compute the spatial and spectral attention from the branch of LrHSI and HrMSI, since they can provide with more faithful spatial and spectral guidance. Next, we multiply the original features with the attention maps from another branch to transfer the significant information. Lastly, we concatenate the original features with the above cross-multiplications in each branch, to construct the input of next layer in the form of such preserved and refined representation.

Formally, the output features $\mathbf{F}_l \in \mathbb{R}^{h \times w}$ of the l-th layer in the encoder part, take f_{en} for example, are formulated as

$$\mathbf{F}_l = f_l(\mathbf{F}_{l-1}) = f_l(f_{l-1}(\cdots f_1(\mathbf{X})\cdots)), \tag{6}$$

which is similar for obtaining $\mathbf{G}_l \in \mathbb{R}^{H \times W}$ from g_{en}. To gather the spatial and spectral significant information, we adopt global and local convolution to generate channel-wise and spatial statistics respectively as

$$o_c = \mathbf{u}_c \odot \mathbf{F}_l^{(c)}, \ \mathbf{S} = \sum_{c=1}^{C} \mathbf{v}^{(c)} \odot \mathbf{G}_l^{(c)}, \tag{7}$$

where $\mathbf{u} = [\mathbf{u}_1, \cdots, \mathbf{u}_C]$ is a set of convolution filters with size $h \times w$, $\mathbf{v}^{(c)}$ is the c-th channel of a 3D convolution filter with spatial size as $p \times p$. Then we apply a softmax layer to the above statistics to get the attention maps $\delta(\mathbf{o}) \in \mathbb{R}^C$, and $\delta(\mathbf{S}) \in \mathbb{R}^{H \times W}$, where $\delta(\cdot)$ denotes the softmax activation function. The original features are finally fused into the input of next layer as $concat(\mathbf{F}_l; \mathbf{F}_l \odot \delta(\mathbf{S}))$, and $concat(\mathbf{G}_l; \mathbf{G}_l \odot \delta(\mathbf{o}))$, where $concat(\cdot)$ denotes the concatenation, and $\odot$ denotes the point-wise multiplication.

Spatial-Spectral Consistency. An essential part that tends to be ignored is related to the coupled factors caused by PSFs and SRFs. Previous researches typically assume an ideal average spatial downsampling and the prior knowledge of SRFs, which rarely exist in reality. Unlike them, we introduce a spatial-spectral consistency module into networks in order to better simulate the to-be-estimated PSF and SRF, which is performed by simple yet effective convolution layers.

We can rewrite the spectral resampling from the HS sensor to the MS sensor by revisiting the left part of Eq. (3) more accurately as follows. Given the spectrum of i-th pixel in HrHSI $\mathbf{z}_i$, for the j-th channel in corresponding LrHSI, the radiance $y_{i,j}$ is defined as

$$y_{i,j} = \int_\phi \mathbf{z}_i(\mu)\mathbf{r}_j(\mu)d\mu/N_r, \tag{8}$$

where ϕ denotes the support set that the wavelength μ belongs to, N_r denotes the normalization constant $\int \mathbf{r}_j(\mu)d\mu$. We directly replace $\mathbf{r}_j$ with a set of L

1×1 convolution kernels with the weights being collected in $\mathbf{w}_j$. Therefore, the SRF layer f_r can be well defined as follows,

$$y_{i,j} = f_r(\mathbf{z}_i; \mathbf{w}_j) = \sum_\phi \mathbf{z}_i(\mu) \mathbf{w}_j(\mu) / N_w, \tag{9}$$

where N_w corresponds to an additional normalization with $\sum_\phi \mathbf{w}_j$. The PSF layer for spatial downsampling is more straightforward. Note that PSF generally indicates that each pixel in LrHSI is produced by combining neighboring pixels in HrHSI with unknown weights in a disjoint manner [30]. To simulate this process, we propose f_s by the means of a channel-wise convolution layer with kernel size and stride both same as the scaling ratio.

To sum up, multiple consistency constraints derived from the statements in Sect. 3.2, either spectrally or spatially, can be defined in our networks as

$$\hat{\mathbf{Y}} = f_r(\hat{\mathbf{Z}}), \ \hat{\mathbf{X}} = f_s(\hat{\mathbf{Z}}), \ f_s(\mathbf{Y}) = f_r(\mathbf{X}), \tag{10}$$

which enables the whole networks to be trained within a closed loop.

3.4 Network Training

Loss Function. As shown in Fig. 1, our CUCaNet mainly consists of two autoencoders for hyperspectral and multispectral data, respectively, thus leading to the following reconstruction loss:

$$\mathcal{L}_R = \|f(\mathbf{X}) - \mathbf{X}\|_1 + \|g(\mathbf{Y}) - \mathbf{Y}\|_1, \tag{11}$$

in which the ℓ_1-norm is selected as the loss criterion for its perceptually satisfying performance in the low-level image processing tasks [39].

The important physically meaningful constraints in spectral unmixing are considered, building on Eq. (4), we then derive the second ASC loss as

$$\mathcal{L}_{ASC} = \|\mathbf{1}_{hw} - f_{en}(\mathbf{X})\mathbf{1}_K\|_1 + \|\mathbf{1}_{HW} - g_{en}(\mathbf{Y})\mathbf{1}_K\|_1, \tag{12}$$

and the ANC is reflected through the activation layer used behind the encoders.

To promote the sparsity of abundances of both stream, we adopt the Kullback-Leibler (KL) divergence-based sparsity loss term by penalizing the discrepancies between them and a tiny scalar ϵ,

$$\mathcal{L}_S = \sum_n \text{KL}(\epsilon \| (f_{en}(\mathbf{X}))_n) + \sum_m \text{KL}(\epsilon \| (g_{en}(\mathbf{Y}))_m), \tag{13}$$

where $\text{KL}(\rho \| \hat{\rho}) = \rho \log \frac{\rho}{\hat{\rho}} + (1 - \rho) \log \frac{1-\rho}{1-\hat{\rho}}$ is the standard KL divergence [24].

Last but not least, we adopt the ℓ_1-norm to define the spatial-spectral consistency loss based on Eq. (10) as follows,

$$\mathcal{L}_C = \|f_s(\mathbf{Y}) - f_r(\mathbf{X})\|_1 + \|\hat{\mathbf{X}} - \mathbf{X}\|_1 + \|\hat{\mathbf{Y}} - \mathbf{Y}\|_1. \tag{14}$$

By integrating all the above-mentioned loss terms, the final objective function for the training of CUCaNet is given by

$$\mathcal{L} = \mathcal{L}_R + \alpha \mathcal{L}_{ASC} + \beta \mathcal{L}_S + \gamma \mathcal{L}_C, \tag{15}$$

where we use (α, β, γ) to trade-off the effects of different constituents.

Implementation Details. Our network is implemented on PyTorch framework. We choose Adam optimizer under default parameters setting for training with the training batch parameterized by 1 [18]. The learning rate is initialized with 0.005 and a linear decay from 2000 to 10000 epochs drop-step schedule is applied [23]. We adopt Kaiming's initialization for the convolutional layers [12]. The hyperparameters are determined using a grid search on the validation set and training will be early stopped before validation loss fails to decrease.

4   Experimental Results

In this section, we first review the HSI-MSI datasets and setup adopted in our experiments. Then, we provide an ablation study to verify the effectiveness of the proposed modules. Extensive comparisons with the state-of-the-art methods on indoor and remotely sensed images are reported at last.

Dataset and Experimental Setting. Three widely used HSI-MSI datasets are investigated in this section, including CAVE dataset [36][1], Pavia University dataset, and Chikusei dataset [37][2]. The CAVE dataset captures 32 different indoor scenes. Each image consists of 512×512 pixels with 31 spectral bands uniformly measured in the wavelength ranging 400 nm to 700 nm. In our experiments, 16 scenes are randomly selected to report performance. The Pavia dataset was acquired by ROSIS airborne sensor over the University of Pavia, Italy, in 2003. The original HSI comprises 610×340 pixels and 115 spectral bands. We use the top-left corner of the HSI with 336×336 pixels and 103 bands (after removing 12 noisy bands), covering the spectral range 430 nm to 838 nm. The Chikusei dataset was taken by a Visible and Near-Infrared (VNIR) imaging sensor over Chikusei, Japan, in 2014. The original HSI consists of 2,517×2,335 pixels and 128 bands with a spectral range 363 nm to 1,018 nm. We crop 6 non-overlapped parts with size of 576×448 pixels from the bottom part for test.

Considering the diversity of MS sensors in generating the HrMS images, we employ the SRFs of Nikon D700 camera[25] and Landsat-8 spaceborne MS sensor[3][3] for the CAVE dataset and two remotely sensed datasets[4], respectively. We adopt the Gaussian filter to obtain the LrHS images, by constructing the filter with the width same as SR ratio and 0.5 valued deviations. The SR ratios are set as 16 for the Pavia University dataset and 32 for the other two datasets.

Evaluation Metrics. We use the following five complementary and widely-used picture quality indices (PQIs) for the quantitative HSI-SR assessment, including peak signal-to-noise ratio (PSNR), spectral angle mapper (SAM) [19], erreur

---

[1] http://www.cs.columbia.edu/CAVE/databases/multispectral.

[2] http://naotoyokoya.com/Download.html.

[3] http://landsat.gsfc.nasa.gov/?p=5779.

[4] We select the spectral radiance responses of blue-green-red(BGR) bands and BGR-NIR bands for the experiments on Pavia and Chikusei datasets, respectively.

relative globale adimensionnellede synthèse (ERGAS) [29], structure similarity (SSIM) [32], and universal image quality index (UIQI) [31]. SAM reflects the spectral similarity by calculating the average angle between two vectors of the estimated and reference spectra at each pixel. PSNR, ERGAS, and SSIM are mean square error (MSE)-based band-wise PQIs indicating spatial fidelity, global quality, and perceptual consistency, respectively. UIQI is also band-wisely used to measure complex distortions among monochromatic images.

4.1 Ablation Study

Our CUCaNet consists of a baseline network – coupled convolutional autoencoder networks – and two newly-proposed modules, i.e., the spatial-spectral consistency module (SSC) and the cross-attention module (CA). To investigate the performance gain of different components in networks, we perform ablation analysis on the Pavia University dataset. We also study the effect of replacing clamp function with conventional softmax activation function at the end of each encoder. Table 1 details the quantitative results, in which CNMF is adopted as the baseline method.

Table 1. Ablation study on the Pavia University dataset by our CUCaNet with different modules and a baseline CNMF. The best results are shown in bold.

Method	Module			Metric				
	Clamp	SSC	CA	PSNR	SAM	ERGAS	SSIM	UQI
CNMF	–	–	–	32.73	7.05	1.18	0.830	0.973
CUCaNet	✗	✗	✗	34.25	6.58	1.01	0.862	0.975
CUCaNet	✓	✗	✗	35.67	5.51	0.92	0.897	0.981
CUCaNet	✓	✓	✗	36.55	4.76	0.85	0.904	**0.991**
CUCaNet	✓	✗	✓	36.49	4.63	0.86	0.902	0.989
CUCaNet	✓	✓	✓	**37.22**	**4.43**	**0.82**	**0.914**	**0.991**

As shown in Table 1, single CUCaNet can outperform CNMF in all metrics owing to its benefit from employing deep networks. We find that the performance is further improved remarkably by the use of clamp function. Meanwhile, single SSC module performs better than single CA module except in SAM, which means that CA module tend to favor spectral consistency. By jointly employing the two modules, the proposed CUCaNet achieves the best results in HSI-SR tasks, demonstrating the effectiveness of our whole network architecture.

4.2 Comparative Experiments

Compared Methods. Here, we make comprehensive comparison with the following eleven state-of-the-art (SOTA) methods in HSI-RS tasks: pioneer work,

Table 2. The ability of learning unkonwn SRF and PSF of competing methods.

Functions	GSA	CNMF	CSU	FUSE	HySure	NSSR	STEREO	CSTF	LTTR	uSDN	MHFnet	CUCaNet
SRF	✗	✗	✗	✗	✓	✗	✗	✗	✗	✗	✓	✓
PSF	✗	✗	✗	✗	✓	✗	✗	✗	✗	–	✓	✓

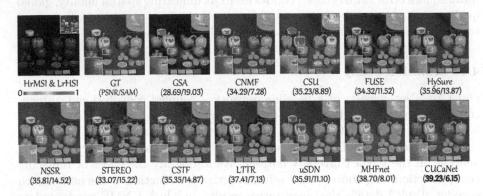

Fig. 3. The HSI-SR performance on the CAVE dataset (*fake and real food*) of CUCaNet in comparison with SOTA methods. For each HSI, the 20th (590 nm) band image is displayed with two demarcated areas zoomed in 3 times for better visual assessment, and two main scores (PSNR/SAM) are reported with the best results in bold.

GSA [1][5], NMF-based approaches, CNMF [38] (see footnote 5) and CSU [20][6], Bayesian-based approaches, FUSE [33][7] and HySure [27][8], dictionary learning-based approach, NSSR [7][9], tensor-based approaches, STEREO [17][10], CSTF [21][11], and LTTR [5] (see footnote 11), and DL-based methods, unsupervised uSDN [25][12] and supervised MHFnet [34][13]. As for the supervised deep method MHFnet, we use the remaining part of each dataset for the training following the strategies in [34].

Note that most of the above methods rely on the prior knowledge of SRFs and PSFs. We summarize the properties of all compared methods in learning SRFs and PSFs (see Table 2), where only HySure and MHFnet are capable of learning the two unknown functions. More specifically, HySure adopts a multi-stage method and MHFnet models them as convolution layers under a supervised framework. Hence our CUCaNet serves as the first unsupervised method that can simultaneously learn SRFs and PSFs in an end-to-end fashion.

---

[5] http://naotoyokoya.com/Download.html.

[6] https://github.com/lanha/SupResPALM.

[7] https://github.com/qw245/BlindFuse.

[8] https://github.com/alfaiate/HySure.

[9] http://see.xidian.edu.cn/faculty/wsdong.

[10] https://github.com/marhar19/HSR_via_tensor_decomposition.

[11] https://sites.google.com/view/renweidian.

[12] https://github.com/aicip/uSDN.

[13] https://github.com/XieQi2015/MHF-net.

Table 3. Quantitative performance comparison with the investigated methods on the CAVE dataset. The best results are shown in bold.

Metric	Method											
	GSA	CNMF	CSU	FUSE	HySure	NSSR	STEREO	CSTF	LTTR	uSDN	MHFnet	CUCaNet
PSNR	27.89	30.11	30.26	29.87	31.26	33.52	30.88	32.74	35.45	34.67	37.30	**37.51**
SAM	19.71	9.98	11.03	16.05	14.59	12.09	15.87	13.13	9.69	10.02	7.75	**7.49**
ERGAS	1.11	0.69	0.65	0.77	0.72	0.69	0.75	0.64	0.53	0.52	0.49	**0.47**
SSIM	0.713	0.919	0.911	0.876	0.905	0.912	0.896	0.914	0.949	0.921	**0.961**	0.959
UQI	0.757	0.911	0.898	0.860	0.891	0.904	0.873	0.902	0.942	0.905	0.949	**0.955**

Fig. 4. The HSI-SR performance on the CAVE dataset (*chart and staffed toy*) of CUCaNet in comparison with SOTA methods. For each HSI, the 7th (460 nm) band image is displayed with two demarcated areas zoomed in 3.5 times for better visual assessment, and two main scores (PSNR/SAM) are reported with the best results in bold.

Indoor Dataset. We first conduct experiments on indoor images of the CAVE dataset. The average quantitative results over 16 testing images are summarized in Table 3 with the best ones highlighted in bold. From the table, we can observe that LTTR and CSTF can obtain better reconstruction results than other conventional methods, mainly by virtue of their complex regularizations under tensorial framework. Note that the SAM values of earlier methods CNMF and CSU are still relatively lower because they consider the coupled unmixing mechanism. As for the DL-based methods, supervised MHFnet outperforms unsupervised uSDN evidently, while our proposed CUCaNet achieves the best results in terms of four major metrics. Only the SSIM value of ours is slightly worse than that of the most powerful competing method MHFnet, due to its extra exploitation of supervised information.

The visual comparison on two selected scenes demonstrated in Fig. 3 and Fig. 4 exhibits a consistent tendency. From the figures, we can conclude that the results of CUCaNet maintain the highest fidelity to the groundtruth (GT) compared to other methods. For certain bands, our method can not only estimate background more accurately, but also maintain the texture details on different objects. The SAM values of CUCaNet on two images are obviously less than

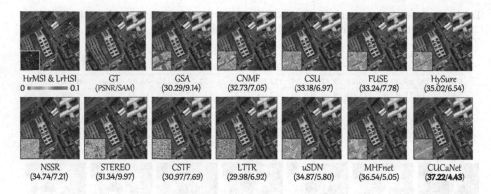

Fig. 5. The HSI-SR performance on the Pavia University dataset (cropped area) of all competing methods. The false-color image with bands 61-36-10 as R-G-B channels is displayed. One demarcated area (red frame) as well as its RMSE-based residual image (blue frame) with respect to GT are zoomed in 3 times for better visual assessment. (Color figure online)

Table 4. Quantitative performance comparison with the investigated methods on the Pavia University dataset. The best results are shown in bold.

Metric	Method											
	GSA	CNMF	CSU	FUSE	HySure	NSSR	STEREO	CSTF	LTTR	uSDN	MHFnet	CUCaNet
PSNR	30.29	32.73	33.18	33.24	35.02	34.74	31.34	30.97	29.98	34.87	36.34	**37.22**
SAM	9.14	7.05	6.97	7.78	6.54	7.21	9.97	7.69	6.92	5.80	5.15	**4.43**
ERGAS	1.31	1.18	1.17	1.27	1.10	1.06	1.35	1.23	1.30	1.02	0.89	**0.82**
SSIM	0.784	0.830	0.815	0.828	0.861	0.831	0.751	0.782	0.775	0.871	**0.919**	0.914
UQI	0.965	0.973	0.972	0.969	0.975	0.966	0.938	0.969	0.967	0.982	0.987	**0.991**

Table 5. Quantitative performance comparison with the investigated methods on the Chikusei dataset. The best results are shown in bold.

Metric	Method											
	GSA	CNMF	CSU	FUSE	HySure	NSSR	STEREO	CSTF	LTTR	uSDN	MHFnet	CUCaNet
PSNR	32.07	38.03	37.89	39.25	39.97	38.35	32.40	36.52	35.54	38.32	**43.71**	42.70
SAM	10.44	4.81	5.03	4.50	4.35	4.97	8.52	6.33	7.31	3.89	3.51	**3.13**
ERGAS	0.98	0.58	0.61	0.47	0.45	0.63	0.74	0.66	0.70	0.51	0.42	**0.40**
SSIM	0.903	0.961	0.945	0.970	0.974	0.961	0.897	0.929	0.918	0.964	0.985	**0.988**
UQI	0.909	0.976	0.977	0.977	0.976	0.914	0.902	0.915	0.917	0.976	**0.992**	0.990

others, which validates the superiority in capturing the spectral characteristics via joint coupled unmixing and degrading functions learning.

Remotely Sensed Dataset. We then carry out more experiments using airborne HS data to further evaluate the generality of our method. The quantitative evaluation results on the Pavia University and Chikusei datasets are provided in Table 4 and Table 5, respectively. Generally, we can observe a significant performance improvements than on CAVE, since more spectral information can be used as the number of HS bands increases. For the same reason, NMF-based and Bayesian-based methods show competitive performance owing to their accurate estimation of high-resolution subspace coefficients [37]. The limited performance of tensor-based methods suggests they may lack robustness to the spectral distortions in real cases. The multi-stage unsupervised training of uSDN makes it easily trapped into local minima, which results in only comparable performance to state-of-the-art conventional methods such as HySure and FUSE. It is particularly evident that MHFnet performs better on Chikusei rather than Pavia University. This can be explained by the fact that training data is relatively adequate on Chikusei so that the tested patterns are more likely to be well learned. We have to admit, however that MHFnet requires extremely rich training samples, which restricts its practical applicability to a great extent. Remarkably, our CUCaNet can achieve better performance in most cases, especially showing advantage in the spectral quality measured by SAM, which confirms that our method is good at capturing the spectral properties and hence attaining a better reconstruction of HrHSI.

Figure 5 and Fig. 6 show the HSI-SR results demonstrated in false-color on these two datasets. Since it is hard to visually discern the differences of most fused results, we display the RMSE-based residual images of local windows compared with GT for better visual evaluation. For both datasets, we can observe that GSA and STEREO yield bad results with relatively higher errors. CNMF and CSU show evident patterns in residuals that are similar to the original image, which indicates that their results are missing actual details. The block pattern-like errors included in CSTF and LTTR make their reconstruction unsmooth.

Fig. 6. The HSI-SR performance on the Chikusei dataset (cropped area) of all competing methods. The false-color image with bands 61-36-10 as R-G-B channels is displayed. One demarcated area (red frame) as well as its RMSE-based residual image (blue frame) with respect to GT are zoomed in 3 times for better visual assessment. (Color figure online)

Note that residual images of CUCaNet and MHFnet exhibit more dark blue areas than other methods. This means that the errors are small and the fused results are more reliable.

5   Conclusion

In this paper, we put forth CUCaNet for the task of HSI-SR by integrating the advantage of coupled spectral unmixing and deep learning techniques. For the first time, the learning of unknown SRFs and PSFs across MS-HS sensors is introduced into an unsupervised coupled unmixing network. Meanwhile, a cross-attention module and reasonable consistency enforcement are employed jointly to enrich feature extraction and guarantee a faithful production. Extensive experiments on both indoor and airborne HS datasets utilizing diverse simulations validate the superiority of proposed CUCaNet with evident performance improvements over competitive methods, both quantitatively and perceptually. Finally, we will investigate more theoretical insights on explaining the effectiveness of the proposed network in our future work.

Acknowledgements. This work has been supported in part by projects of the National Natural Science Foundation of China (No. 61721002, No. U1811461, and No. 11690011) and the China Scholarship Council.

References

1. Aiazzi, B., Baronti, S., Selva, M.: Improving component substitution pansharpening through multivariate regression of ms + pan data. IEEE Trans. Geosci. Remote Sens. **45**(10), 3230–3239 (2007)
2. Akhtar, N., Shafait, F., Mian, A.: Sparse spatio-spectral representation for hyperspectral image super-resolution. In: Fleet, D., Pajdla, T., Schiele, B., Tuytelaars, T. (eds.) ECCV 2014. LNCS, vol. 8695, pp. 63–78. Springer, Cham (2014). https://doi.org/10.1007/978-3-319-10584-0_5
3. Barsi, J.A., Lee, K., Kvaran, G., Markham, B.L., Pedelty, J.A.: The spectral response of the landsat-8 operational land imager. Remote Sens. **6**(10), 10232–10251 (2014)
4. Bieniarz, J., Cerra, D., Avbelj, J., Reinartz, P., Müller, R.: Hyperspectral image resolution enhancement based on spectral unmixing and information fusion. In: ISPRS Hannover Workshop 2011 (2011)
5. Dian, R., Li, S., Fang, L.: Learning a low tensor-train rank representation for hyperspectral image super-resolution. IEEE Trans. Neural Netw. Learn. Syst. **30**(9), 2672–2683 (2019)
6. Dian, R., Li, S., Guo, A., Fang, L.: Deep hyperspectral image sharpening. IEEE Trans. Neural Netw. Learn. Syst. **29**(99), 1–11 (2018)
7. Dong, W., et al.: Hyperspectral image super-resolution via non-negative structured sparse representation. IEEE Trans. Image Process. **25**(5), 2337–2352 (2016)
8. Eismann, M.T.: Resolution enhancement of hyperspectral imagery using maximum a posteriori estimation with a stochastic mixing model. Ph.D. thesis, University of Dayton (2004)

9. Fu, Y., Zhang, T., Zheng, Y., Zhang, D., Huang, H.: Joint camera spectral sensitivity selection and hyperspectral image recovery. In: Proceedings of the European Conference on Computer Vision (ECCV), pp. 788–804 (2018)
10. Fu, Y., Zhang, T., Zheng, Y., Zhang, D., Huang, H.: Hyperspectral image super-resolution with optimized RGB guidance. In: Proceedings of the IEEE Conference on Computer Vision and Pattern Recognition (CVPR), pp. 11661–11670 (2019)
11. Gao, L., Hong, D., Yao, J., Zhang, B., Gamba, P., Chanussot, J.: Spectral super-resolution of multispectral imagery with joint sparse and low-rank learning. IEEE Trans. Geosci. Remote Sens. (2020). https://doi.org/10.1109/TGRS.2020.3000684
12. He, K., Zhang, X., Ren, S., Sun, J.: Delving deep into rectifiers: surpassing human-level performance on imagenet classification. In: Proceedings of the IEEE International Conference on Computer Vision, pp. 1026–1034 (2015)
13. Hong, D., Liu, W., Su, J., Pan, Z., Wang, G.: A novel hierarchical approach for multispectral palmprint recognition. Neurocomputing **151**, 511–521 (2015)
14. Hong, D., Yokoya, N., Chanussot, J., Zhu, X.X.: An augmented linear mixing model to address spectral variability for hyperspectral unmixing. IEEE Trans. Image Process. **28**(4), 1923–1938 (2019)
15. Hong, D., Yokoya, N., Chanussot, J., Zhu, X.X.: Cospace: common subspace learning from hyperspectral-multispectral correspondences. IEEE Trans. Geosci. Remote Sens. **57**(7), 4349–4359 (2019)
16. Hong, D., Yokoya, N., Ge, N., Chanussot, J., Zhu, X.X.: Learnable manifold alignment (lema): a semi-supervised cross-modality learning framework for land cover and land use classification. ISPRS J. Photogramm. Remote Sens. **147**, 193–205 (2019)
17. Kanatsoulis, C.I., Fu, X., Sidiropoulos, N.D., Ma, W.K.: Hyperspectral super-resolution: a coupled tensor factorization approach. IEEE Trans. Signal Process. **66**(24), 6503–6517 (2018)
18. Kingma, D.P., Ba, J.: Adam: a method for stochastic optimization. In: International Conference on Learning Representations (ICLR) (2015)
19. Kruse, F.A.: The spectral image processing system (sips)-interactive visualization and analysis of imaging spectrometer data. Remote Sens. Environ. **44**(2–3), 145–163 (1993)
20. Lanaras, C., Baltsavias, E., Schindler, K.: Hyperspectral super-resolution by coupled spectral unmixing. In: Proceedings of the IEEE International Conference on Computer Vision (ICCV), pp. 3586–3594 (2015)
21. Li, S., Dian, R., Fang, L., Bioucas-Dias, J.M.: Fusing hyperspectral and multispectral images via coupled sparse tensor factorization. IEEE Trans. Image Process. **27**(8), 4118–4130 (2018)
22. Loncan, L., et al.: Hyperspectral pansharpening: a review. IEEE Geosci. Remote Sens. Mag. **3**(3), 27–46 (2015)
23. Loshchilov, I., Hutter, F.: Decoupled weight decay regularization. In: International Conference on Learning Representations (ICLR) (2019)
24. Ng, A., et al.: Sparse autoencoder. CS294A Lect. Notes **72**(2011), 1–19 (2011)
25. Qu, Y., Qi, H., Kwan, C.: Unsupervised sparse dirichlet-net for hyperspectral image super-resolution. In: Proceedings of the IEEE Conference on Computer Vision and Pattern Recognition (CVPR), pp. 2511–2520 (2018)
26. Rasti, B., et al.: Feature extraction for hyperspectral imagery: the evolution from shallow to deep (overview and toolbox). IEEE Geosci. Remote Sens. Mag. (2020). https://doi.org/10.1109/MGRS.2020.2979764

27. Simoes, M., Bioucas-Dias, J., Almeida, L.B., Chanussot, J.: A convex formulation for hyperspectral image superresolution via subspace-based regularization. IEEE Trans. Geosci. Remote Sens. **53**(6), 3373–3388 (2014)
28. Vivone, G., et al.: A critical comparison among pansharpening algorithms. IEEE Trans. Geosci. Remote Sens. **53**(5), 2565–2586 (2014)
29. Wald, L.: Quality of high resolution synthesised images: is there a simple criterion? In: 3rd Conference Fusion Earth Data: Merging Point Measurements, Raster Maps, and Remotely Sensed Images (2000)
30. Wang, Q., Atkinson, P.M.: The effect of the point spread function on sub-pixel mapping. Remote Sens. Environ. **193**, 127–137 (2017)
31. Wang, Z., Bovik, A.C.: A universal image quality index. IEEE Signal Process. Lett. **9**(3), 81–84 (2002)
32. Wang, Z., Bovik, A.C., Sheikh, H.R., Simoncelli, E.P.: Image quality assessment: from error visibility to structural similarity. IEEE Trans. Image Process. **13**(4), 600–612 (2004)
33. Wei, Q., Dobigeon, N., Tourneret, J.Y.: Fast fusion of multi-band images based on solving a sylvester equation. IEEE Trans. Image Process. **24**(11), 4109–4121 (2015)
34. Xie, Q., Zhou, M., Zhao, Q., Meng, D., Zuo, W., Xu, Z.: Multispectral and hyperspectral image fusion by MS/HS fusion net. In: Proceedings of the IEEE Conference on Computer Vision and Pattern Recognition (CVPR), pp. 1585–1594 (2019)
35. Yao, J., Meng, D., Zhao, Q., Cao, W., Xu, Z.: Nonconvex-sparsity and nonlocal-smoothness-based blind hyperspectral unmixing. IEEE Trans. Image Process. **28**(6), 2991–3006 (2019)
36. Yasuma, F., Mitsunaga, T., Iso, D., Nayar, S.K.: Generalized assorted pixel camera: postcapture control of resolution, dynamic range, and spectrum. IEEE Trans. Image Process. **19**(9), 2241–2253 (2010)
37. Yokoya, N., Grohnfeldt, C., Chanussot, J.: Hyperspectral and multispectral data fusion: a comparative review of the recent literature. IEEE Geosci. Remote Sens. Mag. **5**(2), 29–56 (2017)
38. Yokoya, N., Yairi, T., Iwasaki, A.: Coupled nonnegative matrix factorization unmixing for hyperspectral and multispectral data fusion. IEEE Trans. Geosci. Remote Sens. **50**(2), 528–537 (2011)
39. Zhao, H., Gallo, O., Frosio, I., Kautz, J.: Loss functions for image restoration with neural networks. IEEE Trans. Comput. Imaging **3**(1), 47–57 (2016)
40. Zheng, K., Zheng, K., et al.: Coupled convolutional neural network with adaptive response function learning for unsupervised hyperspectral super-resolution. IEEE Trans. Geosci. Remote Sens. (2020). https://doi.org/10.1109/TGRS.2020.3006534

SimPose: Effectively Learning DensePose and Surface Normals of People from Simulated Data

Tyler Zhu[1]($\boxtimes$), Per Karlsson[2], and Christoph Bregler[2]

[1] Google Research, Montreal, Canada
tylerzhu@google.com
[2] Google Research, Mountain View, US
perk@google.com, bregler@google.com

Abstract. With a proliferation of generic domain-adaptation approaches, we report a simple yet effective technique for learning difficult per-pixel 2.5D and 3D regression representations of articulated people. We obtained strong sim-to-real domain generalization for the 2.5D DensePose estimation task and the 3D human surface normal estimation task. On the multi-person DensePose MSCOCO benchmark, our approach outperforms the state-of-the-art methods which are trained on real images that are densely labelled. This is an important result since obtaining human manifold's intrinsic uv coordinates on real images is time consuming and prone to labeling noise. Additionally, we present our model's 3D surface normal predictions on the MSCOCO dataset that lacks any real 3D surface normal labels. The key to our approach is to mitigate the "Inter-domain Covariate Shift" with a carefully selected training batch from a mixture of domain samples, a deep batch-normalized residual network, and a modified multi-task learning objective. Our approach is complementary to existing domain-adaptation techniques and can be applied to other dense per-pixel pose estimation problems.

Keywords: Person pose estimation · Simulated data · Dense pose estimation · 3D surface normal · Multi-task objective · Sim-real mixture

1 Introduction

Robustly estimating multi-person human body pose and shape remains an important research problem. Humans, especially well trained artists, can reconstruct densely three dimensional multi-person sculptures with high accuracy given a RGB reference image. Only recently [2,18,19,40,42,48,53,59] it was demonstrated that this challenging task can also be done using convolutional neural networks with modest in-the-wild accuracy. Improving the accuracy may lead to new applications such as telepresence in virtual reality, large-scale sports video analysis, and 3D MoCap on consumer devices. The challenges presented in

© Springer Nature Switzerland AG 2020
A. Vedaldi et al. (Eds.): ECCV 2020, LNCS 12374, pp. 225–242, 2020.
https://doi.org/10.1007/978-3-030-58526-6_14

natural images of people include mixed clothing and apparel, individual fat-to-muscle ratio, diverse skeletal structure, high-DoF body articulation, soft-tissue deformation, and ubiquitous human-object human-human occlusion.

We explore the possibility of training convolutional neural networks to learn the 2.5D and 3D human representations directly from renderings of animated dressed 3D human figures (non-statistical), instead of designing an increasingly more sophisticated labeling software or building yet another statistical 3D human shape model with simplifications (e.g., ignoring hair or shoes). However it's well-known that "deep neural networks easily fit random labels" [60]. As our simulated 3D human figures are far from being photorealistic, it's easy for the model to overfit the simulated dataset without generalizing to natural images. Therefore the challenge lies in carefully designing a practical mechanism that allows training the difficult 2.5D and 3D human representations on the simulated people domain and generalize well to the in-the-wild domain where there are more details present, e.g., clothing variance (e.g. shoes), occlusion, diverse human skin tone, and hair.

Fig. 1. Visualization of SimPose's 3D surface normal predictions on MSCOCO dataset.

Deep Convolutional Neural Networks (ConvNets) have been successfully applied to human pose estimation [8,23,44,45], person instance segmentation [9,20,37], and dense pose estimation [18] (the representation consists of human mesh's continuous intrinsic uv coordinates). To reach the best task accuracy, they all require collecting real human annotated labels for the corresponding task [18,28]. As the research frontier for human pose and shape estimation expands into predicting new 2.5D and 3D human representations, it becomes self-evident that labelling large-scale precise groundtruth 2.5D and 3D annotations becomes significantly more time-consuming or perhaps even impossible.

Several generic domain-adaptation approaches have been proposed to leverage computer graphics generated precise annotations for in-the-wild tasks: 1) Using Generative Adversarial Network (GAN) framework [7] to jointly train a generator styled prediction network and a discriminator to classify whether the prediction is made on real or simulated domain. 2) Using Domain Randomization [43] to enhance the input image and thus expand the diversity and horizon of the input space of the model. Our approach is complementary to them. For instance, the GRU-domain-classifier [17] or the output space discriminator [46] can be added as additional auxiliary tasks in our multi-task objective, and more uniform, sometimes corrupting, domain randomization can be added on top of our mild data augmentation.

In contrast, we create a new multi-person simulated dataset with accurate dense human pose annotations, and present a simple yet effective technique which mitigates the "Inter-domain Covariate Shift". The closest prior works [48,49] to ours in the human pose estimation field utilize a statistical 3D human shape model SMPL [30] for training (building SMPL requires more 1K 3D scans and geometric simplifications e.g., ignoring hair, shoes). Our approach uses just 17 non-statistic 3D human figures and outperforms SOTA models on DensePose MSCOCO benchmark without additional few-shot real domain fine tuning.

In the following sections of the paper, we discuss related works and describe our SimPose approach in more detail: our simulated dataset, system implementation, experiment setup, the accuracy comparison with SOTA models on DensePose MSCOCO benchmark, and 3D human surface normal results. Our main contributions in the paper are the following:

- Create **simulated multi-person datasets** and present a **simple yet effective batch mixture training strategy with multi-task objective** to learn 2.5D dense pose uv and 3D surface normal estimation model without using any real dense labels.
- Evaluate our proposed approach on the DensePose MSCOCO benchmark [18]. We attained **favourable results using only simulated human uv labels, better than state-of-the-art models** [18,34] which are trained using real DensePose labels. We also present our model's 3D surface normal predictions on MSCOCO dataset which lacks any 3D surface normal labels.
- Show our approach trained using **only 17 non-statistical** 3D human figures can obtain better accuracy on the above benchmark than using SMPL [30] (a statistical 3D human shape model built with more 1K 3D scans). This is a big reduction in number of 3D human scans required.

2   Related Work

Human pose estimation and its extension has been a popular research topic over the years. Early approaches study the structure of the human skeleton with a Pictorial Structure model (PS model) [16]. The method is later improved by combining it with a probabilistic graphics model [3,57]. With the advent of deep learning, the methods are being simplified and the outputs are becoming richer and finer, as we discuss below.

ConvNet Human Pose Estimation. Toshev and Szegedy [45] applies convolutional neural networks on the 2D human pose estimation task by regression to the (x, y) locations of each body joint in a cascaded manner. Tompson et al. [44] proposes a fully convolutional structure by replacing the training target from two scalar (x/y) coordinates to a 2D probabilistic map (or heatmap). The robust heatmap based pose representation became popular and has been extended in many different ways. Wei et al. [51] trains a multi-stage fully convolutional network where the prediction from previous stages are fed into the next stage. Newell et al. [36] proposes an encoder-decoder structured fully convolutional network (called HourglassNet) with long-range skip connection to avoid the dilemma between the output heatmap resolution and the network receptive field. Papandreou et al. [38] augments the heatmap with an offset map for a more accurate pixel location. Chen et al. [11] proposes a cascaded pyramid structured network with hard keypoint mining. Recently, Bin et al. [54] shows that state-of-the-art 2D human pose estimation performance can be achieved with a very simple de-convolutional model.

3D Human Pose and Shape Estimation Tasks. The progress on 2D human pose estimation motivates the community to move forward to more high-level human understanding tasks and take advantage of the 2D pose results. Bottom-up multi-person pose estimation [8,35,37] simultaneously estimates the pose of all persons in the image, by learning the instance-agnostic keypoints and a grouping feature. 3D human pose estimation [32,62,63] learns the additional depth dimension for the 2D skeleton, by learning a 3D pose dictionary [62] or adding a 3D bone length constraint [63]. Human parsing [14,27] segments a human foreground mask into more fine-grained parts. DensePose [18,34] estimation aims to align each human pixel to a canonical 3D mesh. Kanazawa et al.[25] predicts plausible SMPL coefficients using both 2D reprojection loss and 3D adversarial loss. Kolotouros et al. [26] improves it by incorporating SMPL fitting during training. Xu et al. [56] shows that using pre-trained DensePose-RCNN's predictions can further improve 3D pose and shape estimation accuracy.

Simulated Synthetic Datasets. When collecting training data for supervised learning is not feasible, e.g., for dense per-pixel annotation or 3D annotation, simulated training data can be a more useful solution. Chen et al. [10] synthesizes 3D human pose data by deforming a parametric human model (SCAPE [4]) followed by adding texture and background. Similarly, the SURREAL dataset [49] applies the CMU mocap motion data to the SMPL human model [31] on SUN [55] background, with 3D pose, human parsing, depth, and optical flow annotations. Zheng et al. [61] utilizes a Kinect sensor and DoubleFusion algorithm to capture posed people and synthesizes training images which contain one person per image. Compared to the previous simulated human datasets, our datasets contain multi-person images and two 3D human model sources (non-statistical and statistical) for comparison.

Domain Adaptation. As the simulated synthetic data is from a different distribution compared with the real testing data, domain adaptation is often applied to improve the performance. A popular approach is domain adversarial training [47], which trains a domain discriminator and encourages the network to fool this domain discriminator. This results in domain invariant features. Domain randomization [43] generates an aggressively diverse training set which varies all factors instead of the training label, and causes the network to learn the inherited information and ignore the distracting randomness. Other ideas incorporate task specific constraints. Contrained CNN [39] iteratively optimizes a label prior for weakly supervised semantic segmentation in a new domain, and Zhou et al. [63] enforce bone length constraint for 3D human pose estimation in the wild. In this paper, we propose a simpler alternative that generalizes well.

3   Simulated Multi-person Dense Pose Dataset

Fig. 2. Visualization of the 3D human model sources used to create our two simulated people datasets. Renderpeople 3D figures to the left and SMPL with SURREAL textures to the right. For SMPL, we only visualize its two base shapes here, but we sample shapes from its continuous shape space in our experiments. The human manifold's intrinsic uv coordinates labels are visualized as isocontour lines where the background is the dense value of v.

Collecting dense continuous labels on real images is not only expensive and time-consuming but also hard to label accurately. By creating a simulated synthetic dataset, we are able to get consistent per pixel 2.5D and 3D human pose and shape labels (body part segmentation, body part intrinsic uv coordinates, and 3D surface normals), in comparison to [18] where the annotations are not per-pixel labelled. This section describes how we prepare our human 3D models with the non trivial task of generating correct uv mapping (Fig. 2).

3.1   Human 3D Models

We create and compare two simulated datasets with two different sources of human 3D models, one consisting of only 17 rigged high resolution 3D scans of humans acquired from Renderpeople [41] and one using the statistical SMPL body model [31] (built from 1785 scans) together with SURREAL surface textures [49]. Other differences between the sources are the visual quality of the meshes and textures. The Renderpeople models include clothing and hair while the SMPL model does not. The Renderpeople textures are more realistic while the SURREAL textures may have artifacts. 3D reference points are attached to the skeleton of each rigged human model. We align these points with the MSCOCO 2D human keypoint definition for rendering simulated 2D keypoint.

Render People. We ensure that all our human 3D models have the same uv and body part segmentation mapping as the references in DensePose labels [18] so that our approach can be evaluated on the MSCOCO DensePose benchmark. This is not easy to achieve as there are no available tools out there. We propose Algorithm 1, which takes 40 min per Renderpeople model (total 17 3D human model) for a person with previous experience in 3D modeling software.

Algorithm 1. Transfer DensePose UV to Renderpeople meshes

---

for each bodypart submesh bp_{ref} in reference mesh **do**
    Mark a subset of the boundary vertices in bp_{ref} as landmarks L_{ref}.
    Store UV coordinates UV_{ref} for each vertex in L_{ref}.
end for
for each Renderpeople mesh rp **do**
    for each bodypart submesh bp_{ref} in reference mesh **do**
        Cut a submesh bp_{rp} from rp that is similar to bp_{ref}.
        Mark the same landmark vertices L_{rp} as in L_{ref}.
        Copy UV coordinates from UV_{ref} to UV_{rp}.
        Linearly interpolate UV coordinates along the boundary vertices in bp_{rp}.
        Mark boundary vertices in bp_{rp} as static.
        Unwrap UV on the inner non-static vertices, similar to [13].
    end for
end for

SMPL. It is trivial to populate the DensePose UV coordinates from [18] since they also use the SMPL model as the reference mesh.

To make it easier to train and compare our approach on the datasets separately, we make sure that each image only has humans from one of the sources. Figure 3 shows example images from both sources. In total, we generated 100,000 images for each source (Fig. 4).

4   Our Approach

Given a two-stage multi-person pose estimation meta-architecture, our approach, as illustrated by Fig. 5, anchors the domain adaptation of the 2nd-stage model's new 2.5D and 3D human pose & shape estimation tasks on a set of easier 2D tasks. Inspired by the recent works of multi-task & meta-learing [5,15,58], a modified domain-label-based multi-task objective trains a deep normalized network end-to-end on a batch mixture of sim-real samples. We show that within a deep network, batch normalization, together with convolutions, can permit whitening and aligning bi-modal distributions via Batch Mixture Normalization. From an empirical perspective, our approach: i. generalizes from only 17 non-statistical 3D human figures, which we show in Table 1, and hence suffices as a functional

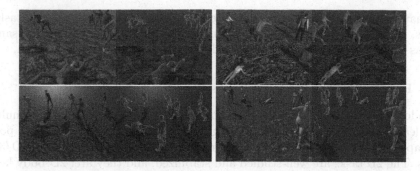

Fig. 3. Examples from our simulated datasets. Top row shows the Renderpeople dataset and bottom row shows the SMPL dataset. Only the *uv* label is shown.

Fig. 4. Examples from our Renderpeople simulated datasets. From left to right: 2D keypoints, instance segmentation, body parts, *uv*, and 3D surface normal.

replacement of using a statistical human shape model built from >1k 3D scans; **ii.** is able to directly domain-adapt continuous intrinsic *uv* coordinates and 3D surface normals.

4.1 Human Pose Estimation Meta-architectures

Among two widely used meta-architectures for multi-person human pose & shape estimation tasks, one is the bottom-up meta-architecture and the other is the two-stage meta-architecture. In the bottom-up meta-architecture, a neural network is applied fully convolutionally to predict instance-agnostic body parts for all people, as well as additional auxiliary representations for grouping atomic parts into individual person instances (e.g., part affinity field [8] and mid-range offsets [37]). The two-stage meta-architecture relies on a separate detection network or an RPN (region proposal network) whose outputs, ROIs (Region of Interest) of the input image or of the intermediate feature maps, are then cropped, resized, and fed into a 2nd stage pose estimation network.

We adopt the two-stage meta-architecture. The crop-and-resize operation is used as spatial attention of the second-stage network to restrict the divergence between real and simulation distributions within the ROIs. More specifically, our two-stage meta-architecture is the following: the 1st stage is a Faster RCNN person detector with ResNet101 [21] backbone (following [38], we make the same surgery to the original Faster RCNN including using dilated convolutions and training only on MSCOCO [29] person bounding box alone); The 2nd stage pose

estimation network is shown in Fig. 5. Before applying the corresponding tasks'
losses, the outputs from the 2nd stage model are upsampled to be of the same
size as the input image crops using the differentiable bilinear interpolation.

4.2 Human Pose Estimation Tasks

We adopt the term "task" that is used unanimously in the meta-learning, multi-
task learning, and reinforcement learning literature. We cluster the human pose
& shape estimation tasks into two categories: **i.** the 2D tasks and **ii.** the 2.5D/3D
tasks. The 2D tasks are well-studied and stabilized, and they are: 2D body keypoint localization and 2D person instance segmentation. The 2.5D/3D tasks are
new, and we select the following two tasks: human mesh intrinsic uv coordinates
regression and 3D surface normal vector regression. Below we describe each task
and the task formulation in details.

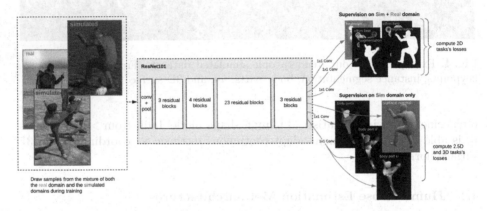

Fig. 5. The ResNet101 network takes image crops batch mixture from both real and
simulated domains as input and generates predictions for both 2.5D 3D tasks and 2D
tasks which consist of sparse 2D body keypoints and instance segmentation prediction
tasks. We compute the 2D tasks' losses for both the real and the simulated domains
and compute the 2.5D and 3D tasks' losses only for the simulated domain.

2D Human Keypoints Estimation Task

The sparse 2D human keypoint estimation task requires precisely localizing the
2D subpixel positions of a set of keypoints on the human body (e.g., 17 keypoints
defined by MSCOCO: "left shoulder", "right elbow", etc). We follow [38] and
let the 2nd stage network predict heatmaps and offset fields. We use Hough-
voting [38] to get more concentrated score maps. The final 2D coordinates of
each keypoint are extracted via the spatial argmax operation on the score maps.
The heatmap predictions are supervised using per-pixel per-keypoint sigmoid
cross entropy loss and the offset field is supervised using per-pixel per-keypoint
Huber loss.

2D Person Instance Segmentation Estimation Task
The task of instance segmentation requires classifying each pixel into foreground pixels that belong to the input crop's center person, and background pixels that don't. We follow [9] and let the 2nd stage network predict per-pixel probability of whether it belongs to the center person or not, and use standard per-pixel sigmoid cross entropy loss to supervise the prediction.

2.5D Body Part Intrinsic UV Coordinates Regression Task
For the 2.5D uv coordinates regression task, following the work of [18], we predict K body part segmentation masks and $2K$ body uv maps in the DensePose convention (e.g., "right face", "front torso", "upper right leg", and etc). We use per-pixel per-part sigmoid cross entropy loss to supervise the body part segmentation and used per-pixel per-part smooth l_1 loss to supervise the uv predictions. We only backprop the l_1 loss through the groundtruth body part segmentation region. For the uv maps we subtract 0.5 from the groundtruth value to center the range of regression.

3D Person Surface Normal Regression Task
For the 3D person surface normal task, the 2nd stage network directly regresses per-pixel 3D coordinates of the unit surface normal vectors. During post-training inference, we l_2 renormalize the 2nd-stage networks's per-pixel 3D surface normal predictions: $\hat{\mathbf{n}} = \frac{\mathbf{n}}{||\mathbf{n}||}$. We use per-pixel smooth l_1 loss to supervise the surface normal predictions. We only backprop the loss L through the groundtruth person instance segmentation region S using a discrete variance of:

$$\frac{\partial L(\theta)}{\partial \theta} = \frac{\partial}{\partial \theta} \sum_i^m \iint_{S_i} ||\mathbf{n} - \mathbf{f}_\theta(I_i)|| \, dx \, dy \tag{1}$$

where θ is all the trainable parameters of the network, m is the batch size, I_i is the i-th image in the batch, (x, y) are the coordinates in the image space, $\mathbf{n} = (n_x, n_y, n_z)$ is the groundtruth 3D surface normal vector at (x, y), $\mathbf{f}_\theta(\cdot)$ is the network predicted surface normal vector at (x, y), S_i is the support of the groundtruth person segmentation in image I_i, and $|| \cdot ||$ is the smooth l_1 norm.

4.3   Multi-task Learning and Batch Mixture Normalization

Learning to estimate people's 2.5D intrinsic uv coordinates and 3D surface normals from renderings of 17 simulated people alone, and generalizing to natural image domain of people, requires a deep neural network to learn domain-invariant representations of the human body. We utilize a shared backbone trained end-to-end to achieve this. The network needs to learn domain-invariant representations of a 3D human body at its higher layers that are robust to the inevitably diverged representations at its lower layers. We refer to this underlying issue as the Inter-domain Covariate Shifts.

Batch Normalization [22] was designed to reduce the "Internal Covariate Shift" [6,12] for supervised image classification tasks. It assumes similar activation statistics within the batch and whitens the input activation distribution by its batch mean μ and batch variance σ. When training a shared backbone using a batch mixture of images sampled from the two domains of a) simulated people and b) natural images of people, the two input activation distributions almost certainly don't share the same mean μ and variance σ. Below we show one non-trivial realization of multi-modal whitening using: 1. Batch Normalization, 2. Fuzzy AND ($\wedge$) OR ($\vee$) implemented as convolution followed by activation, and 3. implicitly learnt domain classifier within the network. We call it Batch Mixture Normalization (BMN), Algorithm 2. During training, we believe the deep batch-normalized shared backbone can whiten bi-modal activations from the sim-real batch mixture using a fuzzy variance of BMN, when it's the optimal thing to do:

Algorithm 2. Batch Mixture Normalization

Require: $s^k(\cdot)$: a k-th layer featuremap (useful only on sim images)
Require: $r^k(\cdot)$: a k-th layer featuremap (useful only on real images)
Require: $z^k(\cdot)$: a k-th layer featuremap implicitly learnt as domain classifier
Require: Fuzzy AND ($\wedge$) OR ($\vee$) implemented as convolution followed by activation

Sample a batch of m image $\{I_i\}_i^m \sim \mathcal{D}_{sim} \cap \mathcal{D}_{real}$
Evaluate k layers to get: $\{z\}_i^m, \{s\}_i^m, \{r\}_i^m = z^+(\{I_i\}_i^m), s^+(\{I_i\}_i^m), r^+(\{I_i\}_i^m)$
Mask domain outliers (the k+1 layer): $[...\{s\}, \{r\}, ...] = [..., \boxed{\{z \wedge s\}, \{\bar{z} \wedge r\}} ...]$
Batch Normalize (the k+2 layer): $[..., \{s\}, \{r\}, ...] = BatchNorm([..., \{s\}, \{r\}, ...])$
Bring into the same channel (the k+3 layer): $[..., \{y\}, ...] = [..., \boxed{\{s \vee r\}}, ...]$

$\triangleright$ For succeeding layers ($\geqslant$ k + 4), feature $\{y\}_i^m$ is the whitened and aligned version of $s^k(\{I_i\}_i^m)$ and $r^k(\{I_i\}_i^m)$.

We incorporate domain labels in the multi-task objective to cope with the constraint that 2.5D human intrinsic uv and 3D human body surface normal labels are only available on the simulated domain. Intuitively each task demands that the network learns more about the human body structure and shape in a complementary way. Due to mixing and normalizing, the network's forward and backward passes depend on the composition of the sim-real batch mixture.

$$L(\{I_i\}, \theta) = \sum_{l=1}^{m} L^{2D}(f_\theta(\{I_i\})) + \sum_{\substack{j=1 \\ I_j \in \mathcal{D}_{sim}}}^{m} [L^{2.5D}(f_\theta(\{I_i\}) + L^{3D}(f_\theta(\{I_i\}))] \quad (2)$$

where L^{2D} is the loss term for the 2D keypoint estimation task and the 2D instance segmentation task, $L^{2.5D}$ is the loss term for the human intrinsic uv coordinates regression task, and L^{3D} is the loss term of the 3D human body surface normal regression task. $f_\theta(\cdot)$ is the shared human pose and shape backbone parameterized by θ, m is the batch size, and $\{I_i\}_i^m$ is the batch mixture of simulated image crops and real image crops. In practice, we adopt ResNet101 as our backbone and apply a set of 1×1 convolutions at the last layer of ResNet101 to predict: keypoint heatmaps, keypoint offset field, person instance segmentation, body part segmentations, body part intrinsic uv coordinates, and 3D human surface normals (see Sect. 4.2). We apply a per task weight to balance the loss for each task. To avoid exhaustive grid-search and task gradient interference [58], we adopt a greedy hyper-parameter search for individual task weights: 1) search for 2D task weights: $w_{heatmap}$, $w_{offsets}$, $w_{segment}$. 2) fix the 2D task weights and sequentially search for the weights for uv and normal tasks: w_{parts}, w_{uv}, and w_{normal}. We used the following weights for each task: $w_{heatmap} = 4.0$, $w_{offsets} = 1.0$, $w_{segment} = 2.0$, $w_{parts} = 0.5$, $w_{uv} = 0.25$, and $w_{normal} = 1.00$. During training we apply an additional gradient multiplier ($10\times$) to the 1×1 convolution weights used to predict the heatmap, offset field, and person instance segmentation.

Compared with [52], our main differences are: 1) we have a new multi-person 3D surface normal prediction (Sect. 4.2) that Detectron2 doesn't have; 2) we obtained competitive DensePose accuracy, learnt from synthetic labels (Sect. 5.2); 3) our modified multi-task objective induces a richer label set and gradients for the simulated part (add. 3D surface normal) of the mini-batch, vs a reduced label set and gradients for the real part of the mini-batch; 4) we optimize only the 2nd-stage pose estimator instead of jointly optimizing RPN and multi-heads.

5   Evaluation

Table 1. UV performance on DensePose COCO **minival** split (higher is better).

	AP	AP$_{50}$	AP$_{75}$	AP$_M$	AP$_L$	AR	AR$_{50}$	A$_{75}$	AR$_M$	AR$_L$
Prior works (use labelled IUV)										
DP-RCNN-cascade [18]	51.6	83.9	55.2	41.9	53.4	60.4	88.9	65.3	43.3	61.6
DP-RCNN-cascade + masks [18]	52.8	85.5	56.1	40.3	54.6	62.0	89.7	67.0	42.4	63.3
DP-RCNN-cascade + keypoints [18]	55.8	87.5	61.2	48.4	57.1	63.9	91.0	69.7	50.3	64.8
Slim DP: DP-RCNN (ResNeXt101) [34]	55.5	**89.1**	60.8	50.7	56.8	63.2	92.6	69.6	51.8	64.0
Slim DP: DP-RCNN + Hourglass [34]	**57.3**	88.4	63.9	57.6	58.2	65.8	92.6	73.0	59.6	66.2
Ours (use only simulated IUV)										
SimPose (3D model source: Renderpeople)	**57.3**	88.4	**67.3**	60.1	**59.3**	**66.4**	**95.1**	**77.8**	62.4	**66.7**
SimPose (3D model source: SMPL)	56.2	87.9	65.3	**61.0**	58.0	65.2	95.1	75.2	**63.2**	65.3

5.1   Experimental Setup

Our SimPose system is implemented in the TensorFlow framework [1]. The 2nd-stage model is trained using the proposed approach. We use 4 P100 GPUs on one single machine and use synchronous stochastic gradient descent optimizer (learning rate is set to 0.005, momentum value is set to 0.9, batch size is set to 8 for each GPU). We train the model for 240K steps. The ResNet101 backbone has been pre-trained for 1M steps on the same sparse 2D 17 human keypoints and instance segmentation annotations mentioned above (no real MSCOCO Dense-Pose labels are used) from an ImageNet classification checkpoint. The 1st-stage Faster RCNN person detector is trained on the MSCOCO person bounding box labels using asynchronous distributed training with 9 K40 GPUs. Stochastic gradient descent with momentum is used as optimizer (learning rate is set to 0.0003, momentum value is set to 0.9, learning rate is decayed by 10× at the 800K step of the total 1M training steps). The ResNet101 backbone has been pretrained on the ImageNet classification task.

We use the Unity game engine to generate our simulated people datasets. We render up to 12 humans per image. Each image is 800 × 800 pixels with 60° field of view and with motion blur disabled. All shaders use the default PBR [33] shading in Unity with low metallic and smoothness values. The shadows use a high resolution shadow map with four shadow cascades. The position of each human on the ground plane is random given the constraint that it can not intersect another human. We pose the human models with a random pose from one of 2,000 different Mixamo animations, similar to [50]. The 17 Renderpeople models do not have blend shapes or different textures. Instead we randomly augment the hue [24] of the Renderpeople clothing to add variation.

Fig. 6. SimPose's UV predictions on DensePose MSCOCO minival split.

5.2   Evaluation of SimPose's UV Prediction

Table 1 shows our SimPose approach's accuracy on the DensePose MSCOCO minival split (1.5K images). Despite the fact that our model has not used any real DensePose labels, it achieves 57.3 average precision measured by the GPS (Geodesic Point Similarity) metrics on the challenging multi-person DensePose MSCOCO benchmark, which is better than the DensePose-RCNN [18] model's average precision of 55.8, and better than the state-of-the-art Slim DensePose [34] on most of the breakdown AP & AR metrics. Both DensePose-RCNN and Slim DensePose have been trained using real DensePose labels.

Table 2. Ablation study of simulated data mixing ratio. Higher mixing ratio of simulated data is detrimental to model's 2D tasks accuracy on MSCOCO and has diminishing returns for the 2.5D and 3D tasks on the Renderpeople validation set.

Task	Metric	Percentage of sim data				
		100%	75%	50%	25%	0%
Normal	ADD (lower is better)	19.3°	19.5°	19.9°	21.5°	74.6°
UV	L2 (lower is better)	0.075	0.075	0.077	0.082	0.386
Segmentation	IOU (higher is better)	0.26	0.68	0.69	0.69	0.70
Keypoint	OKS (higher is better)	0.32	0.74	0.76	0.75	0.77

Table 3. Ablation study of UV task weight. Higher value doesn't affect 2D tasks on COCO and has diminishing returns for UV accuracy on Renderpeople validation set.

Task	Metric	UV task weight				
		0.00	0.25	0.50	1.00	2.00
UV	L2 (lower is better)	0.384	0.077	0.076	0.074	0.071
Segmentation	IOU (higher is better)	0.698	0.692	0.697	0.695	0.697
Keypoint	OKS (higher is better)	0.761	0.756	0.760	0.760	0.756

Furthermore, in Table 1 we also compare the accuracy of our system trained from Renderpeople and SMPL separately. We found using only 17 Renderpeople 3D human models and our proposed approach, the system achieves better performance than using SMPL (a statistical 3D human shape model built from more 1 K 3D human scans). We conduct ablation studies of the simulated data mixing ratio (Table 2) and the UV task weight (Table 3) for the 2nd stage model. In Fig. 6, we visualize our SimPose system's UV predictions on the DensePose MSCOCO minival split.

Table 4. Ablation study of 3D normal task weight. Higher value doesn't affect 2D tasks on COCO and has diminishing returns on the Renderpeople validation set.

Task	Metric	3D normal task weight				
		0.00	0.25	0.50	1.00	2.00
Normal	ADD (lower is better)	76.1°	22.1°	21.1°	19.9°	18.6°
UV	L2 (lower is better)	0.077	0.077	0.076	0.077	0.077
Segmentation	IOU (higher is better)	0.699	0.697	0.700	0.696	0.698
Keypoint	OKS (higher is better)	0.763	0.759	0.760	0.756	0.760

Fig. 7. SimPose's 3D surface normal predictions on MSCOCO dataset.

SMPL   Renderpeople   SMPL   Renderpeople

Fig. 8. Comparison of normal predictions trained from different 3D model sources.

5.3   Evaluation of SimPose's Surface Normal Prediction

In Fig. 1, 7, and 8, we qualitatively visualize SimPose's 3D surface normal predictions on a withheld MSCOCO validation set which the model hasn't been trained on. Our model generalizes well to challenging scenarios: crowded multi-person scenes, occlusions, various poses and skin-tones. We evaluate the 3D surface normal predictions on the Renderpeople validation set using ADD (Average Degree Difference). We conduct ablation studies of the simulated data mixing ratio (Table 2) and the 3D normal task weight (Table 4).

6   Conclusion

We have shown that our SimPose approach achieves more accurate in-the-wild multi-person dense pose prediction without using any real DensePose labels for training. It also learns 3D human surface normal estimation from only simulated labels and can predict 3D human surface normal on the in-the-wild MSCOCO dataset that currently lacks any surface normal labels. Our SimPose approach achieves this using only 17 non-statistical 3D human figures. We hope the creation process of our simulated dataset and the proposed training scheme opens doors for training other accurate 2.5D and 3D human pose and shape estimation models without manually collecting real world annotations, and still generalizes for accurate multi-person prediction in-the-wild usage.

Acknowledgements. We would like to thank Nori Kanazawa for reviewing our implementation, Fabian Pedregosa and Avneesh Sud for proofreading our paper.

References

1. Abadi, M., Agarwal, A., Barham, P., Brevdo, E., et al.: TensorFlow: Large-scale machine learning on heterogeneous systems (2015). https://www.tensorflow.org/
2. Alldieck, T., Magnor, M., Bhatnagar, B.L., Theobalt, C., Pons-Moll, G.: Learning to reconstruct people in clothing from a single RGB camera. In: CVPR (2019)
3. Andriluka, M., Roth, S., Schiele, B.: Pictorial structures revisited: people detection and articulated pose estimation (2009)
4. Anguelov, D., Srinivasan, P., Koller, D., Thrun, S., Rodgers, J., Davis, J.: Scape: shape completion and animation of people. In: ACM Transactions on Graphics (TOG) (2005)
5. Balaji, Y., Sankaranarayanan, S., Chellappa, R.: Metareg: towards domain generalization using meta-regularization. In: NeurIPS (2018)
6. Bjorck, N., Gomes, C.P., Selman, B., Weinberger, K.Q.: Understanding batch normalization. In: NeurIPS (2018)
7. Bousmalis, K., Silberman, N., Dohan, D., Erhan, D., Krishnan, D.: Unsupervised pixel-level domain adaptation with generative adversarial networks. In: CVPR (2017)
8. Cao, Z., Wei, S.E., Simon, T., Sheikh, Y.: Realtime multi-person 2D pose estimation using part affinity fields. In: CVPR (2017)

9. Chen, L.C., Papandreou, G., Kokkinos, I., Murphy, K., Yuille, A.L.: Deeplab: semantic image segmentation with deep convolutional nets, atrous convolution, and fully connected crfs. TPAMI **40**, 834–848 (2017)
10. Chen, W., et al.: Synthesizing training images for boosting human 3D pose estimation. In: 3DV (2016)
11. Chen, Y., Wang, Z., Peng, Y., Zhang, Z., Yu, G., Sun, J.: Cascaded pyramid network for multi-person pose estimation (2017). arXiv preprint arXiv:1711.07319
12. De, S., Smith, S.L.: Batch normalization biases deep residual networks towards shallow paths. ArXiv (2020)
13. Eck, M., DeRose, T., Duchamp, T., Hoppe, H., Lounsbery, M., Stuetzle, W.: Multiresolution analysis of arbitrary meshes. In: Proceedings of the 22nd Annual Conference on Computer Graphics and Interactive Techniques. SIGGRAPH (1995)
14. Fang, H.S., Lu, G., Fang, X., Xie, J., Tai, Y.W., Lu, C.: Weakly and semi supervised human body part parsing via pose-guided knowledge transfer. In: CVPR (2018)
15. Finn, C., Abbeel, P., Levine, S.: Model-agnostic meta-learning for fast adaptation of deep networks. In: ICML (2017)
16. Fischler, M.A., Elschlager, R.: The representation and matching of pictorial structures. In: IEEE TOC (1973)
17. Ganin, Y., et al.: Domain-adversarial training of neural networks. JMLR **17**, 2030–2096 (2015)
18. Güler, R.A., Neverova, N., Kokkinos, I.: Densepose: dense human pose estimation in the wild. In: CVPR (2018)
19. Habermann, M., Xu, W., Zollhöfer, M., Pons-Moll, G., Theobalt, C.: Livecap: real-time human performance capture from monocular video. ACM Trans. Graph. **38**, 1–17 (2019)
20. He, K., Gkioxari, G., Dollár, P., Girshick, R.B.: Mask r-cnn. In: ICCV (2017)
21. He, K., Zhang, X., Ren, S., Sun, J.: Deep residual learning for image recognition. In: CVPR (2016)
22. Ioffe, S., Szegedy, C.: Batch normalization: accelerating deep network training by reducing internal covariate shift. In: ICML (2015)
23. Jain, A., Tompson, J., Andriluka, M., Taylor, G., Bregler, C.: Learning human pose estimation features with convolutional networks. In: ICLR (2014)
24. Joblove, G.H., Greenberg, D.: Color spaces for computer graphics. In: Proceedings of the 5th Annual Conference on Computer Graphics and Interactive Techniques. SIGGRAPH (1978)
25. Kanazawa, A., Black, M.J., Jacobs, D.W., Malik, J.: End-to-end recovery of human shape and pose. In: CVPR (2018)
26. Kolotouros, N., Pavlakos, G., Black, M.J., Daniilidis, K.: Learning to reconstruct 3D human pose and shape via model-fitting in the loop. In: ICCV (2019)
27. Liang, X., Shen, X., Feng, J., Lin, L., Yan, S.: Semantic object parsing with graph LSTM. In: ECCV (2016)
28. Lin, T.Y., et al.: Coco 2016 keypoint challenge. In: ECCV (2016)
29. Lin, T.Y., et al.: Microsoft COCO: Common Objects in Context. In: Fleet, D., Pajdla, T., Schiele, B., Tuytelaars, T. (eds.) ECCV 2014. LNCS, vol. 8693, pp. 740–755. Springer, Cham (2014). https://doi.org/10.1007/978-3-319-10602-1_48
30. Loper, M., Mahmood, N., Romero, J., Pons-Moll, G., Black, M.J.: SMPL: a skinned multi-person linear model. ACM Trans. Graph (Proc. SIGGRAPH Asia) (2015)
31. Loper, M., Mahmood, N., Romero, J., Pons-Moll, G., Black, M.J.: Smpl: a skinned multi-person linear model. ACM Trans. Graph. (TOG) **34**, 1–16 (2015)
32. Martinez, J., Hossain, R., Romero, J., Little, J.J.: A simple yet effective baseline for 3D human pose estimation. In: ICCV (2017)

33. McAuley, S., et al.: Practical physically-based shading in film and game production. In: ACM SIGGRAPH 2012 Courses, SIGGRAPH 2012 (2012)
34. Neverova, N., Thewlis, J., Guler, R.A., Kokkinos, I., Vedaldi, A.: Slim densepose: thrifty learning from sparse annotations and motion cues. In: CVPR (2019)
35. Newell, A., Huang, Z., Deng, J.: Associative embedding: end-to-end learning for joint detection and grouping. In: NeurIPS (2017)
36. Newell, A., Yang, K., Deng, J.: Stacked hourglass networks for human pose estimation. In: Leibe, B., Matas, J., Sebe, N., Welling, M. (eds.) ECCV 2016. LNCS, vol. 9912, pp. 483–499. Springer, Cham (2016). https://doi.org/10.1007/978-3-319-46484-8_29
37. Papandreou, G., Zhu, T., Chen, L.C., Gidaris, S., Tompson, J., Murphy, K.: Personlab: person pose estimation and instance segmentation with a bottom-up, part-based, geometric embedding model. In: ECCV (2018)
38. Papandreou, G., et al.: Towards accurate multi-person pose estimation in the wild. In: CVPR (2017)
39. Pathak, D., Krahenbuhl, P., Darrell, T.: Constrained convolutional neural networks for weakly supervised segmentation. In: ICCV (2015)
40. Pavlakos, G., et al.: Expressive body capture: 3D hands, face, and body from a single image. In: CVPR (2019)
41. RenderPeople: Renderpeople: 3D people for renderings. https://renderpeople.com
42. Saito, S., Huang, Z., Natsume, R., Morishima, S., Kanazawa, A., Li, H.: Pifu: pixel-aligned implicit function for high-resolution clothed human digitization. In: ICCV (2019)
43. Tobin, J., Fong, R., Ray, A., Schneider, J., Zaremba, W., Abbeel, P.: Domain randomization for transferring deep neural networks from simulation to the real world. In: IROS (2017)
44. Tompson, J., Jain, A., LeCun, Y., Bregler, C.: Join training of a convolutional network and a graphical model for human pose estimation. In: NIPS (2014)
45. Toshev, A., Szegedy, C.: Deeppose: human pose estimation via deep neural networks. In: CVPR (2014)
46. Tsai, Y.H., Sohn, K., Schulter, S., Chandraker, M.K.: Domain adaptation for structured output via discriminative patch representations. In: ICCV (2019)
47. Tzeng, E., Hoffman, J., Saenko, K., Darrell, T.: Adversarial discriminative domain adaptation. In: CVPR (2017)
48. Varol, G., et al.: BodyNet: volumetric inference of 3D human body shapes. In: ECCV (2018)
49. Varol, G., et al.: Learning from synthetic humans. In: CVPR (2017)
50. Villegas, R., Yang, J., Ceylan, D., Lee, H.: Neural kinematic networks for unsupervised motion retargetting. In: CVPR (2018)
51. Wei, S.E., Ramakrishna, V., Kanade, T., Sheikh, Y.: Convolutional pose machines. In: CVPR (2016)
52. Wu, Y., Kirillov, A., Massa, F., Lo, W.Y., Girshick, R.: Detectron2 (2019). https://github.com/facebookresearch/detectron2
53. Xiang, D., Joo, H., Sheikh, Y.: Monocular total capture: posing face, body, and hands in the wild. In: CVPR (2019)
54. Xiao, B., Wu, H., Wei, Y.: Simple baselines for human pose estimation and tracking. In: ECCV (2018)
55. Xiao, J., Hays, J., Ehinger, K.A., Oliva, A., Torralba, A.: Sun database: large-scale scene recognition from abbey to zoo. In: CVPR (2010)
56. Xu, Y., Zhu, S.C., Tung, T.: Denserac: joint 3D pose and shape estimation by dense render-and-compare. In: ICCV (2019)

57. Yang, Y., Ramanan, D.: Articulated pose estimation with flexible mixtures of parts. In: CVPR (2011)
58. Yu, T., Kumar, S., Gupta, A., Levine, S., Hausman, K., Finn, C.: Gradient surgery for multi-task learning (2020)
59. Zanfir, A., Marinoiu, E., Zanfir, M., Popa, A.I., Sminchisescu, C.: Deep network for the integrated 3D sensing of multiple people in natural images. In: NeurIPS (2018)
60. Zhang, C., Bengio, S., Hardt, M., Recht, B., Vinyals, O.: Understanding deep learning requires rethinking generalization. In: ICLR (2017)
61. Zheng, Z., Yu, T., Wei, Y., Dai, Q., Liu, Y.: Deephuman: 3D human reconstruction from a single image. In: ICCV (2019)
62. Zhou, X., Zhu, M., Leonardos, S., Derpanis, K.G., Daniilidis, K.: Sparseness meets deepness: 3D human pose estimation from monocular video. In: CVPR (2016)
63. Zhou, X., Huang, Q., Sun, X., Xue, X., Wei, Y.: Towards 3D human pose estimation in the wild: a weakly-supervised approach. In: ICCV (2017)

ByeGlassesGAN: Identity Preserving Eyeglasses Removal for Face Images

Yu-Hui Lee[1](✉) [iD] and Shang-Hong Lai[1,2]

[1] Department of Computer Science, National Tsing Hua University, Hsinchu, Taiwan
s106062508@m106.nthu.edu.tw
[2] Microsoft AI R&D Center, Taipei, Taiwan
shlai@microsoft.com

Abstract. In this paper, we propose a novel image-to-image GAN framework for eyeglasses removal, called ByeGlassesGAN, which is used to automatically detect the position of eyeglasses and then remove them from face images. Our ByeGlassesGAN consists of an encoder, a face decoder, and a segmentation decoder. The encoder is responsible for extracting information from the source face image, and the face decoder utilizes this information to generate glasses-removed images. The segmentation decoder is included to predict the segmentation mask of eyeglasses and completed face region. The feature vectors generated by the segmentation decoder are shared with the face decoder, which facilitates better reconstruction results. Our experiments show that Bye-GlassesGAN can provide visually appealing results in the eyeglasses-removed face images even for semi-transparent color eyeglasses or glasses with glare. Furthermore, we demonstrate significant improvement in face recognition accuracy for face images with glasses by applying our method as a pre-processing step in our face recognition experiment.

Keywords: Generative Adversarial Networks · Face Attributes Manipulation · Face recognition

1 Introduction

Face recognition has been researched extensively and widely used in our daily lives. Although state-of-the-art face recognition systems are capable of recognizing faces for practical applications, their accuracies are degraded when the face images are partially occluded, such as wearing eyeglasses. An obvious reason causes this problem is that the eyeglasses may occlude some important information on faces, leading to discrepancies in facial feature values. For example, the thick frame of glasses may block the eyes. Hence, in the past, researchers

Electronic supplementary material The online version of this chapter (https://doi.org/10.1007/978-3-030-58526-6_15) contains supplementary material, which is available to authorized users.

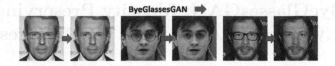

Fig. 1. Examples of glasses removal by ByeGlassesGAN.

proposed to apply the PCA-based methods [20,24] to remove eyeglasses from face images. However, the PCA-based method can only provide approximate glasses removal image via face subspace projection. In addition, they did not really evaluate their methods on diverse face recognition tasks.

Another reason for the degradation of face recognition accuracy with eyeglasses is that face images with eyeglasses are considerably fewer than glasses-free images. It is hard to make the recognition model learn the feature of various kinds of eyeglasses. Recently, alongside with the popularity of face attributes manipulation, some GAN based methods, such as [6] and [30], improved the capability of recognizing faces with eyeglasses by synthesizing a large amount of images of faces with eyeglasses for training a face recognition model.

Different from the previous works, we aim at improving face recognition accuracy by removing eyeglasses with the proposed GAN model before face recognition. With the proposed GAN-based method, we can not only improve face recognition accuracy, the visually appealing glasses-removed images can also be used for some interesting applications, like applying virtual makeup.

The main contributions of this work are listed as follows:

1. We propose a novel glasses removal framework, which can automatically detect and remove eyeglasses from a face image.
2. Our proposed framework combines the mechanisms of the feature sharing between 2 decoders to acquire better visual results, and an identity classifier to make sure the identity in the glasses-removed face image is well preserved.
3. We come up with a new data synthesis method to train a glasses removal network, which effectively simulates color lens, glare of reflection as well as the refraction on eyeglasses.
4. In the experiment, we demonstrate that the face recognition accuracy is significantly improved for faces with eyeglasses after applying the proposed eyeglasses removal method as a pre-processing step.

2   Related Works

2.1   Face Attributes Manipulation

Face attributes manipulation is a research topic that attracts a lot of attention. Along with the popularity of GAN, there are many impressive GAN-based methods proposed for editing face attributes. [16] and [30] edit face attributes through an attribute transformation network and a mask network. Both of them preserve the identity of the source images by using the predicted mask to constrain the editing area. AttGAN [7] edits face images through the attribute classification constraint and reconstruction learning. ELEGANT [25] can not only manipulate face images but also manipulate images according to the attributes of reference images. ERGAN [9] removes eyeglasses by switching features extracted from a face appearance encoder and an eye region encoder. Besides, there are several face attributes editing methods which are not GAN-based. For example, DFI [23] manipulated face images through linear interpolation of the feature vectors of different attributes. [2] achieved identity preserving face attributes editing by disentangling the identity and attributes vectors of face images with the mechanisms of Variational Autoencoder and GAN. However, these face attributes manipulation methods suffer from the instability problem. For example, the identity may not be preserved or some artifacts may be generated.

2.2 Image Completion

Eyeglasses removal can also be seen as a face image completion problem. Recently there are many deep learning works [11, 13, 17, 21, 26, 29] focusing on image completion. Context Encoder [21] is the first deep learning and GAN based inpainting method. After that, [11] significantly improved the quality of inpainting results by using both a global and a local discriminators, with one of them focusing on the whole image and the other focusing on the edited region. Partial-Conv [17] masks the convolution to reduce the discrepancy (e.g. color) between the inpainted part and the non-corrupted part. Recently, there are also some interesting and interactive completion methods [13, 28] which support free-form input. Users can easily control the inpainted result by adding desired sketches on the corrupted regions.

The main difference between the proposed ByeGlassesGAN and existing image completion methods is that our method does not require a predefined mask for the completion. In fact, the eyeglasses removal problem is not the same as image completion because the glasses region could be either transparent or semi-transparent. Our method can exploit the original image in the glasses region to provide better glasses-removed result. Besides, compared with the face attributes manipulation methods described above, our method can automatically remove the glasses and better preserve the face identity in the glasses-removed images.

3   ByeGlassesGAN

In this paper, we propose a multi-task learning method which aims at predicting the position of eyeglasses and removing them from the source image. Since we expect the eyeglasses-removed images can improve the performance of face recognition, the generated results of ByeGlassesGAN must look realistic and well preserve the identities of the source images.

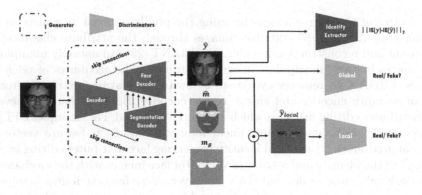

Fig. 2. Framework of ByeGlassesGAN. Each input image (x) is first fed into the Encoder to encode the feature vector. The Face Decoder and the Segmentation Decoder then manipulate the glasses-removed image ($\hat{y}$) and the segmentation mask ($\hat{m}$) of eyeglasses and face shape with the extracted vectors. Two discriminators are included to make sure both the whole generated image ($\hat{y}$) and the edited part ($\hat{y}_{Local}$) look realistic. An Identity Extractor is also applied to minimize the distance between the identity feature vectors computed from the output image ($\hat{y}$) and the ground truth image.

3.1 Proposed Framework

Figure 2 illustrates the overall framework of our ByeGlassesGAN, which contains a generator, an identity extractor, and two discriminators. The generator (G) can be separated into 3 deep neural networks, encoder (E), face decoder (FD), and segmentation decoder (SD). Here we assume the training data contains a set of face images (x) associated with the corresponding glasses-removed images (y) and the corresponding masks (m) of eyeglasses region and the completed face shape. Given a source face image x, which first goes through the encoder to encode the feature vector of image x. After that, we synthesize the glasses-removed image $\hat{y}$ with face decoder using the feature vector mentioned above. Meanwhile, a segmentation decoder is there for generating the binary mask $\hat{m}$ of the glasses region. However, after testing with this baseline model, we found that although there are many good removal results, when the eyeglasses are special or the face is not frontal, the removal effect may degrade. Hence, we were wondering whether there exists a good representation for face that can help remove eyeglasses. Since eyeglasses removal can be regarded as a kind of inpainting task on the face region, we can include semantic segmentation mask of face and eyeglasses regions into the framework. The segmentation of face shape is an excellent hint to guide FD to know the characteristics of each pixel in the face region and should maintain consistency with the neighboring pixels. After the experiment, we found that making SD predict the binary mask of face shape as well greatly improves the glasses-removed results. Hence, we let SD not only predict the binary mask of eyeglasses, but also the mask of face shape. Besides, the information obtained from SD is shared with FD with the skip connections to guide FD synthesizing images. Thus, we have

$$\hat{y} = FD(E(x)) \tag{1}$$

and

$$\hat{m} = SD(E(x)) \tag{2}$$

where $\hat{m}$ is a 2-channel mask, one of the channels indicates the position of the glasses region, and the other is for the face shape. Furthermore, in order to ensure the quality of synthetic output $\hat{y}$, we adopt a global and a local discriminator [11] to make sure both the synthetic image $\hat{y}$ and the inpainted frame area $\hat{y}_{Local}$ look realistic. Besides, we also include an Identity Extractor to minimize the distance between the identity feature vectors computed from the output image ($\hat{y}$) and the ground truth image (y).

3.2   Objective Function

The proposed ByeGlassesGAN is trained with the objective function consisting of four different types of loss functions, i.e. the adversarial loss, per-pixel loss, segmentation loss, and identity preserving loss. They are described in details subsequently.

Adversarial Loss. In order to make the generated images as realistic as possible, we adopt the strategy of adversarial learning. Here we apply the objective function of LSGAN [19,31] since it can make the training process of GAN more stable than the standard adversarial loss. Here we adopt 2 kinds of GAN loss, L_{GAN}^{Global} and L_{GAN}^{Local} for training the discriminators. Equation 3 shows the global adversarial loss L_{D}^{Global}.

$$L_{D}^{Global} = \mathbb{E}_{y \sim P_y}[(D_{Global}(y) - 1)^2] + \mathbb{E}_{x \sim P_x}[(D_{Global}(\hat{y}))^2] \tag{3}$$

When computing L_{D}^{Local}, we replace y, $\hat{y}$, D_{Global} by y_{Local}, $\hat{y}_{Local}$, and D_{Local} in Eq. 3. $y_{Local} = y \odot m_g$, and $\hat{y}_{Local} = \hat{y} \odot m_g$. $\odot$ denotes the element-wise product operator, and m_g is the ground truth binary mask of eyeglasses region. For training the generator, the GAN loss is shown below (Eq. 4). When computing L_{G}^{Local}, we also replace $\hat{y}$ and D_{Global} by $\hat{y}_{Local}$ and D_{Local} in Eq. 4.

$$L_{G}^{Global} = \mathbb{E}_{x \sim P_x}[(D_{Global}(\hat{y}) - 1)^2] \tag{4}$$

Per-Pixel Loss. We compute the L_1 distance between the generated image $\hat{y}$ and the ground truth image y. Per-pixel loss enforces the output of generator to be similar to the ground truth. We adopt two kinds of L_1 loss, $L_{L_1}^{Global}$ and $L_{L_1}^{Local}$. $L_{L_1}^{Local}$ is used for enhancing the removal ability of the generator in the edited region. The global L1 loss is given by

$$L_{L_1}^{Global} = L1(\hat{y}, y) = \mathbb{E}_{x \sim P_x}[\|y - \hat{y}\|_1] \tag{5}$$

When computing $L_{L_1}^{Local}$, we replace $\hat{y}$ and y by $\hat{y}_{Local}$ and y_{Local} in Eq. 5.

Segmentation Loss. Since we expect ByeGlassesGAN to predict the segmentation mask which facilitates eyeglasses removal, here we adopt binary cross entropy loss for generating the segmentation mask of the eyeglasses region and the face shape. It is given by

$$L_{Seg} = \mathbb{E}_{x \sim P_x} - (m \cdot log(\hat{m}) + (1 - m) \cdot log(1 - \hat{m})) \tag{6}$$

where $\hat{m}$ is the generated mask, and m denotes the ground truth segmentation mask.

Identity Preserving. In order to preserve the identity information in the glasses-removed images, we employ an Identity Extractor (IE), which is in fact a face classifier.

The identity distance loss is introduced to our generator, which is used to minimize the distance between $IE(y)$ and $IE(\hat{y})$. Similar to the concept of perceptual loss, after extracting the feature of y and $\hat{y}$ through the identity extractor, we compute the mean square error between these two feature vectors, given by

$$L_{ID} = \mathbb{E}_{x \sim P_x, y \sim P_y}[\|IE(\hat{y}) - IE(y)\|_2] \tag{7}$$

This loss encourages the eyeglasses-removed image $\hat{y}$ shares the same identity information with ground truth image y in the feature space of the identity extractor model.

Note that IE is a ResNet34 classifier pretrained on UMDFaces dataset. When training the Identity Classifier, we transpose the output feature vector of layer4 in ResNet into a 512-dimensional vector, and adopt the ArcFace [5] loss.

Finally, the overall loss function of the generator is given as follows:

$$L_G = \lambda_1 L_G^{Global} + \lambda_2 L_G^{Local} + \lambda_3 L_{L_1}^{Global} + \lambda_4 L_{L_1}^{Local} + \lambda_5 L_{Seg} + \lambda_6 L_{ID} \tag{8}$$

3.3   Network Architecture

Our GAN-based eyeglasses removal framework contains a generator, two discriminators, and an identity extractor. There are one encoder (E) and two decoders (face decoder, FD, and segmentation decoder, SD) in our generator. Following ELEGANT [25], the encoder (E) consists of 5 convolutional blocks, and each block contains a convolutional layer followed by an instance normalization layers and LeakyReLU activation. Both of the face decoder (FD) and the segmentation decoder (SD) consist of 5 deconvolutional blocks and an output block. Each deconvolutional block contains a deconvolutional layer followed by an instance normalization layers and ReLU activation. The output block of the FD is a deconvolutional layer followed by Tanh activation, while the output block of the SD is a deconvolutional layer followed by Sigmoid activation. Since the only area expected to be modified in the source image is the region of eyeglasses, other parts of the image should be kept unchanged. Here we adopt U-NET [22] architecture to be the generator of our ByeGlasess-GAN. Skip connections are

added to the corresponding layers between E-FD and E-SD. U-Net can considerably reduce the information loss compared with the common encoder-decoder.
Besides, skip connections are also added between the corresponding layers of SD
and FD, which are used for making the information acquired from SD guide the
FD to reconstruct images. The network architecture used for the two discriminators are PatchGAN proposed in pix2pix [12].

4   Synthesis of Face Images with Eyeglasses

Since we expect the proposed method can not only remove glasses but also help
improve face recognition performance for images with eyeglasses, we need to
make sure the detailed attributes (eye shape, eye color, skin color etc.) of the
glasses-removed face remain the same as those of the source image. To deal with
this problem, the best solution is to collect a large scale of well-aligned images
of subjects wearing and without wearing glasses. This kind of paired data is
difficult to collect, hence, here we generate the well-aligned paired image data
via synthesizing adding eyeglasses onto real face images.

We use CelebA [18] dataset to train the proposed ByeGlassesGAN in our
experiments. CelebA is a dataset containing 202,599 images of 10,177 celebrities
and annotated with 40 face attribute labels for each image. First, we align all
images according to 5 facial landmarks (left/right eye, nose, left/right mouth)
into the size of 256×256, and then roughly classify all images into 3 kinds of head
pose, frontal, left-front, and right-front using dlib and OpenCV. We manually
label the binary masks of 1,000 images with eyeglasses in CelebA as our glasses
pool (S_G), and use the rest of the images with glasses as our testing set. These
binary masks precisely locate the position of eyeglasses on each image, so we can
make use of them to easily extract 1,000 different styles of glasses. After that, we
randomly put glasses from glasses pool onto each glasses-free images according
to the head pose (Fig. 3, top). In order to make the synthetic glasses images look
more realistic to the real one, we randomly apply different levels of deformation
around the outer side lens to simulate the refraction of eyeglasses. After that,
we dye various colors on the glasses lenses. In addition, we generate many semi-
transparent light spot images in different shape. These spots are used to apply
on the glasses lenses to simulate the reflected glare on real eyeglasses. This step
much improves the ability of our ByeGlassesGAN to manipulate realistic glasses
images (Fig. 3, bottom). Besides, to generate segmentation mask for the face
shape, we used the pre-trained BiSeNet [27] which is trained on CelebAMask-
HQ [15] dataset to obtain the face shape mask of the glasses-free images. Finally,
we obtain 184,862 pairs of data in total as the training dataset. We will release
the glasses pool described above as a new dataset for future research related
eyeglasses detection, synthesis, or removal tasks.

Fig. 3. How to put on glasses? We first label the head pose of all the images in our training set, and label 1,000 eyeglasses segmentation mask of the glasses-images to form glasses pool(S_G). Each glasses-free image in the training set can be randomly put on glasses from S_G according to the head pose label and the binary mask. After that, we've also applied several photorealism steps to our synthetic images with eyeglasses.

5   Experimental Results

5.1   Implementation Details

We implement our ByeGlasses GAN with PyTorch. We use Adam for the optimizer, setting $\beta_1 = 0.5$, $\beta_2 = 0.999$, and the learning rate is 0.0002. For the hyperparameters in loss function, we set $\lambda_1 = 1$, $\lambda_2 = 1$, $\lambda_3 = 100$, $\lambda_4 = 200$, $\lambda_5 = 3$, and $\lambda_6 = 5$. Here we train ByeGlassesGAN on a GTX1080 with the batch size set to 16.

5.2   Qualitative Results

Figure 4, Upper shows the visual results of our method on CelebA [18] dataset. All samples are real glasses images in the testing set. The identity of each sample visually remains the same. The generated segmentation masks are also able to point out the accurate region of the face shape. Here we also show some visual results of the face images not in CelebA dataset. Figure 4, Bottom shows the visual results of our method testing on the wild data. Our method can not only remove eyeglasses from delicate portraits of celebrities, but also images taken from ordinary camera on mobile phones or laptops. Besides, when we synthesize the training data, we take the head pose of each face image into consideration and generate training image pairs for faces of different poses. Hence, our method is able to deal with non-frontal face images as well. As shown in the bottom row, our method can remove not only the glasses frame but also the tinted lenses.

Besides, here we also perform experiments on the other two kinds of models which are (A) Baseline model: Segmentation decoder only predicts the binary mask of eyeglasses, and there is no skip connection between FD and SD. (B) Segmentation decoder predicts the binary mask of both eyeglasses and face shape, but there is still no skip connection between FD and SD. As shown in Fig. 5, when we predict the face shape mask in Experiment B, the removal results improve a lot since the face shape mask shares some similar features with the removed result comparing to only predicting glasses mask in Experiment A. Besides, after

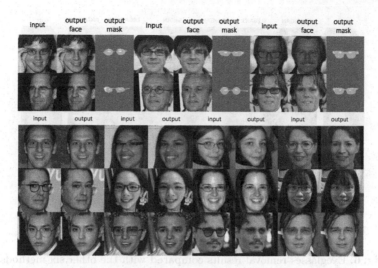

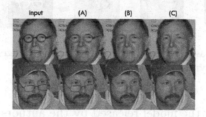

Fig. 4. Eyeglasses-removed results. Upper: CelebA results. Bottom: Wild data results. Images credit: MeGlass [6] dataset, and photos taken by ourselves.

Fig. 5. Some glasses removal results under different combinations: (A) baseline model, (B) baseline model with predicting face shape, and (C) complete model in the proposed network.

we add skip connections between the 2 decoders, the segmentation decoder can better guide the face decoder. Sharing the features of the segmentation mask with the face decoder helps the edited region keep consistency with the neighboring skin region, especially when the face is not-frontal or the glasses may not locate in a normal way.

We compare the visual results of the proposed method to the other state-of-the-art methods as well, including Pix2pix [12], CycleGan [31], StarGAN [4], ELEGANT [25], ERGAN [9], and SaGAN [30]. For comparison, we simply utilize the source code released by the first 5 previous works without any change. For SaGAN, since there is no source code, we carefully implement it ourselves. Pix2pix is a method that needs paired training data, so here we train the pix2pix model using the same data as we mentioned in Sect. 4. CycleGAN, Star-GAN, ELEGANT, and SaGAN are methods adopt unsupervised learning, so we directly use the original CelebA dataset for training the eyeglasses removal networks. ERGAN is an unsupervised method developed for removing eyeglasses.

Fig. 6. Eyeglasses removal results compared with the other six methods.

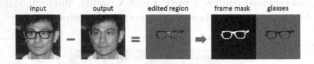

Fig. 7. Extracting eyeglasses from the glasses-removed images. These extracted eyeglasses can be used for synthesizing training pairs.

Here we directly apply the model released by the authors to obtain the results. Figure 6 shows the removal results of different methods. As shown in the figure, there are many artifacts pop out in pix2pix. For the other 5 unsupervised methods, even there exist visually appealing results, it is still difficult for them to directly remove the glasses stably without generating any artifacts. For ELEGANT and ERGAN, both methods need an additional glasses-free image to guide the removal, so the removal results depend on similarity between the input face image with glasses and the reference image without glasses. Besides, it is worth mentioning that in the last 3 rows in Fig. 6, our data synthesis method can effectively strengthen the ability of ByeGlassesGAN to remove reflected glare on the lenses.

Since our method is able to produce high quality glasses-removed result in which the only edited part is the glasses area, we can easily extract the eyeglasses on the input image by applying thresholding operation to the edited region. The edited region is the difference between the input and the output images. Thus, these extracted eyeglasses can also be used for synthesizing training pairs for glasses removal, synthesis, or detection tasks in the future (Fig. 7).

5.3 Quantitative Results

Following [25], here we utilize Fréchet Inception Distance(FID) [8] to measure how well our glasses removal method performs. FID represents the distance between the Inception embedding of the real and the generated images, which reveals how close the embeddings of images in two domains are distributed in feature space. Table 1 shows the FID distances of different methods. Here we perform experiment on 2 different datasets, CelebA [18] and MeGlass [6]. The real images set contains real glasses-free images. For the generated images set, it consists of the glasses removed images after applying each of the 6 different models to some real face images with glasses. As shown in Table 1, our Bye-GlassesGAN outperforms the others in both datasets. The FIDs of celebA with CycleGan, StarGAN and ELEGANT methods were reported in [25]. Besides, we also do the ablation study of removing the segmentation decoder branch, and the perceptual quality of the glasses removal results is not as good as those with the segmentation decoder as shown in Table 1.

Table 1. FID distances of different methods applied on MeGlass and CelebA datasets.

	Pix2pix	CycleGAN	StarGAN	ELEGANT	SaGAN	ERGAN	ours w/o SD	Ours
MeGlass	39.93	29.40	NULL	41.09	44.94	38.25	28.26	**27.14**
CelebA	50.38	48.82	142.35	60.71	50.06	NULL	44.76	**42.97**

To further compare our image synthesis method with the others, we conduct a user study. We randomly select 7 portraits with glasses from the testing set and apply different glasses removal methods on all of them, and there are 42 glasses-removed results in total. We invite 49 subjects to evaluate these images and compute the mean opinion score (MOS). As shown in Table 2, apparently, the glasses-removal results by our method are the most preferred since it receives the highest MOS score.

Table 2. Mean opinion scores of the glasses-removed results of different methods. It is obvious that our ByeGlassesGAN has the highest score.

Methods	Pix2pix	CycleGAN	StarGAN	ELEGANT	SaGAN	Ours
MOS	2.43	3.23	2.06	1.82	2.65	**4.31**

6 Face Recognition Evaluation

In this section, we demonstrate the effect of using our glasses removal GAN as a pre-processing step for the face recognition task.

First, we train a face recognition model on the whole UMDFaces dataset. UMDFaces [1] is a well-annotated dataset containing 367,888 images of 8,277 subjects, there are both faces with and without eyeglasses. The face recognition model we use here is the PyTorch implementation of MobileFaceNets [3]. All the training images are resized to 112 × 112, and the embedding features extracted from the face recognition module is 128-dimensional, following the original setting in MobileFaceNets. We train the face recognition model on a GTX1080 for 40 epochs, and the batchsize is 128. The recognition model achieves the accuracy of 98.3% on LFW [10].

For testing, we use MeGlass [6] dataset, which is a subset of MegaFace [14]. In the testing set of MeGlass, there are images of 1,710 subjects. For each subject, there are 2 images with eyeglasses, one for gallery and the other for probe. There are also 2 images without glasses of each subject, still, one for gallery and the other for probe. Here we show 7 kinds of experimental protocols below. No matter which one of the protocols, all images in gallery are glasses-free faces.

- All images in probe are glasses-free images.
- All images in probe are images with glasses.
- All images in probe are images with glasses, but we remove the glasses with different methods including CycleGAN, SaGAN, ELEGANT, pix2pix, and our ByeGlassesGAN before face recognition.

As shown in Table 3, when images in gallery and probe are all glasses-free, the face recognition model(Experiment M) described above can achieve high accuracy on both verification and identification task. However, if we change the probe into images with glasses, the accuracy degrades a lot.

Table 3. The effect of eyeglasses in face recognition: all the images in gallery are glasses-free images. The first column denotes which kind of images are there in probe. **Experiment M**: The face recognition model used is MobileFaceNet with 112 × 112 input image size.

Experiment M	TAR@FAR=10^{-3}	TAR@FAR=10^{-4}	TAR@FAR=10^{-5}	Rank-1
no glasses	0.9129	0.8567	0.7673	0.9018
with glasses	0.8509	0.7374	0.5708	0.8275

Due to the accuracy degradation for face images with eyeglasses as shown in Table 3, we then apply glasses removal methods to remove the eyeglasses in probe before face recognition. The quantitative results are shown in Table 4. As shown in row 2, row 3, row 4, and row 5 of Table 4, removing glasses with Cycle-GAN, SaGAN, ELEGANT, and pix2pix degrades the accuracy of face recognition. However, removing eyeglasses with our ByeGlassesGAN can improve the accuracy. Especially when FAR is small, the improvement in TAR is more evident. Comparing the unpaired training ones with our work may not be fair, but to utilize glasses removal into face recognition task, paired training is a better strategy. Besides, we also train the proposed GAN model without considering L_{ID}, as shown in row 7, without the Identity Extractor, the improvement of face recognition decrease since there is no mechanism to constrain the generator from producing artifacts which may be seen as noise for face recognition. Here we also demonstrate face recognition experiments on our Identity Extractor used for training our ByeGlassesGAN. As shown in Table 5, when FAR is 10^{-5}, we can improve TAR even more obviously by about 6%.

Table 4. Accuracy of face recognition: all the images in gallery are glasses-free images. The first column denotes the type of images with or without applying a specific glasses removal pre-processing method in the probe.

Experiment M	TAR@FAR=10^{-3}	TAR@FAR=10^{-4}	TAR@FAR=10^{-5}	Rank-1
no removal	0.8509	0.7374	0.5708	0.8275
CycleGAN	0.8298	0.7205	0.5329	0.7994
SaGAN	0.8386	0.7257	0.5684	0.8088
ELEGANT	0.7497	0.5977	0.3719	0.6994
pix2pix	0.8444	0.7327	0.5251	0.8216
ours without IE	0.8573	0.7626	0.5813	0.8358
ours	**0.8632**	**0.7719**	**0.6076**	**0.8415**

Table 5. Accuracy of face recognition: all the images in gallery are glasses-free images. The first column denotes the type of images with or without applying our glasses removal pre-processing method in the probe. **Experiment R**: The face recognition model used here is the Identity Extractor used for training ByeGlassesGAN with 256 × 256 input image size.

Experiment R	TAR@FAR=10^{-3}	TAR@FAR=10^{-4}	TAR@FAR=10^{-5}	Rank-1
no removal	0.8801	0.7830	0.6292	0.8538
ours	**0.8836**	**0.7906**	**0.6819**	**0.8602**

However, for practical applications, there might not be only glasses-portraits in probe and only glasses-free-portraits in gallery. Hence, here we do another face recognition experiment described as follows:

- In gallery: 1 glasses-free image and 1 image with glasses for each person.
- In probe: 1 glasses-free image and 1 image with glasses for each person.
- For the *no removal* experiment, no matter there are eyeglasses on the images or not, we use the original images.
- For the *with removal* experiment, no matter there are eyeglasses on the images or not, we do eyeglasses removal with ByeGlassesGAN for all the images in both probe and gallery before face recognition.

As shown in Table 6, we can see applying glasses removal as a pre-processing step can still benefit face recognition even when there are glasses-free images. When FAR is 10^{-5}, we evidently improve TAR by about 7%. This experiment not only demonstrates the effectiveness of our glasses removal method, but also reveals that when applying our method to the glasses-free images, images remains almost the same, and the feature and identity embedding of the pre-processed face images are still well preserved.

Table 6. Face recognition accuracy when both gallery and probe sets contain face images with and without glasses.

Experiment M	TAR@FAR=10^{-3}	TAR@FAR=10^{-4}	TAR@FAR=10^{-5}	Rank-1
no removal	0.8507	0.7516	0.5927	0.9175
with removal	**0.8646**	**0.7868**	**0.6539**	**0.9289**

Besides, to make sure applying image synthesis before recognition does not harm features of faces, we have done an experiment of computing the cosine distance between features of with-glasses portrait and real glasses-free portrait of same person in the feature space of the recognition model, and the cosine distance between features of glasses-removed image and real glasses-free image. We found that our image synthesis method can effectively shorten the cosine distance for 1,335 out of 1,710 image pairs in MeGlass dataset after applying our glasses removal method. Due to the improvement of face recognition and the reduction in the cosine distance for almost 80% image pairs, we are confident that our method cannot only manipulate visually appealing glasses-removed results, but it's also worth removing eyeglasses with our method as a preprocessing step for face recognition.

7 Conclusions

In this paper, we propose a novel multi-task framework to automatically detect the eyeglasses area and remove them from a face image. We adopt the mechanism of identity extractor to make sure the output of the proposed ByeGlassesGAN model preserves the same identity as that of the source image. As our GAN-based glasses removal framework can predict the binary mask of face shape as well, this spatial information is exploited to remove the eyeglasses from face images and achieve very realistic result. In the face recognition experiment, we showed that our method can significantly enhance the accuracy of face recognition by about 7%TAR@FAR=10^{-5}. However, there are still some limitations of our work, for example, we cannot generate convincing glasses removal results for some special glasses or when the lighting condition is very extreme.

With the advancement of face parsing methods, we believe that combining face parsing can effectively extend this work to other attributes removal tasks, such as removing beard or hat. In the future, we will aim to make our Bye-GlassesGAN more robust under special or extreme conditions, and extend the proposed framework to other face attributes removal tasks.

References

1. Bansal, A., Nanduri, A., Castillo, C.D., Ranjan, R., Chellappa, R.: Umdfaces: an annotated face dataset for training deep networks (2016). arXiv preprint arXiv:1611.01484v2
2. Bao, J., Chen, D., Wen, F., Li, H., Hua, G.: Towards open-set identity preserving face synthesis. In: Proceedings of the IEEE Conference on Computer Vision and Pattern Recognition, pp. 6713–6722 (2018)
3. Chen, S., Liu, Y., Gao, X., Han, Z.: MobileFaceNets: efficient CNNs for accurate real-time face verification on mobile devices. In: Zhou, J., et al. (eds.) CCBR 2018. LNCS, vol. 10996, pp. 428–438. Springer, Cham (2018). https://doi.org/10.1007/978-3-319-97909-0_46
4. Choi, Y., Choi, M., Kim, M., Ha, J.W., Kim, S., Choo, J.: Stargan: unified generative adversarial networks for multi-domain image-to-image translation. In: Proceedings of the IEEE Conference on Computer Vision and Pattern Recognition, pp. 8789–8797 (2018)
5. Deng, J., Guo, J., Xue, N., Zafeiriou, S.: Arcface: additive angular margin loss for deep face recognition (2018). arXiv preprint arXiv:1801.07698
6. Guo, J., Zhu, X., Lei, Z., Li, S.Z.: Face synthesis for eyeglass-robust face recognition. In: Zhou, J., et al. (eds.) CCBR 2018. LNCS, vol. 10996, pp. 275–284. Springer, Cham (2018). https://doi.org/10.1007/978-3-319-97909-0_30
7. He, Z., Zuo, W., Kan, M., Shan, S., Chen, X.: Arbitrary facial attribute editing: only change what you want 1(3) (2017). arXiv preprint arXiv:1711.10678
8. Heusel, M., Ramsauer, H., Unterthiner, T., Nessler, B., Hochreiter, S.: Gans trained by a two time-scale update rule converge to a local nash equilibrium. In: Advances in Neural Information Processing Systems, pp. 6626–6637 (2017)
9. Hu, B., Yang, W., Ren, M.: Unsupervised eyeglasses removal in the wild (2019). arXiv preprint arXiv:1909.06989
10. Huang, G.B., Ramesh, M., Berg, T., Learned-Miller, E.: Labeled faces in the wild: a database for studying face recognition in unconstrained environments. Technical report. 07–49, University of Massachusetts, Amherst (2007)
11. Iizuka, S., Simo-Serra, E., Ishikawa, H.: Globally and locally consistent image completion. ACM Trans. Graph. (ToG) 36(4), 107 (2017)
12. Isola, P., Zhu, J.Y., Zhou, T., Efros, A.A.: Image-to-image translation with conditional adversarial networks. In: Proceedings of the IEEE Conference on Computer Vision and Pattern Recognition, pp. 1125–1134 (2017)
13. Jo, Y., Park, J.: Sc-fegan: face editing generative adversarial network with user's sketch and color (2019). arXiv preprint arXiv:1902.06838
14. Kemelmacher-Shlizerman, I., Seitz, S.M., Miller, D., Brossard, E.: The megaface benchmark: 1 million faces for recognition at scale. In: Proceedings of the IEEE Conference on Computer Vision and Pattern Recognition, pp. 4873–4882 (2016)
15. Lee, C.H., Liu, Z., Wu, L., Luo, P.: Maskgan: towards diverse and interactive facial image manipulation (2019). arXiv preprint arXiv:1907.11922
16. Li, M., Zuo, W., Zhang, D.: Deep identity-aware transfer of facial attributes (2016). arXiv preprint arXiv:1610.05586

17. Liu, G., Reda, F.A., Shih, K.J., Wang, T.C., Tao, A., Catanzaro, B.: Image inpainting for irregular holes using partial convolutions. In: Proceedings of the European Conference on Computer Vision (ECCV), pp. 85–100 (2018)
18. Liu, Z., Luo, P., Wang, X., Tang, X.: Deep learning face attributes in the wild. In: Proceedings of the IEEE International Conference on Computer Vision, pp. 3730–3738 (2015)
19. Mao, X., Li, Q., Xie, H., Lau, R.Y., Wang, Z., Paul Smolley, S.: Least squares generative adversarial networks. In: Proceedings of the IEEE International Conference on Computer Vision, pp. 2794–2802 (2017)
20. Park, J.S., Oh, Y.H., Ahn, S.C., Lee, S.W.: Glasses removal from facial image using recursive error compensation. IEEE Trans. Pattern Anal. Mach. Intell. **27**(5), 805–811 (2005)
21. Pathak, D., Krahenbuhl, P., Donahue, J., Darrell, T., Efros, A.A.: Context encoders: feature learning by inpainting. In: Proceedings of the IEEE Conference on Computer Vision and Pattern Recognition, pp. 2536–2544 (2016)
22. Ronneberger, O., Fischer, P., Brox, T.: U-Net: convolutional networks for biomedical image segmentation. In: Navab, N., Hornegger, J., Wells, W.M., Frangi, A.F. (eds.) MICCAI 2015. LNCS, vol. 9351, pp. 234–241. Springer, Cham (2015). https://doi.org/10.1007/978-3-319-24574-4_28
23. Upchurch, P., et al.: Deep feature interpolation for image content changes. In: Proceedings of the IEEE Conference on Computer Vision and Pattern Recognition, pp. 7064–7073 (2017)
24. Wu, C., Liu, C., Shum, H.Y., Xy, Y.Q., Zhang, Z.: Automatic eyeglasses removal from face images. IEEE Trans. Pattern Anal. Mach. Intell. **26**(3), 322–336 (2004)
25. Xiao, T., Hong, J., Ma, J.: Elegant: Exchanging latent encodings with gan for transferring multiple face attributes. In: Proceedings of the European Conference on Computer Vision (ECCV), pp. 168–184 (2018)
26. Yang, C., Lu, X., Lin, Z., Shechtman, E., Wang, O., Li, H.: High-resolution image inpainting using multi-scale neural patch synthesis. In: Proceedings of the IEEE Conference on Computer Vision and Pattern Recognition, pp. 6721–6729 (2017)
27. Yu, C., Wang, J., Peng, C., Gao, C., Yu, G., Sang, N.: Bisenet: bilateral segmentation network for real-time semantic segmentation. In: Proceedings of the European Conference on Computer Vision (ECCV), pp. 325–341 (2018)
28. Yu, J., Lin, Z., Yang, J., Shen, X., Lu, X., Huang, T.S.: Free-form image inpainting with gated convolution (2018). arXiv preprint arXiv:1806.03589
29. Yu, J., Lin, Z., Yang, J., Shen, X., Lu, X., Huang, T.S.: Generative image inpainting with contextual attention (2018). arXiv preprint arXiv:1801.07892
30. Zhang, G., Kan, M., Shan, S., Chen, X.: Generative adversarial network with spatial attention for face attribute editing. In: Proceedings of the European Conference on Computer Vision (ECCV), pp. 417–432 (2018)
31. Zhu, J.Y., Park, T., Isola, P., Efros, A.A.: Unpaired image-to-image translation using cycle-consistent adversarial networks. In: Proceedings of the IEEE International Conference on Computer Vision, pp. 2223–2232 (2017)

Differentiable Joint Pruning and Quantization for Hardware Efficiency

Ying Wang[1]([✉]) [iD], Yadong Lu[2] [iD], and Tijmen Blankevoort[3]

[1] Qualcomm AI Research, San Diego, California, USA
yinwan@qti.qualcomm.com
[2] University of California, Irvine, USA
yadongl1@uci.edu
[3] Qualcomm AI Research, Amsterdam, The Netherlands
tijmen@qti.qualcomm.com

Abstract. We present a differentiable joint pruning and quantization (DJPQ) scheme. We frame neural network compression as a joint gradient-based optimization problem, trading off between model pruning and quantization automatically for hardware efficiency. DJPQ incorporates variational information bottleneck based structured pruning and mixed-bit precision quantization into a single differentiable loss function. In contrast to previous works which consider pruning and quantization separately, our method enables users to find the optimal trade-off between both in a single training procedure. To utilize the method for more efficient hardware inference, we extend DJPQ to integrate structured pruning with power-of-two bit-restricted quantization. We show that DJPQ significantly reduces the number of Bit-Operations (BOPs) for several networks while maintaining the top-1 accuracy of original floating-point models (e.g., 53× BOPs reduction in ResNet18 on ImageNet, 43× in MobileNetV2). Compared to the conventional two-stage approach, which optimizes pruning and quantization independently, our scheme outperforms in terms of both accuracy and BOPs. Even when considering bit-restricted quantization, DJPQ achieves larger compression ratios and better accuracy than the two-stage approach.

Keywords: Joint optimization · Model compression · Mixed precision · Bit-restriction · Variational information bottleneck · Quantization

1 Introduction

There has been an increasing interest in the deep learning community to neural network model compression, driven by the need to deploy large deep neural networks (DNNs) onto resource-constrained mobile or edge devices. So far, pruning and quantization are two of the most successful techniques in compressing

Y. Lu—Work done during internship at Qualcomm AI Research.
Qualcomm AI Research is an initiative of Qualcomm Technologies, Inc.

© Springer Nature Switzerland AG 2020
A. Vedaldi et al. (Eds.): ECCV 2020, LNCS 12374, pp. 259–277, 2020.
https://doi.org/10.1007/978-3-030-58526-6_16

a DNN for efficient inference [9,12,13,15]. Combining the two is often done in practice. However, almost no research has been published in combining the two in a principled manner, even though finding the optimal trade-off between both methods is non-trivial. There are two challenges with the currently proposed approaches:

First, most of the previous works apply a two-stage compression strategy: pruning and quantization are applied independently to the model [9,20,35]. We argue that the two-stage compression scheme is inefficient since it does not consider the trade-off between sparsity and quantization resolution. For example, if a model is pruned significantly, it is likely that the quantization bit-width has to be large since there is little redundancy left. Further, the heavily pruned model is expected to be more sensitive to input or quantization noise. Since different layers have different sensitivity to pruning and quantization, it is challenging to optimize these holistically, and manually chosen heuristics are probably not optimal.

Second, the joint optimization of pruning and quantization should take into account practical hardware constraints. Although unstructured pruning is more likely to result in higher sparsity, structured pruning is often preferable as it can be more easily exploited on general-purpose devices. Also, typical quantization methods use a fixed bit-width for all layers of a neural network. However, since several hardware platforms can efficiently support mixed-bit precision, joint optimization could automatically support learning the bit-width. The caveat is that only power-of-two bit-widths are often efficiently implemented and supported in typical digital hardware; other bit-widths are rounded up to the nearest power-of-two-two values, resulting in inefficiencies in computation and storage [14]. Arbitrary bit-widths used in some literature [29,30] are more difficult to exploit for power or computational savings unless dedicated silicon or FPGAs are employed.

To address the above challenges, we propose a differentiable joint pruning and quantization (DJPQ) scheme. It first combines the variational information bottleneck [4] approach to structured pruning and mixed-bit precision quantization into a single differentiable loss function. Thus, model training or fine-tuning can be done only once for end-to-end model compression. Accordingly, we show that on a practical surrogate measure Bit-Operations (BOPs) we employ, the complexity of the models is significantly reduced without degradation of accuracy (e.g., 53× BOPs reduction in ResNet18 on ImageNet, 43× in MobileNetV2).

The contributions of this work are:

- We propose a differentiable joint optimization scheme that balances pruning and quantization holistically.
- We show state-of-the-art BOPs reduction ratio on a diverse set of neural networks (i.e., VGG, ResNet, and MobileNet families), outperforming the two-stage approach with independent pruning and quantization.
- The joint scheme is fully end-to-end and requires training only once, reducing the efforts for iterative training and finetuning.

- We extend the DJPQ scheme to the bit restricted case for improved hardware efficiency with little extra overhead. The proposed scheme can learn mixed-precision for a power-of-two bit restriction.

2   Related Work

Both quantization and compression are often considered separately to optimize neural networks, and only a few papers remark on the combination of the two. We can thus relate our work to three lines of papers: mixed-precision quantization, structured pruning, and joint optimization of the two.

Quantization approaches can be summarized into two categories: fixed-bit and mixed-precision quantization. Most of the existing works fall into the first category, which set the same bit-width for all layers beforehand. Many works in this category suffer from lower compression ratio. For works using fixed 4 or 8-bit quantization that are hardware friendly such as [1,8,19], either the compression ratio or performance is not as competitive as the mixed-precision opponents [29]. In [17] 2-bit quantization is utilized for weights, but the scheme suffers from heavy performance loss even with unquantized activations. For mixed-precision quantization, a second-order quantization method is proposed in [6] and [5], which automatically selects quantization bits based on the largest eigenvalue of the Hessian of each layer and the trace respectively. In [30,34] reinforcement-based schemes are proposed to learn the bit-width. Their methods determine the quantization policy by taking the hardware accelerator's feedback into the design loop. In [20], the bit-width is chosen to be proportional to the total variance of that layer, which is heuristic and likely too coarsely estimated. A differentiable quantization (DQ) scheme proposed in [29] can learn the bit-width of both weights and activations. The bit-width is estimated by defining a continuous relaxation of it and using straight-through estimator for the gradients. We employ a similar technique in this paper. For a general overview of quantization, please refer to [15,31].

Structured pruning approaches can also be summarized into two categories: the one with fixed pruning ratio and with learned pruning ratio for each layer. For the first category, [13] proposed an iterative two-step channel pruning scheme, which first selects pruned channels with Lasso regression given a target pruning ratio, and then finetunes weights. In [18] and [11], the pruned channels are determined with L_1 and L_2 norm, respectively for each layer under the same pruning ratio. The approach proposed in [22] is based on first-order Taylor expansion for the loss function. To explore inter-channel dependency, many works explore second-order Taylor expansion, such as [24,26]. For the second category, pruning ratio for each layer is jointly optimized across all the layers. Sparsity learned during training with L_0 [21] regularization has been explored by several works. In [12] the pruning ratio for each layer is learned through reinforcement learning. The pruning approach in [4] relates redundancy of channels to the variational information bottleneck (VIB) [27], adding gates to the network and employing a suitable regularization term to train for sparsity. As mentioned in [27], the VIB-Net scheme has a certain advantage over Bayesian-type of compression [20], as the loss function does not require additional parameters to describe the priors. We also found this method to work better in practice. In our DJPQ scheme, we utilize this VIBNet pruning while optimizing for quantization jointly to achieve a flexible trade-off between the two. For an overview of structured pruning methods, please refer to [16].

To jointly optimize pruning and quantization, two research works have been proposed [28,35]. Unfortunately, neither of them support activation quantization. Thus, the resulting networks cannot be deployed efficiently on edge hardware. Recently, an Alternating Direction Method of Multipliers (ADMM)-based approach was proposed in [33]. It formulates the automated compression problem to a constrained optimization problem and solves it iteratively using ADMM. One key limitation is the lack of support for hardware friendliness. First, it is based on unstructured pruning and thus no performance benefit in typical hardware. Second, they use a non-uniform quantization which has the same issue. In addition, the iterative nature of ADMM imposes excessive training time for end-to-end optimization while our DPJQ method does not, since we train in only a single training pass.

3   Differentiable Joint Pruning and Quantization

We propose a novel differentiable joint pruning and quantization (DJPQ) scheme for compressing DNNs. Due to practical hardware limitations, only uniform quantization and structured pruning are considered. We will first discuss the quantization and pruning method, and then introduce the evaluation metric, and finally go over joint optimization details.

3.1   Quantization with Learnable Mapping

It has been observed that weights and activations of pre-trained neural networks typically have bell-shaped distributions with long tails [9]. Consequently, uniform quantization is sub-optimal for such distributions. We use a non-linear function to map any weight input x to $\tilde{x}$. Let q_s and q_m be the minimum and maximum value to be mapped, where $0 < q_s < q_m$. Let $t > 0$ be the exponent controlling the shape of mapping (c.f., Appendix A). q_m and t are learnable parameters, and q_s is fixed to a small value. The non-linear mapping is defined as

$$\tilde{x} = \text{sign}(x) \cdot \begin{cases} 0, & |x| < q_s \\ (|x| - q_s)^t, & q_s \leq |x| \leq q_m \\ (q_m - q_s)^t, & |x| > q_m \end{cases} \tag{1}$$

For activation quantization, we do not use non-linear mapping, i.e., for any input x, $\tilde{x}$ is derived from (1) with t fixed to 1. After mapping, a uniform quantization is applied to $\tilde{x}$. Let d be quantization step-size, and let x_q be the quantized value of x. The quantized version is given by

$$x_q = \text{sign}(x) \cdot \begin{cases} 0, & |x| < q_s \\ d\lfloor \frac{(|x|-q_s)^t}{d} \rceil, & q_s \leq |x| \leq q_m \\ d\lfloor \frac{(q_m-q_s)^t}{d} \rceil, & |x| > q_m \end{cases} \tag{2}$$

where $\lfloor \cdot \rceil$ is the rounding operation. We note that the quantization grid for weights is symmetric and the bit-width b is given by

$$b = \log_2\lceil \frac{(q_m-q_s)^t}{d} + 1\rceil + 1. \tag{3}$$

For ReLU activations, we use a symmetric unsigned grid since the resulting values are always non-negative. Here, the activation bit-width b is given by

$$b = \log_2\lceil \frac{(q_m-q_s)^t}{d} \rceil. \tag{4}$$

Finally, we use the straight through estimation (STE) method [2] for back-propagating gradients through the quantizers, which [29] has shown to converge fast. The gradients of the quantizer output with respect to d, q_m and t are given by

$$\nabla_d x_q = \begin{cases} \text{sign}(x)\left(\lfloor \frac{(|x|-q_s)^t}{d} \rceil - \frac{(|x|-q_s)^t}{d}\right), & q_s \leq |x| \leq q_m \\ \text{sign}(x)\left(\lfloor \frac{(q_m-q_s)^t}{d} \rceil - \frac{(q_m-q_s)^t}{d}\right), & |x| > q_m, \\ 0 & \text{otherwise} \end{cases} \tag{5}$$

$$\nabla_{q_m} x_q = \begin{cases} 0, & |x| \leq q_m \\ \text{sign}(x)t(q_m-q_s)^{t-1}, & \text{otherwise} \end{cases} \tag{6}$$

$$\nabla_t x_q = \begin{cases} \text{sign}(x)(|x|-q_s)^t \log(|x|-q_s), & q_s \leq |x| \leq q_m \\ \text{sign}(x)(q_m-q_s)^t \log(q_m-q_s), & |x| > q_m \\ 0, & \text{otherwise} \end{cases} \tag{7}$$

3.2 Structured Pruning via VIBNet Gates

Our structured pruning scheme is based on the variational information bottle-neck (VIBNet) [4] approach. Specifically, we add multiplicative Gaussian gates to all channels in a layer and learn a variational posterior of the weight distri-bution aiming at minimizing the mutual information between current layer and next layer's outputs, while maximizing the mutual information between current layer and network outputs. The learned distributions are then used to determine which channels need to be pruned. Assuming the network has L layers, and that there are c_l output channels in the l-th layer, $l = 1, \cdots, L$. Let $h_l \in \mathcal{R}^{c_l}$ be the output of l-th layer after activation. Let $y \in \mathcal{Y}^{c_L}$ be the target output or data labels. Let $I(\boldsymbol{x}; \boldsymbol{y})$ denote the mutual information between $\boldsymbol{x}$ and $\boldsymbol{y}$. The VIB loss function $\mathcal{L}_{\mathrm{VIB}}$ is defined as

$$\mathcal{L}_{\mathrm{VIB}} = \gamma \sum_{l=1}^{L} I(\boldsymbol{h}_l; \boldsymbol{h}_{l-1}) - I(\boldsymbol{h}_l; \boldsymbol{y}), \tag{8}$$

where $\gamma > 0$ is a scaling factor. It then follows that

$$\boldsymbol{h}_l = \boldsymbol{z}_l \odot \boldsymbol{f}_l(\boldsymbol{h}_{l-1}), \tag{9}$$

where $\boldsymbol{z}_l = \{z_{l,1}, \cdots, z_{l,c_l}\}$ is a vector of gates for the l-th layer. The $\odot$ represents the element-wise multiplication operator. $\boldsymbol{f}_l$ represents l-th layer mapping function, i.e., concatenation of a linear or convolutional transformation, batch normalization and some nonlinear activation. Let us assume that $\boldsymbol{z}_l$ follows a Gaussian distribution with mean $\boldsymbol{\mu}_l$ and variance $\boldsymbol{\sigma}_l^2$, which can be re-parameterized as

$$\boldsymbol{z}_l = \boldsymbol{\mu}_l + \boldsymbol{\epsilon}_l \odot \boldsymbol{\sigma}_l \tag{10}$$

where $\boldsymbol{\epsilon}_l \sim \mathcal{N}(0, I)$. It follows that the prior distribution of $\boldsymbol{h}_l$ conditioned on $\boldsymbol{h}_{l-1}$ is

$$p(\boldsymbol{h}_l | \boldsymbol{h}_{l-1}) \sim \mathcal{N}\big(\boldsymbol{\mu}_l \odot \boldsymbol{f}_l(\boldsymbol{h}_{l-1}), \mathrm{diag}[\boldsymbol{\sigma}_l^2 \odot \boldsymbol{f}_l(\boldsymbol{h}_{l-1})^2]\big) \tag{11}$$

We further assume that the prior distribution of $\boldsymbol{h}_l$ is also Gaussian

$$q(\boldsymbol{h}_l) \sim \mathcal{N}(0, \mathrm{diag}[\boldsymbol{\xi}_l]) \tag{12}$$

where $\boldsymbol{\xi}_l$ is a vector of variances chosen to minimize an upper bound of $\mathcal{L}_{\mathrm{VIB}}$ (c.f., (8)), as given in [4]

$$\tilde{\mathcal{L}}_{\mathrm{VIB}} \triangleq \mathrm{CE}(\boldsymbol{y}, \boldsymbol{h}_L) + \gamma \sum_{l=1}^{L} \sum_{i=1}^{c_l} \log\left(1 + \frac{\mu_{l,i}^2}{\sigma_{l,i}^2}\right). \tag{13}$$

In (13), $\mathrm{CE}(\boldsymbol{y}, \boldsymbol{h}_L)$ is the cross-entropy loss. The optimal $\boldsymbol{\xi}_l$ is then derived as

$$\boldsymbol{\xi}_l^* = (\boldsymbol{\mu}_l^2 + \boldsymbol{\sigma}_l^2) \mathbf{E}_{h_{i-1} \sim p(h_{i-1})}[\boldsymbol{f}_i(\boldsymbol{h}_{i-1})^2].$$

Next, we give the condition for pruning a channel. Let α_l be defined as $\alpha_l = \mu_l^2/\sigma_l^2$. If $\alpha_{li} < \alpha_{th}$, where α_{th} is a small pruning threshold, the i-th channel in the l-th layer is pruned. The pruning parameters $\{\boldsymbol{\mu}_l, \boldsymbol{\sigma}_l\}$ are learned during training to minimize the loss function.

3.3   Evaluation Metrics

Actual hardware measurement is generally a poor indicator for the usefulness of these methods. There are many possible target-devices, e.g. GPUs, TPUs, several phone-chips, IoT devices, FPGAs, and dedicated silicon, each with their own quirks and trade-offs. Efficiency of networks also greatly depends on specific kernel-implementations, and currently many devices do not have multiple bit-width kernels implemented to even compare on in a live setting. Therefore, we choose a more general metric to measure the compression performance, namely Bit-Operations (BOPs) count, which assumes computations can be done optimally in an ideal world. The BOPs metric has been used by several papers such as [19] and [3]. Our method can be easily modified to optimize for specific hardware by weighting the contributions of different settings.

We also report results on layerwise pruning ratio and mac counts, for purposes of comparison of the pruning methods. Let p_l be the pruning ratio of l-th layer output channels. We define P_l to be the layerwise pruning ratio, which is the ratio of the number of weights between the uncompressed and compressed models. It can be derived as

$$P_l = 1 - (1 - p_{l-1})(1 - p_l). \tag{14}$$

The l-th layer Multiply-And-Accumulate (MAC) operations is defined as follows. Let l-th layer output feature map have width, height and number of channels of $m_{w,l}$, $m_{h,l}$ and c_l, respectively. Let k_w and k_h be the kernel width and height. The MAC count is computed as

$$\mathrm{MACs}_l \triangleq (1 - p_{l-1})c_{l-1} \cdot (1 - p_l)c_l \cdot m_{w,l} \cdot m_{h,l} \cdot k_w \cdot k_h. \tag{15}$$

The BOP count in the l-th layer BOPs_l is defined as

$$\mathrm{BOPs}_l \triangleq \mathrm{MACs}_l \cdot b_{w,l} \cdot b_{a,l-1}, \tag{16}$$

where $b_{w,l}$ and $b_{a,l}$ denote l-th layer weight and activation bit-width.

The BOP compression ratio is defined as the ratio between total BOPs of the uncompressed and compressed models. MAC compression ratio is similarly defined. Without loss of generality, we use the MAC compression ratio to measure only the pruning effect, and use the BOP compression ratio to measure the overall effect from pruning and quantization. As seen from (16), BOP count is a function of both channel pruning ratio p_l and bit-width $b_{w,l}$ and $b_{a,l}$, hence BOP compression ratio is a suitable metric to measure a DNN's overall compression.

3.4   Joint Optimization of Pruning and Quantization

As described in Sect. 3.3, the BOP count is a function of channel pruning ratio p_l and bit-widths $b_{w,l}$ and $b_{a,l}$. As such, incorporating the BOP count in the loss function allows for joint optimization of pruning and quantization parameters. When combing the two methods for pruning and quantization, we need to define a loss function that combines both in a sensible fashion. So we define the DJPQ loss function as

$$\mathcal{L}_{\mathrm{DJPQ}} \triangleq \tilde{\mathcal{L}}_{\mathrm{VIB}} + \beta \sum_{l=1}^{L} \mathrm{BOPs}_l \tag{17}$$

Algorithm 1: Differentiable joint pruning and quantization

Input: A neural network with weight w_l, $l \in [1, L]$; training data
Output: $b_{w,l}$, $b_{a,l}$, P_l, $l \in [1, L]$
Parameters: w_l, (μ_l, σ_l), $(q_{m,wl}, d_{w,l}, t_{w,l}, q_{m,al}, d_{a,l})$, $l \in [1, L]$
Forward pass:
Anneal strength γ and β after each epoch
for $l \in \{1, \cdots, L\}$ **do**

    Draw samples $z_l \sim \mathcal{N}(\mu_l, \sigma_l^2)$ and multiply to the channel output
    Compute p_l according to (19)
    Compute the layerwise pruning ratio P_l according to (14)
    Compute bit-width $b_{w,l}$ and $b_{a,l}$ according to (3) and (4)
    if *bit-restricted* **then**
        Adjust $b_{w,l}$, $d_{w,l}$, $b_{a,l}$ and $d_{a,l}$ according to Algorithm 2
    Quantize w_l and h_l

Compute $\mathcal{L}_{\text{DJPQ}}$ according to (17)
Backward pass:
for $l \in \{1, \cdots, L\}$ **do**

    Compute the gradients of $(d_{w,l}, d_{a,l})$, $(q_{m,wl}, q_{m,al})$, and $t_{w,l}$ according to (5), (6) and (7), respectively
    Compute the gradients of μ_l and σ_l
    Compute the gradients of w_l
    Update w_l, (μ_l, σ_l), $(q_{m,wl}, d_{w,l}, t_{w,l}, q_{m,al}, d_{a,l})$ with corresponding learning rate

---

where $\tilde{\mathcal{L}}_{\text{VIB}}$ is the upper bound to the VIB loss given by (13), β is a scalar, and BOPs$_l$ is defined in (16). To compute BOPs$_l$, $b_{w,l}$ and $b_{a,l}$ are given by (3) and (4), respectively, and $b_{a,0}$ denotes DNN's input bit-width. For hard pruning, p_l can be computed as

$$p_l = \frac{\sum_{i=1}^{c_l} 1\{\alpha_{l,i} < \alpha_{th}\}}{c_l} \tag{18}$$

where $1\{\cdot\}$ is the indicator function. For soft pruning during training, the indicator function in (18) is relaxed by a sigmoid function $\sigma(\cdot)$, i.e.,

$$p_l = \frac{\sum_{i=1}^{c_l} \sigma\left(\frac{\alpha_{l,i} - \alpha_{th}}{\tau}\right)}{c_l} \tag{19}$$

where τ is a temperature parameter. Note that soft pruning makes p_l, hence the BOPs$_l$, differentiable with respect to μ_l and σ_l.

The parameters of joint optimization of pruning and quantization include w_l, (μ_l, σ_l) and $(q_{m,wl}, d_{w,l}, t_{w,l}, q_{m,al}, d_{a,l})$ for $l = 1, \cdots, L$. In our experimentation we have found that a proper scaling of the learning rates for different pruning and quantization parameters plays an important role in achieving a good performance. Our experiments also showed that using the ADAM optimizer to automatically adapt the learning rates does not give as good a performance as using the SGD optimizer, provided that suitable scaling factors are set manually.

Please refer to Appendix B.3 for detailed setting of the scaling factors and other hyper-parameters.

3.5 Power-of-Two Quantization

To further align this method with what is feasible in common hardware, in this section we extend DJPQ to the scenario where quantization bit-width b is restricted to power-of-two values, i.e., $b \in \{2, 4, 8, 16, 32\}$. As reflected in Algorithm 1, this extension involves an added step where bit-widths are rounded to the nearest power-of-two representable value, and the quantizer stepsizes are updated accordingly without changing q_m, c.f. Algorithm 2 for details.

Algorithm 2: Adjust bit-width to power of two integers

Input: stepsize d_l, max. range $q_{m,l}$, exponent t_l
Output: Adjusted bit-width b'_l and stepsize d'_l
Compute b_l according to (3) or (4)
Compute s_l: $s_l = \log_2 b_l$
Adjust s_l to s'_l: $s'_l = \lceil s_l \rceil$
Adjust b_l to b'_l with s'_l: $b'_l = 2^{s'_l}$
Adjust d_l to d'_l with b'_l: $d'_l = \frac{(q_m - q_s)^t}{2^{b'-1}-1}$ or $d'_l = \frac{(q_m - q_s)^t}{2^{b'}}$

This extension only involves a small overhead to the training phase. More specifically, while there is no change in the backward pass, in the forward pass the quantizer bit-width and stepsize are adjusted. The adjustment of the stepsize would result in larger variance in gradients, however, the scheme is still able to converge quickly in practice.

4   Experiments

We run experiments on different models for image classification, including VGG7 [17] on CIFAR10, ResNet18 [10] and MobileNetV2 [25] on ImageNet. Performance is measured by top-1 accuracy and the BOP count. The proposed DJPQ scheme is always applied to pretrained models. We compare with several other schemes from literature including LSQ [7], TWN [17], RQ [19], WAGE [32] and DQ [29]. We also conduct experiments for the two-stage compression approach, to see how much gain is achieved by co-optimizing pruning and quantization. Furthermore, we conduct experiments of DJPQ with power-of-two quantization, denoted as DJPQ-restrict. We also modify the DQ scheme to do restricted mixed-bit quantization to compare with DJPQ-restrict.

4.1 Comparison of DJPQ with Quantization only Schemes

CIFAR10 Results. For the VGG7 model on CIFAR10 classification, DJPQ performance along with its baseline floating-point model are provided in Table 1. Figure 1 shows the weight and activation bit-widths for each layer. The pruning ratio P_l for each layer is given in Fig. 2. Compared to the uncompressed model, we see that DJPQ reduces the amount of BOPs by 210x with less than a 1.5% accuracy drop. DJPQ is able to achieve a larger compression ratio compared to the other schemes. Comparing e.g. to DQ, which also learns mixed precision, DJPQ has a very similar BOPs reduction.

Table 1. VGG7 results on CIFAR10. If weight and activation bit-width are fixed, they are represented in the format of weight/activation bit-width in the table. If weight and activation bit-width have different values in different layers, they are denoted as 'mixed'. Baseline is the floating point model. 'BOP comp. ratio' denotes the BOP compression ratio defined in Sect. 3.3. 'DJPQ-restrict' denotes DJPQ with power-of-two bit-restricted quantization as presented in Sect. 3.5.

	Bit-width	Test Acc.	MACs(G)	BOPs(G)	BOP comp. ratio
Baseline	32/32	93.0%	0.613	629	–
TWN [17]	2/32	92.56%	0.613	39.23	16.03
RQ [19]	8/8	93.30%	0.613	39.23	16.03
RQ [19]	4/4	92.04%	0.613	9.81	64.12
WAGE [32]	2/8	93.22%	0.613	9.81	64.12
DQ[a] [29]	Mixed	91.59%	0.613	3.03	207.59
DQ-restrict[b] [29]	Mixed	91.59%	0.613	3.40	185.00
DJPQ	Mixed	**91.54%**	0.367	2.99	**210.37**
DJPQ-restrict	Mixed	**91.43%**	0.372	2.92	**215.41**

[a] For a fair comparison, we replaced the memory regularization term in DQ [29] with BOPs regularization.
[b] DQ-restrict [29] refers to the scheme where bit-width learned by DQ are upper rounded to the nearest power-of-two integers.

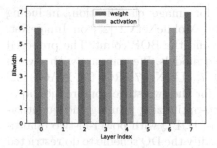

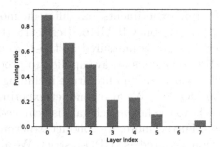

Fig. 1. Bit-width of weights and activations for VGG7 with DJPQ scheme

Fig. 2. Pruning ratio for VGG7 with DJPQ scheme

ImageNet Results. For experiments on the ImageNet dataset, we applied our DJPQ scheme to ResNet18 and MobileNetV2. Table 2 provides a comparison of DJPQ with state-of-art and other works. The learned bit-width and pruning ratio distributions for DJPQ with 69.27% accuracy are shown in Figs. 3 and 4, respectively. From the table we see that DJPQ achieves a 53x BOPs reduction while the top-1 accuracy only drops by 0.47%. The results showed a smooth trade-off achieved by DJPQ between accuracy and BOPs reduction. Compared with other fixed-bit quantization schemes, DJPQ achieves a significantly larger BOPs reduction. Particularly compared with DQ, DJPQ achieves around a 25% reduction in BOP counts (40.71G vs 30.87G BOPs) with a 0.3% accuracy improvement. A more detailed comparison of DJPQ and DQ has been given in Appendix B.1, which includes weight and activation bit-width distribution, along with pruning ratio distribution. Through the comparison, we show that DJPQ can flexibly trade-off pruning and activation to achieve a larger compression ratio.

Table 2. Comparison of ResNet18 compression results on ImageNet. Baseline is the floating point model. 'DJPQ-restrict' denotes DJPQ with bit-restricted quantization.

	Bit-width	Test Acc.	MACs(G)	BOPs(G)	BOP comp. ratio
Baseline	32/32	69.74%	1.81	1853.44	–
LSQ[c] [7]	3/3	70.00%	1.81	136.63	13.56
RQ [19]	8/8	69.97%	1.81	115.84	16.00
DFQ [23]	4/8	69.3%	1.81	57.92	32.00
SR+DR [8]	8/8	68.17%	1.81	115.84	16.00
UNIQ [1]	4/8	67.02%	1.81	57.92	32.00
DFQ [23]	4/4	65.80%	1.81	28.96	64.00
TWN [17]	2/32	61.80%	1.81	115.84	16.00
RQ [19]	4/4	61.52%	1.81	28.96	64.00
DQ[a] [29]	Mixed	68.49%	1.81	40.71	45.53
DQ-restrict[b] [29]	Mixed	68.49%	1.81	58.68	31.59
VIBNet [4]+fixed quant.	8/8	69.24%	1.36	87.04	21.29
VIBNet [4]+DQ[a] [29]	Mixed	68.52%	1.36	39.83	46.53
DJPQ	Mixed	**69.27%**	1.39	35.01	**52.94**
DJPQ	Mixed	68.80%	1.39	30.87	60.04
DJPQ-restrict	Mixed	**69.12%**	1.46	35.45	**52.28**

[a] For a fair comparison, we replaced the memory regularization term in DQ [29] with BOPs regularization.

[b] DQ-restrict [29] refers to the scheme where bit-width learned by DQ are upper rounded to the nearest power-of-two integers.

[c] LSQ [23] does not quantize the first and last layers.

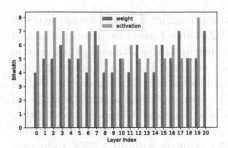

Fig. 3. Bit-width for ResNet18 on ImageNet with DJPQ scheme

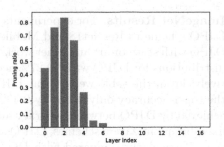

Fig. 4. Pruning ratio for ResNet18 on ImageNet with DJPQ scheme

MobileNetV2 has been shown to be very sensitive to quantization [23]. Despite this, we show in Table 3 that the DJPQ scheme is able to compress MobileNetV2 with a large compression ratio. The optimized bit-width and pruning ratio distributions are provided in Appendix B.2. It is observed that DJPQ achieves a 43x BOPs reduction within 2.4% accuracy drop. Compared with DQ, DJPQ achieves around 25% reduction in BOP counts over DQ on MobileNetV2 (175.24G vs 132.28G BOPs) with 0.5% accuracy improvement. Comparisons of BOPs for different schemes are plotted in Figs. 5 and 6 for ResNet18 and MobileNetV2, respectively. DJPQ provides superior results for ResNet18 and MobileNetV2, and can be easily extended to other architectures.

Table 3. Comparison of MobileNetV2 compression results on ImageNet. Baseline is the uncompressed floating point model.

	Bit-width	Test Acc.	MACs(G)	BOPs(G)	BOP comp. ratio
Baseline	32/32	71.72%	5.55	5682.32	–
SR+DR [8]	8/8	61.30%	5.55	355.14	16.00
DFQ [23]	8/8	70.43%	5.55	355.14	16.00
RQ [19]	8/8	71.43%	5.55	355.14	16.00
RQ [19]	6/6	68.02%	5.55	199.80	28.44
UNIQ [1]	4/8	66.00%	5.55	177.57	32.00
DQ[a] [29]	Mixed	68.81%	4.81	175.24	32.43
DJPQ	Mixed	**69.30%**	4.76	132.28	**42.96**

[a] For a fair comparison, we replaced the memory regularization term in DQ [29] with BOPs regularization.

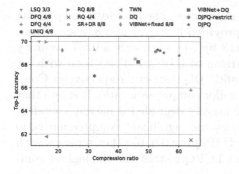
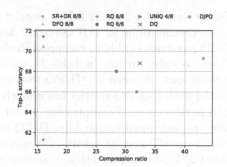

Fig. 5. Comparison of BOPs reduction for ResNet18 on ImageNet

Fig. 6. Comparison of BOPs reduction for MobileNetV2 on ImageNet

4.2 Comparison of DJPQ with Two-Stage Optimization

In this section we provide a comparison of DJPQ and the two-stage approach - pruning first, then quantizing, and show that DJPQ outperforms both in BOPs reduction and accuracy. To ensure a fair comparison, pruning and quantization are optimized independently in the two-stage approach. We have done extensive experiments to find a pruning ratio that results in high accuracy. The two-stage results in Table 2 are derived by first pruning a model with VIBNet gates to a 1.33× MAC compression ratio. The accuracy of the pruned model is 69.54%. Then the pruned model is quantized with both fixed-bit and mixed-precision quantization. For fixed 8-bit quantization, the two-stage scheme obtains a 21.29x BOP reduction with 69.24% accuracy. For the VIBNet+DQ approach, we achieve a 46.53× BOP reduction with 68.52% accuracy.

By comparing the resulting MAC counts of the compressed model of DJPQ with the two-stage approach, we see a very close MAC compression ratio from pruning between the two schemes (1.24× vs 1.33×). They also have comparable BOP counts, however, DJPQ achieves 0.75% higher accuracy (69.27% vs 68.52%) than the two-stage scheme. The results provide good evidence that even under similar pruning ratio and BOP reduction, DJPQ is able to achieve a higher accuracy. This is likely due to the pruned channel distribution of DJPQ being dynamically adapted and optimized jointly with quantization resulting in a higher accuracy.

4.3 Comparison of DJPQ with Others Under Bit Restriction

We further run experiments of the DJPQ scheme for power-of-two bit-restricted quantization. For VGG7 on CIFAR10, the DJPQ result with bit-restricted quantization is provided in Table 1 named 'DJPQ-restrict'. With DJPQ compression, VGG7 is compressed by 215× with a 91.43% accuracy. The degradation of accuracy compared to DJPQ without bit restriction is negligble, showing that our scheme also works well for power-of-two restricted bit-widths. For ResNet18 on

ImageNet, Table 2 shows DJPQ-restrict performance. The bit-width and pruning ratio distributions for DJPQ-restrict are provided in Figs. 7 and 8, respectively. DJPQ-restrict is able to compress ResNet18 by 53× with only a 0.5% accuracy drop. We also provide in Table 2 a comparison of DJPQ and DQ both with bit restrictions, denoted as 'DJPQ-restrict' and 'DQ-restrict', respectively. Compared with DQ, DQ-restrict has a much smaller compression ratio. It shows that the performance of compression would be largely degraded if naively rounding the trained quantization bits. DJPQ-restrict has significant compression ratio gain over DQ-restrict results. Comparing DJPQ-restrict to other schemes with fixed 2/4/8-bit quantization, it is clear that DJPQ-restrict has the highest compression ratio.

4.4   Analysis of Learned Distributions

For ResNet18, the learned bit-width and pruning ratio distributions for DJPQ with 69.27% accuracy are shown in Figs. 3 and 4, respectively. It is observed that pruning occurs more frequently at earlier layers. The pruning ratio can be very large, indicating heavy over-parameterization in that layer. Layers {7,12,17} are residual connections. From Fig. 3 we see that all the three residual connections require larger bits than their corresponding regular branches. Regarding to the distribution of t in the nonlinear mapping, it is observed that for layers with heavy pruning, t is generally smaller and $t < 1$; for layers with no pruning, t is close to 1. This gives a good reflection of interaction between pruning and quantization. For MobileNetV2, we found that point-wise convolution layers require larger bit-width than depth-wise ones, indicating that point-wise convolutional layers are more sensitive than depth-wise ones.

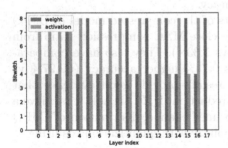
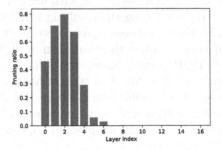

Fig. 7. Bit-width for ResNet18 on ImageNet. Bits are restricted to power of 2.

Fig. 8. Pruning ratio for ResNet18 on ImageNet. Bits are restricted to power of 2.

5   Conclusion

We proposed a differentiable joint pruning and quantization (DJPQ) scheme that optimizes bit-width and pruning ratio simultaneously. The scheme integrates variational information bottleneck, structured pruning and mixed-precision quantization, achieving a flexible trade-off between sparsity and bit precision. We show that DJPQ is able to achieve larger bit-operations (BOPs) reduction over conventional two-stage compression while maintaining the state-of-art performance. Specifically, DJPQ achieves 53× BOPs reduction in ResNet18 and 43× reduction in MobileNetV2 on ImageNet. We further extend DJPQ to support power-of-two bit-restricted quantization with a small overhead. The extended scheme is able to reduce BOPs by 52× on ResNet18 with almost no accuracy loss.

Acknowledgments. We would like to thank Jilei Hou and Joseph Soriaga for consistent support, and thank Jinwon Lee, Kambiz Azarian and Nojun Kwak for their great help in revising this paper and providing valuable feedback.

A   Quantization scheme in DJPQ

Figure 9 illustrates the proposed quantization scheme. First, a non-linear function is applied to map any weight input x to $\tilde{x}$ (shown in blue curve). Then a uniform quantization is applied to $\tilde{x}$. The quantized value x_q is shown in red.

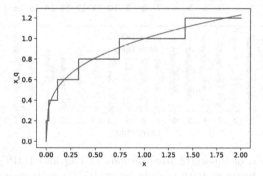

Fig. 9. Illustration of quantization scheme. The blue curve gives the nonlinear mapping function. The red curve corresponds to the quantization value. (Color figure online)

B   Experimental details

B.1   Comparison of DJPQ with DQ

To show that joint optimization of pruning and quantization outperforms quantization only scheme such as DQ, we plot in Fig. 10 a comparison of weight and activation bit-width for DQ and DJPQ. The results are for ResNet18 on ImageNet. We plot the pruning ratio curve of DJPQ in both figures to better show the pruning effect in the joint optimization scheme. As seen in the figure, there is no big difference between weight bit-width for DQ and DJPQ. However, the difference between activation bit-width for the two schemes is significant. Layer 0 to 6 in DJPQ has much larger activation bit-width than those in DQ, while those layers correspond to high pruning ratios in DJPQ. It provides a clear evidence that pruning and quantization can well tradeoff between each other in DJPQ, resulting in lower redundancy in the compressed model.

B.2   DJPQ Results for MobileNetV2

Figures 11 and 12 show the optimized bit-width and pruning ratio distributions, respectively. It is observed that earlier layers tend to have larger pruning ratios than the following layers. And for many layers in MobileNetV2, DJPQ is able to quantize the layers into to small bit-width.

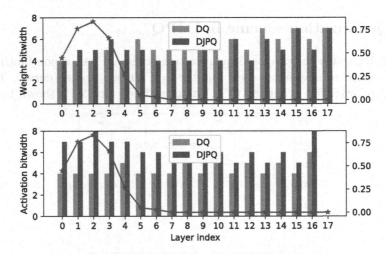

Fig. 10. Comparison of bit-width distributions of DQ and DJPQ for ResNet18 on ImageNet. The top figure plots the weight bit-width for each layer, while the bottom plots the corresponding activation bit-width. The green curve is the pruning ratio of DJPQ in each layer.(Color figure online)

B.3   Experimental Setup

The input for all experiments are uniformly quantized to 8 bits. For the two-stage scheme including 'VIBNet+fixed quant.' and 'VIBNet+DQ', VIBNet pruning is firstly optimized at learning rate 1e–3 with SGD, with a pruning learning rate scaling of 5. The strength γ is set to 5e–6. In DQ for the pruned model, β is chosen to 1e–11 and the learning rate is 5e–4. The learning rate scaling for quantization is 0.05. All the pruning threshold α_{th} is chosen to 1e–3. The number of epochs is 20 for each of the stage.

For DJPQ experiments on VGG7, the learning rate is set to 1e–3 with an ADAM optimizer. The initial bit-width is 6. The strength γ and β are 1e–6

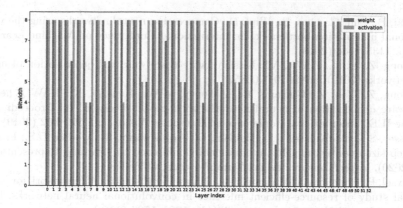

Fig. 11. Bit-width for MobileNetV2 on ImageNet with DJPQ scheme

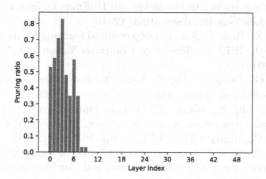

Fig. 12. Pruning ratio for MobileNetV2 on ImageNet with DJPQ scheme

and 1e–9, respectively. The scaling of learning rate for pruning and quantization is 10 and 0.05, respectively. For DJPQ on ResNet18, we chose a learning rate 1e–3 with SGD optimization. The initial bit-width for weights and activations are 6 and 8. The scaling of learning rate for pruning and quantization are 5 and 0.05, respectively. The strength γ and β are 1e–5 and 1e–10. For DJPQ on MobileNetV2, the learning rate is set to 1e–4 for SGD optimization. The initial bit-width is 8. The strength γ and β are set to 1e–8 and 1e–11. The learning rate scaling for pruning and quantization are selected to be 1 and 0.005, respectively.

References

1. Baskin, C., et al.: UNIQ: uniform noise injection for the quantization of neural networks (2018). CoRR abs/1804.10969. http://arxiv.org/abs/1804.10969
2. Bengio, Y., Léonard, N., Courville, A.C.: Estimating or propagating gradients through stochastic neurons for conditional computation (2013). CoRR abs/1308.3432. http://arxiv.org/abs/1308.3432
3. Bethge, J., Bartz, C., Yang, H., Chen, Y., Meinel, C.: MeliusNet: can binary neural networks achieve mobilenet-level accuracy? (2020). arXiv:2001.05936

4. Dai, B., Zhu, C., Guo, B., Wipf, D.: Compressing neural networks using the variational information bottleneck. In: International Conference on Machine Learning, pp. 1135–1144 (2018)

5. Dong, Z., et al.: HAWQ-V2: hessian aware trace-weighted quantization of neural networks (2019). arXiv:1911.03852

6. Dong, Z., Yao, Z., Gholami, A., Mahoney, M.W., Keutzer, K.: HAWQ: hessian aware quantization of neural networks with mixed-precision. In: Proceedings of the IEEE International Conference on Computer Vision, pp. 293–302 (2019)

7. Esser, S.K., McKinstry, J.L., Bablani, D., Appuswamy, R., Modha, D.S.: Learned step size quantization. In: International Conference on Learning Representations (2020). https://openreview.net/forum?id=rkgO66VKDS

8. Gysel, P., Pimentel, J., Motamedi, M., Ghiasi, S.: Ristretto: A framework for empirical study of resource-efficient inference in convolutional neural networks. IEEE Trans. Neural Netw. Learn. Syst. **29**(11), 5784–5789 (2018)

9. Han, S., Mao, H., Dally, W.J.: Deep compression: compressing deep neural network with pruning, trained quantization and Huffman coding. In: 4th International Conference on Learning Representations (2016)

10. He, K., Zhang, X., Ren, S., Sun, J.: Deep residual learning for image recognition. In: Proceedings of the IEEE Conference on Computer Vision and Pattern Recognition, pp. 770–778 (2016)

11. He, Y., Kang, G., Dong, X., Fu, Y., Yang, Y.: Soft filter pruning for accelerating deep convolutional neural networks (2018). arXiv:1808.06866

12. He, Y., Lin, J., Liu, Z., Wang, H., Li, L.J., Han, S.: AMC: AutoML for model compression and acceleration on mobile devices. In: Proceedings of the European Conference on Computer Vision (ECCV), pp. 784–800 (2018)

13. He, Y., Zhang, X., Sun, J.: Channel pruning for accelerating very deep neural networks. In: Proceedings of the IEEE International Conference on Computer Vision, pp. 1389–1397 (2017)

14. Ignatov, A., et al.: AI benchmark: all about deep learning on smartphones in 2019 (2019). arXiv:1910.06663

15. Krishnamoorthi, R.: Quantizing deep convolutional networks for efficient inference: a whitepaper (2018). arXiv:1806.08342

16. Kuzmin, A., Nagel, M., Pitre, S., Pendyam, S., Blankevoort, T., Welling, M.: Taxonomy and evaluation of structured compression of convolutional neural networks (2019). arXiv:1912.09802

17. Li, F., Zhang, B., Liu, B.: Ternary weight networks (2016). arXiv:1605.04711

18. Li, H., Kadav, A., Durdanovic, I., Samet, H., Graf, H.P.: Pruning filters for efficient ConvNets. In: International Conference on Learning Representations (2017). https://openreview.net/pdf?id=rJqFGTslg

19. Louizos, C., Reisser, M., Blankevoort, T., Gavves, E., Welling, M.: Relaxed quantization for discretized neural networks. In: International Conference on Learning Representations (2019). https://openreview.net/forum?id=HkxjYoCqKX

20. Louizos, C., Ullrich, K., Welling, M.: Bayesian compression for deep learning. In: Advances in Neural Information Processing Systems, pp. 3288–3298 (2017)

21. Louizos, C., Welling, M., Kingma, D.P.: Learning sparse neural networks through L_0 regularization. In: International Conference on Learning Representations (2018). https://openreview.net/forum?id=H1Y8hhg0b

22. Molchanov, P., Tyree, S., Karras, T., Aila, T., Kautz, J.: Pruning convolutional neural networks for resource efficient inference. In: International Conference on Learning Representations (2017). https://openreview.net/pdf?id=SJGCiw5gl

23. Nagel, M., Baalen, M.v., Blankevoort, T., Welling, M.: Data-free quantization through weight equalization and bias correction. In: Proceedings of the IEEE International Conference on Computer Vision, pp. 1325–1334 (2019)

24. Peng, H., Wu, J., Chen, S., Huang, J.: Collaborative channel pruning for deep networks. In: International Conference on Machine Learning, pp. 5113–5122 (2019)

25. Sandler, M., Howard, A., Zhu, M., Zhmoginov, A., Chen, L.C.: MobileNetV2: inverted residuals and linear bottlenecks. In: Proceedings of the IEEE Conference on Computer Vision and Pattern Recognition, pp. 4510–4520 (2018)

26. Theis, L., Korshunova, I., Tejani, A., Huszár, F.: Faster gaze prediction with dense networks and fisher pruning (2018). arXiv:1801.05787

27. Tishby, N., Pereira, F., Bialek, W.: The information bottleneck method. In: Proceedings of the 37th Allerton Conference on Communication, Control and Computation, vol. 49 (2001)

28. Tung, F., Mori, G.: CLIP-Q: Deep network compression learning by in-parallel pruning-quantization. In: Proceedings of the IEEE Conference on Computer Vision and Pattern Recognition, pp. 7873–7882 (2018)

29. Uhlich, S., et al.: Differentiable quantization of deep neural networks (2019). CoRR abs/1905.11452. http://arxiv.org/abs/1905.11452

30. Wang, K., Liu, Z., Lin, Y., Lin, J., Han, S.: HAQ: hardware-aware automated quantization with mixed precision. In: Proceedings of the IEEE Conference on Computer Vision and Pattern Recognition, pp. 8612–8620 (2019)

31. Wu, H., Judd, P., Zhang, X., Isaev, M., Micikevicius, P.: Integer quantization for deep learning inference: Principles and empirical evaluation (2020). arXiv:2004.09602

32. Wu, S., Li, G., Chen, F., Shi, L.: Training and inference with integers in deep neural networks. In: International Conference on Learning Representations (2018). https://openreview.net/forum?id=HJGXzmspb

33. Yang, H., Gui, S., Zhu, Y., Liu, J.: Automatic neural network compression by sparsity-quantization joint learning: a constrained optimization-based approach. In: Proceedings of the IEEE/CVF Conference on Computer Vision and Pattern Recognition, pp. 2178–2188 (2020)

34. Yazdanbakhsh, A., Elthakeb, A.T., Pilligundla, P., Mireshghallah, F., Esmaeilzadeh, H.: ReLeQ: an automatic reinforcement learning approach for deep quantization of neural networks (2018). arXiv:1811.01704

35. Ye, S., et al.: A unified framework of DNN weight pruning and weight clustering/quantization using ADMM (2018). arXiv:1811.01907

Learning to Generate Customized Dynamic 3D Facial Expressions

Rolandos Alexandros Potamias$^{(\boxtimes)}$, Jiali Zheng, Stylianos Ploumpis, Giorgos Bouritsas, Evangelos Ververas, and Stefanos Zafeiriou

Department of Computing, Imperial College London, London, UK
{r.potamias,jiali.zheng18,s.ploumpis,g.bouritsas,e.ververas16,
s.zafeiriou}@imperial.ac.uk

Abstract. Recent advances in deep learning have significantly pushed the state-of-the-art in photorealistic video animation given a single image. In this paper, we extrapolate those advances to the 3D domain, by studying 3D image-to-video translation with a particular focus on 4D facial expressions. Although 3D facial generative models have been widely explored during the past years, 4D animation remains relatively unexplored. To this end, in this study we employ a deep mesh encoder-decoder like architecture to synthesize realistic high resolution facial expressions by using a single neutral frame along with an expression identification. In addition, processing 3D meshes remains a non-trivial task compared to data that live on grid-like structures, such as images. Given the recent progress in mesh processing with graph convolutions, we make use of a recently introduced learnable operator which acts directly on the mesh structure by taking advantage of local vertex orderings. In order to generalize to 4D facial expressions across subjects, we trained our model using a high resolution dataset with 4D scans of six facial expressions from 180 subjects. Experimental results demonstrate that our approach preserves the subject's identity information even for unseen subjects and generates high quality expressions. To the best of our knowledge, this is the first study tackling the problem of 4D facial expression synthesis.

Keywords: Expression generation · Facial animation · 4D synthesis · 4DFAB · Graph neural networks

1 Introduction

Recently, facial animation has received attention from the industrial graphics, gaming and filming communities. Face is capable to impart a wide range of information not only about the subject's emotional state but also about the tension of the moment in general. An engaged and crucial task is 3D avatar

Electronic supplementary material The online version of this chapter (https://doi.org/10.1007/978-3-030-58526-6_17) contains supplementary material, which is available to authorized users.

animation, which has lately become feasible [34]. With modern technology, a 3D avatar can be generated by a single uncalibrated camera [9] or even by a self portrait image [33]. At the same time, capturing facial expression is an important task in order to perceive behaviour and emotions of people. To tackle the problem of facial expression generation, it is essential to understand and model facial muscle activations that are related to various emotions. Several studies have attempted to decompose facial expressions on two dimensional spaces such as images and videos [15,31,44,50]. However, modeling facial expressions on high resolution 3D meshes remains unexplored.

In contrast, few studies have attempted 3D speech-driven facial animation exclusively based on vocal audio and identity information [13,21]. Nevertheless, emotional reactions of a subject are not always expressed vocally and speech-driven facial animation approaches neglect the importance of facial expressions. For instance, sadness and happiness are two very common emotions that can be voiced, mainly, through facial deformations. To this end, facial expressions are a major component of entertainment industry, and can convey emotional state of both scene and identity.

People signify their emotions using facial expressions in similar manners. For instance, people express their happiness by mouth and cheek deformations, that vary according to the subject's emotional state and characteristics. Thus, one can describe expressions as "unimodal" distributions [15], with gradual changes from the neutral model till the apex state. Similarly to speech signals, emotion expressions are highly correlated to facial motion, but lie in two different domains. Modeling the relation between those two domains is essential for the task of realistic facial animation. However, in order to disentangle identity information and facial expression it is essential to have a sufficient amount of data. Although most of the publicly available 3D datasets contain a large variety of facial expression, they are captured only from a few subjects. Due to this difficulty, prior work has only focused on generating expressions in 2D.

Our aim is to generate realistic 3D facial animation given a target expression and a static neutral face. Synthesis of facial expression generation on new subjects can be achieved by expression transfer of generalized deformations [39,50]. In order to produce realistic expressions, we map and model facial animation directly on the mesh space, avoiding to focus on specific face landmarks. Specifically, the proposed method comprises two parts: (a) a recurrent LSTM encoder to project the expected expression motion to an expression latent space, and (b) a mesh decoder to decode each latent time-sample to a mesh deformation, which is added to the neutral expression identity mesh. The mesh decoder utilizes intrinsic lightweight mesh convolutions, introduced in [6], along with unpooling operations that act directly on the mesh space [35]. We train our model in an end-to-end fashion on a large scale 4D face dataset. The devised methodology tackles a novel and unexplored problem, i.e. the generation of 4D expressions given a single neutral expression mesh. Both the desired length and the target expression are fully defined and controlled by the user. Our work considerably deviates from methods in the literature as it can be used to generate 4D full-face customised expressions on real-time. Finally, our study is the first 3D facial animation framework that utilizes an intrinsic encoder-decoder architecture that operates directly on mesh space using mesh convolutions instead of fully connected layers, as opposed to [13,21].

2 Related Work

Facial Animation Generation. Following the progress of 3DMMs, several approaches have attempted to decouple expression and identity subspaces and built linear [2,5,45] and nonlinear [6,25,35] expression morphable models. However, all of the aforementioned studies are focused on static 3D meshes and they cannot model 3D facial motion. Recently, a few studies attempted to model the relation between speech and facial deformation for the task of 3D facial motion synthesis. Karras et al. [21] modeled speech formant relationships with 5K vertex positions, generating facial motion from LPC audio features. While this was the first approach to tackle facial motion directly on 3D meshes, their model is subject specific and cannot be generalized across different subjects. Towards the same direction, in [13], facial animation was generated using a static neutral template of the identity and a speech signal, used along with DeepSpeech [20] to generate more robust speech features. A different approach was utilized in [32], where 3D facial motion is generated by regressing on a set of action units, given MFCC audio features processed by RNN units. However, their model is trained on parameters extracted from 2D videos instead of 3D scans. Tzirakis et al. [41] combined predicted blendshape coefficients with a mean face to synthesize 3D facial motion from speech, replacing also fully connected layers, utilized in previous studies, with an LSTM. Blendshape coefficients are also predicted from audio, using attentive LSTMs in [40]. In contrast with the aforementioned studies, the proposed method aims to model facial animations directly on 3D meshes. Furthermore, although blendshape coefficients might be easily modeled, they rely on predefined face rigs, a factor that limits their generalization to new unseen subjects.

Geometric Deep Learning. Recently, the enormous amount of applications related to data residing in non-Euclidean domains motivated the need for the generalization of several popular deep learning operations, such as convolution, to graphs and manifolds. The main efforts include the reformulation of regular convolution operators in order to be applied on structures that lack consistent ordering or directions, as well as the invention of pooling techniques for graph downsampling. All relevant endeavours lie within the new research area of Geometric Deep Learning (GDL) [7]. The first attempts defined convolution in the spectral domain, by applying filters inspired from graph signal processing techniques [37]. These methods mainly boil down to either an eigendecomposition of a Graph Shift Operator (GSO) [8], such as the graph Laplacian, or to approximations thereof, by using polynomials [14,23] or rational complex functions [24] of the GSO in order to obtain strict spatial localization and reduced computational complexity. Subsequent attempts generalize conventional CNNs by introducing patch operators that extract spatial relations of adjacent nodes within a local patch. To this end, several approaches generalized local patches to graph data, using geodesic polar charts [27] anisotropic diffusion operators [4] on manifolds or graphs [3]. MoNet [28], generalized previous spatial approaches by learning the patches themselves with Gaussian kernels. In the same direction, SplineCNN [17] replaced Gaussian kernels with B-spline functions with significant speed advantage. Recent studies focused on soft-attention methods to weight adjacent nodes [42,43]. However, in contrast to regular convolutions, the way permutation invariance is enforced in most of the aforementioned operators, inevitably renders them unaware of vertex correspondences. To tackle that, Bouritsas et al. [6] defined local node orderings, instantiated with the spiral operator of [26], by exploiting the fixed underlying topology of certain deformable shapes, and built correspondence-aware anisotropic operators.

Facial Expression Datasets. Another major reason that 4D generative models have not been widely exploited is due to the limited amount of 3D datasets. During the past decade, several 3D face databases have been published. However, most of them are static [19,29,36,38,49], consisted of few subjects [10,12,35,46], and have limited [1,16,36] or spontaneous expressions [47,48], making them inappropriate for tasks such as facial expression synthesis. On the other hand, the recently proposed 4DFAB dataset [11] consists of six 3D dynamic facial expressions (from 180 subjects), which is ideal for subject independent facial expression generation. In contrast with all previously mentioned datasets, 4DFAB, due to the high range of subjects, can be a promising resource towards disentangling facial expression from the identity information.

3   Learnable Mesh Operators: Background

3.1   Spiral Convolution Networks

We define a 3D facial surface discretized as triangular mesh $\mathcal{M} = (\mathcal{V}, \mathcal{E}, \mathcal{F})$ with $\mathcal{V}$ the set of N vertices, $\mathcal{E}$ and $\mathcal{F}$ the sets of edges and faces, respectively. Let also, $X \in \mathbb{R}^{N \times d}$ denote the feature matrix of the mesh. In contrast to regular domains, when attempting to apply convolution operators on graph-based structures, there does not exist a consistent way to order the input coordinates. However, in a *fixed topology* setting, such an ordering is beneficial so as to be able to keep track of the existing correspondences. In [6], the authors identified this problem and intuitively order the vertices by using spiral trajectories [26]. In particular, given a vertex $v \in \mathcal{V}$, we can define a *k-ring* and *k-disk* as:

$$
\begin{aligned}
ring^{(0)}(v) &= v, \\
ring^{(k+1)}(v) &= \mathcal{N}(ring^{(k)}(v)) - disk^{(k)}(v), \\
disk^{(k)}(v) &= \bigcup_{i=0,\dots,k} ring^{(i)}(v)
\end{aligned}
\tag{1}
$$

where $\mathcal{N}(S)$ is the set of all vertices adjacent to at least one vertex $\in S$.

Once the $ring^{(k)}$ is defined, the spiral trajectory centered around vertex v can be defined as:

$$S(v,k) = \{ring^{(0)}(v), ring^{(1)}(v), ..., ring^{(k)}(v)\} \tag{2}$$

To be consistent across all vertices, one can pad or truncate S(v, k) to a fixed length L. To fully define the spiral ordering, we have to declare the starting direction and the orientation of a spiral sequence. In the current study, we adopt the settings followed in [6], by selecting the initial vertex of $S(v,k)$ to be in the direction of the shortest geodesic distance between a static reference vertex. Given that all 3D faces share the same topology, spiral ordering $S(v,k)$ will be the same across all meshes and so, their calculation is done only once. With all the above mentioned, *Spiral Convolution* can be defined as:

$$\mathbf{f}_v^* = \sum_{j=0}^{|S(v,k)|-1} \mathbf{f}(S_j(v,k))\mathbf{W}_j \tag{3}$$

where $|S(v,k)|$ amounts to the total length of the spiral trajectory, $\mathbf{f}(S_j(v,k))$ are the d-dimensional input features of the *jth* vertex of the spiral trajectory, $\mathbf{f}^*$ the respective output, and $\mathbf{W}_j$ are the filter weights.

3.2   Mesh Unpooling Operations

In order to let our graph convolution decoder to generate faces sampled from a latent space, it is essential to use unpooling operations in analogy with transposed convolutions in regular settings. Each graph convolution is followed by an upsampling layer which acts directly on the mesh space, by increasing the number of vertices. We use sampling operations introduced in [35], based on sparse matrix multiplications with upsampling matrices $Q_u \in \{0,1\}^{n \times m}$, where $m > n$. Since upsampling operation changes the topology of the mesh, and in order to retain the face structure, upsampling matrices Q_u are defined on the basis of down-sampling matrices. The barycentric coordinates of the vertices that were discarded during downsampling procedure are stored and used as the new vertex coordinates of the upsampling matrices.

4   Model

The overall architecture of our model is structured by two major components (see Fig. 1). The first one contains a temporal encoder, using an LSTM layer that encodes the expected facial motion of the target expression. It takes as input a temporal signal $e \in R^{6 \times T}$ with length T, equal to the target facial expression, also equipped with information about the time-stamps that show when the generated facial expression should reach onset, apex and offset modes. Each time-frame of signal e can be characterised as a one-hot encoding of one of the six expressions, with amplitude that indicates the scale of the expression.

The second component of our network consists of a frame decoder, with four layers of mesh convolutions, where each one is followed by an upsampling layer. Each upsampling layer increases the number of vertices by five times, and every mesh convolution is followed by a ReLU activation [30]. Finally, the output of the decoder is added to the identity neutral face. Given a time sample from the latent space the frame decoder network models the expected deformations on the neutral face. Each output time frame can be expressed as:

$$\hat{x}_t = D(z_t) + x_{id},$$
$$z_t = E(e_t) \tag{4}$$

where $D(\cdot)$ denotes the mesh decoder network, $E(\cdot)$ the LSTM encoder, e_t the facial motion information for time-frame t and x_{id} the neutral face of the identity. The network details can be found in Table 1. We trained our model for 100 epochs with learning rate of 0.001 and a weight decay of 0.99 on every epoch. We used Adam optimizer [22] with a 5e−5 weight decay.

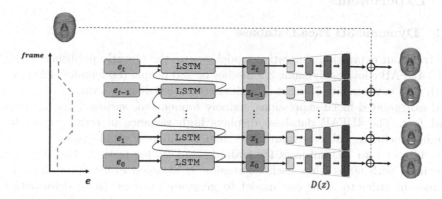

Fig. 1. Network architecture of the proposed method.

Loss Function. The mesh decoder network outputs motion deformation for each time-frame with respect to the expected facial animation. To train our model we minimize both the reconstruction error L_r and the temporal coherence L_c, as proposed in [21]. Specifically, we define our loss function between the generated time frame $\hat{x}_t$ and its ground truth x_t value as:

$$L_r(\hat{x}_t, x_t) = \|\hat{x}_t - x_t\|_1$$
$$L_c(\hat{x}_t, x_t) = \|(\hat{x}_t - \hat{x}_{t-1}) - (x_t - x_{t-1})\|_1 \tag{5}$$
$$L(\hat{x}_t, x_t) = L_r(\hat{x}_t, x_t) + L_c(\hat{x}_t, x_t)$$

Although reconstruction loss L_r term can be sufficient to encourage model to match ground truth vertices at each time step, it does not produce high-quality realistic animation. On the contrary, temporal coherence loss L_c term ensures temporal stability of the generated frames by matching the distances between consecutive frames on ground truth and generated expressions.

Table 1. Mesh decoder architecture

Layer	Input dimension	Output dimension
Fully connected	64	46 × 64
Upsampling	46 × 64	228 × 64
Convolution	228 × 64	228 × 32
Upsampling	228 × 32	1138 × 32
Convolution	1138 × 32	1138 × 16
Upsampling	1138 × 16	5687 × 16
Convolution	5687 × 16	5687 × 8
Upsampling	5687 × 8	28431 × 8
Convolution	28431 × 8	28431 × 3

5   Experiments

5.1   Dynamic 3D Face Database

To train our expression generative model we use the recently published 4DFAB [11]. 4DFAB contains dynamic 3D meshes of 180 people (60 females, 120 males) with ages between 5 to 75 years. The devised meshes display a variety of complex and exaggerated facial expressions, namely *happy, sad, surprise, angry, disgust* and *fear*. The 4DFAB database displays high variance in terms of ethnicity origins, including subjects from more than 30 different ethnic groups. We split the dataset into 153 subjects for training and 27 for testing. The data were captured with 60fps, thus each expression is sampled every approximately 5 frames in order to allow our model to generate extreme facial deformations. Given the high quality of the data (each mesh is composed by 28K vertices) as well as the relatively big number of subjects, 4DFAB presents a rich and rather challenging choice for training generative models.

Expression Motion Labels. In this study, we rely on the assumption that each expression can be characterised by four phases of its evolution (see Fig. 2). First, the subject starts from a neutral pose and at a certain point their face starts to deform, in order to express their emotional state. We call this the *onset phase*. After the subject's expression reaches its *apex state*, it will start again its deformation from the peak emotional state until it becomes neutral again. We call this the *offset phase*. Thus, each time frame is assigned a label that reflects its emotional state phase. We consider the emotional state as a value ranging from 0 to 1 assigned to each frame, with 0 representing the neutral phase and 1 the apex phase. Onset and offset phases are represented via a linear interpolation between the apex and neutral phases (see Fig. 2). However, expressions may also range in terms of extremeness, i.e. the level of intensity in subject's expression. To let our model learn diverse extremeness levels for each expression, it is essential to scale each expression motion label from $[0, 1]$ to $[0, s_i]$, where $s_i \in (0, 1]$ represents the scaled value of the apex state according to the intensity of the expression. Intuitively, the extremeness of each expression is proportional to the absolute mean deformation of the expression, we can thus calculate scaling factor s_i as:

$$s_i = \frac{clip(\frac{m_i - \mu_e}{\sigma_e}) + 1}{2} \tag{6}$$

where m_i is a scalar value representing the absolute value of the mean deformation of the sequence from neutral frame and μ_e, σ_e the mean and standart deviation of the deformation of the respective expression. Clip() function is used to clip values to $[-1, 1]$.

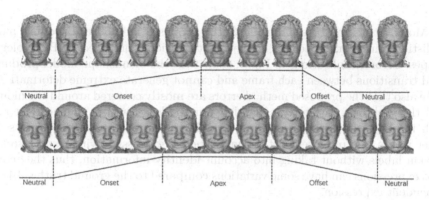

Fig. 2. Sample subjects from the 4DFAB database posing an expression along with expression motion labels.

5.2 Dynamic Facial Expressions

The proposed model for the generation of facial expressions is assessed both qualitatively and quantitatively. The model is trained by feeding the neutral frame of each subject and the manifested motion (i.e. the time-frames where the expression reaches onset, apex and offset modes) of the target expression. We evaluated the performance of the proposed model by its ability to generate expressions of 27 unobserved test subjects. To this end, we calculated the reconstruction loss as the per-vertex Euclidean distance between each generated sample and its corresponding ground truth.

Baseline. As a comparison baseline we implemented an expression blendshape decoder that transforms the latent representation z_t of each time-frame, i.e. the LSTM outputs, to an output mesh. In particular, expression blendshapes were modeled by first subtracting the neutral face of each subject to its corresponding expression, for all the corresponding video frames. With this operation we are able to capture and model just the motion deformation of each expression. Then, we applied Principal Component Analysis (PCA) to the motion deformations to reduce each expression to a latent vector. For a fair comparison, we use the same latent size for both the baseline and the proposed method.

The results presented in Table 2 show that the proposed model outperforms the baseline with regards to all the expressions, as well as on the entire dataset (0.39 mm vs 0.44 mm).

Table 2. Generalization per-vertex loss over all expressions, along with the total loss.

Model	Happy	Angry	Sad	Surprise	Fear	Disgust	Total
Baseline	0.49	0.37	0.37	0.48	0.43	0.45	0.44
Proposed	**0.37**	**0.35**	**0.36**	**0.43**	**0.42**	**0.42**	**0.39**

Moreover, as can be seen in Fig. 3, the proposed method can produce more realistic animation, especially in the mouth region, compared to PCA blend-shapes. Error visualizations in Fig. 3, show that the blendshape model produces mild transitions between each frame and cannot generate extreme deformations. Note also that the proposed method errors are mostly centered around the mouth and the eyebrows, due to the fact that our model is subject independent and each identity expresses its emotions in different ways and varying extents. In other words, the proposed method models each expression with respect to the motion labels without taking into account identity information, thus the generated expressions can have some variations compared to the ground truth subject-dependent expression.

For a subjective qualitative assessment, Fig. 4 shows several expressions that were generated by the proposed method. Due to the fact that several expressions, such as angry and sad expression, mainly relate to eyebrow deformations can not be easily visualised by still images we encourage the reader to check our supplementary material for more qualitative results.

5.3 Classification of Generated 4D Expressions

To further assess the quality of the generated expressions we implemented a classifier trained to identify expressions. The architecture of the classifier is based on a projection on a PCA space followed by three fully connected layers. The sequence classification is performed in two steps. In particular, first we computed 64-PCA coefficients to represent all expression and deformation variations of the training set and then we used them as a frame encoder to map the unseen test data to a latent space. Following that, the latent representations of each frame are concatenated and processed by fully-connected layers in order to predict the expression of the given sequence. The network was trained on the same training set that was originally used for the generation of expressions, with Adam optimizer and 5e−3 weight decay for 13 epochs. Table 3 shows the achieved classification performance for the ground truth test data and the generated data from the proposed model and from the baseline model, respectively.

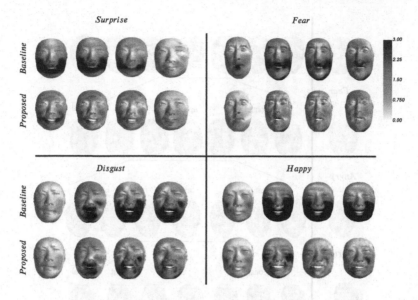

Fig. 3. Color heatmap visualization of error metric of both baseline (top rows) and proposed (bottom rows) model against the ground truth test data for four different expressions.

Table 3. Constructing classification performance between ground truth (test) data, generated (by the proposed method) data, and the blendshape baseline.

	Baseline			Proposed			Ground Truth		
	Pre	Rec	F1	Pre	Rec	F1	Pre	Rec	F1
Surprise	0.69	0.62	0.65	**0.74**	**0.90**	**0.81**	0.69	0.83	0.75
Angry	0.67	0.55	0.60	0.75	0.52	**0.61**	**0.81**	**0.59**	0.60
Disgust	0.51	0.72	0.60	**0.65**	**0.83**	**0.73**	0.58	0.72	0.65
Fear	0.59	0.36	0.44	**0.67**	**0.43**	**0.52**	0.64	0.32	0.43
Happy	0.63	0.59	0.60	0.71	0.86	0.78	**0.73**	**0.93**	**0.82**
Sad	0.43	0.57	0.49	0.62	0.60	0.61	**0.74**	**0.77**	**0.75**
Total	0.58	0.57	0.57	0.69	0.69	0.68	**0.70**	**0.70**	**0.68**

As can be seen in Table 3, the generated data from the proposed model achieve similar classification performance with ground truth data across almost every expression. In particular, the generated *surprise, disgust* and *fear* expressions from the proposed model can be even easier classified compared to the ground truth test data. Note also that both ground truth data and the data generated by the proposed model achieve 0.68 F1-score.

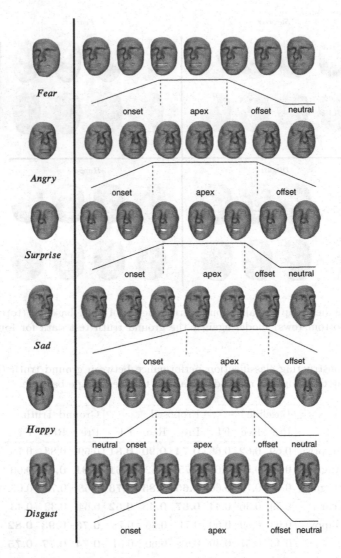

Fig. 4. Frames of generated expressions along with their expected motion labels: Fear, Angry, Surprise, Sad, Happy, Disgust (from top to bottom).

5.4 Loss per Frame

Since the overall loss is calculated for all frames of the generated expression, we cannot assess the ability of the proposed model to generate with low error-rate the onset, apex and offset of each expression. To evaluate the performance of the model on each expression phase we calculated the average L_1 distance between the generated and the corresponding ground truth frame for each of the frames of the evolved expression. Figure 5 shows that apex phase, which is usually taking place between time-frames 30–80, has an increased L_1 error for both models. However, the proposed method exhibits a stable loss around 0.45 mm across the apex phase, compared to the blendshape baseline that struggles to model the extreme deformations that take place and characterize the apex phase of the expression.

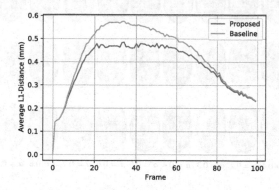

Fig. 5. Average per-frame L_1 error between the proposed method and the PCA-based blendshape baseline.

5.5 Interpolation on the Latent Space

To qualitatively evaluate the representation power of the proposed LSTM encoder we applied linear interpolation to the expression latent space. Specifically, we choose two different apex expression labels from our test set, we encode them using our LSTM encoder to two latent variables z_0 and z_1, each one of size 64. We then produce all intermediate encodings by linearly interpolating the line between them, i.e. $z_\alpha = \alpha z_1 + (1 - \alpha)z_0$, $\alpha \in (0, 1)$. The latent samples z_α are then fed to our mesh decoder network. We visualize the interpolations between different expressions in Fig. 6.

5.6 Expression Generation In-the-Wild

Since expression generation is an essential task in graphics and film industries, we propose a real-world application of our 3D facial expression generator. In particular, we have collected several image pairs with neutral and various expressions of the same identity and we have attempted to realistically synthesize the 4D animation of the target expression solely relying on our proposed approach. In order to acquire a neutral 3D template for our animation purposes, we applied a fitting methodology in the neutral image as proposed in [18]. By utilizing the fitted neutral mesh, we applied our proposed model in order to generate several target expressions as shown in Fig. 7. Our method is able to synthesize/generate a series of realistic 3D facial expressions which demonstrate the ability of our framework to animate a template mesh given a desired expression. We can qualitatively evaluate the similarity of the generated expression with the target expression of the same identity by comparing the generated mesh with the fitted 3D face.

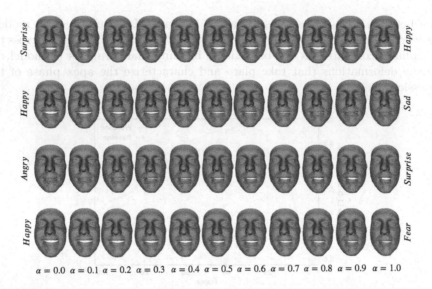

Fig. 6. Interpolation on the latent space between different expressions.

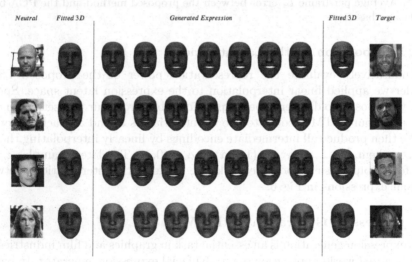

Fig. 7. Generation of expressions in-the-wild form 2D images.

6   Limitations and Future Work

Although the proposed framework is able to model and generate a wide range
of realistic expressions, it cannot model thoroughly extremeness variations. As
mentioned in Sect. 5.1, we attempted to adapt the extremeness variations of each
subject into the training procedure by using an intuitive scaling trick. However,
it's not certain that the mean absolute deformation of each mesh always represents the extremeness of the conducted expression. In addition, in order to

synthesize realistic facial animation, it is essential to model along with shape deformations also facial wrinkles of each expression. Thus, we will attempt to generalize facial expression animation to both shape and texture, by extrapolating our model to texture prediction.

7   Conclusion

In this paper, we propose the first generative model to synthesize 3D dynamic facial expressions from a still neutral 3D mesh. Our model captures both local and global expression deformations using graph based upsampling and convolution operators. Given a neutral expression mesh of a subject and a time signal that conditions the expected expression motion, the proposed model generates dynamic facial expressions for the same subject that respects the time conditions and the anticipated expression. The proposed method models the animation of each expression and deforms the neutral face of each subject according to a desired motion. Both expression and motion can be fully defined by the user. Results show that the proposed method outperforms expression blendshapes and creates motion-consistent deformations, validated both qualitatively and quantitatively. In addition, we assessed whether the generated expressions can be correctly classified and identified by a classifier trained on the same dataset. Classification results endorse our qualitative results showing that the generated data can be similarly classified, compared to the ones created by blendshapes model. In summary, the proposed model is the first that attempts and manages to synthesize realistic and high-quality facial expressions from a single neutral face input.

Acknowledgements. S. Zafeiriou and J. Zheng acknowledge funding from the EPSRC Fellowship DEFORM: Large Scale Shape Analysis of Deformable Models of Humans (EP/S010203/1). R.A. Potamias and G. Bouritsas were funded by the Imperial College London, Department of Computing, PhD scholarship.

References

1. Alashkar, T., Ben Amor, B., Daoudi, M., Berretti, S.: A 3D dynamic database for unconstrained face recognition. In: 5th International Conference and Exhibition on 3D Body Scanning Technologies. Lugano, Switzerland, October 2014
2. Amberg, B., Knothe, R., Vetter, T.: Expression invariant 3D face recognition with a morphable model. In: 2008 8th IEEE International Conference on Automatic Face Gesture Recognition, pp. 1–6 (2008)
3. Atwood, J., Towsley, D.: Diffusion-convolutional neural networks. In: Advances in Neural Information Processing Systems, pp. 1993–2001 (2016)
4. Boscaini, D., Masci, J., Rodolà, E., Bronstein, M.: Learning shape correspondence with anisotropic convolutional neural networks. In: Advances in Neural Information Processing Systems, pp. 3189–3197 (2016)
5. Bouaziz, S., Wang, Y., Pauly, M.: Online modeling for realtime facial animation. ACM Trans. Graph. (ToG) 32(4), 1–10 (2013)

6. Bouritsas, G., Bokhnyak, S., Ploumpis, S., Bronstein, M., Zafeiriou, S.: Neural 3D morphable models: spiral convolutional networks for 3D shape representation learning and generation. In: The IEEE International Conference on Computer Vision (ICCV), October 2019

7. Bronstein, M.M., Bruna, J., LeCun, Y., Szlam, A., Vandergheynst, P.: Geometric deep learning: going beyond Euclidean data. IEEE Signal Process. Mag. **34**(4), 18–42 (2017)

8. Bruna, J., Zaremba, W., Szlam, A., Lecun, Y.: Spectral networks and locally connected networks on graphs. In: International Conference on Learning Representations (ICLR2014), CBLS, April 2014 (2014)

9. Cao, C., Hou, Q., Zhou, K.: Displaced dynamic expression regression for real-time facial tracking and animation. ACM Trans. Graph. (TOG) **33**(4), 1–10 (2014)

10. Chang, Y., Vieira, M., Turk, M., Velho, L.: Automatic 3D facial expression analysis in videos. In: Zhao, W., Gong, S., Tang, X. (eds.) AMFG 2005. LNCS, vol. 3723, pp. 293–307. Springer, Heidelberg (2005). https://doi.org/10.1007/11564386_23

11. Cheng, S., Kotsia, I., Pantic, M., Zafeiriou, S.: 4DFAB: a large scale 4D database for facial expression analysis and biometric applications. In: Proceedings of the IEEE Conference on Computer Vision and Pattern Recognition, pp. 5117–5126 (2018)

12. Cosker, D., Krumhuber, E., Hilton, A.: A FACS valid 3D dynamic action unit database with applications to 3D dynamic morphable facial modeling. In: 2011 International Conference on Computer Vision, pp. 2296–2303. IEEE (2011)

13. Cudeiro, D., Bolkart, T., Laidlaw, C., Ranjan, A., Black, M.J.: Capture, learning, and synthesis of 3D speaking styles. In: Proceedings of the IEEE Conference on Computer Vision and Pattern Recognition, pp. 10101–10111 (2019)

14. Defferrard, M., Bresson, X., Vandergheynst, P.: Convolutional neural networks on graphs with fast localized spectral filtering. In: Advances in Neural Information Processing Systems, pp. 3844–3852 (2016)

15. Fan, L., Huang, W., Gan, C., Huang, J., Gong, B.: Controllable image-to-video translation: a case study on facial expression generation. In: Proceedings of the AAAI Conference on Artificial Intelligence, pp. 3510–3517 (2019)

16. Fanelli, G., Gall, J., Romsdorfer, H., Weise, T., Van Gool, L.: A 3-D audio-visual corpus of affective communication. IEEE Trans. Multimedia **12**(6), 591–598 (2010)

17. Fey, M., Eric Lenssen, J., Weichert, F., Müller, H.: SplineCNN: fast geometric deep learning with continuous B-spline kernels. In: The IEEE Conference on Computer Vision and Pattern Recognition (CVPR), June 2018

18. Gecer, B., Ploumpis, S., Kotsia, I., Zafeiriou, S.: GANFIT: generative adversarial network fitting for high fidelity 3D face reconstruction. In: Proceedings of the IEEE Conference on Computer Vision and Pattern Recognition, pp. 1155–1164 (2019)

19. Gupta, S., Markey, M.K., Bovik, A.C.: Anthropometric 3D face recognition. Int. J. Comput. Vision **90**(3), 331–349 (2010)

20. Hannun, A., et al.: Deep speech: scaling up end-to-end speech recognition. ArXiv (2014)

21. Karras, T., Aila, T., Laine, S., Herva, A., Lehtinen, J.: Audio-driven facial animation by joint end-to-end learning of pose and emotion. ACM Trans. Graph. (TOG) **36**(4), 94 (2017)

22. Kingma, D.P., Ba, J.: Adam: a method for stochastic optimization. ArXiv (2014)

23. Kipf, T.N., Welling, M.: Semi-supervised classification with graph convolutional networks. In: International Conference on Learning Representations (ICLR) (2017)

24. Levie, R., Monti, F., Bresson, X., Bronstein, M.M.: Cayleynets: graph convolutional neural networks with complex rational spectral filters. IEEE Trans. Signal Process. **67**(1), 97–109 (2018)
25. Li, Y., Min, M.R., Shen, D., Carlson, D., Carin, L.: Video generation from text. ArXiv (2017)
26. Lim, I., Dielen, A., Campen, M., Kobbelt, L.: A simple approach to intrinsic correspondence learning on unstructured 3D meshes. In: Proceedings of the European Conference on Computer Vision (ECCV), pp. 349–362 (2018)
27. Masci, J., Boscaini, D., Bronstein, M., Vandergheynst, P.: Geodesic convolutional neural networks on riemannian manifolds. In: Proceedings of the IEEE International Conference on Computer Vision Workshops, pp. 37–45 (2015)
28. Monti, F., Boscaini, D., Masci, J., Rodola, E., Svoboda, J., Bronstein, M.M.: Geometric deep learning on graphs and manifolds using mixture model CNNS. In: Proceedings of the IEEE Conference on Computer Vision and Pattern Recognition, pp. 5115–5124 (2017)
29. Moreno, A.: GavabDB: a 3D face database. In: Proceedings of 2nd COST275 Workshop on Biometrics on the Internet, 2004, pp. 75–80 (2004)
30. Nair, V., Hinton, G.E.: Rectified linear units improve restricted Boltzmann machines. In: Proceedings of the 27th International Conference on Machine Learning (ICML-10), pp. 807–814 (2010)
31. Otberdout, N., Daoudi, M., Kacem, A., Ballihi, L., Berretti, S.: Dynamic facial expression generation on hilbert hypersphere with conditional wasserstein generative adversarial nets. ArXiv (2019)
32. Pham, H.X., Cheung, S., Pavlovic, V.: Speech-driven 3D facial animation with implicit emotional awareness: a deep learning approach. In: 30th IEEE Conference on Computer Vision and Pattern Recognition Workshops, CVPRW 2017, pp. 2328–2336. IEEE Computer Society Conference on Computer Vision and Pattern Recognition Workshops. IEEE Computer Society, United States (2017). https://doi.org/10.1109/CVPRW.2017.287
33. Ploumpis, S., et al.: Towards a complete 3D morphable model of the human head. ArXiv (2019)
34. Ploumpis, S., Wang, H., Pears, N., Smith, W.A., Zafeiriou, S.: Combining 3D morphable models: a large scale face-and-head model. In: Proceedings of the IEEE Conference on Computer Vision and Pattern Recognition, pp. 10934–10943 (2019)
35. Ranjan, A., Bolkart, T., Sanyal, S., Black, M.J.: Generating 3D faces using convolutional mesh autoencoders. In: Proceedings of the European Conference on Computer Vision (ECCV), pp. 704–720 (2018)
36. Savran, A., et al.: Bosphorus database for 3D face analysis. In: Schouten, B., Juul, N.C., Drygajlo, A., Tistarelli, M. (eds.) BioID 2008. LNCS, vol. 5372, pp. 47–56. Springer, Heidelberg (2008). https://doi.org/10.1007/978-3-540-89991-4_6
37. Shuman, D.I., Narang, S.K., Frossard, P., Ortega, A., Vandergheynst, P.: The emerging field of signal processing on graphs: extending high-dimensional data analysis to networks and other irregular domains. IEEE Signal Process. Mag. **30**(3), 83–98 (2013)
38. Stratou, G., Ghosh, A., Debevec, P., Morency, L.P.: Effect of illumination on automatic expression recognition: a novel 3D relightable facial database. In: Face and Gesture 2011, pp. 611–618. IEEE (2011)
39. Thies, J., Zollhöfer, M., Nießner, M., Valgaerts, L., Stamminger, M., Theobalt, C.: Real-time expression transfer for facial reenactment. ACM Trans. Graph. **34**(6), 183–1 (2015)

40. Tian, G., Yuan, Y., Liu, Y.: Audio2Face: generating speech/face animation from single audio with attention-based bidirectional LSTM networks. In: 2019 IEEE International Conference on Multimedia & Expo Workshops (ICMEW), pp. 366–371. IEEE (2019)
41. Tzirakis, P., Papaioannou, A., Lattas, A., Tarasiou, M., Schuller, B.W., Zafeiriou, S.: Synthesising 3D facial motion from "In-the-Wild" speech. CoRR (2019)
42. Veličković, P., Cucurull, G., Casanova, A., Romero, A., Liò, P., Bengio, Y.: Graph attention networks. In: International Conference on Learning Representations (2018)
43. Verma, N., Boyer, E., Verbeek, J.: FeaStNet: feature-steered graph convolutions for 3D shape analysis. In: Proceedings of the IEEE Conference on Computer Vision and Pattern Recognition, pp. 2598–2606 (2018)
44. Yang, C.K., Chiang, W.T.: An interactive facial expression generation system. Multimedia Tools Appl. **40**, 41–60 (2008)
45. Yang, F., Wang, J., Shechtman, E., Bourdev, L., Metaxas, D.: Expression flow for 3D-aware face component transfer. In: ACM SIGGRAPH 2011 Papers. SIGGRAPH 2011. Association for Computing Machinery, New York (2011). https://doi.org/10.1145/1964921.1964955
46. Yin, L., Chen, X., Sun, Y., Worm, T., Reale, M.: A high-resolution 3D dynamic facial expression database. In: 2008 8th IEEE International Conference on Automatic Face Gesture Recognition, pp. 1–6 (2008)
47. Zhang, X., et al.: A high-resolution spontaneous 3D dynamic facial expression database. In: 2013 10th IEEE International Conference and Workshops on Automatic Face and Gesture Recognition (FG), pp. 1–6. IEEE (2013)
48. Zhang, Z., et al.: Multimodal spontaneous emotion corpus for human behavior analysis. In: Proceedings of the IEEE Conference on Computer Vision and Pattern Recognition, pp. 3438–3446 (2016)
49. Zhong, C., Sun, Z., Tan, T.: Robust 3D face recognition using learned visual codebook. In: 2007 IEEE Conference on Computer Vision and Pattern Recognition, pp. 1–6. IEEE (2007)
50. Zhou, Y., Shi, B.E.: Photorealistic facial expression synthesis by the conditional difference adversarial autoencoder. In: 2017 Seventh International Conference on Affective Computing and Intelligent Interaction (ACII), pp. 370–376. IEEE (2017)

LandscapeAR: Large Scale Outdoor Augmented Reality by Matching Photographs with Terrain Models Using Learned Descriptors

Jan Brejcha[1,2]([✉]) [iD], Michal Lukáč[2] [iD], Yannick Hold-Geoffroy[2] [iD],
Oliver Wang[2] [iD], and Martin Čadík[1] [iD]

[1] Faculty of Information Technology, CPhoto@FIT, Brno University of Technology,
Božetěchova 2, 61200 Brno, Czech Republic
ibrejcha@fit.vutbr.cz
http://cphoto.fit.vutbr.cz
[2] Adobe Inc., 345 Park Ave, San Jose, CA 95110-2704, USA

Abstract. We introduce a solution to *large scale* Augmented Reality for outdoor scenes by registering camera images to textured Digital Elevation Models (DEMs). To accommodate the inherent differences in appearance between real images and DEMs, we train a cross-domain feature descriptor using Structure From Motion (SFM) guided reconstructions to acquire training data. Our method runs efficiently on a mobile device and outperforms existing learned and hand-designed feature descriptors for this task.

1 Introduction

Augmented reality systems rely on some approximate knowledge of physical geometry to facilitate the interaction of virtual objects with the physical scene, and tracking of the camera pose in order to render the virtual content correctly. In practice, a suitable scene is tracked with the help of active depth sensors, stereo cameras, or multiview geometry from monocular video (e.g. SLAM). All of these approaches are limited in their *operational range*, due to constraints related to light falloff for active illumination, and stereo baselines and camera parallax for multiview methods.

In this work, we propose a solution for outdoor *landscape-scale* augmented reality applications by registering the user's camera feed to large scale textured Digital Elevation Models (DEMs). As there is significant appearance variation between the DEM and the camera feed, we train a data driven cross-domain feature descriptor that allows us to perform efficient and accurate feature matching.

Electronic supplementary material The online version of this chapter (https://doi.org/10.1007/978-3-030-58526-6_18) contains supplementary material, which is available to authorized users.

A. Vedaldi et al. (Eds.): ECCV 2020, LNCS 12374, pp. 295–312, 2020.
https://doi.org/10.1007/978-3-030-58526-6_18

Fig. 1. Our method matches a query photograph to a rendered digital elevation model (DEM). For clarity, we visualize only four matches (dashed orange). The matches produced by our system can then be used for localization, which is a key component for augmented reality applications. In the right image (zoomed-in for clarity), we render countour lines (white), gravel roads (red), and trails (black) using the estimated camera pose. (Color figue online)

Using this approach, we are able to localize photos based on long-distance cues, allowing us to display large scale augmented reality overlays such as altitude contour lines, map features (roads and trails), or 3D created content, such as educational geographic-focused features. We can also augment long-distance scene content in images with DEM derived features, such as semantic segmentation labels, depth values, and normals.

Since modern mobile devices as well as many cameras come with built-in GPS, compass and accelerometer, we could attempt to compute alignment from this data. Unfortunately, all of these sensors are subject to various sources of imprecision; e.g., the compass suffers from magnetic variation (irregularities of the terrestrial magnetic field) as well as deviation (unpredictable irregularities caused by deposits of ferrous minerals, or even by random small metal objects around the sensor itself). This means that while the computed alignment is usually close enough for rough localization, the accumulated error over geographical distances results in visible mismatches in places such as the horizon line.

The key insight of our approach is that we can take advantage of a robust and readily available source of data, with near-global coverage, that is DEM models, in order to compute camera location using reliable, 3D feature matching based methods. However, registering photographs to DEMs is challenging, as both domains are substantially different. For example, even high-quality DEMs tend to have resolution too rough to capture local high-frequency features like mountain peaks, leading to horizon mismatches. In addition, photographs have (often) unknown camera intrinsics such as focal length, exhibit seasonal and weather variations, foreground occluders like trees or people, and objects not present in the DEM itself, like buildings.

Our method works by learning a data-driven cross-domain feature embedding. We first use Structure From Motion (SFM) to reconstruct a robust 3D model from internet photographs, aligning it to a known terrain model. We then render views at similar poses as photographs, which lets us extract cross-domain patches in correspondence, which we use as supervision for training. At test time, no 3D reconstruction is needed, and features from the query image can be matched directly to renderings of the DEM.

Registration to DEMs only makes sense for images that observe a significant amount of content *farther* away than ca 100 m. For this reason, we focus on mountainous regions, where distant terrain is often visible. While buildings would also provide a reasonable source for registration, in this work we do not test on buildings, as building geometry is diverse, and 3D data and textures for urban areas are not freely available.

Our method is efficient and runs on a mobile device. As a demonstration, we developed a mobile application that performs large-scale visual localization to landscape features locally on a recent iPhone, and show that our approach can be used to refine localization when embedded device sensors are inaccurate.

In summary, we present the following contributions:

- A novel data-driven cross-domain embedding technique suitable for computing similarity between patches from photographs and a textured terrain model.
- A novel approach to Structure-from-Motion using terrain reference to align internet photographs with the terrain model (using D2Net detector & descriptor). Using our technique, a dataset of 16k images has been built and was used for training our method; it is by far the largest dataset of single image precise camera poses in mountainous regions. The dataset and source is available on our project website[1].
- A novel weakly supervised training scheme for positive/negative patch generation from the SfM reconstruction aligned with a DEM.
- We show that our novel embedding can be used for matching photographs to the terrain model to estimate respective camera position and orientation.
- We implement our system on the iPhone, showing that mobile large scale localization is possible on-device.

2   Related Work

2.1   Visual Localization

Localizing cameras in a 3D world is a fundamental component of computer vision and is used in a wide variety of applications. Classic solutions involve computing absolute pose between camera images and a known set of 3D points, typically solving the Perspective-n-Point [14] algorithm, or computing relative pose between two cameras observing the same scene, which can be computed solving the 5-point problem [28]. These approaches are founded in 3D projective geometry and can yield very accurate results when dealing with reliable correspondence measurements.

Recently, deep learning has been proposed as a solution to directly try to predict the camera location from scene observations using a forward pass through a CNN [21,38]. However, recent analysis has shown that these methods operate by image retrieval, computing the pose based on similarity to known images, and

---

[1] http://cphoto.fit.vutbr.cz/LandscapeAR/.

still do not exceed those from classic approaches to this problem [31]. Additionally, such approaches require the whole scene geometry to be represented within the network weights, and can only work on scenes that were seen during training. Our method leverages 3D geometric assumptions external to the model, making it more generalizable and accurate.

Existing approaches to outdoor camera orientation assessment [4,8,27], on the other hand, require a precise camera position. Accordingly, these works are insufficient in our scenario where the camera location is often inaccurate.

2.2   Local Descriptors

A key part of camera localization is correspondence finding. Most classical solutions to this problem involve using descriptors computed from local windows around feature points. These descriptors can be either hand-designed, e.g., SIFT [25], SURF [6], ORB [30], or learned end-to-end [12,15,26,35,39]. While our method is also a local descriptor, it is designed to deal with additional appearance and geometry differences, which is not the case for these methods.

Of these, HardNet++ [26] and D2Net [12] have been trained on outdoor images (HardNet on Brown dataset and HPatches, D2Net on Megadepth which contains 3D reconstructed models in the European Alps and Yosemite). Since it is possible that a powerful enough single-domain method might be able to bridge the domain gap (as demonstrated for D2Net and sketches), and these two methods are compatible with our use-case, we chose them as baselines to compare with our method.

2.3   Cross-Domain Matching

A large body of research work has been devoted to alignment of multi-sensor images [19,20,36] and to modality-invariant descriptors [11,18,23,32,33]. These efforts often focus on optical image alignment with e.g., its infra-red counterpart. However, our scenario is much more challenging, because we are matching an image with a *rendered* DEM where the change in appearance is considerable.

With the advent of deep-learning, several CNN-based works on matching multimodal patches emerged and outperformed previous multimodal descriptors [1,2,5,13,16]. However, cross-spectral approaches [1,2,5,13] need to account only for rapid visual appearance change, compared to our scenario, which needs to cover also the differences in scene geometry, caused by limited DEM resolution. On the other hand, RGB to depth matching approaches, such as Georgakis et al. [16] lack the texture information and need to focus only on geometry, which is not our case.

3   Method

Our goal is to estimate the camera pose of a query image with respect to the synthetic globe, which can be cast as a standard Perspective-n-Point problem [14]

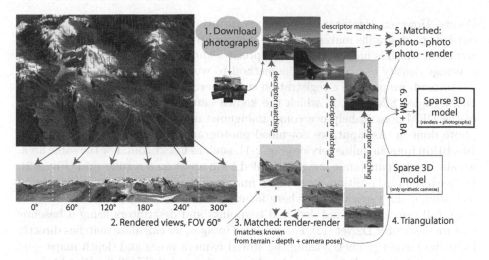

Fig. 2. Structure-from-motion with a terrain reference for automatic cross-domain dataset generation. In the area of interest, camera positions are sampled on a regular grid (red markers). At each position, 6 views covering the full panorama are rendered. A sparse 3D model is created from the synthetic data using known camera poses and scene geometry. Each photograph is localized to the synthetic sparse 3D model. Image credit, photographs left to right: John Bohlmeyer (https://flic.kr/p/gm3zRQ), Tony Tsang (https://flic.kr/p/gWmPbU), distantranges (https://flic.kr/p/gJCPui). (Color figure online)

given accurate correspondences. The main challenge is therefore, to establish correspondences between keypoints in the query photograph and a rendered synthetic frame. We bridge this appearance gap by training an embedding function which projects local neighborhoods of keypoints from either domain into a unified descriptor space.

3.1   Dataset Generation

The central difficulty of training a robust cross-domain embedding function is obtaining accurately aligned pairs of photographs and DEM renders. Manually annotating camera poses is tedious and prone to errors, and capturing diverse enough data with accurate pose information is challenging. Instead, we use internet photo collections, which are highly diverse, but contain unreliable location annotations. For each training photograph, we therefore need to retrieve precise camera pose $P = K[R|t]$, which defines the camera translation t, rotation R, and intrinsic parameters K with respect to the reference frame of the virtual globe.

In previous work [9,37], Structure-from-Motion (SfM) techniques have been used in a two-step process to align the photographs into the terrain. These methods reconstruct a sparse 3D model from photographs and then align it to the terrain model using point cloud alignment methods, such as Iterative Closest

Points. However, significant appearance variation and relatively low density of outdoor photographs makes photo-to-photo matching difficult, leading to reconstruction which is highly unstable, imprecise, and prone to drift. In many areas, coverage density is too low for the method to work at all.

Instead, we propose a registration step where photographs are aligned via a DEM-guided SfM step, in which the known camera parameters and geometry of the DEM domain help overcome ambiguous matches and lack of data in the photo domain. As input, we download photographs within a given rectangle of 10×10 km from an online service (Fig. 2-1), such as Flickr.com. For the same area, we also render panoramic images sampled 1 km apart on a regular grid (Fig. 2-2). For each sampled position, we render 6 images with 60° field-of-view each rotated 60° around the vertical axis, where for each rendered image, we store a depth map, full camera pose and detected keypoints and descriptors using a baseline feature descriptor D2Net [12]. For rendered images, we calculate matches directly from the terrain geometry using the stored camera poses and depth maps – no descriptor matching between rendered images is needed (Fig. 2-3). We obtain an initial sparse 3D model directly from the synthetic data (Fig. 2-4).

In the next step, we extract keypoints and descriptors from the input photographs using D2Net. The input photographs are matched to every other photograph *and* to rendered images using descriptor matching (Fig. 2-5), and localized to the terrain model using Structure-from-Motion (Fig. 2-6). Global bundle adjustment is used to refine camera parameters belonging to photographs and 3D points, while the rendered cameras have fixed all parameters, since they are known precisely.

Importantly, while existing single-domain feature descriptors are not robust to the photo-DEM domain gap, we can overcome this limitation by sheer volume of synthetic data. Most of the matches will be within the same domain (e.g., photo to photo), and only a small handful need to successfully match to DEM images for the entire photo domain model to be accurately registered. This procedure relies on having a collection of photos from diverse views and extensive processing, therefore doing so at inference time would be prohibitive. However, we can use this technique to build a dataset for training, after which our learned descriptor can be used to efficiently register a *single* photograph.

Finally, we check the location for each reconstructed photograph from the terrain model and prune photographs that are located below, or more than 100 m above the terrain since they are unlikely to be localized precisely. This approach proved to be much more robust and drift-free, and was able to georegister photographs in every area we tested. To illustrate this, we reconstructed 6 areas accross the European Alps region, and 1 area in South American Andes. In total, we localized 16,611 photographs using this approach.

3.2 Weakly Supervised Cross-domain Patch Sampling

While the rendered image is assumed to contain a similar view as the photograph, it is not exact. Therefore, our embedding function should be robust to

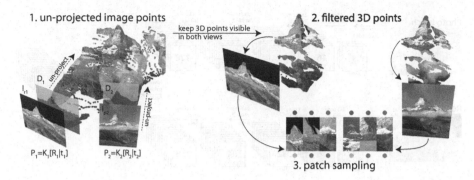

Fig. 3. 1. For a pair of images I_{r1} (render), I_{p2} (photograph), 2D image points are un-projected into 3D using the rendered depth maps D_1, D_2, and the ground truth camera poses P_1, P_2, respectively. 2. Only points visible from both views are kept. 3. A randomly selected subset of 3D points is used to form patch centers, and corresponding patches are extracted. Image credit: John Bohlmeyer (https://flic.kr/p/gm3xwP).

slight geometric deformations caused by viewpoint change, weather and seasonal changes, and different illumination. Note that these phenomena do not occur only in the photograph, but also in the ortho-photo textures. Previous work on wide baseline stereo matching, patch verification and instance retrieval illustrate that these properties could be learned directly from data [3,12,26,29]. For efficient training process, an automatic selection of corresponding (positive) and negative examples is crucial. In contrast with other methods, which rely on the reconstructed 3D points [12,26] dependent on a keypoint detector, we instead propose a weakly supervised patch sampling method completely independent of a preexisting keypoint detector to avoid any bias that might incur. This is an important and desirable property for our cross-domain approach, since (I) the accuracy of existing keypoint detectors in the cross domain matching task is unknown, (II) our embedding function may be used with any keypoint detector in the future without the need for re-training.

Each photograph in our dataset contains ground truth camera pose $P = K[R|t]$ transforming the synthetic world coordinates into the camera space. For each photograph I_{p1}, we render a synthetic image I_{r1} and a depth map D_1, see Fig. 3. We pick all pairs of cameras which have at least 30 corresponding 3D points in the SfM reconstruction described in Sect. 3.1. For each pair, the camera pose and depth map are used to un-project all image pixels into a dense 3D model (Fig. 3-1). Next, for each domain, we keep only the 3D points visible in both views (Fig. 3-2). Finally, we uniformly sample N random correspondences (Fig. 3-3), each defining the center of a local image patch.

3.3   Architecture

In order to account for the appearance gap between our domains, we employ a branched network with one branch for each of the input domains followed

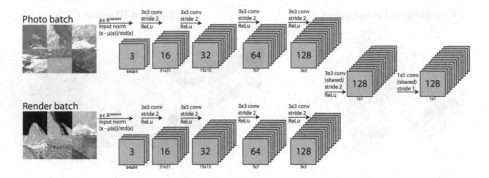

Fig. 4. Architecture of our two branch network with partially shared weights for cross-domain descriptor extraction. Photo and render branches contain four 3 × 3 2D convolutions with stride 2; weights are not shared between branches. The last two convolutions form a trunk of the network with shared weights to embed both domains into a single space. Output is 128-d descriptor. Either one or the other branch is used, each branch is specific for its own domain. Image credit: John Bohlmeyer (https://flic.kr/p/gm3xwP).

by a shared trunk. A description of the architecture is shown in Fig. 4. The proposed architecture is fully convolutional and has a receptive field of 63 px. To get a single descriptor, we use an input patch of size 64 × 64 px. We use neither pooling nor batch normalization layers. Similarly to HardNet [26], we normalize each input patch by subtracting its mean and dividing by its standard deviation. Thanks to the structure of our task formulation and the simplicity of the chosen architecture, our network is quite compact and contains only 261,536 trainable parameters, compared to VGG-16 [34] used by D2Net [12] which contains more than 7.6 million of trainable parameters. The small size allows our architecture to be easily deployed to a mobile device like the iPhone, enabling a wider scale of applications.

3.4   Training

We use a standard triplet loss function adjusted to our cross-domain scenario:

$$L(a^h, p^r, n^r) = \sum_i \max(||f^h(a_i^h) - f^r(p_i^r)||_2 - ||f^h(a_i^h) - f^r(n_i^r)||_2 + \alpha, 0)),$$

(1)

where a, p, n denotes a mini-batch of anchor, positive, and negative patches, respectively, superscript denotes photograph (h), or render (r), f^h and f^r denotes our embedding functions for *photograph* and *render* branches respectively, and α denotes the margin.

Previous work on descriptor learning using the triplet loss function [26] illustrated the importance of sampling strategy for selecting negative examples. In this solution, for each patch in a mini-batch, we know its 3D coordinate in an euclidean world space $x(p_j) \in \mathbf{R}^3$. Given a mini-batch of anchor and positive descriptors $f^h(a_i^h), f^r(p_i^r), i \in [0, N]$ where N is a batch size, we first select subset of *possible* negatives n^r from all positive samples within a current batch, which are farther than m meters from the anchor: $n^r = \{p_j^r | (\|x(p_j^r) - x(a_i^h)\|_2) > m\}$. In HardNet [26], for each positive only a hardest negative from the subset of possible negatives should be selected. However, we found that this strategy led the embedding function to collapse into a singular point. Therefore, we propose an adaptive variant of hard negative sampling inspired by a prior off-line mining strategy [17], modified to operate on-line.

We introduce a curriculum to increase the difficulty of the randomly sampled negatives during training. In classic hard negative mining, for each anchor descriptor a_i we randomly choose descriptor p_j as a negative example n_j, if and only if the triplet loss criterion is violated:

$$\|a_i - p_j\|_2 < \|a_i - p_i\|_2 + \alpha, \tag{2}$$

where we denote $a_i = f^h(a_i^h)$ as an anchor descriptor calculated from a photo patch using the photo encoder, and similarly for $p_j = f^r(p_j^r)$, and $p_i = f^r(p_i^r)$. We build on this, and for each anchor descriptor a_i, randomly choose a descriptor p_j as a negative example n_j iff:

$$\|a_i - p_j\|_2 < d^+ - (d^+ - (n_{\min} + \epsilon)) \cdot \lambda, \tag{3}$$

where λ is a parameter in $[0, 1]$ defining the difficulty of the negative mining, $\epsilon \to 0^+$ is a small positive constant, d^+ is the distance between anchor and positive plus margin: $d^+ = \|a_i - p_i\|_2 + \alpha$, and $n_{\min}$ is the distance between the anchor and the hardest negative: $n_{\min} = \min_{p_j} \|a_i - p_j\|_2$. Intuitively, when $\lambda = 0$, Eq. 3 is reduced to random hard negative sampling defined in Eq. 2, and when $\lambda = 1$, the Eq. 3 is forced to select p_j as a negative only if it is equal to the hardest negative $n_{\min}$, reducing the sampling method to HardNet [26]. Thus, λ allows us to select harder negatives throughout the training. For details, please see the supplementary material.

So far, we defined our loss function to be a cross-domain triplet loss, having an anchor as a *photograph*, and the positive and negative patches as *renders*. However, this loss function optimizes only the distance between the *photograph* and *render* descriptors. As a result, we use a variant with auxiliary loss functions optimizing also the distances between *photo-photo* and *render-render* descriptors:

$$L_{\text{aux}} = L(a^h, p^r, n^r) + L(a^h, p^h, n^h) + L(a^r, p^r, n^r). \tag{4}$$

As we illustrate by our experiments, this variant performs the best in the cross-domain matching scenario.

3.5 Pose Estimation

We illustrate the performance of our descriptor on a camera pose estimation task from a single query image. For each query image, we render a fan of 12 images from the initial position estimate (using GPS in our application and using ground truth position in our experiments) with FOV = 60° rotated by 30° around the vertical axis, similarly to Fig. 2-2. The input photograph is scaled by a factor s proportional to its FOV f: $s = (f \cdot M)/(\pi \cdot I_w)$, where M is the maximum resolution corresponding to FOV = 180° and I_w is the width of the image. We use the SIFT keypoint detector (although any detector could be used), take a 64×64 px patch around each keypoint, and calculate a descriptor using our method.

We start by finding the top candidates from the rendered fan using a simple voting strategy: for each rendered image we calculate the number of mutual nearest neighbor matches with the input photograph. We use the top-3 candidates, since the photograph is unlikely to span more than three consecutive renders, covering a FOV of 120°. For each top candidate, we un-project the 2D points from the rendered image to 3D using rendered camera parameters and a depth map; then we compute full camera pose of the photograph with respect to the 3D coordinates using OpenCV implementation of EPnP [24] algorithm with RANSAC. From the three output camera poses, we select the *best pose* which minimizes the reprojection error while having reasonable number of inliers; if any candidate poses have more than $N = 60$ inliers, we select the one with the lowest reprojection error. If none are found, we lower the threshold N and check for the *best pose* in a new iteration. If there is no candidate pose with at least $N = 20$ inliers, we end the algorithm as unsuccessfull. Finally, we reproject all the matches – not only inliers – into the camera plane using the *best pose*, and select those that are within frame. We repeat the matching proces and EPnP to obtain the *refined pose*.

4   Experiments

We present majority of the results as cumulative error plots, where we count the fraction of images localized below some distance or rotation error threshold. An ideal system is located at the top-left corner, where all the images are localized with zero distance and rotation errors. Througout the experiments section, we denote our architecture and its variants trained on our training dataset as **Ours-***. In addition, we report results for a larger single-branch architecture based on VGG-16 fine-tuned on our data (denoted as **VGG-16-D2-FT**). Similarly as D2Net, we cut the VGG-16 at conv 4-3, load the D2Net weights, and add two more convolutional layers to subsample the result descriptor to 128 dimensions. The newly added layers as well as the conv 4-3 were fine-tuned using our training method and data.

Our methods are compared with state-of-the-art deep local descriptors or matchers: HardNet++ [26], D2Net [12] and NCNet [29], which we use with original weights. Initially, we tried to train the HardNet and D2Net methods on our training dataset using their original training algorithms, but the results did not exhibit any improvements. We did not try to train the NCNet, since this method outputs directly matches and consumes a lot of computational resources, which is undesirable with our target applications capable of running on a mobile device.

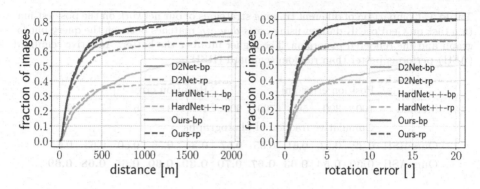

Fig. 5. Comparison between the *best pose* (bp) and the *refined pose* (rp) using different descriptors on GeoPose3K using *cross-domain* matches between the query photograph and synthetically rendered panorama. **Left:** translation error, **right:** rotation error.

4.1 Test Datasets

For evaluation of our method in a cross-domain scenario, we use the publicly available dataset GeoPose3K [7] spanning an area of the European Alps. We used the standard publicly available test split of 516 images [8]. We note that we were very careful while constructing our training dataset *not* to overlap with the test area of the GeoPose3K dataset. To illustrate that our method generalizes over the borders of the European Alps, on which it was trained, we also introduce three more test sets: *Nepal* (244 images), *Andes Huascaran* (126 images), and *Yosemite* (644 images). The *Nepal* and *Yosemite* datasets were constructed using SfM reconstruction using SIFT keypoints aligned to the terrain model with the iterative closest points algoritm as described by Brejcha et al. [9]. The *Huascaran* dataset has been constructed using our novel approach, as described in Sec. 3.1. Please note that this particular dataset may therefore be biased towards D2Net [12] matchable points, while *Nepal* and *Yosemite* datasets might be biased towards SIFT matchable points. Unlike the training images, camera poses in the test sets were manually inspected and outliers were removed.

4.2 Ablation Studies

Best Pose and Refined Pose. We study the behavior of our cross-domain pose estimation approach on the GeoPose3K dataset, on which we evaluate the *best pose* (solid) and the *refined pose* (dashed) for three different embedding algorithms as illustrated in Fig. 5. In the left plot, we can see that the *refined pose* improves over the *best pose* for both HardNet++ and our method for well registered images (up to distance error around 300 m), whereas it decreases result quality with D2Net. We hypothesize that this is because in the pose refinement step, the descriptor needs to disambiguate between more distractors compared

Table 1. Comparison of different training strategies of our network on the pose estimation task on GeoPose3K dataset using *cross-domain* matches between the query photograph and the rendered panorama. The higher number the better. Adaptive semihard (ASH) performs better than random semihard (RSH).

Method	Position error [m]					Rotation error [°]				
	100	300	500	700	900	1	3	5	7	9
	Cumulative fraction of photographs									
Ours-RSH	0.29	0.53	0.61	0.65	0.67	0.34	0.56	0.60	0.63	0.64
Ours-ASH	**0.30**	**0.54**	**0.63**	**0.67**	**0.70**	**0.39**	**0.60**	**0.65**	**0.68**	**0.69**

to the case of the best pose, where a single photograph is matched with a single rendered image, and D2Net seems to be more sensitive to these distractors than other approaches. Furthermore, the right plot of the Fig. 5 shows that the rotation error is improved on the refined pose for all three methods up to the threshold of 5°. Since points from multiple rendered views are already matched, the subsequent matching step covers a wider FOV, and thus a more reliable rotation can be found. For the following experiments, we use the *refined pose*, which seems to estimate camera poses with slightly better accuracy in the low-error regime.

Random Semi-hard and Adaptive Semi-hard Negative Mining. We analyze the difference between the baseline random semi-hard negative mining and adaptive semi-hard negative mining in Table 1. The experiment illustrates that adaptive semi-hard negative mining improves the random semi-hard negative mining baseline in both position and orientation errors, so we use it in all experiments.

Auxiliary Loss. Our network trained with the auxiliary loss function performs the best in the cross-domain scenario evaluated on the GeoPose3K dataset (Fig. 6, see Ours-aux). On this task, it outperforms the cross-domain variant of our network trained with the basic loss function (Ours). We also report the result of our network using a single encoder for both domains (Ours-render) which is consistently worse than the cross-domain variant. Furthermore, we see here that our network significantly outperforms both D2Net and HardNet++ in this task.

Stability with Respect to DEM Sampling Density. One question is how close does our DEM render have to be to the true photo location, for us to still find a correct pose estimate. To evaluate this, for each query photograph (with known ground truth location), we render a synthetic reference panorama offset from the photo location by a random amount (the "baseline"), sampled from a gaussian distribution with parameters $\mathcal{N}(0\,\text{m}, 1000\,\text{m})$. We then estimate

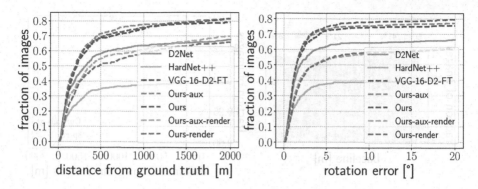

Fig. 6. Comparison of variants of our network with HardNet++ and D2Net for pose estimation task on GeoPose3K using *cross-domain* matches between query photograph and synthetically rendered panorama. **Left:** translation error, **right:** rotation error.

the pose of the query photograph by registering it with the render, and compare the predicted location to the known ground truth location. In Fig. 7-left we show the percentage of cases where the distance from ground truth to the predicted location was predicted to be less than the baseline. This gives us a measure for example, of how incorrect the GPS signal from a photo could be such that our approach improves localization. With low baselines, we see that the geometry mismatch to the DEM dominates and the position is difficult to improve on. With baselines over 200 m, we are able to register the photo, and then performance slowly degrades with increased baselines as matching becomes more difficult. Figure 7-right shows that the cross-over point where the position no longer improves over reference is around 700 m.

4.3 Comparison with State-of-the-Art

We compare our two-branch method and single-branch method based on VGG-16 with three state-of-the-art descriptors and matchers: HardNet [26], D2Net [12], and NCNet [29] in four different locations across the Earth. According to the results in Fig. 8, our two-branch method trained with auxiliary loss function (Ours-aux) exhibits the best performance on *GeoPose3K*, *Nepal*, and *Yosemite* datasets. The only dataset where our two-branch architecture is on-par with D2Net is *Andes Huascaran* (where the ground truth was created by D2Net matching), and where the single-branch VGG-16 architecture trained using our method and data performs the best. This is most probably due to differences in the ortho-photo texture used to render synthetic images. As the larger, pre-trained VGG-16 backbone has most likely learned more general filters than our two-branch network, which was trained solely on our dataset.

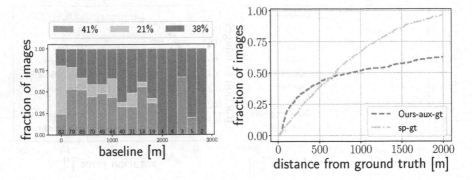

Fig. 7. Evaluation of robustness to baseline. **Left:** Fraction of improved (green), worsen (yellow), and failed (red) positions when matching query photo to a synthetic panorama as a function of baseline. The baseline is the distance between the ground truth position and a *reference position* generated by adding a gaussian noise $\mathcal{N}(0\,\text{m}, 1000\,\text{m})$ to the ground truth position. Position is considered improved when the estimated distance to ground truth is less than the baseline. The numbers at the bottom of each bar give the total number of images within each bar. **Right:** Cumulative fraction of query photos with an estimated position less than a given distance from ground truth (Ours-aux in pink) versus the cumulative fraction of *reference positions* within a given distance of ground truth (sp-gt in yellow). Pink line above yellow line means our method improves over the sampled *reference position* at that baseline. (Color figue online)

5   Applications

Mobile Application. To demonstrate the practicality of our method, we implemented it in an iPhone application. The application takes a camera stream, an initial rotation and position derived from on-board device sensors, and renders synthetic views from the local DEM and ortho-photo textures. It then computes SIFT keypoints on both a still image from the camera stream and the synthetically rendered image and uses our trained CNN to extract local features on the detected keypoints. These features are matched across domains and are then unprojected from the rendered image using the camera parameters and the depth map. Finally, matches between the 2D still keypoints and 3D rendered keypoints are used to estimate the camera pose using PnP method with RANSAC. This estimated camera pose is used to update the camera position and rotation to improve the alignment of the input camera stream with the terrain model (see Fig. 9).

Automatic Photo Augmentation. Furthermore, we demonstrate another use-case of our camera pose estimation approach by augmenting pictures from the internet for which the prior orientation is unknown and GPS position imprecise, see Fig. 9. Please note that many further applications of our method are possible, e.g., image annotation [4,22], dehazing, relighting [22], or refocusing and depth-of-field simulation [10].

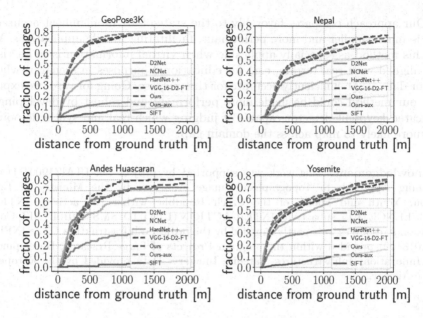

Fig. 8. Comparison of our method with state-of-the-art descriptors in four different locations across the Earth. Our method (dashed red and blue) outperforms Hard-Net [26] on all datasets and D2Net [12] on GeoPose3K, Nepal and Yosemite. Our method seems to be on par with D2Net on Andes Huascaran dataset which has significantly less precise textures (from ESA RapidEye satellite) in comparison to other datasets. (Color figure online)

Fig. 9. An iPhone application (in the left) is used to capture the photograph (in the middle) for which precise camera pose is estimated using our method. The estimated camera pose (in the right) is used to augment the query photograph with contour lines (white) and rivers (blue). (Color figure online)

6   Conclusion and Future Work

We have presented a method for photo-to-terrain alignment for use in augmented reality applications. By training a network on a cross-domain feature embedding, we were able to bridge the domain gap between rendered and real images. This embedding allows for accurate alignment of a photo, or camera view, to the terrain for applications in mobile AR and photo augmentation.

Our approach compares favorably to the state-of-art in alignment accuracy, and is much smaller and more performant, facilitating mobile applications. We see this method as especially applicable when virtual information is to be visually aligned with real terrain, e.g., for educational purposes in scenarios where sensor data is not sufficiently accurate for the purpose. Going forward, we expect that our method could be made more performant and robust by developing a dedicated keypoint detector capable of judging which real and synthetic points are more likely to map across the domain gap.

Acknowledgement. This work was supported by project no. LTAIZ19004 Deep-Learning Approach to Topographical Image Analysis; by the Ministry of Education, Youth and Sports of the Czech Republic within the activity INTER-EXCELENCE (LT), subactivity INTER-ACTION (LTA), ID: SMSM2019LTAIZ. Computational resources were partly supplied by the project e-Infrastruktura CZ (e-INFRA LM2018140) provided within the program Projects of Large Research, Development and Innovations Infrastructures. Satellite Imagery: Data provided by the European Space Agency.

References

1. Aguilera, C.A., Aguilera, F.J., Sappa, A.D., Toledo, R.: Learning cross-spectral similarity measures with deep convolutional neural networks. In: IEEE Computer Society Conference on Computer Vision and Pattern Recognition Workshops, pp. 267–275 (2016). https://doi.org/10.1109/CVPRW.2016.40
2. Aguilera, C.A., Sappa, A.D., Aguilera, C., Toledo, R.: Cross-spectral local descriptors via quadruplet network. Sensors (Switzerland) **17**(4), 1–14 (2017). https://doi.org/10.3390/s17040873
3. Arandjelović, R., Gronat, P., Torii, A., Pajdla, T., Sivic, J.: NetVLAD: CNN architecture for weakly supervised place recognition. Arxiv (2015). http://arxiv.org/abs/1511.07247
4. Baboud, L., Čadík, M., Eisemann, E., Seidel, H.P.: Automatic photo-to-terrain alignment for the annotation of mountain pictures. In: Proceedings of the 2011 IEEE Conference on Computer Vision and Pattern Recognition, CVPR 2011, pp. 41–48. IEEE Computer Society, Washington (2011). https://doi.org/10.1109/CVPR.2011.5995727
5. Baruch, E.B., Keller, Y.: Multimodal matching using a hybrid convolutional neural network. CoRR abs/1810.12941 (2018). http://arxiv.org/abs/1810.12941
6. Bay, H., Tuytelaars, T., Van Gool, L.: SURF: speeded up robust features. In: Leonardis, A., Bischof, H., Pinz, A. (eds.) ECCV 2006. LNCS, vol. 3951, pp. 404–417. Springer, Heidelberg (2006). https://doi.org/10.1007/11744023_32
7. Brejcha, J., Čadík, M.: GeoPose3K: mountain landscape dataset for camera pose estimation in outdoor environments. Image Vis. Comput. **66**, 1–14 (2017). https://doi.org/10.1016/j.imavis.2017.05.009
8. Brejcha, J., Čadík, M.: Camera orientation estimation in natural scenes using semantic cues. In: 2018 International Conference on 3D Vision (3DV), pp. 208–217, September 2018. https://doi.org/10.1109/3DV.2018.00033

9. Brejcha, J., Lukáč, M., Chen, Z., DiVerdi, S., Čadík, M.: Immersive trip reports. In: Proceedings of the 31st Annual ACM Symposium on User Interface Software and Technology, UIST 2018, pp. 389–401. Association for Computing Machinery, New York (2018). https://doi.org/10.1145/3242587.3242653

10. Čadík, M., Sýkora, D., Lee, S.: Automated outdoor depth-map generation and alignment. Elsevier Comput. Graph. **74**, 109–118 (2018)

11. Chen, J., Tian, J.: Real-time multi-modal rigid registration based on a novel symmetric-SIFT descriptor. Prog. Nat. Sci. **19**(5), 643–651 (2009). https://doi.org/10.1016/j.pnsc.2008.06.029

12. Dusmanu, M., et al.: D2-Net: a trainable CNN for joint detection and description of local features. In: The IEEE Conference on Computer Vision and Pattern Recognition (CVPR), June 2019. http://arxiv.org/abs/1905.03561

13. En, S., Lechervy, A., Jurie, F.: TS-NET: Combining modality specific and common features for multimodal patch matching. In: Proceedings - International Conference on Image Processing, ICIP, pp. 3024–3028 (2018). https://doi.org/10.1109/ICIP.2018.8451804

14. Fischler, M.A., Bolles, R.C.: Random sample consensus: a paradigm for model fitting with applications to image analysis and automated cartography. Commun. ACM **24**(6), 381–395 (1981)

15. Georgakis, G., Karanam, S., Wu, Z., Ernst, J., Kosecka, J.: End-to-end learning of keypoint detector and descriptor for pose invariant 3D matching, February 2018. http://arxiv.org/abs/1802.07869

16. Georgakis, G., Karanam, S., Wu, Z., Kosecka, J.: Learning local RGB-to-CAD correspondences for object pose estimation. In: The IEEE International Conference on Computer Vision (ICCV), October 2019

17. Harwood, B., Vijay Kumar, B.G., Carneiro, G., Reid, I., Drummond, T.: Smart mining for deep metric learning. In: Proceedings of the IEEE International Conference on Computer Vision (2017). https://doi.org/10.1109/ICCV.2017.307

18. Hasan, M., Pickering, M.R., Jia, X.: Modified sift for multi-modal remote sensing image registration. In: 2012 IEEE International Geoscience and Remote Sensing Symposium, pp. 2348–2351, July 2012. https://doi.org/10.1109/IGARSS.2012.6351023

19. Irani, M., Anandan, P.: Robust multi-sensor image alignment. In: Sixth International Conference on Computer Vision (IEEE Cat. No.98CH36271), pp. 959–966, January 1998. https://doi.org/10.1109/ICCV.1998.710832

20. Keller, Y., Averbuch, A.: Multisensor image registration via implicit similarity. IEEE Trans. Pattern Anal. Mach. Intell. **28**(5), 794–801 (2006). https://doi.org/10.1109/TPAMI.2006.100

21. Kendall, A., Grimes, M., Cipolla, R.: PoseNet: a convolutional network for real-time 6-DOF camera relocalization. In: Proceedings of the IEEE International Conference on Computer Vision, pp. 2938–2946 (2015)

22. Kopf, J., et al.: Deep photo: model-based photograph enhancement and viewing. In: Transactions on Graphics (Proceedings of SIGGRAPH Asia), vol. 27, no. 6, article no. 116 (2008)

23. Kwon, Y.P., Kim, H., Konjevod, G., McMains, S.: Dude (duality descriptor): a robust descriptor for disparate images using line segment duality. In: 2016 IEEE International Conference on Image Processing (ICIP), pp. 310–314, September 2016. https://doi.org/10.1109/ICIP.2016.7532369

24. Lepetit, V., Moreno-Noguer, F., Fua, P.: EPnP: an accurate O(n) solution to the PnP problem. Int. J. Comput. Vision (2009). https://doi.org/10.1007/s11263-008-0152-6

25. Lowe, D.G., et al.: Object recognition from local scale-invariant features. In: ICCV, vol. 99, pp. 1150–1157 (1999)
26. Mishchuk, A., Mishkin, D., Radenović, F., Matas, J.: Working hard to know your neighbor's margins: local descriptor learning loss. In: Advances in Neural Information Processing Systems, NIPS 2017, vol. 2017-Decem, pp. 4827–4838. Curran Associates Inc., Red Hook (2017)
27. Nagy, B.: A new method of improving the azimuth in mountainous terrain by skyline matching. PFG – J. Photogrammetry Remote Sens. Geoinform. Sci. **88**(2), 121–131 (2020). https://doi.org/10.1007/s41064-020-00093-1
28. Nistér, D.: An efficient solution to the five-point relative pose problem. IEEE Trans. Pattern Anal. Mach. Intell. **26**(6), 0756–777 (2004)
29. Rocco, I., Cimpoi, M., Arandjelović, R., Torii, A., Pajdla, T., Sivic, J.: Neighbourhood consensus networks. In: Advances in Neural Information Processing Systems, vol. 2018-Decem, pp. 1651–1662 (2018)
30. Rublee, E., Rabaud, V., Konolige, K., Bradski, G.: ORB: an efficient alternative to SIFT or SURF. In: Proceedings of the IEEE International Conference on Computer Vision (2011). https://doi.org/10.1109/ICCV.2011.6126544
31. Sattler, T., Zhou, Q., Pollefeys, M., Leal-Taixe, L.: Understanding the limitations of CNN-based absolute camera pose regression. In: Proceedings of the IEEE Conference on Computer Vision and Pattern Recognition, pp. 3302–3312 (2019)
32. Kim, S., Min, D., Ham, B., Ryu, S., Do, M.N., Sohn, K.: DASC: dense adaptive self-correlation descriptor for multi-modal and multi-spectral correspondence. In: 2015 IEEE Conference on Computer Vision and Pattern Recognition (CVPR), pp. 2103–2112, June 2015. https://doi.org/10.1109/CVPR.2015.7298822
33. Shechtman, E., Irani, M.: Matching local self-similarities across images and videos. In: 2007 IEEE Conference on Computer Vision and Pattern Recognition, pp. 1–8, June 2007. https://doi.org/10.1109/CVPR.2007.383198
34. Simonyan, K., Zisserman, A.: Very deep convolutional networks for large-scale image recognition. In: 3rd International Conference on Learning Representations, ICLR 2015 - Conference Track Proceedings (2015)
35. Tian, Y., Fan, B., Wu, F.: L2-Net: deep learning of discriminative patch descriptor in euclidean space. In: 2017 IEEE Conference on Computer Vision and Pattern Recognition (CVPR), pp. 6128–6136, July 2017. https://doi.org/10.1109/CVPR.2017.649
36. Viola, P., Wells, W.M.: Alignment by maximization of mutual information. Int. J. Comput. Vision **24**(2), 137–154 (1997). https://doi.org/10.1023/A:1007958904918
37. Wang, C.P., Wilson, K., Snavely, N.: Accurate georegistration of point clouds using geographic data. In: 2013 International Conference on 3DTV-Conference, pp. 33–40 (2013). https://doi.org/10.1109/3DV.2013.13
38. Weyand, T., Kostrikov, I., Philbin, J.: PlaNet - photo geolocation with convolutional neural networks. In: Leibe, B., Matas, J., Sebe, N., Welling, M. (eds.) ECCV 2016. LNCS, vol. 9912, pp. 37–55. Springer, Cham (2016). https://doi.org/10.1007/978-3-319-46484-8_3
39. Yi, K.M., Trulls, E., Lepetit, V., Fua, P.: LIFT: learned invariant feature transform. In: Leibe, B., Matas, J., Sebe, N., Welling, M. (eds.) ECCV 2016. LNCS, vol. 9910, pp. 467–483. Springer, Cham (2016). https://doi.org/10.1007/978-3-319-46466-4_28

Learning Disentangled Feature Representation for Hybrid-Distorted Image Restoration

Xin Li ⓘ, Xin Jin, Jianxin Lin, Sen Liu ⓘ, Yaojun Wu ⓘ, Tao Yu ⓘ, Wei Zhou, and Zhibo Chen (✉) ⓘ

CAS Key Laboratory of Technology in Geo-Spatial Information Processing and Application System, University of Science and Technology of China, Hefei 230027, China
{lixin666,jinxustc,linjx,yaojunwu,yutao666,weichou}@mail.ustc.edu.cn, elsen@iat.ustc.edu.cn, chenzhibo@ustc.edu.cn

Abstract. Hybrid-distorted image restoration (HD-IR) is dedicated to restore real distorted image that is degraded by multiple distortions. Existing HD-IR approaches usually ignore the inherent interference among hybrid distortions which compromises the restoration performance. To decompose such interference, we introduce the concept of Disentangled Feature Learning to achieve the feature-level divide-and-conquer of hybrid distortions. Specifically, we propose the feature disentanglement module (FDM) to distribute feature representations of different distortions into different channels by revising gain-control-based normalization. We also propose a feature aggregation module (FAM) with channel-wise attention to adaptively filter out the distortion representations and aggregate useful content information from different channels for the construction of raw image. The effectiveness of the proposed scheme is verified by visualizing the correlation matrix of features and channel responses of different distortions. Extensive experimental results also prove superior performance of our approach compared with the latest HD-IR schemes.

Keywords: Hybird-distorted image restoration · Feature disentanglement · Feature aggregation

1 Introduction

Nowadays, Image restoration techniques have been applied in various fields, including streaming media, photo processing, video surveillance, and cloud storage, etc. In the process of image acquisition and transmission, raw images

---

Electronic supplementary material The online version of this chapter (https://doi.org/10.1007/978-3-030-58526-6_19) contains supplementary material, which is available to authorized users.

A. Vedaldi et al. (Eds.): ECCV 2020, LNCS 12374, pp. 313–329, 2020.
https://doi.org/10.1007/978-3-030-58526-6_19

(I)   (II)   (III)Ours

Fig. 1. Examples of hybrid-distorted image restoration. (I) Hybrid-distorted image including noise, blur, and jpeg artifacts. (II) Processed with the cascading single distortion restoration networks including dejpeg, denoise and deblurring. (III) Processed with our FDR-Net.

are usually contaminated with various distortions due to capturing devices, high ratio compression, transmission, post-processing, etc. Previous image restoration methods focusing on single distortion have been extensively studied [5,6,12,14,17,18,25,27,35,36] and achieved satisfactory performance on the field of super resolution [7,21,22], deblurring [24,30,38], denoising [4,43,44], deraining [9,10,34], dehazing [1,41,45] and so on. However, these works are usually designed for solving one specific distortion, which makes them difficult to be applied to real world applications as shown in Fig. 1.

Real world images are typically affected by multiple distortions simultaneously. In addition, different distortions might be interfered with each other, which makes it difficult to restore images. More details about interference between different distortions can be seen in the supplementary material. Recently, there have been proposed some pioneering works for hybrid distortion. For example, Yu et al. [37] pre-train several light-weight CNNs as tools for different distortions. Then they utilize the Reinforcement Learning (RL) agent to learn to choose the best tool-chain for unknown hybrid distortions. Then Suganuma et al. [28] propose to use the attention mechanism to implement the adaptive selection of different operations for hybrid distortions. However, these methods are designed regardless of the interference between hybrid distortions.

Previous literature [2] suggests that deep feature representations could be employed to efficiently characterize various image distortions. In other words, the feature representation extracted from hybrid distortions images might be disentangled to characterize different distortions respectively. Based on the above, we propose to implement the feature-level divide-and-conquer of the hybrid distortions by learning disentangled feature representations. On a separate note, Schwartz et al. [26] point out that a series of filters and gain-control-based normalization could achieve the decomposition of different filter responses. And the convolution layer of CNN is also composed of a series of basic filters/kernels. Inspired by this, we expand this theory and design a feature disentanglement module (FDM) to implement the channel-wise feature decorrelation. By such feature decorrelation, the feature representations of different distortions could be distributed across different channels respectively as shown in Fig. 2.

Different from high-level tasks such as classification [8,23], person re-identification [3,11,19], the processed feature representations are not clustered into several classes. It is crucial for low-level image restoration to aggregate the useful content information from processed feature representations to reconstruct the image. Therefore, we design an adaptive feature aggregation module (FAM) based on channel-wise attention mechanism and inverse transform of gain-control-based normalization [26].

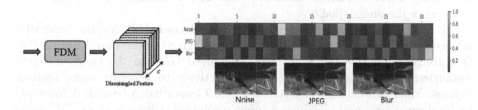

Fig. 2. Illustration of feature disentangle module (FDM) works. The FDM disentangles the input feature representation which characterizes the hybrid-distorted image into disentangled features. The responses of different distortions are distributed across different channels as shown in brighter region. The visualizations of distortions are also displayed.

Extensive experiments are conducted on hybrid distortion datasets and our model achieves the state-of-the-art results. Furthermore, we give a reasonable interpretation for our framework through visualizing the correlation matrix of features and channel response of different distortions. Finally, we verify the robustness of our network on single distortion tasks such as image deblurring and image deraining.

Our main contributions can be summarized as follows:

- We implement the feature-level divide-and-conquer of hybrid-distortion by feature disentanglement. And we design the Feature disentangled module (FDM) by revising gain-control-based normalization to distribute feature representations of different distortions into different channels.
- We propose a Feature Disentanglement Network by incorporating FDM and FAM for hybrid-distorted image restoration named FDR-Net.
- Extensive experiments demonstrate that our FDR-Net has achieved the state-of-the-art results on hybrid-distorted image restoration. We also verify the efficiency of our modules through ablation study and visualization.

2   Related Work

In this section, we briefly summarize the researches on single-distorted image restoration and hybrid-distorted image restoration.

2.1   Image Restoration on Single Distortion

Deep learning has promoted the development of computer vision tasks, especially on image restoration. As the pioneers in image restoration, some early works on special distortion removal have been proposed. Dong et al. first employ a deep convolutional neural network (CNN) in image super-resolution in [5]. Zhang et al. propose Dncnn [43] to handle noise and JPEG compression distortion. And Sun et al. utilize CNN to estimate the motion blur kernels for image deblur in [29]. With the advance of CNN and the improvement of computing resources, the depth and the number of parameters of the neural network have been increased significantly which facilitates a series of excellent works on special tasks such as deblurring [24,30,38], super-resolution [7,21,22], deraining [9,10,34] and so on to improve the image quality.

To further improve the subjective quality of degraded images, the generative adversarial network has been utilized to restore image restoration. SRGAN [16], DeblurGAN [14], ID-CGAN [42] have employed GAN in the tasks of super-resolution, deblurring, and deraining respectively to generate more realistic images. Yuan et al. [39] use Cycle-in-Cycle Generative Adversarial Networks to implement unsupervised image super-resolution. Ulyanov et al. [31] prove that the structure of generator network is sufficient to capture low-level image statistics prior and then utilize it to implement unsupervised image restoration.

2.2   Image Restoration on Hybrid Distortion

Recently, with the wide applications of image restoration, the special image restoration on single distortion cannot meet the need of real world application. Then, some works on hybrid distortion restoration have been proposed. Among them, RL-Restore [37] trains a policy to select the appropriate tools for single distortion from the pre-trained toolbox to restore the hybrid-distorted image. Then Suganuma et al. [28] propose a simple framework which can select the proper operation with attention mechanism. However, these works don't consider the interference between different distortions. In this paper, we propose the FDR-Net for hybrid distortions image restoration by reducing the interference with feature disentanglement, which achieves more superior performance by disentangling the feature representation for hybird-distorted image restoration.

3   Approach

In this section, we start with introducing the gain control based signal decomposition as primary knowledge for feature disentanglement module (FDM), and then elaborate the components of our FDR-Net and its overall architecture.

3.1   Primary Knowledge

To derive the models of sensory processing, gain-control-based normalization has been proposed to implement the nonlinear decomposition of natural signals

[26]. Given a set of signals $\boldsymbol{X}$, the signal decomposition first uses a set of filters $\langle f_1, f_2, ..., f_n \rangle$ to extract different representations of signals $\langle L_1, L_2, ...L_n \rangle$ as

$$L_i = f_i(\boldsymbol{X}), i = 1, 2, ..., n, \tag{1}$$

and then uses gain-control-based normalizaiton as Eq. 2 to further eliminate the dependency between $\langle L_1, L_2, ...L_n \rangle$.

$$R_i = \frac{L_i^2}{\sum_j w_{ji} L_j^2 + \sigma^2}, \tag{2}$$

where independent response R_i can be generated based on the suitable weights w_{ji} and offsets σ^2 with corresponding inputs L_i. In this way, the input signals $\boldsymbol{X}$ can be decomposed into $R_1, R_2, ...R_n$ according to their statistical property.

3.2 Feature Disentanglement

Previous literature [2] has proved that different distortions have different deep feature representation and could be disentangled at feature representation. Based on this analysis, we design the FDM by expanding the signal decomposition into channel-wise feature disentanglement to reduce the interference between hybrid distortions. As shown in [26], the combination of diverse linear filters and divisive normalization has the capability to decompose the responses of filters. And the feature representation of CNN is also composed of responses from a series of basic filters/kernels. Based on such observation, we implement the adaptive channel-wise feature decomposition by applying such algorithm in learning based framework.   Specifically, we use the convolution layer $(C_{in} \times$

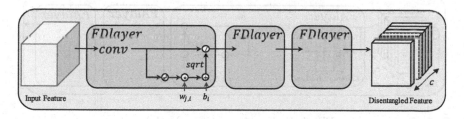

Fig. 3. Feature Disentanglement Module.

$C_{out} \times k \times k)$ in neural network to replace traditional filters $\langle f_1, f_2, ..., f_n \rangle$ in signal decomposition as Sect. 3.1. Here, the number of input channel C_{in} represents the channel dimension of input feature $\boldsymbol{F}_{in}$, the number of output channel C_{out} represents the channel dimension of output feature $\boldsymbol{F}_{out}$, and k is the kernel size of filters. In this way, the extraction results $S_1, S_2, ...S_{C_{out}}$ of convolution layer will be distributed in different channels as:

$$\langle S_1, S_2, ..., S_{C_{out}} \rangle = conv(\boldsymbol{F}_{in}), \tag{3}$$

where S_i represents the ith channel in output feature and *conv* represents the convolution layer.

To introduce the gain-control-based normalization as Eq. 2 into CNN, we modified the Eq. 2 as Eq. 4.

$$D_i = \frac{S_i}{\sqrt{\sum_j w_{ji} S_j^2 + b_i}}, \tag{4}$$

where w_{ji} and b_i can be learned by gradient descent. S_i and D_i represent the ith channel components of features before and after gain control. In formula 4, we make two major improvements to make it applicable to our task. One improvement is that the denominator and numerator is the square root of the original one in Eq. 2 which makes it is easy to implement the gradient propagation. Another improvement is to replace the response of filters L_i with channel components of features S_i, which is proper for channel-wise feature disentanglement.

In order to guide the study of parameters from convolution layer, w_{ji} and b_i, we introduce the spectral value difference orthogonality regularization (SVDO) from [3] as a loss constraint. As a novel method to reduce feature correlations, SVDO can be expressed as Eq. 5.

$$Loss_F = \left\| \lambda_1 \left(FF^T \right) - \lambda_2 \left(FF^T \right) \right\|_2^2 \tag{5}$$

where $\lambda_1 \left(FF^T \right)$ and $\lambda_2 \left(FF^T \right)$ denote the largest and smallest eigenvalues of FF^T, respectively. F is feature maps and T expresses the transposition.

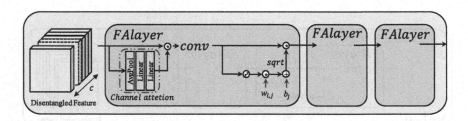

Fig. 4. Feature Aggregation Module.

According to above analysis, we design the corresponding feature disentanglement module (FDM). As seen in the Fig. 3, Each FDM have multiple FDlayers and each FDlayer is composed of one convolution layer and a revised gain-control-based normalization. Experiments demonstrate that three FDlayers for each FDM module are saturated to achieve the best performance as shown in Table 6.

3.3   Feature Aggregation Module

For hybrid-distorted image restoration, we need to restore the image instead of only getting processed feature representation. To further filter out the distortion representation and maintain the raw image content details, we utilize channel-wise attention mechanism to adaptively select useful feature representation from processed disentangled feature as Eq. 6.

$$F_{p\_i} = PM(D_i), F_{p\_i} = CA(F_{p\_i}), \tag{6}$$

where PM represents the process module and CA presents channel attention. D_i is the ith channel of disentangled feature. F_{p_i} represents the ith channel of output feature. To construct the image, we get the inversion formula from Eq. 4 as:

$$F_{c\_i} = F_{p\_i}\sqrt{\sum_j w_{ji}F_{p\_j}^2 + b_i}, \tag{7}$$

where F_{c_i} represents the output feature corresponding to the distribution of clean image. With this module, the processed image information could be aggregated to original feature space, which is proper for reconstructing restored image. Feature aggregation module (FAM) is designed as shown in Fig. 4

3.4   Auxiliary Module

In the processes of feature disentanglement, the mutual information of different channels of features is reduced, which will result in some loss of image information. In order to make up for the weakness of the feature disentanglement branch, we used the existing ResBlock to enhance the transmission of image information in parallel.

3.5   Overview of Whole Framework

Figure. 5To make full use of feature information of different levels, we use multi-phases structure to deal with hybrid distortion as Fig. 5. The whole structure is composed of several same phases. Every phase is composed by two branches, where the lower branch is used for distortion-aware feature disentangle and feature processing, the upper branch is used as auxiliary branch which is used to make up for information lost and enhance the removal of hybird distortion. The details of lower branch are shown in Fig. 5. In order to ensure the matching between FDlayer and FAlayer, we use skip connection to connect the output of FDlayer and the input of FAlayer. To fuse the information of outputs from two branches, we first concatenate the two outputs and then use channel attention and convolution to fuse the outputs to a feature. Furthermore, we use dual residual to relate the different phases to enhance the mutual utilization of information between different phases.

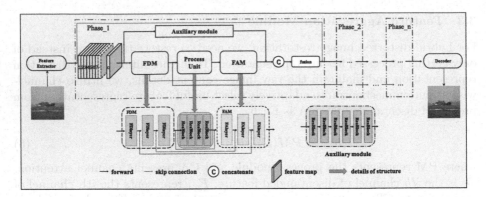

Fig. 5. Architecture of our proposed FDR-Net. It consists of multiple phases and each phase consists of feature disentangle module (FDM) and feature aggregation module (FAM).

3.6 Loss Function

In this paper, the loss function is composed of L1 loss and feature orthogonality loss with the weight β 0.00001 as Eq. 5. The loss function can be expressed as:

$$Loss = L1loss + \beta loss_F \tag{8}$$

4   Experiments

In this section, we first conduct extensive experiments on DIV2K dataset of hybrid distortions [37] and results show that our FDR-Net achieves superior performance over the state-of-the-art methods. Moreover, in order to increase the difficulty of hybrid-distorted image restoration, we construct a more complex hybrid distortion dataset DID-HY based on DID-MDN dataset [40] and test our approach on it. Then we validate the robustness of our FDR-Net by conducting extensive experiments on single distortion. To make our network more explanatory, the correlation maps of features are visualized. Finally, we conduct a series of ablation study to verify the effectiveness of our modules in FDR-Net.

4.1   Dataset for Hybrid-Distorted Image Restoration

DIV2K Dataset. The standard dataset on hybrid distortion are produced in [37] based on DIV2K dataset. The first 750 images are used for training and remaining 50 images are used for testing. They first process the dataset by adding multiple types of distortions, including Gaussian noise, Gaussian blur, and JPEG compression artifacts. The standard deviations of Gaussian blur and Gaussian noise are randomly chosen from the range of [0, 10] and [0, 50]. The quality of JPEG compression is chosen from [10, 100]. According to the degradation

Table 1. Quantitative results on the DIV2K dataset comparison to the state-of-the-art methods in terms of PSNR and SSIM. OP-Attention represents the operation-wise attention network in [28]. The SSIM is computed in the same way with OP-Attention.

Test set	Mild(unseen)		Moderate		Severe(unseen)	
Metric	PSNR	SSIM	PSNR	SSIM	PSNR	SSIM
DnCNN [43]	27.51	0.7315	26.50	0.6650	25.26	0.5974
RL-Restore [37]	28.04	0.7313	26.45	0.6557	25.20	0.5915
OP-Attention [28]	28.33	0.7455	27.07	0.6787	25.88	0.6167
Ours	**28.66**	**0.7568**	**27.24**	**0.6846**	**26.05**	**0.6218**

level, the dataset is divided into three parts, respectively are mild, moderate, and severe which represent that the degradation level from low to high. Then they crop the patches of size 63×63 from these processed images as training set and testing set. The training set consists of 249344 patches of moderate class and testing sets consists of 3584 patches of three classes.

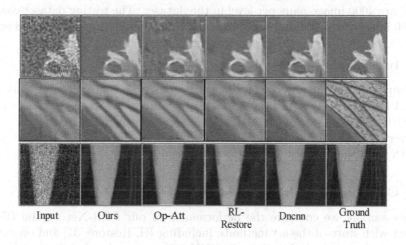

| Input | Ours | Op-Att | RL-Restore | Dncnn | Ground Truth |

Fig. 6. The performance comparison of our algorithm with the stat-of-the-art methods performed on DIV2K dataset. The three groups are corresponding to severe, moderate and mild from top to bottom.Fig. 7

DID-HY Dataset. Since the resolution of test image in DIV2K is too low, which is not practical in real world, we design a more complicated dataset which contains four kinds of distortions and higher resolution. The new dataset is built by adding Gaussian blur, Gaussian noise and JPEG compression artifacts based on DID-MDN dataset [40]. The training set contains 12,000 distortion/clean image pairs, which have resolutions of 512×512. And the testing set contains 1,200 distorted/clean image pairs.

Table 2. Quantization results on DID-HY dataset compared with the state-of-the-art methods. The SSIM is computed in the same way with OP-Attention.

Method	Dncnn [43]	OP-Attention [28]	Ours
PSNR	22.1315	23.1768	**24.9700**
SSIM	0.5953	0.6269	**0.6799**

4.2   Datasets for Single Distortion Restoration

Gopro Dataset for Deblurring. As the standard dataset for image deblurring, GOPRO dataset is produced by [14], which contains 3214 blurry/clean image pairs. The training set and testing set consists of 2103 pairs and 1111 pairs respectively.

DID-MDN Dataset for Deraining. The DID-MDN dataset is produced by [40]. Its training dataset consists of 12000 rain/clean image pairs which are classified to three level (light, medium, and heavy) based on its rain density. There are 4000 image pairs per level in the dataset. The testing dataset consists of 1200 images, which contains rain streaks with different orientations and scales.

4.3   Implementation Details

The implement of the proposed framework is based on PyTorch framework with one NVIDIA 1080Ti GPU. In the process of training, we adopt Adam optimizer [13] with a learning rate of 0.0001. The learning rate will decay by a factor valued 0.8 every 9000 iterations and the number of mini-batches is 28. It takes about 24 h to train the network and get the best results.

4.4   Comparison with State-of-the-Arts

In this section, we compare the performance of our FDR-Net on the DIV2K dataset with state-of-the-art methods, including RL-Restore [37] and operation-wise attention network [28]. Since DnCNN [43] was used as baseline in previous work, we compared with it too. Table 1 have demonstrated the superiority of our method. As shown in Table 1, our algorithm outperforms previous work in mild, moderate and severe level of hybrid distortion just training with moderate level dataset. And the comparison of subjective quality is shown in the Fig. 6.

Table 3. Quantitative results on Gopro dataset compared with the state-of-the-art methods for image deblur. The SSIM is computed in the same way with SRN [30].

Method	DeepDeblur [32]	DeblurGan [14]	SRN [30]	Inception_ResNet [15]	Ours
PSNR	29.23	28.70	30.10	29.55	**30.55**
SSIM	0.916	0.927	0.932	0.934	**0.938**

| Input | Dncnn | OP-Att | Ours | Ground truth |

Fig. 7. Examples of hybrid distortion removal on DID-HY dataset.

Since the size of the image in above dataset is just 63 × 63 and contains fewer distortion types, which is not practical in real world, we synthesize a more complex hybrid distortion dataset named DID-HY based on DID-MDN dataset. This dataset contains four kinds of distortions including blur, noise, compression artifacts and rain, in which the resolution of image is 512 × 512. Then we retrained the DnCNN [43], operation-wise attention network [28] and our FDRNet. The results in Table 2 have demonstrated that the Dncnn [43], operation-wise attention network [28] can't work well for more complicated hybrid distortion with higher resolution. However, our FDRNet can obtain the best results in objective quality as shown in Table 2 and subjective quality as shown in Fig. 7 and Fig. 8.

4.5 Interpretative Experiment

To visualize the channel-wise feature disentanglement, we study the correlation matrix between the channel of features before and after FDM. The feature before FDM reveal high correlations (0.272 in average) as shown in Fig. 8 (a). By disentangling the feature with our FDM, the feature correlations are suppressed as Fig. 8 (b) (0.164 in average). Moreover, we visualize the channel response of different distortion types after FDM in Fig. 8 (c). We take three kinds of distortions including blur, noise, and jpeg artifacts. And Each kind of distortion contains ten kinds of distortion levels. The higher level indicates the larger distortion. As shown in Fig. 8(c), different kinds of distortions have different channel responses regardless of the distortion levels, which means that the distortions have been divided across different channels by feature disentanglement. Besides, the different distortion levels only bring the changes of response strength.

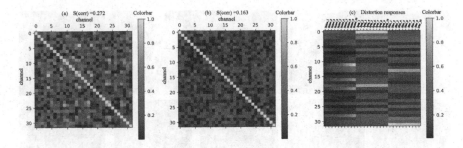

Fig. 8. (a) Visualization of correlation matrix between channels from the feature before FDM. (b) Visualization of correlation matrix between channels from feature after FDM. (c) Channels responses corresponding to different distortion types after FDM. As shown in (a) (b), FDM reduces the channel-wise correlations by disentangling the feature. From (c), the different distortions are distributed in different channels regardless of the levels of distortion by feature disentanglement.

4.6 Experiments on Single Distortion

In order to verify the robustness of our method, we separately apply our FDR-Net on image deblurring and image deraining tasks and achieved the stat-of-the-art performance.

Inputs   Deblurgan   SRN   Ours   Groud Truth

Fig. 9. Visual comparisons for Gopro-test dataset between our FDR-Net and the state-of-art methods for image deblurring.

Comparison to Image Deblurring. On the task of image deblur, we compared our FDR-Net with DeepDeblur [32], Deblurgan [14], SRN [30] and Deblurganv2 [15] with Inception ResNet respectively. The Table 3 have demonstrateTable 4 that our FDR-Net surpasses previous works for image deblurring. As seen in the Fig. 9, our algorithm is better than other state-of-the-art methods in generating rich texture.

Comparison to Image Deraining. In the field of image derain, we compared our FDR-Net with DID-MDN [40], RESCAN [20] and SPANET [33]. The quantization quality as shown in Table 4 have demonstrated the superiority of our method for image deraining task. The subjective quality is compared with the state-of-the-art methods in Fig. 10.

Table 4. Quantitative results on DID-MDN dataset compared with the state-of-the-art methods for image derain. The SSIM is computed in the same way with SPANET [33].

Method	DID-MDN [40]	RESCAN [20]	SPANET [33]	Ours
PSNR	27.95	29.95	30.05	**32.96**
SSIM	0.909	0.884	0.934	**0.951**

Inputs   DID-MDN   SPA-Net   Ours   Groud Truth

Fig. 10. Visual comparisons for DID-MDN dataset between our FDR-Net and the state-of-art methods for image deraining.

4.7 Ablation Studies

Effect of FDM. Distortion-aware feature disentangle method is a key factor to the success of FDR-Net for hybrid-distorted image restoration. To verify its importances and effectiveness, we substitute the FDlayer and FAlayer with a series of ResBlocks [8] as our baseline. Then we retrained the baseline network with DID-HY dataset. The parameters of two networks are almost the same. However, the performance of our FDR-Net is far higher than the baseline without FDlayer and FAlayer. The quantitative results are shown in Table 5.

Table 5. Ablation study on different combinations of FDM and Auxiliary module. Tested on DID-HY dataset.

FDM	Aux	Params	PSNR	SSIM
✓	✓	1.66M	24.97	0.680
×	✓	1.62M	23.91	0.646
✓	×	1.21M	24.27	0.653

Effect of Multiple FDlayers and Multiple Channels. To investigate the effect of multiple FDlayers in FDM, we set the number of FDlayers as 1 to 5 respectively. As shown in Table 6, three FDlayers are saturated to achieve the best performance. Then we set the number of FDlayers as 3 and set the number of channels as 16, 32, 48 and 64 respectively. As shown in Table 6, increasing the number of channels could increase the ability to represent the distortions for Network, which could bring the improvement of performance and 32 is saturated to achieve the best performance.

Table 6. Comparisons between different number of FDlayers in each FDM and comparisons between different number of channels with 3 FDlayers.

DFlayers	1	2	3	4	5	Channels	16	32	48	64
PSNR	24.63	24.70	24.97	24.98	24.99	PSNR	24.47	24.97	24.98	25.00
SSIM	0.666	0.668	0.680	0.680	0.681	SSIM	0.659	0.680	0.682	0.683

Effect of Auxiliary Module. As discussed in Sect. 3, auxiliary module would make up for the weakness of the distortion-aware feature disentangle branch. We demonstrate the effectiveness of auxiliary module by removing it. The performance reduction is shown as Table 5.

5   Conclusion

In this paper, we implement the channel-wise feature disentanglement by FDM to reduce the interference between hybrid distortions, which achieves the best performance for hybrid-distorted image restoration. Furthermore, we also validate the effectiveness of our modules with the visualizations of correlation maps and channel responses for different distortions. Extensive experiments demonstrate that our FDR-Net has stronger robustness for single distortion removal.

Acknowledgement. This work was supported in part by NSFC under Grant U1908209, 61632001 and the National Key Research and Development Program of China 2018AAA0101400.

References

1. Berman, D., Avidan, S., et al.: Non-local image dehazing. In: Proceedings of the IEEE Conference on Computer Vision and Pattern Recognition, pp. 1674–1682 (2016)
2. Bianco, S., Celona, L., Napoletano, P., Schettini, R.: Disentangling image distortions in deep feature space (2020)

3. Chen, T., et al.: ABD-Net: attentive but diverse person re-identification. In: Proceedings of the IEEE International Conference on Computer Vision, pp. 8351–8361 (2019)
4. Dabov, K., Foi, A., Katkovnik, V., Egiazarian, K.: Image denoising by sparse 3-D transform-domain collaborative filtering. IEEE Trans. Image Process. **16**(8), 2080–2095 (2007)
5. Dong, C., Loy, C.C., He, K., Tang, X.: Image super-resolution using deep convolutional networks. IEEE Trans. Pattern Anal. Mach. Intell. **38**(2), 295–307 (2015)
6. Fergus, R., Singh, B., Hertzmann, A., Roweis, S.T., Freeman, W.T.: Removing camera shake from a single photograph. In: ACM Transactions on Graphics (TOG), vol. 25, pp. 787–794. ACM (2006)
7. Haris, M., Shakhnarovich, G., Ukita, N.: Deep back-projection networks for super-resolution. In: Proceedings of the IEEE Conference on Computer Vision and Pattern Recognition, pp. 1664–1673 (2018)
8. He, K., Zhang, X., Ren, S., Sun, J.: Deep residual learning for image recognition. In: Proceedings of the IEEE Conference on Computer Vision and Pattern Recognition, pp. 770–778 (2016)
9. Jin, X., Chen, Z., Lin, J., Chen, J., Zhou, W., Shan, C.: A decomposed dual-cross generative adversarial network for image rain removal. In: BMVC, p. 119 (2018)
10. Jin, X., Chen, Z., Lin, J., Chen, Z., Zhou, W.: Unsupervised single image deraining with self-supervised constraints. In: 2019 IEEE International Conference on Image Processing (ICIP), pp. 2761–2765. IEEE (2019)
11. Jin, X., Lan, C., Zeng, W., Chen, Z., Zhang, L.: Style normalization and restitution for generalizable person re-identification. In: Proceedings of the IEEE/CVF Conference on Computer Vision and Pattern Recognition, pp. 3143–3152 (2020)
12. Kim, J., Kwon Lee, J., Mu Lee, K.: Accurate image super-resolution using very deep convolutional networks. In: Proceedings of the IEEE Conference on Computer Vision and Pattern Recognition, pp. 1646–1654 (2016)
13. Kingma, D.P., Ba, J.: Adam: a method for stochastic optimization. arXiv preprint arXiv:1412.6980 (2014)
14. Kupyn, O., Budzan, V., Mykhailych, M., Mishkin, D., Matas, J.: Deblurgan: blind motion deblurring using conditional adversarial networks. In: Proceedings of the IEEE Conference on Computer Vision and Pattern Recognition, pp. 8183–8192 (2018)
15. Kupyn, O., Martyniuk, T., Wu, J., Wang, Z.: Deblurgan-v2: deblurring (orders-of-magnitude) faster and better. In: Proceedings of the IEEE International Conference on Computer Vision, pp. 8878–8887 (2019)
16. Ledig, C., et al.: Photo-realistic single image super-resolution using a generative adversarial network. In: Proceedings of the IEEE Conference on Computer Vision and Pattern Recognition, pp. 4681–4690 (2017)
17. Lehtinen, J., et al.: Noise2noise: learning image restoration without clean data. arXiv preprint arXiv:1803.04189 (2018)
18. Li, G., He, X., Zhang, W., Chang, H., Dong, L., Lin, L.: Non-locally enhanced encoder-decoder network for single image de-raining. arXiv preprint arXiv:1808.01491 (2018)
19. Li, M., Zhu, X., Gong, S.: Unsupervised person re-identification by deep learning tracklet association. In: Proceedings of the European Conference on Computer Vision (ECCV), pp. 737–753 (2018)
20. Li, X., Wu, J., Lin, Z., Liu, H., Zha, H.: Recurrent squeeze-and-excitation context aggregation net for single image deraining. In: Proceedings of the European Conference on Computer Vision (ECCV), pp. 254–269 (2018)

21. Li, Z., Yang, J., Liu, Z., Yang, X., Jeon, G., Wu, W.: Feedback network for image super-resolution. In: Proceedings of the IEEE Conference on Computer Vision and Pattern Recognition, pp. 3867–3876 (2019)
22. Lim, B., Son, S., Kim, H., Nah, S., Mu Lee, K.: Enhanced deep residual networks for single image super-resolution. In: Proceedings of the IEEE Conference on Computer Vision and Pattern Recognition Workshops, pp. 136–144 (2017)
23. Liu, Z., Wu, B., Luo, W., Yang, X., Liu, W., Cheng, K.T.: Bi-Real Net: enhancing the performance of 1-bit CNNs with improved representational capability and advanced training algorithm. In: Proceedings of the European Conference on Computer Vision (ECCV), pp. 722–737 (2018)
24. Lu, B., Chen, J.C., Chellappa, R.: Unsupervised domain-specific deblurring via disentangled representations. In: Proceedings of the IEEE Conference on Computer Vision and Pattern Recognition, pp. 10225–10234 (2019)
25. Nazeri, K., Ng, E., Joseph, T., Qureshi, F., Ebrahimi, M.: Edgeconnect: generative image inpainting with adversarial edge learning. arXiv preprint arXiv:1901.00212 (2019)
26. Schwartz, O., Simoncelli, E.P.: Natural signal statistics and sensory gain control. Nat. Neurosci. 4(8), 819 (2001)
27. Shi, Y., Wu, X., Zhu, M.: Low-light image enhancement algorithm based on Retinex and generative adversarial network. arXiv preprint arXiv:1906.06027 (2019)
28. Suganuma, M., Liu, X., Okatani, T.: Attention-based adaptive selection of operations for image restoration in the presence of unknown combined distortions. In: Proceedings of the IEEE Conference on Computer Vision and Pattern Recognition, pp. 9039–9048 (2019)
29. Sun, J., Cao, W., Xu, Z., Ponce, J.: Learning a convolutional neural network for non-uniform motion blur removal. In: Proceedings of the IEEE Conference on Computer Vision and Pattern Recognition, pp. 769–777 (2015)
30. Tao, X., Gao, H., Shen, X., Wang, J., Jia, J.: Scale-recurrent network for deep image deblurring. In: Proceedings of the IEEE Conference on Computer Vision and Pattern Recognition, pp. 8174–8182 (2018)
31. Ulyanov, D., Vedaldi, A., Lempitsky, V.: Deep image prior. In: Proceedings of the IEEE Conference on Computer Vision and Pattern Recognition, pp. 9446–9454 (2018)
32. Wang, L., Li, Y., Wang, S.: Deepdeblur: fast one-step blurry face images restoration. arXiv preprint arXiv:1711.09515 (2017)
33. Wang, T., Yang, X., Xu, K., Chen, S., Zhang, Q., Lau, R.W.: Spatial attentive single-image deraining with a high quality real rain dataset. In: Proceedings of the IEEE Conference on Computer Vision and Pattern Recognition, pp. 12270–12279 (2019)
34. Yang, W., Tan, R.T., Feng, J., Liu, J., Guo, Z., Yan, S.: Deep joint rain detection and removal from a single image. In: Proceedings of the IEEE Conference on Computer Vision and Pattern Recognition, pp. 1357–1366 (2017)
35. Yang, X., Xu, Z., Luo, J.: Towards perceptual image dehazing by physics-based disentanglement and adversarial training. In: Thirty-Second AAAI Conference on Artificial Intelligence (2018)
36. Yu, J., Lin, Z., Yang, J., Shen, X., Lu, X., Huang, T.S.: Generative image inpainting with contextual attention. In: Proceedings of the IEEE Conference on Computer Vision and Pattern Recognition, pp. 5505–5514 (2018)
37. Yu, K., Dong, C., Lin, L., Change Loy, C.: Crafting a toolchain for image restoration by deep reinforcement learning. In: Proceedings of the IEEE Conference on Computer Vision and Pattern Recognition, pp. 2443–2452 (2018)

38. Yuan, Q., Li, J., Zhang, L., Wu, Z., Liu, G.: Blind motion deblurring with cycle generative adversarial networks. arXiv preprint arXiv:1901.01641 (2019)
39. Yuan, Y., Liu, S., Zhang, J., Zhang, Y., Dong, C., Lin, L.: Unsupervised image super-resolution using cycle-in-cycle generative adversarial networks. In: Proceedings of the IEEE Conference on Computer Vision and Pattern Recognition Workshops, pp. 701–710 (2018)
40. Zhang, H., Patel, V.M.: Density-aware single image de-raining using a multi-stream dense network. In: Proceedings of the IEEE Conference on Computer Vision and Pattern Recognition, pp. 695–704 (2018)
41. Zhang, H., Sindagi, V., Patel, V.M.: Joint transmission map estimation and dehazing using deep networks. arXiv preprint arXiv:1708.00581 (2017)
42. Zhang, H., Sindagi, V., Patel, V.M.: Image de-raining using a conditional generative adversarial network. IEEE Trans. Circuits Syst. Video Technol. (2019)
43. Zhang, K., Zuo, W., Chen, Y., Meng, D., Zhang, L.: Beyond a Gaussian denoiser: residual learning of deep cnn for image denoising. IEEE Trans. Image Process. **26**(7), 3142–3155 (2017)
44. Zhang, K., Zuo, W., Zhang, L.: FFDNet: toward a fast and flexible solution for CNN-based image denoising. IEEE Trans. Image Process. **27**(9), 4608–4622 (2018)
45. Zhao, J., et al.: DD-CycleGAN: unpaired image dehazing via double-discriminator cycle-consistent generative adversarial network. Eng. Appl. Artif. Intell. **82**, 263–271 (2019)

Jointly De-Biasing Face Recognition and Demographic Attribute Estimation

Sixue Gong$^{(\boxtimes)}$, Xiaoming Liu, and Anil K. Jain

Michigan State University, East Lansing, USA
{gongsixu,liuxm,jain}@msu.edu

Abstract. We address the problem of bias in automated face recognition and demographic attribute estimation algorithms, where errors are lower on certain cohorts belonging to specific demographic groups. We present a novel de-biasing adversarial network (DebFace) that learns to extract disentangled feature representations for both unbiased face recognition and demographics estimation. The proposed network consists of one identity classifier and three demographic classifiers (for gender, age, and race) that are trained to distinguish identity and demographic attributes, respectively. Adversarial learning is adopted to minimize correlation among feature factors so as to abate bias influence from other factors. We also design a new scheme to combine demographics with identity features to strengthen robustness of face representation in different demographic groups. The experimental results show that our approach is able to reduce bias in face recognition as well as demographics estimation while achieving state-of-the-art performance.

Keywords: Bias · Feature disentanglement · Face recognition · Fairness

1 Introduction

Automated face recognition has achieved remarkable success with the rapid developments of deep learning algorithms. Despite the improvement in the accuracy of face recognition, one topic is of significance. Does a face recognition system perform equally well in different demographic groups? In fact, it has been observed that many face recognition systems have lower performance in certain demographic groups than others [23,29,42]. Such face recognition systems are said to be *biased* in terms of demographics.

Electronic supplementary material The online version of this chapter (https://doi.org/10.1007/978-3-030-58526-6_20) contains supplementary material, which is available to authorized users.

© Springer Nature Switzerland AG 2020
A. Vedaldi et al. (Eds.): ECCV 2020, LNCS 12374, pp. 330–347, 2020.
https://doi.org/10.1007/978-3-030-58526-6_20

In a time when face recognition systems are being deployed in the real world for societal benefit, this type of bias[1] is not acceptable. Why does the bias problem exist in face recognition systems? First, state-of-the-art (SOTA) face recognition methods are based on deep learning which requires a large collection of face images for training. Inevitably the distribution of training data has a great impact on the performance of the resultant deep learning models. It is well understood that face datasets exhibit imbalanced demographic distributions where the number of faces in each cohort is unequal. Previous studies have shown that models trained with imbalanced datasets lead to biased discrimination [5, 49]. Secondly, the goal of deep face recognition is to map the input face image to a target feature vector with high discriminative power. The bias in the mapping function will result in feature vectors with lower discriminability for certain demographic groups. Moreover, Klare *et al.* [29] show the errors that are inherent to some demographics by studying non-trainable face recognition algorithms.

To address the bias issue, data re-sampling methods have been exploited to balance the data distribution by under-sampling the majority [16] or over-sampling the minority classes [8,39]. Despite its simplicity, valuable information may be removed by under-sampling, and over-sampling may introduce noisy samples. Naively training on a balanced dataset can still lead to bias [56]. Another common option for imbalanced data training is cost-sensitive learning that assigns weights to different classes based on (i) their frequency or (ii) the effective number of samples [6,12]. To eschew the overfitting of Deep Neural Network (DNN) to minority classes, hinge loss is often used to increase margins among classification decision boundaries [21,27]. The aforementioned methods have also been adopted for face recognition and attribute prediction on imbalanced datasets [24,58]. However, such face recognition studies only concern bias in terms of *identity*, rather than our focus of *demographic bias*.

In this paper, we propose a framework to address the influence of bias on face recognition and demographic attribute estimation. In typical deep learning based face recognition frameworks, the large capacity of DNN enables the face representations to embed demographic details, including gender, race, and age [3,17]. Thus, the biased demographic information is transmitted from the training dataset to the output representations. To tackle this issue, we assume that if the face representation does not carry discriminative information of demographic attributes, it would be unbiased in terms of demographics. Given this assumption, one common way to remove demographic information from face representations is to perform feature disentanglement via adversarial learning (Fig. 1b). That is, the classifier of demographic attributes can be used to encourage the identity representation to *not* carry demographic information. However, one issue of this common approach is that, the demographic classifier itself could be biased (*e.g.*, the race classifier could be biased on gender), and hence it will

---

[1] This is different from the notion of machine learning bias, defined as "any basis for choosing one generalization [hypothesis] over another, other than strict consistency with the observed training instances" [15].

act differently while disentangling faces of different cohorts. This is clearly undesirable as it leads to demographic biased identity representation.

To resolve the chicken-and-egg problem, we propose to *jointly* learn unbiased representations for both the identity and demographic attributes. Specifically, starting from a multi-task learning framework that learns disentangled feature representations of gender, age, race, and identity, respectively, we request the classifier of each task to act as adversarial supervision for the other tasks (*e.g.*, the dash arrows in Fig. 1c). These four classifiers help each other to achieve better feature disentanglement, resulting in unbiased feature representations for both the identity and demographic attributes. As shown in Fig. 1, our framework is in sharp contrast to either multi-task learning or adversarial learning.

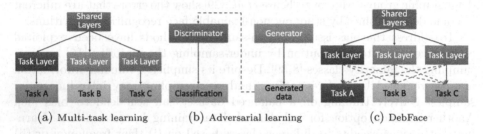

(a) Multi-task learning   (b) Adversarial learning   (c) DebFace

Fig. 1. Methods to learn different tasks simultaneously. Solid lines are typical feature flow in CNN, while dash lines are adversarial losses.

Moreover, since the features are disentangled into the demographic and identity, our face representations also contribute to privacy-preserving applications. It is worth noticing that such identity representations contain little demographic information, which could undermine the recognition competence since demographic features are *part* of identity-related facial appearance. To retain the recognition accuracy on demographic biased face datasets, we propose another network that combines the demographic features with the demographic-free identity features to generate a new identity representation for face recognition.

The key contributions and findings of the paper are:

⋄ A thorough analysis of deep learning based face recognition performance on three different demographics: (i) gender, (ii) age, and (iii) race.

⋄ A de-biasing face recognition framework, called DebFace, that generates disentangled representations for both identity and demographics recognition while jointly removing discriminative information from other counterparts.

⋄ The identity representation from DebFace (DebFace-ID) shows lower bias on different demographic cohorts and also achieves SOTA face verification results on demographic-unbiased face recognition.

⋄ The demographic attribute estimations via DebFace are less biased across other demographic cohorts.

⋄ Combining ID with demographics results in more discriminative features for face recognition on biased datasets.

2   Related Work

Face Recognition on Imbalanced Training Data. Previous efforts on face recognition aim to tackle class imbalance problem on training data. For example, in prior-DNN era, Zhang *et al.* [66] propose a cost-sensitive learning framework to reduce misclassification rate of face identification. To correct the skew of separating hyperplanes of SVM on imbalanced data, Liu *et al.* [33] propose Margin-Based Adaptive Fuzzy SVM that obtains a lower generalization error bound. In the DNN era, face recognition models are trained on large-scale face datasets with highly-imbalanced class distribution [63,65]. Range Loss [65] learns a robust face representation that makes the most use of every training sample. To mitigate the impact of insufficient class samples, center-based feature transfer learning [63] and large margin feature augmentation [58] are proposed to augment features of minority identities and equalize class distribution. Despite their effectiveness, these studies ignore the influence of demographic imbalance on the dataset, which may lead to demographic bias. For instance, The FRVT 2019 [42] shows the demographic bias of over 100 face recognition algorithms. To uncover deep learning bias, Alexander *et al.* [4] develop an algorithm to mitigate the hidden biases within training data. Wang *et al.* [57] propose a domain adaptation network to reduce racial bias in face recognition. They recently extended their work using reinforcement learning to find optimal margins of additive angular margin based loss functions for different races [56]. To our knowledge, no studies have tackled the challenge of de-biasing demographic bias in DNN-based face recognition and demographic attribute estimation algorithms.

Adversarial Learning and Disentangled Representation. Adversarial learning [44] has been well explored in many computer vision applications. For example, Generative Adversarial Networks (GANs) [18] employ adversarial learning to train a generator by competing with a discriminator that distinguishes real images from synthetic ones. Adversarial learning has also been applied to domain adaptation [36,48,52,53]. A problem of current interest is to learn interpretable representations with semantic meaning [60]. Many studies have been learning factors of variations in the data by supervised learning [31,32,34,50,51], or semi-supervised/unsupervised learning [28,35,40,68], referred to as disentangled representation. For supervised disentangled feature learning, adversarial networks are utilized to extract features that only contain discriminative information of a target task. For face recognition, Liu *et al.* [34] propose a disentangled representation by training an adversarial autoencoder to extract features that can capture identity discrimination and its complementary knowledge. In contrast, our proposed DebFace differs from prior works in that each branch of a multi-task network acts as both a generator and discriminators of other branches (Fig. 1c).

3   Methodology

3.1   Problem Definition

The ultimate goal of unbiased face recognition is that, given a face recognition system, no statistically significant difference among the performance in different categories of face images. Despite the research on pose-invariant face recognition that aims for equal performance on all poses [51,62], we believe that it is inappropriate to define variations like pose, illumination, or resolution, as the categories. These are instantaneous *image-related* variations with intrinsic bias. E.g., large-pose or low-resolution faces are inherently harder to be recognized than frontal-view high-resolution faces.

Rather, we would like to define *subject-related* properties such as demographic attributes as the categories. *A face recognition system is **biased** if it performs worse on certain demographic cohorts.* For practical applications, it is important to consider what demographic biases may exist, and whether these are intrinsic biases across demographic cohorts or algorithmic biases derived from the algorithm itself. This motivates us to analyze the demographic influence on face recognition performance and strive to reduce algorithmic bias for face recognition systems. One may achieve this by training on a dataset containing uniform samples over the cohort space. However, the demographic distribution of a dataset is often imbalanced and under represents demographic minorities while overrepresenting majorities. Naively re-sampling a balanced training dataset may still induce bias since the diversity of latent variables is different across cohorts and the instances cannot be treated fairly during training. To mitigate demographic bias, we propose a face de-biasing framework that jointly reduces mutual bias over all demographics and identities while disentangling face representations into gender, age, race, and demographic-free identity in the mean time.

3.2   Algorithm Design

The proposed network takes advantage of the relationship between demographics and face identities. On one hand, demographic characteristics are highly correlated to face features. On the other hand, demographic attributes are heterogeneous in terms of data type and semantics [20]. A male person, for example, is not necessarily of a certain age or of a certain race. Accordingly, we present a framework that jointly generates demographic features and identity features from a single face image by considering both the aforementioned attribute correlation and attribute heterogeneity in a DNN.

While our main goal is to mitigate demographic bias from face representation, we observe that demographic estimations are biased as well (see Fig. 5). How can we remove the bias of face recognition when demographic estimations themselves are biased? Cook *et al.* [11] investigated this effect and found the performance of face recognition is affected by multiple demographic covariates. We propose a de-biasing network, DebFace, that disentangles the representation into gender (DebFace-G), age (DebFace-A), race (DebFace-R), and identity (DebFace-ID), to decrease bias of both face recognition and demographic estimations. Using adversarial learning, the proposed method is capable of jointly learning multiple discriminative representations while ensuring that each classifier cannot distinguish among classes through non-corresponding representations.

Though less biased, DebFace-ID loses demographic cues that are useful for identification. In particular, race and gender are two critical components that constitute face patterns. Hence, we desire to incorporate race and gender with DebFace-ID to obtain a more integrated face representation. We employ a lightweight fully-connected network to aggregate the representations into a face representation (DemoID) with the same dimensionality as DebFace-ID.

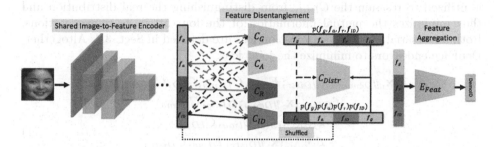

Fig. 2. Overview of the proposed De-biasing face (DebFace) network. DebFace is composed of three major blocks, *i.e.*, a shared feature encoding block, a feature disentangling block, and a feature aggregation block. The solid arrows represent the forward inference, and the dashed arrows stand for adversarial training. During inference, either DebFace-ID (*i.e.*, $\mathbf{f}_{ID}$) or DemoID can be used for face matching given the desired trade-off between biasness and accuracy.

3.3   Network Architecture

Figure 2 gives an overview of the proposed DebFace network. It consists of four components: the shared image-to-feature encoder E_{Img}, the four attribute classifiers (including gender C_G, age C_A, race C_R, and identity C_{ID}), the distribution classifier C_{Distr}, and the feature aggregation network E_{Feat}. We assume access to N labeled training samples $\{(\mathbf{x}^{(i)}, y_g^{(i)}, y_a^{(i)}, y_r^{(i)}, y_{id}^{(i)})\}_{i=1}^N$. Our approach takes an image $\mathbf{x}^{(i)}$ as the input of E_{Img}. The encoder projects $\mathbf{x}^{(i)}$ to its feature representation $E_{Img}(\mathbf{x}^{(i)})$. The feature representation is then decoupled into four D-dimensional feature vectors, gender $\mathbf{f}_g^{(i)}$, age $\mathbf{f}_a^{(i)}$, race $\mathbf{f}_r^{(i)}$, and identity $\mathbf{f}_{ID}^{(i)}$, respectively. Next, each attribute classifier operates the corresponding feature vector to correctly classify the target attribute by optimizing parameters of both E_{Img} and the respective classifier C_*. For a demographic attribute with K categories, the learning objective $\mathcal{L}_{C_{Demo}}(\mathbf{x}, y_{Demo}; E_{Img}, C_{Demo})$ is the standard cross entropy loss function. For the $n-$identity classification, we adopt AM-Softmax [54] as the objective function $\mathcal{L}_{C_{ID}}(\mathbf{x}, y_{id}; E_{Img}, C_{ID})$. To de-bias all of the feature representations, adversarial loss $\mathcal{L}_{Adv}(\mathbf{x}, y_{Demo}, y_{id}; E_{Img}, C_{Demo}, C_{ID})$ is applied to the above four classifiers such that each of them will not be able to predict correct labels when operating irrelevant feature vectors. Specifically, given a classifier, the remaining three attribute feature vectors are imposed on it and attempt to mislead the classifier by only optimizing the parameters of E_{Img}. To further improve the disentanglement, we also reduce the mutual information among the attribute features by introducing a distribution classifier C_{Distr}. C_{Distr} is trained to identify whether an input representation is sampled from the joint distribution $p(\mathbf{f}_g, \mathbf{f}_a, \mathbf{f}_r, \mathbf{f}_{ID})$ or the multiplication of margin distributions $p(\mathbf{f}_g)p(\mathbf{f}_a)p(\mathbf{f}_r)p(\mathbf{f}_{ID})$ via a binary cross entropy loss $\mathcal{L}_{C_{Distr}}(\mathbf{x}, y_{Distr}; E_{Img}, C_{Distr})$, where y_{Distr} is the distribution label. Similar to adversarial loss, a factorization objective function $\mathcal{L}_{Fact}(\mathbf{x}, y_{Distr}; E_{Img}, C_{Distr})$ is utilized to restrain the C_{Distr} from distinguishing the real distribution and thus minimizes the mutual information of the four attribute representations. Both adversarial loss and factorization loss are detailed in Sect. 3.4. Altogether, DebFace endeavors to minimize the joint loss:

$$
\begin{aligned}
\mathcal{L}(\mathbf{x}, y_{Demo}, y_{id}, y_{Distr}; & E_{Img}, C_{Demo}, C_{ID}, C_{Distr}) = \\
& \mathcal{L}_{C_{Demo}}(\mathbf{x}, y_{Demo}; E_{Img}, C_{Demo}) \\
& + \mathcal{L}_{C_{ID}}(\mathbf{x}, y_{id}; E_{Img}, C_{ID}) \\
& + \mathcal{L}_{C_{Distr}}(\mathbf{x}, y_{Distr}; E_{Img}, C_{Distr}) \\
& + \lambda \mathcal{L}_{Adv}(\mathbf{x}, y_{Demo}, y_{id}; E_{Img}, C_{Demo}, C_{ID}) \\
& + \nu \mathcal{L}_{Fact}(\mathbf{x}, y_{Distr}; E_{Img}, C_{Distr}),
\end{aligned}
\tag{1}
$$

where λ and ν are hyper-parameters determining how much the representation is decomposed and decorrelated in each training iteration.

The discriminative demographic features in DebFace-ID are weakened by removing demographic information. Fortunately, our de-biasing network preserves all pertinent demographic features in a disentangled way. Basically, we train another multilayer perceptron (MLP) E_{Feat} to aggregate DebFace-ID and the demographic embeddings into a unified face representation DemoID. Since age generally does not pertain to a person's identity, we only consider gender and race as the identity-informative attributes. The aggregated embedding, $\mathbf{f}_{DemoID} = E_{feat}(\mathbf{f}_{ID}, \mathbf{f}_g, \mathbf{f}_r)$, is supervised by an identity-based triplet loss:

$$
\mathcal{L}_{E_{Feat}} = \frac{1}{M} \sum_{i=1}^{M} [\|\mathbf{f}_{DemoID^a}^{(i)} - \mathbf{f}_{DemoID^p}^{(i)}\|_2^2 - \|\mathbf{f}_{DemoID^a}^{(i)} - \mathbf{f}_{DemoID^n}^{(i)}\|_2^2 + \alpha]_+ (2)
$$

where $\{\mathbf{f}_{DemoID^a}^{(i)}, \mathbf{f}_{DemoID^p}^{(i)}, \mathbf{f}_{DemoID^n}^{(i)}\}$ is the ith triplet consisting of an anchor, a positive, and a negative DemoID representation, M is the number of hard triplets in a mini-batch. $[x]_+ = \max(0, x)$, and α is the margin.

3.4   Adversarial Training and Disentanglement

As discussed in Sect. 3.3, the adversarial loss aims to minimize the task-independent information semantically, while the factorization loss strives to dwindle the interfering information statistically. We employ both losses to disentangle the representation extracted by E_{Img}. We introduce the adversarial loss as a means to learn a representation that is invariant in terms of certain attributes, where a classifier trained on it cannot correctly classify those attributes using that representation. We take one of the attributes, *e.g.*, gender, as an example to illustrate the adversarial objective. First of all, for a demographic representation $\mathbf{f}_{Demo}$, we learn a gender classifier on $\mathbf{f}_{Demo}$ by optimizing the classification loss $\mathcal{L}_{C_G}(\mathbf{x}, y_{Demo}; E_{Img}, C_G)$. Secondly, for the same gender classifier, we intend to maximize the chaos of the predicted distribution [26]. It is well known that a uniform distribution has the highest entropy and presents the most randomness. Hence, we train the classifier to predict the probability distribution as close as possible to a uniform distribution over the category space by minimizing the cross entropy:

$$\mathcal{L}_{Adv}^{G}(\mathbf{x}, y_{Demo}, y_{id}; E_{Img}, C_G) = -\sum_{k=1}^{K_G} \frac{1}{K_G} \cdot (\log \frac{e^{C_G(\mathbf{f}_{Demo})_k}}{\sum_{j=1}^{K_G} e^{C_G(\mathbf{f}_{Demo})_j}} + \log \frac{e^{C_G(\mathbf{f}_{ID})_k}}{\sum_{j=1}^{K_G} e^{C_G(\mathbf{f}_{ID})_j}}),$$

(3)

where K_G is the number of categories in gender[2], and the ground-truth label is no longer an one-hot vector, but a K_G-dimensional vector with all elements being $\frac{1}{K_G}$. The above loss function corresponds to the dash lines in Fig. 2. It strives for gender-invariance by finding a representation that makes the gender classifier C_G perform poorly. We minimize the adversarial loss by only updating parameters in E_{Img}.

We further decorrelate the representations by reducing the mutual information across attributes. By definition, the mutual information is the relative entropy (KL divergence) between the joint distribution and the product distribution. To increase uncorrelation, we add a distribution classifier C_{Distr} that is trained to simply perform a binary classification using $\mathcal{L}_{C_{Distr}}(\mathbf{x}, y_{Distr}; E_{Img}, C_{Distr})$ on samples $\mathbf{f}_{Distr}$ from both the joint distribution and dot product distribution. Similar to adversarial learning, we factorize the representations by tricking the classifier via the same samples so that the predictions are close to random guesses,

$$\mathcal{L}_{Fact}(\mathbf{x}, y_{Distr}; E_{Img}, C_{Distr}) = -\sum_{i=1}^{2} \frac{1}{2} \log \frac{e^{C_{Distr}(\mathbf{f}_{Distr})_i}}{\sum_{j=1}^{2} e^{C_{Distr}(\mathbf{f}_{Distr})_j}}.$$

(4)

In each mini-batch, we consider $E_{Img}(\mathbf{x})$ as samples of the joint distribution $p(\mathbf{f}_g, \mathbf{f}_a, \mathbf{f}_r, \mathbf{f}_{ID})$. We randomly shuffle feature vectors of each attribute, and re-concatenate them into $4D$-dimension, which are approximated as samples of the product distribution $p(\mathbf{f}_g)p(\mathbf{f}_a)p(\mathbf{f}_r)p(\mathbf{f}_{ID})$. During factorization, we only update E_{Img} to minimize mutual information between decomposed features.

---

[2] In our case, $K_G = 2$, *i.e.*, male and female.

4   Experiments

4.1   Datasets and Pre-processing

We utilize 15 total face datasets in this work, for learning the demographic estimation models, the baseline face recognition model, DebFace model as well as their evaluation. To be specific, CACD [9], IMDB [43], UTKFace [67], AgeDB [38], AFAD [41], AAF [10], FG-NET [1], RFW [57], IMFDB-CVIT [45], Asian-DeepGlint [2], and PCSO [13] are the datasets for training and testing demographic estimation models; and MS-Celeb-1M [19], LFW [25], IJB-A [30], and IJB-C [37] are for learning and evaluating face verification models. All faces are detected by MTCNN [64]. Each face image is cropped and resized to 112×112 pixels using a similarity transformation based on the detected landmarks.

4.2   Implementation Details

DebFace is trained on a cleaned version of MS-Celeb-1M [14], using the Arc-Face architecture [14] with 50 layers for the encoder E_{Img}. Since there are no demographic labels in MS-Celeb-1M, we first train three demographic attribute estimation models for gender, age, and race, respectively. For age estimation, the model is trained on the combination of CACD, IMDB, UTKFace, AgeDB, AFAD, and AAF datasets. The gender estimation model is trained on the same datasets except CACD which contains no gender labels. We combine AFAD, RFW, IMFDB-CVIT, and PCSO for race estimation training. All three models use ResNet [22] with 34 layers for age, 18 layers for gender and race.

We predict the demographic labels of MS-Celeb-1M with the well-trained demographic models. Our DebFace is then trained on the re-labeled MS-Celeb-1M using SGD with a momentum of 0.9, a weight decay of 0.01, and a batch size of 256. The learning rate starts from 0.1 and drops to 0.0001 following the schedule at 8, 13, and 15 epochs. The dimensionality of the embedding layer of E_{Img} is 4×512, *i.e.*, each attribute representation (gender, age, race, ID) is a $512\text{-}dim$ vector. We keep the hyper-parameter setting of AM-Softmax as [14]: $s = 64$ and $m = 0.5$. The feature aggregation network E_{Feat} comprises of two linear residual units with P-ReLU and BatchNorm in between. E_{Feat} is trained on MS-Celeb-1M by SGD with a learning rate of 0.01. The triplet loss margin α is 1.0. The disentangled features of gender, race, and identity are concatenated into a $3 \times 512\text{-}dim$ vector, which inputs to E_{Feat}. The network is then trained to output a $512\text{-}dim$ representation for face recognition on biased datasets. Our source code is available at https://github.com/gongsixue/DebFace.git.

4.3   De-Biasing Face Verification

Baseline: We compare DebFace-ID with a regular face representation model which has the same architecture as the shared feature encoder of DebFace. Referred to as BaseFace, this baseline model is also trained on MS-Celeb-1M, with the representation dimension of 512.

To show the efficacy of DebFace-ID on bias mitigation, we evaluate the verification performance of DebFace-ID and BaseFace on faces from each demographic cohort. There are 48 total cohorts given the combination of demographic attributes including 2 gender (male, female), 4 race[3] (Black, White, East Asian, Indian), and 6 age group (0 − 12, 13 − 18, 19 − 34, 35 − 44, 45 − 54, 55 − 100). We combine CACD, AgeDB, CVIT, and a subset of Asian-DeepGlint as the testing set. Overlapping identities among these datasets are removed. IMDB is excluded from the testing set due to its massive number of wrong ID labels. For the dataset without certain demographic labels, we simply use the corresponding models to predict the labels. We report the Area Under the Curve (AUC) of the Receiver Operating Characteristics (ROC). We define the degree of bias, termed *biasness*, as the standard deviation of performance across cohorts.

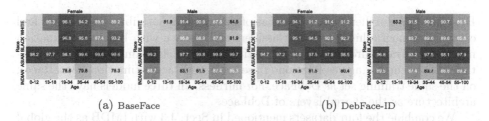

(a) BaseFace   (b) DebFace-ID

Fig. 3. Face Verification AUC (%) on each demographic cohort. The cohorts are chosen based on the three attributes, *i.e.*, gender, age, and race. To fit the results into a 2D plot, we show the performance of male and female separately. Due to the limited number of face images in some cohorts, their results are gray cells.

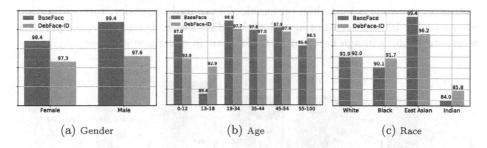

(a) Gender   (b) Age   (c) Race

Fig. 4. The overall performance of face verification AUC (%) on gender, age, and race.

Figure 3 shows the face verification results of BaseFace and DebFace-ID on each cohort. That is, for a particular face representation (*e.g.*, DebFace-ID), we report its AUC on each cohort by putting the number in the corresponding cell.

---

[3] To clarify, we consider two race groups, Black and White; and two ethnicity groups, East Asian and Indian. The word race denotes both race and ethnicity in this paper.

From these heatmaps, we observe that both DebFace-ID and BaseFace present bias in face verification, where the performance on some cohorts are significantly worse, especially the cohorts of Indian female and elderly people. Compared to BaseFace, DebFace-ID suggests less bias and the difference of AUC is smaller, where the heatmap exhibits smoother edges. Figure 4 shows the performance of face verification on 12 demographic cohorts. Both DebFace-ID and BaseFace present similar relative accuracies across cohorts. For example, both algorithms perform worse on the younger age cohorts than on adults; and the performance on the Indian is significantly lower than on the other races. DebFace-ID decreases the bias by gaining discriminative face features for cohorts with less images in spite of the reduction in the performance on cohorts with more samples.

4.4   De-Biasing Demographic Attribute Estimation

Baseline: We further explore the bias of demographic attribute estimation and compare demographic attribute classifiers of DebFace with baseline estimation models. We train three demographic estimation models, namely, gender estimation (BaseGender), age estimation (BaseAge), and race estimation (BaseRace), on the same training set as DebFace. For fairness, all three models have the same architecture as the shared layers of DebFace.

We combine the four datasets mentioned in Sect. 4.3 with IMDB as the global testing set. As all demographic estimations are treated as classification problems, the classification accuracy is used as the performance metric. As shown in Fig. 5, all demographic attribute estimations present significant bias. For gender estimation, both algorithms perform worse on the White and Black cohorts than on East Asian and Indian. In addition, the performance on young children is significantly worse than on adults. In general, the race estimation models perform better on the male cohort than on female. Compared to gender, race estimation shows higher bias in terms of age. Both baseline methods and DebFace perform worse on cohorts in age between 13 to 44 than in other age groups.

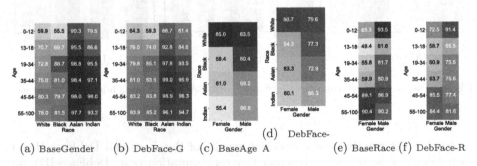

(a) BaseGender   (b) DebFace-G   (c) BaseAge A   (d) DebFace-   (e) BaseRace (f) DebFace-R

Fig. 5. Classification accuracy (%) of demographic attribute estimations on faces of different cohorts, by DebFace and the baselines. For simplicity, we use DebFace-G, DebFace-A, and DebFace-R to represent the gender, age, and race classifier of DebFace.

Table 1. Biasness of face recognition and demographic attribute estimation.

Method	Face verification				Demographic estimation		
	All	Gender	Age	Race	Gender	Age	Race
Baseline	6.83	0.50	3.13	5.49	12.38	10.83	14.58
DebFace	**5.07**	**0.15**	**1.83**	**3.70**	**10.22**	**7.61**	**10.00**

Similar to race, age estimation still achieves better performance on male than on female. Moreover, the white cohort shows dominant advantages over other races in age estimation. In spite of the existing bias in demographic attribute estimations, the proposed DebFace is still able to mitigate bias derived from algorithms. Compared to Fig. 5a, e, c, cells in Fig. 5b, f, d present more uniform colors. We summarize the biasness of DebFace and baseline models for both face recognition and demographic attribute estimations in Table 1. In general, we observe DebFace substantially reduces biasness for both tasks. For the task with larger biasness, the reduction of biasness is larger.

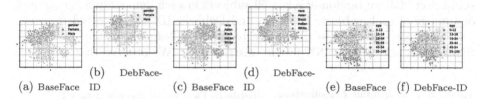

(a) BaseFace (b) DebFace-ID (c) BaseFace (d) DebFace-ID (e) BaseFace (f) DebFace-ID

Fig. 6. The distribution of face identity representations of BaseFace and DebFace. Both collections of feature vectors are extracted from images of the same dataset. Different colors and shapes represent different demographic attributes. Zoom in for details.

Fig. 7. Reconstructed Images using Face and Demographic Representations. The first row is the original face images. From the second row to the bottom, the face images are reconstructed from 2) BaseFace; 3) DebFace-ID; 4) DebFace-G; 5) DebFace-R; 6) DebFace-A. Zoom in for details.

4.5   Analysis of Disentanglement

We notice that DebFace still suffers unequal performance in different demographic groups. It is because there are other latent variables besides the demographics, such as image quality or capture conditions that could lead to biased performance. Such variables are difficult to control in pre-collected large face datasets. In the framework of DebFace, it is also related to the degree of feature disentanglement. A fully disentangling is supposed to completely remove the factors of bias from demographic information. To illustrate the feature disentanglement of DebFace, we show the demographic discriminative ability of face representations by using these features to estimate gender, age, and race. Specifically, we first extract identity features of images from the testing set in Sect. 4.1 and split them into training and testing sets. Given demographic labels, the face features are fed into a two-layer fully-connected network, learning to classify one of the demographic attributes. Table 2 reports the demographic classification accuracy on the testing set. For all three demographic estimations, DebFace-ID presents much lower accuracies than BaseFace, indicating the decline of demographic information in DebFace-ID. We also plot the distribution of identity representations in the feature space of BaseFace and DebFace-ID. From the testing set in Sect. 4.3, we randomly select 50 subjects in each demographic group and one image of each subject. BaseFace and DebFace-ID are extracted from the selected image set and are then projected from 512-dim to 2-dim by T-SNE. Figure 6 shows their T-SNE feature distributions. We observe that BaseFace

Table 2. Demographic classification accuracy (%) by face features.

Method	Gender	Race	Age
BaseFace	95.27	89.82	78.14
DebFace-ID	73.36	61.79	49.91

Table 3. Face verification accuracy (%) on RFW dataset.

Method	White	Black	Asian	Indian	Biasness
[56]	96.27	94.68	94.82	95.00	0.93
DebFace-ID	95.95	93.67	94.33	94.78	0.83

Table 4. Verification Performance on LFW, IJB-A, and IJB-C.

Method	LFW (%)	Method	IJB-A (%) 0.1% FAR	IJB-C @ FAR (%) 0.001%	0.01%	0.1%
DeepFace+ [47]	97.35	Yin et al. [61]	73.9 ± 4.2	–	–	69.3
CosFace [55]	99.73	Cao et al. [7]	90.4 ± 1.4	74.7	84.0	91.0
ArcFace [14]	99.83	Multicolumn [59]	92.0 ± 1.3	77.1	86.2	92.7
PFE [46]	99.82	PFE [46]	95.3 ± 0.9	89.6	93.3	95.5
BaseFace	99.38	BaseFace	90.2 ± 1.1	80.2	88.0	92.9
DebFace-ID	98.97	DebFace-ID	87.6 ± 0.9	82.0	88.1	89.5
DemoID	99.50	DemoID	92.2 ± 0.8	83.2	89.4	92.9

presents clear demographic clusters, while the demographic clusters of DebFace-ID, as a result of disentanglement, mostly overlap with each other.

To visualize the disentangled feature representations of DebFace, we train a decoder that reconstructs face images from the representations. Four face decoders are trained separately for each disentangled component, i.e., gender, age, race, and ID. In addition, we train another decoder to reconstruct faces from BaseFace for comparison. As shown in Fig. 7, both BaseFace and DebFace-ID maintain the identify features of the original faces, while DebFace-ID presents less demographic characteristics. No race or age, but gender features can be observed on faces reconstructed from DebFace-G. Meanwhile, we can still recognize race and age attributes on faces generated from DebFace-R and DebFace-A.

4.6 Face Verification on Public Testing Datasets

We report the performance of three different settings, using 1) BaseFace, the same baseline in Sect. 4.3, 2) DebFace-ID, and 3) the fused representation DemoID. Table 4 reports face verification results on three public benchmarks: LFW, IJB-A, and IJB-C. On LFW, DemoID outperforms BaseFace while maintaining similar accuracy compared to SOTA algorithms. On IJB-A/C, DemoID outperforms all prior works except PFE [46]. Although DebFace-ID shows lower discrimination, TAR at lower FAR on IJB-C is higher than that of BaseFace. To evaluate DebFace on a racially balanced testing dataset RFW [57] and compare with the work [56], we train a DebFace model on BUPT-Balancedface [56] dataset. The new model is trained to reduce racial bias by disentangling ID and race. Table 3 reports the verification results on RFW. While DebFace-ID gives a slightly lower face verification accuracy, it improves the biasness over [56].

We observe that DebFace-ID is less discriminative than BaseFace, or DemoID, since demographics are essential components of face features. To understand the deterioration of DebFace, we analyse the effect of demographic heterogeneity on face verification by showing the tendency for one demographic group to experience a false accept error relative to another group. For any two demographic cohorts, we check the number of falsely accepted pairs that are from different groups at 1% FAR. Figure 8 shows the percentage of such falsely accepted demographic-heterogeneous pairs. Compared to BaseFace, DebFace exhibits more cross-demographic pairs that are falsely accepted, resulting in the performance decline on demographically biased datasets. Due to the demographic information reduction, DebFace-ID is more susceptible to errors between demographic groups. In the sense of de-biasing, it is preferable to decouple demographic information from identity features. However, if we prefer to maintain the overall performance across all demographics, we can still aggregate all the relevant information. It is an application-dependent trade-off between accuracy and de-biasing. DebFace balances the accuracy vs. bias trade-off by generating both debiased identity and debiased demographic representations, which may be aggregated into DemoID if bias is less of a concern.

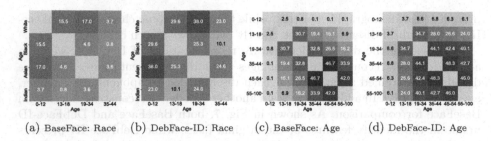

(a) BaseFace: Race   (b) DebFace-ID: Race   (c) BaseFace: Age   (d) DebFace-ID: Age

Fig. 8. The percentage of false accepted cross race or age pairs at 1% FAR.

5   Conclusion

We present a de-biasing face recognition network (DebFace) to mitigate demographic bias in face recognition. DebFace adversarially learns the disentangled representation for gender, race, and age estimation, and face recognition simultaneously. We empirically demonstrate that DebFace can not only reduce bias in face recognition but in demographic attribute estimation as well. Our future work will explore an aggregation scheme to combine race, gender, and identity without introducing algorithmic and dataset bias.

Acknowledgement. This work is supported by U.S. Department of Commerce (#60NANB19D154), National Institute of Standards and Technology. The authors thank reviewers, area chairs, Dr. John J. Howard, and Dr. Yevgeniy Sirotin for offering constructive comments.

References

1. https://yanweifu.github.io/FG_NET_data
2. http://trillionpairs.deepglint.com/overview
3. Abadi, M., et al.: Deep learning with differential privacy. In: Proceedings of the ACM SIGSAC Conference on Computer and Communications Security, pp. 308–318 (2016)
4. Amini, A., Soleimany, A., Schwarting, W., Bhatia, S., Rus, D.: Uncovering and mitigating algorithmic bias through learned latent structure. In: AAAI/ACM Conference on AI, Ethics, and Society (2019)
5. Bolukbasi, T., Chang, K.W., Zou, J.Y., Saligrama, V., Kalai, A.T.: Man is to computer programmer as woman is to homemaker? debiasing word embeddings. In: Advances in Neural Information Processing Systems, pp. 4349–4357 (2016)
6. Cao, K., Wei, C., Gaidon, A., Arechiga, N., Ma, T.: Learning imbalanced datasets with label-distribution-aware margin loss. arXiv preprint arXiv:1906.07413 (2019)
7. Cao, Q., Shen, L., Xie, W., Parkhi, O.M., Zisserman, A.: Vggface2: a dataset for recognising faces across pose and age. In: IEEE International Conference on Automatic Face & Gesture Recognition. IEEE (2018)
8. Chawla, N.V., Bowyer, K.W., Hall, L.O., Kegelmeyer, W.P.: Smote: synthetic minority over-sampling technique. J. Artif. Intell. Res. **16**, 321–357 (2002)

9. Chen, B.C., Chen, C.S., Hsu, W.H.: Cross-age reference coding for age-invariant face recognition and retrieval. In: ECCV (2014)
10. Cheng, J., Li, Y., Wang, J., Yu, L., Wang, S.: Exploiting effective facial patches for robust gender recognition. Tsinghua Sci. Technol. **24**(3), 333–345 (2019)
11. Cook, C.M., Howard, J.J., Sirotin, Y.B., Tipton, J.L., Vemury, A.R.: Demographic effects in facial recognition and their dependence on image acquisition: an evaluation of eleven commercial systems. IEEE Trans. Biometrics Behav. Identity Sci. **1**, 32–41 (2019)
12. Cui, Y., Jia, M., Lin, T.Y., Song, Y., Belongie, S.: Class-balanced loss based on effective number of samples. In: CVPR (2019)
13. Deb, D., Best-Rowden, L., Jain, A.K.: Face recognition performance under aging. In: CVPRW (2017)
14. Deng, J., Guo, J., Xue, N., Zafeiriou, S.: ArcFace: additive angular margin loss for deep face recognition. In: CVPR (2019)
15. Dietterich, T.G., Kong, E.B.: Machine learning bias, statistical bias, and statistical variance of decision tree algorithms. Tech. rep. (1995)
16. Drummond, C., Holte, R.C., et al.: C4. 5, class imbalance, and cost sensitivity: why under-sampling beats over-sampling. In: Workshop on Learning from Imbalanced Datasets II. Citeseer (2003)
17. Fredrikson, M., Jha, S., Ristenpart, T.: Model inversion attacks that exploit confidence information and basic countermeasures. In: The 22nd ACM SIGSAC (2015)
18. Goodfellow, I., et al.: Generative adversarial nets. In: NIPS (2014)
19. Guo, Y., Zhang, L., Hu, Y., He, X., Gao, J.: MS-Celeb-1M: a dataset and benchmark for large-scale face recognition. In: Leibe, B., Matas, J., Sebe, N., Welling, M. (eds.) ECCV 2016. LNCS, vol. 9907, pp. 87–102. Springer, Cham (2016). https://doi.org/10.1007/978-3-319-46487-9_6
20. Han, H., A, K.J., Shan, S., Chen, X.: Heterogeneous face attribute estimation: a deep multi-task learning approach. IEEE Trans. Pattern Anal. Mach. Intelli. PP(99), 1–1 (2017)
21. Hayat, M., Khan, S., Zamir, W., Shen, J., Shao, L.: Max-margin class imbalanced learning with Gaussian affinity. arXiv preprint arXiv:1901.07711 (2019)
22. He, K., Zhang, X., Ren, S., Sun, J.: Deep residual learning for image recognition. In: CVPR (2016)
23. Howard, J., Sirotin, Y., Vemury, A.: The effect of broad and specific demographic homogeneity on the imposter distributions and false match rates in face recognition algorithm performance. In: IEEE BTAS (2019)
24. Huang, C., Li, Y., Chen, C.L., Tang, X.: Deep imbalanced learning for face recognition and attribute prediction. IEEE Trans Pattern Anal. Mach. Intell. (2019)
25. Huang, G.B., Mattar, M., Berg, T., Learned-Miller, E.: Labeled faces in the wild: a database for studying face recognition in unconstrained environments (2008)
26. Jourabloo, A., Yin, X., Liu, X.: Attribute preserved face de-identification. In: ICB (2015)
27. Khan, S., Hayat, M., Zamir, S.W., Shen, J., Shao, L.: Striking the right balance with uncertainty. In: CVPR (2019)
28. Kim, H., Mnih, A.: Disentangling by factorising. arXiv preprint arXiv:1802.05983 (2018)
29. Klare, B.F., Burge, M.J., Klontz, J.C., Bruegge, R.W.V., Jain, A.K.: Face recognition performance: Role of demographic information. IEEE Trans. Inform. Forensics Secur. **7**(6), 1789–1801 (2012)
30. Klare, B.F., et al.: Pushing the frontiers of unconstrained face detection and recognition: Iarpa janus benchmark a. In: CVPR (2015)

31. Liu, F., Zeng, D., Zhao, Q., Liu, X.: Disentangling features in 3D face shapes for joint face reconstruction and recognition. In: CVPR (2018)
32. Liu, Y., Wang, Z., Jin, H., Wassell, I.: Multi-task adversarial network for disentangled feature learning. In: CVPR (2018)
33. Liu, Y.H., Chen, Y.T.: Face recognition using total margin-based adaptive fuzzy support vector machines. IEEE Trans. Neural Networks 18(1), 178–192 (2007)
34. Liu, Y., Wei, F., Shao, J., Sheng, L., Yan, J., Wang, X.: Exploring disentangled feature representation beyond face identification. In: CVPR (2018)
35. Locatello, F., Bauer, S., Lucic, M., Gelly, S., Schölkopf, B., Bachem, O.: Challenging common assumptions in the unsupervised learning of disentangled representations. arXiv preprint arXiv:1811.12359 (2018)
36. Long, M., Cao, Z., Wang, J., Jordan, M.I.: Conditional adversarial domain adaptation. In: NIPS (2018)
37. Maze, B., et al.: Iarpa janus benchmark-c: face dataset and protocol. In: 2018 ICB (2018)
38. Moschoglou, S., Papaioannou, A., Sagonas, C., Deng, J., Kotsia, I., Zafeiriou, S.: AgeDB: the first manually collected, in-the-wild age database. In: CVPRW (2017)
39. Mullick, S.S., Datta, S., Das, S.: Generative adversarial minority oversampling. arXiv preprint arXiv:1903.09730 (2019)
40. Narayanaswamy, S., et al.: Learning disentangled representations with semi-supervised deep generative models. In: NIPS (2017)
41. Niu, Z., Zhou, M., Wang, L., Gao, X., Hua, G.: Ordinal regression with multiple output CNN for age estimation. In: CVPR (2016)
42. Patrick Grother, M.N., Hanaoka, K.: Face recognition vendor test (FRVT) part 3: demographic effects. Tech. rep., National Institute of Standards and Technology (2019)
43. Rothe, R., Timofte, R., Van Gool, L.: Deep expectation of real and apparent age from a single image without facial landmarks. IJCV 126(2–4), 144–157 (2018)
44. Schmidhuber, J.: Learning factorial codes by predictability minimization. Neural Comput. 4(6), 863–879 (1992)
45. Setty, S., et al.: Indian movie face database: a benchmark for face recognition under wide variations. In: NCVPRIPG (2013)
46. Shi, Y., Jain, A.K., Kalka, N.D.: Probabilistic face embeddings. arXiv preprint arXiv:1904.09658 (2019)
47. Taigman, Y., Yang, M., Ranzato, M., Wolf, L.: Deepface: Closing the gap to human-level performance in face verification. In: CVPR (2014)
48. Tao, C., Lv, F., Duan, L., Wu, M.: Minimax entropy network: learning category-invariant features for domain adaptation. arXiv preprint arXiv:1904.09601 (2019)
49. Torralba, A., Efros, A.A., et al.: Unbiased look at dataset bias. In: CVPR (2011)
50. Tran, L., Yin, X., Liu, X.: Disentangled representation learning GAN for pose-invariant face recognition. In: CVPR (2017)
51. Tran, L., Yin, X., Liu, X.: Representation learning by rotating your faces. IEEE Trans. Pattern Anal. Mach. Intell. 41(12), 3007–3021 (2019)
52. Tzeng, E., Hoffman, J., Darrell, T., Saenko, K.: Simultaneous deep transfer across domains and tasks. In: CVPR (2015)
53. Tzeng, E., Hoffman, J., Saenko, K., Darrell, T.: Adversarial discriminative domain adaptation. In: CVPR (2017)
54. Wang, F., Cheng, J., Liu, W., Liu, H.: Additive margin softmax for face verification. IEEE Signal Process. Lett. 25(7), 926–930 (2018)
55. Wang, H., et al.: CosFace: large margin cosine loss for deep face recognition. In: CVPR (2018)

56. Wang, M., Deng, W.: Mitigating bias in face recognition using skewness-aware reinforcement learning. In: CVPR (2020)
57. Wang, M., Deng, W., Hu, J., Tao, X., Huang, Y.: Racial faces in the wild: reducing racial bias by information maximization adaptation network. In: ICCV (2019)
58. Wang, P., Su, F., Zhao, Z., Guo, Y., Zhao, Y., Zhuang, B.: Deep class-skewed learning for face recognition. Neurocomputing **363**, 35–45 (2019)
59. Xie, W., Zisserman, A.: Multicolumn networks for face recognition. arXiv preprint arXiv:1807.09192 (2018)
60. Yin, B., Tran, L., Li, H., Shen, X., Liu, X.: Towards interpretable face recognition. In: ICCV (2019)
61. Yin, X., Liu, X.: Multi-task convolutional neural network for pose-invariant face recognition. IEEE Trans. Image Process. **27**(2), 964–975 (2017)
62. Yin, X., Yu, X., Sohn, K., Liu, X., Chandraker, M.: Towards large-pose face frontalization in the wild. In: ICCV (2017)
63. Yin, X., Yu, X., Sohn, K., Liu, X., Chandraker, M.: Feature transfer learning for face recognition with under-represented data. In: CVPR (2019)
64. Zhang, K., Zhang, Z., Li, Z., Qiao, Y.: Joint face detection and alignment using multitask cascaded convolutional networks. IEEE Signal Process. Lett. **23**(10), 1499–1503 (2016)
65. Zhang, X., Fang, Z., Wen, Y., Li, Z., Qiao, Y.: Range loss for deep face recognition with long-tailed training data. In: CVPR (2017)
66. Zhang, Y., Zhou, Z.H.: Cost-sensitive face recognition. IEEE Trans. Pattern Anal. Mach. Intell. **32**(10), 1758–1769 (2009)
67. Zhang, Z., Song, Y., Qi, H.: Age progression/regression by conditional adversarial autoencoder. In: CVPR. IEEE (2017)
68. Zhang, Z., et al.: Gait recognition via disentangled representation learning. In: CVPR (2019)

Regularized Loss for Weakly Supervised Single Class Semantic Segmentation

Olga Veksler[✉]

University of Waterloo, Waterloo, ON N2L3G1, Canada
oveksler@uwaterloo.ca, https://cs.uwaterloo.ca/~oveksler/

Abstract. Fully supervised semantic segmentation is highly successful, but obtaining dense ground truth is expensive. Thus there is an increasing interest in weakly supervised approaches. We propose a new weakly supervised method for training CNNs to segment an object of a single class of interest. Instead of ground truth, we guide training with a regularized loss function. Regularized loss models prior knowledge about the likely object shape properties and thus guides segmentation towards the more plausible shapes. Training CNNs with regularized loss is difficult. We develop an annealing strategy that is crucial for successful training. The advantage of our method is simplicity: we use standard CNN architectures and intuitive and computationally efficient loss function. Furthermore, we apply the same loss function for any task/dataset, without any tailoring. We first evaluate our approach for salient object segmentation and co-segmentation. These tasks naturally involve one object class of interest. In some cases, our results are only a few points of standard performance measure behind those obtained training the same CNN with full supervision, and state-of-the art results in weakly supervised setting. Then we adapt our approach to weakly supervised multi-class semantic segmentation and obtain state-of-the-art results.

1 Introduction

Convolutional Neural Networks (CNNs) [20,22] lead to a breakthrough in semantic segmentation [4,5,11,35,57]. Each year new architectures push the state of the art in on fully labeled datasets [6,39,49]. However, labeling datasets is expensive. Thus there is an interest in segmentation without fully labeled data.

We consider a new approach to train CNNs in the absence of pixel precise ground truth. Our approach is inspired by [40,41] who demonstrate the utility of regularized loss[1] for semantic segmentation with partially labeled data. Regularized losses are used in computer vision in the context of Conditional Random

---

[1] In the context of CNNs, regularization is a term often used to refer to norm regularization on network weights [12], or other techniques to prevent overfitting. In this work, regularized loss refers to the loss function on the output of CNN.

Electronic supplementary material The online version of this chapter (https://doi.org/10.1007/978-3-030-58526-6_21) contains supplementary material, which is available to authorized users.

A. Vedaldi et al. (Eds.): ECCV 2020, LNCS 12374, pp. 348–365, 2020.
https://doi.org/10.1007/978-3-030-58526-6_21

Fields (CRFs) [21,25] for modelling the likely properties of an object, such as short boundary [2], connectivity [45], shape convexity [13], etc.

Our main idea is to supervise training with a regularized loss. Instead of encouraging CNN output to match to ground truth with cross entropy loss, we encourage CNN output to have desired properties with regularized loss. We use *sparse-CRF* loss [2] that encourages shorter object boundaries aligning to intensity edges. We show that sparse-CRF has a high correlation with segmentation accuracy, and thus is a good candidate to use for training. While it is possible to design more complex regularized losses, it is interesting to evaluate the utility of sparse-CRF loss, widely used in classical computer vision, now for CNN training.

Although regularized loss has been used before [40,41] for settings with partially labeled pixels, our task is significantly more difficult, as we assume no labeled pixels. Consider that in [41] the difference in performance with and without regularized loss is only about 4%, i.e. most of the learning is done by cross entropy on partially labeled pixels. To enable learning in absence of pixel labels, we make an assumption that the training dataset contains a single object class of interest, making our approach a weakly supervised single class segmentation.

Regularized loss by itself is insufficient, since an empty object would be a trivial optimal solution. We need additional priors implemented as "helper" loss functions. Assuming a single object class lets us use the *positional* loss. Positional loss encourages the image border to be assigned to the background, and the image center to the foreground. This prior is reasonable for single class datasets. We also use *volumetric loss* [46] to counteract the well known shrinking bias of sparse-CRF loss. It penalizes segmentations that are either too large or too small in size. Both volumetric and positional loss have a significant weight only in the initial training stage, to push CNN training in the right direction. Their weight is significantly reduced after the initial stage, to allow segmentation of objects that are not centered and have a diverse range of sizes.

Training CNN with regularized loss is hard. We develop an *annealing* strategy [17] that is essential for successful training. The goal of our annealing is different from the traditional one. Our annealing helps CNN to first identify an easy hypothesis for the appearance of the object, and then to slowly refine it.

Training only on the class of interest can result in a CNN that has a good response to similar ("confuser") classes. We can add "confuser" class images to the training set as negative examples, and add a loss function that encourages labeling them as background. This is still in the realm of weakly supervised segmentation since negative examples need only an image-level "negative" tag.

Our method results in a simple end-to-end trainable CNN system. We use standard architectures and just add regularized, volumetric, and positional loss instead of cross entropy. We do not adapt our loss function for any task. Our loss function is easy to interpret, intuitive, and efficient.

First, we evaluate our approach on the salient object segmentation [27] and co-segmentation [32]. These tasks naturally involve a single class of interest. In many cases, our CNN, trained without ground truth is only a few points behind the same CNN trained with full ground truth. For some saliency and cosegmentation datasets we obtain state-of-the art weakly supervised results.

Next we extend our approach to weakly supervised semantic segmentation in a naive fashion: we train on datasets containing a single semantic class and use the other classes as negative examples. The resulting pseudo-ground truth is then used to train a standard multi-class semantic segmentation CNN. We obtain state-of-the art results on Pascal VOC 2012 benchmark in weakly supervised setting. Our code for single class segmentation is publicly available[2].

2   Related Work

In this section, we review prior work on CNN segmentation without full supervision for salient object, co-segmentation and semantic segmentation.

There is some prior works for CNN based salient object segmentation without human annotation [28,55,56]. All three are based on exploiting multiple saliency segmentation methods that are not based on machine learning, i.e. "weak methods". These approaches are very specific to saliency segmentation, whereas our method is generic an applies to any single object segmentation task.

There is a prior work on unsupervised co-segmentation in [14]. Their co-attention loss function is expensive to compute, it relies on all pairs of images in the training dataset. Again, the advantage of our method is generic architecture and generic computationally efficient loss function.

The prior work on weakly and partially supervised semantic segmentation can be divided in two groups, depending on the supervision type. In the first group are weakly supervised approaches based on image-level tags [1,18,23,29, 30]. Most successful of these methods use CAM (class activation maps) [34, 58] to produce pseudo-ground truth which is then used to train a semantic segmentation CNN. The drawback of these approaches is their reliance on CAMs, which can vary in accuracy. Our approach is does not require CAMs.

In the second group are methods based on imprecise labeling (i.e. bounding boxes) or partial labeling of the data. Approaches in [7,15,29,51] assume a bounding box is placed around semantic objects. Approaches in [26,40,41,44,51] are based on partially labeled data or "scribbles". The earlier of these approaches first use scribbles or boxes to generate full pseudo ground truth, which is then used to train CNNs. The later approaches [40,41] argue against using pseudo ground truth and show more accurate results by using cross entropy loss only on the labeled pixels and regularized loss for the rest of the pixels. The advantage of our approach is that it does not require any labeling or annotation of the data. Note that our approach is different from training with bounding boxes. A bounding box annotation is typically placed relatively tight around the object. In the datasets we use, objects vary in size significantly. Thus if we regard image as a "box", it is anything but tight for the majority of dataset images.

---

[2] https://github.com/morduspordus/SingleClassRL

3   Our Approach

We now describe our approach. We explain regularized losses in Sect. 3.1, evaluate correlation between regularized loss and segmentation accuracy in Sect. 3.2, give our complete loss function in Sect. 3.3, and describe training strategy in Sect. 3.4.

3.1   Regularized Losses

A variety of regularized loss functions are used in CRFs [21,25]. The main task in CRF is to assign some label x_p, for example, a class label, to each image pixel p. Here we assume $x_p \in \{0,1\}$. If $x_p = 1$ then pixel p is assigned the object class, and if $x_p = 0$ then p is assigned the background. A regularized loss is computed for $\boldsymbol{x} = (x_p \mid p \in \mathcal{P})$, an assignment of labels to all pixels in $\mathcal{P}$, where $\mathcal{P}$ is the set of image pixels. We use color edge sensitive Potts [2] regularized loss:

$$L_{reg}(\boldsymbol{x}) = \sum_{p,q \in \mathcal{P}} w_{pq} \cdot [x_p \neq x_q], \tag{1}$$

The pairwise term $w_{pq} \cdot [x_p \neq x_q]$ in Eq. (1) gives a penalty of w_{pq} whenever pixels p, q are not assigned to the same label. To align object boundaries to image edges, we use:

$$w_{pq} = e^{-\frac{||C_p - C_q||^2}{2\sigma^2}}, \tag{2}$$

where σ controls "edge detection". Decreasing σ leads to an increased number of weights w_{pq} that are small enough to be considered a color edge.

The sum in Eq. (1) is over all neighboring pixels p, q. We assume a 4-connected neighborhood. The loss function in Eq. (1) with 4-connected neighborhood is called sparse-CRF. Sparse-CRF regularizes a labeling by encouraging shorter object boundaries that align to image edges [3].

We also test dense-CRF regularized loss from [19], which has the same form as in Eq. (1) but the summation is over all (thus dense) pixel pairs and w_{pq} is a Gaussian weight that depends both on the color difference and the distance between pixels p, q. Unlike sparse-CRF, the regularization properties of dense-CRF are not well understood, although there is some work [43] that analyzes the properties of an approximation to a dense-CRF.

In the context of CNNs, the output label x_p is no longer discrete but in continuous range $(0, 1)$, assuming last layer is softmax. We use the absolute value relaxation of Eq. (1):

$$L_{reg}^{abs}(\boldsymbol{x}) = \sum_{p,q \in \mathcal{P}} w_{pq} \cdot |x_p - x_q|. \tag{3}$$

3.2   Loss and Segmentation Accuracy Correlation

If regularized loss is a good criterion for single class segmentation, then the accuracy of segmentation should be negatively correlated with the value of the loss. While this may not be a sufficient condition, it is a necessary one. We measure correlation experimentally, using four single-object datasets: OxfordPet [42], MSRA-B [27], ECSSD [52], and DUT-OMRON [53]. The first dataset has ground truth segmentation for two object classes: *cat* and *dog*. We merge them into a single class *pet*. The last three datasets are for salient object segmentation and contain one class. We use F-measure as the accuracy metric.

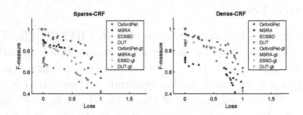

Fig. 1. Scatter plots of regularized loss vs. F-measure on four datasets (different colors): sparse-CRF (left), dense-CRF (right). See text for details.

Table 1. Correlation coefficients between F-measure and regularized loss for sparse-CRF and dense-CRF.

	OxfordPet	MSRA	ECSSD	DUT-O
Sparse-CRF	−0.862	−0.895	−0.956	−0.957
Dense-CRF	−0.830	−0.758	−0.772	−0.896

To measure correlation between regularized loss and F-measure, for each dataset, we need several segmentation sets of varying accuracy. We are interested in segmentations that can be learned with a CNN, not just any random segmentations, since we assume a dataset has a concept that can be learned with a CNN. Therefore we train a CNN with ground truth and obtain several early/intermediate segmentation results in addition to the final one. We use 20 time steps to obtain 20 different accuracy segmentation sets for each dataset.

We average sparse-CRF losses and F-measures over segmentations of each dataset, separately for each time step. We do the same for the dense-CRF. The scatter plots of regularized loss vs. F-measure are in Fig. 1. Each dataset is in different color. We linearly normalize (no effect on correlation) the loss into range $(0, 1)$ for visualization. The correlation coefficients for sparse-CRF and dense-CRF are in Table 1. Both losses show significant negative correlation, however sparse-CRF correlation is better for all datasets. Given high negative correlation, the task of performing CNN training with regularized loss is promising.

We also measure regularized loss of ground truth, shown with asterisks in Fig. 1. For dense-CRF only one asterisk is visible because the loss value is almost the same for all datasets. Except OxfordPet, for all datasets the ground truth has the lowest value of the regularized loss. For the OxfordPet, the ground truth has a slightly higher value than the best solution obtained through training CNN.

3.3 Complete Loss Function

Sparse-CRF has a shrinking bias [46], it favors segments with a shorter boundary. To counteract, we add volumetric loss, also frequently used in MRFs [46]. Let x be an output of the network, and let $\bar{x} = \frac{1}{|\mathcal{P}|} \sum_{p \in \mathcal{P}} x_p$, i.e. normalized object size. Our volumetric loss is defined for a batch of m images with outputs $x_1, \ldots, x_m$:

$$L_v(x_1, \ldots, x_m) = \left(\frac{1}{m}\sum_i \bar{x}_i - 0.5\right)^2 + \lambda_s \sum_i \left[\bar{x}_i < 0.15\right] \cdot \left(\bar{x}_i - 0.15\right)^2, \quad (4)$$

The first sum in Eq. (4) encourages the average size of the object in a batch to be 0.5, i.e. roughly half of the image size. We average object size over a batch to allow for a large range of object sizes in a batch: some objects in can be small, some can be large, but the average size is expected to be 0.5. We could encourage any object/background size ratio by replacing 0.5 appropriately. However, we found no need for more precise modeling as volumetric loss is essential only in the first, *annealing* stage of training, see Sect. 3.4. Batch size is set to $m = 16$.

The second sum in Eq. (4) encourages each object inside the batch to have normalized size of at least 0.15. This is a conservative measure of the minimum object size to prevent collapse to an empty object. Here $[\cdot]$ is 1 if the argument is true and 0 otherwise. The weight λ_s balancing these two sums is set to 5 in all experiments. In practice, this term is important only during the second training stage, when regularization loss has a much higher weight than volumetric loss.

In the beginning of the training, it is important to guide the network in the right direction. For a dataset that contains only one object class of interest, for many images the border pixels do not contain the object, and the center of the image contains the object. We incorporate this prior through *positional* loss. Let $\mathcal{B}$ be the set of pixels on the image border of width w, and $\mathcal{C}$ be the set of image pixels in the center box of side size c. We formulate positional loss $L_p(x)$ as:

$$L_p(x) = \left(\frac{1}{|\mathcal{B}|}\sum_{p \in \mathcal{B}} x_p\right)^2 + \left(\frac{1}{|\mathcal{C}|}\sum_{p \in \mathcal{C}} x_p - 1\right)^2. \quad (5)$$

For all experiments, we set $w = c = 3$, so the number of effected pixels is small. However, positional loss is enough to push the network in the right direction in the beginning of training, seeking object segments that are preferably in the image center and do not overlap the border. Just as with volumetric loss, positional loss is important only in the *annealing* stage. In the *normal* stage, the weight of positional loss is low, thus allowing segmenting objects that overlap with the border and do not contain the central four pixels of the image.

Our complete loss function for training is a weighted sum of regularized, volumetric, and positional losses, and is applied to $\boldsymbol{x}$, the output of CNN[3]:

$$L(\boldsymbol{x}) = \lambda_{reg}L_{reg}(\boldsymbol{x}) + \lambda_v L_v(\boldsymbol{x}) + \lambda_p L_p(\boldsymbol{x}). \tag{6}$$

All parts of the loss in Eq. (6) are important. Training fails if any part is omitted. When we say we train with "regularized loss", we mean training with the complete loss in Eq. (6). Saying "train with regularized loss" emphasizes the largest important component of our complete loss function.

Sometimes we have images containing a class similar to the class of interest but not the class we want to segment. It helps to include these images as negative examples. Negative images are assumed not to contain the object of interest, and thus no ground truth is required for training. Therefore, including negative images is, again, a form of a weak supervision. If negative images are available, we apply the following *negative* loss to $\boldsymbol{x}$, the output of CNN on negative images:

$$L_{neg}(\boldsymbol{x}) = \lambda_{neg}\Big(\frac{1}{|\mathcal{P}|}\sum_{p\in\mathcal{P}} x_p - 1\Big)^2 \tag{7}$$

3.4 Training

Training CNN with the loss function in Eq. (6) is difficult. Out-of-the-box training produces results with either all pixels assigned to the foreground or background. We found that *annealing* [17] (i.e. slowly increasing) parameter σ in Eq. (2) is essential for obtaining good results. Thus there are two training stages in our approach: the *annealing* stage and the *normal* stage.

The annealing stage has $n = 15$ iterations during which parameter σ changes according to the following schedule, where i is the iteration number:

$$\sigma(i) = 0.05 + \frac{i}{n} * 0.1, \quad 1 \le i \le 15 \tag{8}$$

The other parameters are not changed and are set to: $\lambda_{reg} = \lambda_v = \lambda_p = 1$.

Traditionally, annealing is used to increase parameter λ_{reg}, increasing the difficulty of loss function optimization. The larger is λ_{reg}, the more difficult the loss function tends to be. We anneal parameter σ that controls edge sensitivity, and its annealing seems to serve a different purpose. Intuitively, annealing σ helps CNN to find an easy object hypothesis first, and then refine it.

Consider Fig. 2. The input image is shown on the left. In the middle, for several annealing iterations, we show the segmentation result together with w_{pq} weights (below the result). The lower weights w_{pq} are illustrated with darker intensities. The last column in the figure is the final result after the normal training stage, and the ground truth is below it. Consider the result at iteration one. The value of σ is low, and, therefore, there are many areas in the image with

---

[3] Note that our volumetric prior is actually defined on batches of images. However for the simplicity of notation, we write $L_v(\boldsymbol{x})$ in this equation

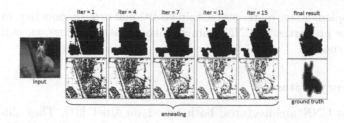

Fig. 2. Illustration of annealing. Left: input image. Middle: segmentation results for several iterations of annealing together with w_{pq} weights at these iterations, shown below each segmentation result; darker intensities correspond to lower w_{pq}. Right, top: the final segmentation after the normal training stage. Right, bottom: the ground truth, with "void" class in orange. (Color figure online)

small w_{pq}. At this first iteration, CNN learns to classify most of the areas with low w_{pq} as the object. This happens because placing object/background boundaries around pixels with lower w_{pq} is cheap. It also happens to be a reasonable start as objects usually have edges on their boundary, and often they have texture edges in the interior. Of course, many of the pixels included in the object mask in the first iteration are errors. As annealing proceeds, σ is raised, and there are fewer areas with small w_{pq} weights, making the old placements of boundaries on the weaker edges more expensive. Increasingly more erroneous object areas are discarded as CNN learns the common object appearance from the "surviving" areas. The final annealing result (iter $= 15$) is a crude approximation to the actual object, and it gets refined further during the normal training stage (last column, right), as the weight of regularized loss is increased significantly.

After the annealing stage, we have the *normal* training stage. The parameters are set to: $\lambda_{reg} = 100$, $\lambda_v = \lambda_p = 1$, $\sigma = 0.15$. Thus during normal training stage, regularized loss is the major contributing component. After annealing, we have a good approximate solution and can relax the penalty for objects that deviate widely from half of the image size, do not overlap the image center, and overlap image border. We do not set λ_v and λ_p to zero because then the optimal solution is a trivial one, everything assigned to the object (or background).

If negative loss in Eq. 7 is used, we set $\lambda_{neg} = 2$. Also, we first train in the annealing and normal stages without negative loss, and then train with the negative loss in the normal stage for an additional 200 epochs.

4   Experimental Results

This section is organized as follows. We start with saliency datasets and Oxford-Pet dataset in Sect. 4.1. We use these datasets to make choices about CNN architecture and training protocol and then use these choices for all the other experiments. In Sect. 4.2 we test our approach for cosegmentation. In Sect. 4.3 we extend our approach to weakly supervised multi-class semantic segmentation.

The supplementary materials contain additional experiments that experimentaly analyze properties of CNN trained with regularized loss, as well as other minor experiments.

4.1 Saliency Datasets

We use two CNN architectures, both based on Unet [31]. They differ in the module for feature extraction. The first one is based on MobileNetV2 [33] and the second on ResNeXt [50]. The feature extraction modules are pretrained on Imagenet [8]. The pretrained features are used in the encoder part Unet, and they stay fixed during training. We refer to these networks, respectively, as *UMobV2* and *UResNext*. *UResNext* is much bigger than *UMobV2*. For both networks, we add softmax as the last layer to convert the output to the range of $(0, 1)$. We train each network with regularized loss in Eq. (6) and with cross-entropy loss on ground truth for comparison. To distinguish between these training types, we add "gt" to the name of any network that is trained with ground truth, and "reg" if it is trained with regularized loss.

For the experiments in this section, we use three saliency datasets: MSRA-B [27], ECSSD [52], and DUT-OMRON [53]. We split MSRA-B into $3,000$ training and $2,000$ test images, ECSSD into 700 training and 300 test images, and DUT-OMRON into $3,678$ training and $1,490$ test images.

We also use OxfordPet [42] dataset that has segmentations of cats and dogs. To make it appropriate for single class segmentation, we combine the cat and dog classes into one *pet* class. OxfordPet has $3,680$ training and $3,669$ test images.

We use F-measure as an accuracy metric, $F = \frac{(1+\beta^2) precision \times recall}{\beta^2 \times precision + recall}$, with $\beta^2 = 0.3$.

Table 2. F-measure for all networks and datasets; for each network and dataset, we also show the gap between training with ground truth and regularized loss.

	OxfordPet	MSRA-B	ECSSD	DUT-O
UMobV2-reg	94.62	86.20	57.96	53.79
UMobV2-gt	96.07	89.35	84.78	80.86
gap	**1.45**	3.15	26.82	27.07
UResNext-reg	**94.77**	**88.36**	**73.47**	**64.48**
UResNext-gt	96.53	89.68	86.11	81.47
gap	1.76	**1.32**	**12.64**	**16.99**

F-measures are computed on the test fold. However, since our method does not use ground truth, the test and training folds have a similar F-measure.

For efficiency, we decrease resolution of all images on all experiments in this section to 128×128. We use Adam optimizer [16] with a fixed learning rate 0.001 and train all networks for 50 epochs.

The F-measures of *UMobV2* and *UResNext* on the OxfordPet, MSRA-B, ECSSD, and DUT-OMRON are in Table 2, both in the case of training with regularized loss and training with ground truth. We also give the gap between training with ground truth and regularized loss. Whether training with ground truth or regularized loss, *UResNext*, a bigger network, performs better than *UMobV2*, and has a smaller gap from the network trained with ground truth on 3 out of 4 datasets. OxfordPet dataset is the easiest to fit dataset.

Replacing the Annealing Stage. The purpose of the annealing training stage is to find a good start for the normal training stage. We found that instead of annealing, using weights from the same network trained on another dataset also gives a good start for the normal training stage. Our training consists of 15 annealing and 50 normal epochs. So skipping 15 epochs of annealing gives only a modest time saving. However, we found that transferring weights from a network trained on an easy-to-fit dataset can lead to a much better result. We use OxfordPet for weight transfer because it is the easiest dataset to fit. Specifically, we train a network on OxfordPet both in the annealing and normal training stage, and then use the resulting weights to start training in the normal stage (skipping annealing) on another dataset.

The results are summarized in Table 3. Most F-measures improve, in some cases by a large margin. *UResNext* is, again, the best, and is less than 3 points of F-measure behind the same network trained with ground truth on all datasets except DUT-OMRON. We also show the new gap between the same networks trained with ground truth and regularized loss. The new best gaps are in bold. For the rest of the experiments in this paper, we always start with the weights pretrained on OxfordPet, skipping the annealing stage.

Table 3. Replacing annealing stage by transferring the weights from OxfordPet.

	MSRA-B	ECSSD	DUT-O
UMobV2-reg	85.51	79.20	54.49
UMobV2-gt	89.35	84.78	80.86
gap	3.84	5.58	26.37
UResNext-reg	**88.64**	**83.49**	**73.42**
UResNext-gt	89.68	86.11	81.47
gap	**1.04**	**2.61**	**8.05**

Comparison to Saliency CNNs Trained Without Ground Truth. There is prior work that train saliency CNN without ground truth [28,55,56]. The F-measure comparison is in Table 4. For fairness of comparison, as done in [28, 55,56], we use the model trained on MSRA-B to test on ECSSD and DUT-OMRON.

Our method outperforms other methods on DUT-OMRON, and is not far from the best other methods on MSRA-B and ECSSD. Note that in addition, our

Table 4. Comparison to other saliency-CNNs that do not use ground truth for training.

	MSRA-B	ECSSD	DUT-O
[55]	NA	78.70	58.30
[56]	87.70	**87.83**	71.56
[28]	**90.31**	87.42	73.58
Ours *UMobV2-reg*	85.51	83.50	70.47
Ours *UResNext-reg*	88.64	86.10	**74.77**

method is generic and not tuned in architecture or loss function to the saliency segmentation task, where as all three methods in [55] are saliency specific.

4.2 Cosegmentation

The goal of cosegmentation [32] is to segment an object common to all images in a dataset. The setting of this problem is somewhat different from the setting we assume. We assume that the object of interest is the main subject of the image. For co-segmentation datasets, there maybe several salient objects of different classes, and the task is to find the object class that occurs in all images.

We use Pascal-VOC cosegmentation dataset [10] for evaluation, the most challenging cosegmentation benchmark. It has 20 separate subsets, one for each Pascal category. This dataset is particularly challenging for our approach because the number of images per class is only around 50, and our method needs more images for accurate training, compared to training with ground truth, see supplementary materials. We use UnetX network and train at resolution 256×256.

There is one prior works that trains CNN for cosegmentation without full supervision [14]. The comparison with our method is in Table 5. Our performance, marked as $ours_1$, is slightly inferior. However, our method is generic and unchanged between various applications/datasets. The method in [14] is handcrafted for co-segmentation, has a costly loss function that depends on pairs of images, and thus has quadratic complexity, which is infeasible for large datasets. Furthermore, in [14] they use post-processing with dense-CRF and do not report results without post-processing. We do not use any post-processing. For comparison, in this table, we also include the most recent prior work [24] on cosegmentation that does use full ground truth for training. It performs only slightly better than [14] and $ours_1$, despite using full supervision.

Next we add negative loss from Eq. 7. Namely, when training for, say, 'airplane' category, images in all the other categories are used as negative examples. Note that negative images can also have objects of class 'airplane', since the categories in cosegmentation dataset are not "pure". That is the category "person" has an object of type person in all examples, but may also have "airplane" objects in some examples. Still, negative loss improves results, see column $ours_2$ in Table 5. We now outperform [14]. Although we are using the same dataset

Table 5. Results on Pascal VOC Cosegmentation, evaluation metric is Jaccard. Here '*' indicates that a method uses full ground truth for training.

	[14]	[24]*	$ours_1$	$ours_2$	$ours_3$
Airplane	0.77	0.78	0.78	0.82	**0.83**
Bike	0.27	0.29	0.27	0.28	**0.32**
Bird	0.70	**0.71**	0.68	0.70	0.67
Boat	0.61	0.66	0.53	0.61	**0.68**
Bottle	0.58	0.58	0.58	0.61	**0.72**
Bus	0.79	0.82	0.76	0.84	**0.84**
Car	0.76	0.79	0.76	0.79	**0.82**
Cat	0.79	0.81	0.81	0.82	**0.86**
Chair	0.29	0.35	0.34	0.35	**0.38**
Cow	0.75	0.78	0.77	0.80	**0.83**
Dining table	**0.28**	0.26	0.15	0.16	0.21
Dog	0.63	0.65	0.70	0.70	**0.76**
Horse	0.66	0.78	0.75	0.77	**0.79**
Motorbike	0.65	0.69	0.69	0.71	**0.73**
Person	0.37	0.39	0.44	0.46	**0.60**
Potted plant	0.42	0.45	0.41	0.45	**0.60**
Sheep	0.75	0.77	0.77	0.80	**0.81**
Sofa	**0.67**	0.70	0.55	0.60	**0.67**
Train	0.68	0.73	0.68	0.75	**0.82**
Tv	0.51	**0.55**	0.33	0.40	0.50
Mean	0.60	0.63	0.59	0.62	**0.67**

as [14], this comparison is somewhat unfair since [14] does not use any loss term that depends on data across different object categories. Still, it is reasonable to assume that the user provides examples of images without the object of interest as negative examples. Figure 3 illustrates how negative loss improves performance.

Cosegmentation dataset has a small number of images, a problem for our method. If we use Web dataset from Sect. 4.3 for training and evaluate on Pascal cosegmentation, then our results are much improved, under $ours_3$ in Table 5. Training with more images does give us an advantage over [14]. But the method in [14], with its quadratic loss function, is infeasible for large datasets.

4.3   Semantic Segmentation

We extend our approach to handle weakly supervised semantic segmentation following a simpler version of [36]. We use Web image search on 20 class names in Pascal VOC [9] to automatically collect about 800 images per class. Let us call

Fig. 3. Sample results on Pascal VOC cosegmentation dataset. The columns are: input image, ground truth, our results without negative loss, our results with negative loss. Negative loss helps to rule out the salient objects that are not of the class of interest.

these datasets as 'airplane'-Web, 'bicycle'-Web, etc. Then we train our *UMobV2-reg* on each class separately. We train at resolution 256×256. Let the resulting trained CNNs be called 'airplane'-CNN, 'bicycle'-CNN, etc.

We use the obtained segmentations to generate pseudo-ground truth. We run airplane-CNN on airplane-Web. Each pixel that is labelled 'salient' is assigned to label 'airplane'. Each pixel that is labeled 'background' stays assigned 'background'. There are no ambiguity issues as images in airplane-Web dataset go only through airplane-CNN. We process all other classes similarly. Note that our pseudo-ground truth is multi-label, but each individual image has only 2 labels: background and one of the object classes. Similarly to cosegmentation, we train both with and without negative loss in Eq. (7).

Table 6. Comparison (mIoU metric) to other weakly supervised semantic segmentation methods on Pascal VOC 2012 validation fold.

Method	[36]*	[47]#	[48]#	[1]#	[37]*	[54]	[23]	[38]	$ours_1^*$	$ours_2^*$	$ours_3^*$
mIoU	56.4	60.3	60.4	61.7	63.0	63.3	64.9	64.9	64.0	66.7	67.1

Our Web dataset has around 16 K images and thus is larger than the augmented Pascal training dataset. However, our Web images are of lower similarity to Pascal validation images. Therefore we also train on the subset of augmented Pascal training dataset that has only one class per image. We call this dataset PascalSingle, it has around 5 K images. We generate pseudo-ground truth for PascalSingle using the same procedure as for our Web dataset, described above.

Next we train a multi-class CNN on our pseudo-ground truth. We use DeepLab-ResNet101 [5] pretrained on ImageNet [8]. We follow the same training strategy as in [5], except our initial learning rate is 0.001. DeepLab is trained on images of size 513 × 513. We also use random horizontal flip, Gaussian blur, and rescaling for data augmentation. We train for 100 epochs.

Fig. 4. Sample results on Pascal VOC 2012 validation fold.

Table 6 compares the performance of our method with the most related or most recent work. Methods marked with "*" use additional data for training (with image-level tags), usually obtained by Web search. Methods marked with "#" use saliency datasets together with their pixel precise ground truth. We compare three versions of our method. $Ours_1$ is trained on Web images. $Ours_2$ and $Ours_3$ are trained on PascalSingle dataset, without and with negative loss, respectively. Our method trained with negative loss outperforms all prior work.

Figure 4 shows some sample segmentations. Despite being trained on images each containing only one semantic class, we are able to segment images containing multiple semantic classes. Note that we do not use any post-processing, and therefore our results are likely to improve with CRF-based refinement.

5 Conclusions

We presented a new method of training segmentation CNNs using regularized loss instead of ground truth. Our approach is a simple end-to-end trainable system with no need to adapt architecture or loss for a specific dataset/task. In some cases, our method obtains results that are only a few points behind the results when training the same CNN with pixel precise ground truth. There are several future research directions. First our regularized loss is rather simple and cannot work well for all object classes. It will be interesting to research other loss functions that can be useful for training. Second research direction is extending our approach to a multi-class setting in a unified manner, not as a separate pseudo-ground truth generating stage, as we do currently.

References

1. Ahn, J., Kwak, S.: Learning pixel-level semantic affinity with image-level supervision for weakly supervised semantic segmentation. In: Conference on Computer Vision and Pattern Recognition, pp. 4981–4990 (2018)
2. Boykov, Y., Veksler, O., Zabih, R.: Fast approximate energy minimization via graph cuts. IEEE Trans. Pattern Anal. Mach. Intell. **23**(11), 1222–1239 (2001)
3. Boykov, Y., Kolmogorov, V.: Computing geodesics and minimal surfaces via graph cuts. In: 9th IEEE International Conference on Computer Vision (ICCV 2003), 14–17 October 2003, Nice, France, pp. 26–33 (2003)
4. Chen, L., Papandreou, G., Kokkinos, I., Murphy, K., Yuille, A.L.: Semantic image segmentation with deep convolutional nets and fully connected CRFs. In: International Conference on Learning Research (2015)
5. Chen, L., Papandreou, G., Kokkinos, I., Murphy, K., Yuille, A.L.: Deeplab: semantic image segmentation with deep convolutional nets, atrous convolution, and fully connected CRFs. IEEE Trans. Pattern Anal. Mach. Intell. **40**(4), 834–848 (2018)
6. Chen, L.-C., Zhu, Y., Papandreou, G., Schroff, F., Adam, H.: Encoder-decoder with atrous separable convolution for semantic image segmentation. In: Ferrari, V., Hebert, M., Sminchisescu, C., Weiss, Y. (eds.) ECCV 2018. LNCS, vol. 11211, pp. 833–851. Springer, Cham (2018). https://doi.org/10.1007/978-3-030-01234-2_49
7. Dai, J., He, K., Sun, J.: Boxsup: exploiting bounding boxes to supervise convolutional networks for semantic segmentation. In: International Conference on Computer Vision, pp. 1635–1643 (2015)
8. Deng, J., Dong, W., Socher, R., Li, L.J., Li, K., Fei-Fei, L.: ImageNet: a large-scale hierarchical image database. In: CVPR 2009 (2009)
9. Everingham, M., Van Gool, L., Williams, C.K.I., Winn, J., Zisserman, A.: The PASCAL visual object classes (VOC) challenge. Int. J. Comput. Vision **88**(2), 303–338 (2010)
10. Faktor, A., Irani, M.: Co-segmentation by composition. In: International Conference on Computer Vision, pp. 1297–1304 (2013)
11. Farabet, C., Couprie, C., Najman, L., LeCun, Y.: Learning hierarchical features for scene labeling. IEEE Trans. Pattern Anal. Mach. Intell. (2013). (in press)
12. Goodfellow, I., Bengio, Y., Courville, A.: Deep Learning. MIT Press, Cambridge (2016)
13. Gorelick, L., Veksler, O., Boykov, Y., Nieuwenhuis, C.: Convexity shape prior for segmentation. In: Fleet, D., Pajdla, T., Schiele, B., Tuytelaars, T. (eds.) ECCV 2014. LNCS, vol. 8693, pp. 675–690. Springer, Cham (2014). https://doi.org/10.1007/978-3-319-10602-1_44
14. Hsu, K.J., Lin, Y.Y., Chuang, Y.Y.: Co-attention CNNs for unsupervised object co-segmentation. In: IJCAI, pp. 748–756 (2018)
15. Khoreva, A., Benenson, R., Hosang, J.H., Hein, M., Schiele, B.: Simple does it: weakly supervised instance and semantic segmentation. In: Conference on Computer Vision and Pattern Recognition, pp. 1665–1674 (2017)

16. Kingma, D.P., Ba, J.: Adam: a method for stochastic optimization. arXiv preprint arXiv:1412.6980 (2014)
17. Kirkpatrick, S., Gelatt, C., Vecchi, M.: Optimization by simulated annealing. Science **220**(4598), 671–680 (1983)
18. Kolesnikov, A., Lampert, C.H.: Seed, expand and constrain: three principles for weakly-supervised image segmentation. In: Leibe, B., Matas, J., Sebe, N., Welling, M. (eds.) ECCV 2016. LNCS, vol. 9908, pp. 695–711. Springer, Cham (2016). https://doi.org/10.1007/978-3-319-46493-0_42
19. Krähenbühl, P., Koltun, V.: Efficient inference in fully connected CRFs with gaussian edge potentials. In: Neural Information Processing Systems, pp. 109–117 (2011)
20. Krizhevsky, A., Sutskever, I., Hinton, G.E.: Imagenet classification with deep convolutional neural networks. In: Neural Information Processing Systems (2012)
21. Lafferty, J., McCallum, A., Pereira, F.: Conditional random fields. In: International Conference on Machine Learning (2001)
22. LeCun, Y., et al.: Backpropagation applied to handwritten zip code recognition. Neural Comput. **1**(4), 541–551 (1989)
23. Lee, J., Kim, E., Lee, S., Lee, J., Yoon, S.: Ficklenet: weakly and semi-supervised semantic image segmentation using stochastic inference. In: Conference on Computer Vision and Pattern Recognition, pp. 5267–5276 (2019)
24. Li, B., Sun, Z., Li, Q., Wu, Y., Hu, A.: Group-wise deep object co-segmentation with co-attention recurrent neural network. In: International Conference on Computer Vision, October 2019
25. Li, S.: Markov Random Field Modeling in Computer Vision. Springer, London (1995). https://doi.org/10.1007/978-4-431-66933-3
26. Lin, D., Dai, J., Jia, J., He, K., Sun, J.: Scribblesup: scribble-supervised convolutional networks for semantic segmentation. In: Conference on Computer Vision and Pattern Recognition, pp. 3159–3167 (2016)
27. Liu, T., Sun, J., Zheng, N.N., Tang, X., Shum, H.Y.: Learning to detect a salient object. In: Conference on Computer Vision and Pattern Recognition, pp. 1–8. IEEE (2007)
28. Nguyen, T., et al.: Deepusps: deep robust unsupervised saliency prediction via self-supervision. In: Advances in Neural Information Processing Systems, pp. 204–214 (2019)
29. Papandreou, G., Chen, L.C., Murphy, K.P., Yuille, A.L.: Weakly and semi supervised learning of a deep convolutional network for semantic image segmentation. In: International Conference on Computer Vision, pp. 1742–1750 (2015)
30. Pinheiro, P.H.O., Collobert, R.: From image-level to pixel-level labeling with convolutional networks. In: Conference on Computer Vision and Pattern Recognition, pp. 1713–1721 (2015)
31. Ronneberger, O., Fischer, P., Brox, T.: U-net: convolutional networks for biomedical image segmentation. In: Medical Image Computing and Computer-Assisted Intervention, pp. 234–241 (2015)
32. Rother, C., Minka, T., Blake, A., Kolmogorov, V.: Cosegmentation of image pairs by histogram matching-incorporating a global constraint into MRFs. In: Conference on Computer Vision and Pattern Recognition, vol. 1, pp. 993–1000 (2006)
33. Sandler, M., Howard, A.G., Zhu, M., Zhmoginov, A., Chen, L.: Mobilenetv 2: inverted residuals and linear bottlenecks. In: Conference on Computer Vision and Pattern Recognition, pp. 4510–4520 (2018)

34. Selvaraju, R.R., Cogswell, M., Das, A., Vedantam, R., Parikh, D., Batra, D.: Grad-CAM: visual explanations from deep networks via gradient-based localization. In: International Conference on Computer Vision, pp. 618–626 (2017)
35. Shelhamer, E., Long, J., Darrell, T.: Fully convolutional networks for semantic segmentation. Trans. Pattern Anal. Mach. Intell. **39**(4), 640–651 (2017)
36. Shen, T., Lin, G., Liu, L., Shen, C., Reid, I.D.: Weakly supervised semantic segmentation based on co-segmentation. In: BMVC (2017)
37. Shen, T., Lin, G., Shen, C., Reid, I.: Bootstrapping the performance of webly supervised semantic segmentation. In: Proceedings of the IEEE Conference on Computer Vision and Pattern Recognition, pp. 1363–1371 (2018)
38. Shimoda, W., Yanai, K.: Self-supervised difference detection for weakly-supervised semantic segmentation. In: Proceedings of the IEEE International Conference on Computer Vision, pp. 5208–5217 (2019)
39. Takikawa, T., Acuna, D., Jampani, V., Fidler, S.: Gated-SCNN: gated shape CNNs for semantic segmentation. In: International Conference on Computer Vision (2019)
40. Tang, M., Djelouah, A., Perazzi, F., Boykov, Y., Schroers, C.: Normalized cut loss for weakly-supervised CNN segmentation. In: Conference on Computer Vision and Pattern Recognition, pp. 1818–1827 (2018)
41. Tang, M., Perazzi, F., Djelouah, A., Ayed, I.B., Schroers, C., Boykov, Y.: On regularized losses for weakly-supervised CNN segmentation. In: European Conference on Computer Vision. pp. 524–540 (2018)
42. Vedaldi, A.: Cats and dogs. In: Conference on Computer Vision and Pattern Recognition (2012)
43. Veksler, O.: Efficient graph cut optimization for full CRFs with quantized edges. IEEE Trans. Pattern Anal. Mach. Intell. **42**(4), 1005–1012 (2020)
44. Vernaza, P., Chandraker, M.: Learning random-walk label propagation for weakly-supervised semantic segmentation. In: Conference on Computer Vision and Pattern Recognition, pp. 2953–2961 (2017)
45. Vicente, S., Kolmogorov, V., Rother, C.: Graph cut based image segmentation with connectivity priors. In: Conference on Computer Vision and Pattern Recognition, pp. 1–8 (2008)
46. Vogiatzis, G., Torr, P., Cipolla, R.: Multi-view stereo via volumetric graph-cuts. In: Conference on Computer Vision and Pattern Recognition, pp. II: 391–398 (2005)
47. Wang, X., You, S., Li, X., Ma, H.: Weakly-supervised semantic segmentation by iteratively mining common object features. In: Proceedings of the IEEE Conference on Computer Vision and Pattern Recognition, pp. 1354–1362 (2018)
48. Wei, Y., Xiao, H., Shi, H., Jie, Z., Feng, J., Huang, T.S.: Revisiting dilated convolution: a simple approach for weakly-and semi-supervised semantic segmentation. In: Proceedings of the IEEE Conference on Computer Vision and Pattern Recognition, pp. 7268–7277 (2018)
49. Wu, H., Zhang, J., Huang, K., Liang, K., Yu, Y.: Fastfcn: Rethinking dilated convolution in the backbone for semantic segmentation. CoRR abs/1903.11816 (2019)
50. Xie, S., Girshick, R.B., Dollár, P., Tu, Z., He, K.: Aggregated residual transformations for deep neural networks. In: Conference on Computer Vision and Pattern Recognition, pp. 5987–5995 (2017)
51. Xu, J., Schwing, A.G., Urtasun, R.: Learning to segment under various forms of weak supervision. In: Conference on Computer Vision and Pattern Recognition, pp. 3781–3790 (2015)
52. Yan, Q., Xu, L., Shi, J., Jia, J.: Hierarchical saliency detection. In: Conference on Computer Vision and Pattern Recognition, pp. 1155–1162. IEEE (2013)

53. Yang, C., Zhang, L., Lu, H., Ruan, X., Yang, M.H.: Saliency detection via graph-based manifold ranking. In: Conference on Computer Vision and Pattern Recognition, pp. 3166–3173. IEEE (2013)
54. Zeng, Y., Zhuge, Y., Lu, H., Zhang, L.: Joint learning of saliency detection and weakly supervised semantic segmentation. In: Proceedings of the IEEE International Conference on Computer Vision, pp. 7223–7233 (2019)
55. Zhang, D., Han, J., Zhang, Y.: Supervision by fusion: Towards unsupervised learning of deep salient object detector. In: Proceedings of the IEEE International Conference on Computer Vision, pp. 4048–4056 (2017)
56. Zhang, J., Zhang, T., Dai, Y., Harandi, M., Hartley, R.: Deep unsupervised saliency detection: A multiple noisy labeling perspective. In: Proceedings of the IEEE Conference on Computer Vision and Pattern Recognition, pp. 9029–9038 (2018)
57. Zheng, S., Jayasumana, S., Romera-Paredes, B., Vineet, V., Su, Z., Du, D., Huang, C., Torr, P.: Conditional random fields as recurrent neural networks. In: International Conference on Computer Vision, pp. 1529–1537 (2015)
58. Zhou, B., Khosla, A., Lapedriza, À., Oliva, A., Torralba, A.: Learning deep features for discriminative localization. In: Conference on Computer Vision and Pattern Recognition, pp. 2921–2929 (2016)

Spike-FlowNet: Event-Based Optical Flow Estimation with Energy-Efficient Hybrid Neural Networks

Chankyu Lee[1]([✉]) [iD], Adarsh Kumar Kosta[1] [iD], Alex Zihao Zhu[2] [iD],
Kenneth Chaney[2] [iD], Kostas Daniilidis[2] [iD], and Kaushik Roy[1] [iD]

[1] Purdue University, 47907 West Lafayette, IN, USA
{lee2216,akosta,kaushik}@purdue.edu
[2] University of Pennsylvania, 19104 Philadelphia, PA, USA
{alexzhu,chaneyk,kostas}@seas.upenn.edu

Abstract. Event-based cameras display great potential for a variety of tasks such as high-speed motion detection and navigation in low-light environments where conventional frame-based cameras suffer critically. This is attributed to their high temporal resolution, high dynamic range, and low-power consumption. However, conventional computer vision methods as well as deep Analog Neural Networks (ANNs) are not suited to work well with the asynchronous and discrete nature of event camera outputs. Spiking Neural Networks (SNNs) serve as ideal paradigms to handle event camera outputs, but deep SNNs suffer in terms of performance due to the spike vanishing phenomenon. To overcome these issues, we present Spike-FlowNet, a deep hybrid neural network architecture integrating SNNs and ANNs for efficiently estimating optical flow from sparse event camera outputs without sacrificing the performance. The network is end-to-end trained with self-supervised learning on Multi-Vehicle Stereo Event Camera (MVSEC) dataset. Spike-FlowNet outperforms its corresponding ANN-based method in terms of the optical flow prediction capability while providing significant computational efficiency.

Keywords: Event-based vision · Optical flow estimation · Hybrid network · Spiking neural network · Self-supervised learning

1 Introduction

The dynamics of biological species such as winged insects serve as prime sources of inspiration for researchers in the field of neuroscience, machine learning as well as robotics. The ability of winged insects to perform complex, high-speed

---

The code is publicly available at: https://github.com/chan8972/Spike-FlowNet.

Electronic supplementary material The online version of this chapter (https://doi.org/10.1007/978-3-030-58526-6_22) contains supplementary material, which is available to authorized users.

A. Vedaldi et al. (Eds.): ECCV 2020, LNCS 12374, pp. 366–382, 2020.
https://doi.org/10.1007/978-3-030-58526-6_22

maneuvers effortlessly in cluttered environments clearly highlights the efficiency of these resource-constrained biological systems [5]. The estimation of motion patterns corresponding to spatio-temporal variations of structured illumination - commonly referred to as optical flow, provides vital information for estimating ego-motion and perceiving the environment. Modern deep Analog Neural Networks (ANNs) aim to achieve this at the cost of being computationally intensive, placing significant overheads on current hardware platforms. A competent methodology to replicate such energy efficient biological systems would greatly benefit edge-devices with computational and memory constraints (Note, we will be referring to standard deep learning networks as Analog Neural Networks (ANNs) due to their analog nature of inputs and computations. This would help to distinguish them from Spiking Neural Networks (SNNs), which involve discrete spike-based computations).

Over the past years, the majority of optical flow estimation techniques relied on images from traditional frame-based cameras, where the input data is obtained by sampling intensities on the entire frame at fixed time intervals irrespective of the scene dynamics. Although sufficient for certain computer vision applications, frame-based cameras suffer from issues such as motion blur during high speed motion, inability to capture information in low-light conditions, and over- or under-saturation in high dynamic range environments.

Event-based cameras, often referred to as bio-inspired silicon retinas, overcome these challenges by detecting log-scale brightness changes asynchronously and independently on each pixel-array element [20], similar to retinal ganglion cells. Having a high temporal resolution (in the order of microseconds) and a fraction of power consumption compared to frame-based cameras make event cameras suitable for estimating high-speed and low-light visual motion in an energy-efficient manner. However, because of their fundamentally different working principle, conventional computer vision as well as ANN-based methods become no longer effective for event camera outputs. This is mainly because these methods are typically designed for pixel-based images relying on photo-consistency constraints, assuming the color and brightness of object remain the same in all image sequences. Thus, the need for development of handcrafted-algorithms for handling event camera outputs is paramount.

SNNs, inspired by the biological neuron model, have emerged as a promising candidate for this purpose, offering asynchronous computations and exploiting the inherent sparsity of spatio-temporal events (spikes). The Integrate and Fire (IF) neuron is one spiking neuron model [8], which can be characterized by an internal state, known as the membrane potential. The membrane potential accumulates the inputs over time and emits an output spike whenever it exceeds a set threshold. This mechanism naturally encapsulates the event-based asynchronous processing capability across SNN layers, leading to energy-efficient computing on specialized neuromorphic hardware such as IBM's TrueNorth [24] and Intel's Loihi [9]. However, recent works have shown that the number of spikes drastically vanish at deeper layers, leading to performance degradations in deep SNNs [18]. Thus, there is a need for an efficient hybrid architecture, with SNNs in the initial layers, to exploit their compatability with event camera outputs while having ANNs in the deeper layers in order to retain performance.

In regard to this, we propose a deep hybrid neural network architecture, accommodating SNNs and ANNs in different layers, for energy efficient optical flow estimation using sparse event camera data. To the best of our knowledge, this is the first SNN demonstration to report the state-of-art performance on event-based optical flow estimation, outperforming its corresponding fully-fledged ANN counterpart.

The main contributions of this work can be summarized as:

- We present an input representation that efficiently encodes the sequences of sparse outputs from event cameras over time to preserve the spatio-temporal nature of spike events.
- We introduce a deep hybrid architecture for event-based optical flow estimation referred to as Spike-FlowNet, integrating SNNs and ANNs in different layers, to efficiently process the sparse spatio-temporal event inputs.
- We evaluate the optical flow prediction capability and computational efficiency of Spike-FlowNet on the Multi-Vehicle Stereo Event Camera dataset (MVSEC) [33] and provide comparison results with current state-of-the-art approaches.

The following contents are structured as follows. In Sect. 2, we elucidate the related works. In Sect. 3, we present the methodology, covering essential backgrounds on the spiking neuron model followed by our proposed input event (spike) representation. This section also discusses the self-supervised loss, Spike-FlowNet architecture, and the approximate backpropagation algorithm used for training. Section 4 covers the experimental results, including training details and evaluation metrics. It also discusses the comparison results with the latest works in terms of performance and computational efficiency.

2   Related Work

In recent years, there have been an increasing number of works on estimating optical flow by exploiting the high temporal resolution of event cameras. In general, these approaches have either been adaptations of conventional computer vision methods or modified versions of deep ANNs to encompass discrete outputs from event cameras.

For computer vision based solutions to estimate optical flow, gradient-based approaches using the Lucas-Kanade algorithm [22] have been highlighted in [4, 7]. Further, plane fitting approaches by computing the slope of the plane for estimating optical flow have been presented in [1,3]. In addition, bio-inspired frequency-based approaches have been discussed in [2]. Finally, correlation-based approaches are presented in [12,32] employing convex optimization over events. In addition, [21] interestingly uses an adaptive block matching technique to estimate sparse optical flow.

For deep ANN-based solutions, optical flow estimation from frame-based images has been discussed in Unflow [23], which utilizes a U-Net [28] architecture and computes a bidirectional census loss in an unsupervised manner with an added smoothness term. This strategy is modified for event camera outputs in EV-FlowNet [34] incorporating a self-supervised loss based on gray images as a replacement for ground truth. Other previous works employ various modifications to the training methodology, such as [15], which imposes certain brightness constancy and smoothness constraints to train a network and [17] which adds an adversarial loss over the standard photometric loss. In contrast, [35] presents an unsupervised learning approach using only event camera data to estimate optical flow by accounting for and then learning to rectify the motion blur.

All the above strategies employ ANN architectures to predict the optical flow. However, event cameras produce asynchronous and discrete outputs over time, and SNNs can naturally capture their spatio-temporal dynamics, which are embedded in the precise spike timings. Hence, we posit that SNNs are suitable for handling event camera outputs. Recent SNN-based approaches for event-based optical flow estimation include [13, 25, 27]. Researchers in [25] presented visual motion estimation using SNNs, which accounts for synaptic delays in generating motion-sensitive receptive fields. In addition, [13] demonstrated real-time model-based optical flow computations on TrueNorth hardware for evaluating patterns including rotating spirals and pipes. Authors of [27] presented a methodology for optical flow estimation using convolutional SNNs based on Spike-Time-Dependent-Plasticity (STDP) learning [11]. The main limitation of these works is that they employ shallow SNN architectures, because deep SNNs suffer in terms of performance. Besides, the presented results are only evaluated on relatively simple tasks. In practice, they do not generally scale well to complex and real-world data, such as that presented in MVSEC dataset [33]. In view of these, a hybrid approach becomes an attractive option for constructing deep network architectures, leveraging the benefits of both SNNs and ANNs.

3 Method

3.1 Spiking Neuron Model

The spiking neurons, inspired by biological models [10], are computational primitives in SNNs. We employ a simple IF neuron model, which transmits the output signals in the form of spike events over time. The behavior of IF neuron at the lth layer is illustrated in Fig. 1. The input spikes are weighted to produce an influx current that integrates into neuronal membrane potential (V^l).

$$V^l[n+1] = V^l[n] + w^l * o^{l-1}[n] \tag{1}$$

where $V^l[n]$ represents the membrane potential at discrete time-step n, w^l represents the synaptic weights and $o^{l-1}[n]$ represents the spike events from the previous layer at discrete time-step n. When the membrane potential overcomes the firing threshold, the neuron emits an output spike and resets the membrane potential to the initial state (zero). Over time, these mechanisms are repeatedly carried out in each IF neuron, enabling event-based computations throughout the SNN layers.

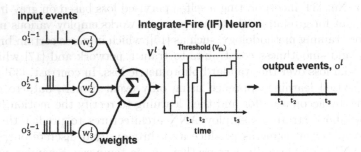

Fig. 1. The dynamics of an Integrate and Fire (IF) neuron. The input events are modulated by the synaptic weight to be integrated as the current influx in the membrane potential. Whenever the membrane potential crosses the threshold, the neuron fires an output spike and resets the membrane potential.

3.2   Spiking Input Event Representation

An event-based camera tracks the changes in log-scale intensity (I) at every element in the pixel-array independently and generates a discrete event whenever the change exceeds a threshold (θ):

$$\| \log(I_{t+1}) - \log(I_t) \| \geq \theta \tag{2}$$

A discrete event contains a 4-tuple $\{x, y, t, p\}$, consisting of the coordinates: x, y; timestamp: t; and polarity (direction) of brightness change: p. This input representation is called Address Event Representation (AER), and is the standard format used by event-based sensors.

There are prior works that have modified the representations of asynchronous event camera outputs to be compatible with ANN-based methods. To overcome the asynchronous nature, event outputs are typically recorded for a certain time period and transformed into a synchronous image-like representation. In EV-FlowNet [34], the most recent pixel-wise timestamps and the event counts encoded the motion information (within a time window) in an image. However, fast motions and dense events (in local regions of the image) can vastly overlap per-pixel timestamp information, and temporal information can be lost. In addition, [35] proposed a discretized event volume that deals with the time domain as a channel to retain the spatio-temporal event distributions. However, the number of input channels increases significantly as the time dimensions are finely discretized, further aggravating the computation and parameter overheads.

In this work, we propose a discretized input representation (fine-grained in time) that preserves the spatial and temporal information of events for SNNs.

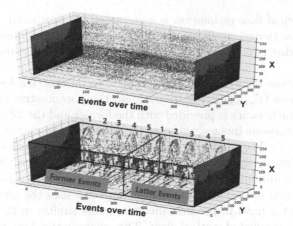

Fig. 2. Input event representation. (*Top*) Continuous raw events between two consecutive grayscale images from an event camera. (*Bottom*) Accumulated event frames between two consecutive grayscale images to form the former and the latter event groups, serving as inputs to the network.

Our proposed input encoding scheme discretizes the time dimension within a time window into two groups (former and latter). Each group contains N number of event frames obtained by accumulating raw events from the timestamp of the previous frame till the current timestamp. Each of these event frames is also composed of two channels for ON/OFF polarity of events. Hence, the input to the network consists of a sequence of N frames with four channels (one frame each from the former and the latter groups having two channels each). The proposed input representation is displayed in Fig. 2 for one channel (assuming the number of event frames in each group equals to five). The main characteristic of our proposed input event representation (compared to ANN-based methods) are as follows:

– Our spatio-temporal input representations encode only the presence of events over time, allowing asynchronous and event-based computations in SNNs. In contrast, ANN-based input representation often requires the timestamp and the event count images in separate channels.
– In Spike-FlowNet, each event frame from the former and the latter groups sequentially passes through the network, thereby preserving and utilizing the spatial and temporal information over time. On the contrary, ANN-based methods feed-forward all input information to the network at once.

3.3 Self-Supervised Loss

The DAVIS camera [6] is a commercially available event-camera, which simultaneously provides synchronous grayscale images and asynchronous event streams. The number of available event-based camera datasets with annotated labels suitable for optical flow estimation is quite small, as compared to frame-based camera datasets. Hence, a self-supervised learning method that uses proxy labels from the recorded grayscale images [15,34] is employed for training our Spike-FlowNet.

The overall loss incorporates a photometric reconstruction loss ($\mathcal{L}_{\text{photo}}$) and a smoothness loss ($\mathcal{L}_{\text{smooth}}$) [15]. To evaluate the photometric loss within each time window, the network is provided with the former and the latter event groups and a pair of grayscale images, taken at the start and the end of the event time window (I_t, I_{t+dt}). The predicted optical flow from the network is used to warp the second grayscale image to the first grayscale image. The photometric loss ($\mathcal{L}_{\text{photo}}$) aims to minimize the discrepancy between the first grayscale image and the inverse warped second grayscale image. This loss uses the photo-consistency assumption that a pixel in the first image remains similar in the second frame mapped by the predicted optical flow. The photometric loss is computed as follows:

$$\mathcal{L}_{\text{photo}}(u, v; I_t, I_{t+dt}) = \sum_{x,y} \rho(I_t(x, y) - I_{t+dt}(x + u(x, y),\ y + v(x, y))) \quad (3)$$

where, I_t, I_{t+dt} indicate the pixel intensity of the first and second grayscale images, u, v are the flow estimates in the horizontal and vertical directions, ρ is the Charbonnier loss $\rho(x) = (x^2 + \eta^2)^r$, which is a generic loss used for outlier rejection in optical flow estimation [30]. For our work, $r = 0.45$ and $\eta =$1e-3 show the optimum results for the computation of photometric loss.

Furthermore, a smoothness loss ($\mathcal{L}_{\text{smooth}}$) is applied for enhancing the spatial collinearity of neighboring optical flow. The smoothness loss minimizes the difference in optical flow between neighboring pixels and acts as a regularizer on the predicted flow. It is computed as follows:

$$\mathcal{L}_{\text{smooth}}(u, v) = \frac{1}{HD} \sum_{j}^{H} \sum_{i}^{D} (\|u_{i,j} - u_{i+1,j}\| + \|u_{i,j} - u_{i,j+1}\|$$

$$+ \|v_{i,j} - v_{i+1,j}\| + \|v_{i,j} - v_{i,j+1}\|) \quad (4)$$

where H is the height and D is the width of the predicted flow output. The overall loss is computed as the weighted sum of the photometric and smoothness loss:

$$\mathcal{L}_{\text{total}} = \mathcal{L}_{\text{photo}} + \lambda \mathcal{L}_{\text{smooth}} \quad (5)$$

where λ is the weight factor.

3.4   Spike-FlowNet Architecture

Spike-FlowNet employs a deep hybrid architecture that accommodates SNNs and ANNs in different layers, enabling the benefits of SNNs for sparse event data processing and ANNs for maintaining the performance. The use of a hybrid architecture is attributed to the fact that spike activities reduce drastically with

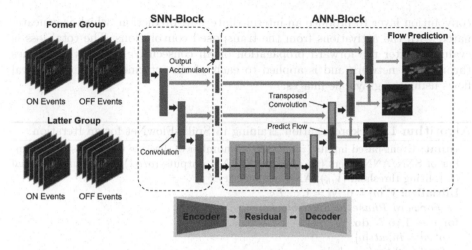

Fig. 3. Spike-FlowNet architecture. The four-channeled input images, comprised of ON/OFF polarity events for former and latter groups, are sequentially passed through the hybrid network. The SNN-block contains the encoder layers followed by output accumulators, while the ANN-block contains the residual and decoder layers. The loss is evaluated after forward propagating all consecutive input event frames (a total of N inputs, sequentially taken in time from the former and the latter event groups) within the time window. The black arrows denote the forward path, green arrows represent residual connections, and blue arrows indicate the flow predictions. (Color figure online)

growing the network depth in the case of full-fledged SNNs. This is commonly referred to as the vanishing spike phenomenon [26], and potentially leads to performance degradation in deep SNNs. Furthermore, high numerical precision is essentially required for estimating the accurate pixel-wise network outputs, namely the regression tasks. Hence, very rare and binary precision spike signals (in input and intermediate layers) pose a crucial issue for predicting the accurate flow displacements. To resolve these issues, only the encoder block is built as an SNN, while the residual and decoder blocks maintain an ANN architecture.

Spike-FlowNet's network topology resembles the U-Net [28] architecture, containing four encoder layers, two residual blocks, and four decoder layers as shown in Fig. 3. The events are represented as the four-channeled input frames as presented in Sect. 3.2, and are sequentially passed through the SNN-based encoder layers over time (while being downsampled at each layer). Convolutions with a stride of two are employed for incorporating the functionality of dimensionality reduction in the encoder layers. The outputs from encoder layers are collected in their corresponding output accumulators until all consecutive event images have passed. Next, the accumulated outputs from final encoder layer are passed through two residual blocks and four decoder layers. The decoder layers upsample the activations using transposed convolution. At each decoder layer, there is a skip connection from the corresponding encoder layer, as well as another convolution layer to produce an intermediate flow prediction, which is concatenated with the activations from the transposed convolutions. The total loss is evaluated after the forward propagation of all consecutive input event frames through the network and is applied to each of the intermediate dense optical flows using the grayscale images.

Algorithm 1. Backpropagation Training in Spike-FlowNet for an Iteration.

Input: Event-based inputs ($inputs$), total number of discrete time-steps (N), number of SNN/ANN layers (L_S/L_A), SNN/ANN outputs (o/o_A) membrane potential (V), firing threshold (V_{th}), ANN nonlinearity (f)

Initialize: $V^l[n] = 0$, $\forall l = 1, ..., L_S$

// *Forward Phase in SNN-blocks*

for $n \leftarrow 1$ **to** N **do**

　$o^1[n] = inputs[n]$

　for $l \leftarrow 2$ **to** $L_S - 1$ **do**

　　$V^l[n] = V^l[n-1] + w^l o^{l-1}[n]$ //*weighted spike-inputs are integrated to V*

　　if $V^l[n] > V_{th}$ **then**

　　　$o^l[n] = 1$, $V^l[n] = 0$ //*if V exceeds V_{th}, a neuron emits a spike and reset V*

　　end if

　end for

　$o_A^{L_S} = V^{L_S}[n] = V^{L_S}[n-1] + w^{L_S} o^{L_S-1}[n]$ //*final SNN layer does not fire*

end for

// *Forward Phase in ANN-blocks*

for $l \leftarrow L_S + 1$ **to** $L_S + L_A$ **do**

　$o_A^l = f(w^l o_A^{l-1})$

end for

// *Backward Phase in ANN-blocks*

for $l \leftarrow L_S + L_A$ **to** L_S **do**

　$\triangle w^l = \frac{\partial \mathcal{L}_{total}}{\partial o_A^l} \frac{\partial o_A^l}{\partial w^l}$

end for

// *Backward Phase in SNN-blocks*

for $n \leftarrow N$ **to** 1 **do**

　for $l \leftarrow L_S - 1$ **to** 1 **do**

　　//*evaluate partial derivatives of loss w.r.t. w_S by unrolling the SNN over time*

　　$\triangle w^l[n] = \frac{\partial \mathcal{L}_{total}}{\partial o^l[n]} \frac{\partial o^l[n]}{\partial V^l[n]} \frac{\partial V^l[n]}{\partial w^l[n]}$

　end for

end for

3.5   Backpropagation Training in Spike-FlowNet

The spike generation function of an IF neuron is a hard threshold function that emits a spike when the membrane potential exceeds a firing threshold. Due to this discontinuous and non-differentiable neuron model, standard backpropagation algorithms cannot be applied to SNNs in their native form. Hence, several approximate methods have been proposed to estimate the surrogate gradient of spike generation function. In this work, we adopt the approximate gradient method proposed in [18,19] for back-propagating errors through SNN layers. The approximate IF gradient is computed as $\frac{1}{V_{th}}$, where the threshold value accounts for the change of the spiking output with respect to the input. Algorithm 1 illustrates the forward and backward pass in ANN-block and SNN-block.

In the forward phase, neurons in the SNN layers accumulate the weighted sum of the spike inputs in membrane potential. If the membrane potential exceeds a threshold, a neuron emits a spike at its output and resets. The final SNN layer neurons just integrate the weighted sum of spike inputs in the output accumulator, while not producing any spikes at the output. At the last time-step, the integrated outputs of SNN layers propagate to the ANN layers to predict the optical flow. After the forward pass, the final loss ($\mathcal{L}_{total}$) is evaluated, followed by backpropagation of gradients through the ANN layers using standard backpropagation.

Next, the backpropagated errors ($\frac{\partial \mathcal{L}_{total}}{\partial o^L{}_S}$) pass through the SNN layers using the approximate IF gradient method and BackPropagation Through Time (BPTT) [31]. In BPTT, the network is unrolled for all discrete time-steps, and the weight update is computed as the sum of gradients from each time-step. This procedure is displayed in Fig. 4 where the final loss is back-propagated through an ANN-block and a simple SNN-block consisting of a single input IF neuron. The parameter updates of the lth SNN layers are described as follows:

$$\triangle w^l = \sum_n \frac{\partial \mathcal{L}_{total}}{\partial o^l[n]} \frac{\partial o^l[n]}{\partial V^l[n]} \frac{\partial V^l[n]}{\partial w^l}, \text{ where } \frac{\partial o^l[n]}{\partial V^l[n]} = \frac{1}{V_{th}}(o^l[n] > 0) \qquad (6)$$

where o^l represents the output of spike generation function. This method enables the end-to-end self-supervised training in the proposed hybrid architecture.

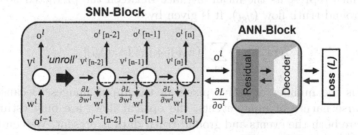

Fig. 4. Error backpropagation in Spike-FlowNet. After the forward pass, the gradients are back-propagated through the ANN block using standard backpropagation whereas the backpropagated errors ($\frac{\partial \mathcal{L}}{\partial o^l}$) pass through the SNN layers using the approximate IF gradient method and BPTT technique.

4   Experimental Results

4.1   Dataset and Training Details

We use the MVSEC dataset [33] for training and evaluating the optical flow predictions. MVSEC contains stereo event-based camera data for a variety of environments (e.g., indoor flying and outdoor driving) and also provides the corresponding ground truth optical flow. In particular, the indoor and outdoor sequences are recorded in dissimilar environments where the indoor sequences (indoor_flying) have been captured in a lab environment and the outdoor sequences (outdoor_day) have been recorded while driving on public roads.

Even though the indoor_flying and outdoor_day scenes are quite different, we only use outdoor_day2 sequence for training Spike-FlowNet. This is done to provide fair comparisons with prior works [34,35] which utilized only outdoor_day2 sequence for training. During training, input images are randomly flipped horizontally and vertically (with 0.5 probability) and randomly cropped to 256×256 size. Adam optimizer [16] is used, with the initial learning rate of 5e-5, and scaled by 0.7 every 5 epochs until 10 epoch, and every 10 epochs thereafter. The model is trained on the left event camera data of outdoor_day2 sequence for 100 epochs with a mini-batch size 8. Training is done for two different time windows lengths (i.e, 1 grayscale image frame apart ($dt = 1$) and 4 grayscale image frames apart ($dt = 4$)). The number of event frame (N) and weight factor for the smoothness loss (λ) are set to 5, 10 for a $dt = 1$ case and 20, 1 for a $dt = 4$ case, respectively. The threshold of the IF neurons are set to 0.5 ($dt = 4$) and 0.75 ($dt = 1$) in SNN layers.

4.2   Algorithm Evaluation Metric

The evaluation metric for optical flow prediction is the Average End-point Error (AEE), which represents the mean distance between the predicted flow (y_{pred}) and the ground truth flow (y_{gt}). It is given by:

$$\text{AEE} = \frac{1}{m} \sum_{m} \|(u, v)_{\text{pred}} - (u, v)_{\text{gt}}\|_2 \qquad (7)$$

where m is the number of active pixels in the input images. Because of the highly sparse nature of input events, the optical flows are only estimated at pixels where both the events and ground truth data is present. We compute the AEE for $dt = 1$ and $dt = 4$ cases.

4.3   Average End-Point Error (AEE) Results

During testing, optical flow is estimated on the center cropped 256×256 left camera images of the indoor_flying 1,2,3 and outdoor_day 1 sequences. We use all events for the indoor_flying sequences, but we take events within 800 grayscale frames for the outdoor_day1 sequence, similar to [34]. Table 1 provides the AEE

Grayscale	Spike Image	Ground Truth Flow	Masked Ground Truth Flow	EV-FlowNet$_{2R}$ Flow	Spike-FlowNet Flow	Masked Spike-FlowNet Flow

Fig. 5. Optical flow evaluation and comparison with EV-FlowNet. The samples are taken from (*top*) *outdoor_day*1 and (*bottom*) *indoor_day*1. The Masked Spike-FlowNet Flow is basically a sparse optical flow computed at pixels at which events occurred. It is computed by masking the predicted optical flow with the spike image.

evaluation results in comparison with the prior event camera based optical flow estimation works. Overall, our results show that Spike-FlowNet can accurately predict the optical flow in both the indoor_flying and outdoor_day1 sequences. This demonstrates that the proposed Spike-FlowNet can generalize well to distinctly different environments. The grayscale, spike event, ground truth flow and the corresponding predicted flow images are visualized in Fig. 5 where the images are taken from (*top*) *outdoor_day*1 and (*bottom*) *indoor_day*1, respectively. Since event cameras work based on changing light intensity at pixels, the regions having low texture produce very sparse events due to minimal intensity changes, resulting in scarce optical flow predictions in the corresponding areas such as the flat surfaces. Practically, the useful flows are extracted by using flow estimations at points where significant events exist in the input frames.

Moreover, we compare our quantitative results with the recent works [34,35] on event-based optical flow estimation, as listed in Table 1. We observe that Spike-FlowNet outperforms EV-FlowNet [34] in terms of AEE results in both the $dt = 1$ and $dt = 4$ cases. It is worth noting here that EV-FlowNet employs a similar network architecture and self-supervised learning method, providing a

Table 1. Average Endpoint Error (AEE) comparisons with Zhu et al. [35] and EV-FlowNet [34].

	dt = 1 frame				dt = 4 frame			
	indoor1	indoor2	indoor3	outdoor1	indoor1	indoor2	indoor3	outdoor1
Zhu et al. [35]	**0.58**	**1.02**	**0.87**	**0.32**	**2.18**	3.85	**3.18**	1.30
EV-FlowNet [34]	1.03	1.72	1.53	0.49	2.25	4.05	3.45	1.23
This work	0.84	1.28	1.11	0.49	2.24	**3.83**	**3.18**	**1.09**

* EV-FlowNet also uses a self-supervised learning method, providing the fair comparison baseline compared to Spike-FlowNet.

fair comparison baseline for fully ANN architectures. In addition, Spike-FlowNet attains AEE results slightly better or comparable to [35] in the $dt = 4$ case, while underperforming in the $dt = 1$ case. [35] presented an image deblurring based unsupervised learning that employed only the event streams. Hence, it seems to not suffer from the issues related to grayscale images such as motion blur or aperture problems during training. In view of these comparisons, Spike-FlowNet (with presented spatio-temporal event representation) is more suitable for motion detection when the input events have a certain minimum level of spike density. We further provide the ablation studies for exploring the optimal design choices in the supplementary material.

4.4 Computational Efficiency

To further analyze the benefits of Spike-FlowNet, we estimate the gain in computational costs compared to a fully ANN architecture. Typically, the number of synaptic operations is used as a metric for benchmarking the computational energy of neuromorphic hardware [18,24,29]. Also, the required energy consumption per synaptic operation needs to be considered. Now, we describe the procedures for measuring the computational costs in SNN and ANN layers.

In a neuromorphic hardware, SNNs carry out event-based computations only at the arrival of input spikes. Hence, we first measure the mean spike activities at each time-step in the SNN layers. As presented in the first row of Table 2, the mean spiking activities (averaged over indoor1,2,3 and outdoor1 sequences) are

Table 2. Analysis for Spike-FlowNet in terms of the mean spike activity, the total and normalized number of SNN operations in an encoder-block, the encoder-block and overall computational energy benefits.

	indoor1		indoor2		indoor3		outdoor1	
	dt = 1	dt = 4	dt = 1	dt = 4	dt = 1	dt = 4	dt = 1	dt = 4
Encoder spike activity (%)	**0.33**	0.87	0.65	1.27	0.53	1.11	0.41	0.78
Encoder SNN # operation ($\times 10^8$)	**0.16**	1.69	0.32	2.47	0.26	2.15	0.21	1.53
Encoder normalized # operation (%)	**1.68**	17.87	3.49	26.21	2.81	22.78	2.29	16.23
Encoder compute-energy benefit ($\times$)	**305**	28.6	146.5	19.5	182.1	22.44	223.2	31.5
Overall compute-energy reduction (%)	**17.57**	17.01	17.51	16.72	17.53	16.84	17.55	17.07

* For an ANN, the number of synaptic operations is 9.44×10^8 for the encoder-block and 5.35×10^9 for overall network.

0.48% and 1.01% for $dt = 1$ and $dt = 4$ cases, respectively. Note that the neuronal threshold is set to a higher value in $dt = 1$ case; hence the average spiking activity becomes sparser compared to $dt = 4$ case. The extremely rare mean input spiking activities are mainly due to the fact that event camera outputs are highly sparse in nature. This sparse firing rate is essential for exploiting efficient event-based computations in SNN layers. In contrast, ANNs execute dense matrix-vector multiplication operations without considering the sparsity of inputs. In other words, ANNs simply feed-forward the inputs at once, and the total number of operations are fixed. This leads to the high energy requirements (compared to SNNs) by computing both zero and non-zero entities, especially when inputs are very sparse.

Essentially, SNNs need to compute the spatio-temporal spike images over a number of time-steps. Given M is the number of neurons, C is number of synaptic connections and F indicates the mean firing activity, the number of synaptic operations at each time-step in the lth layer is calculated as $M_l \times C_l \times F_l$. The total number of SNN operations is the summation of synaptic operations in SNN layers during the N time-steps. Hence, the total number of SNN and ANN operations become $\sum_l (M_l \times C_l \times F_l) \times N$ and $\sum_l M_l \times C_l$, respectively. Based on these, we estimate and compare the average number of synaptic operations on Spike-FlowNet and a fully ANN architecture. The total and the normalized number of SNN operations compared to ANN operations on the encoder-block are provided in the second and the third row of Table 2, respectively.

Due to the binary nature of spike events, SNNs perform only accumulation (AC) per synaptic operation. On the other hand, ANNs perform the multiply-accumulate (MAC) computations since the inputs consist of analog-valued entities. In general, AC computation is considered to be significantly more energy-efficient than MAC. For example, AC is reported to be $5.1\times$ more energy-efficient than a MAC in the case of 32-bit floating-point numbers (45 nm CMOS process) [14]. Based on this principle, the computational energy benefits of encoder-block and overall Spike-FlowNet are obtained, as provided in the fourth and the fifth rows of Table 2, respectively. These results reveal that the SNN-based encoder-block is $214.2\times$ and $25.51\times$ more computationally efficient compared to ANN-based one (averaged over indoor1,2,3 and outdoor1 sequences) for $dt = 1$ and $dt = 4$ cases, respectively. The number of time-steps (N) is four times less in $dt = 1$ case than in $dt = 4$ case; hence, the computational energy benefit is much higher in $dt = 1$ case.

From our analysis, the proportion of required computations in encoder-block compared to the overall architecture is 17.6%. This reduces the overall energy benefits of Spike-FlowNet. In such a case, an approach of interest would be to perform a distributed edge-cloud implementation where the SNN- and ANN-blocks are administered on the edge device and the cloud, respectively. This would lead to high energy benefits on edge devices, which are limited by resource constraints while not compromising on algorithmic performance.

5   Conclusion

In this work, we propose Spike-FlowNet, a deep hybrid architecture for energy-efficient optical flow estimations using event camera data. To leverage the benefits of both SNNs and ANNs, we integrate them in different layers for resolving the spike vanishing issue in deep SNNs. Moreover, we present a novel input encoding strategy for handling outputs from event cameras, preserving the spatial and temporal information over time. Spike-FlowNet is trained with a self-supervised learning method, bypassing expensive labeling. The experimental results show that the proposed architecture accurately predicts the optical flow from discrete and asynchronous event streams along with substantial benefits in terms of computational efficiency compared to the corresponding ANN architecture.

Acknowledgment. This work was supported in part by C-BRIC, one of six centers in JUMP, a Semiconductor Research Corporation (SRC) program sponsored by DARPA, the National Science Foundation, Sandia National Laboratory, and the DoD Vannevar Bush Fellowship.

References

1. Aung, M.T., Teo, R., Orchard, G.: Event-based plane-fitting optical flow for dynamic vision sensors in FPGA. In: 2018 IEEE International Symposium on Circuits and Systems (ISCAS), pp. 1–5, May 2018. https://doi.org/10.1109/ISCAS.2018.8351588
2. Barranco, F., Fermuller, C., Aloimonos, Y.: Bio-inspired motion estimation with event-driven sensors. In: Rojas, I., Joya, G., Catala, A. (eds.) IWANN 2015. LNCS, vol. 9094, pp. 309–321. Springer, Cham (2015). https://doi.org/10.1007/978-3-319-19258-1_27
3. Benosman, R., Clercq, C., Lagorce, X., Ieng, S., Bartolozzi, C.: Event-based visual flow. IEEE Transa. Neural Networks Learn. Syst. **25**(2), 407–417 (2014). https://doi.org/10.1109/TNNLS.2013.2273537
4. Benosman, R., Ieng, S.H., Clercq, C., Bartolozzi, C., Srinivasan, M.: Asynchronous frameless event-based optical flow. Neural Networks **27**, 32–37 (2012). https://doi.org/10.1016/j.neunet.2011.11.001. http://www.sciencedirect.com/science/article/pii/S0893608011002930
5. Borst, A., Haag, J., Reiff, D.F.: Fly motion vision. Ann. Rev. Neurosci. **33**(1), 49–70 (2010). https://doi.org/10.1146/annurev-neuro-060909-153155. https://doi.org/10.1146/annurev-neuro-060909-153155, pMID: 20225934
6. Brandli, C., Berner, R., Yang, M., Liu, S., Delbruck, T.: A 240 × 180 130 db 3 μs latency global shutter spatiotemporal vision sensor. IEEE J. Solid-State Circuits **49**(10), 2333–2341 (2014). https://doi.org/10.1109/JSSC.2014.2342715
7. Brosch, T., Tschechne, S., Neumann, H.: On event-based optical flow detection. Front. Neurosci. **9**, 137 (2015). https://doi.org/10.3389/fnins.2015.00137
8. Burkitt, A.N.: A review of the integrate-and-fire neuron model: I. homogeneous synaptic input. Biol. Cybern. **95**(1), 1–19 (2006)
9. Davies, M., et al.: Loihi: a neuromorphic manycore processor with on-chip learning. IEEE Micro **38**(1), 82–99 (2018). https://doi.org/10.1109/MM.2018.112130359

10. Dayan, P., Abbott, L.F.: Theoretical Neurosci., vol. 806. MIT Press, Cambridge (2001)

11. Diehl, P.U., Cook, M.: Unsupervised learning of digit recognition using spike-timing-dependent plasticity. Front. Comput. Neurosci. **9**, 99 (2015)

12. Gallego, G., Rebecq, H., Scaramuzza, D.: A unifying contrast maximization framework for event cameras, with applications to motion, depth, and optical flow estimation. CoRR abs/1804.01306 (2018). http://arxiv.org/abs/1804.01306

13. Haessig, G., Cassidy, A., Alvarez, R., Benosman, R., Orchard, G.: Spiking optical flow for event-based sensors using ibm's truenorth neurosynaptic system. IEEE Trans. Biomed. Circuits Syst. **12**(4), 860–870 (2018)

14. Horowitz, M.: 1.1 computing's energy problem (and what we can do about it). In: 2014 IEEE International Solid-State Circuits Conference Digest of Technical Papers (ISSCC), pp. 10–14. IEEE (2014)

15. Yu, J.J., Harley, A.W., Derpanis, K.G.: Back to basics: unsupervised learning of optical flow via brightness constancy and motion smoothness. In: Hua, G., Jégou, H. (eds.) ECCV 2016. LNCS, vol. 9915, pp. 3–10. Springer, Cham (2016). https://doi.org/10.1007/978-3-319-49409-8_1

16. Kingma, D.P., Ba, J.: Adam: a method for stochastic optimization. arXiv preprint arXiv:1412.6980 (2014)

17. Lai, W.S., Huang, J.B., Yang, M.H.: Semi-supervised learning for optical flow with generative adversarial networks. In: Guyon, I., et al. (eds.) Advances in Neural Information Processing Systems 30, pp. 354–364. Curran Associates, Inc. (2017). http://papers.nips.cc/paper/6639-semi-supervised-learning-for-optical-flow-with-generative-adversarial-networks.pdf

18. Lee, C., Sarwar, S.S., Panda, P., Srinivasan, G., Roy, K.: Enabling spike-based backpropagation for training deep neural network architectures. Front. Neurosci. **14**, 119 (2020)

19. Lee, J.H., Delbruck, T., Pfeiffer, M.: Training deep spiking neural networks using backpropagation. Front. Neurosci. **10**, 508 (2016)

20. Lichtsteiner, P., Posch, C., Delbruck, T.: A 128× 128 120 db 15 μs latency asynchronous temporal contrast vision sensor. IEEE J. Solid-State Circuits **43**(2), 566–576 (2008). https://doi.org/10.1109/JSSC.2007.914337

21. Liu, M., Delbrück, T.: ABMOF: A novel optical flow algorithm for dynamic vision sensors. CoRR abs/1805.03988 (2018). http://arxiv.org/abs/1805.03988

22. Lucas, B.D., Kanade, T.: An iterative image registration technique with an application to stereo vision. In: Proceedings of the 7th International Joint Conference on Artificial Intelligence - Volume 2, IJCAI 1981, pp. 674–679. Morgan Kaufmann Publishers Inc., San Francisco (1981). http://dl.acm.org/citation.cfm?id=1623264.1623280

23. Meister, S., Hur, J., Roth, S.: Unflow: unsupervised learning of optical flow with a bidirectional census loss. In: Thirty-Second AAAI Conference on Artificial Intelligence (2018)

24. Merolla, P.A., et al.: A million spiking-neuron integrated circuit with a scalable communication network and interface. Science **345**(6197), 668–673 (2014)

25. Orchard, G., Benosman, R.B., Etienne-Cummings, R., Thakor, N.V.: A spiking neural network architecture for visual motion estimation. In: 2013 IEEE Biomedical Circuits and Systems Conference (BioCAS), pp. 298–301 (2013)

26. Panda, P., Aketi, S.A., Roy, K.: Toward scalable, efficient, and accurate deep spiking neural networks with backward residual connections, stochastic softmax, and hybridization. Front. Neurosci. **14**, 653 (2020)

27. Paredes-Vallés, F., Scheper, K.Y.W., De Croon, G.C.H.E.: Unsupervised learning of a hierarchical spiking neural network for optical flow estimation: from events to global motion perception. IEEE Trans. Pattern Anal. Mach. Intell. **42**, 2051–2064 (2019). https://ieeexplore.ieee.org/abstract/document/8660483

28. Ronneberger, O., Fischer, P., Brox, T.: U-net: convolutional networks for biomedical image segmentation. CoRR abs/1505.04597 (2015). http://arxiv.org/abs/1505.04597

29. Rueckauer, B., Lungu, I.A., Hu, Y., Pfeiffer, M., Liu, S.C.: Conversion of continuous-valued deep networks to efficient event-driven networks for image classification. Front. Neurosci. **11**, 682 (2017)

30. Sun, D., Roth, S., Black, M.J.: A quantitative analysis of current practices in optical flow estimation and the principles behind them. Int. J. Comput. Vision **106**(2), 115–137 (2014). https://doi.org/10.1007/s11263-013-0644-x. http://dx.doi.org/10.1007/s11263-013-0644-x

31. Werbos, P.J.: Backpropagation through time: what it does and how to do it. Proc. IEEE **78**(10), 1550–1560 (1990)

32. Zhu, A.Z., Atanasov, N., Daniilidis, K.: Event-based feature tracking with probabilistic data association. In: 2017 IEEE International Conference on Robotics and Automation (ICRA), pp. 4465–4470, May 2017. https://doi.org/10.1109/ICRA.2017.7989517

33. Zhu, A.Z., Thakur, D., Özaslan, T., Pfrommer, B., Kumar, V., Daniilidis, K.: The multivehicle stereo event camera dataset: an event camera dataset for 3D perception. IEEE Robot. Autom. Lett. **3**(3), 2032–2039 (2018)

34. Zhu, A.Z., Yuan, L., Chaney, K., Daniilidis, K.: Ev-flownet: self-supervised optical flow estimation for event-based cameras. arXiv preprint arXiv:1802.06898 (2018)

35. Zhu, A.Z., Yuan, L., Chaney, K., Daniilidis, K.: Unsupervised event-based learning of optical flow, depth, and egomotion. In: Proceedings of the IEEE Conference on Computer Vision and Pattern Recognition, pp. 989–997 (2019)

Forgetting Outside the Box: Scrubbing Deep Networks of Information Accessible from Input-Output Observations

Aditya Golatkar[(✉)], Alessandro Achille, and Stefano Soatto

Department of Computer Science, University of California, Los Angeles,
Los Angeles, USA
{aditya29,achille,soatto}@cs.ucla.edu

Abstract. We describe a procedure for removing dependency on a cohort of training data from a trained deep network that improves upon and generalizes previous methods to different readout functions, and can be extended to ensure forgetting in the final activations of the network. We introduce a new bound on how much information can be extracted per query about the forgotten cohort from a black-box network for which only the input-output behavior is observed. The proposed forgetting procedure has a deterministic part derived from the differential equations of a linearized version of the model, and a stochastic part that ensures information destruction by adding noise tailored to the geometry of the loss landscape. We exploit the connections between the final activations and weight dynamics of a DNN inspired by Neural Tangent Kernels to compute the information in the final activations.

Keywords: Forgetting · Data removal · Neural tangent kernel · Information theory

1 Introduction

We study the problem of removing information pertaining to a given set of data points from the weights of a trained network. After removal, a potential attacker should not be able to recover information about the forgotten cohort. We consider both the cases in which the attacker has full access to the weights of the trained model, and the less-studied case where the attacker can only query the model by observing some input data and the corresponding output, for instance through a web Application Programming Interface (API). We show that we can quantify the maximum amount of information that an attacker can extract from observing inputs and outputs (black-box attack), as well as from direct

Electronic supplementary material The online version of this chapter (https://doi.org/10.1007/978-3-030-58526-6_23) contains supplementary material, which is available to authorized users.

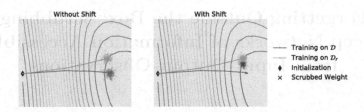

Fig. 1. Scrubbing procedure: PCA-projection of training paths on $\mathcal{D}$ (blue), $\mathcal{D}_r$ (orange) and the weights after scrubbing, using **(Left)** The Fisher method of [16], and **(Right)** the proposed scrubbing method. Our proposed scrubbing procedure (red cross) moves the model towards $w(\mathcal{D}_r)$, which reduces the amount of noise (point cloud) that needs to be added to achieve forgetting. (Color figure online)

knowledge of the weights (white-box), and propose tailored procedures for removing such information from the trained model in *one shot*. That is, assuming the model has been obtained by fine-tuning a pre-trained generic backbone, we compute a single perturbation of the weights that, in one go, can erase information about a cohort to be forgotten in such a way that an attacker cannot access it.

More formally, let a dataset $\mathcal{D}$ be partitioned into a subset $\mathcal{D}_f$ to be forgotten, and its complement $\mathcal{D}_r$ to be retained, $\mathcal{D} = \mathcal{D}_f \sqcup \mathcal{D}_r$. A (possibly stochastic) training algorithm A takes $\mathcal{D}_f$ and $\mathcal{D}_r$ and outputs a weight vector $w \in \mathbb{R}^p$:

$$A(\mathcal{D}_f, \mathcal{D}_r) \to w.$$

Assuming an attacker knows the training algorithm A (*e.g.*, stochastic gradient descent, or SGD), the weights w, and the retainable data $\mathcal{D}_r$, she can exploit their relationship to recover information about $\mathcal{D}_f$, at least for state-of-the-art deep neural networks (DNNs) using a "readout" function [16], that is, a function $R(w)$ that an attacker can apply on the weights of a DNN to extract information about the dataset. For example, an attacker may measure the confidence of the classifier on an input, or measure the time that it takes to fine-tune the network on a given subset of samples to decide whether that input or set was used to train the classifier (membership attack). We discuss additional examples in Sect. 5.4. Ideally, the forgetting procedure should be robust to different choices of readout functions. Recent work [16,17], introduces a "scrubbing procedure" (forgetting procedure or data deletion/removal) $S_{\mathcal{D}_f}(w) : \mathbb{R}^p \to \mathbb{R}^p$, that attempts to remove information about $\mathcal{D}_f$ from the weights, *i.e.*,

$$A(\mathcal{D}_f, \mathcal{D}_r) \to w \to S_{\mathcal{D}_f}(w)$$

with an upper-bound on the amount of information about $\mathcal{D}_f$ that can be extracted after the forgetting procedure, provided the attack has access to the scrubbed weights $S_{\mathcal{D}_f}(w)$, a process called "white-box attack."

Bounding the information that can be extracted from a white-box attack is often complex and may be overly restrictive: Deep networks have large sets of equivalent solutions – a "null space" – that would give the same activations on all test samples. Changes in $\mathcal{D}_f$ may change the position of the weights in the null space. Hence, the position of the weight in the null-space, even if irrelevant for the input-output behavior, may be exploited to recover information about $\mathcal{D}_f$.

This suggests that the study of forgetting should be approached from the perspective of the *final activations*, rather than the weights, since there could be infinitely many different models that produce the same input-output behavior, and we are interested in preventing attacks that affect the behavior of the network, rather than the specific solution to which the training process converged. More precisely, denote by $f_w(x)$ the final activations of a network on a sample x (for example the softmax or pre-softmax vector). We assume that an attacker makes queries on n images $\mathbf{x} = (x_1, \ldots, x_n)$, and obtains the activations $f_w(\mathbf{x})$. The pipeline then is described by the Markov Chain

$$A(\mathcal{D}_f, \mathcal{D}_r) \to w \to S_{\mathcal{D}_f}(w) \to f_{S_{\mathcal{D}_f}(w)}(\mathbf{x}).$$

The key question now is to determine how much information can an attacker recover about $\mathcal{D}_f$, starting from the final activations $f_{S_{\mathcal{D}_f}(w)}(\mathbf{x})$? We provide a new set of bounds that quantifies the average information per query an attacker can extract from the model.

The forgetting procedure we propose is obtained using the Neural Tangent Kernel (NTK). We show that this forgetting procedure is able to handle the null space of the weights better than previous approaches when using overparametrized models such as DNNs. In experiments, we confirm that it works uniformly better than previous proposals on all forgetting metrics introduced [16], *both in the white-box and black-box case* (Fig. 1).

Note that one may think of forgetting in a black-box setting as just changing the activations (*e.g.*, adding noise or hiding one class output) so that less information can be extracted. This, however, is not proper forgetting as the model still contains information, it is just not visible outside. We refer to forgetting as removing information from the weights, but we provide bounds for how much information can be extracted after scrubbing in the black-box case, and show that they are order of magnitudes smaller than the corresponding bounds for white boxes for the same target accuracy.

Key Contributions: To summarize, our contributions are as follow:

1. We introduce methods to scrub information from, and analyze the content of deep networks from their final activations (black-box attacks).
2. We introduce a "one-shot" forgetting algorithms that work better than the previous method [16] for both white-box and black-box attacks for DNNs.
3. This is possible thanks to an elegant connection between activations and weights dynamics inspired by the neural tangent kernel (NTK), which allows us to better deal with the null-space of the network weights. Unlike the NTK formalism, we do not need infinite-width DNNs.
4. We show that better bounds can be obtained against black-box attacks than white-box, which gives a better forgetting vs error trade-off curve.

2 Related Work

Differential Privacy: [12] aims to learn the parameters of a model in such a way that no information about any particular training sample can be recovered. This is a much stronger requirement than forgetting, where we only want to remove – *after* training is done – information about a given subset of samples. Given the stronger requirements, enforcing differential privacy is difficult for deep networks and often results in significant loss of accuracy [1,10].

Forgetting: The term "machine unlearning" was introduced by [9], who shows an efficient forgetting algorithm in the restricted setting of statistical query learning, where the learning algorithm cannot access individual samples. [30] provided the first framework for instantaneous data summarization with data deletion using robust streaming submodular optimization. [14] formalizes the problem of efficient data elimination, and provides engineering principles for designing forgetting algorithms. However, they only provide a data deletion algorithms for k-means clustering. [6] propose a forgetting procedure based on sharding the dataset and training multiple models. Aside from the storage cost, they need to retrain subset of the models, while we aim for one-shot forgetting. [5] proposed a forgetting method for logit-based classification models by applying linear transformation to the output logits, but do not remove information from the weights. [17] formulates data removal mechanisms using differential privacy, and provides an algorithm for convex problems based on a second order Newton update. They suggest applying this method on top of the features learned by a differentially private DNN, but do not provide a method to remove information from a DNN itself. [22] provides a projective residual update based method using synthetic data points to delete data points for linear regression based models. [15,24] provide a newton based method for computing the influence of a training point on the model predictions in the context of model interpretation and cross-validation, however, such an approach can also be used for data removal. Closer to us, [16] proposed a selective forgetting procedure for DNNs trained with SGD, using an information theoretic formulation and exploiting the stability of SGD [18]. They proposed a forgetting mechanism which involves a shift in weight space, and addition of noise to the weights to destroy information. They also provide an upper bound on the amount of remaining information in the weights of the network after applying the forgetting procedure. We extend this framework to activations, and show that using an NTK based scrubbing procedure uniformly improves the scrubbing procedure in all metrics that they consider.

Membership Inference Attacks: [19,21,31,32,34,35,37] try to guess if a particular sample was used for training a model. Since a model has forgotten only if an attacker cannot guess at better than chance level, these attacks serve as a good metric for measuring the quality of forgetting. In Fig. 2 we construct a black-box membership inference attack similar to the shadow model training approach in [34]. Such methods relate to model inversion methods [13] which aim to gain information about the training data from the model output.

Neural Tangent Kernel: [23,27] show that the training dynamics of a linearized version of a Deep Network — which are described by the so called NTK matrix — approximate increasingly better the actual training dynamics as the network width goes to infinity. [4,28] extend the framework to convolutional networks. [33] compute information-theoretic quantities using the closed form expressions for various quantities that can be derived in this settings. While we do use the infinite width assumption, we show that the same linearization framework and solutions are a good approximation of the network dynamics during fine-tuning, and use them to compute an optimal scrubbing procedure.

3   Out of the Box Forgetting

In this section, we derive an upper-bound for how much information can be extracted by an attacker that has black-box access to the model, that is, they can query the model with an image, and obtain the corresponding output. We will then use this to design a forgetting procedure (Sect. 4).

While the problem itself may seem trivial — can the relation $A(\mathcal{D}_f, \mathcal{D}_r) = w$ be inverted to extract $\mathcal{D}_f$? — it is made more complex by the fact that the algorithm is stochastic, and that the map may not be invertible, but still partially invertible, that is, only a subset of information about $\mathcal{D}_f$ can be recovered. Hence, we employ a more formal information-theoretic framework, inspired by [16] and that in turns generalizes Differential Privacy [12]. We provide a-posteriori bounds which provide tighter answers, and allow one to design and benchmark scrubbing procedures even for very complex models such as deep networks, for which a-priori bounds would be impossible or vacuous. In this work, we focus on a-posteriori bounds for Deep Networks and use them to design a scrubbing procedure.

3.1   Information Theoretic Formalism

We start by modifying the framework of [16], developed for the weights of a network, to the final activations. We expect an adversary to use a readout function applied to the final activations. Given a set of images $\mathbf{x} = (x_0, \ldots, x_n)$, we denote by $f_w(\mathbf{x}) = (f(x_0), \ldots, f(x_m))$ the concatenation of their respective final activations. Let $\mathcal{D}_f$ be the set of training data to forget, and let y be some function of $\mathcal{D}_f$ that an attacker wants to reconstruct (*i.e.*, y is some piece of information regarding the samples). To keep the notation uncluttered, we write $S_{\mathcal{D}_f}(w) = S(w)$ for the scrubbing procedure to forget $\mathcal{D}_f$. We then have the following Markov chain

$$y \longleftarrow \mathcal{D}_f \longrightarrow w \longrightarrow S(w) \longrightarrow f_{S(w)}(\mathbf{x})$$

connecting all quantities. Using the Data Processing Inequality [11] we have the following inequalities:

$$\underbrace{I(y; f_{S(w)}(\mathbf{x}))}_{\text{Recovered information}} \leq \underbrace{I(\mathcal{D}_f; f_{S(w)}(\mathbf{x}))}_{\text{Black-box upper bound}} \leq \underbrace{I(\mathcal{D}_f; S(w))}_{\text{White box upper-bound}} \tag{1}$$

where $I(x; y)$ denotes the Shannon Mutual Information between random variables x and y. Bounding the last term — which is a general bound on how much information an attacker with full access to the weights could extract — is the focus of [16]. In this work, we also consider the case where the attacker can only access the final activations, and hence focus on the central term. As we will show, if the number of queries is bounded then the central term provides a sharper bound compared to the black-box case.

3.2 Bound for Activations

The mutual information in the central term is difficult to compute,[1] but in our case has a simple upper-bound:

Lemma 1 (Computable bound on mutual information). *We have the following upper bound:*

$$I(\mathcal{D}_f; f_{S(w)}(\mathbf{x})) \leq \mathbb{E}_{\mathcal{D}_f}\big[\, \mathrm{KL}\left(p(f_{S(w)}(\mathbf{x})|\mathcal{D}_f \cup \mathcal{D}_r) \,\|\, p(f_{S_0(w)}(\mathbf{x})|\mathcal{D}_r)\right)\big], \quad (2)$$

where $p(f_{S(w)}(\mathbf{x})|\mathcal{D} = \mathcal{D}_f \cup \mathcal{D}_r)$ is the distribution of activations after training on the complete dataset $\mathcal{D}_f \sqcup \mathcal{D}_r$ and scrubbing. Similarly, $p(f_{S_0(w)}(\mathbf{x})|\mathcal{D} = \mathcal{D}_r)$ is the distribution of possible activations after training only on the data to retain $\mathcal{D}_r$ and applying a function S_0 (that does not depend on $\mathcal{D}_f$) to the weights.

The lemma introduces the important notion that we can estimate how much information we erased by comparing the activations of our model with the activations of a reference model that was trained in the same setting, but without $\mathcal{D}_f$. Clearly, if the activations after scrubbing are identical to the activations of a model that has never seen $\mathcal{D}_f$, they cannot contain information about $\mathcal{D}_f$.

We now want to convert this bound in a more practical expected information gain per query. This is not yet trivial due to them stochastic dependency of w on $\mathcal{D}$: based on the random seed ϵ used to train, we may obtain very different weights for the same dataset. Reasoning in a way similar to that used to obtain the "local forgetting bound" of [16], we have:

Lemma 2. *Write a stochastic training algorithm A as $A(\mathcal{D}, \epsilon)$, where ϵ is the random seed and $A(\mathcal{D}, \epsilon)$ is a deterministic function. Then,*

$$I(\mathcal{D}_f; f_{S(w)}(\mathbf{x})) \leq \mathbb{E}_{\mathcal{D}_f, \epsilon}\Big[\, \mathrm{KL}\left(p(f_{S(w_\mathcal{D})}(\mathbf{x})) \,\|\, p(f_{S_0(w_{\mathcal{D}_r})}(\mathbf{x}))\right)\Big] \quad (3)$$

where we call $w_\mathcal{D} = A(\mathcal{D}, \epsilon)$ the deterministic result of training on the dataset $\mathcal{D}$ using random seed ϵ. The probability distribution inside the KL accounts only for the stochasticity of the scrubbing map $S(w_\mathcal{D})$ and the baseline $S_0(w_{\mathcal{D}_r})$.

The expression above is general. To gain some insight, it is useful to write it for a special case where scrubbing is performed by adding Gaussian noise.

---

[1] Indeed, it is even difficult to define, as the quantities at play, $\mathcal{D}_f$ and $\mathbf{x}$, are fixed values, but that problem has been addressed by [2,3] and we do not consider it here.

3.3   Close Form Bound for Gaussian Scrubbing

We start by considering a particular class of scrubbing functions $S(w)$ in the form

$$S(w) = h(w) + n, \quad n \sim N(0, \Sigma(w, \mathcal{D})) \tag{4}$$

where $h(w)$ is a deterministic shift (that depends on $\mathcal{D}$ and w) and n is Gaussian noise with a given covariance (which may also depend on w and $\mathcal{D}$). We consider a baseline $S_0(w) = w + n'$ in a similar form, where $n' \sim N(0, \Sigma_0(w, \mathcal{D}_r))$.

Assuming that the covariance of the noise is relatively small, so that $f_{h(w)}(\mathbf{x})$ is approximately linear in w in a neighborhood of $h(w)$ (we drop $\mathcal{D}$ without loss of generality), we can easily derive the following approximation for the distribution of the final activations after scrubbing for a given random seed ϵ:

$$f_{S(w_\mathcal{D})}(\mathbf{x}) \sim N(f_{h(w_\mathcal{D})}(\mathbf{x}), \nabla_w f_{h(w_\mathcal{D})}(\mathbf{x}) \Sigma \nabla_w f_{h(w_\mathcal{D})}(\mathbf{x})^T) \tag{5}$$

where $\nabla_w f_{h(w_\mathcal{D})}(\mathbf{x})$ is the matrix whose row is the gradient of the activations with respect to the weights, for each sample in $\mathbf{x}$. Having an explicit (Gaussian) distribution for the activations, we can plug it in Lemma 2 and obtain:

Proposition 1. *For a Gaussian scrubbing procedure, we have the bounds:*

$$I(\mathcal{D}_f; S(w)) \leq \mathbb{E}_{\mathcal{D}_f, \epsilon}[\Delta w^T \Sigma_0^{-1} \Delta w + d(\Sigma, \Sigma_0)] \quad \textit{(white-box)} \tag{6}$$

$$I(\mathcal{D}_f; f_{S(w)}(\mathbf{x})) \leq \mathbb{E}_{\mathcal{D}_f, \epsilon}[\Delta f^T \Sigma_\mathbf{x}'^{-1} \Delta f + d(\Sigma_\mathbf{x}, \Sigma_\mathbf{x}')] \quad \textit{(black-box)} \tag{7}$$

where for a $k \times k$ matrix Σ we set $d(\Sigma, \Sigma_0) := \mathrm{tr}(\Sigma \Sigma_0^{-1}) + \log |\Sigma \Sigma_0^{-1}| - k$, and

$$\Delta w := h(w_\mathcal{D}) - w_{\mathcal{D}_r}, \quad \Delta f := f_{h(w_\mathcal{D})}(\mathbf{x}) - f_{w_{\mathcal{D}_r}}(\mathbf{x})$$

$$\Sigma_\mathbf{x} := \nabla_w f_{h(w_\mathcal{D})}(\mathbf{x}) \Sigma \nabla_w f_{h(w_\mathcal{D})}(\mathbf{x})^T, \quad \Sigma_\mathbf{x}' := \nabla_w f_{w_{\mathcal{D}_r}}(\mathbf{x}) \Sigma_0 \nabla_w f_{w_{\mathcal{D}_r}}(\mathbf{x})^T.$$

Avoiding Curse of Dimensionality: Proposition 1. Comparing Eq. 6 and Eq. 7, we see that the bound in Eq. 6 involves variables of the same dimension as the number of weights, while Eq. 7 scales with the number of query points. Hence, for highly overparametrized models such as DNNs, we expect that the black-box bound in Eq. 7 will be much smaller if the number of queries is bounded, which indeed is what we observe in the experiments (Fig. 3).

Blessing of the Null-Space: The white-box bound depends on the difference Δw in weight space between the scrubbed model and the reference model $w_{\mathcal{D}_r}$, while the black-box bound depends on the distance Δf in the activations. As we mentioned in Sect. 1, over-parametrized models such as deep networks have a large null-space of weights with similar final activations. It may hence happen that even if Δw is large, Δf may still be small (and hence the bound in Eq. 7 tighter) as long as Δw lives in the null-space. Indeed, we often observe this to be the case in our experiments.

Adversarial Queries: Finally, this should not lead us to think that whenever the activations are similar, little information can be extracted. Notice that the relevant quantity for the black box bound is $\Delta f(J_{\mathbf{x}} \Sigma_0 J_{\mathbf{x}}^T)^{-1} \Delta f^T$, which involves also the gradient $J_{\mathbf{x}} = \nabla_w f_{w_{\mathcal{D}_r}}(\mathbf{x})$. Hence, if an attacker crafts an adversarial query $\mathbf{x}$ such that its gradient $J_{\mathbf{x}}$ is small, they may be able to extract a large amount of information even if the activations are close to each other. In particular, this happens if the gradient of the samples lives in the null-space of the reference model, but not in that of the scrubbed model. In Fig. 3 (right), we show that indeed different images can extract different amount of information.

4   An NTK-Inspired Forgetting Procedure

We now introduce a new scrubbing procedure, which aims to minimize both the white-box and black-box bounds of Proposition 1. It relates to the one introduced in [16,17], but it enjoys better numerical properties and can be computed without approximations (Sect. 4.1). In Sect. 5 we show that it gives better results under all commonly used metrics.

The main intuition we exploit is that most networks commonly used are fine-tuned from pre-trained networks (*e.g.*, on ImageNet), and that the weights do not move much during fine-tuning on $\mathcal{D} = \mathcal{D}_r \cup \mathcal{D}_f$ will remain close to the pre-trained values. In this regime, the network activations may be approximated as a linear function of the weights. This is inspired by a growing literature on the so called Neural Tangent Kernel, which posits that large networks during training evolve in the same way as their linear approximation [27]. Using the linearized model we can derive an analytical expression for the optimal forgetting function, which we validate empirically. However, we observe this to be misaligned with weights actually learned by SGD, and introduce a very simple "isoceles trapezium" trick to realign the solutions (see Supplementary Material).

Using the same notation as [27], we linearize the final activations around the pre-trained weights θ_0 as:

$$f_t^{\mathrm{lin}}(x) \equiv f_0(x) + \nabla_\theta f_0(x)|_{\theta=\theta_0} w_t$$

where $w_t = \theta_t - \theta_0$ and gives the following expected training dynamics for respectively the weights and the final activations:

$$\dot{w}_t = -\eta \nabla_\theta f_0(\mathcal{D})^T \nabla_{f_t^{\mathrm{lin}}(\mathcal{D})} \mathcal{L} \tag{8}$$

$$\dot{f}_t^{\mathrm{lin}}(x) = -\eta \Theta_0(x, \mathcal{D}) \nabla_{f_t^{\mathrm{lin}}(\mathcal{D})} \mathcal{L} \tag{9}$$

The matrix $\Theta_0 = \nabla_\theta f(\mathcal{D}) \nabla_\theta f(\mathcal{D})^T$ of size $c|\mathcal{D}| \times c|\mathcal{D}|$, where c the number of classes, is called the Neural Tangent Kernel (NTK) matrix [23,27]. Using this dynamics, we can approximate in closed form the final training point when training with $\mathcal{D}$ and $\mathcal{D}_r$, and compute the optimal "one-shot forgetting" vector to jump from the weights $w_{\mathcal{D}}$ that have been obtained by training on $\mathcal{D}$ to the weights $w_{\mathcal{D}_r}$ that would have been obtained training on $\mathcal{D}_r$ alone:

Proposition 2. *Assuming an L_2 regression loss,[2] the optimal scrubbing procedure under the NTK approximation is given by*

$$\boxed{h_{NTK}(w) = w + P\nabla f_0(\mathcal{D}_f)^T MV} \tag{10}$$

where $\nabla f_0(\mathcal{D}_f)^T$ is the matrix whose columns are the gradients of the sample to forget, computed at θ_0 and $w = A(\mathcal{D}_r \cup \mathcal{D}_f)$. $P = I - \nabla f_0(\mathcal{D}_r)^T \Theta_{rr}^{-1} \nabla f_0(\mathcal{D}_r)$ is a projection matrix, that projects the gradients of the samples to forget $\nabla f_0(\mathcal{D}_f)$ onto the orthogonal space to the space spanned by the gradients of all samples to retain. The terms $M = [\Theta_{ff} - \Theta_{rf}^T \Theta_{rr}^{-1} \Theta_{rf}]^{-1}$ and $V = [(Y_f - f_0(\mathcal{D}_f)) + \Theta_{rf}^T \Theta_{rr}^{-1}(Y_r - f_0(\mathcal{D}_r))]$ re-weight each direction before summing them together.

Given this result, our proposed scrubbing procedure is:

$$\boxed{S_{\text{NTK}}(w) = h_{\text{NTK}}(w) + n} \tag{11}$$

where $h_{\text{NTK}}(w)$ is as in Eq. 10, and we use noise $n \sim N(0, \lambda F^{-1})$ where $F = F(w)$ is the Fisher Information Matrix computed at $h_{\text{NTK}}(w)$ using $\mathcal{D}_r$. The noise model is as in [16], and is designed to increase robustness to mistakes due to the linear approximation.

4.1 Relation Between NTK and Fisher Forgetting

In [16] and [17], a different forgetting approach is suggested based on either the Hessian or the Fisher Matrix at the final point: assuming that the solutions $w(\mathcal{D}_r)$ and $w(\mathcal{D})$ of training with and without the data to forget are close and that they are both minima of their respective loss, one may compute the shift to jump from one minimum to the other of a slightly perturbed loss landscape. The resulting "scrubbing shift" $w \mapsto h(w)$ relates to the newton update:

$$h(w) = w - H(\mathcal{D}_r)^{-1} \nabla_w L_{\mathcal{D}_r}(w). \tag{12}$$

In the case of an L_2 loss, and using the NTK model, the Hessian is given by $H(w) = \nabla f_0(\mathcal{D}_r)^T \nabla f_0(\mathcal{D}_r)$ which in this case also coincides with the Fisher Matrix [29]. To see how this relates to the NTK matrix, consider determining the convergence point of the linearized NTK model, that for an L_2 regression is given by $w^* = w_0 + \nabla f_0(\mathcal{D}_r)^+ \mathcal{Y}$, where $\nabla f_0(\mathcal{D}_r)^+$ denotes the matrix pseudo-inverse, and $\mathcal{Y}$ denotes the regression targets. If $\nabla f_0(\mathcal{D}_r)$ is a tall matrix (more samples in the dataset than parameters in the network), then the pseudo-inverse is $\nabla f_0(\mathcal{D}_r)^+ = H^{-1} \nabla f_0(\mathcal{D}_r)^T$, recovering the scrubbing procedure considered by [16,17]. However, if the matrix is wide (more parameters than samples in the network, as is often the case in Deep Learning), the Hessian is not invertible, and the pseudo-inverse is instead given by $\nabla f_0(\mathcal{D}_r)^+ = \nabla f_0(\mathcal{D}_r)^T \Theta^{-1}$, leading

---

[2] This assumption is to keep the expression simple, in the Supplementary Material we show the corresponding expression for a softmax classification loss.

to our proposed procedure. In general, when the model is over-parametrized there is a large null-space of weights that do not change the final activations or the loss. The degenerate Hessian is not informative of where the network will converge in this null-space, while the NTK matrix gives the exact point.

5   Experiments

5.1   Datasets

We report experiments on smaller versions of CIFAR-10 [26] and Lacuna-10 [16], a dataset derived from the VGG-Faces [8] dataset. We obtain the small datasets using the following procedure: we randomly sample 500 images (100 images from each of the first 5 classes) from the training/test set of CIFAR-10 and Lacuna-10 to obtain the small-training/test respectively. We also sample 125 images from the training set (5 classes × 25 images) to get the validation set. So, in short, we have 500 (5 × 100) examples for training and testing respectively, and 125 (5 × 25) examples for validation. On both the datasets we choose to forget 25 random samples (5% of the dataset). Without loss of generality we choose to forget samples from class 0.

5.2   Models and Training

We use All-CNN [36] (to which we add batch-normalization before non-linearity) and ResNet-18 [20] as the deep neural networks for our experiments. We pre-train the models on CIFAR-100/Lacuna-100 and then fine-tune them (all the weights) on CIFAR-10/Lacuna-10. We pre-train using SGD for 30 epochs with a learning rate of 0.1, momentum 0.9 and weight decay 0.0005. Pre-training helps in improving the stability of SGD while fine-tuning. For fine-tuning we use a learning rate of 0.01 and weight decay 0.1. While applying weight decay we bias the weights with respect to the initialization. During training we always use a batch-size of 128 and fine-tune the models till zero training error. Also, during fine-tuning we do not update the running mean and variance of the batch-normalization parameters to simplify the training dynamics. We perform each experiment 3 times and report the mean and standard deviation.

5.3   Baselines

We consider three baselines for comparison: (i) **Fine-tune**, we fine-tune $w(\mathcal{D})$ on $\mathcal{D}_r$ (similar to catastrophic forgetting) and (ii) **Fisher forgetting** [16], we scrubs the weights by adding Gaussian noise using the inverse of the Fisher Information Matrix as covariance matrix, (iii) **Original** corresponds to the original model trained on the complete dataset ($w(\mathcal{D})$) without any forgetting. We compare those, and our proposal, with optimal reference the model $w(\mathcal{D}_r)$ trained from scratch on the retain set, that is, without using $\mathcal{D}_f$ in the first place. Values read from this reference model corresponds to the green region in Fig. 2 and represent the gold standard for forgetting: In those plots, an optimal algorithm should lie inside the green area.

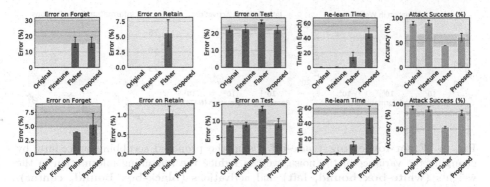

Fig. 2. Comparison of different models baselines (original, finetune) and forgetting methods (Fisher [16] and our NTK proposed method), using several readout functions (**(Top)** CIFAR and (**Bottom**) Lacuna). We benchmark them against a model that has never seen the data (the gold reference for forgetting): values (mean and standard deviation) measured from this models corresponds to the green region. Optimal scrubbing procedure should lie in the green region, or they will leak information about $\mathcal{D}_f$. We compute three read-out functions: **(a)** Error on forget set $\mathcal{D}_f$, **(b)** Error on retain set $\mathcal{D}_r$, **(c)** Error on test set $\mathcal{D}_{\text{test}}$. **(d)** Black-box membership inference attack: We construct a simple yet effective membership attack using the entropy of the output probabilities. We measures how often the attack model (using the activations of the scrubbed network) classify a sample belonging $\mathcal{D}_f$ as a training sample rather than being fooled by the scrubbing. **(e)** Re-learn time for different scrubbing methods: How fast a scrubbed model learns the forgotten cohort when fine-tuned on the complete dataset. We measure the re-learn time as the first epoch when the loss on $\mathcal{D}_f$ goes below a certain threshold. (Color figure online)

5.4 Readout Functions

We use multiple readout functions similar to [16]: (i) **Error on residual** (should be small), (ii) **Error on cohort to forget** (should be similar to the model re-trained from scratch on $\mathcal{D}_r$), (iii) **Error on test set** (should be small), (iv) **Re-learn time**, measures how quickly a scrubbed model learns the cohort to forget, when fine-tuned on the complete data. Re-learn time (measured in epochs) is the first epoch when the loss during fine-tuning (the scrubbed model) falls below a certain threshold (loss of the original model (model trained on $\mathcal{D}$) on $\mathcal{D}_f$). (v) **Blackbox membership inference attack**: We construct a simple yet effective blackbox membership inference attack using the entropy of the output probabilities of the scrubbed model. Similar to the method in [34], we formulate the attack as a binary classification problem (class 1 - belongs to training set and class 0 - belongs to test set). For training the attack model (Support Vector Classifier with Radial Basis Function Kernel) we use the retain set ($\mathcal{D}_r$) as class 1 and the test set as class 0. We test the success of the attack on the cohort to forget ($\mathcal{D}_f$). Ideally, the attack accuracy for an optimally scrubbed model should be the same as a model re-trained from scratch on $\mathcal{D}_f$, having a higher value implies incorrect (or no) scrubbing, while a lower value may result in

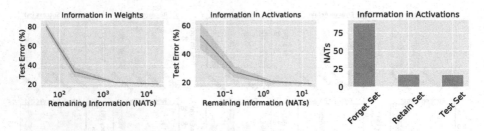

Fig. 3. Error-forgetting trade-off. Using the proposed scrubbing procedure, by changing the variance of the noise, we can reduce the remaining information in the weights **(white-box bound, left)** and activations **(black-box bound, center)**. However, it comes at the cost of increasing the test error. Notice that the bound on activation is much sharper than the bound on error at the same accuracy. **(Right) Different samples leak different information.** An attacker querying samples from $\mathcal{D}_f$ can gain much more information than querying unrelated images. This suggest that adversarial samples may be created to leak even more information.

Streisand Effect, (vi) **Remaining information in the weights** [16] and (vii) **Remaining information in the activations**: We compute an upper bound on the information the activations contain about the cohort to forget $(\mathcal{D}_f)$ (after scrubbing) when queried with images from different subsets of the data $(\mathcal{D}_r, \mathcal{D}_r)$.

5.5 Results

Error Readouts: In Fig. 2(a–c), we compare error based readout functions for different forgetting methods. Our proposed method outperforms Fisher forgetting which incurs high error on the retain $(\mathcal{D}_r)$ and test $(\mathcal{D}_{\text{Test}})$ set to attain the same level of forgetting. This is due the large distance between $w(\mathcal{D})$ and $w(\mathcal{D}_r)$ in weight space, which forces it to add too much noise to erase information about $\mathcal{D}_f$, and ends up also erasing information about the retain set $\mathcal{D}_r$ (high error on $\mathcal{D}_r$ in Fig. 2). Instead, our proposed method first moves $w(\mathcal{D})$ in the direction of $w(\mathcal{D}_r)$, thus minimizing the amount of noise to be added (Fig. 1). Fine-tuning the model on $\mathcal{D}_r$ (catastrophic forgetting) does not actually remove information from the weights and performs poorly on all the readout functions.

Relearn Time: In Fig. 2(d), we compare the re-learn time for different methods. Re-learn time can be considered as a proxy for the information remaining in the weights about the cohort to forget $(\mathcal{D}_f)$ after scrubbing. We observe that the proposed method outperforms all the baselines which is in accordance with the previous observations (in Fig. 2(a–c)).

Membership Attacks: In Fig. 2(e), we compare the robustness of different scrubbed models against blackbox membership inference attacks (attack aims to identify if the scrubbed model was ever trained on $\mathcal{D}_f$). This can be considered as a proxy for the remaining information (about $\mathcal{D}_f$) in the activations. We observe that attack accuracy for the proposed method lies in the optimal region (green),

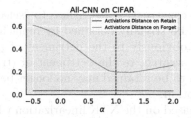
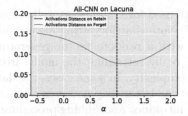

Fig. 4. Scrubbing brings activations closer to the target. We plot the L_1 norm of the difference between the final activations (post-softmax) of the target model trained only on $\mathcal{D}_r$, and models sampled along the line joining the original model $w(D)$ ($\alpha = 0$) and the proposed scrubbed model ($\alpha = 1$). The distance between the activations decreases as we move along the scrubbing direction. The L_1 distance is already low on the retain set ($\mathcal{D}_r$) (red) as it corresponds to the data common to $w(\mathcal{D})$ and $w(\mathcal{D}_r)$. However, the two models differ on the forget set ($\mathcal{D}_f$) (blue) and we observe that the L_1 distance decreases as move along the proposed scrubbing direction. (Color figure online)

while Fine-tune does not forget ($\mathcal{D}_f$), Fisher forgetting may result in Streisand effect which is undesireable.

Closeness of Activations: In Fig. 4, we show that the proposed scrubbing method brings the activations of the scrubbed model closer to retrain model (model retrained from scratch on $\mathcal{D}_r$). We measure the closeness by computing: $\mathbb{E}_{x \sim \mathcal{D}_f / \mathcal{D}_r}[\|f_{\text{scrubbed}}(x) - f_{\text{retrain}}(x)\|_1]$ along the scrubbing direction, where $f_{\text{scrubbed}}(x)$, $f_{\text{retrain}}(x)$ are the activations (post soft-max) of the proposed scrubbed and retrain model respectively. The distance between the activations on the cohort to forget ($\mathcal{D}_f$) decreases as we move along the scrubbing direction and achieves a minimum value at the scrubbed model, while it almost remains constant on the retain set. Thus, the activations of the scrubbed model shows desirable behaviour on both $\mathcal{D}_r$ and $\mathcal{D}_f$.

Error-Forgetting Trade-off: In Fig. 3, we plot the trade-off between the test error and the remaining information in the weights and activations respectively by changing the scale of the variance of Fisher noise. We can reduce the remaining information but this comes at the cost of increasing the test error. We observe that the black-box bound on the information accessible with one query is much tighter than the white box bound at the same accuracy (compare left and center plot x-axes). Finally, in (right), we show that query samples belonging to the cohort to be forgotten ($\mathcal{D}_f$) leaks more information about the $\mathcal{D}_f$ rather than the retain/test set, proving that indeed carefully selected samples are more informative to an attacker than random samples.

6   Discussion

Recent work [3,16,17] has started providing insights on both the amount of information that can be extracted from the weights about a particular cohort of the data used for training, as well as give constructive algorithms to "selectively forget." Note that forgetting alone could be trivially obtained by replacing the model with a random vector generator, obviously to the detriment of performance, or by retraining the model from scratch, to the detriment of (training time) complexity. In some cases, the data to be retained may no longer be available, so the latter may not even be an option.

We introduce a scrubbing procedure based on the NTK linearization which is designed to minimize both a white-box bound (which assumes the attacker has the weights), and a newly introduced black-box bound. The latter is a bound on the information that can be obtained about a cohort using only the observed input-output behavior of the network. This is relevant when the attacker performs a bounded number of queries. If the attacker is allowed infinitely many observations, the matter of whether the black-box and white-box attack are equivalent remains open: Can an attacker always craft sufficiently exciting inputs so that the exact values of all the weights can be inferred? An answer would be akin to a generalized "Kalman Decomposition" for deep networks. This alone is an interesting open problem, as it has been pointed out recently that good "clone models" can be created by copying the response of a black-box model on a relatively small set of exciting inputs, at least in the restricted cases where every model is fine-tuned from a common pre-trained models [25].

While our bounds are tighter that others proposed in the literature, our model has limitations. Chiefly, computational complexity. The main source of computational complexity is computing and storing the matrix P in Eq. 10, which naively would require $O(c^2|\mathcal{D}_r|^2)$ memory and $O(|w| \cdot c^2|\mathcal{D}_r|^2)$ time. For this reason, we conduct experiments on a relatively small scale, sufficient to validate the theoretical results, but our method is not yet scalable to production level. However, we notice that P is the projection matrix on the orthogonal of the subspace spanned by the training samples, and there is a long history [7] of numerical methods to incrementally compute such operator incrementally and without storing it fully in memory. We leave these promising option to scale the method as subject of future investigation.

Acknowledgement. We would like to thank the anonymous reviewers for their feedback and suggestions. This work is supported by ARO W911NF-17-1-0304 and ONR N00014-19-1-2229.

References

1. Abadi, M., et al.: Deep learning with differential privacy. In: Proceedings of the 2016 ACM SIGSAC Conference on Computer and Communications Security, pp. 308–318. ACM (2016)
2. Achille, A., Soatto, S.: Emergence of invariance and disentanglement in deep representations. J. Mach. Learn. Res. **19**(1), 1947–1980 (2018)
3. Achille, A., Soatto, S.: Where is the Information in a Deep Neural Network? arXiv e-prints arXiv:1905.12213, May 2019

4. Arora, S., Du, S.S., Hu, W., Li, Z., Salakhutdinov, R.R., Wang, R.: On exact computation with an infinitely wide neural net. In: Advances in Neural Information Processing Systems, pp. 8139–8148 (2019)
5. Baumhauer, T., Schöttle, P., Zeppelzauer, M.: Machine unlearning: linear filtration for logit-based classifiers. arXiv preprint arXiv:2002.02730 (2020)
6. Bourtoule, L., et al.: Machine unlearning. arXiv preprint arXiv:1912.03817 (2019)
7. Bunch, J.R., Nielsen, C.P., Sorensen, D.C.: Rank-one modification of the symmetric eigenproblem. Numer. Math. **31**(1), 31–48 (1978)
8. Cao, Q., Shen, L., Xie, W., Parkhi, O.M., Zisserman, A.: VGGFace2: a dataset for recognising faces across pose and age. In: International Conference on Automatic Face and Gesture Recognition (2018)
9. Cao, Y., Yang, J.: Towards making systems forget with machine unlearning. In: 2015 IEEE Symposium on Security and Privacy, pp. 463–480. IEEE (2015)
10. Chaudhuri, K., Monteleoni, C., Sarwate, A.D.: Differentially private empirical risk minimization. J. Mach. Learn. Res. **12**, 1069–1109 (2011)
11. Cover, T.M., Thomas, J.A.: Elements of Information Theory. Wiley, New York (2012)
12. Dwork, C., Roth, A., et al.: The algorithmic foundations of differential privacy. Found. Trends Theor. Comput. Sci. **9**(3–4), 211–407 (2014)
13. Fredrikson, M., Jha, S., Ristenpart, T.: Model inversion attacks that exploit confidence information and basic countermeasures. In: Proceedings of the 22nd ACM SIGSAC Conference on Computer and Communications Security, pp. 1322–1333. ACM (2015)
14. Ginart, A., Guan, M., Valiant, G., Zou, J.Y.: Making AI forget you: data deletion in machine learning. In: Advances in Neural Information Processing Systems, pp. 3513–3526 (2019)
15. Giordano, R., Stephenson, W., Liu, R., Jordan, M., Broderick, T.: A swiss army infinitesimal jackknife. In: The 22nd International Conference on Artificial Intelligence and Statistics, pp. 1139–1147 (2019)
16. Golatkar, A., Achille, A., Soatto, S.: Eternal sunshine of the spotless net: selective forgetting in deep networks (2019)
17. Guo, C., Goldstein, T., Hannun, A., van der Maaten, L.: Certified data removal from machine learning models. arXiv preprint arXiv:1911.03030 (2019)
18. Hardt, M., Recht, B., Singer, Y.: Train faster, generalize better: stability of stochastic gradient descent. arXiv preprint arXiv:1509.01240 (2015)
19. Hayes, J., Melis, L., Danezis, G., De Cristofaro, E.: Logan: membership inference attacks against generative models. Proc. Priv. Enhancing Technol. **2019**(1), 133–152 (2019)
20. He, K., Zhang, X., Ren, S., Sun, J.: Deep residual learning for image recognition. In: Proceedings of the IEEE Conference on Computer Vision and Pattern Recognition, pp. 770–778 (2016)
21. Hitaj, B., Ateniese, G., Perez-Cruz, F.: Deep models under the GAN: information leakage from collaborative deep learning. In: Proceedings of the 2017 ACM SIGSAC Conference on Computer and Communications Security, pp. 603–618. ACM (2017)
22. Izzo, Z., Smart, M.A., Chaudhuri, K., Zou, J.: Approximate data deletion from machine learning models: algorithms and evaluations. arXiv preprint arXiv:2002.10077 (2020)
23. Jacot, A., Gabriel, F., Hongler, C.: Neural tangent kernel: convergence and generalization in neural networks. In: Advances in Neural Information Processing Systems, pp. 8571–8580 (2018)

24. Koh, P.W., Liang, P.: Understanding black-box predictions via influence functions. arXiv preprint arXiv:1703.04730 (2017)
25. Krishna, K., Tomar, G.S., Parikh, A.P., Papernot, N., Iyyer, M.: Thieves on sesame street! model extraction of bert-based APIS (2019)
26. Krizhevsky, A., et al.: Learning multiple layers of features from tiny images. Technical report, Citeseer (2009)
27. Lee, J., et al.: Wide neural networks of any depth evolve as linear models under gradient descent. In: Advances in Neural Information Processing Systems, pp. 8570–8581 (2019)
28. Li, Z., et al.: Enhanced convolutional neural tangent kernels. arXiv preprint arXiv:1911.00809 (2019)
29. Martens, J.: New insights and perspectives on the natural gradient method. arXiv preprint arXiv:1412.1193 (2014)
30. Mirzasoleiman, B., Karbasi, A., Krause, A.: Deletion-robust submodular maximization: data summarization with "the right to be forgotten". In: International Conference on Machine Learning, pp. 2449–2458 (2017)
31. Pyrgelis, A., Troncoso, C., De Cristofaro, E.: Knock knock, who's there? membership inference on aggregate location data. arXiv preprint arXiv:1708.06145 (2017)
32. Sablayrolles, A., Douze, M., Ollivier, Y., Schmid, C., Jégou, H.: White-box vs black-box: Bayes optimal strategies for membership inference. arXiv preprint arXiv:1908.11229 (2019)
33. Schwartz-Ziv, R., Alemi, A.A.: Information in infinite ensembles of infinitely-wide neural networks. arXiv preprint arXiv:1911.09189 (2019)
34. Shokri, R., Shmatikov, V.: Privacy-preserving deep learning. In: Proceedings of the 22nd ACM SIGSAC Conference on Computer and Communications Security, pp. 1310–1321. ACM (2015)
35. Song, C., Ristenpart, T., Shmatikov, V.: Machine learning models that remember too much. In: Proceedings of the 2017 ACM SIGSAC Conference on Computer and Communications Security, pp. 587–601. ACM (2017)
36. Springenberg, J.T., Dosovitskiy, A., Brox, T., Riedmiller, M.: Striving for simplicity: the all convolutional net. arXiv preprint arXiv:1412.6806 (2014)
37. Truex, S., Liu, L., Gursoy, M.E., Yu, L., Wei, W.: Demystifying membership inference attacks in machine learning as a service. IEEE Trans. Serv. Comput. (2019)

Inherent Adversarial Robustness of Deep Spiking Neural Networks: Effects of Discrete Input Encoding and Non-linear Activations

Saima Sharmin[1]([✉]) [iD], Nitin Rathi[1] [iD], Priyadarshini Panda[2] [iD],
and Kaushik Roy[1] [iD]

[1] Purdue University, West Lafayette, IN 47907, USA
{ssharmin,rathi2,kaushik}@purdue.edu
[2] Yale University, New Haven, CT 06520, USA
priya.panda@yale.edu

Abstract. In the recent quest for trustworthy neural networks, we present Spiking Neural Network (SNN) as a potential candidate for inherent robustness against adversarial attacks. In this work, we demonstrate that adversarial accuracy of SNNs under gradient-based attacks is higher than their non-spiking counterparts for CIFAR datasets on deep VGG and ResNet architectures, particularly in blackbox attack scenario. We attribute this robustness to two fundamental characteristics of SNNs and analyze their effects. First, we exhibit that input discretization introduced by the Poisson encoder improves adversarial robustness with reduced number of timesteps. Second, we quantify the amount of adversarial accuracy with increased leak rate in Leaky-Integrate-Fire (LIF) neurons. Our results suggest that SNNs trained with LIF neurons and smaller number of timesteps are more robust than the ones with IF (Integrate-Fire) neurons and larger number of timesteps. Also we overcome the bottleneck of creating gradient-based adversarial inputs in temporal domain by proposing a technique for crafting attacks from SNN (https://github.com/ssharmin/spikingNN-adversarial-attack).

Keywords: Spiking neural networks · Adversarial attack ·
Leaky-integrate-fire neuron · Input discretization

1 Introduction

Adversarial attack is one of the biggest challenges against the success of today's deep neural networks in mission critical applications [10,16,32]. The underlying concept of an adversarial attack is to purposefully modulate the input to a neural

Electronic supplementary material The online version of this chapter (https://doi.org/10.1007/978-3-030-58526-6_24) contains supplementary material, which is available to authorized users.

network such that it is subtle enough to remain undetectable to human eyes, yet capable of fooling the network into incorrect decisions. This malicious behavior was first demonstrated in 2013 by Szegedy *et al.* [27] and Biggio *et al.* [4] in the field of computer vision and malware detection, respectively. Since then, numerous defense mechanisms have been proposed to address this issue. One category of defense includes fine-tuning the network parameters like adversarial training [11,18], network distillation [22], stochastic activation pruning [8] etc. Another category focuses on preprocessing the input before passing through the network like thermometer encoding [5], input quantization [21,31], compression [12] etc. Unfortunately, most of these defense mechanisms have been proved futile by many counter-attack techniques. For example, an ensemble of defenses based on "gradient-masking" collapsed under the attack proposed in [1]. Defensive distillation was broken by Carlini-Wagner method [6,7]. Adversarial training has the tendency to overfit to the training samples and remain vulnerable to transfer attacks [28]. Hence, the threat of adversarial attack continues to persist.

In the absence of adversarial robustness in the existing state-of-the-art networks, we feel there is a need for a network with *inherent* susceptibility against adversarial attacks. In this work, we present Spiking Neural Network (SNN) as a potential candidate due to two of its fundamental distinctions from the non-spiking networks:

1. SNNs operate based on discrete binary data (0/1), whereas their non-spiking counterparts, referred as Analog Neural Network (ANN), take in continuous-valued analog signals. Since SNN is a binary spike-based model, input discretization is a constituent element of the network, most commonly done by Poisson encoding.
2. SNNs employ nonlinear activation function of the biologically inspired Integrate-Fire (IF) or Leaky-Integrate-Fire (LIF) neurons, in contrast to the piecewise-linear ReLU activations used in ANNs.

Among the handful of works done in the field of SNN adversarial attacks [2, 19], most of them are restricted to either simple datasets (MNIST) or shallow networks. However, this work extends to complex datasets (CIFAR) as well as deep SNNs which can achieve comparable accuracy to the state-of-the-art ANNs [23,24]. For robustness comparison with non-spiking networks, we analyze two different types of spiking networks: (1) converted SNN (trained by ANN-SNN conversion [24]) and (2) backpropagated SNN (an ANN-SNN converted network, further incrementally trained by surrogate gradient backpropagation [23]). We identify that converted SNNs fail to demonstrate more robustness than ANNs. Although authors in [25] show similar analysis, we explain with experiments the reason behind this discrepancy and, thereby, establish the necessary criteria for an SNN to become adversarially robust. Moreover, we propose an SNN-crafted attack generation technique with the help of the surrogate gradient method. We summarize our contribution as follows:

- We show that the adversarial accuracies of SNNs are higher than ANNs under a gradient-based blackbox attack scenario, where the respective clean accuracies are comparable to each other. The attacks were performed on deep VGG

and ResNet architectures trained on CIFAR10 and CIFAR100 datasets. For whitebox attacks, the comparison is dependent on the relative strengths of the adversary.
- The increased robustness of SNN is attributed to two fundamental characteristics: input discretization through Poisson encoding and non-linear activations of LIF (or IF) neurons.
  - We investigate how adversarial accuracy changes with the number of timesteps (inference latency) used in SNN for different levels of input pre-quantization[1]. In case of backpropagated SNNs (trained with smaller number of timesteps), the amount of discretization as well as adversarial robustness increases as we reduce the number of timesteps. Pre-quantization of the analog input brings about further improvement. However, converted SNNs appear to depend only on the input pre-quantization, but invariant to the variation in the number of timesteps. Since these converted SNNs operate under larger number of timesteps [24], discretization effect is minimized by input averaging, and hence, the observed invariance.
  - We show that piecewise-linear activation (ReLU) in ANN linearly propagates the adversarial perturbation throughout the network, whereas LIF (or IF) neurons diminish the effect of perturbation at every layer. Additionally, the leak factor in LIF neurons offers an extra knob to control the adversarial perturbation. We perform a quantitative analysis to demonstrate the effect of leak on the adversarial robustness of SNNs.

  Overall, we show that SNNs employing LIF neurons, trained with surrogate gradient-based backpropagation, and operating at less number of timesteps are more robust than SNNs trained with ANN-SNN conversion that requires IF neurons and more number of timesteps. Hence, the training technique plays a crucial role in fulfilling the prerequisites for an adversarially robust SNN.
- Gradient-based attack generation in SNNs is non-trivial due to the discontinuous gradient of the LIF (or IF) neurons. We propose a methodology to generate attacks based on the approximate surrogate gradients.

2 Background

2.1 Adversarial Attack

Given a clean image x belonging to class i and a trained neural network M, an adversarial image x_{adv} needs to meet two criteria:

1. x_{adv} is visually "similar" to x i.e. $|x - x_{adv}| = \epsilon$, where ϵ is a small number.
2. x_{adv} is misclassified by the neural network, i.e. $M(x_{adv}) \neq i$

The choice of the distance metric $|.|$ depends on the method used to create x_{adv} and ϵ is a hyper-parameter. In most methods, l_2 or l_∞ norm is used to measure

---

[1] Full-precision analog inputs are quantized to lower bit precision values before undergoing the discretization process by the Poisson encoder of SNN.

the similarity and the value of ϵ is limited to $\leq \frac{8}{255}$ where normalized pixel intensity $x \in [0, 1]$ and original $x \in [0, 255]$.

In this work, we construct adversarial examples using the following two methods:

Fast Gradient Sign Method (FGSM). This is one of the simplest methods for constructing adversarial examples, introduced in [11]. For a given instance x, true label y_{true} and the corresponding cost function of the network $J(x, y_{true})$, this method aims to search for a perturbation δ such that it maximizes the cost function for the perturbed input $J(x + \delta, y_{true})$, subject to the constraint $|\delta|_\infty < \epsilon$. In closed form, the attack is formulated as,

$$x_{adv} = x + \epsilon \times sign(\nabla_x J(x, y_{true}))$$ (1)

Here, ϵ denotes the strength of the attack.

Projected Gradient Descent (PGD). This method, proposed in [18], produces more powerful adversary. PGD is basically a *k-step* variant of FGSM computed as,

$$x_{adv}^{(k+1)} = \Pi_{x+\epsilon} \left\{ \left(x_{adv}^{(k)} + \alpha \times sign\left(\nabla_x \left(J(x_{adv}^{(k)}, y_{true}) \right) \right) \right) \right\}$$ (2)

where $x_{adv}^{(0)} = x$ and $\alpha(\leq \epsilon)$ refers to the amount of perturbation used per iteration or step, k is the total number of iterations. $\Pi_{x+\epsilon}\{.\}$ performs a projection of its operand on an ϵ-ball around x, *i.e.*, the operand is clipped between $x + \epsilon$ and $x - \epsilon$. Another variant of this method adds a random perturbation of strength ϵ to x before performing PGD operation.

2.2   Spiking Neural Network (SNN)

The main difference between an SNN and an ANN is the concept of time. The incoming signals as well as the intermediate node inputs/outputs in an ANN are static analog values, whereas in an SNN, they are binary spikes with a value of 0 or 1, which are also functions of time. In the input layer, a Poisson event generation process is used to convert the continuous valued analog signals into binary spike train. Suppose the input image is a 3-D matrix of dimension $h \times w \times l$ with pixel intensity in the range $[0, 1]$. At every time step of the SNN operation, a random number (from the normal distribution $\mathcal{N}(0, 1)$) is generated for each of these pixels. A spike is triggered at that particular time step if the corresponding pixel intensity is greater than the generated random number. This process continues for a total of T timesteps to produce a spike train for each pixel. Hence, the size of the input spike train is $T \times h \times w \times l$. For a large enough T, the timed average of the spike train will be proportional to its analog value. Every node of the SNN is accompanied with a neuron. Among many neuron models, the most commonly used ones are Integrate-Fire (IF) or Leaky-Integrate-Fire (LIF) neurons. The dynamics of the neuron membrane potential at time $t+1$ is described by,

$$V(t + 1) = \lambda V(t) + \sum_i w_i x_i(t) \tag{3}$$

Here $\lambda = 1$ for IF neurons and < 1 for LIF neurons. w_i denotes the synaptic weight between current neuron and i-th pre-neuron. $x_i(t)$ is the input spike from the i-th pre-neuron at time t. When $V(t + 1)$ reaches the threshold voltage V_{th}, an output spike is generated and the membrane potential is reset to 0, or in case of soft reset, reduced by the amount of the threshold voltage. At the output layer, inference is performed based on the cumulative membrane potential of the output neurons after the total time T has elapsed.

One of the main shortcomings of SNNs is that they are difficult to train, especially for deeper networks. Since the neurons in an SNN have discontinuous gradients, standard gradient-descent techniques do not directly apply. In this work, we use two of the supervised training algorithms [23, 24], which achieve ANN-like accuracy even for deep neural networks and on complex datasets.

ANN-SNN Conversion. This training algorithm was originally outlined in [9] and subsequently improved in [24] for deep networks. Note that the algorithm is suited for training SNNs with IF neurons only. They propose a threshold-balancing technique that sequentially tunes the threshold voltage of each layer. Given a trained ANN, the first step is to generate the input Poisson spike train for the network over the training set for a large enough time-window so that the timed average accurately describes the input. Next, the maximum value of $\sum_i w_i x_i$ (term 2 in Eq. 3) received by layer 1 is recorded over the entire time range for several minibatches of the training data. This is referred as the maximum activation for layer 1. The threshold value of layer 1 neuron is replaced by this maximum activation keeping the synaptic weights unchanged. Such threshold tuning operation ensures that the IF neuron activity precisely mimics the ReLU function in the corresponding ANN. After balancing layer 1 threshold, the method is continued for all subsequent layers sequentially.

Conversion and Surrogate-Gradient Backpropagation. In order to take advantage of the standard backpropagation-based optimization procedures, authors in [3, 30, 33] introduced the surrogate gradient technique. The input-output characteristics of an LIF (or IF) neuron is a step function, the gradient of which is discontinuous at the threshold point (Fig. 1). In surrogate gradient technique, the gradient is approximated by pseudo-derivatives like linear or exponential functions. Authors in [23] proposed a novel approximation function for these

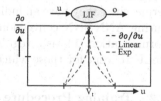

Fig. 1. Surrogate gradient approximation of an LIF neuron.

gradients by utilizing the spike time information in the derivative. The gradient at timestep t is computed as follows:

$$\frac{\partial o^t}{\partial u^t} = \alpha e^{-\beta \Delta t} \tag{4}$$

Here, o^t is the output spike at time t, u^t is the membrane potential at t, Δt is the difference between current timestep and the last timestep post-neuron generated a spike. α and β are hyperparameters. Once the neuron gradients are approximated, backpropagation through time (BPTT) [29] is performed using the chain rule of derivatives. In BPTT, the network is unrolled over all timesteps. The final output is computed as the cumulation of outputs at every timestep and eventually, loss is defined on the summed output. During backward propagation of the loss, the gradients are accumulated over time and used in gradient-descent optimization. Authors in [23] proposed a hybrid training procedure in which the surrogate-gradient training is preceded by an ANN-SNN conversion to initialize the weights and thresholds of the network. The advantage of this method over ANN-SNN conversion is twofold: one can train a network with both IF and LIF neurons and the number of timesteps required for training is reduced by a factor of 10 without losing accuracy.

3 Experiments

3.1 Dataset and Models

We conduct our experiments on VGG5 and ResNet20 for CIFAR10 dataset and VGG11 for CIFAR100. The network topology for VGG5 consists of conv3,64-avgpool-conv3,128 ($\times$2)-avgpool-fc1024 ($\times$2)-fc10. Here conv3,64 refers to a convolutional layer with 64 output filters and 3$\times$3 kernel size. fc1024 is a fully-connected layer with 1024 output neurons. VGG11 contains 11 weight layers corresponding to the configuration A in [26] with maxpool layers replaced by average pooling. ResNet20 follows the proposed architecture for CIFAR10 in [13], except the initial 7×7 non-residual convolutional layer is replaced by a series of two 3×3 convolutional layers. For ANN-SNN conversion of ResNet20, threshold balancing is performed only on these initial non-residual units (as demonstrated by [24]). The neurons (in both ANN and SNN) contain no bias terms, since they have an indirect effect on the computation of threshold voltages during ANN-SNN conversion. The absence of bias eliminates the use of batch normalization [14] as a regularizer. Instead, a dropout layer is used after every ReLU (except for those which are followed by a pooling layer).

3.2 Training Procedure

The aim of our experiment is to compare adversarial attack on 3 networks: 1) ANN, 2) SNN trained by ANN-SNN conversion and 3) SNN trained by back-propagation, with initial conversion. These networks will be referred as ANN, SNN-conv and SNN-BP, respectively from this point onward.

For both CIFAR10 and CIFAR100 datasets, we follow the data augmentation techniques in [17]: 4 pixels are padded on each side, and a 32×32 crop is randomly sampled from the padded image or its horizontal flip. Testing is performed on the original 32 × 32 images. Both training and testing data are normalized to [0, 1]. For training the ANNs, we use cross-entropy loss with stochastic gradient descent optimization (weight decay = 0.0001, mometum = 0.9). VGG5 (and ResNet20) are trained for a total of 200 epochs, with an initial learning rate of 0.1 (0.05), which is divided by 10 at 100-th (80-th) and 150-th (120-th) epoch. VGG11 with CIFAR100 is trained for 250 epochs with similar learning schedule. During training SNN-conv networks, a total of 2500 timesteps are used for all VGG and ResNet architectures. SNN-BP networks are trained for 15 epochs with cross-entropy loss and adam [15] optimizer (weight decay = 0.0005). Initial learning rate is 0.0001, which is halved every 5 epochs. A total of 100 timesteps is used for VGG5 and 200 timesteps for ResNet20 & VGG11. Training is performed with either linear surrogate gradient approximation [3] or spike time dependent approximation [23] with $\alpha = 0.3$, β=0.01 (in Eq. 4). Both techniques yield approximately similar results. Leak factor λ is kept at 0.99 in all cases, except in the analysis for the leak effect.

In order to analyze the effect of input quantization (with varying number of timesteps) and leak factors, only VGG5 networks with CIFAR10 dataset is used.

3.3   Adversarial Input Generation Methodology

For the purpose of whitebox attacks, we need to construct adversarial samples from all three networks (ANN, SNN-conv, SNN-BP). The ANN-crafted FGSM and PGD attacks are generated using the standard techniques described in Eq. 1 and 2, respectively. We carry out non-targeted attacks with $\epsilon = 8/255$. PGD attacks are performed with iteration steps $k = 7$ and per-step perturbation $\alpha = 2/255$. FGSM or PGD method cannot be directly applied to SNN due to its discontinuous gradient problem (described in Sect. 2.2). To that end, we outline a surrogate-gradient based FGSM (and PGD) technique. In SNN, analog input X is converted to Poisson spike train X_{spike} which is fed into the 1st convolutional layer. If X_{rate} is the timed average of X_{spike}, the membrane potential of the 1st convolutional layer X_{conv1} can be approximated as

$$X_{conv1} \approx Conv(X_{rate}, W_{conv1}) \tag{5}$$

W_{conv1} is the weight of the 1st convolutional layer. From this equation, the sign of the gradient of the network loss function J w.r.t. X_{rate} or X is described by (detailed derivation is provided in *supplementary*),

$$sign\left(\frac{\partial J}{\partial X}\right) \approx sign\left(\frac{\partial J}{\partial X_{rate}}\right) = sign\left(Conv\left(\frac{\partial J}{\partial X_{conv1}}, W_{conv1}^{180rotated}\right)\right) \tag{6}$$

Surrogate gradient technique yields $\frac{\partial J}{\partial X_{conv1}}$ from SNN, which is plugged into Eq. 6 to calculate the sign of the input gradient. This sign matrix is later used to compute X_{adv} according to standard FGSM or PGD method. The algorithm is summarized in Algorithm 1.

Algorithm 1. SNN-crafted X_{adv} : $FGSM$

Require: Input (X, y_{true}), Trained SNN (N) with loss function J.

Ensure: $\frac{\partial J}{\partial X_{conv1}} \leftarrow 0$

    for timestep t in total time T **do**

        forward: Loss $J(X, y_{true})$

        backward : Accumulate gradient $\frac{\partial J}{\partial X_{conv1}} + = X_{conv1}.grad$

    end for

    post-processing: $sign(\frac{\partial J}{\partial X}) = sign\left(Conv(\frac{\partial J}{\partial X_{conv1}}, W_{conv1}^{180rotated})\right)$

    SNN-crafted adversary: $X_{adv}^{SNN} = X + \epsilon \times sign(\frac{\partial J}{\partial X})$

---

4   Results

4.1   ANN vs SNN

Table 1 summarizes our results for CIFAR10 (VGG5 & ResNet20) and CIFAR100 (VGG11) datasets in whitebox and blackbox settings. For each architecture, we start with three networks: ANN, SNN-conv and SNN-BP, trained to achieve comparable baseline clean accuracy. Let us refer them as M_{ANN}, $M_{SNN-conv}$ and M_{SNN-BP}, respectively. During blackbox attack, we generate an adversarial test dataset x_{adv} from a separately trained ANN of the same network topology as the target model but different initialization. It is clear that the adversarial accuracy of SNN-BP during FGSM and PGD blackbox attacks is higher than the corresponding ANN and SNN-conv models, irrespective of the size of the dataset or network architecture (the highest value of the accuracy for each attack case is highlighted by *orange text*). The amount of improvement in adversarial accuracy, compared to ANN, is listed as Δ in the Table. If M_{ANN} and M_{SNN-BP} yield adversarial accuracy of $p_{ANN}\%$ and $p_{SNN-BP}\%$, respectively, the value of Δ amounts to $p_{SNN-BP}\% - p_{ANN}\%$. On the other hand, during whitebox attack, we generate three sets of adversarial test dataset: $x_{adv,ANN}$ (generated from M_{ANN}), $x_{adv,SNN-conv}$ (generated from $M_{SNN-conv}$) and so on. Since ANN and SNN have widely different operating dynamics and constituent elements, the strength of the constructed adversary varies significantly from ANN to SNN during whitebox attack (demonstrated in Sect. 4.2). SNN-BP shows significant improvement in whitebox adversarial accuracy (Δ ranging from 2% to 4.6%) for both VGG and Resnet architectures with CIFAR10 dataset. In contrast, VGG11 ANN with CIFAR100 manifests higher whitebox accuracy than SNN-BP. We attribute this discrepancy to the difference in adversary-strength of ANN & SNN for different dataset and network architectures. From Table 1, it is evident that SNN-BP networks exhibit the highest amount of adversarial accuracy (*orange text*) among the three networks in all blackbox attack cases (attacked by a common adversary), whereas SNN-conv and ANN demonstrate comparable accuracy, irrespective of the dataset, network topology or attack generation method. Hence, we conclude that SNN-BP is inherently more robust compared to their non-spiking counterpart as well as SNN-conv models, when all three networks are attacked by identical adversarial inputs. It is important to men-

Table 1. A comparison of the clean and adversarial (FGSM and PGD) test accuracy (%) among ANN, SNN-conv and SNN-BP networks. Highest value of the accuracy for each attack case is marked in *orange text*. FGSM accuracy is calculated at $\epsilon = 8/255$. For PGD, $\epsilon = 8/255$, α (per-step perturbation) $= 2/255$, k (number of steps) $= 7$. The blackbox attacks are generated from a separately trained ANN of the same network topology as the target model but different initialization

		Whitebox				Blackbox			
		ANN	SNN-conv	SNN-BP	$\Delta^\dagger$	ANN	SNN-conv	SNN-BP	$\Delta^\dagger$
					CIFAR10				
VGG5	Clean	90%	89.9%	89.3%	—	90%	89.9%	89.3%	—
	FGSM	10.4%	7.7%	15%	**4.6%**	18.9%	19.3%	21.5%	**2.6%**
	PGD	1.8%	1.7%	3.8%	**2.0%**	9.3%	9.6%	16.0%	**6.7%**
ResNet20	Clean	88.0%	87.5%	86.1%	—	88.0%	87.5%	86.1%	—
	FGSM	28.9%	28.8%	31.3%	**2.4%**	56.7%	56.8%	56.8%	**0.1%**
	PGD	1.9%	1.4%	4.9%	**3.0%**	41.5%	41.6%	46.5%	**5.0%**
					CIFAR100				
VGG11	Clean	67.1%	66.8%	64.4%	—	67.1%	66.8%	64.4%	—
	FGSM	17.1%	10.5%	15.5%	**-1.6%**	21.2%	21.4%	21.4%	**0.2%**
	PGD	8.5%	4.1%	6.3%	**-2.2%**	15.6%	15.8%	16.5%	**0.9%**

$^\dagger \Delta$ = Adversarial accuracy (SNN-BP) - Adversarial accuracy (ANN)

tion here that our conclusion is validated for VGG & ResNet architectures and gradient-based attacks only.

In the next two subsections, we explain two characteristics of SNNs contributing towards this robustness, as well as the reason for SNN-conv not being able to show similar behavior.

Effect of Input Quantization and Number of Timesteps. The main idea behind non-linear input pre-processing as a defense mechanism is to discretize continuous-valued input signals so that the network becomes non-transparent to adversarial perturbations, as long as they lie within the discretization bin. SNN is a binary spike-based network, which demands encoding any analog valued input signal into binary spike train, and hence, we believe it has the inherent robustness. In our SNN models, we employ Poisson rate encoding, where the output spike rate is proportional to the input pixel intensity. However, the amount of discretization introduced by the Poisson encoder varies with the number of timesteps used. Hence, the adversarial accuracy of the network can be controlled by varying the number of timesteps as long as the clean accuracy remains within reasonable limit. This effect can be further enhanced by quantizing the analog input before feeding into the Poisson encoder. In Fig. 2(a), we demonstrate the

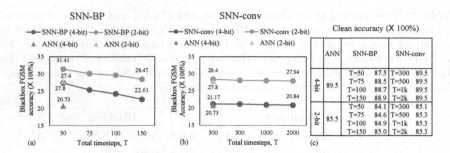

Fig. 2. Blackbox FGSM accuracy(%) versus total number of timesteps (T) plot with 4-bit (*blue*) and 2-bit (*orange*) input quantization for (a) SNN-BP and (b) SNN-conv networks. SNN-BP adversarial accuracy increases drastically with decreased number of timesteps, whereas SNN-conv is insensitive to it. (c) A table summarizing the clean accuracy of ANN, SNN-conv and SNN-BP for different input quantizations and number of timesteps. (Color figure online)

FGSM adversarial accuracy of an SNN-BP network (VGG5) trained for 50, 75, 100 and 150 timesteps with CIFAR10 dataset. As number of timesteps drop from 150 to 50, accuracy increases by ~5% (*blue line*) for 4-bit input quantization. Note that clean accuracy drops by only 1.4% within this range, from 88.9% (150 timesteps) to 87.5% (50 timesteps), as showed in the table in Fig. 2(c). Additional reduction of the number of timesteps leads to larger degradation of clean accuracy. The adversarial accuracy for corresponding ANN (with 4-bit input quantization) is showed in *gray triangle* in the same plot for comparison. Further increase in adversarial accuracy is obtained by pre-quantizing the analog inputs to 2-bits (*orange line*) and it follows the same trend with number of timesteps. Thus varying the number of timesteps introduces an extra knob for controlling the level of discretization in SNN-BP in addition to the input pre-quantizations. In contrast, in Fig. 2(b), similar experiments performed on SNN-conv network demonstrate little increase in adversarial accuracy with the number of timesteps. Only pre-quantization of input signal causes improvement of accuracy from ~21% to ~28%. Note that the range of the number of timesteps used for SNN-conv (300 to 2000) is much higher than SNN-BP, because converted networks have higher inference latency. The reason behind the invariance of SNN-conv towards the number of timesteps is explained in Fig. 3. We plot the input-output characteristics of the Poisson-encoder for 4 cases: (b) 4-bit input quantization with smaller number of timesteps (50 and 150), (c) 4-bit quantization, larger number of timesteps (300 and 2000) and their 2-bit counterparts in (d) and (e), respectively. It is evident from (c) and (e) that larger number of timesteps introduces more of an averaging effect, than quantization, and hence, varying the number of timesteps has negligible effect on the transfer plots (*solid dotted* and *dashed* lines coincide), which is not true for (b) and (d). Due to this averaging effect, Poisson output y for SNN-conv tends to follow the trajectory of x_{quant} (quantized ANN input), leading to comparable adversarial accuracy

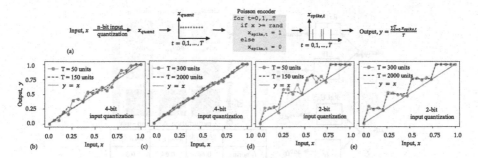

Fig. 3. The input-output characteristics of Poisson encoder to demonstrate the effect of the total number of timseteps T used to generate the spike train with different levels of pre-quantization of the analog input. When T is in the low value (between 50 to 150) regime (subplots (b) and (d)), the amount of quantization significantly changes for varying the number of timesteps (*solid dotted and dashed lines*). But in the high value regime of T (plots (c) and (e)), *solid dotted* and *dashed* lines almost coincide due to the averaging effect. The flow of data from output y to input x is showed in the schematic in (a)

to the corresponding ANN over the entire range of timesteps in Fig. 2(b). Note that, in these plots input x refers to the analog input signal, whereas output y is the timed average of the spike train (as showed in the schematic in Fig. 3(a)).

Effect of LIF (or IF) Neurons and the Leak Factor. Another major contributing factor towards SNN robustness is their highly nonlinear neuron activations (Integrate-Fire or Leaky-Integrate-Fire), whereas ANNs use mostly piecewise linear activations like ReLU. In order to explain the effect of this non-linearity, we perform a proof of concept experiment. We feed a clean and corresponding adversarial input to a ReLU and an LIF neuron ($\lambda = 0.99$ in Eq. 3). Both of the inputs are 32×32 images with pixel intensity normalized to $[0, 1]$. Row 1, 2 and 3 in Fig. 4 present the clean image, corresponding adversarial image and their absolute difference (amount of perturbation), respectively. Note, the outputs of the LIF neurons are binary at each timestep, hence, we take an average over the entire time-window to obtain corresponding pixel intensity. ReLU passes both clean and adversarial inputs without any transformation, hence the l_2-norm of the perturbation is same at the input and ReLU output (bottom table of the figure). However, the non-linear transformation in LIF reduces the perturbation of 0.6 at input layer to 0.3 at its output. Basically, the output images of LIF (*column 3*) neurons is a low pixel version of the input images, due to the translation of continuous analog values into a binary spike representation. This behavior helps diminish the propagation of adversarial perturbation through the network. IF neurons also demonstrate this non-linear transformation. However, the quantization effect is minimized due to their operation over longer time-window (as explained in the previous section).

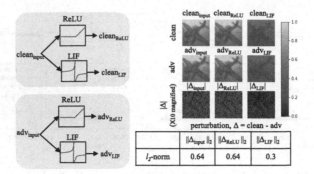

Fig. 4. The input and output of ReLU and LIF neurons for each of clean and adversarial image. Column 1 shows clean image, adversarial image and the absolute value of the adversarial perturbation before passing through the neurons. Column 2 and 3 depict the corresponding images after passing through a ReLU and an LIF, respectively. The bottom table contains the l_2-*norm* of the perturbation at the input, ReLU output and LIF output

Unlike SNN-conv, SNN-BP networks can be trained with LIF neurons. The leak factor in an LIF neuron provides an extra knob to manipulate the adversarial robustness of these networks. In order to investigate the effect of leak on the amount of robustness, we develop a simple expression relating the leak factor with neuron spike rate in an LIF neuron. In this case, the membrane

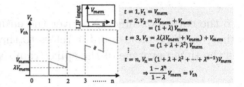

Fig. 5. Output of an LIF neuron for constant input voltage.

potential V_t at timestep t is updated as $V_t = \lambda V_{t-1} + V_{input,t}$ given the membrane potential has not reached threshold yet, and hence, reset signal = 0. Here, $\lambda \, (< 1)$ is the leak factor and $V_{input,t}$ is the input to the neuron at timestep t. Let us consider the scenario, where a constant voltage V_{mem} is fed into the neuron at every timestep and the membrane potential reaches the threshold voltage V_{th} after n timesteps. As explained in Fig. 5, membrane potential follows a geometric progression with time. After replacing $\frac{V_{th}}{V_{mem}}$ with a constant r, we obtain the following relation between the rate of spike $(1/n)$ and leak factor (λ):

$$\text{Spike rate,} \quad \frac{1}{n} = \frac{log\lambda}{log[r\lambda - (r-1)]}, \lambda < 1 \tag{7}$$

In Fig. 6(a), we plot the spike rate as a function of leak factor λ for different values of V_{mem} according to Eq. 7, where λ is varied from 0.9999 to 0.75. In every case, spike rate decreases (*i.e.* sparsity and hence, robustness increases) with increased amount of leak (smaller λ). The plot in Fig. 6(b) justifies this idea where we show the adversarial accuracy of an SNN-BP network (VGG5 with CIFAR10) trained with different values of leak. For both FGSM and PGD

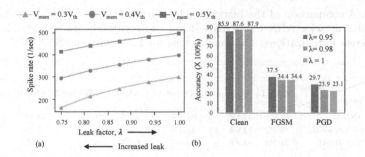

Fig. 6. (a) Spike rate versus leak factor λ for different values of $\frac{V_{mem}}{V_{th}}$. Smaller value of λ corresponds to more leak. (b) A bar plot showing the comparison of clean, FGSM and PGD($\epsilon = 8/255$) accuracy for a VGG5 SNN-BP network trained on CIFAR10 for different values of λ. These are blackbox attacks crafted from a VGG5 ANN model

attacks, adversarial accuracy increases by 3–6% as λ is decreased to 0.95. Note, the clean accuracy of the trained SNNs with different leak factors lies within a range of $\sim$2%. In addition to sparsity, leak makes the membrane potential (in turn, the output spike rate) dependent on the temporal information of the incoming spike train [20]. Therefore, for a given input, while the IF neuron produces a deterministic spike pattern, the input-output spike mapping is non-deterministic in an LIF neuron. This effect gets enhanced with increased leak. We assume that this phenomenon is also responsible to some extent for the increased robustness of backpropagated SNNs with increased leak. It is worth mentioning here that Eq. 7 holds when input to the neuron remains unchanged with the leak factor. In our experiments, we train SNN-BP with different values of λ starting from the same initialized ANN-SNN converted network. Hence, the parameters of SNN-BP trained with different leak factors do not vary much from one another. Therefore, the assumption in the equation applies to our results.

4.2   ANN-Crafted vs SNN-Crafted Attack

Lastly, we propose an attack-crafting technique from SNN with the aid of the surrogate gradient calculation. The details of the method is explained in Sect. 3.3. Table 2 summarizes a comparison between ANN-crafted and SNN-crafted (our proposed technique) attacks. Note, these are blackbox attacks, *i.e.*, we train two separate and independently initialized models for each of the 3 networks (ANN, SNN-conv, SNN-BP). One of them is used as the source (*marked as ANN-I, SNN-conv-I etc.*) and the other ones as the target (*marked as ANN-II, SNN-conv-II etc.*). It is clear that SNN-BP adversarial accuracy (last row) is the highest for both SNN-crafted and ANN-crafted inputs. Moreover, let us analyze row 1 of Table 2 for FGSM attack. When ANN-II is attacked by ANN-I, FGSM accuracy is 18.9%, whereas, if attacked by an SNN-conv-I (or SNN-BP-I), the accuracy is 32.7% (or 31.3%). Hence, these results suggest that ANN-crafted attacks are stronger than the corresponding SNN counterparts.

412 S. Sharmin et al.

Table 2. A comparison of the blackbox adversarial accuracy for ANN-crafted *versus* SNN-crafted attacks. ANN-I and ANN-II are two separately trained VGG5 networks with different initializations. The same is true for SNN-conv and SNN-BP.

Source / Target	FGSM			PGD		
	ANN-I	SNN-conv-I	SNN-BP-I	ANN-I	SNN-conv-I	SNN-BP-I
ANN-II	18.9%	32.7%	31.3%	4.7%	31.7%	13.8%
SNN-conv-II	19.2%	33.0%	31.4%	11.6%	32.4%	14.3%
SNN-BP-II	21.5%	38.8%	32.9%	9.7%	43.6%	17.0%

5 Conclusions

The current defense mechanisms in ANNs are incapable of preventing a range of adversarial attacks. In this work, we show that SNNs are inherently resilient to gradient-based adversarial attacks due to the discrete nature of input encoding and non-linear activation functions of LIF (or IF) neurons. The resiliency can be further improved by reducing the number of timesteps in the input-spike generation process and increasing the amount of leak of the LIF neurons. SNNs trained using ANN-SNN conversion technique (with IF neurons) require larger number of timesteps for inference than the corresponding SNNs trained with spike-based backpropagation (with LIF neurons). Hence, the latter technique leads to more robust SNNs. Our conclusion is validated only for gradient-based attacks on deep VGG and ResNet networks with CIFAR datasets. Future analysis on more diverse attack methods and architectures is necessary. We also propose a method to generate gradient-based attacks from SNNs by using the surrogate gradients.

Acknowledgement. The work was supported in part by, Center for Brain-inspired Computing (C-BRIC), a DARPA sponsored JUMP center, Semiconductor Research Corporation, National Science Foundation, Intel Corporation, the DoD Vannevar Bush Fellowship and U.S. Army Research Laboratory.

References

1. Athalye, A., Carlini, N., Wagner, D.: Obfuscated gradients give a false sense of security: circumventing defenses to adversarial examples. In: Proceedings of the International Conference on Machine Learning (ICML) (2018)
2. Bagheri, A., Simeone, O., Rajendran, B.: Adversarial training for probabilistic spiking neural networks. In: 2018 IEEE 19th International Workshop on Signal Processing Advances in Wireless Communications (SPAWC) (2018)
3. Bellec, G., Salaj, D., Subramoney, A., Legenstein, R., Maass, W.: Long short-term memory and learning-to-learn in networks of spiking neurons. In: Advances in Neural Information Processing Systems, pp. 787–797 (2018)
4. Biggio, B., et al.: Evasion attacks against machine learning at test time. In: Blockeel, H., Kersting, K., Nijssen, S., Železný, F. (eds.) ECML PKDD 2013. LNCS (LNAI), vol. 8190, pp. 387–402. Springer, Heidelberg (2013). https://doi.org/10.1007/978-3-642-40994-3_25

5. Buckman, J., Roy, A., Raffel, C., Goodfellow, I.: Thermometer encoding: one hot way to resist adversarial examples. In: International Conference on Learning Representations (ICLR) (2018)
6. Carlini, N., Wagner, D.: Defensive distillation is not robust to adversarial examples (2016), arXiv:1607.04311
7. Carlini, N., Wagner, D.: Towards evaluating the robustness of neural networks. In: 2017 IEEE Symposium on Security and Privacy (SP). pp. 39–57 (2017)
8. Dhillon, G.S., et al.: Stochastic activation pruning for robust adversarial defense. In: International Conference on Learning Representations (ICLR) (2018)
9. Diehl, P.U., Neil, D., Binas, J., Cook, M., Liu, S.C., Pfeiffer, M.: Fast-classifying, high-accuracy spiking deep networks through weight and threshold balancing. In: 2015 International Joint Conference on Neural Networks (IJCNN), pp. 1–8 (2015)
10. Eykholt, K., et al.: Robust physical world attacks on deep learning visual classifications. In: CVPR (2018)
11. Goodfellow, I.J., Shlens, J., Szegedy, C.: Explaining and harnessing adversarial examples. In: International Conference on Learning Representations (ICLR) (2015)
12. Guo, C., Rana, M., Cisse, M., van der Maaten, L.: Countering adversarial images using input transformations (2018)
13. He, K., Zhang, X., Ren, S., Sun, J.: Deep residual learning for image recognition. Technical report, Microsoft Research, arXiv: 1512.03385 (2015)
14. Ioffe, S., Szegedy, C.: Batch normalization: accelerating deep network training by reducing internal covariate shift. In: Proceedings of International Conference on Machine Learning (ICML), pp. 448–456 (2015)
15. Kingma, D.P., Ba., J.L.: Adam: a method for stochastic optimization. In: International Conference on Learning Representations (ICLR) (2014)
16. Kurakin, A., Goodfellow, I., Bengio, S.: Adversarial examples in the physical world. In: International Conference on Learning Representations (ICLR) (2017)
17. Lee, C.Y., Xie, S., Gallagher, P., Zhang, Z., Tu, Z.: Deeply-supervised nets (2014)
18. Madry, A., Makelov, A., Schmidt, L., Tsipras, D., Vladu, A.: Towards deep learning models resistant to adversarial attacks. In: International Conference on Learning Representations (ICLR) (2018)
19. Marchisio, A., Nanfa, G., Khalid, F., Hanif, M.A., Martina, M., Shafique, M.: SNN under attack: are spiking deep belief networks vulnerable to adversarial examples? (2019)
20. Olin-Ammentorp, W., Beckmann, K., Schuman, C.D., Plank, J.S., Cady, N.C.: Stochasticity and robustness in spiking neural networks (2019)
21. Panda, P., Chakraborty, I., Roy, K.: Discretization based solutions for secure machine learning against adversarial attacks. IEEE Access 7, 70157–70168 (2019)
22. Papernot, N., McDaniel, P., Wu, X., Jha, S., Swami, A.: Distillation as a defense to adversarial perturbations against deep neural networks. In: 2016 IEEE Symposium on Security and Privacy (SP), pp. 582–597 (2016)
23. Rathi, N., Srinivasan, G., Panda, P., Roy, K.: Enabling deep spiking neural networks with hybrid conversion and spike timing dependent backpropagation. In: International Conference on Learning Representations (ICLR) (2020)
24. Sengupta, A., Ye, Y., Wang, R., Liu, C., Roy, K.: Going deeper in spiking neural networks: VGG and residual architectures. Front. Neurosci. 13, 95 (2019)
25. Sharmin, S., Panda, P., Sarwar, S.S., Lee, C., Ponghiran, W., Roy, K.: A comprehensive analysis on adversarial robustness of spiking neural networks. In: 2019 International Joint Conference on Neural Networks (IJCNN) (2019)

26. Simonyan, K., Zisserman, A.: Very deep convolutional networks for large-scale image recognition. In: International Conference on Learning Representations (ICLR) (2015)
27. Szegedy, C., et al.: Intriguing properties of neural networks. In: International Conference on Learning Representations (ICLR) (2014)
28. Tramèr, F., Kurakin, A., Papernot, N., Goodfellow, I., Boneh, D., McDaniel, P.: Ensemble adversarial training: attacks and defenses. In: International Conference on Learning Representations (ICLR) (2018)
29. Werbos, P.J.: Backpropagation through time: what it does and how to do it. Proc. IEEE **78**(10), 1550–1560 (1990)
30. Wu, Y., Deng, L., Li, G., Zhu, J., Shi, L.: Spatio-temporal backpropagation for training high-performance spiking neural networks. Front. Neurosci. **12**, 331 (2018)
31. Xu, W., Evans, D., Qi, Y.: Feature squeezing: Detecting adversarial examples in deep neural networks. In: Network and Distributed Systems Security Symposium (NDSS) (2018)
32. Xu, W., Qi, Y., Evans, D.: Automatically evading classifiers. In: Network and Distributed Systems Security Symposium (NDSS) (2016)
33. Zenke, F., Ganguli, S.: Superspike: Supervised learning in multilayer spiking neural networks. Neural Comput. **30**(6), 1514–1541 (2018)

Synthesizing Coupled 3D Face Modalities by Trunk-Branch Generative Adversarial Networks

Baris Gecer[1,2]($\boxtimes$) (iD), Alexandros Lattas[1,2] (iD), Stylianos Ploumpis[1,2] (iD),
Jiankang Deng[1,2] (iD), Athanasios Papaioannou[1,2] (iD), Stylianos Moschoglou[1,2] (iD),
and Stefanos Zafeiriou[1,2] (iD)

[1] Imperial College, London, UK
{b.gecer,a.lattas,s.ploumpis,j.deng16,a.papaioannou11,
stylianos.moschoglou15,s.zafeiriou}@imperial.ac.uk,
https://ibug.doc.ic.ac.uk/
[2] FaceSoft.io, London, UK

Abstract. Generating realistic 3D faces is of high importance for computer graphics and computer vision applications. Generally, research on 3D face generation revolves around linear statistical models of the facial surface. Nevertheless, these models cannot represent faithfully either the facial texture or the normals of the face, which are very crucial for photo-realistic face synthesis. Recently, it was demonstrated that Generative Adversarial Networks (GANs) can be used for generating high-quality textures of faces. Nevertheless, the generation process either omits the geometry and normals, or independent processes are used to produce 3D shape information. In this paper, we present the first methodology that generates high-quality texture, shape, and normals jointly, which can be used for photo-realistic synthesis. To do so, we propose a novel GAN that can generate data from different modalities while exploiting their correlations. Furthermore, we demonstrate how we can condition the generation on the expression and create faces with various facial expressions. The qualitative results shown in this paper are compressed due to size limitations, full-resolution results and the accompanying video can be found in the supplementary documents. The code and models are available at the project page: https://github.com/barisgecer/TBGAN.

Keywords: Synthetic 3D Face · Face generation · Generative Adversarial Networks · 3D morphable models · Facial expression generation

1 Introduction

Generating 3D faces with high-quality texture, shape, and normals is of paramount importance in computer graphics, movie post-production, computer

Electronic supplementary material The online version of this chapter (https://doi.org/10.1007/978-3-030-58526-6_25) contains supplementary material, which is available to authorized users.

© Springer Nature Switzerland AG 2020
A. Vedaldi et al. (Eds.): ECCV 2020, LNCS 12374, pp. 415–433, 2020.
https://doi.org/10.1007/978-3-030-58526-6_25

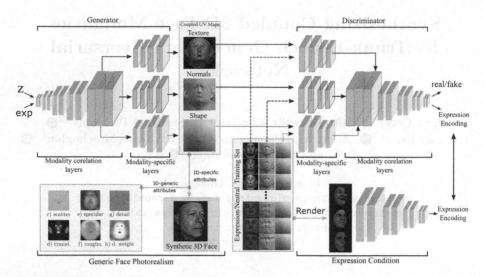

Fig. 1. We propose a novel GAN that can synthesize high-quality texture, shape, and normals jointly for realistic and coherent 3D faces of novel identities. The separation of branch networks allows the specialization of the characteristicofeach one of the modalities while the trunk network maintains the local correspondences among them. Moreover, we demonstrate how we can condition the generation on the expression and create faces with various facial expressions. We annotate the training dataset automatically by an expression recognition network to couple those expression encodings to the texture, shape, and normals UV maps.

games, etc. Other applications of such approaches include generating synthetic training data for face recognition [23] and modeling the face manifold for 3D face reconstruction [24]. Currently, 3D face generation in computer games and movies is performed by expensive capturing systems or by professional technical artists. The current state-of-the-art methods generate faces, which can be suitable for applications such as caricature avatar creation in mobile devices [29] but do not generate high-quality shape and normals that can be used for photo-realistic face synthesis. In this paper, we propose the first methodology for high-quality face generation that can be used for photo-realistic face synthesis (i.e., joint generation of texture, shape, and normals) by capitalizing on the recent developments on Generative Adversarial Networks (GANs).

The early face models, such as [6], represent 3D face by disentangled PCA models of geometry, expression [13], and colored texture, called 3D morphable models (3DMM). 3DMMs and its variants were the most popular method for modeling shape and texture separately. However, the linear nature of PCA is often unable to capture high-frequency signals properly, thus the quality of generation and reconstruction by PCA is sub-optimal.

GANs is a recently introduced family of techniques that train samplers of high-dimensional distributions [25]. It has been demonstrated that when a GAN

is trained on facial images, it can generate images that have realistic characteristics. In particular, the recently introduced GANs [11,32,33] can generate
photo-realistic high-resolution faces. Nevertheless, because they are trained on
partially-aligned 2D images, they cannot properly model the manifold of faces
and thus (a) inevitably create many unrealistic instances and (b) it is not clear
how they can be used to generate photo-realistic 3D faces.

Recently, GANs have been applied for generating facial texture for various
applications. In particular, [54] and [23] utilize style transfer GANs to generate
photorealistic images of 3DMM-sampled novel identities. [57] directly generates
high-quality 3D facial textures by GANs and [24] replaces 3D Morphable Models
(3DMMs) with GAN models for 3D texture reconstruction while the shape is
still maintained by statistical models. [35] propose to generate 4K diffuse and
specular albedo and normals from a texture map by an image-to-image GAN.
On the other hand, [44] model 3D shape by GANs in a parametric UV map and
[53] utilize mesh convolutions with variational autoencoders to model shape in its
original structure. Although one can model 3D faces with such shape and texture
GAN approaches, these studies omit the correlation between shape, normals,
and texture which is very important for photorealism in identity space. The
significance of such correlation is most visible with inconsistent facial attributes
such as age, gender, and ethnicity (i.e. old-aged texture on a baby-face geometry).

In order to address these gaps, we propose a novel multi-branch GAN architecture that preserves the correlation between different 3D modalities (such as
texture, shape, normals, and expression). After converting all modalities into
UV space and concatenate over channels, we train a GAN that generates all
modalities in a meaningful local and global correspondence. In order to prevent
incompatibility issues due to the intensity distribution of different modalities,
we propose a trunk-branch architecture that can synthesize photorealistic 3D
faces with coupled texture and geometry. Further, we condition this GAN by
expression labels to generate faces in any desired expression.

From a computer graphics point of view, a photorealistic face rendering
requires a number of elements to be tailored, i.e. shape, normals and albedo
maps, some of which should or can be specific to a particular identity. However,
the cost of hand-crafting novel identities limits their usage on large-scale applications. The proposed approach tackles this down with reasonable photorealism
with a massively generalized identity space. Although the results in this paper
are limited to aforementioned modalities by the dataset at hand, the proposed
method allows adding more identity-specific modalities (i.e. cavity, gloss, scatter)
once such a dataset becomes available.

The contributions of this paper can be summarized as follows:

– We propose to model and synthesize coherent 3D faces by jointly training a
  novel Trunk-branch based GAN (TBGAN) architecture for shape, texture,
  and normals modalities. TBGAN is designed to maintain correlation while
  tolerating domain-specific differences of these three modalities and can be
  easily extended to other modalities and domains.

- In the domain of identity-generic face modeling, we believe this is the first study that utilizes normals as an additional source of information.
- We propose the first methodology for face generation that correlates expression and identity geometries (i.e. modeling personalized expression) and also the first attempt to model expression in texture and normals space.

2   Related Work

2.1   3D Face Modeling

There is an underlying assumption that human faces lie on a manifold with respect to the appearance and geometry. As a result, one can model the geometry and appearance of the human face analytically based upon the identity and expression space of all individuals. Two of the first attempts in the history of face modeling were [1], which proposes part-based 3D face reconstruction from frontal and profile images, and [48], which represents expression action units by a set of muscle fibers.

Twenty years ago methods that generated 3D faces revolved around parametric generative models that are driven by a small number of anthropometric statistics (e.g., sparse face measurements in a population) which act as constraints [18]. The seminal work of 3D morphable models (3DMMs) [6] demonstrated for the first time that is possible to learn a linear statistical model from a population of 3D faces [12,46]. 3DMMs are often constructed by using a Principal Component Analysis (PCA) based on a dataset of registered 3D scans of hundreds [47] or thousands [7] subjects. Similarly, facial expressions are also modeled by applying PCA [2,10,38,62], or are manually defined using linear blendshapes [9,36,58]. 3DMMs, despite their advantages, are bounded by the capacity of linear space that under-represents the high-frequency information and often result in overly-smoothed geometry and texture models. [14] and [59] attempt to address this issue by using local displacement maps. Furthermore, the 3DMM line of research assumes that texture and shape are uncorrelated, hence they can only be produced by separate models (i.e., separate PCA models for texture and shape). Early attempts in correlated shape and texture have been made in Active Appearance Models (AAMs) by computing joint PCA models of sparse shape and texture [16]. Nevertheless, due to the inherent limitations of PCA to model high-frequency texture, it is rarely used to correlate shape and texture for 3D face generation.

Recent progress in generative models [25,34] is being utilized in 3D face modeling to tackle this issue. [44] trained a GAN that models face geometry based on UV representations for neutral faces, and likewise, [53] modeled identity and expression geometry by variational autoencoders with mesh convolutions. [24] proposed a GAN-based texture modeling for 3D face reconstruction while modeling geometry by PCA and [57] trained a GAN to synthesize facial textures. To the best of our knowledge, these methodologies totally omit the correlation between geometry and texture and moreover, they ignore identity-specific expression modeling by decoupling them into separate models. In order to address this issue, we propose a trunk-branch GAN that is trained jointly for texture, shape, normals, and expression in order to leverage non-linear generative networks for capturing the correlation between these modalities.

2.2   Photorealistic Face Synthesis

Although most of the aforementioned 3D face models can synthesize 2D face images, there are also some dedicated 2D face generation studies. [42] combines non-parametric local and parametric global models to generate various set of face images. Recent family of GAN approaches [11,32,33,52] offers the state-of-the-art high quality random face generation without constraints.

Some other GAN-based studies allow to condition synthetic faces by rendered 3DMM images [23], by landmarks [5] or by another face image [4] (i.e. by disentangling identity and certain facial attributes). Similarly, facial expression is also conditionally synthesized by an audio input [31], by action unit codes [51], by predefined 3D geometry [65] or by expression of an another face image [37].

In this work, we jointly synthesize the aforementioned modalities for coherent photorealistic face synthesis by leveraging high-frequency generation by GANs. Unlike many of its 2D and 3D alternatives, the resulting generator models provide absolute control over disentangled identity, pose, expression and illumination spaces. Unlike many other GAN works that are struggling due to misalignments among the training data, our entire latent space correspond to realistic 3D faces as the data representation is naturally aligned on UV space.

2.3   Boosting Face Recognition by Synthetic Training Data

There have been also some works to synthesize face images to be used as synthetic training data for face recognition methods either by directly using GAN-generated images [61] or by controlling pose-space with a conditional-GAN [30,56,60]. [41] propose many augmentation techniques, such as rotation, expression, and shape, based on 3DMMs. Other GAN-based approaches that capitalize 3D facial priors include [66], which rotates faces by fitting 3DMM and preserves photorealism by translation GANs and [64], which frontalize face images by a GAN and 3DMM regression network. [19] complete missing parts of UV texture representations of 2D images after 3DMM fitting by a translation GAN. [23] first synthesizes face images of novel identities by sampling from 3DMM and then removes the photorealistic domain gap by an image-to-image translation GAN.

All of these studies show the significance of photorealistic and identity-generic face synthesization for the next generation of facial recognition algorithms. Although this study focuses more on the graphical aspect of face synthesization, we show that synthetic images can also improve face recognition performance.

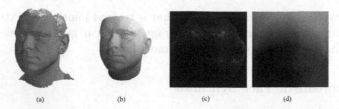

(a)   (b)   (c)   (d)

Fig. 2. UV extraction process. In (a) we present a raw mesh, in (b) the registered mesh using the Large Scale Face Model (LSFM) template [8], in (c) the unwrapped 3D mesh in the 2D UV space, and (d) the interpolated 2D UV map. Interpolation is carried out using the barycentric coordinates of each pixel in the registered 3D mesh.

3   Approach

3.1   UV Maps for Shape, Texture and Normals

In order to feed the shape, the texture, and the normals of the facial meshes into a deep network we need to reparameterize them into an image-like tensor format to apply 2D-convolutions[1]. We begin by describing all the raw 3D facial scans with the same topology and number of vertices (dense correspondence). This is achieved by morphing non-rigidly a template mesh to each one of the raw scans. We employ a standard non-rigid iterative closest point algorithm as described in [3,17] and we deform our chosen template so that it captures correctly the facial surface of the raw scans. As a template mesh, we choose the mean face of the LSFM model proposed in [8], which consists approximately of $54K$ vertices that are sufficient enough to depict non-linear, high facial details.

After reparameterizing all the meshes into the LSFM [8] topology, we cylindrically unwrap the mean face of the LSFM [8] to create a UV representation for that specific mesh topology. In the literature, a UV map is commonly utilized for storing only the RGB texture values. Apart from storing the texture values of the 3D meshes, we utilize the UV space to store the 3D coordinates of each vertex (x, y, z) and the normal orientation (n_x, n_y, n_z). Before storing the 3D coordinates into the UV space, all meshes are aligned in the 3D spaces by performing General Procrustes Analysis (GPA) [26] and are normalized to be in the scale of $[1, -1]$. Moreover, we store each 3D coordinate and normals in the UV space given the respective UV pixel coordinate. Finally, we perform a barycentric interpolation based on the barycentric coordinates of each pixel on the registered mesh to fill out the missing areas in order to produce a dense illustration of the UV map. In Fig. 2, we illustrate a raw 3D scan, the registered 3D scan on the LSFM [8] template, the sparse UV map of 3D coordinates and finally the interpolated one.

---

[1] Another line of research is mesh convolutional networks [15,39,53] which cannot preserve high-frequency details of the texture and normals at the current state-of-the-art.

3.2 Trunk-Branch GAN to Generate Coupled Texture, Shape and Normals

In order to train a model that handles multiple modalities, we propose a novel trunk-branch GAN architecture to generate entangled modalities of the 3D face such as texture, shape, and normals as UV maps. For this task, we exploit the MeIn3D dataset [8] which consists of approximately 10,000 neutral 3D facial scans with wide diversity in age, gender, and ethnicity.

Given a generator network $\mathcal{G}^L$ with a total of L convolutional upsampling layers and gaussian noise $\mathbf{z} \sim \mathcal{N}(\mathbf{0}, \mathbf{I})$ as input, the activation at the end of layer d (i.e., $\mathcal{G}^d(\mathbf{z})$) is split into three branch networks $\mathcal{G}_T^{L-d}$, $\mathcal{G}_N^{L-d}$, $\mathcal{G}_S^{L-d}$ each of which consists of $L-d$ upsampling convolutional layers that generate texture, normals and shape UV maps respectively. The discriminator $\mathcal{D}^L$ starts with the branch networks $\mathcal{D}_T^{L-d}$, $\mathcal{D}_N^{L-d}$, $\mathcal{D}_S^{L-d}$ whose activations are concatenated before fed into trunk network $\mathcal{D}^d$. The output of $\mathcal{D}^L$ is regression of real/fake score.

Although the proposed approach is compatible with most of the GAN architectures and loss functions, in our experiments, we base TBGAN on progressive growing GAN architecture [32] train it by WGAN-GP loss [27] as following:

$$\mathcal{L}_{\mathcal{G}^L} = \mathbb{E}_{\mathbf{z} \sim \mathcal{N}(\mathbf{0},\mathbf{I})} \left[-\mathcal{D}^L \left(\mathcal{G}^L(\mathbf{z}) \right) \right] \tag{1}$$

$$\mathcal{L}_{\mathcal{D}^L} = \mathbb{E}_{x \sim p_{\text{data}}, \; \mathbf{z} \sim \mathcal{N}(\mathbf{0},\mathbf{I})} \left[\mathcal{D}^L \left(\mathcal{G}^L(\mathbf{z}) \right) - \mathcal{D}^L(x) + \lambda * GP(x, \mathcal{G}^L(\mathbf{z})) \right] \tag{2}$$

where gradient penalty calculated by $GP(x, \hat{x}) = (\|\nabla \mathcal{D}^L (\alpha \hat{x} + (1 - \alpha)x))\|_2 - 1)^2$ and α denotes uniform random numbers between 0 and 1. λ is a balancing factor which is typically $\lambda = 10$. An overview of this trunk-branch architecture is illustrated in Fig. 1

3.3 Expression Augmentation by Conditional GAN

Further, we modify our GAN in order to generate 3D faces with expression by conditioning it with expression annotations ($\mathbf{p_e}$). Similar to the MeIn3D dataset, we have captured approximately 35,000 facial scans of around 5,000 distinct identities during a special exhibition in the Science Museum, London. All subjects were recorded in various guided expressions with a 3dMD face capturing apparatus. All of the subjects were asked to provide meta-data regarding their age, gender, and ethnicity. The database consists of 46% male, 54% female, 85% White, 7% Asian, 4% Mixed Heritage, 3% Black, and 1% other.

In order to avoid the cost and potential inconsistency of manual annotation, we render those scans and automatically annotate them by an expression recognition network. The resulting expression encodings $((*, \mathbf{p_e}) \sim p_{\text{data}})$ are used as label vector during the training of our trunk-branch conditional GAN. This training scheme is illustrated in Fig. 1. $\mathbf{p_e}$ is basically a vector of 7 for universal expressions (neutral, happy, angry etc.), randomly drawn from our dataset. During the training, Eq. 1 and 2 are updated to condition expression encodings by AC-GAN [45] as following:

$$\mathcal{L}_{\mathcal{G}^L} \mathrel{+}= \mathbb{E}_{(*,\mathbf{p_e})\sim p_{\text{data}},\ \mathbf{z}\sim\mathcal{N}(\mathbf{0},\mathbf{I})}\left[\sum_e \mathbf{p_e}\log(\mathcal{D}_e^L(\mathcal{G}^L(\mathbf{z},\mathbf{p_e})))\right] \qquad (3)$$

$$\mathcal{L}_{\mathcal{D}^L} \mathrel{+}= \mathbb{E}_{(x,\mathbf{p_e})\sim p_{\text{data}},\ \mathbf{z}\sim\mathcal{N}(\mathbf{0},\mathbf{I})}\left[\sum_e \mathbf{p_e}\log(\mathcal{D}_e^L(x)) + \mathbf{p_e}\log(\mathcal{D}_e^L(\mathcal{G}^L(\mathbf{z},\mathbf{p_e})))\right] \qquad (4)$$

which performs softmax cross entropy between expression prediction of the discriminator $(\mathcal{D}_e^L(x))$ and the random expression vector input $(\mathbf{p_e})$ for real (x) and generated samples $(\mathcal{G}^L(\mathbf{z},\mathbf{p_e}))$.

Unlike previous expression models that omit the effect of the expression on textures, the resulting generator is capable of generating coupled texture, shape, and normals map of a face with controlled expression. Similarly, our generator respects the identity-expression correlation thanks to correlated supervision provided by the training data. This is in contrast to the traditional statistical expression models which decouples expression and identity models into two separate entities.

3.4   Photorealistic Rendering with Generated UV Maps

For the renderings to appear photorealistic, we use the generated identity-specific mesh, texture, and normals, in combination with the generic reflectance properties, and employ a commercial rendering application: *Marmoset Toolbag* [40].

In order to extract the 3D representation from the UV domain we employ the inverse procedure explained in Sect. 3.1 based on the UV pixel coordinates of each vertex of the 3D mesh. Figure 3 shows the rendering results, under a single light source, when using the generated geometry (Fig. 3(a)) and the generated texture (Fig. 3(b)). Here the specular reflection is calculated on the per-face normals of the mesh and exhibits steep changes between on the face's edges. By interpolating the generated normals on each face (Fig. 3(c)), we are able to smooth the specular highlights and correct any high-frequency noise on the geometry of the mesh. However, these results do not correctly model the human skin and resemble a metallic surface. In reality, the human skin is rough and as a body tissue, it both reflects and absorbs light, thus exhibiting specular reflection, diffuse reflection, and subsurface scattering.

Although we can add such modalities as additional branches with the availability of such data, we find that rendering can be still improved by adding some identity-generic maps. Using our training data, we create maps that define certain reflectance properties per-pixel, which will match the features of the average generated identity, as shown in bottom-left of Fig. 1. *Scattering* (c) defines the intensity of subsurface scattering of the skin. *Translucency* (d) defines the amount of light, that travels inside the skin and gets emitted in different directions. *Specular albedo* (e) gives the intensity of the specular highlights, which differ between hair-covered areas, the eyes, and the teeth. *Roughness* (f) describes the scattering of specular highlights and controls the glossiness of the skin.

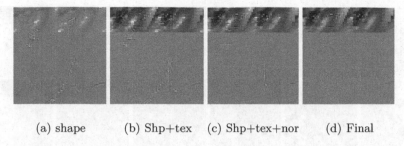

(a) shape   (b) Shp+tex   (c) Shp+tex+nor   (d) Final

Fig. 3. Zoom-in on rendering results with (a) only the shape, (b) adding the albedo texture, (c) adding the generated normals, and (d) using identity-generic detail normal, specular albedo, roughness, scatter and translucency maps.

A *detail normal map* (g) is also tilled and added on the generated normal maps, to mimic the skin pores and a *detail weight map* (h) controls the appearance of the detail normals, so that they do not appear on the eyes, lips, and hair. The final result (Fig. 3(d)) properly models the skin surface and reflection, by adding plausible high-frequency specularity and subsurface scattering, both weighted by the area of the face where they appear.

4   Results

In this section, we give qualitative and quantitative results of our method for generating 3D faces with novel identities and various expressions. In our experiments, there are total $L = 8$ up- and down-sampling layers where $d = 6$ of them in the trunk and 2 layers in each branch. These choices are empirically validated to ensure sufficient correlation among modalities without incompatibility artifacts. Running time is a few milliseconds to generate UV images from a latent code on a high-end GPU. Transforming from UV image to mesh is just sampling with UV coordinates and can be considered free of cost. Renderings in this paper take a few seconds due to high resolution but this cost depends on the application. The memory needed for the generator network is 1.25 GB compared to the 6 GB PCA model of the same resolution and %95 of the total variance.

In the following sections, we first visualize generated UV maps and their contributions to the final renderings on several generated faces. Next, we show the generalization ability of the identity and expression generators on some facial characteristics. We also demonstrate its well-generalization latent space by interpolating between different identities. Additionally, we perform full-head completion to the interpolated faces. Finally, we perform face recognition experiments by using the generated face images as additional training data.

4.1   Qualitative Results

Combining Coupled Modalities: Fig. 4 presents the generated shape, normals, and texture maps by the proposed GAN and their additive contributions

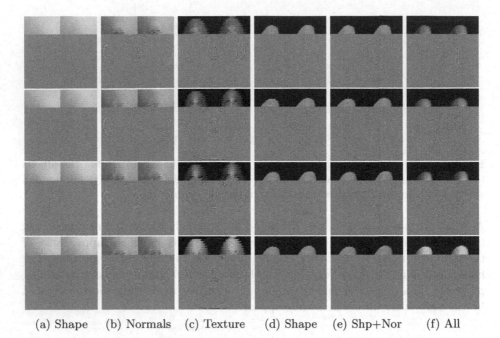

(a) Shape   (b) Normals   (c) Texture   (d) Shape   (e) Shp+Nor   (f) All

Fig. 4. Generated UV representations and their corresponding additive renderings. Please note the strong correlation between UV maps, high fidelity and photorealistic renderings. The figure is best viewed in zoom.

to the final renderings. As can be seen from local and global correspondences, the generated UV maps are highly correlated and coherent. Attributes like age, gender, race, etc. can be easily grasped from all of the UV maps and rendered images. Please also note that some of the minor artifacts of the generated geometry in Fig. 4(d) are compensated by the normals in Fig. 4(e).

Diversity: Our model is well-generalized with different age, gender, ethnicity groups and many facial attributes. Although Fig. 5 shows diversity in some of those categories, the reader is encouraged to see identity variation throughout the paper and the supplementary video.

Expression: We also show that our expression generator is capable of synthesizing quite a diverse set of expressions. Moreover, the expressions can be controlled by the input label as can be seen in Fig. 6. The reader is encouraged to see more expression generations in the supplementary video.

Interpolation Between Identities: As shown in the supplementary video and in Fig. 7, our model can easily interpolate between any generation in a visually continuous set of identities which is another indication that the model is free

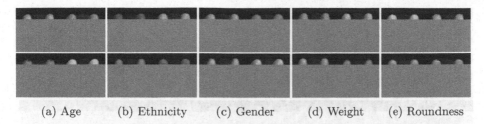

(a) Age   (b) Ethnicity   (c) Gender   (d) Weight   (e) Roundness

Fig. 5. Variation of generated 3D faces by our model. Each block shows diversity in a different aspect. Readers are encouraged to zoom in on a digital version.

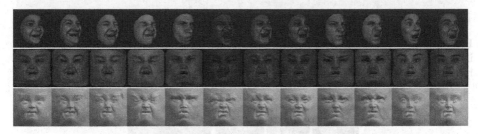

Fig. 6. (Top) generations of six universal expressions (i.e. each two columns respective the following expressions: Happiness, Sadness, Anger, Fear, Disgust, Surprise). (Middle) texture and (Bottom) normals maps are used to generate the corresponding 3D faces. Please note how expressions are represented and correlated in the texture and normals space.

from mode collapse. Interpolation is done by randomly generating two identities and generates faces by evenly spaced samples in latent space between the two.

Full Head Completion: We also extend our facial 3D meshes to full head representations by employing the framework proposed in [50]. We achieve this by regressing from a latent space that represents only the 3D face to the PCA latent space of the Universal Head Model (UHM) [49,50]. We begin by building a PCA model of the inner face based on the 10,000 neutral scans of the MeIn3D dataset. Similarly, we exploit the extended full head meshes of the same identities utilized by UHM model and project them to the UHM subspace to acquire the latent shape parameters of the entire head topology. Finally, we learn a regression matrix by solving a linear least-square optimization problem as proposed in [50], which maps the latent space of the face shape to the full head representation. Figure 7 demonstrates the extended head representations of our approach in conjunction with the synthesized crop faces.

Comparison to Decoupled Modalities and PCA: Results in Fig. 8 reveal a set of advantages of such unified 3D face modeling over separate GAN and statistical models. Clearly, the figure shows that the correlation among texture,

Fig. 7. Interpolation between pair of identities in the latent space. Smooth transition indicates generalization of our GAN model. The last two rows show complete full head representations respective to the first two rows.

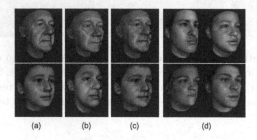

Fig. 8. Comparison with separate GAN models and PCA model. (a) Generation by our model. (b) Same texture with random shape and normals. (c) Same texture and shape with random normals (i.e. beard). (d) Generation by a PCA model constructed by the same training data and the same identity-generic rendering tools as explained in Sec. 3.4.

shape, and normals is an important component for realistic face synthesis. Also, generations by PCA models are missing photorealism and details significantly.

4.2 Pose-Invariant Face Recognition

In this section, we present an experiment that demonstrates that the proposed methodology can generate faces of different and diverse identities. That is, we use the generated faces to train one of the most recent state-of-the-art face recognition method, ArcFace [20], and show that the proposed shape and texture generation model can boost the performance of pose-invariant face recognition. **Training Data:** We randomly synthesize 10 K new identities from the proposed model and render 50 images per identity with a random camera and illumination parameters from the Gaussian distribution of the 300W-LP dataset [23,67]. For clarity, we call this dataset "Gen" in the rest of the text. Figure 9 illustrates

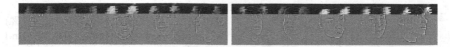

Fig. 9. Examples of generated data ("Gen") by the proposed method.

some examples of "Gen" dataset which show larger pose variations than the real-world collected data. We augment "Gen" with an in-the-wild training data, CASIA dataset [63], which consists of 10,575 identities with 494,414 images.
Test Data: For evaluation, we employ Celebrities in Frontal Profile (CFP) [55] and Age Database (AgeDB) [43]. **CFP** [55] consists of 500 subjects, each with 10 frontal and 4 profile images. The evaluation protocol includes frontal-frontal (FF) and frontal-profile (FP) face verification. In this paper, we focus on the most challenging subset, CFP-FP, to investigate the performance of pose-invariant face recognition. There are 3,500 same-person pairs and 3,500 different-person pairs in CFP-FP for the verification test. **AgeDB** [21,43] contains 12,240 images of 440 distinct subjects. The minimum and maximum ages are 3 and 101, respectively. The average age range for each subject is 49 years. There are four groups of test data with different year gaps (5 years, 10 years, 20 years and 30 years, respectively) [21]. In this paper, we only use the most challenging subset, AgeDB-30, to report the performance. There are 3,000 positive pairs and 3,000 negative pairs in AgeDB-30 for the verification test.

Data Prepossessing: We follow the baseline [20] to generate the normalized face crops (112×112) by utilizing five facial points.

Training and Testing Details: For the embedding networks, we employ the widely used ResNet50 architecture [28]. After the last convolutional layer, we also use the BN-Dropout-FC-BN [20] structure to get the final 512-D embedding feature. For the hyper-parameter setting and loss functions, we follow [20-22]. The overlapping identities between the CASIA data set and the test set are removed for strict evaluations, and we only use a single crop for all testing.

Result Analysis: In Table 1, we show the contribution of the generated data on pose-invariant face recognition. We take UV-GAN [19] as the baseline method, which attaches the completed UV texture map onto the fitted mesh and generates instances of arbitrary poses to increase pose variation during training and minimize pose discrepancy during testing. As we can see from Table 1, generated data significantly boost the verification performance on CFP-FP from 95.56% to 97.12%, decreasing the verification error by 51.2% compared to the result of UV-GAN [19]. On AgeDB-30, combining CASIA and generated data achieves similar performance compared to using single CASIA because we only include intra-variance from pose instead of age.

In Fig. 10, we show the angle distributions of all positive pairs and negative pairs from CFP-FP. By incorporating generation data, the overlap indistinguishable area between the positive histogram and the negative histogram is obviously decreased, which confirms that ArcFace can learn pose-invariant

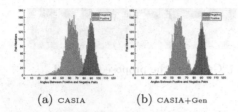

(a) CASIA   (b) CASIA+Gen

Fig. 10. Angle distributions of CFP-FP positive (red) and negative (blue) pairs in the 512-D feature space

Table 1. Verification performance (%) of different models on CFP-FP and AgeDB-30.

Methods	CFP-FP	AgeDB-30
UVGAN [19]	94.05	94.18
Ours (CASIA)	95.56	95.15
Ours (CASIA+Gen)	**97.12**	**95.18**

Table 2. The angles between face pairs from CFP-FP predicted by different models trained from the CASIA and combined data. The generated data can obviously enhance the pose-invariant feature embedding.

Training Data					
CASIA	84.06°	82.39°	84.72°	88.06°	84.37°
CASIA+Gen	57.60°	63.12°	66.10°	59.72°	60.25°

feature embedding from the generated data. In Table 2, we select some verification pairs from CFP-FP and calculate the cosine distance (*angle*) between these pairs predicted by different models trained from the CASIA and combined data. Intuitively, the angles between these challenging pairs are significantly reduced when generated data are used for the model training.

5 Conclusion

We presented the first 3D face model for joint texture, shape, and normal generation based on Generative Adversarial Networks (GANs). The proposed GAN model implements a new architecture for exploiting the correlation between different modalities and can synthesize different facial expressions in accordance with the embeddings of an expression recognition network. We demonstrate that randomly synthesized images of our unified generator show strong relations between texture, shape, and normals and that rendering with normals provides excellent shading and overall visual quality. Finally, in order to demonstrate the generalization of our model, we have used a set of generated images to train a deep face recognition network.

Acknowledgement. Baris Gecer is supported by the Turkish Ministry of National Education, Stylianos Ploumpis by the EPSRC Project EP/N007743/1 (FACER2VM), and Stefanos Zafeiriou by EPSRC Fellowship DEFORM (EP/S010203/1).

References

1. Akimoto, T., Suenaga, Y., Wallace, R.S.: Automatic creation of 3D facial models. IEEE Comput. Graphics Appl. **13**(5), 16–22 (1993). https://doi.org/10.1109/38. 232096
2. Amberg, B., Knothe, R., Vetter, T.: Expression invariant 3D face recognition with a morphable model. In: 2008 8th IEEE International Conference on Automatic Face and Gesture Recognition, FG 2008, pp. 1–6. IEEE (2008). https://doi.org/10.1109/AFGR.2008.4813376
3. Amberg, B., Romdhani, S., Vetter, T.: Optimal step nonrigid ICP algorithms for surface registration. In: Proceedings of the IEEE Computer Society Conference on Computer Vision and Pattern Recognition, pp. 1–8 (2007). https://doi.org/10.1109/CVPR.2007.383165
4. Bao, J., Chen, D., Wen, F., Li, H., Hua, G.: Towards open-set identity preserving face synthesis. In: Proceedings of the IEEE Computer Society Conference on Computer Vision and Pattern Recognition, pp. 6713–6722 (2018). https://doi.org/10.1109/CVPR.2018.00702
5. Bazrafkan, S., Javidnia, H., Corcoran, P.: Face synthesis with landmark points from generative adversarial networks and inverse latent space mapping. arXiv preprint arXiv:1802.00390 (2018)
6. Blanz, V., Vetter, T.: A morphable model for the synthesis of 3D faces. In: Proceedings of the 26th Annual Conference on Computer Graphics and Interactive Techniques, SIGGRAPH 1999, pp. 187–194. ACM Press/Addison-Wesley Publishing Co. (1999). https://doi.org/10.1145/311535.311556
7. Booth, J., Roussos, A., Ponniah, A., Dunaway, D., Zafeiriou, S.: Large scale 3D morphable models. Int. J. Comput. Vision **126**(2–4), 233–254 (2018). https://doi.org/10.1007/s11263-017-1009-7
8. Booth, J., Roussos, A., Zafeiriou, S., Ponniahy, A., Dunaway, D.: A 3D morphable model learnt from 10,000 faces. In: Proceedings of the IEEE Computer Society Conference on Computer Vision and Pattern Recognition, vol. 2016-December, pp. 5543–5552 (2016). https://doi.org/10.1109/CVPR.2016.598
9. Bouaziz, S., Wang, Y., Pauly, M.: Online modeling for realtime facial animation. ACM Trans. Graph. **32**(4), 40 (2013). https://doi.org/10.1145/2461912.2461976
10. Breidt, M., Bülthoff, H.H., Curio, C.: Robust semantic analysis by synthesis of 3D facial motion. In: 2011 IEEE International Conference on Automatic Face and Gesture Recognition and Workshops, FG 2011, pp. 713–719. IEEE (2011). https://doi.org/10.1109/FG.2011.5771336
11. Brock, A., Donahue, J., Simonyan, K.: Large scale GaN training for high fidelity natural image synthesis. In: 7th International Conference on Learning Representations, ICLR 2019 (2019)
12. Brunton, A., Salazar, A., Bolkart, T., Wuhrer, S.: Review of statistical shape spaces for 3D data with comparative analysis for human faces. Comput. Vis. Image Underst. **128**, 1–17 (2014). https://doi.org/10.1016/j.cviu.2014.05.005
13. Cao, C., Weng, Y., Zhou, S., Tong, Y., Zhou, K.: FaceWarehouse: a 3D facial expression database for visual computing. IEEE Trans. Visual Comput. Graphics **20**(3), 413–425 (2014). https://doi.org/10.1109/TVCG.2013.249
14. Chen, A., Chen, Z., Zhang, G., Mitchell, K., Yu, J.: Photo-realistic facial details synthesis from single image. In: Proceedings of the IEEE International Conference on Computer Vision, vol. 2019-October, pp. 9428–9438, October 2019. https://doi.org/10.1109/ICCV.2019.00952

15. Cheng, S., Bronstein, M., Zhou, Y., Kotsia, I., Pantic, M., Zafeiriou, S.: MeshGAN: non-linear 3D morphable models of faces. arXiv preprint arXiv:1903.10384 (2019)
16. Cootes, T.F., Edwards, G.J., Taylor, C.J.: Active appearance models. In: Burkhardt, H., Neumann, B. (eds.) ECCV 1998. LNCS, vol. 1407, pp. 484–498. Springer, Heidelberg (1998). https://doi.org/10.1007/BFb0054760
17. De Smet, M., Van Gool, L.: Optimal regions for linear model-based 3D face reconstruction. In: Kimmel, R., Klette, R., Sugimoto, A. (eds.) ACCV 2010. LNCS, vol. 6494, pp. 276–289. Springer, Heidelberg (2011). https://doi.org/10.1007/978-3-642-19318-7_22
18. DeCarlo, D., Metaxas, D., Stone, M.: An anthropometric face model using variational techniques. In: Proceedings of the 25th Annual Conference on Computer Graphics and Interactive Techniques, SIGGRAPH 1998, vol. 98, pp. 67–74 (1998). https://doi.org/10.1145/280814.280823
19. Deng, J., Cheng, S., Xue, N., Zhou, Y., Zafeiriou, S.: UV-GAN: adversarial facial UV map completion for pose-invariant face recognition. In: Proceedings of the IEEE Computer Society Conference on Computer Vision and Pattern Recognition, pp. 7093–7102 (2018). https://doi.org/10.1109/CVPR.2018.00741
20. Deng, J., Guo, J., Xue, N., Zafeiriou, S.: ArcFace: additive angular margin loss for deep face recognition. In: Proceedings of the IEEE Computer Society Conference on Computer Vision and Pattern Recognition, June 2019, pp. 4685–4694 (2019). https://doi.org/10.1109/CVPR.2019.00482
21. Deng, J., Zhou, Y., Zafeiriou, S.: Marginal loss for deep face recognition. In: IEEE Computer Society Conference on Computer Vision and Pattern Recognition Workshops, vol. 2017-July, pp. 2006–2014 (2017). https://doi.org/10.1109/CVPRW.2017.251
22. Gecer, B., Balntas, V., Kim, T.K.: Learning deep convolutional embeddings for face representation using joint sample- and set-based supervision. In: 2017 IEEE International Conference on Computer Vision Workshops (ICCVW), pp. 1665–1672, October 2017. https://doi.org/10.1109/ICCVW.2017.195
23. Gecer, B., Bhattarai, B., Kittler, J., Kim, T.-K.: Semi-supervised adversarial learning to generate photorealistic face images of new identities from 3D morphable model. In: Ferrari, V., Hebert, M., Sminchisescu, C., Weiss, Y. (eds.) ECCV 2018. LNCS, vol. 11215, pp. 230–248. Springer, Cham (2018). https://doi.org/10.1007/978-3-030-01252-6_14
24. Gecer, B., Ploumpis, S., Kotsia, I., Zafeiriou, S.: GANFIT: generative adversarial network fitting for high fidelity 3D face reconstruction. In: Conference on Computer Vision and Pattern Recognition (CVPR), pp. 1155–1164, June 2019. https://doi.org/10.1109/CVPR.2019.00125
25. Goodfellow, I.J., et al.: Generative adversarial nets. In: Advances in Neural Information Processing Systems, vol. 3, pp. 2672–2680 (2014). https://doi.org/10.3156/jsoft.29.5_177_2
26. Gower, J.C.: Generalized procrustes analysis. Psychometrika 40(1), 33–51 (1975). https://doi.org/10.1007/BF02291478
27. Gulrajani, I., Ahmed, F., Arjovsky, M., Dumoulin, V., Courville, A.: Improved training of wasserstein GANs. In: Advances in Neural Information Processing Systems, vol. 2017-December, pp. 5768–5778 (2017)

28. He, K., Zhang, X., Ren, S., Sun, J.: Deep residual learning for image recognition. In: Proceedings of the IEEE Computer Society Conference on Computer Vision and Pattern Recognition, vol. 2016-December, pp. 770–778 (2016). https://doi.org/10.1109/CVPR.2016.90

29. Hu, L., et al.: Avatar digitization from a single image for real-time rendering. ACM Trans. Graph. 36(6), 195 (2017). https://doi.org/10.1145/3130800.3130887

30. Hu, Y., Wu, X., Yu, B., He, R., Sun, Z.: Pose-guided photorealistic face rotation. In: Proceedings of the IEEE Computer Society Conference on Computer Vision and Pattern Recognition, pp. 8398–8406 (2018). https://doi.org/10.1109/CVPR.2018.00876

31. Jamaludin, A., Chung, J.S., Zisserman, A.: You said that?: synthesising talking faces from audio. Int. J. Comput. Vision 127(11–12), 1767–1779 (2019). https://doi.org/10.1007/s11263-019-01150-y

32. Karras, T., Aila, T., Laine, S., Lehtinen, J.: Progressive growing of GANs for improved quality, stability, and variation. In: 6th International Conference on Learning Representations, ICLR 2018 - Conference Track Proceedings (2018)

33. Karras, T., Laine, S., Aila, T.: A style-based generator architecture for generative adversarial networks. In: Proceedings of the IEEE Computer Society Conference on Computer Vision and Pattern Recognition 2019-June, pp. 4396–4405 (2019). https://doi.org/10.1109/CVPR.2019.00453

34. Kingma, D.P., Welling, M.: Auto-encoding variational bayes. In: 2nd International Conference on Learning Representations, ICLR 2014 - Conference Track Proceedings (2014)

35. Lattas, A., et al.: AvatarMe: realistically renderable 3D facial reconstruction "in-the-wild". In: Conference on Computer Vision and Pattern Recognition (CVPR), pp. 760–769 (2020)

36. Li, H., Weise, T., Pauly, M.: Example-based facial rigging. ACM SIGGRAPH 2010 Papers, SIGGRAPH 2010 29(4), 32 (2010). https://doi.org/10.1145/1778765.1778769

37. Li, K., Dai, Q., Wang, R., Liu, Y., Xu, F., Wang, J.: A data-driven approach for facial expression retargeting in video. IEEE Trans. Multimedia 16, 299–310 (2014). https://doi.org/10.1109/TMM.2013.2293064

38. Li, T., Bolkart, T., Black, M.J., Li, H., Romero, J.: Learning a model of facial shape and expression from 4D scans. ACM Trans. Graph. 36(6), 194 (2017). https://doi.org/10.1145/3130800.3130813

39. Litany, O., Bronstein, A., Bronstein, M., Makadia, A.: Deformable shape completion with graph convolutional autoencoders. In: Proceedings of the IEEE Computer Society Conference on Computer Vision and Pattern Recognition, pp. 1886–1895 (2018). https://doi.org/10.1109/CVPR.2018.00202

40. Marmoset LLC: Marmoset toolbag (2019)

41. Masi, I., Tran, A.T., Hassner, T., Leksut, J.T., Medioni, G.: Do we really need to collect millions of faces for effective face recognition? In: Leibe, B., Matas, J., Sebe, N., Welling, M. (eds.) ECCV 2016. LNCS, vol. 9909, pp. 579–596. Springer, Cham (2016). https://doi.org/10.1007/978-3-319-46454-1_35

42. Mohammed, U., Prince, S.J., Kautz, J.: Visio-lization: generating novel facial images. ACM Trans. Graph. 28(3), 57 (2009). https://doi.org/10.1145/1531326.1531363

432   B. Gecer et al.

43. Moschoglou, S., Papaioannou, A., Sagonas, C., Deng, J., Kotsia, I., Zafeiriou, S.: AgeDB: the first manually collected, in-the-wild age database. In: IEEE Computer Society Conference on Computer Vision and Pattern Recognition Workshops, vol. 2017-July, pp. 1997–2005 (2017). https://doi.org/10.1109/CVPRW.2017.250
44. Moschoglou, S., Ploumpis, S., Nicolaou, M., Papaioannou, A., Zafeiriou, S.: 3DFaceGAN: adversarial nets for 3D face representation, generation, and translation. arXiv preprint arXiv:1905.00307 (2019)
45. Odena, A., Olah, C., Shlens, J.: Conditional image synthesis with auxiliary classifier GANs. In: Proceedings of the 34th International Conference on Machine Learning, Sydney, NSW, Australia, vol. 70, pp. 2642–2651. ICML 2017, JMLR.org, August 2017
46. Patel, A., Smith, W.A.: 3D morphable face models revisited. In: 2009 IEEE Computer Society Conference on Computer Vision and Pattern Recognition Workshops, CVPR Workshops 2009, vol. 2009 IEEE Computer Society Conference on Computer Vision and Pattern Recognition, pp. 1327–1334. IEEE (2009). https://doi.org/10.1109/CVPRW.2009.5206522
47. Paysan, P., Knothe, R., Amberg, B., Romdhani, S., Vetter, T.: A 3D face model for pose and illumination invariant face recognition. In: 2009 Sixth IEEE International Conference on Advanced Video and Signal Based Surveillance, pp. 296–301 (Sep 2009). https://doi.org/10.1109/AVSS.2009.58
48. Platt, S.M., Badler, N.I.: Animating facial expressions. In: Proceedings of the 8th Annual Conference on Computer Graphics and Interactive Techniques, SIGGRAPH 1981, vol. 15, pp. 245–252. ACM (1981). https://doi.org/10.1145/800224.806812
49. Ploumpis, S., et al.: Towards a complete 3D morphable model of the human head. IEEE Trans. Pattern Anal. Mach. Intell. (TPAMI), 1–1 (2020). https://doi.org/10.1109/TPAMI.2020.2991150
50. Ploumpis, S., Wang, H., Pears, N., Smith, W.A., Zafeiriou, S.: Combining 3D morphable models: a large scale face-and-head model. In: Proceedings of the IEEE Computer Society Conference on Computer Vision and Pattern Recognition, vol. 2019-June, pp. 10926–10935 (2019). https://doi.org/10.1109/CVPR.2019.01119
51. Pumarola, A., Agudo, A., Martinez, A.M., Sanfeliu, A., Moreno-Noguer, F.: GANimation: anatomically-aware facial animation from a single image. In: Ferrari, V., Hebert, M., Sminchisescu, C., Weiss, Y. (eds.) ECCV 2018. LNCS, vol. 11214, pp. 835–851. Springer, Cham (2018). https://doi.org/10.1007/978-3-030-01249-6_50
52. Radford, A., Metz, L., Chintala, S.: Unsupervised representation learning with deep convolutional generative adversarial networks. In: 4th International Conference on Learning Representations, ICLR 2016 - Conference Track Proceedings (2016)
53. Ranjan, A., Bolkart, T., Sanyal, S., Black, M.J.: Generating 3D faces using convolutional mesh autoencoders. In: Ferrari, V., Hebert, M., Sminchisescu, C., Weiss, Y. (eds.) ECCV 2018. LNCS, vol. 11207, pp. 725–741. Springer, Cham (2018). https://doi.org/10.1007/978-3-030-01219-9_43
54. Sela, M., Richardson, E., Kimmel, R.: Unrestricted facial geometry reconstruction using image-to-image translation. In: Proceedings of the IEEE International Conference on Computer Vision. vol. 2017-October, pp. 1585–1594 (2017). https://doi.org/10.1109/ICCV.2017.175
55. Sengupta, S., Chen, J.C., Castillo, C., Patel, V.M., Chellappa, R., Jacobs, D.W.: Frontal to profile face verification in the wild. In: 2016 IEEE Winter Conference on Applications of Computer Vision, WACV 2016 (2016). https://doi.org/10.1109/WACV.2016.7477558

56. Shen, Y., Luo, P., Yan, J., Wang, X., Tang, X.: FaceID-GAN: learning a symmetry three-player GAN for identity-preserving face synthesis. In: Proceedings of the IEEE Computer Society Conference on Computer Vision and Pattern Recognition, pp. 821–830 (2018). https://doi.org/10.1109/CVPR.2018.00092

57. Slossberg, R., Shamai, G., Kimmel, R.: High quality facial surface and texture synthesis via generative adversarial networks. In: Leal-Taixé, L., Roth, S. (eds.) ECCV 2018. LNCS, vol. 11131, pp. 498–513. Springer, Cham (2019). https://doi.org/10.1007/978-3-030-11015-4_36

58. Thies, J., Zollhöfer, M., Nießner, M., Valgaerts, L., Stamminger, M., Theobalt, C.: Real-time expression transfer for facial reenactment. ACM Trans. Graph. **34**(6), 181–183 (2015). https://doi.org/10.1145/2816795.2818056

59. Tran, A.T., Hassner, T., Masi, I., Paz, E., Nirkin, Y., Medioni, G.: Extreme 3D face reconstruction: seeing through occlusions. In: 2018 IEEE/CVF Conference on Computer Vision and Pattern Recognition, Salt Lake City, UT, pp. 3935–3944. IEEE, June 2018. https://doi.org/10.1109/CVPR.2018.00414

60. Tran, L., Yin, X., Liu, X.: Representation learning by rotating your faces. IEEE Trans. Pattern Anal. Mach. Intell. **41**(12), 3007–3021 (2019). https://doi.org/10.1109/TPAMI.2018.2868350

61. Trigueros, D.S., Meng, L., Hartnett, M.: Generating photo-realistic training data to improve face recognition accuracy. arXiv preprint arXiv:1811.00112 (2018)

62. Yang, F., Metaxas, D., Wang, J., Shechtman, E., Bourdev, L.: Expression flow for 3D-aware face component transfer. ACM Trans. Graph. **30**(4), 1–10 (2011). https://doi.org/10.1145/2010324.1964955

63. Yi, D., Lei, Z., Liao, S., Li, S.Z.: Learning face representation from scratch. arXiv:1411.7923 (2014)

64. Yin, X., Yu, X., Sohn, K., Liu, X., Chandraker, M.: Towards large-pose face frontalization in the wild. In: Proceedings of the IEEE International Conference on Computer Vision, vol. 2017-October, pp. 4010–4019 (2017). https://doi.org/10.1109/ICCV.2017.430

65. Zhang, Q., Liu, Z., Guo, B., Shum, H.: Geometry-driven photorealistic facial expression synthesis. In: Proceedings of the 2003 ACM SIGGRAPH/Eurographics Symposium on Computer Animation, SCA 2003, vol. 12, no. 1, pp. 48–60 (2003)

66. Zhao, J., et al.: Dual-agent GANs for photorealistic and identity preserving profile face synthesis. In: Advances in Neural Information Processing Systems, vol. 2017-December, pp. 66–76 (2017)

67. Zhu, X., Lei, Z., Liu, X., Shi, H., Li, S.Z.: Face alignment across large poses: a 3D solution. In: Proceedings of the IEEE Computer Society Conference on Computer Vision and Pattern Recognition, vol. 2016-December, pp. 146–155 (2016). https://doi.org/10.1109/CVPR.2016.23

Learning to Learn Words from Visual Scenes

Dídac Surís[1]([✉]), Dave Epstein[1], Heng Ji[2], Shih-Fu Chang[1], and Carl Vondrick[1]

[1] Columbia University, New York, USA
ds3819@columbia.edu
[2] UIUC, Champaign, USA
http://expert.cs.columbia.edu

Abstract. Language acquisition is the process of learning words from the surrounding scene. We introduce a meta-learning framework that *learns how to learn* word representations from unconstrained scenes. We leverage the natural compositional structure of language to create training episodes that cause a meta-learner to learn strong policies for language acquisition. Experiments on two datasets show that our approach is able to more rapidly acquire novel words as well as more robustly generalize to unseen compositions, significantly outperforming established baselines. A key advantage of our approach is that it is data efficient, allowing representations to be learned from scratch without language pre-training. Visualizations and analysis suggest visual information helps our approach learn a rich cross-modal representation from minimal examples.

1 Introduction

Language acquisition is the process of learning words from the surrounding environment. Although the sentence in Fig. 1 contains new words, we are able to leverage the visual scene to accurately acquire their meaning. While this process comes naturally to children as young as six months old [54] and represents a major milestone in their development, creating a machine with the same malleability has remained challenging.

The standard approach in vision and language aims to learn a common embedding space [13,27,50], however this approach has a number of key limitations. Firstly, these models are inefficient because they often require millions of examples to learn. Secondly, they consistently generalize poorly to the natural compositional structure of language [16]. Thirdly, fixed embeddings are unable to adapt to novel words at inference time, such as in realistic scenes that are naturally open world. We believe these limitations stem fundamentally from the process that models use to acquire words.

D. Surís and D. Epstein—Equal contribution.

Electronic supplementary material The online version of this chapter (https://doi.org/10.1007/978-3-030-58526-6_26) contains supplementary material, which is available to authorized users.

A. Vedaldi et al. (Eds.): ECCV 2020, LNCS 12374, pp. 434–452, 2020.
https://doi.org/10.1007/978-3-030-58526-6_26

While most approaches learn the word embeddings, we propose to instead learn the *process* for acquiring word embeddings. We believe the language acquisition process is too complex and subtle to handcraft. However, there are large amounts of data available to learn the process. In this paper, we introduce a framework that *learns how to learn* vision and language representations.

We present a model that receives an episode of examples consisting of vision and language pairs, where the model meta-learns word embeddings from the episode. The model is trained to complete a masked word task, however it must do so by copying and pasting words across examples within the episode. Although this is a roundabout way to fill in masked words, this requires the model to learn a robust process for word acquisition. By controlling the types of episodes from which the model learns, we are able to explicitly learn a process to acquire novel words and generalize to novel compositions. Figure 2 illustrates our approach.

"then, I spread the *ghee* on the *roti*"

Fig. 1. What is "ghee" and "roti"? The answer is in the footnote.[1] Although the words "ghee" and "roti" may be unfamiliar, you are able to leverage the structure of the visual world and knowledge of other words to acquire their meaning. In this paper, we propose a model that learns how to learn words from visual context. (Answer: "ghee" is the butter on the knife, and "roti" is the bread in the pan)

Our experiments show that our framework meta-learns a strong policy for word acquisition. We evaluate our approach on two established datasets, Flickr30k [62] and EPIC-Kitchens [8], both of which have a large diversity of natural scenes and a long-tail word distribution. After learning the policy, the model can receive a stream of images and corresponding short phrases containing unfamiliar words. Our model is able to learn the novel words and point to them to describe other scenes. Visualizations of the model suggest strong cross-modal interaction from language to visual inputs and vice versa.

A key advantage of our approach is that it is able to acquire words with orders of magnitude less examples than previous approaches. Although we train our model from scratch without any language pre-training, it either outperforms or matches methods with massive corpora. In addition, the model is able to effectively generalize to compositions outside of the training set, *e.g.* to unseen compositions of nouns and verbs, outperforming the state-of-the-art in visual language models by over fifteen percent when the compositions are new.

Our primary contribution is a framework that meta-learns a policy for visually grounded language acquisition, which is able to robustly generalize to both new words and compositions. The remainder of the paper is organized around this contribution. In Sect. 2, we review related work. In Sect. 3, we present our approach to meta-learn words from visual episodes. In Sect. 4, we analyze the

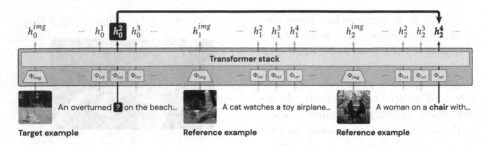

h_0^{img} ... h_0^1 h_0^2 h_0^3 ... h_1^{img} ... h_1^2 h_1^3 h_1^4 ... h_2^{img} ... h_2^2 h_2^3 h_2^4 ...

Transformer stack

Φ_{img} ... Φ_{txt} Φ_{txt} Φ_{txt} ... Φ_{img} ... Φ_{txt} Φ_{txt} Φ_{txt} ... Φ_{img} ... Φ_{txt} Φ_{txt} Φ_{txt} ...

An overturned **?** on the beach...   A cat watches a toy airplane...   A woman on a **chair** with...

Target example   **Reference example**   **Reference example**

Fig. 2. Learning to Learn Words from Scenes: Rather than directly learning word embeddings, we instead learn the *process* to acquire word embeddings. The input to our model is an episode of image and language pairs, and our approach meta-learns a policy to acquire word representations from the episode. Experiments show this produces a representation that is able to acquire novel words at inference time as well as more robustly generalize to novel compositions.

performance of our approach and ablate components with a set of qualitative and quantitative experiments. We will release all code and trained models.

2   Related Work

Visual Language Modeling: Machine learning models have leveraged large text datasets to create strong language models that achieve state-of-the-art results on a variety of tasks [10,38,39]. To improve the representation, a series of papers have tightly integrated vision as well [2,7,26,27,30,40,48–50,52,63]. However, since these approaches directly learn the embedding, they often require large amounts of data, poorly generalize to new compositions, and cannot adapt to an open-world vocabulary. In this paper, we introduce a meta-learning framework that instead learns the language acquisition process itself. Our approach outperforms established vision and language models by a significant margin. Since our goal is word acquisition, we evaluate both our method and baselines on language modeling directly.

Compositional Models: Due to the diversity of the visual world, there has been extensive work in computer vision on learning compositional representations for objects and attributes [20,32,34,35,37] as well as for objects and actions [22,37,58]. Compositions have also been studied in natural language processing [9,12]. Our paper builds on this foundation. The most related is [24], which also develops a meta-learning framework for compositional generalization. However, unlike [24], our approach works for realistic language and natural images.

Out-of-Vocabulary Words: This paper is related but different to models of out-of-vocabulary words (OOV) [18,19,23,25,43–45,45]. Unlike this paper, most of them require extra training, or gradient updates on new words. We compare to the most competitive approach [45], which reduces to regular BERT in our

Episode n-1 (new word)

Target example   Reference set

peel [?]   move broccoli   cut bottom off   clean the
              from chopping   **carrot**   countertop
              board to pan

Episode n (new composition)

Target example   Reference set

[?] [?]   stir pasta   cut **paneer** into
                           cubes

Fig. 3. Episodes for Meta-Learning: We illustrate two examples of training episodes. Each episode consists of several pairs of image and text. During learning, we mask out one or more words, indicated by a [?], and train the model to reconstruct it by pointing to ground truth (in **bold**) among other examples within the episode. By controlling the generalization gaps within an episode, we can explicitly train the model to generalize and learn new words and new compositions. For example, the left episode requires the model to learn how to acquire a new word ("carrot"), and the right episode requires the model to combine known words to form a novel composition ("stir paneer").

setting, as a baseline. Moreover, we incorporate OOV words not just as an input to the system, but also as output. Previous work on captioning [28,31,59,61] produces words never seen in the ground truth captions. However, they use pre-trained object recognition systems to obtain labels and use them to caption the new words. Our paper is different because we instead learn the word acquisition process from vision and text data. Finally, unlike [4], our approach does not require any side information or external information, and instead acquires new words using their surrounding textual and visual context.

Few-Shot Learning: Our paper builds on foundational work in few-shot learning, which aims to generalize with little or no labeled data. Past work has explored a variety of tasks, including image classification [47,51,60], translating between a language pair never seen explicitly during training [21] or understanding text from a completely new language [1,5], among others. In contrast, our approach is designed to acquire *language* from minimal examples. Moreover, our approach is not limited to just few-shot learning. Our method also learns a more robust underlying representation, such as for compositional generalization.

Learning to Learn: Meta-learning is a rapidly growing area of investigation. Different approaches include learning to quickly learn new tasks by finding a good initialization [14,29], learning efficient optimization policies [3,6,29,41,46], learning to select the correct policy or oracle in what is also known as hierarchical learning [15,19], and others [11,33]. In this paper, we apply meta-learning to acquire new words and compositions from visual scenes.

3   Learning to Learn Words

We present a framework that learns how to acquire words from visual context. In this section, we formulate the problem as a meta-learning task and propose a model that leverages self-attention based transformers to learn from episodes.

3.1 Episodes

We aim to learn the word acquisition process. Our key insight is that we can construct training episodes that demonstrate language acquisition, which provides the data to meta-learn this process. We create training *episodes*, each of which contain multiple *examples* of text-image pairs. During meta-learning, we sample episodes and train the model to acquire words from examples within each episode. Figure 3 illustrates some episodes and their constituent examples.

To build an episode, we first sample a *target example*, which is an image and text pair, and mask some of its word tokens. We then sample *reference examples*, some of which contain tokens masked in the target. We build episodes that require overcoming substantial generalization gaps, allowing us to explicitly meta-learn the model to acquire robust word representations. Some episodes may contain new words, requiring the model to learn a policy for acquiring the word from reference examples and using it to describe the target scene in the episode. Other episodes may contain familiar words but novel compositions in the target. In both cases, the model will need to generalize to target examples by using the reference examples in the episode. Since we train our model on a distribution of episodes instead of a distribution of examples, and each episode contains new scenes, words, and compositions, the learned policy will be robust at generalizing to testing episodes from the same distribution. By propagating the gradient from the target scene back to other examples in the episode, we can directly train the model to learn a word acquisition process.

3.2 Model

Let an episode be the set $e_k = \{v_1, \ldots, v_i, w_{i+1}, \ldots, w_j\}$ where v_i is an image and w_i is a word token in the episode. We present a model that receives an episode e_k, and train the model to reconstruct one or more masked words w_i by pointing to other examples within the same episode. Since the model must predict a masked word by drawing upon other examples within the same episode, it will learn a policy to acquire words from one example and use them for another example.

Transformers on Episodes: To parameterize our model, we need a representation that is able to capture pairwise relationships between each example in the episode. We propose to use a stack of transformers based on self-attention [55], which is able to receive multiple image and text pairs, and learn rich contextual outputs for each input [10]. The input to the model is the episode $\{v_1, \ldots, w_j\}$, and the stack of transformers will produce hidden representations $\{h_1, \ldots, h_j\}$ for each image and word in the episode.

Transformer Architecture: We input each image and word into the transformer stack. One transformer consists of a multi-head attention block followed by a linear projection, which outputs a hidden representation at each location, and is passed in series to the next transformer layer. Let $H^z \in \mathbb{R}^{d \times j}$ be the d dimensional hidden vectors at layer z. The transformer first computes vectors for queries $Q = W_q^z H^z$, keys $K = W_k^z H^z$, and values $V = W_v^t H^z$ where each

$W_* \in \mathbb{R}^{d \times d}$ is a matrix of learned parameters. Using these queries, keys, and values, the transformer computes the next layer representation by attending to all elements in the previous layer:

$$H^{z+1} = SV \quad \text{where} \quad S = \text{softmax}\left(\frac{QK^T}{\sqrt{d}}\right). \tag{1}$$

In practice, the transformer uses multi-head attention, which repeats Eq. 1 once for each head, and concatenates the results. The network produces a final representation $\{h_1^Z, \ldots, h_i^Z\}$ for a stack of Z transformers.

Input Encoding: Before inputting each word and image into the transformer, we encode them with a fixed-length vector representation. To embed input words, we use an $N \times d$ word embedding matrix ϕ_w, where N is the size of the vocabulary considered by the tokenizer. To embed visual regions, we use a convolutional network $\phi_v(\cdot)$ over images. We use ResNet-18 initialized on ImageNet [17,42]. Visual regions can be the entire image in addition to any region proposals. Note that the region proposals only contain spatial information without any category information.

To augment the input encoding with both information about the modality and the positional information (word index for text, relative position of region proposal), we translate the encoding by a learned vector:

$$\begin{aligned}
\phi_{\text{img}}(v_i) &= \phi_v(v_i) + \phi_{\text{loc}}(v_i) + \phi_{\text{mod}}(\text{IMG}) + \phi_{\text{id}}(v_i) \\
\phi_{\text{txt}}(w_j) &= \phi_{wj} + \phi_{\text{pos}}(w_j) + \phi_{\text{mod}}(\text{TXT}) + \phi_{\text{id}}(w_j)
\end{aligned} \tag{2}$$

where ϕ_{loc} encodes the spatial position of v_i, ϕ_{pos} encodes the word position of w_j, ϕ_{mod} encodes the modality and ϕ_{id} encodes the example index.

Please see the supplementary material for all implementation details of the model architecture. Code will be released.

3.3 Learning Objectives

To train the model, we mask input elements from the episode, and train the model to reconstruct them. We use three different complementary loss terms.

Pointing to Words: We train the model to "point" [56] to other words within the same episode. Let w_i be the target word that we wish to predict, which is masked out. Furthermore, let $w_{i'}$ be the same word which appears in a reference example in the episode ($i' \neq i$). To fill in the masked position w_i, we would like the model to point to $w_{i'}$, and not any other word in the reference set.

We estimate similarity between the ith element and the jth element in the episode. Pointing to the right word within the episode corresponds to maximizing the similarity between the masked position and the true reference position, which we implement as a cross-entropy loss:

$$\mathcal{L}_{\text{point}} = -\log\left(\frac{A_{ii'}}{\sum_k A_{ik}}\right) \quad \text{where} \quad \log A_{ij} = f(h_i)^T f(h_j) \tag{3}$$

where A is the similarity matrix and $f(h_i) \in \mathbb{R}^d$ is a linear projection of the hidden representation for the ith element. Minimizing the above loss over a large number of episodes will cause the neural network to produce a policy such that a novel reference word $w_{i'}$ is correctly routed to the right position in the target example within the episode.

Other similarity matrices are possible. The similarity matrix A will cause the model to fill in a masked word by pointing to another contextual representation. However, we can also define a similarity matrix that points to the input word embedding instead. To do this, the matrix is defined as $\log A_{ij} = f(h_i)^T \phi_{w_j}$. This prevents the model from solely relying on the context and forces it to specifically attend to the reference word, which our experiments will show helps generalizing to new words.

Word Cloze: We additionally train the model to reconstruct words by directly predicting them. Given the contextual representation of the masked word h_i, the model predicts the missing word by multiplying its contextual representation with the word embedding matrix, $\hat{w}_i = \arg \max \phi_w^T h_i$. We then train with cross-entropy loss between the predicted word $\hat{w}_i$ and true word w_i, which we write as $\mathcal{L}_{\text{cloze}}$. This objective is the same as in the original BERT [10].

Visual Cloze: In addition to training the word representations, we train the visual representations on a cloze task. However, whereas the word cloze task requires predicting the missing word, generating missing pixels is challenging. Instead, we impose a metric loss such that a linear projection of h_i is closer to $\phi_v(v_i)$ than $\phi_v(v_{k \neq i})$. We use the tripet loss [57] with cosine similarity and a margin of one. We write this loss as $\mathcal{L}_{\text{vision}}$. This loss is similar to the visual loss used in state-of-the-art visual language models [7].

Combination: Since each objective is complementary, we train the model by optimizing the neural network parameters to minimize the sum of losses:

$$\min_{\Omega} \mathbb{E} \left[\mathcal{L}_{\text{point}} + \alpha \mathcal{L}_{\text{cloze}} + \beta \mathcal{L}_{\text{vision}} \right] \tag{4}$$

where $\alpha \in \mathbb{R}$ and $\beta \in \mathbb{R}$ are scalar hyper-parameters to balance each loss term, and Ω are all the learned parameters. We sample an episode, compute the gradients with back-propagation, and update the model parameters by stochastic gradient descent.

3.4   Information Flow

We can control how information flows in the model by constraining the attention in different ways. **Isolated attention** implies examples can only attend within themselves. In the **full attention** setting, every element can attend to all other elements. In the **target-to-reference attention** setting we can constrain the attention to only allow the target elements to attend to the reference elements. Finally, in **attention via vision** case, we constrain the attention to only transfer information through vision. See the supplementary material for more details.

3.5   Inference

After learning, we obtain a policy that can acquire words from an episode consisting of vision and language pairs. Since the model produces words by pointing to them, which is a non-parametric mechanism, the model is consequently able to acquire words that were absent from the training set. As image and text pairs are encountered, they are simply inserted into the reference set. When we ultimately input a target example, the model is able to use new words to describe it by pulling from other examples in the reference set.

Moreover, the model is not restricted to only producing words from the reference set. Since the model is also trained on a cloze task, the underlying model is able to perform any standard language modeling task. In this setting, we only give the model a target example without a reference set. As our experiments will show, the meta-learning objective also improves these language modeling tasks.

4   Experiments

The goal of our experiments is to analyze the language acquisition process that is learned by our model. Therefore, we train the model on vision-and-language datasets, without any language pretraining. We call our approach **EXPERT**.[1]

4.1   Datasets

We use two datasets with natural images and realistic textual descriptions.

EPIC-Kitchens is a large dataset consisting of $39,594$ video clips across 32 homes. Each clip has a short text narration, which spans 314 verbs and 678 nouns, as well as other word types. EPIC-Kitchens is challenging due to the complexities of unscripted video. We use object region proposals on EPIC-Kitchens, but discard any class labels for image regions. We sample frames from videos and feed them to our models along with the corresponding narration. Since we aim to analyze generalization in language acquisition, we create a train-test split such that some words and compositions will only appear at test time. We list the full train-test split in the supplementary material.

Flickr30k contains $31,600$ images with five descriptions each. The language in Flickr30k is more varied and syntactically complex than in EPIC-Kitchens, but comes from manual descriptive annotations rather than incidental speech. Images in Flickr30k are not frames from a video, so they do not present the same amount of visual challenges in motion blur, clutter, etc., but they cover a wider range of scene and object categories. We again use region proposals without their labels and create a train-test split that withholds some words and compositions.

Our approach does not require additional image regions as input beyond the full image, and our experiments show that our method outperforms baselines similarly even when trained only with the full image as input, without other cropped regions (see supplementary material).

---

[1] Episodic Cross-Modal Pointing for Encoder Representations from Transformers.

Table 1. Acquiring New Words on EPIC-Kitchens: We test our model's ability to acquire new words at test time by pointing. The difficulty of this task varies with the number of distractor examples in the reference set. We show **top-1 accuracy** results on both 1:1 and 2:1 ratios of distractors to positives. The rightmost column shows computational cost of the attention variant used.

		Ratio		
		1:1	2:1	Cost
	Chance	13.5	8.7	–
	BERT (scratch) [10]	36.5	26.3	
	BERT+Vision [7]	63.4	57.5	
EXPERT	Isolated attention	69.0	57.8	$O(n)$
	Tgt-to-ref attention	71.0	63.2	$O(n)$
	Via-vision attention	72.7	64.5	$O(n)$
	+ Input pointing	**75.0**	**67.4**	$O(n)$
	Full attention	**76.6**	**68.4**	$O(n^2)$
	BERT (pretrained) [10]	53.4	48.8	

4.2 Baselines

We compare to established, state-of-the-art models in vision and language, as well as to ablated versions of our approach.

BERT is a language model that recently obtained state-of-the-art performance across several natural language processing tasks [10]. We consider two variants. Firstly, we download the pre-trained model, which is trained on three billion words, then fine-tune it on our training set. Secondly, we train BERT from scratch on our data. We use BERT as a strong language-only baseline.

BERT+Vision refers to the family of visually grounded language models [2, 7, 26, 27, 36, 48, 50, 63], which adds visual pre-training to BERT. We experimented with several of them on our tasks, and we report the one that performs the best [7]. Same as our model, this baseline does not use language pretraining.

We also compare several different attention mechanisms. **Tgt-to-ref attention, Via-vision attention,** and **Full attention** indicate the choice of attention mask; the base one is **Isolated attention. Input pointing** indicates the use of pointing to the input encodings along with contextual encodings. Unless otherwise noted, EXPERT refers to the variant trained with via-vision attention.

4.3 Acquisition of New Words

Our model learns the word acquisition *process*. We evaluate this learned process at how well it acquires new words not encountered in the training set. At test time, we feed the model an episode containing many examples, which contain previously unseen words. Our model has learned a strong word acquisition policy

Fig. 4. Word Acquisition: We show examples where the model acquires new words. ? in the target example indicates the masked out new word. **Bold** words in the reference set are ground truth. The model makes predictions by pointing into the reference set, and the weight of each pointer is visualized by the shade of the arrows shown (weight <3% is omitted). In bottom right, we show an error where the model predicts the plate is being placed, where the ground truth is "grabbed".

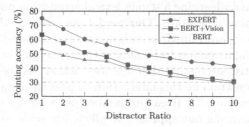

Fig. 5. Word Acquisition versus Distractors: As more distractors are added (testing on EPIC-Kitchens), the problem becomes more difficult, causing performance for all models to go down. However, EXPERT decreases at a lower rate than baselines.

if it can learn a representation for the new words, and correctly use them to fill in the right masked words in the target example.

Specifically, we pass each example in an episode forward through the model and store hidden representations at each location. We then compute hidden representation similarity between the masked location in the target example and every example in the reference set. We experimented with a few similarity metrics, and found dot-product similarity performs the best, as it is a natural extension of the attention mechanism that transformers are composed of.

We compare our meta-learned representations to state-of-the-art vision and language representations, i.e. BERT and BERT with Vision. When testing, baselines use the same pointing mechanism (similarity score between hidden representations) and reference set as our model. Baselines achieve strong performance since they are trained to learn contextual representations that have meaningful similarities under the same dot-product metric used in our evaluation.

We show results on this experiment in Table 1. Our complete model obtains the best performance in word acquisition on both EPIC-Kitchens and Flickr30k.

Table 2. Acquiring New Words on Flickr30k: We run the same experiment as Table 1 (**top-1 accuracy** pointing to new words), except on the Flickr30k dataset, which has more complex textual data. As before, we show results on 1:1 and 2:1 ratios of distractors to positives. By learning the acquisition policy, our model obtains competitive performance with orders of magnitude less training data.

	Ratio	
Method	1:1	2:1
Chance	3.4	2.3
BERT (scratch)	31.6	25.7
BERT with Vision [7]	32.1	26.8
EXPERT	**69.3**	**60.9**
BERT (pretrained)	**69.4**	**60.8**

In the case of EPIC-Kitchens, where linguistic information is scarce and sentence structure simpler, meta-learning a strong lexical acquisition policy is particularly important for learning new words. Our model outperforms the strongest baselines (including those pretrained on enormous text corpora) by up to 13% in this setting. Isolating attention to be only within examples in an episode harms accuracy significantly, suggesting that the interaction between examples is key for performance. Constraining this interaction to pass through the visual modality, the computational cost is linear in number of examples with only a minor drop in accuracy, allowing our approach to efficiently scale to larger episodes.

Figure 4 shows qualitative examples where the model must acquire novel language by learning from its reference set, and use it to describe another scene with both nouns and verbs. In the bottom right of the figure, an incorrect example is shown, in which EXPERT points to *place* and *put* instead of *grab*. However, both incorrect options are plausible guesses given only the static image and textual context "plate". This example suggests that video information would further improve EXPERT's performance.

Figure 5 shows that, even as the size of the reference set (and thus the difficulty of language acquisition) increases, the performance of our model remains relatively robust compared to baselines. EXPERT outperforms baselines by 18% with one distractor example, and by 36% with ten.

In Flickr30k, visual scenes are manually described in text by annotators rather than transcribed from incidental speech, so they present a significant challenge in their complexity of syntactic structure and diversity of subject matter. In this setting, our model significantly outperforms all baselines that train from scratch on Flickr30k, with an increase in accuracy of up to 37% (Table 2). Since text is more prominent, a state-of-the-art language model pretrained on huge (>3 billion token) text datasets performs well, but EXPERT achieves the same accuracy while requiring several orders of magnitude less training data.

Table 3. Acquiring Familiar Words: We report **top-5 accuracy** on masked language modeling of words which appear in training. Our model outperforms all other baselines.

Method	EPIC-Kitchens			Flickr30k		
	Verbs	Nouns	All	Verbs	Nouns	All
Chance	0.1	0.1	0.1	< 0.1	< 0.1	< 0.1
BERT (scratch) [10]	68.2	48.9	57.9	64.8	69.4	66.2
BERT with vision [7]	77.3	63.2	65.6	65.1	70.2	66.5
EXPERT	**81.9**	**73.0**	**74.9**	69.1	79.8	72.0
BERT (pretrained) [10]	71.4	51.5	59.8	**69.5**	79.4	**72.2**

4.4 Acquisition of Familiar Words

By learning a policy for word acquisition, the model also jointly learns a representation for the familiar words in the training set. Since the representation is trained to facilitate the acquisition process, we expect these embeddings to also be robust at standard language modeling tasks. We directly evaluate them on the standard cloze test [53], which all models are trained to complete.

Table 3 shows performance on language modeling. The results suggest that visual information helps learn a more robust language model. Moreover, our approach, which learns the process in addition to the embeddings, outperforms all baselines by between 4 and 9% across both datasets. While a fully pretrained BERT model also obtains strong performance on Flickr30k, our model is able to match its accuracy with orders of magnitude less training data.

Our results suggest that learning a process for word acquisition also collaterally improves standard vision and language modeling. We hypothesize this happens because learning acquisition provides an incentive for the model to generalize, which acts as a regularization for the underlying word embeddings.

4.5 Compositionality

Since natural language is compositional, we quantify how well the representations generalize to novel combinations of verbs and nouns that were absent from the training set. We again use the cloze task to evaluate models, but require the model to predict both a verb and a noun instead of only one word.

We report results on compositions in Table 4 for both datasets. We breakdown results by whether the compositions were seen or not during training. Note that, for all approaches, there is a substantial performance gap between seen and novel compositions. However, since our model is explicitly trained for generalization, the gap is significantly smaller (nearly twice as small). Moreover, our approach also shows substantial gains over baselines for both seen and novel compositions, improving by seven and sixteen points respectively. Additionally, our approach is able to exceed or match the performance of pretrained BERT, even though our model is trained on three orders of magnitude less training data.

Table 4. Compositionality: We show top-5 accuracy at predicting masked compositions of seen nouns and verbs. Both the verb and the noun must be correctly predicted. EXPERT achieves the best performance on both datasets.

Method	EPIC-Kitchens			Flickr30k		
	Seen	New	Diff	Seen	New	Diff
Chance	<0.1	<0.1	–	<0.1	<0.1	–
BERT (scratch) [10]	34.3	17.7	16.6	43.4	39.4	4.0
BERT with vision [7]	56.1	37.6	18.5	45.0	42.0	3.0
EXPERT	**63.5**	**53.0**	**10.5**	48.7	47.1	1.6
BERT (pretrained) [10]	39.8	20.7	19.1	**48.8**	**47.2**	1.6

Table 5. Retrieval: We test the model's top-1 retrieval accuracy (in %) from a 10 sample retrieval set. T→I and I→T represent retrieval from image to text and text to image.

Method	EPIC-Kitchens		Flickr30k	
	T→I	I→T	T→I	I→T
Chance	10.0	10.0	10.0	10.0
BERT with vision [7]	13.8	13.9	54.9	57.4
EXPERT	**32.6**	**25.3**	**57.5**	**60.6**

4.6 Retrieval

Following prior work [7,26,30,48], we evaluate our representation on cross-modal retrieval. We observe significant gains from our approach, outperforming baselines by up to 19%. Specifically, we run an image/text cross-modal retrieval test on both the baseline BERT+Vision model and ours. We freeze model weights and train a classifier on top to decide whether input image and text match, randomly replacing data from one modality to create negative pairs. We then test on samples containing new compositions. Please see Table 5 for results.

4.7 Analysis

In this section, we analyze *why* EXPERT obtains better performance.

How Are New Words Embedded in EXPERT? Fig. 6 shows how EXPERT represents new words in its embedding space at test time. We run sentences which contain previously unseen words through our model. Then, we calculate the nearest neighbor of generated hidden representations of these unseen words in the learned word embedding matrix. Our model learns a representation space such that new words are embedded near semantically similar words (dependent on context), even though we use no such supervisory signal in training.

Does EXPERT Use Vision? We take our complete model, trained with both text and images, and withhold images at test time. Performance drops to nearly

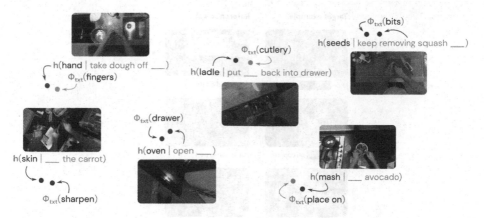

Fig. 6. Embedding New Words with EXPERT: We give EXPERT sentences with unfamiliar language at test time. We show the hidden vectors $h(new\ word\ |\ context, image)$ it produces, conditioned on visual and linguistic context, and their nearest neighbors in word embedding space $\phi_{txt}(known\ word)$. EXPERT can use its learned vision-and-language policy to embed new words near other words that are similar in object category, affordances, and semantic properties.

chance, showing that EXPERT uses visual information to predict words and disambiguate between similar language contexts.

What Visual Information Does EXPERT Use? To study this, we withhold one visual region at a time from the episode and find the regions that cause the largest decrease in prediction confidence. Figure 7 visualizes these regions, showing that removing the object that corresponds to the target word causes the largest drop in performance. This suggests that the model is correlating these words with the right visual region, without direct supervision.

How Does Information Flow Through EXPERT? Our model makes predictions by attending to other elements within its episode. To analyze the learned attention, we take the variant of EXPERT trained with full pairwise attention and measure changes in accuracy as we disable query-key interactions one by one. Figure 8 shows which connections are most important for performance. This reveals a strong dependence on cross-modal attention, where information flows from text to image in the first layer, and back to text in the last layer.

How Does EXPERT Disambiguate Multiple New Words? We evaluate our model on episodes that contain five new words in the reference set, only one of which matches the target token. Our model obtains an accuracy of 56% in this scenario, while randomly picking one of the novel words would give 20%. This shows that our model is able to discriminate between many new words in an episode. We also evaluate the fine-tuned BERT model in this same setting, where it obtains a 37% accuracy, significantly worse than our model. This suggests that vision is important to disambiguate new words.

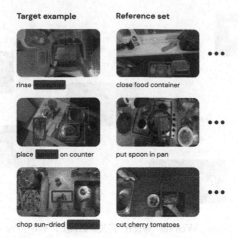

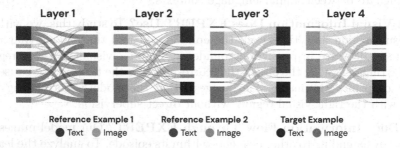

Fig. 7. Visualizing the Attention: We probe how the model uses visual information. We remove various objects from input images in an episode, and evaluate the model's confidence in predicting masked words . Removing image regions with a yellow box causes the greatest drop in confidence (other regions are shown in red). The most important visual regions for the prediction task contain an instance of the target word. These results suggest that our model learns some spatial localization of words automatically. (Color figure online)

Fig. 8. Visualizing the Learned Process: We visualize how information flows through the learned word acquisition process. The width of the pipe indicates the importance of the connection, as estimated by how much performance drops if removed. In the first layer, information tends to flow from the textual nodes to the image nodes. In subsequent layers, information tends to flow from image nodes back to text nodes.

5 Discussion

We believe the language acquisition process is too complex to hand-craft. In this paper, we instead propose to meta-learn a policy for word acquisition from visual scenes. Compared to established baselines, our experiments show significant gains at acquiring novel words, generalizing to novel compositions, and learning more robust word representations. Visualizations and analysis reveal that the learned policy leverages both the visual scene and linguistic context.

Acknowledgements. We thank Alireza Zareian, Bobby Wu, Spencer Whitehead, Parita Pooj and Boyuan Chen for helpful discussion. Funding for this research was provided by DARPA GAILA HR00111990058. We thank NVidia for GPU donations.

References

1. Adams, O., Makarucha, A., Neubig, G., Bird, S., Cohn, T.: Cross-lingual word embeddings for low-resource language modeling. In: Proceedings of the 15th Conference of the European Chapter of the Association for Computational Linguistics: Volume 1, Long Papers, Valencia, Spain, pp. 937–947. Association for Computational Linguistics, April 2017. https://www.aclweb.org/anthology/E17-1088
2. Alberti, C., Ling, J., Collins, M., Reitter, D.: Fusion of detected objects in text for visual question answering (B2T2), August 2019. http://arxiv.org/abs/1908.05054
3. Andrychowicz, M., et al.: Learning to learn by gradient descent by gradient descent. In: Advances in Neural Information Processing Systems, pp. 3981–3989 (2016)
4. Anne Hendricks, L., Venugopalan, S., Rohrbach, M., Mooney, R., Saenko, K., Darrell, T.: Deep compositional captioning: describing novel object categories without paired training data. In: Proceedings of the IEEE Conference on Computer Vision and Pattern Recognition, pp. 1–10 (2016)
5. Artetxe, M., Schwenk, H.: Massively multilingual sentence embeddings for zero-shot cross-lingual transfer and beyond. Tech. rep. (2019)
6. Bengio, S., Bengio, Y., Cloutier, J., Gecsei, J.: On the optimization of a synaptic learning rule (2002)
7. Chen, Y.C., et al.: UNITER: Learning UNiversal Image-TExt Representations. Tech. rep. (2019)
8. Damen, D., et al.: Scaling egocentric vision: the EPIC-KITCHENS Dataset. In: The European Conference on Computer Vision (ECCV) (2018). http://youtu.be/Dj6Y3H0ubDw
9. Dasgupta, I., Guo, D., Stuhlmüller, A., Gershman, S.J., Goodman, N.D.: Evaluating compositionality in sentence embeddings. arXiv preprint arXiv:1802.04302 (2018)
10. Devlin, J., Chang, M.W., Lee, K., Toutanova, K.: BERT: pre-training of deep bidirectional transformers for language understanding. Tech. rep. https://github.com/tensorflow/tensor2tensor
11. Duan, Y., Schulman, J., Chen, X., Bartlett, P.L., Sutskever, I., Abbeel, P.: Rl2: fast reinforcement learning via slow reinforcement learning. arXiv preprint arXiv:1611.02779 (2016)
12. Ettinger, A., Elgohary, A., Phillips, C., Resnik, P.: Assessing composition in sentence vector representations. arXiv preprint arXiv:1809.03992 (2018)
13. Farhadi, A., et al.: Every picture tells a story: generating sentences from images. In: Daniilidis, K., Maragos, P., Paragios, N. (eds.) ECCV 2010. LNCS, vol. 6314, pp. 15–29. Springer, Heidelberg (2010). https://doi.org/10.1007/978-3-642-15561-1_2
14. Finn, C., Abbeel, P., Levine, S.: Model-agnostic meta-learning for fast adaptation of deep networks. In: Proceedings of the 34th International Conference on Machine Learning, vol. 70, pp. 1126–1135. JMLR. org (2017)
15. Frans, K., Ho, J., Chen, X., Abbeel, P., Schulman, J.: Meta learning shared hierarchies. arXiv preprint arXiv:1710.09767 (2017)
16. Gandhi, K., Lake, B.M.: Mutual exclusivity as a challenge for neural networks. arXiv preprint arXiv:1906.10197 (2019)

17. He, K., Zhang, X., Ren, S., Sun, J.: Deep residual learning for image recognition. In: Proceedings of the IEEE Conference on Computer Vision and Pattern Recognition, pp. 770–778 (2016)
18. Herbelot, A., Baroni, M.: High-risk learning: acquiring new word vectors from tiny data. In: Proceedings of the 2017 Conference on Empirical Methods in Natural Language Processing, pp. 304–309 (2017)
19. Hu, Z., Chen, T., Chang, K.W., Sun, Y.: Few-shot representation learning for out-of-vocabulary words. In: Proceedings of the 57th Annual Meeting of the Association for Computational Linguistics, pp. 4102–4112 (2019)
20. Johnson, J., Fei-Fei, L., Hariharan, B., Zitnick, C.L., Van Der Maaten, L., Girshick, R.: CLEVR: a diagnostic dataset for compositional language and elementary visual reasoning. In: Proceedings of the IEEE Conference on Computer Vision and Pattern Recognition (CVPR) (2017). https://arxiv.org/pdf/1612.06890.pdf
21. Johnson, M., et al.: Google's multilingual neural machine translation system: enabling zero-shot translation. Trans. Assoc. Comput. Linguist. **5**, 339–351 (2017)
22. Kato, K., Li, Y., Gupta, A.: Compositional Learning for Human Object Interaction. In: Ferrari, V., Hebert, M., Sminchisescu, C., Weiss, Y. (eds.) Computer Vision – ECCV 2018. LNCS, vol. 11218, pp. 247–264. Springer, Cham (2018). https://doi. org/10.1007/978-3-030-01264-9_15
23. Khodak, M., Saunshi, N., Liang, Y., Ma, T., Stewart, B.M., Arora, S.: A la carte embedding: cheap but effective induction of semantic feature vectors. In: 56th Annual Meeting of the Association for Computational Linguistics, ACL 2018, pp. 12–22. Association for Computational Linguistics (ACL) (2018)
24. Lake, B.M.: Compositional generalization through meta sequence-to-sequence learning. In: NeurIPS (2019)
25. Lazaridou, A., Marelli, M., Baroni, M.: Multimodal word meaning induction from minimal exposure to natural text. Cogn. Sci. **41**, 677–705 (2017)
26. Li, G., Duan, N., Fang, Y., Jiang, D., Zhou, M.: Unicoder-VL: a universal encoder for vision and language by cross-modal pre-training. ArXiv abs/1908.06066 (2019)
27. Li, L.H., Yatskar, M., Yin, D., Hsieh, C.J., Chang, K.W.: VisualBERT: a simple and performant baseline for vision and language. Tech. rep. (2019). http://arxiv. org/abs/1908.03557
28. Li, Y., Yao, T., Pan, Y., Chao, H., Mei, T.: Pointing novel objects in image captioning. In: Proceedings of the IEEE Conference on Computer Vision and Pattern Recognition, pp. 12497–12506 (2019)
29. Li, Z., Zhou, F., Chen, F., Li, H.: Meta-SGD: learning to learn quickly for few-shot learning. arXiv preprint arXiv:1707.09835 (2017)
30. Lu, J., Batra, D., Parikh, D., Lee, S.: ViLBERT: pretraining task-agnostic visiolinguistic representations for vision-and-language tasks. In: Neural Information Processing Systems (NeurIPS) (2019). http://arxiv.org/abs/1908.02265
31. Lu, J., Yang, J., Batra, D., Parikh, D.: Neural baby talk. In: Proceedings of the IEEE Conference on Computer Vision and Pattern Recognition, pp. 7219–7228 (2018)
32. Mao, J., Gan, C., Kohli, P., Tenenbaum, J.B., Wu, J.: The neuro-symbolic concept learner: interpreting scenes, words, and sentences from natural supervision. In: International Conference on Learning Representations (2019). https://openreview. net/forum?id=rJgMlhRctm
33. Mishra, N., Rohaninejad, M., Chen, X., Abbeel, P.: A simple neural attentive meta-learner. arXiv preprint arXiv:1707.03141 (2017)

34. Misra, I., Gupta, A., Hebert, M.: From red wine to red tomato: composition with context. In: Proceedings of the IEEE Conference on Computer Vision and Pattern Recognition, pp. 1792–1801 (2017)
35. Nagarajan, T., Grauman, K.: Attributes as operators: factorizing unseen attribute-object compositions. In: European Conference on Computer Vision (ECCV) (2018). https://arxiv.org/pdf/1803.09851.pdf
36. Nangia, N., Bowman, S.R.: Human vs. muppet: a conservative estimate of human performance on the glue benchmark. arXiv preprint arXiv:1905.10425 (2019)
37. Nikolaus, M., Abdou, M., Lamm, M., Aralikatte, R., Elliott, D.: Compositional generalization in image captioning. In: CoNLL (2018)
38. Peters, M.E., et al.: Deep contextualized word representations. arXiv preprint arXiv:1802.05365 (2018)
39. Radford, A., Narasimhan, K., Salimans, T., Sutskever, I.: Improving language understanding by generative pre-training (2018)
40. Rahman, W., Hasan, M.K., Zadeh, A., Morency, L.P., Hoque, M.E.: M-BERT: injecting multimodal information in the BERT structure. Tech. rep. (2019). http://arxiv.org/abs/1908.05787
41. Ravi, S., Larochelle, H.: Optimization as a model for few-shot learning (2016)
42. Russakovsky, O., et al.: Imagenet large scale visual recognition challenge. Int. J. Comput. Vision **115**(3), 211 252 (2015)
43. Schick, T., Schütze, H.: Attentive mimicking: Better word embeddings by attending to informative contexts. In: Proceedings of the 2019 Conference of the North American Chapter of the Association for Computational Linguistics: Human Language Technologies, Volume 1 (Long and Short Papers), pp. 489–494 (2019)
44. Schick, T., Schütze, H.: Learning semantic representations for novel words: leveraging both form and context. In: Proceedings of the AAAI Conference on Artificial Intelligence, vol. 33, pp. 6965–6973 (2019)
45. Schick, T., Schütze, H.: Rare words: A major problem for contextualized embeddings and how to fix it by attentive mimicking. arXiv preprint arXiv:1904.06707 (2019)
46. Schmidhuber, J.: Evolutionary Principles in Self-Referential Learning. On Learning now to Learn: The Meta-Meta-Meta...-Hook. Diploma thesis, Technische Universitat Munchen, Germany, 14 May 1987. http://www.idsia.ch/~juergen/diploma.html
47. Snell, J., Swersky, K., Zemel, R.: Prototypical networks for few-shot learning. In: Guyon, I., et al. (eds.) Advances in Neural Information Processing Systems, vol. 30, pp. 4077–4087. Curran Associates, Inc. (2017). http://papers.nips.cc/paper/6996-prototypical-networks-for-few-shot-learning.pdf
48. Su, W., et al.: VL-BERT: pre-training of generic visual-linguistic representations. Tech. rep. (2019). http://arxiv.org/abs/1908.08530
49. Sun, C., Baradel, F., Murphy, K., Schmid, C.: Contrastive bidirectional transformer for temporal representation learning. Tech. rep. (2019)
50. Sun, C., Myers, A., Vondrick, C., Murphy, K., Schmid, C.: VideoBERT: a joint model for video and language representation learning, April 2019. http://arxiv.org/abs/1904.01766
51. Sung, F., Yang, Y., Zhang, L., Xiang, T., Torr, P.H., Hospedales, T.M.: Learning to compare: relation network for few-shot learning. In: The IEEE Conference on Computer Vision and Pattern Recognition (CVPR), June 2018
52. Tan, H., Bansal, M.: LXMERT: learning cross-modality encoder representations from transformers. In: Proceedings of the 2019 Conference on Empirical Methods in Natural Language Processing, August 2019. http://arxiv.org/abs/1908.07490

53. Taylor, W.L.: "cloze procedure": a new tool for measuring readability. J. Bull. **30**(4), 415–433 (1953)
54. Tincoff, R., Jusczyk, P.W.: Some beginnings of word comprehension in 6-month-olds. Psychol. Sci. **10**(2), 172–175 (1999)
55. Vaswani, A., et al.: Attention Is All You Need (2017)
56. Vinyals, O., Fortunato, M., Jaitly, N.: Pointer networks. In: Advances in Neural Information Processing Systems, pp. 2692–2700 (2015)
57. Weinberger, K.Q., Saul, L.K.: Distance metric learning for large margin nearest neighbor classification. J. Mach. Learn. Res. **10**(Feb), 207–244 (2009)
58. Wray, M., Larlus, D., Csurka, G., Damen, D.: Fine-grained action retrieval through multiple parts-of-speech embeddings. In: IEEE/CVF International Conference on Computer Vision (ICCV) (2019)
59. Wu, Y., Zhu, L., Jiang, L., Yang, Y.: Decoupled novel object captioner. In: Proceedings of the 26th ACM International Conference on Multimedia, pp. 1029–1037 (2018)
60. Xian, Y., Schiele, B., Akata, Z.: Zero-shot learning - the good, the bad and the ugly. In: The IEEE Conference on Computer Vision and Pattern Recognition (CVPR), July 2017
61. Yao, T., Pan, Y., Li, Y., Mei, T.: Incorporating copying mechanism in image captioning for learning novel objects. In: Proceedings of the IEEE Conference on Computer Vision and Pattern Recognition, pp. 6580–6588 (2017)
62. Young, P., Lai, A., Hodosh, M., Hockenmaier, J.: From image descriptions to visual denotations: new similarity metrics for semantic inference over event descriptions. Trans. Assoc. Comput. Linguist. **2**, 67–78 (2014)
63. Zhou, L., Palangi, H., Zhang, L., Hu, H., Corso, J.J., Gao, J.: Unified vision-language pre-training for image captioning and VQA. Tech. rep. (2019). https://github.com/LuoweiZhou/VLP

On Transferability of Histological Tissue Labels in Computational Pathology

Mahdi S. Hosseini[1]([⊠])(iD), Lyndon Chan[1](iD), Weimin Huang[1](iD),
Yichen Wang[1](iD), Danial Hasan[1](iD), Corwyn Rowsell[2,3](iD), Savvas
Damaskinos[4](iD), and Konstantinos N. Plataniotis[1](iD)

[1] Department of Electrical and Computer Engineering, University of Toronto,
Toronto, Canada
{mahdi.hosseini,lyndon.chan,cheryl.huang,
yichenk.wang,danial.hasan,kostas.plataniotis}@utoronto.ca
[2] Division of Pathology, St. Michaels Hospital, Toronto, ON M4N 1X3, Canada
[3] Department of Laboratory Medicine and Pathobiology, University of Toronto,
Toronto, Canada
[4] Huron Digital Pathology, St. Jacobs, ON N0B 2N0, Canada
https://github.com/mahdihosseini/HistoLabelTransfer/

Abstract. Deep learning tools in computational pathology, unlike natural vision tasks, face with limited histological tissue labels for classification. This is due to expensive procedure of annotation done by expert pathologist. As a result, the current models are limited to particular diagnostic task in mind where the training workflow is repeated for different organ sites and diseases. In this paper, we explore the possibility of transferring diagnostically-relevant histology labels from a source-domain into multiple target-domains to classify similar tissue structures and cancer grades. We achieve this by training a Convolutional Neural Network (CNN) model on a source-domain of diverse histological tissue labels for classification and then transfer them to different target domains for diagnosis without re-training/fine-tuning (zero-shot). We expedite this by an efficient color augmentation to account for color disparity across different tissue scans and conduct thorough experiments for evaluation.

Keywords: Cancer detection · Cancer grade classification · Deep learning · Domain adaptation · Zero-shot transfer · Color augmentation

1 Introduction

Recent deep learning techniques have been achieving competitive (at times even superior) performances compared to medical pathologists when diagnosing disease from Whole Slide Images (WSI). Histology slides are collected from

Electronic supplementary material The online version of this chapter (https://doi.org/10.1007/978-3-030-58526-6_27) contains supplementary material, which is available to authorized users.

particular organs and annotated with a particular disease to solve a particular diagnostic task in mind and deep learning models are trained with these annotated images to produce accurate and diagnostically meaningful predictions [1,9,11,12,19,27,29]. While the latter approach largely solves specific diagnostic problems, it also prevents ready application to other organs and diseases. There has been little research to date on generalizing these state-of-the-art deep learning models to other datasets with different but related organs and diseases. This is problematic, because without the ability to transfer relevant knowledge from other datasets, it becomes prohibitively *expensive* and *time-consuming* to collect the histological annotations needed to train a new deep learning model for each new application [30,46].

There are two main bottlenecks to such knowledge transfer: (1) the lack of annotated labels with diverse and generalizable tissue types from different organs and diseases; and (2) the variation in WSI scanners and staining protocols. Firstly, most openly available datasets were collected to solve a particular diagnostic problem and hence only contain slides focusing on specific organs and diseases: colorectal [22,23,37], breast [2,4,6], brain [13,14], and various organs [26,35]. This restricts the scope of histopathology for representational learning. In particular, most images are provided as single-label patches referring to specific tissue/disease related components. This restricts the ability to train classifiers that can discriminate tissue components smaller than the patch. Secondly, the color fidelity of WSI scans varies considerably: (a) digital pathology scanners follow different optics configurations and camera sensor calibration guidelines, and (b) pathology laboratories adopt different standards for staining the histology slides [31,33,36]. This means the same histology slide prepared by different institutions and digitized by different scanners can vary drastically by color and illumination contrast, which causes enormous challenges for training generalizable computational pathology algorithms [3,8,25,39–43].

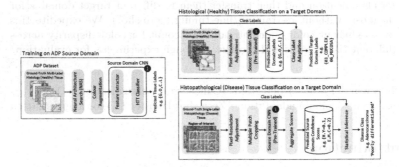

Fig. 1. Transferring HTT labels from source domain CNN (with labels available for training) into target domains for both tissue and disease classification (with labels unavailable for training) without re-training/fine-tuning. Labels are transferred based on prior histological/histopathological knowledge.

In this paper, we address these gaps by proposing a new approach to computational pathology - by training a *"universal"* model to recognize diverse histological tissue types (HTTs) from a source domain dataset of healthy slides (obtained from various organs), we can adapt the model to transfer diagnostically relevant labels for tissue and disease classification in target domain datasets without re-training or fine-tuning (i.e. zero-shot), see Fig. 1 for demonstration. Unlike the existing domain adaptation methods, our approach requires training only once on the source domain dataset followed by a simple label adaptation to the target domain consulted by expert pathologist. For this purpose, we employ the Atlas of Digital Pathology (ADP) database [18] as our source domain dataset, where its diverse multi-label/multi-class set overlaps with many other existing datasets. To account for color disparities in WSI scans, we explore two different color augmentation methods for training using HSV [39, 42, 43] as well as a less complex color space transform i.e. YCbCr. Furthermore, we develop a simple and yet efficient Convolutional Neural Network (CNN) architecture called *"HistoNet"*, guided by the Reinforcement Learning (RL)-based Neural Architecture Search (NAS) [48] as a means to the end goal of domain adaptation. We further study optimum optical resolution that can produce acceptable performance in processing WSI scans. We define two tasks in target domain labels: (1) recognize tissues by matching labels in the source and target domains using prior histological knowledge; and (2) classify cancer grades by employing the confidence scores of diagnostically relevant labels as a surrogate for disease progression. Both tasks can be seen as variants of the "transductive" and "inductive" transfer learning problems without the target domain images or labels available during source domain training [32, 45]. Our results, for the first time, reveal that different but related histopathology datasets can be unified and efficiently represented by a deep learning model trained on a sufficiently diverse label set. Without retraining, our approach can form reasonable predictions of the tissue classes and diseases in unseen images.

The summary of our main contributions is as follows.

1. We introduce a new label transferring solution in computational pathology from the ADP source domain to different target domain based on diagnostically relevant labels for tissue type and cancer grade classification
2. We explore the possibility of efficient CNN training that can be robust and optimized toward color disparities and pixel-resolution for tissue recognition in WSI scans.
3. We provide thorough adaptation experiments on variety of target domain datasets and show the strengths of source domain for generalization

1.1   Related Works

The problem of domain adaptation is widely used in machine learning such as vision and language [15, 16, 32, 45, 47] to study the problem of knowledge transfer from a source domain to perform a predefined target domain task. Most knowledge transfer methods are done in an unsupervised fashion where the target-domain data is unlabeled [15] but its representation is employed during the training such as adversarial training in [16,47]. This concept has been recently investigated in computational pathology to train a classifier using the source domain labels and adversaries from target-domain [7,28,31,34]. A common disadvantage to such methods is the target domain images must be available for training in the source domain. Therefore, retraining must be done independently for each target domain task.

Color augmentation is noticeably becoming a key factor in computational pathology domain adaptation, since the color disparities have strong effects on CNN predictions [3,39,41–43,46]. Methods such as HSV transformation [3,39,43] randomly perturbs the original images during training to expand the color space and hence generalize the model to images with colors not seen in the original training set. Stain normalization is also used to map the color distribution in the target domain to the source domain [5,7,8,21,28,31,34]. The downside of the latter approaches is that HSV performs a non-linear mapping and produces color residues during transformation which tend to produce misleading results [3]. Also, the stain normalization performs a one-to-one mapping between two histological stains (usually H&E). Hence it is unsuitable for training domain adaptation datasets such as ADP which contain multiple stains and using it would limit model generalizability to other stains.

2 Transferring Diagnostically-Relevant Labels

In this section, we describe a useful tissue label mapping (a rule based approach) that can be adapted between the HTT labels from ADP source domain and other target domain labels. This adaptation is shown in Fig. 2 for both healthy and disease tissue labels, where the connections are consulted by expert pathologist for label mapping. If the target task is tissue classification, we only map the source labels to the target label set using its prior histological knowledge. If the target task is disease classification problem, the corresponding source labels correlated to disease level are identified and use their inverse confidence scores for statistical inference of disease class. Note that the primary site in computational pathology (i.e. the type of organ) is usually given as a prior knowledge. Our hypothesis here is if ADP contains such organ tissues, then with good probability, the relevant tissue type and disease class(es) can be well predicted through the inference of HTTs. For instance, in GlaS (Colon tissues) dataset, we employ two HTTs from ADP, i.e. E.M.C and H.Y, which are highly relevant for cancer detection in Colon tissues.

While our approach is indeed rule-based, histological tissues exhibit many superficial visual similarities, so taking a data-driven approach would be counterproductive. Our label mapping was derived through consultation with a medical pathologist and is a quick, one-step procedure. Our approach is more explainable (and hence trustworthy) to pathologists because it mimics their own diagnostic workflow better than a black-box solution. Pathologists cannot exhaustively learn to diagnose all possible cases, so they learn from labeled educational examples (i.e. train on source domain healthy samples), transfer their knowledge to

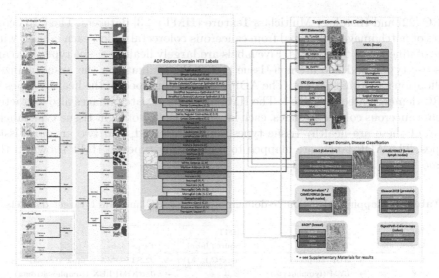

Fig. 2. Transferring Histological Tissue Type (HTT) labels from ADP source domain to target domains consulted by expert pathologist. For the tissue classification case (top), transferable target classes are in solid color and nontransferable classes are in striped color. For the disease classification case (bottom), the normal classes are indicated in magenta and the diseased classes in blue. (Color figure online)

diagnostic task at hand by searching for abnormalities (i.e. label mapping), and diagnose each new case by classifying visual appearance of cancer diseases (i.e. predict on target domain for grading).

2.1   Source Domain: ADP

The Atlas of Digital Pathology (ADP) [18] is a database of patch images extracted from 100 healthy slides from the same medical institution scanned with a TissueScope LE1.2 at 0.25 μm/px resolution. Each patch in the database is annotated with up to 33 hierarchical tissue types (modified from the 42 types in the original release - see the Supplementary Materials for details) in multi-labeled class format, covering a diverse set of morphological and functional types across different organs, such as stomach, colon, and thyroid. For visual presentation of this hierarchy, please refer to Fig. 2. The label set in ADP contains tissue types observed in different organs and hence overlaps with the healthy tissue types in other histopathology databases.

2.2   Task 1: Tissue Classification

We simply map those target labels representing the same healthy tissue types as in the source domain set. Our approach is both simpler (no retraining is required) and requires less information (only the label set) from the target domain. We evaluate our approach for tissue classification in the ColoRectal Cancer dataset

CRC [22] and Histology Multiclass Texture (HMT) [23] datasets. The CRC consists of patch images extracted from cancerous colorectal slides, each labeled with one of nine tissue type labels. Five labels are largely healthy tissue types, two are disease types, and two are non-tissue types. We evaluate on the un-normalized validation set of this dataset. The ADP labels are mapped to the healthy types in CRC demonstrated in Table 1. The HMT consists of patch images also extracted from cancerous colorectal slides, each labeled with one of eight tissue type labels. Five of these are healthy tissue types, one is diseased, and two are non-tissue types. The ADP labels are mapped to the healthy types in HMT shown in the same Table 1.

Table 1. Mapping ADP Source-domain labels to CRC and HMT target datasets.

CRC			HMT		
Source label		Target label	Source label		Target label
A	→	ADI (adipose)	max(C.L, M)	→	02_STROMA (simple stroma)
H.Y	→	LYM (lymphocyte)	C.L	→	03_COMPLEX (complex stroma)
M	→	MUS (muscle)	H.Y	→	04_LYMPHO (lymhocyte)
C.L	→	NORM (normal stroma)	G.O	→	06_MUCOSA (mucosa)
			A	→	07_ADIPOSE (adipose)

2.3 Task 2: Disease Classification

In preliminary experiments, we noticed that if the disease classes are quantified between 0 (normal) and 1 (most diseased), the confidence scores of diagnostically relevant classes would deteriorate with worsening disease. Here, we study three diseased datasets i.e. Gland Segmentation (GlaS) challenge [37], PatchCamelyon [44] extracted from the WSI scans of the original Camelyon16 challenge dataset [6], and Grand Challenge on Breast Cancer Histology (BACH) [2] all listed in Table 2 including the diagnostic features of each dataset. For each dataset, the diagnostically relevant HTT labels are identified from ADP source domain. Note that GlaS is comprised of Colon tissues which already exist in the ADP source domain. However, both PatchCamelyon and BACH are from breast tissues missing from ADP. We will show in experiments that, in fact, classification success is directly related to the type of organ tissues contained in the source domain. Once an organ type is included in ADP, it will be accurately classified through domain adaptation, e.g. GlaS but not for PatchCamelyon/BACH.

3   YCbCr Color Augmentation

ADP contains a wide range of stained tissue colors which extends beyond H&E spectrum. The goal is to develop simple and efficient color augmentation for CNN training that accounts for such wide variations. HSV color augmentation

Table 2. HTT labels from source domain ADP identified as being diagnostically relevant in three different disease datasets i.e. GlaS, PatchCamelyon and BACH. For description of the labels please refer to label adaptation in Fig. 2.

	GlaS	PatchCamelyon	BACH
Type of data	Patches extracted from colorectal slides	Patches extracted from lymph node sections of breast slides	Patches extracted from breast slides
Disease lass(es)	5-classes: 1 normal + 4 progressive cancer grades	2-classes: normal + tumorous	4-classes: 1 normal + 3 progressive cancer grades
Diagnostically Relevant Labels	"E.M.C" and "H.Y"	"E.T.C" and "G.O"	"E.T.C" and "G.O"
Availability of primary organ in ADP	Included	Not included	Not included

from [39, 42, 43] could be an alternative solution since it perturbs wide color range but it is (a) slow for CNN training; and (b) produces color residuals during forward/backward color conversion. Our solution to solve these issues is to switch to YCbCr using a linear transformation.

$$\begin{bmatrix} Y \\ C_b \\ C_r \end{bmatrix} = \begin{bmatrix} 0.2568 & 0.5041 & 0.0979 \\ -0.1482 & -0.2910 & 0.4392 \\ 0.4392 & -0.3678 & -0.0714 \end{bmatrix} \begin{bmatrix} I_R \\ I_G \\ I_B \end{bmatrix} + \begin{bmatrix} 16 \\ 128 \\ 128 \end{bmatrix} \tag{1}$$

Consider the raw input image to the network (without channel normalization) I in RGB color space to be augmented in YCbCr space through random perturbations of both red and blue chroma channels $I_{[R][G][B]} \mapsto I_{[Y][C_b][C_r]} \mapsto I_{[Y][C_b+\mathcal{N}(0,s\sigma_b)][C_r+\mathcal{N}(0,s\sigma_r)]} \mapsto \tilde{I}_{[R][G][B]}$ where σ_b and σ_r are the standard deviation of the chroma channels, and s is the significance of perturbation. This is similar to the ratio defined in [43] where $s = 0.1$ is referred to as "Light" and $s = 1$ as "Strong" augmentation. The main advantages of this transformation are: (a) it separates the color chroma spaces (i.e. tuples of C_b and C_r) from the Luma channel Y, enabling direct manipulation of the stain colors without affecting the tissue illumination stimulated by WSI scanner condenser; and (b) the computational cost of such transform is much lower due to simple multiplication and addition operations.

We demonstrate the effectiveness of YCbCr compared to the HSV augmentations in Table 3 by randomly perturbing several image patches from ADP [18] and visually comparing them to patch examples from other datasets, i.e. HMT [23], GlaS [37], PCam [6, 44], and CRC [22]. Notice how applying different color augmentation methods (using "Light"/"Strong") can match color distribution of other datasets. HSV-Strong method creates random outliers while YCbCr-Strong remains more stable.

4   HistoNet for ADP Source Domain

Here, we design an efficient CNN model optimized on ADP database for HTT classification. This is highly preferable in computational pathology for process-

Table 3. Distribution of color space shown in Hue-Saturation domain for five different datasets: (a) HMT, (b) GlaS, (c) CRC, and (d) ADP. Two different augmentation methods i.e. HSV [43] and YCbCr (Proposed) are considered to perturb ADP pixels shown from (e) to (h) using scaling factors $s = 0.1$ (light) and $s = 1$ (strong).

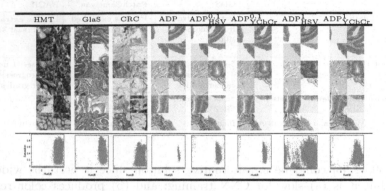

ing thousands of image-patches cropped from GigaPixel WSI scans. Therefore, both computational complexity and precision are equally important for practical considerations.

4.1　Neural Architecture Search (NAS) for HistoNet

We design a CNN model with six sequential convolutional layers and one fully connected layer, followed by a sigmoid layer for multi-label class activation. After each convolutional layer, ReLU activation, Batch Normalization (BN), and max pooling (2×2) are applied. Global max pooling is used at the end of the sixth layer. The parameters we explore include the kernel size $\{w_\ell \times w_\ell\}_{\ell=1}^6$ and the number of filters $\{D_\ell\}_{\ell=1}^6$ of the convolutional weights (tensor) for each layer i.e. $\Phi_\ell \in \mathbb{R}^{w_\ell \times w_\ell \times D_{\ell-1} \times D_\ell}$. We recast the HistoNet design in two phases. First, we seek the optimal configuration for $\{w_\ell\}_{\ell=1}^6$ and $\{D_\ell\}_{i=1}^6$ using the neural architecture search with reinforcement learning algorithm introduced in [48]. A controller RNN is used to generate child CNN architectures with different kernel sizes and kernel number configurations. The child CNN model is trained, and the classification accuracy of child model is used as a reward signal to update the controller to sample configurations for the next step. We explore the search space of the RNN controller using different kernel size $w_\ell \in \{1, 3, 5, 7\}$ and channel depth $D_\ell \in \{32, 64, 128, 192, 256\}$. We run the controller RNN to generate 250 child CNN architectures that have different configurations, and train each child CNN for 20 epochs on ADP patches of size 224×224. The 250 trained CNN models are applied to the ADP test dataset and choose the seven best configurations (by F1 score) for kernel size and number of filters. In the second phase, we trained all the seven HistoNet models for 100 epochs and selected the best configuration to finalize the network. The overall layout of the optimized architecture is shown in Fig. 3(a).

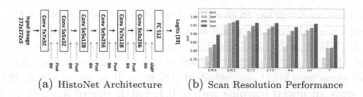

(a) HistoNet Architecture   (b) Scan Resolution Performance

Fig. 3. (a) HistoNet serial architecture. The network consists of six convolutional layers, followed by a fully-connected and sigmoid-activation layers for ADP multil-label HTT classification. The kernel size and channel depth are optimized using the NAS-Reinforcement Learning method in [48]; and (b) Classification performance (AUC) of HistoNet on selected HTTs at different scan resolutions.

4.2 Choice of Pixel Resolution

Although the input image size studied to optimize the HistoNet is 224 × 224 (@1.21 μm/pixel resolution), it is important to understand how the network training corresponds to different scan resolutions. To study this, we downsize the original ADP image from 1088 × 1088 (@0.25 μm/pixel resolution) into four different pixel resolutions of 1 μm, 2 μm, 3 μm, 4 μm. As shown in Fig. 3(b), decreasing the pixel resolution improves the AUC for all classes, especially for smaller tissues such as simple squamous epithelium (E.M.S), leukocytes (H.K), and transport vessels (T) (see Fig. 2 for full HTT names).

The overall HistoNet performance for different scan resolutions and augmentation methods is demonstrated in Table 4. While the network yields better classification on higher resolutions, the choice of color augmentation impacts the overall results. For instance, employing YCbCr-Strong augmentation improves about 0.5% compared to no augmentation at the 1 μm scan resolution. While both YCbCr and HSV provide similar performances, HSV adds about 40% computational overhead per epoch during training.

Table 4. HistoNet test set performance applied to HTT classification on ADP database using different pixel-resolution scan and color augmentation methods. The performances are reported by Area Under the Curve (AUC) of the ROC and F_1 measure.

	Sec/Epoch			AUC					F_1				
	None	HSV	YCbCr	None	HSV		YCbCr		None	HSV		YCbCr	
					Light	Strong	Light	Strong		Light	Strong	Light	Strong
4 μm	72	92	72	0.9136	0.9310	0.9205	0.9277	0.9208	0.7485	0.7780	0.7536	0.7699	0.7658
3 μm	94	129	95	0.9276	0.9367	0.9325	0.9374	0.9369	0.7670	0.7836	0.7790	0.7895	0.7943
2 μm	156	236	155	0.9454	0.9539	0.9526	0.9511	0.9525	0.8085	0.8247	0.8215	0.8203	0.8204
1 μm	567	937	569	0.9594	0.9650	0.9638	0.9646	0.9645	0.8378	0.8534	0.8476	0.8499	0.8537

4.3 Network Performance Comparison

We further compare the performance of HistoNet to three CNNs, i.e. ResNet18 [17], MobileNet [20] and Xception [10]. Note that we trimmed number of middle flow blocks in Xception from eight (baseline) to one to reduce the network parameters - we call this Xception-1. The criteria of our comparison networks selection here is mainly based on the simplicity of architectures for practical implementations. All models are trained at 1 μm scan resolution following the multi-label class weighting suggested in [18] as well as using Cyclical Learning Rate [38] with an initial learning rate of 0.1, batch size of 32, and termination after 100 epochs. Table 5 demonstrates the predictive performance over different networks and color augmentation methods. The rank performance of HistoNet is preserved compared to the other CNNs, while consuming the lowest complexity with 3M parameters.

Table 5. Test set performance of four different networks for HTT classification on ADP database at 1 μm/pixel with 272 μm field-of-view scan (272 × 272 input image).

	# Conv layers	Params	AUC			F_1		
			NA	HSV	YCbCr	NA	HSV	YCbCr
ResNet18 [17]	18	11.20M	0.9533	0.9511	0.9521	0.8244	0.8200	0.8209
Xception-1 [10]	15	9.60M	0.9576	0.9576	0.9567	0.8363	0.8362	0.8339
MobileNet [20]	14	3.26M	0.9521	0.9503	0.9518	0.8233	0.8200	0.8181
HistoNet/NAS [48]	6	3.00M	0.9594	0.9638	0.9645	.8378	0.8476	0.8537

5   Experiments on HTT Transferability

To evaluate the transferability of HTT labels from ADP source domain into different target domains, we adopt the CNN models trained in previous Sect. 4 w- and w/o- color augmentation. Then, we transfer the models to solve two datasets in tissue classification and three datasets in cancer detection and cancer grade classification tasks. For tissue classification we compare the ROC performance of the domain-adapted CNNs; for cancer detection we compare the ROC performance of the relevant inverted source class scores against the target cancer labels; and finally for cancer classification we compare the statistical correlation of inverted source class scores against the target cancer grades. For details on datasets, visual examples, and additional results, see the Supplementary Materials. Codes and trained models are available on GitHub[1].

5.1   Image Modifications and Pixel-Resolution Adjustment

The source-domain CNN is trained on ADP square patch images $\mathbf{X_s}$ of size $W_s \times W_s$ pixels with a ρ_s resolution (in μm/px) and a fixed Field-Of-View (FOV) of 272μm, so to ensure the target-domain images $\mathbf{X_t}$ of size $H_t \times W_t$ and ρ_t resolution have the same pixel resolution and FOV, a few modifications must be performed. To ensure the same pixel resolution, they are resized by a constant

---

[1] https://github.com/mahdihosseini/HistoLabelTransfer/.

factor $\alpha = \rho_t/\rho_s$, such that $(W_t, H_t) \leftarrow \alpha \cdot (W_t, H_t)$. The target image must be also padded (if the FOV is too small) and/or cropped (if the FOV is too large). It is symmetrically padded by $(k_x, k_y) = (\max(W_s - W_t, 0), \max(W_s - H_t, 0))$, such that $(W_t, H_t) \leftarrow (W_t + k_x, H_t + k_y)$. Then, if $W_s < W_t$ or $W_s < H_t$, crops of size $W_s \times W_s$ are extracted and the confidence scores of the cropped patches are later aggregated back to the image level by taking their average.

5.2   Transferability in CRC Dataset for Tissue Classification

The images in CRC are colorectal tissues which partially correlates with primary sites compiled in ADP. The stain color distributions of CRC, however, differs from ADP (see Fig. 3). Figure 4 (top) shows that HistoNet performs well in all adapted classes except for the LYM (Lymphocyte) class ([22] reports a mean 4-class AUC of 0.995). Here, the *"lighter"* forms of color augmentation are better. Furthermore, class performance in CRC is heavily dependent on the network architecture used. Figure 4 (bottom) shows that HistoNet performs best in the large tissue classes (ADI, MUS, and NORM) and that the other networks progressively get better performance with increasing depth.

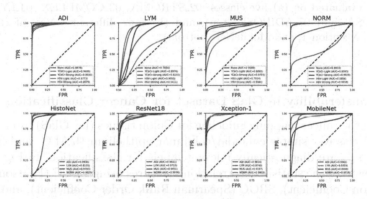

Fig. 4. Network ROC performance on CRC trained with different color augmentations and evaluated on (a) four classes of ADI, LYM, MUS, and NORM (top row); and (b) four different CNNs of HistoNet, ResNet18, Xception-1, and MobileNet (bottom row).

5.3   Transferability in HMT Dataset for Tissue Classification

The HMT images are very similar to CRC in the organ of origin but have a smaller FOV and a different color distribution (see Fig. 3). Figure 5 (top) shows that HistoNet struggles to classify the smaller tissues (02_STROMA, 03_COMPLEX, 04_LYMPHO) and excels in the larger tissues (06_MUCOSA, 07_ADIPOSE). We hypothesize this phenomena due to stain normalization used in original dataset for training where color representation of tissues are deteriorated from their original spectrum. ([23] reports a mean 8-class AUC of 0.976).

YCbCr-Strong color augmentation is the only method to consistently improve upon the unaugmented case. Again, class performance in HMT depends on the architecture used. Figure 5 (bottom) shows that HistoNet is superior in three of five classes, with the other networks performing better with decreasing depth.

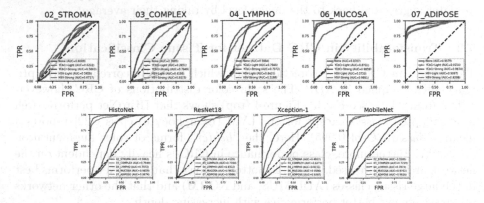

Fig. 5. Network ROC performance on HMT trained with different color augmentations and evaluated on (a) five classes: 02_STROMA, 03_COMPLEX, 04_LYMPHO, 06_MUCOSA, and 07_ADIPOSE (top); and (b) four different CNNs of HistoNet, ResNet18, Xception-1, and MobileNet (bottom).

5.4 Transferability in GlaS Dataset for Cancer Classification

Following the cancer mapping guideline for Colon tissues (i.e. GlaS) from Table 2, two ADP classes - simple cuboidal/columnar epithelium (E.M.C) and lymphocytes (H.Y) - are diagnostically relevant labels for classification. In Fig. 6 (left) we demonstrate the statistical correlation measures using PLCC (Pearson Linear Correlation Coefficient), SROC (Spearman Rank Order Coefficient), and KROC (Kendall Rank Order Coefficient) calculated between predicted confidence scores (inverted) from network and five cancer grades for classification. Both YCbCr-Strong and HSV-Strong yield higher correlation results. This observation is interesting because it matches prior knowledge of colorectal cancer grades: high tumor differentiation distorts epithelial cell boundaries and absence of tumor infiltrating lymphocytes (TILs) has been linked to poor cancer prognosis. We further demonstrate the ROC analysis in Fig. 6 (right) on cancer detection (binarizing healthy versus cancer) in Glas using E.M.C and H.Y. The HistoNet performs excellently here for all color augmentation methods achieving 0.95 AUC as the best result for lighter augmentation.

5.5 Alternative Source Domain Choice

While the ADP is mainly studied here to highlight the importance of transferring pathologists knowledge in the form of annotated labels, we further investigate

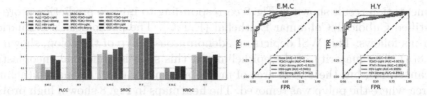

Fig. 6. Cancer classification on GlaS dataset using two HTTs: (a) Correlation measures (PLCC, SROC, and KROC) between select inverse HistoNet confidence scores (E.M.C and H.Y) and five quantified disease classes in GlaS (left plot); and ROC curves of HistoNet in GlaS on classifying normal/cancerous (right plot).

the reproducibility of similar results using other source domain choice. For this purpose, we directly train the HistoNet on CRC dataset using 1 μm resolution, five different color augmentations, and 6:1:3 train-validation-test split. The obtained class AUCs for all healthy labels (see Table 1) achieved almost perfect classification i.e. AUC> 0.99 similar to what original authors reported in [22]. Two labels are selected from CRC as being diagnostically relevant in Colon cancer i.e. LYM (lymphocyte) and NORM (normal stroma) and transfer into GlaS for cancer classification. Note that the cancer cells alter and embed in normal stroma [24]. The results are shown in Fig. 7, where inferior performances are achieved for cancer grade classification comparing to ADP source domain, implying that the annotated labels in CRC is less comprehensive compared to ADP. This is in spite the fact that the train data size in CRC is statistically significant (70K patches) compared to ADP (∼ 14.3K patches).

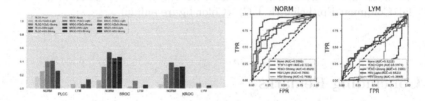

Fig. 7. (Cancer classification on GlaS dataset using two HTTs: (a) Correlation measures (PLCC, SROC, and KROC) between select inverse HistoNet confidence scores (E.M.C and H.Y) and five quantified disease classes in GlaS (left plot); and ROC curves of HistoNet in GlaS on classifying normal/cancerous (right plot).

6   Cancer Detection on WSI Level

In this section, we analyze a Colon tissue organ for cancer detection on the WSI level shown in Fig. 8. The slide is mosaiced into multiple patches and classified by HistoNet with 1μm/pixel resolution. We construct the heatmap for the best color augmentation result shown in Fig. 8 corresponding to inverse prediction score of Stratified Cuboidal/Columnar Epithelial (E.T.C). The pathologist's evaluation reads as follows: This WSI depicts an adenomatous polyp of the colon, shown in Fig. 8(a). The majority of the epithelium in this slide is abnormal (neoplastic, precancerous), but there is an area of muscularis mucosa and normal epithelium at area where the polyp was removed. The heatmaps in Fig. 8 show a high probability of abnormality in the areas of adenomatous epithelium (yellow, orange, and red), and indicate a low probability of abnormality in the regions with muscularis mucosa and normal epithelium (blue). The HSV Strong protocol appears to show the strongest correlation with histologic findings, followed by HSV Light, then YCbCr Light. The YCbCr Strong protocol shows the least correlation (while it still correctly indicates the normal areas, it appears to be less sensitive in identifying areas of abnormality compared to the other methodologies).

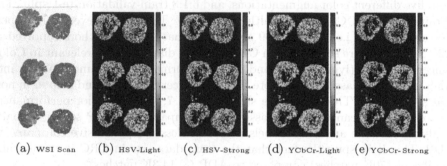

(a) WSI Scan   (b) HSV-Light   (c) HSV-Strong   (d) YCbCr-Light   (e) YCbCr-Strong

Fig. 8. A Colon tissue organ is selected for processing and diagnosed by pathologist. The heatmaps of abnormality is based on the inverse prediction of Stratified Cuboidal/Columnar Epithelial (E.T.C). (Color figure online)

7   Concluding Remarks

In this paper, a new tissue label transferring method is proposed to classify different histological tissue structures and cancer grades across diverse target domains. The method is based on training a CNN model on the source domain dataset (using the Atlas of Digital Pathology's multi-label tissue types) and then transferring those labels to the relevant target labels using prior histological knowledge. The ability of proposed method is demonstrated to produce reasonable predictions in related tissue classification datasets. Furthermore, the confidence prediction scores of diagnostically relevant labels are inferred as a surrogate model for cancer grade progression. The results suggested that the diagnostically relevant labels can be better transferred by adopting an appropriate source domain with broad spectrum of tissue structures for classification.

References

1. Araújo, T., et al.: Classification of breast cancer histology images using convolutional neural networks. PloS ONE **12**(6), e0177544 (2017)
2. Aresta, G., et al.: Bach: grand challenge on breast cancer histology images. Med. Image Anal. **56**, 122–139 (2019)
3. Arvidsson, I., Overgaard, N.C., Åström, K., Heyden, A.: Comparison of different augmentation techniques for improved generalization performance for gleason grading. In: 2019 IEEE 16th International Symposium on Biomedical Imaging (ISBI 2019), pp. 923–927. IEEE (2019)
4. Bandi, P., et al.: From detection of individual metastases to classification of lymph node status at the patient level: the CAMELYON17 challenge. IEEE Trans. Med. Imaging **38**(2), 550–560 (2018)
5. Bejnordi, B.E., et al.: Stain specific standardization of whole-slide histopathological images. IEEE Trans. Med. Imaging **35**(2), 404–415 (2015)
6. Bejnordi, B.E., Veta, M., Van Diest, P.J., Van Ginneken, B., Karssemeijer, N., Litjens, G., Van Der Laak, J.A., Hermsen, M., Manson, Q.F., Balkenhol, M., et al.: Diagnostic assessment of deep learning algorithms for detection of lymph node metastases in women with breast cancer. Jama **318**(22), 2199–2210 (2017)
7. Brieu, N., et al.: Domain adaptation-based augmentation for weakly supervised nuclei detection. In: MICCAI 2019 Computational Pathology Workshop COMPAY (2019)
8. Bug, D., et al.: Context-based normalization of histological stains using deep convolutional features. In: Cardoso, M.J., et al. (eds.) DLMIA/ML-CDS -2017. LNCS, vol. 10553, pp. 135–142. Springer, Cham (2017). https://doi.org/10.1007/978-3-319-67558-9_16
9. Campanella, G., et al.: Clinical-grade computational pathology using weakly supervised deep learning on whole slide images. Nat. Med. **25**(8), 1301–1309 (2019)
10. Chollet, F.: Xception: deep learning with depthwise separable convolutions. In: Proceedings of the IEEE Conference on Computer Vision and Pattern Recognition, pp. 1251–1258 (2017)
11. Coudray, N., et al.: Classification and mutation prediction from non-small cell lung cancer histopathology images using deep learning. Nat. Med. **24**(10), 1559 (2018)
12. Djuric, U., Zadeh, G., Aldape, K., Diamandis, P.: Precision histology: how deep learning is poised to revitalize histomorphology for personalized cancer care. NPJ Precision Oncol. **1**(1), 22 (2017)
13. Faust, K., et al.: Intelligent feature engineering and ontological mapping of brain tumour histomorphologies by deep learning. Nat. Mach. Intell. **1**(7), 316–321 (2019)
14. Faust, K., et al.: Visualizing histopathologic deep learning classification and anomaly detection using nonlinear feature space dimensionality reduction. BMC Bioinformatics **19**(1), 173 (2018)
15. Ganin, Y., et al.: Domain-adversarial training of neural networks. J. Mach. Learn. Res. **17**(1), 2030–2096 (2016)
16. Goodfellow, I., Bengio, Y., Courville, A.: Deep Learning. MIT Press, Cambridge (2016)
17. He, K., Zhang, X., Ren, S., Sun, J.: Deep residual learning for image recognition. In: Proceedings of the IEEE Conference on Computer Vision and Pattern Recognition, pp. 770–778 (2016)

18. Hosseini, M.S., et al.: Atlas of digital pathology: a generalized hierarchical histological tissue type-annotated database for deep learning. In: Proceedings of the IEEE Conference on Computer Vision and Pattern Recognition, pp. 11747–11756 (2019)

19. Hou, L., Agarwal, A., Samaras, D., Kurc, T.M., Gupta, R.R., Saltz, J.H.: Robust histopathology image analysis: to label or to synthesize? In: Proceedings of the IEEE Conference on Computer Vision and Pattern Recognition, pp. 8533–8542 (2019)

20. Howard, A.G., et al.: MobileNets: efficient convolutional neural networks for mobile vision applications. arXiv preprint arXiv:1704.04861 (2017)

21. Karimi, D., Nir, G., Fazli, L., Black, P.C., Goldenberg, L., Salcudean, S.E.: Deep learning-based gleason grading of prostate cancer from histopathology images-role of multiscale decision aggregation and data augmentation. IEEE J. Biomed. Health Inform. **24**(5), 1413–1426 (2019)

22. Kather, J.N., et al.: Predicting survival from colorectal cancer histology slides using deep learning: a retrospective multicenter study. PLoS Med. **16**(1), e1002730 (2019)

23. Kather, J.N., et al.: Multi-class texture analysis in colorectal cancer histology. Sci. rep. **6**, 27988 (2016)

24. Kaukonen, R., et al.: Normal stroma suppresses cancer cell proliferation via mechanosensitive regulation of JMJD1A-mediated transcription. Nat. Commun. **7**(1), 1–15 (2016)

25. Lafarge, M., Pluim, J., Eppenhof, K., Veta, M.: Learning domain-invariant representations of histological images. Front. Med. **6**, 162 (2019)

26. Li, J., et al.: Signet ring cell detection with a semi-supervised learning framework. In: Chung, A.C.S., Gee, J.C., Yushkevich, P.A., Bao, S. (eds.) IPMI 2019. LNCS, vol. 11492, pp. 842–854. Springer, Cham (2019). https://doi.org/10.1007/978-3-030-20351-1_66

27. Litjens, G., et al.: Deep learning as a tool for increased accuracy and efficiency of histopathological diagnosis. Sci. rep. **6**, 26286 (2016)

28. Mahmood, F., et al.: Deep adversarial training for multi-organ nuclei segmentation in histopathology images. IEEE Trans. Med. Imaging, 1 (2019)

29. Mobadersany, P., et al.: Predicting cancer outcomes from histology and genomics using convolutional networks. Proc. Nat. Acad. Sci. **115**(13), E2970–E2979 (2018)

30. Niazi, M.K.K., Parwani, A.V., Gurcan, M.N.: Digital pathology and artificial intelligence. Lancet Oncol. **20**(5), e253–e261 (2019)

31. Otálora, S., Atzori, M., Andrearczyk, V., Khan, A., Müller, H.: Staining invariant features for improving generalization of deep convolutional neural networks in computational pathology. Front. Bioeng. Biotechnol. **7**, 198 (2019)

32. Pan, S.J., Yang, Q.: A survey on transfer learning. IEEE Trans. Knowl. Data Eng. **22**(10), 1345–1359 (2009)

33. Pantanowitz, L., et al.: Validating whole slide imaging for diagnostic purposes in pathology: guideline from the College of American pathologists pathology and laboratory quality center. Arch. Pathol. Lab. Med. **137**(12), 1710–1722 (2013)

34. Ren, J., Hacihaliloglu, I., Singer, E.A., Foran, D.J., Qi, X.: Unsupervised domain adaptation for classification of histopathology whole-slide images. Front. Bioeng. Biotechnol. **7**, 102 (2019)

35. Riordan, D.P., Varma, S., West, R.B., Brown, P.O.: Automated analysis and classification of histological tissue features by multi-dimensional microscopic molecular profiling. PloS ONE **10**(7), e0128975 (2015)

36. Rolls, G., et al.: 101 Steps to Better Histology. Leica Microsystems 7, Melbourne (2008)
37. Sirinukunwattana, K., et al.: Gland segmentation in colon histology images: the GLaS challenge contest. Med. Image Anal. **35**, 489–502 (2017)
38. Smith, L.N.: Cyclical learning rates for training neural networks. In: 2017 IEEE Winter Conference on Applications of Computer Vision (WACV), pp. 464–472. IEEE (2017)
39. Stacke, K., Eilertsen, G., Unger, J., Lundström, C.: A closer look at domain shift for deep learning in histopathology. In: MICCAI 2019 Computational Pathology Workshop COMPAY (2019)
40. Takahama, S., et al.: Multi-stage pathological image classification using semantic segmentation. In: Proceedings of the IEEE International Conference on Computer Vision, pp. 10702–10711 (2019)
41. Tellez, D., Balkenhol, M., Karssemeijer, N., Litjens, G., van der Laak, J., Ciompi, F.: H and E stain augmentation improves generalization of convolutional networks for histopathological mitosis detection. In: Medical Imaging 2018: Digital Pathology, vol. 10581, p. 105810Z. International Society for Optics and Photonics (2018)
42. Tellez, D., et al.: Whole-slide mitosis detection in H&E breast histology using PHH3 as a reference to train distilled stain-invariant convolutional networks. IEEE Trans. Med. Imaging **37**(9), 2126–2136 (2018)
43. Tellez, D., et al.: Quantifying the effects of data augmentation and stain color normalization in convolutional neural networks for computational pathology. Med. Image Anal. **58**, 101544 (2019)
44. Veeling, B.S., Linmans, J., Winkens, J., Cohen, T., Welling, M.: Rotation equivariant CNNs for digital pathology. In: Frangi, A.F., Schnabel, J.A., Davatzikos, C., Alberola-López, C., Fichtinger, G. (eds.) MICCAI 2018. LNCS, vol. 11071, pp. 210–218. Springer, Cham (2018). https://doi.org/10.1007/978-3-030-00934-2_24
45. Wilson, G., Cook, D.J.: A survey of unsupervised deep domain adaptation. arXiv preprint arXiv:1812.02849 (2019)
46. Wu, B., et al.: P3SGD: patient privacy preserving SGD for regularizing deep CNNS in pathological image classification. In: Proceedings of the IEEE Conference on Computer Vision and Pattern Recognition, pp. 2099–2108 (2019)
47. Zhang, Y., Barzilay, R., Jaakkola, T.: Aspect-augmented adversarial networks for domain adaptation. Trans. Assoc. Comput. Linguist. **5**, 515–528 (2017)
48. Zoph, B., Le, Q.V.: Neural architecture search with reinforcement learning. In: International Conference on Learning Representations (2017)

Learning Actionness via Long-Range Temporal Order Verification

Dimitri Zhukov[1,2]([✉]), Jean-Baptiste Alayrac[3], Ivan Laptev[1,2], and Josef Sivic[1,2,4]

[1] Département d'informatique de L'ENS, ENS, CNRS, PSL University, Paris, France
[2] INRIA, Paris, France
dmitry.zhukov@inria.fr
[3] DeepMind, London, UK
[4] CIIRC, Prague, Czech Republic

Abstract. Current methods for action recognition typically rely on supervision provided by manual labeling. Such methods, however, do not scale well given the high burden of manual video annotation and a very large number of possible actions. The annotation is particularly difficult for temporal action localization where large parts of the video present no action, or *background*. To address these challenges, we here propose a self-supervised and generic method to isolate actions from their background. We build on the observation that actions often follow a particular temporal order and, hence, can be predicted by other actions in the same video. As consecutive actions might be separated by minutes, differently to prior work on the arrow of time, we here exploit long-range temporal relations in 10–20 min long videos. To this end, we propose a new model that learns actionness via a self-supervised proxy task of order verification. The model assigns high actionness scores to clips which order is easy to predict from other clips in the video. To obtain a powerful and action-agnostic model, we train it on the large-scale unlabeled HowTo100M dataset with highly diverse actions from instructional videos. We validate our method on the task of action localization and demonstrate consistent improvements when combined with other recent weakly-supervised methods.

Keywords: Temporal order · Action localization · Video recognition

1 Introduction

Learning from web videos is becoming increasingly popular in computer vision as such videos are available in large quantities, and cover diverse activities and scenes. In particular, instructional videos have been recently explored as a rich

Electronic supplementary material The online version of this chapter (https://doi.org/10.1007/978-3-030-58526-6_28) contains supplementary material, which is available to authorized users.

© Springer Nature Switzerland AG 2020
A. Vedaldi et al. (Eds.): ECCV 2020, LNCS 12374, pp. 470–487, 2020.
https://doi.org/10.1007/978-3-030-58526-6_28

Action frames   Background frames

Fig. 1. We show pairs of action frames and background frames from the same video. Can you predict the order of frames within each pair? While the order is relatively easy to guess for actions, the same task is more difficult for the background. We use this observation and exploit the predictability of temporal order as a measure of actionness (**Quiz answers:** Action frames are shown in correct temporal order; Background frames are shown in reverse temporal order.)

source for many tasks and goal-driven sequences of actions [2,13,23,31,40,41]. While the quantity and diversity of video data appears crucial for training current recognition models [4,22,23], the manual annotation of actions in large-scale video data requires large efforts [4] and may not scale well to the large number of possible actions. This is particularly true for the task of temporal action localization where sparse actions should be isolated in video streams from the large portion of "background" with no actions. For example, action frames in the CrossTask dataset [41] with typical instructional videos represent only 25.9% of the total video length.

In our work we address the above challenges and aim to develop a self-supervised approach for separating a large and diverse set of actions from their background. We observe that actions typically do not happen in isolation and are often surrounded by other related actions. Moreover, action sequences often demonstrate a consistent order (taking off a car wheel should be preceded by lifting the car), hence, many actions can be identified by the predictability of their order with respect to other actions in the same video. On the contrary, the order of background frames is often hard to predict, hence, the low predictability for the order could be used to signify the background. We illustrate this idea with a quiz in Fig. 1.

Temporal order has been explored as a supervisory signal by a number of recent works [9,24,34,39]. The goal of such methods, however, is to learn video representations by verifying the order for a short range of consecutive frames. We here address a different task and learn actionness [5] by exploiting long-range relations between video clips on the scale of minutes. To this end, we propose a new model and a method to learn actionness scores via a self-supervised proxy task of order verification. The model assigns high actionness scores to clips which order is easy to predict from other clips in the video. Our method is self-supervised and requires no manual annotation. Given this property, we use a very large HowTo100M dataset [23] with diverse and unlabeled instructional videos to learn an action-agnostic model for actionness. We show interesting insights of our method and demonstrate improved performance of action localization when combining our model with recent weakly-supervised approaches.

Contributions. This work makes the following contributions: (1) We develop a new model for action-agnostic action/background classification and propose to learn it via a self-supervised proxy task of long-range order verification. (2) We demonstrate a successful application of our method to the tasks of frame-wise action/background classification and action proposal generation evaluated on datasets with instructional videos COIN [31] and CrossTask [41]; (3) We further demonstrate the benefit of our model for action localization by combining it with recent weakly-supervised methods of step localization on the CrossTask dataset; and (4) We provide ablation studies that give insights about our approach.

2   Related Work

2.1   Self-supervised Learning

Our work exploits the natural source of supervision in videos: the temporal order between frames. The proposed method is, hence, related to a large body of recent work on self-supervised learning, where the supervisory signal is obtained directly from the data and does not require manual annotation. The variety of recently proposed self-supervised tasks in the image domain include prediction of image rotation [11], spotting artifacts [14], image colorization [17,36], cross-channel prediction [37], inpainting [27] and predicting relative position of patches [6,25]. In the video domain, motion has been used as a cue for learning video representations in [1,7,26,33]. More related to our work, previous methods [9,18,24,34,35] explore the temporal order, either by predicting the exact order of consecutive frames [18,35] or verifying their partial order [9,24,34]. Our work builds on these ideas but brings two important innovations. First, in contrast to previous work that exploits temporal order to learn local video representations, we address a different task of action/background classification. As actions are often separated by minutes, our task requires reasoning about *long-range* temporal order, as opposed to *short-range* frame permutations explored by previous methods. Our second innovation is, hence, a new method that exploits long-range order verification for video clips and enables to model relations between actions in 10–20 min long videos.

2.2   Learning from Instructional Videos

Instructional videos have recently been in the focus of numerous works in the context of action localization [2,28,41], joint learning of object states and actions [3], joint modeling of video and language [23,30] and visual reference resolution [12,13]. Some of this work exploits specific properties of instructional videos, such as the approximate temporal alignment between narrations and the visual content [2,23,30,41], and the order consistency [2,28,41]. Similarly, we rely on the partial order between actions. Our novelty is to use the order verification as a proxy task to discover most relevant parts of the video. To demonstrate the value of our approach, we combine it with the previous methods [22,23,41] for the task of weakly-supervised step localization in instructional videos and demonstrate consistent improvements.

2.3   Action Proposals

We apply our method to generate action proposals. Action proposals is an essential part of many methods for action detection, explored by a number of recent papers [8,10,15,19–21,38]. A popular approach to generate action proposals is to estimate an *actionness* score for each temporal unit and then apply some sort of temporal grouping and non-maxima suppression. The notion of actionness was first introduced in [5] as a confidence measure of intentional bodily movement of biological agents. Most works [10,19–21,38] address actionness with supervised methods based on manual annotation of a known and limited set of action classes. This is done by training a binary classifier for estimating actionness score as first proposed in the context of spatial action detection [32]. Contrary to this approach, we aim to learn an actionness score without manual supervision by relying on generic assumptions about action order. Our definition of actions is narrower than in [5]. In particular, we only consider *goal-oriented actions*, necessary to perform specific manipulation tasks. This definition excludes actions such as gesticulation and conversations.

3   Unsupervised Learning of Actionness Score

Given a large corpus of instructional videos depicting complex tasks, our goal is to automatically discover which segments of the videos are the most relevant for the successful completion of the tasks. We refer to these relevant segments as actions. Example of a complex task would be *"building a shelve"* and a relevant action would be *"drilling a hole"*. Formally, we learn an *actionness* scoring function S that takes as input a video clip $x \in \mathbb{R}^{T \times H \times W \times 3}$ containing T frames of height H and width W and outputs a score $S(x) \in \mathbb{R}$ that is high when x corresponds to a relevant action and is low otherwise, *e.g.* on background scenes that are not relevant for completing the task.

We propose to learn S in a self-supervised manner through the pretext task of long range temporal order verification, which consists in predicting whether or not a set of video clips spanning a long temporal interval are in the correct order. The intuition is that one needs to isolate the relevant segments of the video that allow to best identify the correct ordering of events to be able to solve that task. We use that observation to train our actionness model S.

This section formally describes our method for joint order verification and actionness prediction. Section 3.1 introduces the model used. Section 3.2 details the training procedure, that allows to train the actionness score S via order verification.

3.1   Models for Actionness and Order Verification

We represent each video clip x by a d-dimensional feature vector $h = f(x) \in \mathbb{R}^d$ obtained from a pre-trained video network f that we keep fixed throughout the work. In practice we use averaged pooled I3D representation [4] pretrained

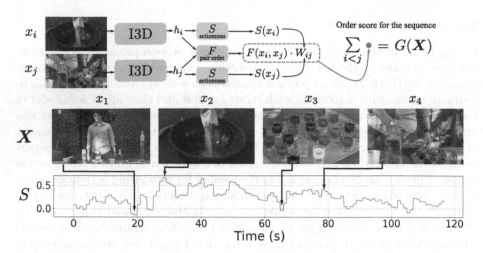

Fig. 2. Given a sequence of clips $X = (x_1, x_2, ..., x_M)$ extracted from the same video as input, our model produces two types of outputs: confidence $S(x_i)$ that video clip x_i displays an action, and confidence $F(x_i, x_j)$ that x_i occurs before x_j in the original video. We combine these scores together to produce an order score $G(\boldsymbol{X})$ that reflects the model confidence that the sequence $\boldsymbol{X}$ is displayed in the correct order. We generate training data for G for *free* by simply maintaining or reverting order of videos. By doing so, the model automatically learns to put more weight $W_{ij} \propto \exp(S(x_i) + S(x_j))$ to clips from which it is easy to predict whether clip x_i has happened before clip x_i. We argue that clips with such properties are more likely to be actions that allow S to become an *actionness* score. **Top.** Architecture for producing the actionness score S and its training. **Bottom.** Evolution of the learnt actionness score S throughout a video. Note how the model learned to put higher weights to frames that correspond to actions such as *adding sugar* (x_2) or *pouring* (x_4) and low score on clips that are not relevant to the completion of the task (*i.e.*, *background*) such as a *man standing still* (x_1) or a clip only showing *empty glasses* (x_3)

in [22], with $d = 1024$. For simplicity, we only refer to the feature vector h in the following (implicitly assuming h is associated with a video clip x). The actionness score $S(x)$ is estimated by a linear function on h, *i.e.*, $S(x) = w^\top h + b$, where $w \in \mathbb{R}^d$ and $b \in \mathbb{R}$ are learnt weights and biases, respectively. To predict the order of clips, we also introduce a model F that takes as input two clips x and x' and outputs the confidence $F(x, x') \in \mathbb{R}$ that x happens before x'. We force F to be antisymmetric (*i.e.*, $F(x, x') = -F(x', x)$), by defining $F(x, x') = a^\top (h - h')$, where $a \in \mathbb{R}^d$ and $h' = f(x')$. Our choice of simple linear models is practically motivated by the fact that in our experiments we did not see improvements from using more complex models. Next, we describe the training strategy used to train S and F using order verification.

3.2   Training with Ordering Verification

Actionness Through Order Verification. Our goal is to learn actionness score in an unsupervised manner. As explained earlier we believe that actions contain more information than background in terms of predicting what happens before or what may come next in instructional videos as they carry more information about the global temporal structure of the video than the background. In this section, we use that observation to automatically differentiate actions from background. In short, the idea is to train a network to predict if a set of clips are in the correct order, a task for which it's trivial to get *free* supervision as correct ordering is naturally present in the video. In order to do so, we allow the model to softly select which pairs of clips from the set are best to perform that prediction, *i.e.*, those that are most informative in terms of their relative order in the video. Hence, by learning to predict order through weighted relative ordering of pairs of clips, we expect the model to pay more attention on important actions and therefore learn a good actionness score S. Details are given next.

Order Verification Task. As illustrated in Fig. 2, given a sequence X of M video clips $X = (x_i)_{i=1}^M$ randomly sampled from the *same* video, the task of order verification is to predict whether or not the clips in X are in the correct order, *i.e.*, the same order as the original video.

Ground Truth Generation. We create positives and negatives for the verification task by simply randomly sampling M clips from a video and either **(i)** create a positive sequence by sorting clips in the order of their appearance in the video (label $y = 1$) or **(ii)** reverse completely the original sequence (negative label $y = 0$).

Order Verification Prediction. We seek to predict if the sequence X is in the correct order through our pairwise model F that can predict the relative ordering of a given pair of clips (x_i, x_j). To make a more accurate prediction, it is better to aggregate scores over many pairs from the entire sequence X. For this reason, we predict the confidence $G(X)$ that $X = (x_i)_{i=1}^M$ is in the correct order as a weighted average of all pair-wise predictions:

$$G(X) = \sum_{i<j} W_{ij} F(x_i, x_j). \tag{1}$$

Note that the sum indices $i < j$ are to make sure that (i) we only compare pairs in the order given by the sequence and (ii) we don't compare a clip to itself. Finally the weights W_{ij} are defined with a softmax over all pairs of clips:

$$W_{ij} = \frac{\exp(S(x_i) + S(x_j))}{\sum_{i'<j'} \exp(S(x_{i'}) + S(x_{j'}))}, \tag{2}$$

where $S(x_i)$ and $S(x_j)$ are the actionness score of our model for clips x_i and x_j, respectively. Because of the softmax (2) we have $\sum_{i<j} W_{ij} = 1$ and $W_{ij} \geq 0$. Hence the weights W_{ij} can be seen as a way to softly select the contribution of every individual pair (x_i, x_j) since they control the contribution of the individual order pairwise prediction $F(x_i, x_j)$ in the global order score $G(X)$ (see (1)). This contribution to $G(X)$ is proportional to the sum $S(x_i)+S(x_j)$ of actionness score of both clips x_i *and* x_j. This is to match our intuition that *both* clips should be depicting a relevant action to facilitate the order prediction. By learning G we will therefore indirectly learn S as described by the training objective below.

Training Objective. Given a sequence X and associated label y indicating whether or not the sequence of clips is in the correct order, we use the binary cross-entropy loss $\mathcal{L}$ as follows

$$\mathcal{L}(X,y) = -y \log(\sigma(G(X))) - (1-y)\log(1 - \sigma(G(X))), \tag{3}$$

where σ is the sigmoid function. This loss will enforce that when X is in the correct order (*i.e.*, $y = 1$), then $G(X)$ should be maximized. To do so, the actionness model S needs to be *high* on clips x_i and x_j from which it's *easier* to predict their correct ordering (meaning high value of $F(x_i, x_j)$) so that their contribution in $G(X)$ will be maximal. Conversely, when X is in the incorrect order (*i.e.*, $y = 0$), then $G(X)$ should be minimized. Again, to do so the actionness model S needs to be *high* on clips x_i and x_j from which it's *easier* to predict their incorrect ordering (meaning low value of $F(x_i, x_j)$) so that their contribution in $G(X)$ will be maximal. Following the standard procedure, we minimize the expected loss $\mathcal{L}$ on our training dataset.

4 Experiments

The experimental setup is described in Sect. 4.1. We then provide in Sect. 4.2 an ablation study of highlighting the most important components of our method. Finally, in Sect. 4.3, we show how we can use our learned actionness score for applications such as action proposals or weakly supervised action localization.

4.1 Experimental Setup

Input Processing. Given a clip x containing T frames, we extract $h(x)$ using the I3D backbone from [22], pre-trained on HowTo100M for the task of joint embedding of videos and subtitles. We extract the features at Mixed_5c in a fully convolutional manner and perform a global average pooling to have a single feature vector $h(x)$ of dimension 1024. Note that this network was trained without requiring manual annotations. To reduce computation cost during training, we pre-extract these features and directly work in feature space. In particular, this means that we do not finetune the I3D backbone for our task.

Training Dataset. Due to the large variety of actions in instructional videos, the variety of their visual appearance as well as the order in which they are performed in videos, we require a large amount of data for our self-supervised task. For this reason, we train our model on HowTo100M [23], a large dataset containing more than 1.2 million videos depicting around 23,000 different tasks and was collected without manual annotation.

Training Details. We optimize the objective (3) using the Adam optimizer [16] on a single GPU with batch size 1024, initial learning rate of 10^{-4} that we decay by a factor 0.9 at every epoch. We train for a total of 15 epochs. In addition, since HowTo100M is biased towards some specific domains (e.g. 40% of videos are cooking), we resample the data, using the task taxonomy of HowTo100M. Precisely, we consider all subcategories of depth 3, i.e., subcategories of the principal categories, such as Food and Cars, and sample equal number of videos from each subcategory. We also remove subcategories with less than 3000 videos.

Evaluation Datasets. We use two instructional video datasets, COIN [31] and CrossTask [41]. Both datasets contain untrimmed videos that have been temporally annotated with action labels corresponding to the different steps of the task. In the following, time intervals without any action labels are considered as background. We use the official COIN test subset of 2797 videos for evaluation. Since there is no official test subset of CrossTask, we randomly split it into a training (2062 videos) and test sets (688 videos). In addition, we made sure to discard all COIN and CrossTask videos from our training set.

Evaluation Tasks. We use three different tasks for evaluation, detailed next.

Background vs. Action Classification. In this task, videos from the respective test sets are split into non overlapping 0.2 s segments. For each segment we assign a binary label: positive (action) if the segment overlaps with an annotated action interval and negative (background) otherwise. The goal is to classify each segment as an action or a background. We use average precision (AP) as an evaluation metric for this task.

Action Temporal Proposal Generation. The task is to generate a set of proposals, that overlap well with ground truth action intervals. To generate proposals from the outputs of the network $S(x)$, we use the Temporal Actionness Grouping (TAG) method [38]. We set the Intersection over Union (IoU) threshold for non-maximum suppression to 0.8. We follow the evaluation protocol of [20] using the implementation provided in [19] to compute Average Recall (AR) at multiple IoU thresholds: [0.5 : 0.05 : 0.95]. Finally, we report AR as a function of the Average Number (AN) of proposals per video AR@AN as done in [20].

Action Step Localization. Third task is step localization on CrossTask. Given an ordered list of actions for a video, the goal is to assign each action to exactly one frame. We use the same evaluation protocol as in [41] and report average recall.

4.2   Ablations Studies

This section ablates the important components of our method. We report performance using *background vs action classification* task on COIN and CrossTask.

Video Trimming. When first training our method, we notice that the model could learn to be good at the self-supervised ordering task by putting high score on intro and outro segments of videos. This can be intuitively explained by the fact that the beginning and end of instructional videos are distinctive: they start with some typical introduction and often finish with credits. This effect is demonstrated on the left part of Fig. 3: when trained on untrimmed videos, the actionness score $S(x)$ is high for beginning and end of videos simply because the model can easily discriminate if a frame is from intro (beginning) or outro (end) and hence can safely select these frames to predict relative ordering of pairs. Top scoring frames illustrate that the model picks up on typical credit frames. This led to bad performance as these segments do not contain relevant actions. To alleviate this effect, we employ a simple but effective strategy: during training we trim off 30% of frames in the beginning and 30% of frames in the end of the video. Note that we only do that at training and do not trim test videos to keep the evaluation protocol consistent. Figure 3 shows that the temporal distribution of scores that we obtain by this technique better matches the ground truth one, as expected. In particular it's interesting to note that our model is still able to assign frames to actions even for frames that are in the beginning or the end when relevant (*i.e.*, this trimming did not handicapped the model for true actions that happen early or late in the video). Finally, the table in Fig. 3 demonstrates the effectiveness of this approach on our two evaluation datasets.

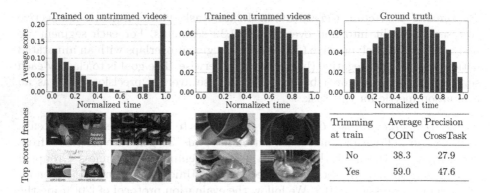

Fig. 3. *Actionness scores at different temporal locations of COIN videos test set.* (**left**) scores of the actionness model S, trained on untrimmed videos. (**center**) scores of the model S, trained on trimmed videos. (**right**) Distribution of actions in the videos. When trained on untrimmed videos, the score concentrates in the beginning and in the end of the video. When trained on untrimmed videos, the score distribution is closer to the ground truth action label distribution which leads to significant increase in background vs. action classification performance

Long vs. Short Range Order Verification. The driving hypothesis of this work is that learning actionness score S is possible thanks to the long range order consistency between actions in instructional videos. We claim that this is not true for short-term ordering between frames as used in previous work [24], as in that case order verification can be done via low level visual cues regardless of whether images depict actions or background. We verify that claim by training our model at different temporal scales. Formally, we set d to be the temporal window length in which we are going to sample $M = 5$ segments of length $d/10$. In other words, d corresponds to the maximum distance we can have between two clips from X, and is a good measure of the *range* at which we perform the order verification task. Figure 4 (left) shows the results for different values of d. For small values of d, the model shows poor performance, compared to larger values. This demonstrates that our method works better on the scale of several minutes (long range), rather than 10–20 s (short range).

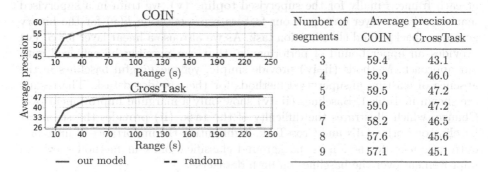

Number of segments	Average precision	
	COIN	CrossTask
3	59.4	43.1
4	59.9	46.0
5	59.5	47.2
6	59.0	47.2
7	58.2	46.5
8	57.6	45.6
9	57.1	45.1

Fig. 4. Left: long vs. short range. Action vs. background average precision of our model on COIN (top) and on CrossTask (bottom) as a function of the range d (in seconds) spanned by the sequence X. Duration of each sampled clip equals $d/10$. Performance increases with the range used for the order prediction task, confirming our hypothesis that long range action dependencies are better to learn about action-ness. **Right: number of segments.** Average precision of our model on COIN and CrossTask for different number of segments M sampled per video

Number of Segments Per Video. In Fig. 4 (right), we study the effect on performance when training our model with different number of segments M per video. Results are given for $M \geq 3$ ($M = 2$ makes training of S impossible since there are no pairs to select from). We observe that overall the method is not too sensitive to M. However, only sampling 3 segments per video may often lead to a situation where all selected segments are background, hence selecting larger value for M leads to better results. Figure 4 also shows a decline in precision for $M > 5$ on both datasets. This can be explained by the fact that for large values of M there is a high probability of having at least one pair of segments, for which the temporal order is easy to guess. This may decrease the ability of our model to learn temporal order for other more subtle pairs. Based on that study, we use $M = 5$ in all of our other experiments.

4.3 Actionness Score for Practical Applications

Action vs. Background Classification. In Table 1, we compare against five methods: **(i)** chance baseline, **(ii)** chance baseline with trimming, **(iii)** hand detector, **(iv)** optical flow and **(v)** a supervised model. **(i)** simply assigns random uniform action scores to segments in the video. Following our observation from Sect. 4.2 about trimming, **(ii)** does the same as **(i)** but also assigns to background the segments that occur in the first and last 30% portion of the video. **(iii)** assigns the actionness score to the maximum score of a hand detector [29] computed for each frame. This baseline is based on the assumption that the actions correlate with the appearance of hands. **(iv)** assumes that the actions correlate with motion and estimates an actionness score as an average magnitude of optical flow at each frame. Finally for the supervised topline **(v)**, we train in a supervised manner a linear layer on top of our feature representation $h(x)$ for the binary action vs. background classification task. As we also use a linear layer for S, **(v)** provides an upper bound of performance, that can possibly be achieved with our approach. Methods **(ii–iv)** provide simple, yet meaningful baselines in the absence of existing unsupervised methods for the considered task. The results are shown in Table 1. Baselines **(ii–iv)** show only a marginal improvement over Chance, which illustrates the difficulty of the task. **(ii)** provides the strongest baseline on both COIN and CrossTask, highlighting the importance of intro and outro segments for action vs. background classification. Our method shows an improvement over the baselines on both datasets.

Table 1. Action vs. Background. Frame-wise average precision of background separation on COIN and CrossTask datasets

Dataset	(i) Chance	(ii) Chance (trim@0.3)	(iii) Hand detector	(iv) Optical flow	Ours	(v) Supervised
COIN	45.6%	53.5%	50.6%	47%	**59.0%**	70.7%
CrossTask	27.5%	32.6%	28.4%	30.3%	**47.6%**	56.2%

Temporal Action Proposals. Following the evaluation protocol described in Sect. 4.1, we compare to 6 different methods: **(i)** chance baseline, **(ii)** chance baseline with trimming, **(iii)** hand detector, **(iv)** optical flow, **(v)** supervised and **(vi)** temporal prior. **(i–v)** are the same as for the Action vs. Background classification task. For **(vi)**, we use a temporal prior to generate action segments: it consists in sampling proposal start and length from a prior distribution, obtained from ground truth action intervals. In details, we compute the empirical distribution of the normalized start time of actions as well as their duration. We then randomly sample segments from that distribution by first sampling a start time and then a duration. Note that this baseline has access to more annotation than our proposed approach, since we don't have access to any temporal annotation. Figure 5 shows the average recall for CrossTask and COIN, as a function of average number of proposals per video (AN). On both datasets we outperform baselines (i–iv) for most values of AN. More interestingly, we also significantly outpeform the temporal prior (vi) approach despite using less annotation information. Finally it is worth noting that the gap between our method and the supervised (v) topline is not large (less than between our method and (vi)).

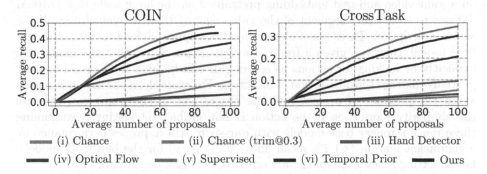

Fig. 5. Action proposal. Average recall versus average number of proposals per video

Step Localization. In this experiment, we explore how our actionness model can improve weakly supervised action localization. This task is particularly relevant since presence of background is one of the main challenges for weakly supervised action localization methods. To do so, we augment various action localization methods from [23,41] and [22] with our actionness score $S(x)$ and evaluate on the CrossTask dataset following the protocol described in [41]. These methods work in a similar way, described next. First, step classifiers are applied to every frame of the video. Then, each step is assigned to exactly one short clip, using a dynamic programming to solve: $(t_1^*, ..., t_K^*) = \arg\max_{t_1 < ... < t_K} \sum_{t=1}^{T} \sum_{k=1}^{K} f_k(x_t)$, where $f_k(x_t)$ is the output of the step classifier k for the t-th input clip x_t, and $t_1, ..., t_K$ are the clips ids assigned to steps $1, ..., K$, respectively. [23,41] and [22] differ only in the form of the classifier $f_k(x)$. We augment these methods with our actionness score by simply adding it to the objective during inference:

$$(t_1^*, ..., t_K^*) = \arg\max_{t_1 < ... < t_K} \sum_{t=1}^{T} \left[(1 - \alpha) \sum_{k=1}^{K} f_k(x_t) + \alpha S(x_t) \right], \qquad (4)$$

where $S(x_t)$ is the actionness score of our model for clip x_t and α is a combination parameter. Intuitively, the role of our score is to lower the confidence of background clips and increase the score on foreground action clips.

Results are provided in Table 2. α is selected independently for each method, which equals 0.1 for [23] and 0.8 for other methods. We use the same value of α for all test tasks. Combining the baseline method with our score improves the performance in every case. The gap in performance is particularly large for the method from Zhukov *et al.* [41]. This can be in part attributed to the fact that this method does not try to model the background for a given frame (indeed a simple constraint imposes that the score should sum to one across time for all actions without trying to explicitly lower score on detected background frames). Adding our score to the outputs of the model resolves this problem and leads to a large improvement (+4.6% recall). Other methods in Table 2 rely instead on a joint video and text embedding pretrained on the large scale HowTo100M dataset to score every segment of the video against the text embeddings of the action description. These text-video embeddings approach lead to much stronger base model. However, given a frame, there is no guarantee that the model will explicitly be looking for actions as scores can still be high if its visual content partially matches the description of an action. For example, if the action is *season steak*, the presence of a *steak* and the object *salt* in the frame can increase the similarity score even if the action is not visible. Interestingly, combining these much stronger base models with our score still improves performance by a significant margin (+1.2% recall and +0.7% recall for the best S3D model), hence setting a new state-of-the-art on the CrossTask benchmark [41].

4.4   Qualitative Results

To provide more insight about the kind of signal captured by our model S, we provide clips from HowTo100M with high and low actionness scores in Fig. 6. In order to obtain these examples, we run our model on 50,000 randomly sampled videos from HowTo100M. To show the variety of different tasks, we illustrate the top 10 highest scoring clips within each of the four largest HowTo100M categories: Food and Entertaining, Home and Garden, Hobbies and Craft and Cars & Other Vehicles. Finally, we also give the 10 lowest scoring clips across all categories to illustrate the type of background discovered by our model.

Table 2. Step localization results on CrossTask with and without our actionness score

	+ actionness	Make kimchi rice	Pickle cucumber	Make banana ice cream	Grill steak	Jack up car	Make jello shots	Change tire	Make lemonade	Add oil to car	Make latte	Build shelves	Make taco salad	Make French toast	Make Irish coffee	Make strawberry cake	Make pancakes	Make meringue	Make fish curry	Average
[41]	✗	13.3	18.0	23.4	23.1	16.9	16.5	30.7	21.6	4.6	19.5	35.3	10.0	32.3	13.8	29.5	37.6	43.0	13.3	22.4
[41]	✓	18.4	24.9	25.6	24.1	19.0	29.6	33.8	30.0	7.7	23.7	45.0	13.4	36.1	23.7	34.3	41.9	42.0	15.8	27.0
[23]	✗	33.5	27.1	36.6	37.9	24.1	35.6	32.7	35.1	30.7	28.5	43.2	19.8	34.7	33.6	40.4	41.6	41.9	27.4	33.6
[23]	✓	33.6	28.6	35.4	38.5	25.0	37.3	35.1	41.2	30.9	30.1	45.1	21.4	33.7	34.3	39.1	41.2	40.3	26.3	34.0
I3D [22]	✗	28.7	37.9	42.8	36.3	22.0	42.9	27.4	43.1	30.8	32.7	42.8	27.5	34.0	33.7	44.3	48.0	46.0	33.9	36.4
I3D [22]	✓	31.1	37.2	42.6	37.4	23.5	43.4	27.1	43.4	32.2	35.9	46.0	29.4	33.9	36.6	45.6	49.7	45.2	37.0	37.6
S3D [22]	✗	31.5	36.0	46.5	38.5	25.2	45.0	33.3	48.1	38.4	37.0	48.1	34.2	38.7	41.9	44.6	48.2	52.2	38.0	40.3
S3D [22]	✓	34.1	40.0	48.7	40.3	30.7	46.1	34.5	45.9	38.1	35.9	50.0	35.4	38.1	42.6	42.6	45.9	51.6	37.8	41.0

Highest scoring clips (Food and Entertainment)

Highest scoring frames (Home and Garden)

Highest scoring frames (Hobbies and Craft)

Highest scoring frames (Cars & Other Vehicles)

Lowest scoring frames (all categories)

Fig. 6. First four rows show highest actionness scoring clips from the top 4 categories of HowTo100M: Food and Entertaining, Home and Garden, Hobbies and Craft and Cars & Other Vehicles. Bottom row illustrates the lowest scoring clips according to our actionness score S (see Supplementary material for more examples)

5   Conclusion

In this paper, we have presented a self-supervised method that can separate actions from background without resorting to any form of manual annotation. It does so by leveraging the assumption that frames that depict key actions are more informative when it comes to predict what may come next or what happens in the past. Equipped with our method, we managed to improve the state-of-the-art on a challenging action localization benchmark. As future work we notably plan to investigate if our method would generalize to broader domains than instructional videos. Another potential direction would be to jointly discriminate background and actions while also clustering similar actions together, thus paving the way for unsupervised action discovery.

Acknowledgements. This work was partially supported by the European Regional Development Fund under project IMPACT (reg. no. CZ.02.1.01/0.0/0.0/15 003/0000468), Louis Vuitton ENS Chair on Artificial Intelligence, the MSR-Inria joint lab, and the French government under management of Agence Nationale de la Recherche as part of the "Investissements d'avenir" program, reference ANR-19-P3IA-0001 (PRAIRIE 3IA Institute).

References

1. Agrawal, P., Carreira, J., Malik, J.: Learning to see by moving. In: The IEEE International Conference on Computer Vision (ICCV), December 2015
2. Alayrac, J.B., Bojanowski, P., Agrawal, N., Laptev, I., Sivic, J., Lacoste Julien, S.: Unsupervised learning from narrated instruction videos. In: CVPR (2016)
3. Alayrac, J.B., Sivic, J., Laptev, I., Lacoste-Julien, S.: Joint discovery of object states and manipulation actions. In: ICCV (2017)
4. Carreira, J., Zisserman, A.: Quo Vadis, action recognition? A new model and the kinetics dataset. In: CVPR (2017)
5. Chen, W., Xiong, C., Xu, R., Corso, J.J.: Actionness ranking with lattice conditional ordinal random fields. In: The IEEE Conference on Computer Vision and Pattern Recognition (CVPR), June 2014
6. Doersch, C., Gupta, A., Efros, A.A.: Unsupervised visual representation learning by context prediction. In: The IEEE International Conference on Computer Vision (ICCV), December 2015
7. Dwibedi, D., Aytar, Y., Tompson, J., Sermanet, P., Zisserman, A.: Temporal cycle-consistency learning. In: Proceedings of the IEEE/CVF Conference on Computer Vision and Pattern Recognition (CVPR), June 2019
8. Escorcia, V., Caba Heilbron, F., Niebles, J.C., Ghanem, B.: DAPs: deep action proposals for action understanding. In: Leibe, B., Matas, J., Sebe, N., Welling, M. (eds.) ECCV 2016. LNCS, vol. 9907, pp. 768–784. Springer, Cham (2016). https://doi.org/10.1007/978-3-319-46487-9_47
9. Fernando, B., Bilen, H., Gavves, E., Gould, S.: Self-supervised video representation learning with odd-one-out networks. In: The IEEE Conference on Computer Vision and Pattern Recognition (CVPR), July 2017

10. Gao, J., Chen, K., Nevatia, R.: CTAP: complementary temporal action proposal generation. In: Ferrari, V., Hebert, M., Sminchisescu, C., Weiss, Y. (eds.) ECCV 2018. LNCS, vol. 11206, pp. 70–85. Springer, Cham (2018). https://doi.org/10.1007/978-3-030-01216-8_5

11. Gidaris, S., Singh, P., Komodakis, N.: Unsupervised representation learning by predicting image rotations. In: ICLR, April 2018

12. Huang, D.A., Lim, J.J., Fei-Fei, L., Niebles, J.C.: Unsupervised visual-linguistic reference resolution in instructional videos. In: CVPR (2017)

13. Huang, D.A., et al.: Finding "it": weakly-supervised reference-aware visual grounding in instructional video. In: CVPR (2018)

14. Jenni, S., Favaro, P.: Self-supervised feature learning by learning to spot artifacts. In: The IEEE Conference on Computer Vision and Pattern Recognition (CVPR), June 2018

15. Ji, J., Cao, K., Niebles, J.C.: Learning temporal action proposals with fewer labels. In: The IEEE International Conference on Computer Vision (ICCV), October 2019

16. Kingma, D., Ba, J.: Adam: a method for stochastic optimization. arXiv preprint arXiv:1412.6980 (2014)

17. Larsson, G., Maire, M., Shakhnarovich, G.: Colorization as a proxy task for visual understanding. In: The IEEE Conference on Computer Vision and Pattern Recognition (CVPR), July 2017

18. Lee, H.Y., Huang, J.B., Singh, M., Yang, M.H.: Unsupervised representation learning by sorting sequences. In: The IEEE International Conference on Computer Vision (ICCV), October 2017

19. Lin, T., Liu, X., Li, X., Ding, E., Wen, S.: BMN: boundary-matching network for temporal action proposal generation. In: The IEEE International Conference on Computer Vision (ICCV), October 2019

20. Lin, T., Zhao, X., Su, H., Wang, C., Yang, M.: BSN: boundary sensitive network for temporal action proposal generation. In: Ferrari, V., Hebert, M., Sminchisescu, C., Weiss, Y. (eds.) ECCV 2018. LNCS, vol. 11208, pp. 3–21. Springer, Cham (2018). https://doi.org/10.1007/978-3-030-01225-0_1

21. Liu, Y., Ma, L., Zhang, Y., Liu, W., Chang, S.F.: Multi-granularity generator for temporal action proposal. In: The IEEE Conference on Computer Vision and Pattern Recognition (CVPR), June 2019

22. Miech, A., Alayrac, J.B., Smaira, L., Laptev, I., Sivic, J., Zisserman, A.: End-to-end learning of visual representations from uncurated instructional videos (2020)

23. Miech, A., Zhukov, D., Alayrac, J.B., Tapaswi, M., Laptev, I., Sivic, J.: HowTo100M: learning a text-video embedding by watching hundred million narrated video clips. In: The IEEE International Conference on Computer Vision (ICCV) (2019)

24. Misra, I., Zitnick, C.L., Hebert, M.: Shuffle and learn: unsupervised learning using temporal order verification. In: Leibe, B., Matas, J., Sebe, N., Welling, M. (eds.) ECCV 2016. LNCS, vol. 9905, pp. 527–544. Springer, Cham (2016). https://doi.org/10.1007/978-3-319-46448-0_32

25. Noroozi, M., Favaro, P.: Unsupervised learning of visual representations by solving jigsaw puzzles. In: Leibe, B., Matas, J., Sebe, N., Welling, M. (eds.) ECCV 2016. LNCS, vol. 9910, pp. 69–84. Springer, Cham (2016). https://doi.org/10.1007/978-3-319-46466-4_5

26. Pathak, D., Girshick, R., Dollar, P., Darrell, T., Hariharan, B.: Learning features by watching objects move. In: The IEEE Conference on Computer Vision and Pattern Recognition (CVPR), July 2017

27. Pathak, D., Krahenbuhl, P., Donahue, J., Darrell, T., Efros, A.A.: Context encoders: feature learning by inpainting. In: The IEEE Conference on Computer Vision and Pattern Recognition (CVPR), June 2016
28. Sener, F., Yao, A.: Unsupervised learning and segmentation of complex activities from video. In: CVPR (2018)
29. Shan, D., Geng, J., Shu, M., Fouhey, D.F.: Understanding human hands in contact at internet scale. In: Proceedings of the IEEE/CVF Conference on Computer Vision and Pattern Recognition (CVPR), June 2020
30. Sun, C., Myers, A., Vondrick, C., Murphy, K., Schmid, C.: VideoBERT: a joint model for video and language representation learning. In: The IEEE International Conference on Computer Vision (ICCV) (2019)
31. Tang, Y., et al.: COIN: a large-scale dataset for comprehensive instructional video analysis. In: CVPR (2019)
32. Wang, L., Qiao, Y., Tang, X., Van Gool, L.: Actionness estimation using hybrid fully convolutional networks. In: The IEEE Conference on Computer Vision and Pattern Recognition (CVPR), June 2016
33. Wang, X., Gupta, A.: Unsupervised learning of visual representations using videos. In: The IEEE International Conference on Computer Vision (ICCV), December 2015
34. Wei, D., Lim, J.J., Zisserman, A., Freeman, W.T.: Learning and using the arrow of time. In: The IEEE Conference on Computer Vision and Pattern Recognition (CVPR), June 2018
35. Xu, D., Xiao, J., Zhao, Z., Shao, J., Xie, D., Zhuang, Y.: Self-supervised spatiotemporal learning via video clip order prediction. In: The IEEE Conference on Computer Vision and Pattern Recognition (CVPR), June 2019
36. Zhang, R., Isola, P., Efros, A.A.: Colorful image colorization. In: Leibe, B., Matas, J., Sebe, N., Welling, M. (eds.) ECCV 2016. LNCS, vol. 9907, pp. 649–666. Springer, Cham (2016). https://doi.org/10.1007/978-3-319-46487-9_40
37. Zhang, R., Isola, P., Efros, A.A.: Split-brain autoencoders: unsupervised learning by cross-channel prediction. In: The IEEE Conference on Computer Vision and Pattern Recognition (CVPR), July 2017
38. Zhao, Y., Xiong, Y., Wang, L., Wu, Z., Tang, X., Lin, D.: Temporal action detection with structured segment networks. In: The IEEE International Conference on Computer Vision (ICCV), October 2017
39. Zhou, B., Andonian, A., Oliva, A., Torralba, A.: Temporal relational reasoning in videos. In: Ferrari, V., Hebert, M., Sminchisescu, C., Weiss, Y. (eds.) ECCV 2018. LNCS, vol. 11205, pp. 831–846. Springer, Cham (2018). https://doi.org/10.1007/978-3-030-01246-5_49
40. Zhou, L., Chenliang, X., Corso, J.J.: Towards automatic learning of procedures from web instructional videos. In: AAAI (2018)
41. Zhukov, D., Alayrac, J.B., Cinbis, R.G., Fouhey, D., Laptev, I., Sivic, J.: Cross-task weakly supervised learning from instructional videos. In: The IEEE Conference on Computer Vision and Pattern Recognition (CVPR), June 2019

Fully Embedding Fast Convolutional Networks on Pixel Processor Arrays

Laurie Bose[1]([✉]), Piotr Dudek[2], Jianing Chen[2], Stephen J. Carey[2],
and Walterio W. Mayol-Cuevas[1]

[1] University of Bristol, Bristol, UK
{lb7943,Walterio.Mayol-Cuevas}@bristol.ac.uk
[2] University of Manchester, Manchester, UK
{pdudek,jianing.chen,stephen.carey}@manchester.ac.uk

Abstract. We present a novel method of CNN inference for pixel processor array (PPA) vision sensors, designed to take advantage of their massive parallelism and analog compute capabilities. PPA sensors consist of an array of processing elements (PEs), with each PE capable of light capture, data storage and computation, allowing various computer vision processes to be executed directly upon the sensor device. The key idea behind our approach is storing network weights "in-pixel" within the PEs of the PPA sensor itself to allow various computations, such as multiple different image convolutions, to be carried out in parallel. Our approach can perform convolutional layers, max pooling, ReLu, and a final fully connected layer entirely upon the PPA sensor, while leaving no untapped computational resources. This is in contrast to previous works that only use a sensor-level processing to sequentially compute image convolutions, and must transfer data to an external digital processor to complete the computation. We demonstrate our approach on the SCAMP-5 vision system, performing inference in a MNIST digit classification network at over 3000 frames per second and over 93% classification accuracy. This is the first work demonstrating CNN inference conducted entirely upon a PPA vision sensor, requiring no external processing.

Keywords: Low-level vision · PPA · CNN · Vision sensor · Edge computing

1 Introduction

Recently, there has been much interest in developing hardware architectures for acceleration of deep learning algorithms. In particular, as Convolutional Neural Networks (CNNs) have become a staple of computer vision applications, there have been many approaches to implementing these efficiently in

Electronic supplementary material The online version of this chapter (https://doi.org/10.1007/978-3-030-58526-6_29) contains supplementary material, which is available to authorized users.

© Springer Nature Switzerland AG 2020
A. Vedaldi et al. (Eds.): ECCV 2020, LNCS 12374, pp. 488–503, 2020.
https://doi.org/10.1007/978-3-030-58526-6_29

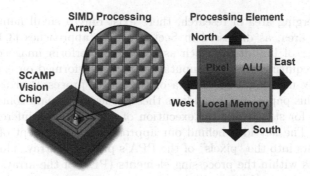

Fig. 1. A Pixel Processor Array device performs computations on the image sensor chip, using a SIMD processor array, with each pixel containing arithmetic logic unit (ALU), local memory circuits, and nearest-neighbour communication links.

hardware [1,7,9,13,15,17]. Some of the most challenging application scenarios involve "edge computing" or "on-device computing", where computations are carried out as close to sensors as possible, to achieve low power operation and minimise bandwidth of downstream communications. Ultimately, the sensing and processing can be integrated in a single device. One approach to such integration is through distribution of photosensors of the image sensor within a massively-parallel fine-grain SIMD cellular processor array [4,12,14], an approach we term Pixel Processor Array (PPA). The PPA concept is illustrated in Fig. 1.

In areas of computer vision and robotics applications, PPA sensors may potentially offer a wealth of benefits over standard camera sensors that are primarily developed with the human viewer in mind, and designed to capture entire high fidelity images for later inspection. The complete image capture, read-out, analog-digital conversion and transfer process in standard sensors introduces a significant time and energy bottleneck in computer vision pipelines, and typically results in low temporal resolution visual information (e.g. typical video-rate of 30 frames per second) that is highly prone to motion blur. A PPA sensor circumvents this scenario by instead performing visual computation directly at the point of light capture, extracting the desired information on-sensor, before transferring it over to a host processor. In many situations this can result in a vast decrease in data bandwidth between the sensor and the external hardware, allowing the system to conduct visual processing at much higher frame-rates, well beyond the capabilities of more standard sensors, while maintaining a low power consumption [2,4].

One application of such PPA sensors is that of neural network inference in which captured visual information is immediately fed through a neural network being executed wholly or partially on-sensor, with the PPA's output then being compressed to simply neuron activations, ideally of the network's final layer. Such an application of future PPA sensors may offer real world network inference at speeds well beyond standard visual pipelines, however implementation on current PPA hardware is a highly challenging area of investigation.

As an emerging area of research, there exist only a small number of prior works in this area, as discussed in Sect. 3. These approaches [3,10,16] suffer from a number of limitations, such as having to perform image convolutions sequentially, requiring certain computation to be performed on external hardware, and only utilizing a small area of the entire processor array. The work presented in this paper aims to address these issues. Our main contribution is a new approach for structuring the execution of CNN network inference on PPA architectures. The key idea behind our approach is the concept of embedding network weights into the "pixels" of the PPA's processor array. This is done by storing weights within the processing elements (PEs) of the array, rather than weights being contained in the instructions transmitted to the processor array during inference as in previous works. This embedding of weights allows different parts of the processor array to perform different computations, upon different local data, simultaneously. As such, our approach can perform many different image convolutions, upon multiple images, spread across the PPA array in parallel, and efficiently perform a final fully connected layer entirely on-sensor. This computation can be structured to make use of the entire processor array at all times, improving the utilisation of available computational resources. To the best of our knowledge this is the first work to present such an approach, and the first to demonstrate multiple convolutional layers, a fully connected layer, and complete network inference upon a PPA. We demonstrate inference of both 2 and 3 layer networks upon the SCAMP-5 PPA performing digit recognition, able to achieve classifications at over 3000 frames per second and over 93% accuracy.

2   SCAMP-5 Overview

The PPA used for this work is the SCAMP-5 vision sensor [4,6] consisting of a 256×256 array of processing elements (PEs), each containing processor circuitry allowing visual data to be stored, and manipulated directly at the point of light capture. The chip architecture, as shown in Fig. 2, has been described in [4]. Briefly, each PE contains 13 digital registers (1-Bit) and 7 analog memory registers. Various operations can be performed between the memory registers of a PE, such as addition and subtraction of analog registers, and standard Boolean logic operations between digital registers. PEs can also exchange data with their neighbours. The array operates as an SIMD computer. The operations on local memory registers are performed across all PEs of the 256×256 array in parallel, using a single instruction. Each PE also contains an Execution Flag register allowing it to ignore received operations and allowing for conditional execution.

The operations performed by the PE array are dictated by a central controller, built upon ARM Cortex M0 processor. This controller executes its own program, primarily for sending instructions to the SCAMP-5 PPA to perform the sequence of operations that will result in some desired computation being performed upon the array.

The near-sensor processing approach of this architecture is very efficient. The SCAMP-5 chip performs up to 535 GOPS/W (Giga Operations Per Second per

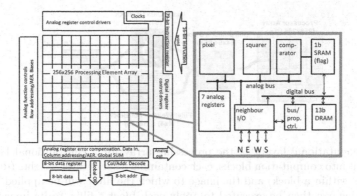

Fig. 2. SCAMP-5 Architecture [4]. Left: The PPA chip contains a 256 × 256 SIMD processor array and associated control, readout and interface peripheral circuits. Right: The processing element shown contains the photosensor, seven analog local registers, supporting arithmetic operations of addition, negation and division, neighbour communications with 4 nearest neighbour, 1 bit activity flag, and 13 bits of digital memory supporting logic operations.

watt). Note that this device is manufactured using two decades old 180nm CMOS silicon technology [4]. Very significant gains can clearly be made on future devices in terms of increasing computing power and decreasing power consumption.

3   Related Work

While previous works exist regarding CNN inference on PPAs [3,10,16] typically these methods perform various parts of the network computation in serial, rely on external hardware for additional computation, and only make use of a small area of the PPA's processor array leaving a great amount of processing power untapped. For example, these approaches are demonstrated upon MNIST/digit classification task on SCAMP-5 in which they load a single small MNIST digit (28×28) into the center of the 256×256 SIMD processor array. These approaches then sequentially compute image convolutions upon this central digit, effectively leaving well over 90% of the processor array unused. These convolution results are passed to the ARM controller connected to the SCAMP-5 PPA, which is used to perform one or more fully connected layers. Therefore, while demonstrating the concept of on-sensor CNN, a significant portion of the neural network computation in these approaches is actually conducted upon the ARM controller in a standard C++ program rather than by making use of the PPA's processing power.

By comparison, the approach proposed in this paper performs complete inference computation, including the fully-connected layer, upon the PPA device, potentially utilizing 100% of the processing array, and efficiently performing convolutional layers by computing many different image convolutions in parallel.

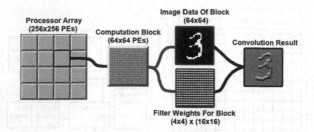

Fig. 3. Computational layout of the processing array for a convolutional layer. The array is split into computation blocks, each containing a set of filter weights (duplicated many times within a block) and the image to which the filter will be applied. A single SIMD routine can then be executed to apply each block's filter to its image data in parallel.

The proposed approach requires all network weights to be stored upon the processing elements (PEs) of the PPA itself. Due to the limited memory (13 bits, 7 analog values) of each PE on current generation PPA hardware, we are restricted to low-bit quantised weights and a limited number of layers. However it should be noted that many tasks have been successfully demonstrated on such low-bit weight networks [11,18,19], and it is likely that next generation PPA hardware will see a significant boost in memory per PE.

4   Parallel Convolutional Layer Computation

In this section we describe our approach for the computation of convolutional layers upon the PPA. The weights of all the various convolutional filters are stored upon the processing array simultaneously, within the registers of the PEs. This enables different convolutional filters to be applied to different areas of the PE array in parallel. This can allow us to perform all the computation required for a convolutional layer in parallel.

For example, in the case of SCAMP-5, up to 64 MNIST digits can be spread across the 256×256 PE array. This allows for 64 different convolutions to be performed simultaneously at no additional time or power cost. In the case of digit classification this can be used to compute 64 different convolutions on the same digit duplicated 64 times in parallel.

4.1   Computational Layout on PE Array

Our convolutional layer approach effectively divides the PE array into multiple rectangular "computation" blocks of processing elements. The PEs of each computation block contain both the weights of a specific convolutional filter and image data to which the filter should be applied as shown in Fig. 3. A sequence of SIMD operations can then be formulated to simultaneously apply each computational block's filter to its stored image data, performing all computation

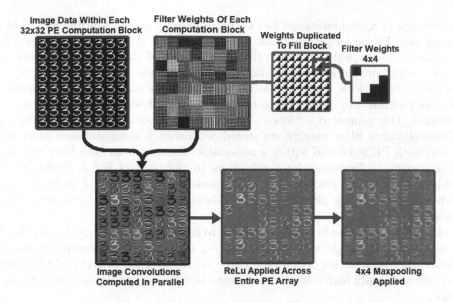

Fig. 4. Examples of convolutional layer computation, illustrating how multiple convolutions are computed in parallel by splitting the PPA array into distinct computation blocks. In this case the Scamp5's processing array is split into 64 computational blocks of 32×32 PEs each. Upon computing image convolutions ReLu and Maxpooling can also be applied across the array at very little computational cost.

required for a convolutional layer. Examples of such computation are illustrated in Fig. 4 for MNIST digits.

Note this approach is flexible in that each computational block may contain different image data and vary in size, however, for convolutional layer computation we use identical square blocks. For digit recognition we demonstrate convolutional layers of both 64 and 16 convolutions, using computational blocks of size 32 and 64 respectively. In both cases, the 28×28 MNIST digits are rescaled to fill these computation blocks.

4.2   In-Pixel Filter Weights

Each computation block stores within it the weights of a specific filter. When the SIMD routine for a convolutional layer is sent to the processor array, each block will use these weights to compute a convolution upon its stored image data. Directly storing filter weights upon the processor array at the locations where they are to be applied is what allows our approach to perform multiple filters simultaneously.

There are many possible layouts for storing a set of filter weights within a computational block of PEs. However, it is generally not possible for each PE to store a complete copy of its block's filter weights due to the limited local memory resources available on current generation PPA devices. The solution is to spread the storage of a computational block's filter weights across multiple PEs. This means each PE no longer has immediate access to every filter weight, however, weights can be copied over from other nearby PEs of the same computational block during convolution computation. To minimize the time transferring filter weights between PEs, its important to use a layout in which each PE is located in close proximity to other PEs storing the weights it will require during computation. This prompted a "checker board" style layout, where multiple copies of convolutional filter weights are stored within each computational block to ensure each PE is located within a reasonable distance from each filter weight. This concept is illustrated for 4×4 filters in the right of Fig. 4. Future PPA devices, with greater resources per PE, should allow each PE to store its own dedicated copy of any filter weights, significantly speeding up the convolution computation.

In our demonstrated networks each PE in a block stores a single binary filter weight, with the weight values of $+1$ and -1 naturally corresponding to image addition and subtraction operations. There are many schemes that could be used to store and apply higher bit-count weights but for now we leave this to future work.

4.3 Parallel ReLU and Max Pooling

After performing a convolutional layer the PE array will hold multiple convolution images such as those shown in Fig. 4. We then turn these images into activation data by applying the ReLU activation function. The SCAMP-5 hardware has the function to flag all PEs whose stored values in a certain analog register are positive or negative. This allows us to simply flag all PEs whose convolution result is negative and input a value of zero into these flagged registers, generating ReLu activations.

We then perform a 4×4 max pooling routine by first making a copy of the activation values image. This copied image is then shifted horizontally right, with each PE then containing both its original activation value and a value from this shifted data. In parallel, every PE then compares these two activation values, replacing the stored activation data with the shifted data whenever it is greater in value. This routine of shifting, comparing activations and replacing with the higher value is repeated horizontally right three times, and three times vertically down. This results in every PE holding the highest activation value in the 4×4 square of which it is in the top left corner. The pixels holding the correct maxpooled values for each 4×4 grid space are then copied back into each PE of their 4×4 block.

4.4 Further Convolutional Layers

After performing an initial convolutional layer, either a final fully connected layer, or an additional convolutional layer is performed. This section describes one possible method to compute such an additional convolution layer, where each feature map is constructed from those of the previous layer as standard.

Note that this approach could in future be used to add multiple additional convolutional layers, however this is difficult to achieve within the limited memory resources of current SCAMP-5 hardware.

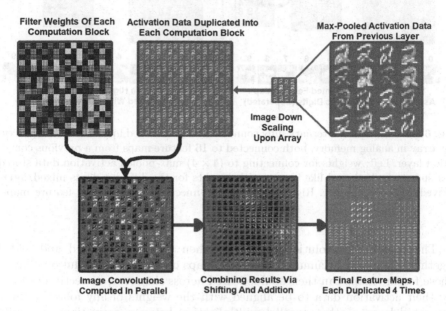

Fig. 5. Computation of feature maps from an additional convolutional Layer. Max-pooled activation data from the previous layer (top right) is shrunk and duplicated across the processor array. Image convolutions for the new layer are them performed upon each duplicated image, in the same manner as network's initial convolution layer. The resulting convolutional images (bottom left) are then combined accordingly and ReLu is applied forming feature maps of this new layer (bottom right). In this example 16 feature maps from the initial layer are duplicated and 256 convolutions computed, before being recombined to form 16 feature maps.

In brief, the feature maps of a previous convolution layer (consisting of max-pooled activation data) are shrunk and duplicated to fill the processor array. Each duplicate of a feature map is then used in computing a feature map in the new convolutional layer. An example of this is shown in Fig. 5, where 256 convolutions are computed in parallel upon the 16 feature maps (each duplicated 16 times) of the previous layer. These convolution results are then added together accordingly to form the 16 feature maps of the new convolutional layer.

Many of the concepts introduced previously are re-used for this computation. The in-pixel storage of filter weights and computational layout is identical to the initial convolutional layer, with the processor array again being split into computation blocks each storing its own set of weights and image data as shown in Fig. 5. The same SIMD routine used to perform the initial convolutional layer can simply be executed again for computing this additional layer, helping to reduce program size.

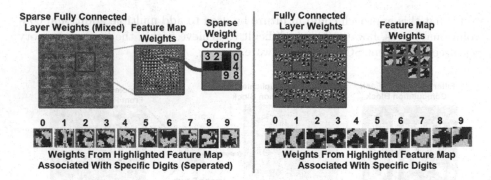

Fig. 6. Two examples of ternary fully connected weights stored upon the PPA's processor array in analog memory, both connected to 16 feature maps from a previous convolution layer. Left: weights for connecting to (4×4) max-pooled activation data stored in a sparse checkerboard like layout, with weights for the different digits mixed/interweaved with one another. Right: weights for connecting to duplicated feature maps.

The resulting convolution results are then repeatedly shifted and added together, iteratively accumulating feature maps of this new convolutional layer. These feature maps can then be duplicated across the array, correctly positioning their activation data to be aligned with the weights of any following fully connected layer, so that parallel multiplication between activations and fully connected weights can be performed as described in Sect. 5 (Fig. 7).

4.5 Feature Map Shrinking and Duplication

The process of shrinking the max-pooled activation data upon the PPA leverages the image transformation methods first introduced in [2] for image scaling. However, conducting such scaling operations using analog memory registers results in the build-up of systematic errors and noise [4], from analog data having to be repeatedly copied from one PE to the next. This would corrupt the activation data beyond use. To avoid this issue we instead convert the analog activation data to a 3-bit digital representation, with each PE's stored analog value being split across 3 digital registers (within the same PE). This creates 3 binary images, one for each bit, which then can then all be scaled and duplicated across the array. Afterwards this digital data can be recombined to once again form a single gray-scale analog image, but devoid of corruption.

5   Parallel Fully Connected Layer Computation

Following on from a convolutional layers, we perform computation of a final ternary weight fully connected layer upon the PPA, again storing weights directly in the PEs of the processor array. The activations of the previous convolution layer are duplicated as shown in Fig. 7, either by max-pooling (which creates blocks of duplicated values) or by duplicating the feature maps multiple times across the array. By correctly arranging the layout of the fully connected weights, each weight's PE can then directly receive the activation data associated with that weight. This layout varies as illustrated in Fig. 7 depending on whether the previous layer produces max-pooled data or duplicated feature maps. All duplicated activations from the previous layer can then be multiplied by their associated fully connected weights simultaneously in parallel, using the native analog image addition (for weights of value 1) and subtraction (weights of value −1) operations of SCAMP-5 (or no operation for weights of value 0). Examples of this process are shown in Fig. 7.

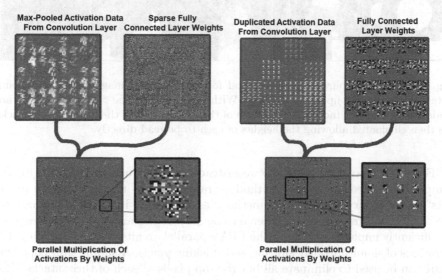

Fig. 7. Layout and application of fully connected layer weights, accepting as input either max-pooled activation data (Left) or duplicated feature maps (right). The arrangement of weights is different in each case to enable the correct parallel in-line multiplication between the weights and activation data. After this multiplication, all resulting data associated with a specific fully connected neuron can be summed in parallel by flagging the appropriate PEs.

The limited memory of the processing elements on SCAMP-5 restricts the programmer to the use of ternary weights for the final fully connected layer, stored within the analog registers of the PEs to save digital resources. Note that the content of analog registers decays over time, drifting away from the stored value [4]. However, with quantized ternary values being stored one can "Refresh" the register's content at set intervals to prevent such decay.

5.1 Activation Value Summation

After the multiplication step each PE contains a synaptic contribution to the activation of one of the final neurons in the fully connected layer. SCAMP-

5 has the capability of performing a global summation of many analog values distributed across the PE array in parallel, which can be used to effectively add all synaptic contributions, however this summation introduces significant noise.

For two layer networks this does not pose a major issue and analog summation can be simply be performed a number times and averaged. However for three layer networks, with more discriminative second layer features, analog summation proves too noisy for accurate classification.

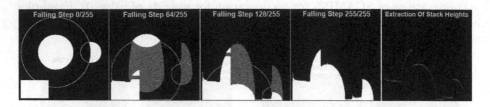

Fig. 8. Example of our proposed method for rapid binary image summation being iteratively performed upon a test image. With each iteration the "1s" of the image are made to fall forming stacks at the bottom of the image. All but the tops of these stacks are then eliminated allowing the heights of each to be read directly.

For such three layer networks we instead turned to using the PPA's digital computation, creating a new method to rapidly count the "set" white pixels ("1s") in a binary image. This method, as visualised in Fig. 8, functions by essentially stacking pixels together on one side of the array. This process can be efficiently implemented upon the PPA's parallel architecture, performing 255 iterations of simultaneously shifting and stacking pixels. A simple shift copy and XOR can be used to eliminate all but the top pixels of each of these stacks. The image coordinates of these remaining pixels (up to 256) can then be read directly (using an address-event readout scheme of SCAMP-5) to give the heights of each stack, which when added together give the total number of white pixels in the original image. This entire process takes $260\,\mu s$ to complete, and while slower than the analog global summation, it provides a perfectly accurate summation result. This method is employed in the fully connected layer of our demonstrated three layer network, converting the analog activations into multi-bit representations, which can then be summed.

6   Results

6.1   MNIST Network Training

We trained networks with a mixture of binary and ternary weights (for convolutional filters and fully connected weights respectively) using an approach similar as [8], whereby real-valued weights are stochastically quantized during every forward pass. The errors obtained from the forward pass are then used to update the real-valued weights using the standard error back propagation algorithm, resulting in these real values converging towards binary/ternary ones over the course of training.

6.2 Inference on SCAMP-5 Hardware

We evaluated our inference approach using both two and three layer networks, trained on MNIST classification. The two layer networks used 32×32 input images, and consisted of one convolutional layer (64 feature maps, 4×4 filters), max pooling (4×4), and a final fully connected layer. Three layer networks used up scaled 64×64 input images, a first convolutional layer (16 feature maps), max pooling (4×4), a second convolutional layer (also 16 feature maps), and a final fully connected layer. Some sample classifications of such networks are shown in Figs. 9 and 10. Training is performed on a standard PC, using the 60,000 samples dataset. The trained weights were then loaded into the PEs of the SCAMP-5's processor array as described in previous sections, and evaluation of inference was performed by directly loading the test set images (one at a time) onto the PE array. With each image the SCAMP-5 then executed the SIMD routines to compute the network layers and output a final classification. Table 1 shows the computation times of the various processes used during inference.

The total computation time of two layer networks was $272\,\mu s$ corresponding to the processing speed of 3676 classifications per second (excluding the time to load a testing set image to the array). In a real-life scenario, digit images are not loaded to the array but captured via the image sensing capabilities of the chip, as

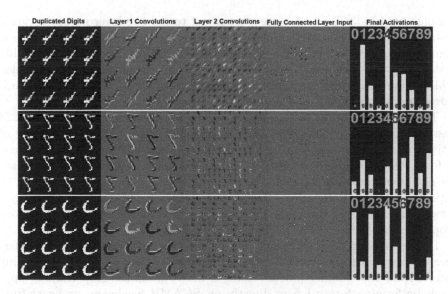

Fig. 9. Example classifications of some ambiguous digits via inference of a three layer network upon SCAMP-5. The convolutions computed from both layers are shown, along with the activations multiplied by the fully connected layer weights, and the final neuron activations for digits 0–9.

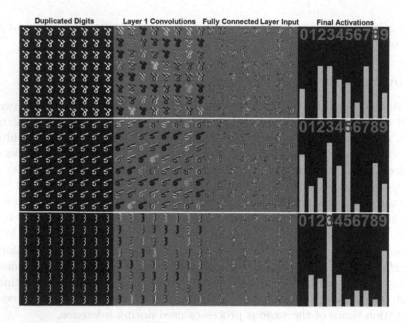

Fig. 10. Examples of digit classification via inference of a two layer network on SCAMP-5. Showing both the 64 convolutions computed, parallel multiplication of fully connected weights and the activation values of the final fully connected layer's neurons.

demonstrated in [3]. The inference accuracy of tested two layer networks varied around 92%–94% classification accuracy with different networks, a reduction from accuracy levels around 95% obtained in training on PC. It is worth noting that due to the nature of the analog computing used during inference, which exhibits noise and systematic errors varying from one device to another [4], such a drop in accuracy is not unexpected.

Three layer networks had a total computation time of 4.46 ms (giving 224 classifications per second), the majority of which was from the shrinking of activation data between convolution layers, and the merging of convolutions to form feature maps. The methods and SIMD routines for these components are not as optimized as those for layer computation, and this is something to improve in future works. The classification accuracy obtained for three layer network inference was also in the range of 92%–94%, but with a more significant drop from the accuracy of 97% obtained in training.

It may be possible to reduce these discrepancies between training and inference accuracy in future work by investigating analog errors [5] and modelling them within the training process, using a hardware-in-the-loop training approach performing forward pass directly on SCAMP hardware, or by shifting certain operations to use digital computation upon the PPA. That said, the PPAs massively parallel analog computing still results in high performance and efficiency. During inference vision sensor itself consumes 1.25 W, with the rest of

Table 1. Computation times of various network components.

Component	Two layer network	Three layer network
Digit duplication	28 μs	28 μs
Convolutional layer/s	160 μs	320 μs (160 × 2)
ReLu	<1 μs	<1 μs
Max pooling	25 μs	25 μs
Feature map shrink and duplicate	–	2095 μs
Feature map creation	–	1055 μs
Fully connected layer	59 μs	901 μs
Total	272 μs (3676 fps)	4464 μs (224 fps)

the current camera system contributing another 750 mW, when operating at the maximum throughput of 3676 classifications per second. It can be extrapolated, that for applications where frame rates in the range of 30 fps are acceptable, the operating power of the system executing a two-layer network model would be in the range of 10–20 mW.

Our approach also compares favorably to existing CNN works using the SCAMP-5 PPA. Against the digital CNN implementation of [3], our presented 2-layer network offers similar accuracy while being capable of frame-rates 10 times higher (200 vs 3000+ FPS). Against the works of [10,16] our approach offers similar speed and accuracy, however a key distinction is that these works make heavy use of traditional computation on the ARM Cortex controller connected to the array. Specifically these works compute 2 fully connected layers on the ARM controller, with then only a single convolutional layer of 3 convolutions computed upon the PPA array. This is in contrast to this work in which every layer is computed directly upon the PPA array, with our weights "in-pixel" approach allowing the entire array to be utilized. As such, we argue that this approach is far better suited to make use of any future PPA devices, as it can immediately make full use of any hardware improvements such as greater memory per PE or increased array size.

7 Conclusions

We have presented a novel approach for conducting CNN inference upon PPA hardware, exploiting analog computations, and storing the weights of the network directly within the array's processor elements rather than in the program running upon the array's controller chip. Unlike previous works, our approach can perform multiple convolution layers, and a final fully connected layer entirely upon the PE array of the device, with the only information read-out being the activations of the final neuron layer. Thus we demonstrate a complete visual CNN on-chip solution, from light sensing, to classification results.

Our experiments considered only small networks using binary filters and ternary fully connected weights. However our approach can be used for deeper more complex networks as PPA hardware improves, and even these smaller networks have found practical applications in edge computing devices. Our inference of digit classification networks on the SCAMP-5 PPA sensor performs at over 93% classification accuracy and at speeds exceeding 3,000 image classifications per second, for the first time executing a complete classification network entirely upon the "focal-plane" of an image sensor. We expect that new PPA hardware utilizing more recent silicon technologies will provide substantial gains in performance and efficiency which our approach is well positioned to make use of. We hope this work also motivates better architectures designed for visual perception.

References

1. Aimar, A., et al.: NullHop: a flexible convolutional neural network accelerator based on sparse representations of feature maps. IEEE Trans. Neural Netw. Learn. Syst. **99**, 1–13 (2018)
2. Bose, L., Chen, J., Carey, S.J., Dudek, P., Mayol-Cuevas, W.: Visual odometry for pixel processor arrays. In: Proceedings of the IEEE International Conference on Computer Vision, pp. 4604–4612 (2017)
3. Bose, L., Chen, J., Carey, S.J., Dudek, P., Mayol-Cuevas, W.: A camera that CNNs: towards embedded neural networks on pixel processor arrays. arXiv preprint arXiv:1909.05647 (2019). (ICCV 2019 Accepted Submission)
4. Carey, S.J., Lopich, A., Barr, D.R., Wang, B., Dudek, P.: A 100,000 fps vision sensor with embedded 535GOPS/W 256 × 256 SIMD processor array. In: 2013 Symposium on VLSI Circuits, pp. C182–C183. IEEE (2013)
5. Carey, S.J., Zarándy, Á., Dudek, P.: Characterization of processing errors on analog fully-programmable cellular sensor-processor arrays. In: 2014 IEEE International Symposium on Circuits and Systems (ISCAS), pp. 1580–1583. IEEE (2014)
6. Chen, J., Carey, S.J., Dudek, P.: Scamp5d vision system and development framework. In: Proceedings of the 12th International Conference on Distributed Smart Cameras, p. 23. ACM (2018)
7. Chen, Y.H., Emer, J., Sze, V.: Eyeriss: a spatial architecture for energy-efficient dataflow for convolutional neural networks. In: ACM SIGARCH Computer Architecture News, vol. 44, pp. 367–379. IEEE Press (2016)
8. Courbariaux, M., Bengio, Y., David, J.P.: Binaryconnect: training deep neural networks with binary weights during propagations. In: Advances in Neural Information Processing Systems, pp. 3123–3131 (2015)
9. Du, Z., et al.: ShiDianNao: shifting vision processing closer to the sensor. In: ACM SIGARCH Computer Architecture News, vol. 43, pp. 92–104. ACM (2015)
10. Guillard, B.: Optimising convolutional neural networks for super fast inference on focal-plane sensor-processor arrays. Master's thesis, Imperial College London (2019)
11. Hubara, I., Courbariaux, M., Soudry, D., El-Yaniv, R., Bengio, Y.: Quantized neural networks: training neural networks with low precision weights and activations. J. Mach. Learn. Res. **18**(1), 6869–6898 (2017)
12. Komuro, T., Kagami, S., Ishikawa, M.: A dynamically reconfigurable SIMD processor for a vision chip. IEEE J. Solid-State Circuits **39**(1), 265–268 (2004)

13. Liang, S., Yin, S., Liu, S., Luk, W., Wei, S.: FP-BNN: binarized neural network on FPGA. Neurocomputing **275**, 1072–1086 (2018)
14. Rodriguez-Vazquez, A., Fernández-Berni, J., Leñero-Bardallo, J.A., Vornicu, I., Carmona-Galán, R.: CMOS vision sensors: embedding computer vision at imaging front-ends. IEEE Circuits Syst. Mag. **18**(2), 90–107 (2018)
15. Sim, J., Park, J.S., Kim, M., Bae, D., Choi, Y., Kim, L.S.: A 1.42 TOPS/W deep convolutional neural network recognition processor for intelligent IoE systems. In: 2016 IEEE International Solid-State Circuits Conference (ISSCC), pp. 264–265. IEEE (2016)
16. Wong, M.: Analog vision - neural network inference acceleration using analog SIMD computation in the focal plane. M.Sc. dissertation, Imperial College London (2018)
17. Zhao, R., et al.: Accelerating binarized convolutional neural networks with software-programmable FPGAs, pp. 15–24 (02 2017)
18. Zhou, A., Yao, A., Guo, Y., Xu, L., Chen, Y.: Incremental network quantization: towards lossless CNNs with low-precision weights. arXiv preprint arXiv:1702.03044 (2017)
19. Zhu, C., Han, S., Mao, H., Dally, W.J.: Trained ternary quantization. arXiv preprint arXiv:1612.01064 (2016)

Character Region Attention
for Text Spotting

Youngmin Baek, Seung Shin, Jeonghun Baek, Sungrae Park, Junyeop Lee,
Daehyun Nam, and Hwalsuk Lee[✉]

Clova AI Research, NAVER Corp., Seongnam-si, South Korea
{youngmin.baek,seung.shin,jh.baek,sungrae.park,junyeop.lee,
daehyun.nam,hwalsuk.lee}@navercorp.com

Abstract. A scene text spotter is composed of text detection and recognition modules. Many studies have been conducted to unify these modules into an end-to-end trainable model to achieve better performance. A typical architecture places detection and recognition modules into separate branches, and a RoI pooling is commonly used to let the branches share a visual feature. However, there still exists a chance of establishing a more complimentary connection between the modules when adopting recognizer that uses attention-based decoder and detector that represents spatial information of the character regions. This is possible since the two modules share a common sub-task which is to find the location of the character regions. Based on the insight, we construct a tightly coupled single pipeline model. This architecture is formed by utilizing detection outputs in the recognizer and propagating the recognition loss through the detection stage. The use of character score map helps the recognizer attend better to the character center points, and the recognition loss propagation to the detector module enhances the localization of the character regions. Also, a strengthened sharing stage allows feature rectification and boundary localization of arbitrary-shaped text regions. Extensive experiments demonstrate state-of-the-art performance in publicly available straight and curved benchmark dataset.

Keywords: Optical character recognition (OCR) · Character Region Attention · Text spotting · Scene text detection · Scene text recognition

1 Introduction

Scene text spotting, including text detection and recognition, has recently attracted much attention because of its variety of applications in instant translation, image retrieval, and scene parsing. Although existing text detectors and recognizers work well on horizontal texts, it still remains as a challenge when it comes to spotting curved text instances in scene images.

To spot curved texts in an image, a classic method is to cascade existing detection and recognition models to manage text instances on each side. The detectors [2,31,32] attempt to capture the geometric attributes of curved texts

© Springer Nature Switzerland AG 2020
A. Vedaldi et al. (Eds.): ECCV 2020, LNCS 12374, pp. 504–521, 2020.
https://doi.org/10.1007/978-3-030-58526-6_30

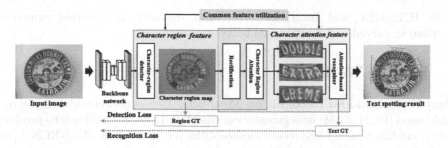

Fig. 1. Concept of the proposed method. The character region feature from the detector is used as an input character attention feature to the recognizer. Having a tightly coupled architecture lets the recognition loss flow through the whole network.

by applying complicated post-processing techniques, and the recognizers apply multi-directional encoding [6] or take rectification modules [11,37,46] to enhance the accuracy of the recognizer on curved texts.

As deep learning advanced, researches have been made to combine detectors and recognizers into a jointly trainable end-to-end network [14,29]. Having a unified model not only provides efficiency in the size and speed of the model, but also helps the model learn a shared feature that pulls up the overall performance. To gain benefit from this property, attempts have also been made to handle curved text instances using an end-to-end model [10,32,34,44]. However, most of the existing works only adopt a RoI pooling to share low-level feature layers between detection and recognition branches. In the training phase, instead of training the whole network, only shared feature layers are trained using both detection and recognition losses.

As shown in Fig. 1, we propose a novel end-to-end *Character Region Attention For Text Spotting* model, referred to as CRAFTS. Instead of isolating detection and recognition modules in two separate branches, we form a single pipeline by establishing a complimentary connection between the modules. We observe that recognizer [1] using attention-based decoder and detector [2] encapsulating character spatial information share a common sub-task which is to localize character regions. By tightly integrating the two modules, the outputs from the detection stage helps recognizer attend better to the character center points, and loss propagated from recognizer to detector stage enhances the localization of the character regions. Furthermore, the network is able to maximize the quality of the feature representation used in the common sub-tasks. To best of our knowledge, this is the first end-to-end work that constructs a tightly coupled loss propagation.

The summary of our contribution follows; (1) We propose an end-to-end network that could detect and recognize arbitrary-shaped texts. (2) We construct a complementary relationship between the modules by utilizing spatial character information from the detector on the rectification and recognition module. (3) We establish a single pipeline by propagating the recognition loss throughout all the features in the network. (4) We achieve the state-of-the-art performances in IC13,

IC15, IC19-MLT, and TotalText [7,19,20,33] datasets that contain numerous horizontal, curved and multilingual texts.

2   Related Work

Text Detection and Recognition Methods. Detection networks use regression based [16,24,25,48] or segmentation based [9,31,43,45] methods to produce text bounding boxes. Some recent methods like [17,26,47] take Mask-RCNN [13] as the base network and gain advantages from both regression and segmentation methods by employing multi-task learning. In terms of units for text detection, all methods could also be sub-categorized depending on the use of word-level or character-level[2,16] predictions.

Text recognizers typically adopt CNN-based feature extractor and RNN based sequence generator, and are categorized by their sequence generators; connectionist temporal classification (CTC) [35] and attention-based sequential decoder [21,36]. Detection model provides information of the text regions, but it is still a challenge for the recognizer to extract useful information in arbitrary-shaped texts. To help recognition networks handle irregular texts, some researches [28,36,37] utilize spatial transformer network (STN)[18]. Also, the papers [11,46] further extend the use of STN by iterative executing the rectification method. These studies show that running STN recursively helps recognizer extract useful features in extremely curved texts. In [27], Recurrent RoIWarp Layer was proposed to crop individual characters before recognizing them. The work proves that the task of finding a character region is closely related to the attention mechanism used in the attention-based decoder.

One way to construct a text spotting model is to sequentially place detection and recognition networks. A well known two-staged architecture couples TextBox++ [24] detector and CRNN [35] recognizer. With its simplicity, the method achieves favorable results.

End-to-End Using RNN-Based Recognizer. EAA [14] and FOTS[29] are end-to-end models based on EAST detector [49]. The difference between these two networks lies in the recognizer. The FOTS model uses CTC decoder [35], and the EAA model uses attention decoder [36]. Both works implement an affine transformation layer to pool the shared feature. The proposed affine transformation works well on horizontal texts, but shows limitations when handling arbitrary-shaped texts. TextNet [42] proposed a spatial-aware text recognizer with perspective-RoI transformation in the feature pooling layer. The network keeps an RNN layer to recognize a sequence of text in the 2D feature map, but due to the lack of expressively of the quadrangles, the network still shows limitations when detecting curved texts.

Qin et al. [34] proposed a Mask-RCNN [13] based end-to-end network. Given the box proposals, features are pooled from the shared layer and the ROI-masking layer is used to filter out the background clutters. The proposed method increases its performance by ensuring attention only in the text region. Busta et al. proposed Deep TextSpotter [3] network and extended their work in E2E-MLT [4]. The network is composed of FPN based detector and a CTC-based recognizer. The model predicts multiple languages in an end-to-end manner.

End-to-End Using CNN-Based Recognizer. Most CNN-based models that recognize texts in character level have advantages when handling arbitrary-shaped texts. MaskTextSpotter [32] is a model that recognizes text using a segmentation approach. Although it has strengths in detecting and recognizing individual characters, it is difficult to train the network since character-level annotations are usually not provided in the public datasets. CharNet [44] is another segmentation-based method that makes character level predictions. The model is trained in a weakly-supervised manner to overcome the lack of character-level annotations. During training, the method performs iterative character detection to create pseudo-ground-truths.

While segmentation-based recognizers have shown great success, the method suffers when the number of target characters increases. Segmentation based models require more output channels as the number of character sets grow, and this increases memory requirements. The journal version of MaskTextSpotter [23] expands the character set to handle multiple languages, but the authors added a RNN-based decoder instead of using their initially proposed CNN-based recognizer. Another limitation of segmentation-based recognizer is the lack of contextual information in the recognition branch. Due to the absence of sequential modeling like RNNs, the accuracy of the model drops under noisy images.

TextDragon [10] is another segmentation-based method that localize and recognize text instances. However, a predicted character segment is not guaranteed to cover a single character region. To solve the issue, the model incorporates CTC to remove overlapping characters. The network shows good detection performance but shows limitations in the recognizer due to the lack of sequential modeling.

3   Methodology

3.1   Overview

Proposed CRAFTS network can be divided into three stages; detection stage, sharing stage, and recognition stage. A detailed pipeline of the network is illustrated in Fig. 2. Detection stage takes an input image and localizes oriented text boxes. Sharing stage then pools backbone high-level features and detector outputs. The pooled features are then rectified using the rectification module, and are concatenated together to form a *character attended feature*. In the recognition stage, attention-based decoder predicts text labels using the *character attended feature*. Finally, a simple post-processing technique is optionally used for better visualization.

3.2 Detection Stage

CRAFT detector [2] is selected as a base network because of its capability of representing semantic information of the character regions. The outputs of the CRAFT network represent center probability of character regions and linkage between them. We contemplate that this character centeredness information can be used to support the attention module in the recognizer since both modules aim to localize the center position of characters. In this work, we make three changes in the original CRAFT model; backbone replacement, link representation, and orientation estimation.

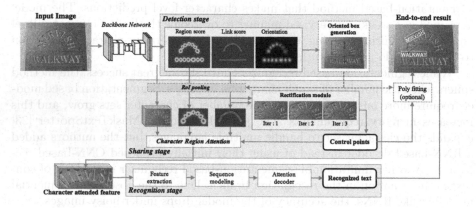

Fig. 2. Schematic overview of CRAFTS pipeline.

Backbone Replacement. Recent studies show that the use of ResNet50 captures well-defined feature representations of both the detector and the recognizer [1,30]. We therefore replace the backbone of the network from VGG-16 [40] to ResNet50 [15].

Link Representation. The occurrence of vertical texts is not common in Latin texts, but it is frequently found in East Asian languages like Chinese, Japanese, and Korean. In this work, a binary center line is used to connect the sequential character regions. This change was made because employing the original affinity maps on vertical texts often produced ill-posed perspective transformation that generated invalid box coordinates. To generate ground truth linkmap, a line segment with thickness t is drawn between adjacent characters. Here, $t = max((d_1 + d_2)/2 * \alpha, 1)$, where d_1 and d_2 are the diagonal lengths of adjacent character boxes and α is the scaling coefficient. Use of the equation lets the width of the center line proportional to the size of the characters. We set α as 0.1 in our implementation.

Orientation Estimation. It is important to obtain the right orientation of text boxes since the recognition stage requires well-defined box coordinates to recognize the text properly. To this end, we add two-channel outputs in the detection stage; channel is used to predict angles of characters along the x-axis, y-axis each. To generate the ground truth of orientation map, the upward angle of the GT character bounding box is represented as θ^*_{box}, the channel predicting x-axis has a value of $S^*_{cos}(p) = (\cos\theta + 1) \times 0.5$, and the channel predicting y-axis has a value of $S^*_{sin}(p) = (\sin\theta + 1) \times 0.5$. The ground truth orientation map is generated by filling the pixels p in the region of the word box with the values of $S^*_{cos}(p)$ and $S^*_{sin}(p)$. The trigonometric function is not directly used to let the channels have the same output range with the region map and the link map; between 0 and 1.

The loss function for orientation map is calculated by Eq. 1.

$$L_\theta = S^*_r(p) \cdot (||S_{sin}(p) - S^*_{sin}(p)||^2_2 + ||S_{cos}(p) - S^*_{cos}(p)||^2_2) \qquad (1)$$

where $S^*_{sin}(p)$ and $S^*_{cos}(p)$ denote the ground truth of text orientation. Here, the character region score $S_r(p)$ is used as a weighting factor because it represents the confidence of the character centeredness. By doing this, the orientation loss is calculated only in the positive character regions.

The final objective function in the detection stage L_{det} is defined as,

$$L_{det} = L_r + L_l + \lambda L_\theta \qquad (2)$$

where L_r and L_l denote character region loss and link loss, which are exactly same in [2]. The L_θ is the orientation loss, and is multiplied with λ to control the weight. In our experiment, we set λ to 0.1.

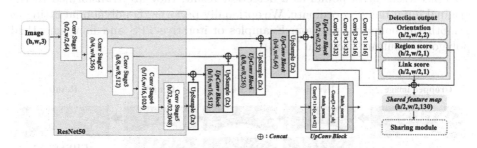

Fig. 3. Schematic illustration of the backbone network and the detection head.

The architecture of the backbone and modified detection head is illustrated in Fig. 3. The final output of the detector has four channels, each representing *character region map* S_r, *character link map* S_l, and two *orientation maps* S_{sin}, S_{cos}.

During inference, we apply the same post-processing as described in [2] to obtain text bounding boxes. First, by using predefined threshold values, we make binary maps of *character region map* S_r and *character link map* S_l. Then, using the two maps, the text blobs are constructed by using connected components labeling (CCL). The final boxes are obtained by finding a minimum bounding box enclosing each text blob. We additionally determine the orientation of the bounding box by utilizing pixel-wise averaging scheme. As shown in the Eq. 3, the angle of the text box is found by taking the arctangent of accumulated sine and cosine values at the predicted orientation map.

$$\theta_{box} = \arctan\left(\frac{\sum(S_r(p) \times (S_{sin}(p) - 0.5)}{\sum(S_r(p) \times (S_{cos}(p) - 0.5)}\right) \tag{3}$$

θ_{box} denotes orientation of the text box, S_{cos} and S_{sin} are the 2-ch orientation outputs. The same character centerdeness-based weighting scheme that used in the loss calculation is applied to predict the orientation as well.

3.3   Sharing Stage

Sharing stage consists of two modules: text rectification module and character region attention(CRA) modules. To rectify arbitrarily-shaped text region, a thin-plate spline (TPS) [37] transformation is used. Inspired by the work of [46], our rectification module incorporates iterative-TPS to acquire a better representation of the text region. By updating the control points attractively, the curved geometry of a text in an image becomes ameliorated. Through empirical studies, we discover that three TPS iterations are sufficient for rectification.

Typical TPS module takes an word image as input, but we feed the character region map and link map since they encapsulate geometric information of the text regions. We use twenty control points to tightly cover the curved text region. To use these control points as a detection result, they are transformed to the original input image coordinate. We optionally perform 2D polynomial fitting to smooth the bounding polygon. Examples of iterative-TPS and final smoothed polygon output are shown in Fig. 4.

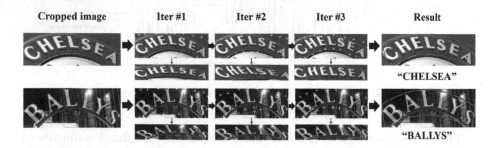

Fig. 4. Example of iterative TPS. The middle rows show TPS control points on each iteration, and the bottom row shows rectified images on each stage. The control points are drawn in the image level for better visualization. Actual rectification is done in the feature space. Final result was smoothed using a 2D polynomial.

CRA module is the key component that tightly couples detection and recognition modules. By simply concatenating rectified character score map with feature representation, the model establishes following advantages. Creating a link between detector and recognizer allows recognition loss to propagate through detection stage, and this improves the quality of character score map. Also, attaching character region map to the feature helps recognizer attend better to the character regions. Ablation study of using this module will be discussed further in the experiment section.

3.4 Recognition Stage

The modules in the recognition stage are formed based on the results reported in [1]. There are three components in the recognition stage: feature extraction, sequence modeling, and prediction. The feature extraction module is made lighter than a solitary recognizer since it takes high-level semantic features as input.

Detailed architecture of the module is shown in Table 1. After extracting the features, a bidirectional LSTM is applied for sequence modeling, and attention-based decoder makes a final text prediction.

Table 1. A simplified ResNet feature extraction module.

Layers	Configurations	Output
Input	Pooled feature	$64 \times 16 \times 130$
Block1	$\begin{bmatrix} c:256, k:3 \times 3 \\ c:256, k:3 \times 3 \end{bmatrix} \times 2$	$64 \times 16 \times 256$
Conv1	c: 256 k: 3×3	$64 \times 16 \times 256$
MaxPool	k: 2×2 s: 1×2 p: 1×0	$65 \times 8 \times 256$
Block2	$\begin{bmatrix} c:512, k:3 \times 3 \\ c:256, k:3 \times 3 \end{bmatrix} \times 5$	$65 \times 8 \times 512$
Conv2	c: 512 k: 3×3	$65 \times 8 \times 512$
Block3	$\begin{bmatrix} c:512, k:3 \times 3 \\ c:512, k:3 \times 3 \end{bmatrix} \times 3$	$65 \times 8 \times 512$
Conv3	c: 512 k: 2×2 s: 1×2 p: 1×0	$65 \times 4 \times 512$
Conv4	c: 512 k: 2×2 s: 1×1 p: 0×0	$65 \times 3 \times 512$
AvgPool	k: 1×3 s: 1×2 p: 1×0	$65 \times 1 \times 512$

At each time step, attention-based recognizer decodes textual information by masking attention outputs to the features. Although attention module works well in most cases, it fails to predict characters when attention points are misaligned or vanished [5,14]. Figure 5 shows the effect of using CRA module. Well-placed attention points allow robust text prediction.

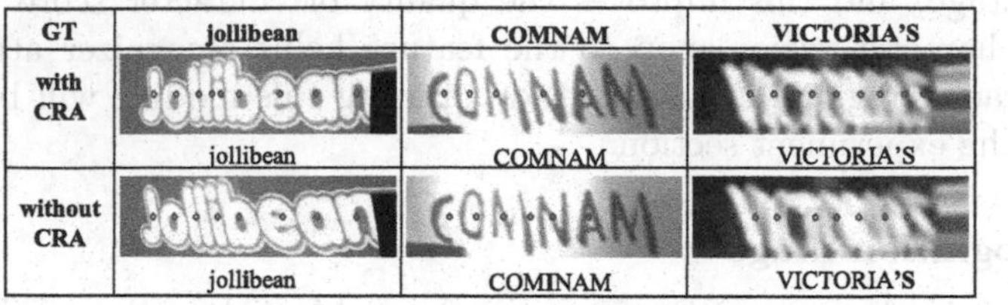

GT	jollibean	COMNAM	VICTORIA'S
with CRA	Jollibean	COMNAM	VICTORIA'S
	jollibean	COMNAM	VICTORIA'S
without CRA	Jollibean	COMNAM	VICTORIA'S
	jollibean	COMINAM	VICTORIA'S

Fig. 5. Attention problems with and without Character Region Attention module. The red dots represent attention points of the decoding characters. Missing characters are colored in blue, and misrecognized characters are colored in red. The cropped images are slightly different since generated control points in each rectification modules are inconsistent. (Color figure online)

The objective function, L_{reg}, in the recognition stage is

$$L_{reg} = -\sum_i \log p(Y_i|X_i) \tag{4}$$

where $p(Y_i|X_i)$ indicates the generation probability of the character sequence, Y_i, from the cropped feature representation, X_i of the i-th word box.

The final loss, L, used for training is composed of detection loss and recognition loss by taking $L = L_{det} + L_{reg}$. The overall flow of the recognition loss is shown in Fig. 6. The loss flows through the weights in the recognition stage, and propagates towards detection stage through *Character Region Attention* module. Detection loss on the other hand is used as an intermediate loss, and thus the weights before detection stage are updated using both detection and recognition losses.

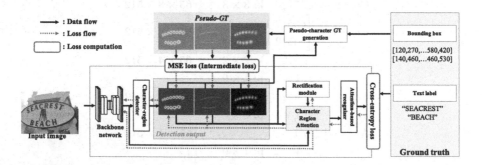

Fig. 6. The entire loss flow of CRAFTS model.

4   Experiment

4.1   Datasets

English Datasets. *IC13* [20] dataset consists of high-resolution images, 229 for training and 233 for testing. A rectangular box is used to annotate word-level text instances. *IC15* [20] consists of 1000 training and 500 testing images. A quadrilateral box is used to annotate word-level text instances. *TotalText* [7] has 1255 training and 300 testing images. Unlike IC13 and IC15 datasets, it contains curved text instances and is annotated using polygon points.

Multi-language Dataset. *IC19* [33] dataset contains 10,000 training and 10,000 testing images. The dataset contains texts in 7 different languages and is annotated using quadrilateral points.

4.2   Training Strategy

We jointly train both the detector and recognizer in the CRAFTS model. To train the detection stage, we follow the weakly-supervised training method described in [2]. The recognition loss is calculated by making a batch of randomly sampled cropped word features in each image. Maximum number of words per image is set to 16 to prevent out-of-memory error. Data augmentations in the detector apply techniques like crops, rotations, and color variations. For the recognizer, the corner points of the ground truth boxes are perturbed in a range between 0 to 10% of the shorter length of the box.

Table 2. Results on horizontal Latin datasets. * denote the results based on multi-scale tests. R, P, and H refer to recall, precision and H-mean, and S, W, and G indicate strongly-, weakly- and generic-contextualization results, respectively. The best score is highlighted in **bold**. The evaluation metric of ICDAR 2013 detection task is DetEval, and IoU metric is used for other three cases. FPS is for reference only due to the different experimental environments.

Method	IC13 (Det)			IC13 (E2E)			IC15 (Det)			IC15 (E2E)			FPS
	R	P	H	S	W	G	R	P	H	S	W	G	
Deep TextSpotter [3]	–	–	–	89	86	77	–	–	–	54	51	47	9
TextBoxes++* [24]	86	92	89	93	92	85	78.5	87.8	82.9	73.3	65.8	51.9	–
TextNet* [42]	89.1	93.6	91.3	89.7	88.8	82.9	80.8	85.7	83.2	78.6	74.9	60.4	2.7
EAA [14]	89	91	90	91	89	86	86	87	87	82	77	63	–
TextDragon [10]	–	–	–	–	–	–	83.7	92.4	87.8	82.5	78.3	65.1	–
FOTS* [29]	–	–	92.8	91.9	90.1	84.7	87.9	91.8	89.8	83.5	79.1	65.3	7.5
Li et al. [22]	80.5	91.4	85.6	92.5	91.2	84.9	–	–	–	84.4	78.9	66.1	1.3
Qin et al. [34]	–	–	–	–	–	–	87.9	91.6	89.7	**85.5**	81.9	69.9	4.7
CharNet* [44]	–	–	–	–	–	–	90.4	**92.6**	**91.5**	85.0	81.2	71.0	–
MaskTextSpotter* [23]	89.5	94.8	92.1	93.3	91.3	88.2	87.3	86.6	87.0	83.0	77.7	73.5	2.0
CRAFTS (ours)	**90.9**	**96.1**	**93.4**	**94.2**	**93.8**	**92.2**	85.3	89.0	87.1	83.1	**82.1**	**74.9**	5.4

The model is first trained on the SynthText dataset [12] for 50k iterations, and we further train the network on target datasets. Adam optimizer is used, and *On-line Hard Negative Mining (OHEM)* [39] is applied to enforce 1:3 ratio of positive and negative pixels in the detection loss. When fine-tuning the model, SynthText dataset is mixed with the ratio of 1:5. We take 94 characters to cover alphabets, numbers, and special characters, and take 4267 characters for the multi-language dataset.

4.3 Experimental Results

Horizontal Datasets (IC13, IC15). To target the IC13 benchmark, we take the model trained on the SynthText dataset and perform finetuning on IC13 and IC19 datasets. During inference, we resize the longer side of the input to 1280. The results show significant increase in performance when compared with the previous state-of-the-art works.

The model trained on IC13 dataset is then fine-tuned on the IC15 dataset. During the evaluation process, the input size of the model is set to 2560 × 1440. Note that we perform generic evaluation without the generic vocabulary set. The quantitative results on IC13 and IC15 datasets are listed in Table 2.

Our method surpasses previous methods in both generic and weakly- contextualization end-to-end tasks, and shows comparable results in other tasks. The generic performance is meaningful because a vocabulary set is not provided in practical scenarios. Note that we get slightly low detection scores on IC15 dataset and also observe low performance in strongly-contextualization results. The relatively low detection performance is obtained mainly due to the granularity difference, and will be discussed further in the later section.

Curved Datasets (TotalText). From the model trained on IC13 dataset, we further train the model on TotalText dataset. During inference, we resize the longer side of the input to 1920, and the control points from rectification module are used for detector evaluation. The qualitative results are shown in Fig. 7. The character region map and the link map are illustrated using a heatmap, and the weighted pixel-wise angle values are visualized in the HSV color space. As it is shown in the figure, the network successfully localizes polygon regions and recognizes characters in the curved text region. Two top-left figures show successful recognition of fully rotated and highly curved text instances.

Quantitative results on TotalText dataset are listed in Table 3. DetEval [7] evaluates the performance of the detector and modified IC15 evaluation scheme measures the end-to-end performance. Our method outperforms previously reported methods by a large margin. Note that even without the vocabulary set, the end-to-end result significantly exceeds the h-mean score by 8.0%.

Table 3. Results on TotalText dataset. None means no lexicon is used for contextualization. The full lexicon contains all words in the test set. * indicates the multi-scale inference, and † denotes models trained on the private datasets.

Method	Detection			E2E (None)			E2E (Full)
	R	P	H	R	P	H	H
TextDragon [10]	75.7	85.6	80.3	–	–	48.4	74.8
TextNet [42]	59.4	68.2	63.5	56.4	51.9	54.0	–
Li et al. [22]	59.8	64.8	62.2	–	–	57.8	–
MaskTextSpotter [23]	75.4	81.8	78.5	–	–	65.3	77.4
CharNet* [44]	85.0	88.0	86.5	–	–	69.2	–
Qin et al.† [34]	85.0	87.8	86.4	–	–	70.7	–
CRAFTS (ours)	**85.4**	**89.5**	**87.4**	**72.2**	**86.5**	**78.7**	–

Multi-language Dataset (IC19). Evaluation on multiple languages is performed using IC19-MLT dataset. The output channel in the prediction layer of the recognizer was expanded to 4267 to handle the characters in Arabic, Latin, Chinese, Japanese, Korean, Bangladesh, and Hindi. However, occurrence of characters in the dataset is not evenly distributed. Among 4267 characters in the training set, 1017 characters occur once in the dataset, and this insufficiency makes it hard for the model to make accurate label predictions. To solve class imbalance problem, we first freeze the weights in the detection stage and pre-train the weights in the recognizer with other publicly available multi-language datasets: SynthMLT, ArT, LSVT, ReCTS and RCTW [4,8,38,41]. We then let the loss flow through the whole network and use IC19 dataset to finetune the model. Since no paper reports performance, we compare our results with E2E-MLT [4,33]. The samples from the IC19 dataset are shown in Fig. 8. We hope our study is set as a baseline for future works on the IC19-MLT benchmark (Table 4).

Table 4. Results on IC19-MLT dataset.

Method	Detection			E2E		
	R	P	H	R	P	H
E2E-MLT [4]	–	–	–	20.5	37.4	26.5
CRAFTS (ours)	**70.1**	**81.7**	**75.5**	**48.5**	**72.9**	**58.2**

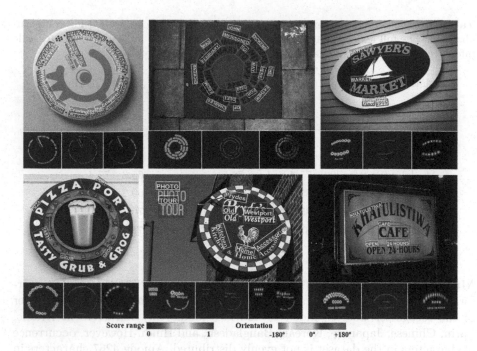

Fig. 7. Results on the TotalText dataset. First row: each column shows the input image (top) with its respective region score map, link map, and orientation map.

4.4 Ablation Study

Attention Assisted by *Character Region Attention*. In this section, we study how *Character Region Attention(CRA)* affects the performance of the recognizer by training a separate network without CRA.

Table 5 shows the effect of using CRA on benchmark datasets. Without CRA, we observe performance drops on all of the datasets. Especially on the perspective dataset (IC15) and the curved dataset (TotalText), we observe a greater

Table 5. The end-to-end performance comparisons of using character attention maps. CRA denotes the use of *Character Region Attention* in recognition stage. R, P, and H refers to Recall, Precision and Hmean values.

Dataset	Type	CRA	R	P	H	Gain
IC13	Horizontal	–	89.2	94.8	91.9	–
		✓	89.5	95.0	92.2	+0.3
IC15	Perspective	–	64.9	84.1	73.2	–
		✓	65.9	86.7	74.9	+1.7
TotalText	Curved	–	71.9	84.0	77.5	–
		✓	72.2	86.5	78.7	+1.2

Fig. 8. Qualitative results on IC19 MLT dataset.

gap when compared with the horizontal dataset (IC13). This implies that feeding character attention information improves the performance of the recognizer when dealing with irregular texts.

Recognition Loss in the Detection Stage. The recognition loss flowing through the detection stage affects the quality of the character region map and character link map. It is expected that the recognition loss helps detector localize character regions more explicitly. However, the improvement of character localization is not clearly presented in word-level evaluation. Therefore, in order to show individual character localization ability of the detector, we take advantage of the pseudo-character box generation process in the CRAFT detector. When generating pseudo-ground-truths, supervision network calculates the difference between the number of generated pseudo characters with the number of ground-truth characters in the word transcription. Table 6 shows number of *character error length* on each dataset measured with fully trained networks.

Table 6. Comparison of *character error length* on each dataset with trained networks.

Dataset	Total lengths	SynthText	w.o R-loss	w. R-loss	Diff.
ICDAR2015	46,107	9,324	1,251	1,147	−104 (−8.3%)
TotalText	53,645	16,385	5,050	4,521	−529 (−10.4%)

When training the network on SynthText dataset, *character error length* on each dataset is large. Error decreases further as training is performed on real datasets, but the value drops further by propagating recognition loss to the detection stage. This implies that the use of *Character Region Attention* improves the quality of the localization ability of the detector.

Importance of Orientation Estimation. The orientation estimation is important because there are many oriented texts in scene text images. Our pixel-wise averaging scheme is very useful for the recognizer to receive well-defined features. We compare the results of our model when the orientation information is not used. On the IC15 dataset, the performance drops from 74.9% to 74.1% (−0.8%), and on TotalText dataset, the h-mean value drops from 78.7% to 77.5% (−1.2%). The results show that the use of accurate angle information escalates performance on rotated texts.

4.5 Discussions

Inference Speed. Since inference speed varies depending on the input image size, we measure the FPS on different input resolutions, each having a longer side of 960, 1280, 1600, and 2560. The test results give FPS of 9.9, 8.3, 6.8, and 5.4, respectively. For all experiments, we use Nvidia P40 GPU with Intel(R) Xeon(R) CPU. When compared with the 8.6 FPS of the VGG based CRAFT detector [2], the ResNet based CRAFTS network achieves higher FPS on the same sized input. Also, directly using the control points from the rectification module alleviates the need of post processing for polygon generation (Fig. 9).

Fig. 9. Failure cases in IC15 dataset due to granularity difference. Red boxes denote detection results, cyan boxes denote ground truths. (Color figure online)

Granularity Difference Issue. We assume that the granularity difference between the ground-truth and prediction box causes relatively low detection performance on the IC15 dataset. Character-level segmentation methods tend to generalize character connectivity based on space and color cues, and not capture the whole feature of word instance. For this reason, the outputs do not follow the annotation style of the boxes required by the benchmark. The Fig. 9 shows the failure cases in the IC15 dataset, which proves that the detection results are marked incorrect while we observe acceptable qualitative results.

5   Conclusion

In this paper, we present an end-to-end trainable single pipeline model that tightly couples detection and recognition modules. *Character region attention* in the sharing stage fully exploit *character region map* to help recognizer rectify and attend better to the text regions. Also, we design the recognition loss propagate through detection stage and enhances the character localization ability of the detector. In addition, the rectification module in the sharing stage enables fine localization of curved texts, and obviates the need of developing hand crafted post-processing. The experimental results validate state-of-the-art performance of CRAFTS on various datasets.

References

1. Baek, J., et al.: What is wrong with scene text recognition model comparisons? Dataset and model analysis. In: The IEEE International Conference on Computer Vision (ICCV), October 2019
2. Baek, Y., Lee, B., Han, D., Yun, S., Lee, H.: Character region awareness for text detection. In: Proceedings of the IEEE Conference on Computer Vision and Pattern Recognition, pp. 9365–9374 (2019)
3. Busta, M., Neumann, L., Matas, J.: Deep TextSpotter: an end-to-end trainable scene text localization and recognition framework. In: Proceedings of the IEEE International Conference on Computer Vision, pp. 2204–2212 (2017)
4. Bušta, M., Patel, Y., Matas, J.: E2E-MLT - an unconstrained end-to-end method for multi-language scene text. In: Carneiro, G., You, S. (eds.) ACCV 2018. LNCS, vol. 11367, pp. 127–143. Springer, Cham (2019). https://doi.org/10.1007/978-3-030-21074-8_11
5. Cheng, Z., Bai, F., Xu, Y., Zheng, G., Pu, S., Zhou, S.: Focusing attention: towards accurate text recognition in natural images. In: Proceedings of the IEEE International Conference on Computer Vision, pp. 5076–5084 (2017)
6. Cheng, Z., Xu, Y., Bai, F., Niu, Y., Pu, S., Zhou, S.: AON: towards arbitrarily-oriented text recognition. In: Proceedings of the IEEE Conference on Computer Vision and Pattern Recognition, pp. 5571–5579 (2018)
7. Ch'ng, C.K., Chan, C.S.: Total-text: a comprehensive dataset for scene text detection and recognition. In: ICDAR, vol. 1, pp. 935–942. IEEE (2017)
8. Chng, C.K., et al.: ICDAR 2019 robust reading challenge on arbitrary-shaped text-RRC-ArT. In: 2019 International Conference on Document Analysis and Recognition (ICDAR), pp. 1571–1576. IEEE (2019)
9. Deng, D., Liu, H., Li, X., Cai, D.: PixelLink: detecting scene text via instance segmentation. In: Thirty-Second AAAI Conference on Artificial Intelligence (2018)
10. Feng, W., He, W., Yin, F., Zhang, X.Y., Liu, C.L.: TextDragon: an end-to-end framework for arbitrary shaped text spotting. In: Proceedings of the IEEE International Conference on Computer Vision, pp. 9076–9085 (2019)
11. Gao, Y., Chen, Y., Wang, J., Lei, Z., Zhang, X.Y., Lu, H.: Recurrent calibration network for irregular text recognition. arXiv preprint arXiv:1812.07145 (2018)
12. Gupta, A., Vedaldi, A., Zisserman, A.: Synthetic data for text localisation in natural images. In: CVPR, pp. 2315–2324 (2016)
13. He, K., Gkioxari, G., Dollár, P., Girshick, R.: Mask R-CNN. In: ICCV, pp. 2980–2988. IEEE (2017)

14. He, T., Tian, Z., Huang, W., Shen, C., Qiao, Y., Sun, C.: An end-to-end textspotter with explicit alignment and attention. In: CVPR, pp. 5020–5029 (2018)
15. He, W., Zhang, X.Y., Yin, F., Liu, C.L.: Deep direct regression for multi-oriented scene text detection. In: CVPR, pp. 745–753 (2017)
16. Hu, H., Zhang, C., Luo, Y., Wang, Y., Han, J., Ding, E.: WordSup: exploiting word annotations for character based text detection. In: ICCV (2017)
17. Huang, Z., Zhong, Z., Sun, L., Huo, Q.: Mask R-CNN with pyramid attention network for scene text detection. In: 2019 IEEE Winter Conference on Applications of Computer Vision (WACV), pp. 764–772. IEEE (2019)
18. Jaderberg, M., Simonyan, K., Zisserman, A., et al.: Spatial transformer networks. In: Advances in Neural Information Processing Systems, pp. 2017–2025 (2015)
19. Karatzas, D., et al.: ICDAR 2015 competition on robust reading. In: ICDAR, pp. 1156–1160. IEEE (2015)
20. Karatzas, D., et al.: ICDAR 2013 robust reading competition. In: ICDAR, pp. 1484–1493. IEEE (2013)
21. Lee, C.Y., Osindero, S.: Recursive recurrent nets with attention modeling for OCR in the wild. In: Proceedings of the IEEE Conference on Computer Vision and Pattern Recognition, pp. 2231–2239 (2016)
22. Li, H., Wang, P., Shen, C.: Towards end-to-end text spotting in natural scenes. arXiv preprint arXiv:1906.06013 (2019)
23. Liao, M., Lyu, P., He, M., Yao, C., Wu, W., Bai, X.: Mask TextSpotter: an end-to-end trainable neural network for spotting text with arbitrary shapes. IEEE Trans. Pattern Anal. Mach. Intell. (2019)
24. Liao, M., Shi, B., Bai, X.: TextBoxes++: a single-shot oriented scene text detector. IEEE Trans. Image Process. **27**(8), 3676–3690 (2018)
25. Liao, M., Zhu, Z., Shi, B., Xia, G.s., Bai, X.: Rotation-sensitive regression for oriented scene text detection. In: CVPR, pp. 5909–5918 (2018)
26. Liu, J., Liu, X., Sheng, J., Liang, D., Li, X., Liu, Q.: Pyramid mask text detector. arXiv preprint arXiv:1903.11800 (2019)
27. Liu, W., Chen, C., Wong, K.Y.K.: Char-Net: a character-aware neural network for distorted scene text recognition. In: Thirty-Second AAAI Conference on Artificial Intelligence (2018)
28. Liu, W., Chen, C., Wong, K.Y.K., Su, Z., Han, J.: STAR-Net: a spatial attention residue network for scene text recognition. In: BMVC, vol. 2, p. 7 (2016)
29. Liu, X., Liang, D., Yan, S., Chen, D., Qiao, Y., Yan, J.: FOTS: fast oriented text spotting with a unified network. In: CVPR, pp. 5676–5685 (2018)
30. Long, S., He, X., Ya, C.: Scene text detection and recognition: the deep learning era. arXiv preprint arXiv:1811.04256 (2018)
31. Long, S., Ruan, J., Zhang, W., He, X., Wu, W., Yao, C.: TextSnake: a flexible representation for detecting text of arbitrary shapes. In: Ferrari, V., Hebert, M., Sminchisescu, C., Weiss, Y. (eds.) ECCV 2018. LNCS, vol. 11206, pp. 19–35. Springer, Cham (2018). https://doi.org/10.1007/978-3-030-01216-8_2
32. Lyu, P., Liao, M., Yao, C., Wu, W., Bai, X.: Mask TextSpotter: an end-to-end trainable neural network for spotting text with arbitrary shapes. In: Ferrari, V., Hebert, M., Sminchisescu, C., Weiss, Y. (eds.) Computer Vision – ECCV 2018. LNCS, vol. 11218, pp. 71–88. Springer, Cham (2018). https://doi.org/10.1007/978-3-030-01264-9_5
33. Nayef, N., et al.: ICDAR 2019 robust reading challenge on multi-lingual scene text detection and recognition–RRC-MLT-2019. In: 2019 International Conference on Document Analysis and Recognition (ICDAR), pp. 1582–1587. IEEE (2019)

34. Qin, S., Bissacco, A., Raptis, M., Fujii, Y., Xiao, Y.: Towards unconstrained end-to-end text spotting. In: Proceedings of the IEEE International Conference on Computer Vision, pp. 4704–4714 (2019)
35. Shi, B., Bai, X., Yao, C.: An end-to-end trainable neural network for image-based sequence recognition and its application to scene text recognition. IEEE Trans. Pattern Anal. Mach. Intell. **39**(11), 2298–2304 (2016)
36. Shi, B., Wang, X., Lyu, P., Yao, C., Bai, X.: Robust scene text recognition with automatic rectification. In: Proceedings of the IEEE Conference on Computer Vision and Pattern Recognition, pp. 4168–4176 (2016)
37. Shi, B., Yang, M., Wang, X., Lyu, P., Yao, C., Bai, X.: ASTER: an attentional scene text recognizer with flexible rectification. IEEE Trans. Pattern Anal. Mach. Intell. **41**(9), 2035–2048 (2018)
38. Shi, B., et al.: ICDAR 2017 competition on reading Chinese text in the wild (RCTW-17). In: 2017 14th IAPR International Conference on Document Analysis and Recognition (ICDAR), vol. 1, pp. 1429–1434. IEEE (2017)
39. Shrivastava, A., Gupta, A., Girshick, R.: Training region-based object detectors with online hard example mining. In: CVPR, pp. 761–769 (2016)
40. Simonyan, K., Zisserman, A.: Very deep convolutional networks for large-scale image recognition. In: ICLR (2015)
41. Sun, Y., et al.: ICDAR 2019 competition on large-scale street view text with partial labeling-RRC-LSVT. In: 2019 International Conference on Document Analysis and Recognition (ICDAR), pp. 1557–1562. IEEE (2019)
42. Sun, Y., Zhang, C., Huang, Z., Liu, J., Han, J., Ding, E.: TextNet: irregular text reading from images with an end-to-end trainable network. In: Jawahar, C.V., Li, H., Mori, G., Schindler, K. (eds.) ACCV 2018. LNCS, vol. 11363, pp. 83–99. Springer, Cham (2019). https://doi.org/10.1007/978-3-030-20893-6_6
43. Wang, W., et al.: Shape robust text detection with progressive scale expansion network. In: Proceedings of the IEEE Conference on Computer Vision and Pattern Recognition, pp. 9336–9345 (2019)
44. Xing, L., Tian, Z., Huang, W., Scott, M.R.: Convolutional character networks. In: Proceedings of the IEEE International Conference on Computer Vision, pp. 9126–9136 (2019)
45. Xu, Y., Wang, Y., Zhou, W., Wang, Y., Yang, Z., Bai, X.: TextField: learning a deep direction field for irregular scene text detection. IEEE Trans. Image Process. **28**, 5566–5579 (2019)
46. Zhan, F., Lu, S.: ESIR: end-to-end scene text recognition via iterative image rectification. In: Proceedings of the IEEE Conference on Computer Vision and Pattern Recognition, pp. 2059–2068 (2019)
47. Zhang, C., et al.: Look more than once: an accurate detector for text of arbitrary shapes. In: Proceedings of the IEEE Conference on Computer Vision and Pattern Recognition, pp. 10552–10561 (2019)
48. Zhong, Z., Sun, L., Huo, Q.: An anchor-free region proposal network for faster R-CNN-based text detection approaches. Int. J. Doc. Anal. Recogn. (IJDAR) **22**(3), 315–327 (2019)
49. Zhou, X., et al.: EAST: an efficient and accurate scene text detector. In: CVPR, pp. 2642–2651 (2017)

Stable Low-Rank Tensor Decomposition for Compression of Convolutional Neural Network

Anh-Huy Phan[1(✉)], Konstantin Sobolev[1], Konstantin Sozykin[1], Dmitry Ermilov[1], Julia Gusak[1], Petr Tichavský[2], Valeriy Glukhov[3], Ivan Oseledets[1], and Andrzej Cichocki[1]

[1] Skolkovo Institute of Science and Technology (Skoltech), Moscow, Russia
{a.phan,konstantin.sobolev,konstantin.sozykin,dmitrii.ermilov,y.gusak,
i.oseledets,a.cichocki}@skoltech.ru
[2] The Czech Academy of Sciences, Institute of Information Theory and Automation,
Prague, Czech Republic
tichavsk@utia.cas.cz
[3] Noah's Ark Lab, Huawei Technologies, Shenzhen, China
glukhov.valery@huawei.com

Abstract. Most state-of-the-art deep neural networks are overparameterized and exhibit a high computational cost. A straightforward approach to this problem is to replace convolutional kernels with its low-rank tensor approximations, whereas the Canonical Polyadic tensor Decomposition is one of the most suited models. However, fitting the convolutional tensors by numerical optimization algorithms often encounters diverging components, i.e., extremely large rank-one tensors but canceling each other. Such degeneracy often causes the non-interpretable result and numerical instability for the neural network ne-tuning. This paper is the first study on degeneracy in the tensor decomposition of convolutional kernels. We present a novel method, which can stabilize the low-rank approximation of convolutional kernels and ensure efficient compression while preserving the high-quality performance of the neural networks. We evaluate our approach on popular CNN architectures for image classification and show that our method results in much lower accuracy degradation and provides consistent performance.

Keywords: Convolutional neural network acceleration · Low-rank tensor decomposition · Sensitivity · Degeneracy correction

1 Introduction

Convolutional neural networks (CNNs) and their recent extensions have significantly increased their ability to solve complex computer vision tasks, such as image classification, object detection, instance segmentation, image generation, etc. Together with big data and fast development of the internet of things, CNNs bring new tools for solving computer science problems, which are intractable using classical approaches.

Despite the great successes and rapid development of CNNs, most modern neural network architectures contain a huge number of parameters in the convolutional

© Springer Nature Switzerland AG 2020
A. Vedaldi et al. (Eds.): ECCV 2020, LNCS 12374, pp. 522–539, 2020.
https://doi.org/10.1007/978-3-030-58526-6_31

and fully connected layers, therefore, demand extremely high computational costs [46], which makes them difficult to deploy on devices with limited computing resources, like PC or mobile devices. Common approaches to reduce redundancy of the neural network parameters are: structural pruning [13,20,21,59], sparsification [12,15,36], quantization [2,44] and low-rank approximation [4,10,14,26,28,33].

The weights of convolutional and fully connected layers are usually overparameterized and known to lie on a low-rank subspace [9]. Hence, it is possible to represent them in low-rank tensor/tensor network formats using e.g., Canonical Polyadic decomposition (CPD) [1,10,33], Tucker decomposition [14,26], Tensor Train decomposition [37,55]. The decomposed layers are represented by a sequence of new layers with much smaller kernel sizes, therefore, reducing the number of parameters and computational cost in the original model.

Various low-rank tensor/matrix decompositions can be straightforwardly applied to compress the kernels. This article intends to promote the simplest tensor decomposition model, the Canonical Polyadic decomposition (CPD).

1.1 Why CPD

In neural network models working with images, the convolutional kernels are usually tensors of order 4 with severely unbalanced dimensions, e.g., $D \times D \times S \times T$, where $D \times D$ represents the filter sizes, S and T denote the number of input and output channels, respectively. The typical convolutional filters are often of relatively small sizes, e.g., 3×3, 7×7, compared to the input (S) and output (T) dimensions, which in total may have hundred of thousands of filters. This leads to excessive redundancy among the kernel filters, which are particularly suited for tensor decomposition methods. Among low-rank tensor decomposition and tensor networks, the Canonical Polyadic tensor decomposition [17,22] is the simplest and elegant model, which represents a tensor by sum of rank-1 tensors[1] or equivalently by factor matrices interconnected through a diagonal tensor (Fig. 1a). The number of parameters for a CP model of rank-R is $R(2D + S + T)$ or $R(D^2 + S + T)$ when we consider kernels as order-4 tensors or their reshaped order-3 versions, respectively. Usually, CPD gains a relatively high compression ratio since the decomposition rank is not very large [14,33].

Representation of the high order convolutional kernels in the form of the CP model is equivalent to the use of separable convolutions. In [28], the authors modeled the high order kernels in the generalized multiway convolution by the CP model.

The Tucker tensor decomposition (TKD) [52] is an alternative tensor decomposition method for convolutional kernel compression [26]. The TKD provides more flexible interaction between the factor matrices through a core tensor, which is often dense in practice (Fig. 1b). Kim et al. [26] investigated low-rank models at the most suited noise level for different unfoldings[2] of the kernel tensor. This heuristic method does not consider a common noise level for multi modes and is not optimal to attain the approximation error bound.

---

[1] Rank-1 tensor of size $n_1 \times n_2 \times \cdots \times n_d$ is an outer product of d vectors with dimensions $n_1, n_1, \ldots, n_d$.

[2] The mode-j unfolding of an order-d tensor of size $n_1 \times n_2 \times \cdots \times n_d$ reorders the elements of the tensor into a matrix with n_j rows and $n_1 \ldots n_{j-1} n_{j+1} \ldots n_d$ columns.

Block tensor decomposition [6] is an extension of the TKD, which models data as the sum of several Tucker or Kruskal terms, i.e., a TKD with block-diagonal core tensor. For the same multilinear rank as in TKD, BTD exhibits a smaller number of parameters; however, there are no available proper criteria for block size selection (rank of BTD).

In addition, the other tensor networks, e.g., Tensor Train [38] or Tensor Chain (TC) [11,25], are not applicable unless the kernel filters are tensorized to higher orders. Besides, the Tensor Chain contains a loop, is not closed and leads to severe numerical instability to nd the best approximation, see Theorem 14.1.2.2 [16,31].

We later show that CPD can achieve much better performance with an even higher compression ratio by further compression the Tucker core tensors by solving a suitably formulated optimization problem.

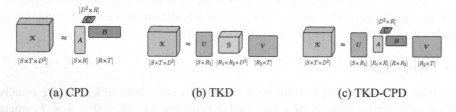

(a) CPD   (b) TKD   (c) TKD-CPD

Fig. 1. Approximation of a third-order tensor using Canonical Polyadic tensor decomposition (CPD), Tucker-2 tensor decomposition (TKD), and their combination (TKD-CPD). CPD and TKD are common methods applied for CNN compression.

1.2 Why Not Standard CPD

In one of the first works applying CPD to convolutional kernels, Denton et al. [10] computed the CPD by sequentially extracting the best rank-1 approximation in a greedy way. This type of deflation procedure is not a proper way to compute CPD unless decomposition of orthogonally decomposable tensors [57] or with a strong assumption, e.g., at least two factor matrices are linearly independent, and the tensor rank must not exceed any dimension of the tensor [41]. The reason is that subtracting the best rank-1 tensor does not guarantee to decrease the rank of the tensor [49].

In [33], the authors approximated the convolution kernel tensors using the Nonlinear Least Squares (NLS) algorithm [54], one of the best existing algorithms for CPD. However, as mentioned in the Ph.D. thesis [32], it is not trivial to optimize a neural network even when weights from a single layer are factorized, and the authors "failed to find a good SGD learning rate" with fine-tuning a classification model on the ILSVRC-12 dataset.

Diverging Component - Degeneracy. Common phenomena when using numerical optimization algorithms to approximate a tensor of relatively high rank by a low-rank model or a tensor, which has nonunique CPD, is that there should exist at least two rank-one tensors such that their Frobenius norms or intensities are relatively high but cancel each other [47], $\|\mathbf{a}_r^{(1)} \circ \mathbf{a}_r^{(2)} \circ \cdots \circ \mathbf{a}_r^{(d)}\|_F \to \infty$.

The degeneracy of CPD is reported in the literature, e.g., in [5,18,29,35,39,45]. Some efforts which impose additional constraints on the factor matrices can improve stability and accelerate convergence, such as, column-wise orthogonality [29,45], positivity or nonnegativity [34]. However, the constraints are not always applicable in some data, and thus prevent the estimator from getting lower approximation error, yielding to the trade-off between estimation stability and good approximation error.[3]

We have applied CPD approximations for various CNNs and confirm that the diverging component occurs for most cases when we used either Alternating Least Squares (ALS) or NLS [54] algorithm. As an example, we approximated one of the last convolutional layers from ResNet-18 with rank-500 CPD and plotted in Fig. 2(left) intensities of CPD components, i.e., Frobenius norm of rank-1 tensors. The ratio between the largest and smallest intensities of rank-1 tensors was greater than 30. Figure 2(right) shows that the sum of squares of intensities for CPD components is (exponentially) higher when the decomposition is with a higher number of components. Another criterion, sensitivity (Definition 1), shows that the standard CPD algorithms are not robust to small perturbations of factor matrices, and sensitivity increases with higher CP rank.

Such degeneracy causes the instability issue when training a CNN with decomposed layers in the CP (or Kruskal) format. More specifically, it causes difficulty for a neural network to perform fine-tuning, selecting a good set of parameters, and maintaining stability in the entire network. This problem has not been investigated thoroughly. To the best of our knowledge, there is no method for handling this problem.

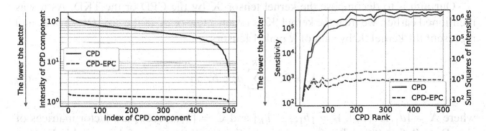

Fig. 2. (Left) Intensity (Frobenius norm) of rank-1 tensors in CPDs of the kernel in the 4th layer of ResNet-18. (Right) Sum of squares of the intensity and Sensitivity vs Rank of CPD. EPC-CPD demonstrates much lower intensity and sensitivity as compared to CPD.

1.3 Contributions

In this paper, we address the problem of CNN stability compressed by CPD. The key advantages and major contributions of our paper are the following:

- We propose a new stable and efficient method to perform neural network compression based on low-rank tensor decompositions.

---

[3] As shown in [53], RMS error is not the only one minimization criterion for a particular computer vision task.

- We demonstrate how to deal with the degeneracy, the most severe problem when approximating convolutional kernels with CPD. Our approach allows finding CPD a reliable representation with minimal sensitivity and intensity.
- We show that the combination of Tucker-2 (TKD) and the proposed stable CPD (Fig. 1c) outperforms CPD in terms of accuracy/compression trade-off.

We provide results of extensive experiments to confirm the efficiency of the proposed algorithms. Particularly, we empirically show that the neural network with weights in factorized CP format obtained using our algorithms is more stable during fine-tuning and recovers faster (close) to initial accuracy.

2   Stable Tensor Decomposition Method

2.1   CP Decomposition of Convolutional Kernel

In CNNs, the convolutional layer performs mapping of an input (source) tensor $\mathcal{X}$ of size $H \times W \times S$ into output (target) tensor $\mathcal{Y}$ of size $H' \times W' \times T$ following the relation

$$\mathcal{Y}_{h',w',t} = \sum_{i=1}^{D} \sum_{j=1}^{D} \sum_{s=1}^{S} \tilde{\mathcal{K}}_{i,j,s,t} \mathcal{X}_{h_i,w_j,s}, \tag{1}$$

where $h_i = (h' - 1)\varDelta + i - P$, and $w_j = (w' - 1)\varDelta + j - P$, and $\tilde{\mathcal{K}}$ is an order-4 kernel tensor of size $D \times D \times S \times T$, $\varDelta$ is stride, and P is zero-padding size.

Our aim is to decompose the kernel tensor $\tilde{\mathcal{K}}$ by the CPD or the TKD. As it was mentioned earlier, we treat the kernel $\tilde{\mathcal{K}}$ as order-3 tensor $\mathcal{K}$ of the size $D^2 \times S \times T$, and represent the kernel $\mathcal{K}$ by sum of R rank-1 tensors

$$\mathcal{K} \simeq \hat{\mathcal{K}} = \sum_{r=1}^{R} a_r \circ b_r \circ c_r, \tag{2}$$

where $\mathbf{A} = [a_1, \ldots, a_R]$, $\mathbf{B} = [b_1, \ldots, b_R]$ and $\mathbf{C} = [c_1, \ldots, c_R]$ are factor matrices of size $D^2 \times R$, $S \times R$ and $T \times R$, respectively. See an illustration of the model in Fig. 1a. The tensor $\hat{\mathcal{K}} = [\![\mathbf{A}, \mathbf{B}, \mathbf{C}]\!]$ in the Kruskal format uses $(D^2 + S + T) \times R$ parameters.

2.2   Degeneracy and Its Effect to CNN Stability

Degeneracy occurs in most CPD of the convolutional kernels. The Error Preserving Correction (EPC) method [42] suggests a correction to the decomposition results in order to get a more stable decomposition with lower sensitivity. There are two possible measures for assessment of the degeneracy degree of the CPD: sum of Frobenius norms of the rank-1 tensors [42]

$$\mathrm{sn}([\![\mathbf{A}, \mathbf{B}, \mathbf{C}]\!]) = \sum_{r=1}^{R} \|a_r \circ b_r \circ c_r\|_F^2 \tag{3}$$

and sensitivity, defined as follows.

Definition 1 (Sensitivity [51]). *Given a tensor* $\mathcal{T} = [\![\mathbf{A}, \mathbf{B}, \mathbf{C}]\!]$, *define the sensitivity as*

$$\text{ss}([\![\mathbf{A}, \mathbf{B}, \mathbf{C}]\!]) = \lim_{\sigma^2 \to 0} \frac{1}{R\sigma^2} E\{\|\mathcal{T} - [\![\mathbf{A} + \delta\mathbf{A}, \mathbf{B} + \delta\mathbf{B}, \mathbf{C} + \delta\mathbf{C}]\!]\|_F^2\} \tag{4}$$

where $\delta\mathbf{A}$, $\delta\mathbf{B}$, $\delta\mathbf{C}$ *have random i.i.d. elements from* $N(0, \sigma^2)$.

The sensitivity of the decomposition can be measured by the expectation ($E\{\cdot\}$) of the normalized squared Frobenius norm of the difference. In other words, sensitivity of the tensor $\mathcal{T} = [\![\mathbf{A}, \mathbf{B}, \mathbf{C}]\!]$ is a measure with respect to perturbations in individual factor matrices. CPDs with high sensitivity are usually useless.

Lemma 1.

$$\text{ss}([\![\mathbf{A}, \mathbf{B}, \mathbf{C}]\!]) = \text{tr}\{(\mathbf{A}^T\mathbf{A}) \circledast (\mathbf{B}^T\mathbf{B}) + (\mathbf{B}^T\mathbf{B}) \circledast (\mathbf{C}^T\mathbf{C}) + (\mathbf{A}^T\mathbf{A}) \circledast (\mathbf{C}^T\mathbf{C})\}. \tag{5}$$

where $\circledast$ *denotes the Hadamard element-wise product.*

Proof. First, the perturbed tensor in (4) can be expressed as sum of 8 Kruskal terms

$$[\![\mathbf{A} + \delta\mathbf{A}, \mathbf{B} + \delta\mathbf{B}, \mathbf{C} + \delta\mathbf{C}]\!] = [\![\mathbf{A}, \mathbf{B}, \mathbf{C}]\!] + [\![\delta\mathbf{A}, \mathbf{B}, \mathbf{C}]\!] + [\![\mathbf{A}, \delta\mathbf{B}, \mathbf{C}]\!] + [\![\mathbf{A}, \mathbf{B}, \delta\mathbf{C}]\!]$$
$$+ [\![\delta\mathbf{A}, \delta\mathbf{B}, \mathbf{C}]\!] + [\![\delta\mathbf{A}, \mathbf{B}, \delta\mathbf{C}]\!] + [\![\mathbf{A}, \delta\mathbf{B}, \delta\mathbf{C}]\!] + [\![\delta\mathbf{A}, \delta\mathbf{B}, \delta\mathbf{C}]\!].$$

Since these Kruskal terms are uncorrelated and expectation of the terms composed by two or three factor matrices $\delta\mathbf{A}$, $\delta\mathbf{B}$ and $\delta\mathbf{C}$ are negligible, the expectation in (4) can be expressed in the form

$$E\{\|\mathcal{T} - [\![\mathbf{A} + \delta\mathbf{A}, \mathbf{B} + \delta\mathbf{B}, \mathbf{C} + \delta\mathbf{C}]\!]\|_F^2\} = E\{\|[\![\delta\mathbf{A}, \mathbf{B}, \mathbf{C}]\!]\|_F^2\}$$
$$+ E\{\|[\![\mathbf{A}, \delta\mathbf{B}, \mathbf{C}]\!]\|_F^2\} + E\{\|[\![\mathbf{A}, \mathbf{B}, \delta\mathbf{C}]\!]\|_F^2\}. \tag{6}$$

Next we expand the Frobenius norm of the three Kruskal tensors

$$\begin{aligned} E\{\|[\![\delta\mathbf{A}, \mathbf{B}, \mathbf{C}]\!]\|_F^2\} &= E\{\|\left((\mathbf{C} \odot \mathbf{B}) \otimes \mathbf{I}\right) \text{vec}(\delta\mathbf{A})\|^2\} \\ &= E\{\text{tr}((\mathbf{C} \odot \mathbf{B}) \otimes \mathbf{I})^T \left((\mathbf{C} \odot \mathbf{B}) \otimes \mathbf{I}\right) \text{vec}(\delta\mathbf{A}) \text{vec}(\delta\mathbf{A})^T)\} \\ &= \sigma^2 \text{tr}((\mathbf{C} \odot \mathbf{B})^T (\mathbf{C} \odot \mathbf{B}) \otimes \mathbf{I}) \\ &= R\sigma^2 \text{tr}((\mathbf{C}^T\mathbf{C}) \circledast (\mathbf{B}^T\mathbf{B})) \end{aligned} \tag{7}$$

$$E\{\|[\![\mathbf{A}, \delta\mathbf{B}, \mathbf{C}]\!]\|_F^2\} = R\sigma^2 \text{tr}((\mathbf{C}^T\mathbf{C}) \circledast (\mathbf{A}^T\mathbf{A})) \tag{8}$$

$$E\{\|[\![\mathbf{A}, \mathbf{B}, \delta\mathbf{C}]\!]\|_F^2\} = R\sigma^2 \text{tr}((\mathbf{B}^T\mathbf{B}) \circledast (\mathbf{A}^T\mathbf{A})) \tag{9}$$

where $\odot$ and $\otimes$ are Khatri-Rao and Kronecker products, respectively.

Finally, we replace these above expressions into (6) to obtain the compact expression of sensitivity.

2.3   Stabilization Method

Sensitivity Minimization. The first method to correct CPD with diverging components proposed in [42] minimizes the sum of Frobenius norms of rank-1 tensors while the approximation error is bounded. In [51]. the Krylov Levenberg-Marquardt algorithm was proposed for the CPD with bounded sensitivity constraint.

In this paper, we propose a variant of the EPC method which minimizes the sensitivity of the decomposition while preserving the approximation error, i.e.,

$$\min_{\{A,B,C\}} \quad \text{ss}([\![A, B, C]\!]) \tag{10}$$
$$\text{s.t.} \quad \|\mathcal{K} - [\![A, B, C]\!]\|_F^2 \le \delta^2.$$

The bound, δ^2, can represent the approximation error of the decomposition with diverging components. Continuing the CPD using a new tensor $\hat{\mathcal{K}} = [\![A, B, C]\!]$ with a lower sensitivity can improve its convergence.

Update Rules. We derive alternating update formulas for the above optimization problem. While B and C are kept fixed, the objective function is rewritten to update A as

$$\min_{A} \quad \text{tr}\{(A^T A) \circledast W\} = \|A \, \text{diag}(w)\|_F^2 \tag{11}$$
$$\text{s.t.} \quad \|K_{(1)} - AZ^T\|_F^2 \le \delta^2,$$

where $K_{(1)}$ is mode-1 unfolding of the kernel tensor $\mathcal{K}$, $Z = C \odot B$ and $W = B^T B + C^T C$ is a symmetric matrix of size $R \times R$, $w = [\sqrt{w_{1,1}}, \ldots, \sqrt{w_{R,R}}]$ is a vector of length R taken from the diagonal of W.

Remark 1. The problem (11) can be reformulated as a regression problem with bound constraint

$$\min_{\widetilde{A}} \quad \|\widetilde{A}\|_F^2 \tag{12}$$
$$\text{s.t.} \quad \|K_{(1)} - \widetilde{A}\widetilde{Z}^T\|_F^2 \le \delta^2,$$

where $\widetilde{A} = A \, \text{diag}(w)$ and $\widetilde{Z} = Z \, \text{diag}(w^{-1})$. This problem can be solved in closed form solution through the quadratic programming over a sphere [43]. We skip the algorithm details and refer to the solver in [43].

Remark 2. If factor matrices B and C are normalized to unit length columns, i.e., $\|b_r\|_2 = \|c_r\|_2 = 1$, $r = 1, \ldots, R$, then all entries of the diagonal of W are identical. The optimization problem in (11) becomes seeking a weight matrix, A, with minimal norm

$$\min_{A} \quad \|A\|_F^2 \tag{13}$$
$$\text{s.t.} \quad \|K_{(1)} - AZ^T\|_F^2 \le \delta^2.$$

This sub-optimization problem is similar to that in the EPC approach [42].

2.4   Tucker Decomposition with Bound Constraint

Another well-known representation of multi-way data is the Tucker Decomposition [52], which decomposes a given tensor into a core tensor and a set of factor matrices (see Fig. 1b for illustration). The Tucker decomposition is particularly suited as prior-compression for CPD. In this case, we compute CPD of the core tensor in TKD, which is of smaller dimensions than the original kernels.

For our problem, we are interested in the Tucker-2 model (see Fig. 1b)

$$\mathcal{K} \simeq \mathcal{G} \times_2 \mathbf{U} \times_3 \mathbf{V}, \tag{14}$$

where $\mathcal{G}$ is the core tensor of size $D^2 \times R_1 \times R_2$, $\mathbf{U}$ and $\mathbf{V}$ are matrices of size $S \times R_1$ and $T \times R_2$, respectively. Because of rotational ambiguity, without loss in generality, the matrices $\mathbf{U}$ and $\mathbf{V}$ can be assumed to have orthonormal columns.

Different from the ordinary TK-2, we seek the smallest TK-2 model which holds the approximation error bound δ^2 [40], i.e.,

$$\min_{\{\mathcal{G},\mathbf{U},\mathbf{V}\}} \quad R_1 S + R_2 T + R_1 R_2 D^2 \tag{15}$$

$$\text{s.t.} \quad \|\mathcal{K} - \mathcal{G} \times_2 \mathbf{U} \times_3 \mathbf{V}\|_F^2 \leq \delta^2$$

$$\mathbf{U}^T \mathbf{U} = \mathbf{I}_{R_1}, \mathbf{V}^T \mathbf{V} = \mathbf{I}_{R_2} .$$

We will show that the core tensor $\mathcal{G}$ has closed-form expression as in the HOOI algorithm for the orthogonal Tucker decomposition [7], and the two-factor matrices, $\mathbf{U}$ and $\mathbf{V}$, can be *sequentially estimated* through Eigenvalue decomposition (EVD).

Lemma 2. *The core tensor $\mathcal{G}$ has closed-form expression* $\mathcal{G}^\star = \mathcal{K} \times_2 \mathbf{U}^T \times_3 \mathbf{V}^T$.

Proof. From the error bound condition, we can derive

$$\delta^2 \geq \|\mathcal{K} - \mathcal{G} \times_2 \mathbf{U} \times_3 \mathbf{V}\|_F^2 = \|\mathcal{K}\|_F^2 - \|\mathcal{G}^\star\|_F^2 + \|\mathcal{G} - \mathcal{G}^\star\|_F^2,$$

which indicates that the core tensor can be expressed as $\mathcal{G} = \mathcal{G}^\star + \mathcal{E}$, where $\mathcal{E}$ is an error tensor such that its norm $\gamma^2 = \|\mathcal{E}\|_F^2 \leq \delta^2 + \|\mathcal{G}^\star\|_F^2 - \|\mathcal{K}\|_F^2$.

Next define a matrix $\mathbf{Q}_1$ of size $S \times S$

$$\mathbf{Q}_1(i, j) = \sum_{r=1}^{R_2} \mathbf{V}(:, r)^T \mathbf{K}(:, i, :)\mathbf{K}(:, j, :)^T \mathbf{V}(:, r). \tag{16}$$

Assume that $\mathbf{V}^\star$ is the optimal factor matrix with the minimal rank $R_2^\star$. The optimization in (15) becomes the rank minimization problem for $\mathbf{U}$

$$\min_{\mathbf{U}} \quad \text{rank}(\mathbf{U}) \tag{17}$$

$$\text{s.t.} \quad \text{tr}(\mathbf{U}^T \mathbf{Q}_1 \mathbf{U}) \geq \|\mathcal{K}\|_F^2 + \gamma^2 - \delta^2 ,$$

$$\mathbf{U}^T \mathbf{U} = \mathbf{I}_{R_1} .$$

The optimal factor matrix $\mathbf{U}^\star$ comprises R_1 principal eigenvectors of $\mathbf{Q}_1$, where R_1 is the smallest number of eigenvalues, $\lambda_1 \geq \lambda_2 \geq \cdots \geq \lambda_{R_1}$ such that their norm exceeds the bound $\|\mathcal{Y}\|_F^2 - \delta^2 + \gamma^2$, that is, $\sum_{r=1}^{R_1} \lambda_r \geq \|\mathcal{K}\|_F^2 - \delta^2 + \gamma^2 > \sum_{r=1}^{R_1-1} \lambda_r$. It is obvious that the minimal number of columns R_1 is achieved, when the bound $\|\mathcal{K}\|_F^2 + \gamma^2 - \delta^2$ is smallest, i.e., $\gamma = 0$. Implying that the optimal $\mathcal{G}$ is $\mathcal{G}^\star$. This completes the proof.

Similar to the update of $\mathbf{U}$, the matrix $\mathbf{V}$ comprises R_2 principal eigenvectors of the matrix $\mathbf{Q}_2$ of size $T \times T$

$$\mathbf{Q}_2(i, j) = \sum_{r=1}^{R_1} \mathbf{U}(:, r)^T \mathbf{K}(:, :, i)^T \mathbf{K}(:, :, k) \mathbf{U}(:, r), \tag{18}$$

where R_2 is either given or determined based on the bound $\|\mathcal{Y}\|_F^2 - \delta^2$. The algorithm for TKD sequentially updates $\mathbf{U}$ and $\mathbf{V}$.

3 Implementation

Our method for neural network compression includes the following main steps (see Fig. 3):

1. Each convolutional kernel is approximated by a tensor decomposition (CPD/TKD-CPD in case of ordinary convolutions and SVD in case of 1×1 convolution) with given rank R.
2. The CP decomposition with diverging components is corrected using the error preserving method. The result is a new CP model with minimal sensitivity.
3. An initial convolutional kernel is replaced with a tensor in CPD/TKD-CPD or SVD format, which is equivalent to replacing one convolutional layer with a sequence of convolutional layers with a smaller total number of parameters.
4. The entire network is then fine-tuned using backpropagation.

CPD Block results in three convolutional layers with shapes ($C_{in} \times R \times 1 \times 1$), depthwise ($R \times R \times D \times D$) and ($R \times C_{out} \times 1 \times 1$), respectively (see Fig. 3a). In obtained structure, all spatial convolutions are performed by central $D \times D$ group convolution with R channels. 1×1 convolutions allow the transfer of input data to a more compact channel space (with R channels) and then return data to initial channel space.

TKD-CPD Block is similar to the CPD block, but has 4 (1×1) convolutional layers with the condition that the CP rank must exceed the multilinear ranks, R_1 and R_2 (see Fig. 3c). This structure allows additionally to reduce the number of parameters and floating operations in a factorized layer. Otherwise, when $R < R_1$ and $R < R_2$, sequential 1×1 convolutions can be merged into one 1×1 convolution, converting the TKD-CPD layer format to CPD block.

SVD Block is a variant of CPD Block but comprises only two-factor layers, computed using SVD. Degeneracy is not considered in this block, and no correction is applied (see Fig. 3b).

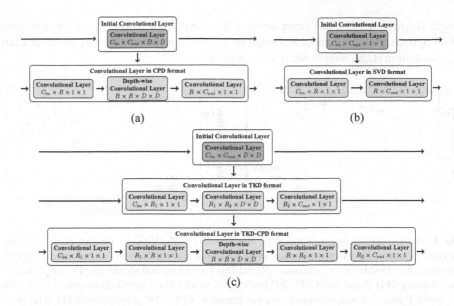

Fig. 3. Graphical illustration to the proposed layer formats that show how decomposed factors are used as new weights of the compressed layer. C_{in}, C_{out} are the number of input of and output channels and D is a kernel size. (a) CPD layer format, R is a CPD rank. (b) SVD layer format, R is a SVD rank. (c) TKD-CPD layer format, R is a CPD rank, R_1 and R_2 are TKD ranks.

Rank Search Procedure. Determination of CP rank is an NP-hard problem [22]. We observe that the drop in accuracy by a factorized layer influences accuracy with fine-tuning of the whole network. In our experiments, we apply a heuristic binary search to find the smallest rank such that drop after single layer fine-tuning does not exceed a predefined accuracy drop threshold EPS.

4 Experiments

We test our algorithms on three representative convolutional neural network architectures for image classification: *VGG-16* [48], *ResNet-18*, *ResNet-50* [19]. We compressed 7×7 and 3×3 convolutional kernels with CPD, CPD with sensitivity correction (CPD-EPC), and Tucker-CPD with the correction (TKD-CPD-EPC). The networks after fine-tuning are evaluated through *top 1* and *top 5* accuracy on *ILSVRC-12* [8] and *CIFAR-100* [30].

We conducted a series of layer-wise compression experiments and measured accuracy recovery and whole model compression of the decomposed architectures. Most of our experiments were devoted to the approximation of single layers when other layers remained intact. In addition, we performed compression of entire networks.

The experiments were conducted with the popular neural networks framework *Pytorch* on GPU server with NVIDIA V-100 GPUs. As a baseline for ILSVRC-12 we used a pre-trained model shipped with *Torchvision*. Baseline CIFAR-100 model was trained using the Cutout method. The fine-tuning process consists of two parts: local or single layer fine-tuning, and entire network fine-tuning. The model was trained with an SGD optimizer with an initial learning rate of 10^{-3} and learning decay of 0.1 at each loss saturation stage, weight decay was set as 10^{-4}.

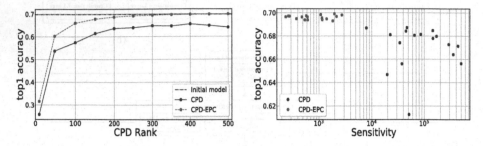

Fig. 4. (Left) Performance evaluation of ResNet-18 on ILSVRC-12 dataset after replacing `layer4.1.conv1` by its approximation using CPD and CPD-EPC with various ranks. The networks are fine-tuned after compression. (Right) Top-1 accuracy and sensitivity of the models estimated using CPD (blue) and CPD-EPC (red). Each model has a single decomposed layer with the best CP rank and was fine-tuned after compression. CPD-EPC outperforms CPD in terms of accuracy/sensitivity trade-off. `layer4.1.conv1` – layer 4, residual block 2 (indexing starts with 0), convolutional layer 1 (Color figure online)

4.1   Layer-Wise Study

CPD-EPC Vs CPD. For this study, we decomposed the kernel filters in 17 convolutional layers of ResNet-18 with different CP ranks, R, ranging from small (10) to relatively high rank (500). The CPDs were run with a sufficiently large number of iterations so that all models converged or there was no significant improvement in approximation errors.

Experiment results show that for all decomposition ranks, the CPD-EPC regularly results in considerably higher *top 1* and *top 5* model accuracy than the standard CPD algorithm. Figure 4 (left) demonstrates an illustrative example for `layer4.1.conv1`. An important observation is that the compressed network using CPD even with the rank of 500 (and fine-tuning) does not achieve the original network's accuracy. However, with EPC, the performances are much better and attain the original accuracy with the rank of 450. Even a much smaller model with the rank of 250 yields a relatively good result, with less than 1% loss of accuracy.

Next, each convolutional layer in ResNet-18 was approximated with different CP ranks and fine-tuned. The best model in terms of top-1 accuracy was then selected. Figure 4 (right) shows relation between the sensitivity and accuracy of the best models. It is straightforward to see that the models estimated using CPD exhibit high sensitivity, and are hard to train. The CPD-EPC suppressed sensitivities of the estimated models and improved the performance of the compressed networks. The CPD-EPC gained the most remarkable accuracy recovery on deeper layers of CNNs.

The effect is significant for some deep convolutional layers of the network with ~ 2% top-1 accuracy difference.

CPD-EPC Vs TKD-EPC. Next, we investigated the proposed compression approach based on the hybrid TKD-CPD model with sensitivity control. Similar experiments were conducted for the CIFAR-100 dataset. The TK multi-linear ranks (R_1, R_2) were kept fixed, while the CP rank varied in a wide range.

In Fig. 5, we compare accuracy of the two considered compressed approaches applied to the layer 4.0.conv1 in ResNet-18. For this case, CPD-EPC still demonstrated a good performance. The obtained accuracy is very consistent, implying that the layer exhibits a low-rank structure. The hybrid TKD-CPD yielded a rather low accuracy for small models, i.e., with small ranks, which are much worse than the CPD-based model with less or approximately the same number of parameters. However, the method quickly attained the original top-1 accuracy and even exceeded the top-5 accuracy when the $R_{CP} \geq 110$.

Comparison of accuracy vs. the number of FLOPs and parameters for the other layers is provided in Fig. 6. Each dot in the figure represents (accuracy, no. FLOPs) for each model. The dots for the same layers are connected by dashed lines. Once again, TKD-EPC achieved higher *top 1* and *top 5* accuracy with a smaller number of parameters and FLOPs, compared to CPD-EPC.

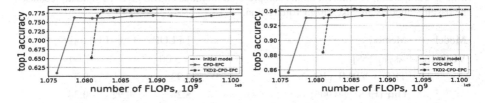

Fig. 5. Performance comparison (top1 accuracy – left, top5 accuracy – right) of CPD-EPC and TKD-CPD-EPC in compression of the layer 4.0.conv1 in the pre-trained ResNet-18 on ILSVRC-12 dataset. TKD-CPD-EPC shows better accuracy recovery with a relatively low number of FLOPs. Initial model has $\approx 1.11 \times 10^9$ FLOPs.

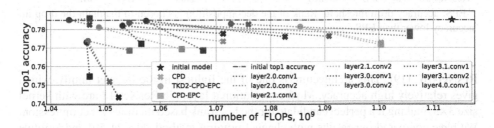

Fig. 6. Accuracy vs FLOPs for models obtained from ResNet-18 (CIFAR-100) via compression of one layer using standard CPD (cross), CPD-EPC (square), or TKD-CPD-EPC (circle) decomposition. Each color corresponds to one layer, which has been compressed using three different methods. For each layer, TKD-CPD-EPC outperforms other decompositions in terms of FLOPs, or accuracy, or both.

4.2 Full Model Compression

In this section, we demonstrate the efficiency of our proposed method in a full model compression of three well-known CNNs *VGG-16* [48], *ResNet-18, ResNet-50* [19] for the ILSVRC-12. We compressed all convolutional layers remaining fully-connected layers intact. The proposed scheme gives ($\times 1.10, \times 5.26$) for VGG-16, ($\times 3.82, \times 3.09$) for ResNet-18 and ($\times 2.51, \times 2.64$) for ResNet-50 reduction in the number of weights and FLOPs respectively. Table 1 shows that our approach yields a high compression ratio while having a moderate accuracy drop.

VGG [48]. We compared our method with other low-rank compression approaches on *VGG-16*. The Asym method [58] is one of the first successful methods on the whole VGG-16 network compression. This method exploits matrix decomposition, which is based on SVD and is able to reduce the number of flops by a factor of 5. Kim et al. [26] applied TKD with ranks selected by VBMF, and achieved a comparable compression ratio but with a smaller accuracy drop. As can be seen from the Table 1, our approach outperformed both Asym and TKD in terms of compression ratio and accuracy drop.

Table 1. Comparison of different model compression methods on ILSVRC-12 validation dataset. The baseline models are taken from Torchvision.

Model	Method	↓ FLOPs	Δ top-1	Δ top-5
VGG-16	Asym. [58]	≈ 5.00	–	−1.00
	TKD + VBMF [26]	4.93	–	−0.50
	Our (EPS[a] = 0.005)	**5.26**	**−0.92**	**−0.34**
ResNet-18	Channel gating NN [24]	1.61	−1.62	−1.03
	Discrimination-aware channel Pruning [59]	1.89	−2.29	−1.38
	FBS [13]	1.98	−2.54	−1.46
	MUSCO [14]	2.42	−0.47	−0.30
	Our (EPS[a] = 0.00325)	**3.09**	**−0.69**	**−0.15**
ResNet-50	**Our (EPS[1] = 0.0028)**	**2.64**	**−1.47**	**−0.71**

[a] EPS: accuracy drop threshold. Rank of the decomposition is chosen to maintain the drop in accuracy lower than EPS.

ResNet-18 [19]. This architecture is one of the lightest in the ResNet family, which gives relatively high accuracy. Most convolutional layers in ResNet-18 are with kernel size 3×3, making it a perfect candidate for the low-rank based methods for compression. We have compared our results with channel pruning methods [13,24,59] and iterative low-rank approximation method [14]. Among all the considered results, our approach has shown the best performance in terms of compression - accuracy drop trade-off.

ResNet-50 [19]. Compared to *ResNet-18, ResNet-50* is a deeper and heavier neural network, which is used as backbone in various modern applications, such as object detection and segmentation. A large number of 1 × 1 convolutions deteriorate performance of low-rank decomposition-based methods. There is not much relevant literature available for compression of this type of ResNet. To the best of our knowledge, the results we obtained can be considered the first attempt to compress the entire *ResNet-50*.

Inference Time for Resnet-50. We briefly compare the inference time of Resnet-50 for the image classification task in Table 2. The measures were taken on 3 platforms: CPU server with Intel® Xeon® Silver 4114 CPU 2.20 GHz, NVIDIA GPU server with ® Tesla® V100 and Qualcomm mobile CPU ® Snapdragon™ 845. The batch size was choosen to yield small variance in inference measurements, e.g., 16 for the measures on CPU server, 128 for the GPU server and 1 for the mobile CPU.

Table 2. Inference time and acceleration for ResNet-50 on different platforms.

Platform	Model inference time	
	Original	Compressed
Intel® Xeon®Silver 4114 CPU 2.20 GHz	3.92 ± 0.02 s	2.84 ± 0.02 s
NVIDIA®Tesla®V100	102.3 ± 0.5 ms	89.5 ± 0.2 ms
Qualcomm®Snapdragon™845	221 ± 4 ms	171 ± 4 ms

5   Discussion and Conclusions

Replacing a large dense kernel in a convolutional or fully-connected layer by its low-rank approximation is equivalent to substituting the initial layer with multiple ones, which in total have fewer parameters. However, as far as we concerned, the sensitivity of the tensor-based models has never been considered before. The closest method proposes to add regularizer on the Frobenius norm of each weight to prevent over-fitting.

In this paper, we have shown a more direct way to control the tensor-based network's sensitivity. Through all the experiments for both ILSVRC-12 and CIFAR-100 dataset, we have demonstrated the validity and reliability of our proposed method for compression of CNNs, which includes a stable decomposition method with minimal sensitivity for both CPD and the hybrid TKD-CPD.

As we can see from recent deep learning literature [23, 28, 50], modern state-of-the-art architectures exploit the CP format when constructing blocks of consecutive layers, which consist of 1 × 1 convolution followed by depth-wise separable convolution. The intuition that stays behind the effectiveness of such representation is that first 1 × 1 convolution maps data to a higher-dimensional subspace, where the features are more separable, so we can apply separate convolutional kernels to preprocess them. Thus, representing weights in CP format using stable and efficient algorithms is the simplest and efficient way of constructing reduced convolutional kernels.

To the best of our knowledge, our paper is the first work solving a problem of building weights in the CP format that is stable and consistent with the fine-tuning procedure.

The ability to control sensitivity and stability of factorized weights might be crucial when approaching incremental learning task [3] or multi-modal tasks, where information fusion across different modalities is performed through shared weight factors.

Our proposed CPD-EPC method can allow more stable fine-tuning of architectures containing higher-order CP convolutional layers [27,28] that are potentially very promising due to the ability to propagate the input structure through the whole network. We leave the mentioned directions for further research.

Acknowledgements. The work of A.-H. Phan, A. Cichocki, I. Oseledets, J. Gusak, K. Sobolev, K. Sozykin and D. Ermilov was supported by the Ministry of Education and Science of the Russian Federation under Grant 14.756.31.0001. The results of this work were achieved during the cooperation project with Noah's Ark Lab, Huawei Technologies. The authors sincerely thank the Referees for very constructive comments which helped to improve the quality and presentation of the paper. The computing for this project was performed on the Zhores CDISE HPC cluster at Skoltech [56].

References

1. Astrid, M., Lee, S.: CP-decomposition with tensor power method for convolutional neural networks compression. In: 2017 IEEE International Conference on Big Data and Smart Computing, BigComp 2017, Jeju Island, South Korea, 13–16 February 2017, pp. 115–118. IEEE (2017). https://doi.org/10.1109/BIGCOMP.2017.7881725
2. Bulat, A., Kossaifi, J., Tzimiropoulos, G., Pantic, M.: Matrix and tensor decompositions for training binary neural networks. arXiv preprint arXiv:1904.07852 (2019)
3. Bulat, A., Kossaifi, J., Tzimiropoulos, G., Pantic, M.: Incremental multi-domain learning with network latent tensor factorization. In: AAAI (2020)
4. Chen, T., Lin, J., Lin, T., Han, S., Wang, C., Zhou, D.: Adaptive mixture of low-rank factorizations for compact neural modeling. In: CDNNRIA Workshop, NIPS (2018)
5. Cichocki, A., Lee, N., Oseledets, I., Phan, A.H., Zhao, Q., Mandic, D.P.: Tensor networks for dimensionality reduction and large-scale optimization: Part 1 low-rank tensor decompositions. Found. Trends® Mach. Learn. **9**(4–5), 249–429 (2016)
6. De Lathauwer, L.: Decompositions of a higher-order tensor in block terms – Part I and II. SIAM J. Matrix Anal. Appl. **30**(3), 1022–1066 (2008). http://publi-etis.ensea.fr/2008/De08e. special Issue on Tensor Decompositions and Applications
7. De Lathauwer, L., De Moor, B., Vandewalle, J.: On the best rank-1 and rank-(R1, R2,., RN) approximation of higher-order tensors. SIAM J. Matrix Anal. Appl. **21**, 1324–1342 (2000)
8. Deng, J., Dong, W., Socher, R., Li, L.J., Li, K., Fei-Fei, L.: ImageNet: a large-scale hierarchical image database. 2009 IEEE Conference on Computer Vision and Pattern Recognition (CVPR), pp. 248–255 (2009)
9. Denil, M., Shakibi, B., Dinh, L., Ranzato, M., de Freitas, N.: Predicting parameters in deep learning. In: Proceedings of the 26th International Conference on Neural Information Processing Systems - Volume 2, NIPS 2013, pp. 2148–2156. Curran Associates Inc. (2013)
10. Denton, E.L., Zaremba, W., Bruna, J., LeCun, Y., Fergus, R.: Exploiting linear structure within convolutional networks for efficient evaluation. In: Advances in Neural Information Processing Systems, vol. 27, pp. 1269–1277. Curran Associates, Inc. (2014)
11. Espig, M., Hackbusch, W., Handschuh, S., Schneider, R.: Optimization problems in contracted tensor networks. Comput. Vis. Sci. **14**(6), 271–285 (2011)
12. Figurnov, M., Ibraimova, A., Vetrov, D.P., Kohli, P.: PerforatedCNNs: acceleration through elimination of redundant convolutions. In: Advances in Neural Information Processing Systems, pp. 947–955 (2016)
13. Gao, X., Zhao, Y., Dudziak, Ł., Mullins, R., Xu, C.Z.: Dynamic channel pruning: feature boosting and suppression. In: International Conference on Learning Representations (2019)

14. Gusak, J., et al.: Automated multi-stage compression of neural networks. In: 2019 IEEE/CVF International Conference on Computer Vision Workshop (ICCVW), pp. 2501–2508 (2019)

15. Han, S., Pool, J., Tran, J., Dally, W.: Learning both weights and connections for efficient neural network. In: Cortes, C., Lawrence, N.D., Lee, D.D., Sugiyama, M., Garnett, R. (eds.) Advances in Neural Information Processing Systems, vol. 28, pp. 1135–1143 (2015)

16. Handschuh, S.: Numerical Methods in Tensor Networks. Ph.D. thesis, Faculty of Mathematics and Informatics, University Leipzig, Germany, Leipzig, Germany (2015)

17. Harshman, R.A.: Foundations of the PARAFAC procedure: models and conditions for an "explanatory" multimodal factor analysis. In: UCLA Working Papers in Phonetics, vol. 16 pp. 1–84 (1970)

18. Harshman, R.A.: The problem and nature of degenerate solutions or decompositions of 3-way arrays. In: Tensor Decomposition Workshop, Palo Alto, CA (2004)

19. He, K., Zhang, X., Ren, S., Sun, J.: Deep residual learning for image recognition. In: 2016 IEEE Conference on Computer Vision and Pattern Recognition (CVPR), pp. 770–778 (2016)

20. He, Y., Kang, G., Dong, X., Fu, Y., Yang, Y.: Soft filter pruning for accelerating deep convolutional neural networks. In: Proceedings of the Twenty-Seventh International Joint Conference on Artificial Intelligence, IJCAI 2018, pp. 2234–2240 (7 2018)

21. He, Y., Lin, J., Liu, Z., Wang, H., Li, L.-J., Han, S.: AMC: AutoML for model compression and acceleration on mobile devices. In: Ferrari, V., Hebert, M., Sminchisescu, C., Weiss, Y. (eds.) ECCV 2018. LNCS, vol. 11211, pp. 815–832. Springer, Cham (2018). https://doi.org/10.1007/978-3-030-01234-2_48

22. Hillar, C.J., Lim, L.H.: Most tensor problems are NP-hard. J. ACM (JACM) **60**(6), 45 (2013)

23. Howard, A., et al.: Searching for MobileNetv3. In: Proceedings of the IEEE International Conference on Computer Vision, pp. 1314–1324 (2019)

24. Hua, W., Zhou, Y., De Sa, C.M., Zhang, Z., Suh, G.E.: Channel gating neural networks. In: Wallach, H., Larochelle, H., Beygelzimer, A., d'Alché Buc, F., Fox, E., Garnett, R. (eds.) Advances in Neural Information Processing Systems, vol. 32, pp. 1886–1896 (2019)

25. Khoromskij, B.: $O(d \log N)$-quantics approximation of N-d tensors in high-dimensional numerical modeling. Constr. Approximation **34**(2), 257–280 (2011)

26. Kim, Y., Park, E., Yoo, S., Choi, T., Yang, L., Shin, D.: Compression of deep convolutional neural networks for fast and low power mobile applications. In: 4th International Conference on Learning Representations, ICLR 2016, San Juan, Puerto Rico, 2–4 May 2016, Conference Track Proceedings (2016). http://arxiv.org/abs/1511.06530

27. Kossaifi, J., Bulat, A., Tzimiropoulos, G., Pantic, M.: T-net: parametrizing fully convolutional nets with a single high-order tensor. In: Proceedings of the IEEE Conference on Computer Vision and Pattern Recognition (CVPR), pp. 7822–7831 (2019)

28. Kossaifi, J., Toisoul, A., Bulat, A., Panagakis, Y., Hospedales, T.M., Pantic, M.: Factorized higher-order CNNs with an application to spatio-temporal emotion estimation. In: Proceedings of the IEEE/CVF Conference on Computer Vision and Pattern Recognition, pp. 6060–6069 (2020)

29. Krijnen, W., Dijkstra, T., Stegeman, A.: On the non-existence of optimal solutions and the occurrence of "degeneracy" in the CANDECOMP/PARAFAC model. Psychometrika **73**, 431–439 (2008)

30. Krizhevsky, A.: Learning multiple layers of features from tiny images. Technical Report TR-2009, University of Toronto, Toronto (2009)

31. Landsberg, J.M.: Tensors: Geometry and Applications, vol. 128. American Mathematical Society, Providence (2012)

32. Lebedev, V.: Algorithms for speeding up convolutional neural networks. Ph.D. thesis, Skoltech, Russia (2018). https://www.skoltech.ru/app/data/uploads/2018/10/Thesis-Final.pdf

33. Lebedev, V., Ganin, Y., Rakhuba, M., Oseledets, I., Lempitsky, V.: Speeding-up convolutional neural networks using fine-tuned CP-decomposition. In: International Conference on Learning Representations (2015)
34. Lim, L.H., Comon, P.: Nonnegative approximations of nonnegative tensors. J. Chemom. **23**(7–8), 432–441 (2009)
35. Mitchell, B.C., Burdick, D.S.: Slowly converging PARAFAC sequences: Swamps and two-factor degeneracies. J. Chemom. **8**, 155–168 (1994)
36. Molchanov, D., Ashukha, A., Vetrov, D.: Variational dropout sparsifies deep neural networks. In: Proceedings of the 34th International Conference on Machine Learning - Volume 70, pp. 2498–2507 (2017). JMLR.org
37. Novikov, A., Podoprikhin, D., Osokin, A., Vetrov, D.: Tensorizing neural networks. In: Proceedings of the 28th International Conference on Neural Information Processing Systems - Volume 1, NIPS 2015, pp. 442–450. MIT Press, Cambridge (2015)
38. Oseledets, I., Tyrtyshnikov, E.: Breaking the curse of dimensionality, or how to use SVD in many dimensions. SIAM J. Sci. Comput. **31**(5), 3744–3759 (2009)
39. Paatero, P.: Construction and analysis of degenerate PARAFAC models. J. Chemometrics **14**(3), 285–299 (2000)
40. Phan, A.H., Cichocki, A., Uschmajew, A., Tichavský, P., Luta, G., Mandic, D.: Tensor networks for latent variable analysis: novel algorithms for tensor train approximation. IEEE Trans. Neural Network Learn. Syst. (2020). https://doi.org/10.1109/TNNLS.2019.2956926
41. Phan, A.H., Tichavský, P., Cichocki, A.: Tensor deflation for CANDECOMP/PARAFAC. Part 1: alternating subspace update algorithm. IEEE Trans. Sig. Process. **63**(12), 5924–5938 (2015)
42. Phan, A.H., Tichavský, P., Cichocki, A.: Error preserving correction: a method for CP decomposition at a target error bound. IEEE Trans. Signal Process. **67**(5), 1175–1190 (2019)
43. Phan, A.H., Yamagishi, M., Mandic, D., Cichocki, A.: Quadratic programming over ellipsoids with applications to constrained linear regression and tensor decomposition. Neural Comput. Appl. (2020). https://doi.org/10.1007/s00521-019-04191-z
44. Rastegari, M., Ordonez, V., Redmon, J., Farhadi, A.: XNOR-Net: ImageNet classification using binary convolutional neural networks. In: Leibe, B., Matas, J., Sebe, N., Welling, M. (eds.) ECCV 2016. LNCS, vol. 9908, pp. 525–542. Springer, Cham (2016). https://doi.org/10.1007/978-3-319-46493-0_32
45. Rayens, W., Mitchell, B.: Two-factor degeneracies and a stabilization of PARAFAC. Chemometr. Intell. Lab. Syst. **38**(2), 173–181 (1997)
46. Rigamonti, R., Sironi, A., Lepetit, V., Fua, P.: Learning separable filters. In: Proceedings of the 2013 IEEE Conference on Computer Vision and Pattern Recognition. pp. 2754–2761. CVPR '13, IEEE Computer Society, Washington, DC, USA (2013)
47. de Silva, V., Lim, L.H.: Tensor rank and the ill-posedness of the best low-rank approximation problem. SIAM J. Matrix Anal. Appl. **30**, 1084–1127 (2008)
48. Simonyan, K., Zisserman, A.: Very deep convolutional networks for large-scale image recognition. In: 3rd International Conference on Learning Representations, ICLR (2015)
49. Stegeman, A., Comon, P.: Subtracting a best rank-1 approximation may increase tensor rank. Linear Algebra Appl. **433**(7), 1276–1300 (2010)
50. Tan, M., Le, Q.V.: EfficientNet: rethinking model scaling for convolutional neural networks. In: ICML (2019)
51. Tichavský, P., Phan, A.H., Cichocki, A.: Sensitivity in tensor decomposition. IEEE Signal Process. Lett. **26**(11), 1653–1657 (2019)
52. Tucker, L.R.: Implications of factor analysis of three-way matrices for measurement of change. Probl. Measuring Change **15**, 122–137 (1963)

53. Vasilescu, M.A.O., Terzopoulos, D.: Multilinear subspace analysis of image ensembles. In: 2003 IEEE Computer Society Conference on Computer Vision and Pattern Recognition (CVPR 2003), Madison, WI, USA, 16–22 June 2003, pp. 93–99. IEEE Computer Society (2003). https://doi.org/10.1109/CVPR.2003.1211457
54. Vervliet, N., Debals, O., Sorber, L., Barel, M.V., Lathauwer, L.D.: Tensorlab 3.0, March 2016. http://www.tensorlab.net
55. Wang, D., Zhao, G., Li, G., Deng, L., Wu, Y.: Lossless compression for 3DCNNs based on tensor train decomposition. CoRR abs/1912.03647 (2019). http://arxiv.org/abs/1912.03647
56. Zacharov, I., et al.: Zhores – petaflops supercomputer for data-driven modeling, machine learning and artificial intelligence installed in Skolkovo Institute of Science and Technology. Open Eng. **9**(1) (2019)
57. Zhang, T., Golub, G.H.: Rank-one approximation to high order tensors. SIAM J. Matrix Anal. Appl. **23**(2), 534–550 (2001). https://doi.org/10.1137/S0895479899352045
58. Zhang, X., Zou, J., He, K., Sun, J.: Accelerating very deep convolutional networks for classification and detection. IEEE Trans. Pattern Anal. Mach. Intell. **38**(10), 1943–1955 (2016)
59. Zhuang, Z., et al.: Discrimination-aware channel pruning for deep neural networks. In: Advances in Neural Information Processing Systems, pp. 883–894 (2018)

Dual Mixup Regularized Learning
for Adversarial Domain Adaptation

Yuan Wu[1(✉)], Diana Inkpen[2], and Ahmed El-Roby[1]

[1] Carleton University, Ottawa, Canada
{yuan.wu3,Ahmed.ElRoby}@carleton.ca
[2] University of Ottawa, Ottawa, Canada
Diana.Inkpen@uottawa.ca

Abstract. Recent advances on unsupervised domain adaptation (UDA) rely on adversarial learning to disentangle the explanatory and transferable features for domain adaptation. However, there are two issues with the existing methods. First, the discriminability of the latent space cannot be fully guaranteed without considering the class-aware information in the target domain. Second, samples from the source and target domains alone are not sufficient for domain-invariant feature extracting in the latent space. In order to alleviate the above issues, we propose a dual mixup regularized learning (DMRL) method for UDA, which not only guides the classifier in enhancing consistent predictions in-between samples, but also enriches the intrinsic structures of the latent space. The DMRL jointly conducts category and domain mixup regularizations on pixel level to improve the effectiveness of models. A series of empirical studies on four domain adaptation benchmarks demonstrate that our approach can achieve the state-of-the-art.

Keywords: Domain adaptation · Mixup · Regularization

1 Introduction

The development of deep neural networks has significantly improved the state of the arts for a wide variety of machine learning tasks, such as computer vision [13], speech recognition [11], and reinforcement learning [23]. However, these advancements often rely on the existence of a large amount of labeled training data. In many real-world applications, collecting sufficient labeled data is often prohibitive due to time, financial, and expertise constraints. Therefore, there is a strong motivation to train an effective predictive model which can leverage knowledge learned from a label-abundant dataset and perform well on another label-scarce domains [17]. However, due to the existence of domain shift, deep neural networks trained on one large scale labeled dataset can be weak at generalizing learned knowledge to new datasets and tasks [17].

To address the above issue, a general strategy called domain adaptation is introduced by transferring knowledge from a label-rich domain, referred as the

© Springer Nature Switzerland AG 2020
A. Vedaldi et al. (Eds.): ECCV 2020, LNCS 12374, pp. 540–555, 2020.
https://doi.org/10.1007/978-3-030-58526-6_32

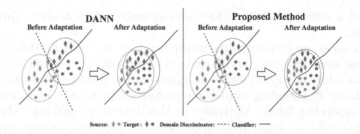

Fig. 1. Comparison of DANN and the proposed method. **Left**: DANN only tries to match the feature distribution by utilizing adversarial learning; it does not consider the class-aware information in the target domain and samples from the source and target domains may not be sufficient to ensure domain-invariance of the latent space. **Right**: Our proposed method uses category mixup regularization to enforce prediction consistency in-between samples and domain mixup regularization to explore more intrinsic structures across domains, resulting in better adaptation performance.

source domain, to a label-scarce domain, referred as the target domain [1]. Unsupervised domain adaptation addresses a more challenging scenario where there is no labeled data in the target domain. A theoretical analysis of domain adaptation is introduced by [1], it suggests that in UDA tasks, the risk on the target domain can be bounded by the risk of a model on the source domain and the discrepancy between distributions of the two domains. Early UDA methods learned to reduce the discrepancy between domains in a shallow regime [7] or to re-weight source instances based on their relevance to the target domain [6]. Later on, Maximum Mean Discrepancy (MMD) [9] was proposed to measure the distribution difference between source and target domains [14,16,25]. More recently, the UDA models are largely built on deep neural networks, and focus on learning domain-invariant features across domains by using adversarial learning [4]. Adversarial domain adaptation models can learn discriminative and domain invariant features across domains by playing a minimax game between a feature extractor and a domain discriminator. The domain discriminator is trained to tell whether the sample comes from the source domain or target domain, while the feature extractor is learned to fool the domain discriminator. Many recent UDA methods based on adversarial learning can achieve the state-of-the-art performance [15,20,30].

Even though adversarial domain adaptation method has shown impressive performance for various tasks, such as image classification [15] and semantic segmentation [20]. This approach still faces two issues: First, the adversarial domain adaptation does not take class-aware information in the target domain into consideration. Second, as we always use mini-batch stochastic gradient descent (SGD) for optimization in practice, if the batch size is small, samples from the source and target domains may not be sufficient to guarantee the domain-invariance in the latent space. Therefore, after adaptation, as shown in the left side of Fig. 1, the classifier may falsely align target samples of one label with samples of a different label in the source domain, which leads to inconsistent predictions.

In this paper, we propose a dual mixup regularized learning (DMRL) method, which implements category and domain mixup regularizations on pixel level to address the aforementioned issues for unsupervised domain adaptation. Mixup has the ability of generating convex combinations of pairs of training samples and their corresponding labels. Motivated by this data augmentation technique, we propose two effective regularization mechanisms including domain-level mixup regularization and category-level mixup regularization, which play crucial roles in reducing the domain discrepancy for unsupervised domain adaptation. In particular, category mixup regularization is used to enforce consistent predictions in the latent space, which is conducted on both source and target domains, achieving a stronger discriminability of the latent space. Domain mixup regularization can reveal more mixed instances within each domain and allows the model to enrich internal feature patterns in the latent space, which can lead to a more continuous domain-invariant latent space and help match the global domain statistics across different domains. By using the two mixup-based regularization mechanisms, our model can effectively generate discriminative and domain-invariant representations. Empirical studies on four benchmarks demonstrate the performance of our approach. The contributions of our paper are summarized as follows:

- We propose a dual mixup regularized learning method which can project the source and target domains to a common latent space, and efficiently transfer the knowledge learned from the labeled source domain to the unlabeled target domain.
- The proposed regularization mechanisms (category-level mixup regularization and domain-level mixup regularization) can learn discriminative and domain-invariant representations effectively to help reduce the distribution discrepancy across different domains.
- We empirically confirm the effectiveness of our proposed method by evaluating it on four benchmark datasets. Conducting ablation studies and parameter sensitivity analysis to validate the contributions of different components in our model and evaluate how each hyperparameter influences the performance of our method.

2   Related Work

This work builds on two threads of research: interpolation-based regularization and domain adaptation. In this section, we briefly overview methods that are related to these two tasks.

2.1   Interpolation-Based Regularization

Interpolation-based regularization has been recently proposed for supervised learning [26,29], it can help models to alleviate issues such as instability in adversarial training and sensitivity to adversarial samples. In particular, Mixup [29] is proposed to train models on virtual examples constructed as the convex combinations of pairs of inputs and labels. It can also encourage models to have a strictly linear behavior between training samples, by smoothing the models' output for a convex combination of two inputs to the convex combination of the outputs of each individual input [2]. Moreover, Mixup can also be used to guarantee consistent predictions in the data distribution [27], which can induce the separation of samples into different labels [3]. More recently, various variants of Mixup have been studied. A manifold extension to Mixup, Manifold-Mixup [26], proposes to perform interpolation in the latent space representations. [2] exploits interpolations in the latent space generated by an autoencoder to improve performance.

2.2 Domain Adaptation

The main goal of domain adaptation is to transfer the knowledge learned from a label-abundant domain to a label-scarce domain. Unsupervised domain adaptation tackles a more challenging scenario where the target domain has no labeled data at all. Deep neural network based methods have been widely studied for UDA. The Deep Domain Confusion (DDC) method leverages Maximum Mean Discrepancy (MMD) [9] metric in the last fully-connected layer to learn representations that are both discriminative and transferable [25]. [14] proposes a Deep Adaptation Network (DAN) to enhance the feature transferability by minimizing the multi-kernel MMD in several task specific layers. Asymmetric Tri-Training (ATT) exploits three different networks to generates pseudo-labels for target domain samples and utilizes these pseudo-labels to train the final classifier [19]. Joint Adaptation Networks (JAN) learn a transfer network by aligning the joint distributions of multiple domain-specific layers across different domains based on a joint maximum mean discrepancy criterion [16]. [31] proposes a Confidence Regularized Self-Training (CRST) framework to construct the soft pseudo-label, smoothing the one-hot pseudo-label to a conservative target distribution.

More recently, unsupervised domain adaptation methods are largely focusing on learning domain-invariant features by using adversarial training [4]. Generative Adversarial Network (GAN) is proposed in [8], which plays a minimax game between two networks: the discriminator is trained by minimizing the binary classification error of distinguishing the real images from the generated ones, while the generator is learned to generate high-quality images that are indistinguishable by the discriminator. Motivated by GAN, [4] proposes a domain adversarial neural network (DANN) that can learn discriminative and domain-invariant features by exploiting adversarial learning between a feature extractor and a domain discriminator. Many recent works have adopted the adversarial learning mechanism and achieved the state-of-the-art performance for unsupervised domain adaptation. The Adversarial Discriminative Domain Adaptation method (ADDA) uses an untied weight sharing strategy to align the feature distributions of source and target domains [24]. The Maximum Classification Discrepancy (MCD) utilizes different task-specific classifiers to learn a feature extractor that can generate category-related discriminative features [20]. [12] proposes Cycle-Consistent Adversarial Domain Adaptation (CyCADA) which implements domain adaptation at both pixel-level and feature-level by using cycle-consistent adversarial training. Multi-Adversarial Domain Adaptation (MADA) [18] can exploit multiplicative interactions between feature representations and category predictions to enforce adversarial learning. Generate to Adapt (GTA) provides an adversarial image generation approach that directly learns a joint feature space in which the distance between the source and target domains can be minimized [21]. Conditional Domain Adversarial Network (CDAN) conditions the domain discriminator on a multilinear map of feature representations and category predictions so as to enable discriminative alignment of multi-mode structures [15]. Consensus Adversarial Domain Adaptation (CADA) [30] enforces the source and target encoders to achieve consensus to ensure the domain-invariance of the latent space.

Our proposed DMRL can be regarded as an extension of this line of research by introducing both category and domain mixup regularizations on pixel level to solve complex, high dimensional unsupervised domain adaptation tasks.

3   Method

In this paper, we consider unsupervised domain adaptation in the following setting. We have a labeled source domain $D^s = \{(x_i^s, y_i^s)\}_{i=1}^{n^s}$ with $x_i^s \in \mathcal{X}$ and $y_i^s \in \mathcal{Y}$, and an unlabeled target domain $D^t = \{x_i^t\}_{i=1}^{n^t}$ with $x_i^t \in \mathcal{X}$. The data in two domains are sampled from two distributions P_S and P_T. P_S and P_T are assumed to be different but related (refereed as covariate shift in the literature [22]). The target task is assumed to be the same with the source task. Our ultimate goal is to utilize the labeled data in the source domain to learn a predictive model $h : \mathcal{X} \mapsto \mathcal{Y}$ which can generalize well on the target domain.

3.1   Adversarial Domain Adaptation

Motivated by the domain adaptation theory [1] and GANs [8], [4] proposes Domain Adversarial Neural Network (DANN), which can learn the domain-invariant features that are generalizable across domains. The standard DANN consists of three components: a feature extractor G, a category classifier C and a domain discriminator D. We consider a feature extractor $G : \mathcal{X} \mapsto \mathbb{R}^m$, which maps any input instance $\mathbf{x} \in \mathcal{X}$ from the input space $\mathcal{X}$ into the latent space $G(\mathbf{x}) \in \mathbb{R}^m$; a category classifier $C : \mathbb{R}^m \mapsto \mathcal{Y}$, which transforms a feature vector in the latent space into the output label space $\mathcal{Y}$; a domain discriminator $D : \mathbb{R}^m \mapsto [0, 1]$, which distinguishes the source domain data (with domain label 1) from the target domain data (with domain label 0). By training G adversarially to confuse D, DANN can learn domain-invariant features to bridge the divergence between domains. Formally, the DANN can be formulated as:

$$\min_{G} \max_{D} \quad \mathcal{L}_c(G, C) + \lambda_d \mathcal{L}_{Adv}(G, D) \tag{1}$$

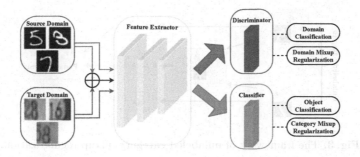

Fig. 2. The architecture of the proposed dual mixup regularized learning (DMRL) method. Our DMRL consists of two mixup-based regularization mechanisms, including category-level mixup regularization and domain-level mixup regularization, which can enhance discriminability and domain-invariance of the latent space. The feature extractor G aims to learn discriminative and domain-invariant features, the domain discriminator D is trained to tell whether the sampled feature comes from the source domain or target domain, and the classifier C is used to conduct object classification.

$$\mathcal{L}_c(G, C) = \mathbb{E}_{(\mathbf{x}^s, y^s) \sim D^s} \ell(C(G(\mathbf{x}^s)), y^s) \tag{2}$$

$$\mathcal{L}_{Adv}(G, D) = \mathbb{E}_{\mathbf{x}^s \sim D^s} \log D(G(\mathbf{x}^s)) + \mathbb{E}_{\mathbf{x}^t \sim D^t} \log(1 - D(G(\mathbf{x}^t))) \tag{3}$$

where $\ell(\cdot, \cdot)$ is the canonical cross-entropy loss, and λ_d is a trade-off hyperparameter.

3.2 Dual Mixup Regularization

In this work, we propose a dual mixup regularized learning (DMRL) method based on adversarial domain adaptation. This method conducts category and domain mixup on pixel level. In general, Mixup performs data augmentation by constructing virtual samples with convex combinations of a pair of samples and their corresponding labels: $(\mathbf{x}_i, y_i)$ and $(\mathbf{x}_j, y_j)$:

$$\widetilde{\mathbf{x}} = \mathcal{M}_\lambda(\mathbf{x}_i, \mathbf{x}_j) = \lambda \mathbf{x}_i + (1 - \lambda)\mathbf{x}_j \tag{4}$$

$$\widetilde{y} = \mathcal{M}_\lambda(y_i, y_j) = \lambda y_i + (1 - \lambda)y_j \tag{5}$$

where λ is randomly sampled from a beta distribution $Beta(\alpha, \alpha)$ for $\alpha \in (0, \infty)$. By encouraging linear interpolation regularization in-between training samples, Mixup has been demonstrated effective in both supervised and semi-supervised learning [27, 29].

We largely enhance the ability of prior adversarial-learning-based domain adaptation methods by our proposed dual mixup regularized learning module, which jointly conducts category and domain mixup on pixel level to guide adversarial learning. Figure 2 depicts the whole framework of our proposed method.

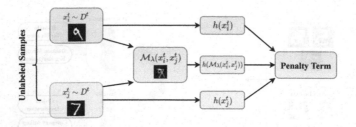

Fig. 3. The framework of unlabeled category mixup regularization.

For the input, images are linearly mixed by pixel-wise addition within each individual domain. Therefore, in the input space, there exists four kinds of samples: source samples, target samples, mixed source samples obtained by mixing two source samples, and mixed target samples obtained by mixing two target samples. After that, there exists two streams. For one stream, features of the source and target domains are used to align the global distribution statistics and conduct domain mixup regularization by the domain discriminator D. For the other stream, object classification and category mixup regularization are implemented by the classifier C. More details are provided in the following parts.

Category Mixup Regularization. The category mixup regularization consists of two components: labeled category mixup regularization and unlabeled category mixup regularization. For the source domain, since we have the labeled samples, we directly use mixed source samples and their corresponding labels to enforce prediction consistency:

$$\mathcal{L}_s^r(G, C) = \mathbb{E}_{(\mathbf{x}_i^s, y_i^s),(\mathbf{x}_j^s, y_j^s) \sim D^s} \ell(h(\widetilde{\mathbf{x}}^s), \widetilde{y}^s) \tag{6}$$

where h denotes the composition of the feature extractor G and the classifier C, $h = G \circ C$, and it can be treated as the classification function in the input space.

For the target domain, we have no access to the label information. Therefore, mixup needs to be applied on the pseudo-labels generated by the classifier C. Specifically, we replace y_i^t and y_j^t with $h(\mathbf{x}_i^t)$ and $h(\mathbf{x}_j^t)$, which are the current predictions of C. Literally, $h(\mathbf{x}_i^t)$ and $h(\mathbf{x}_j^t)$ are termed as pseudo-labels of $(\mathbf{x}_i^t)$ and $(\mathbf{x}_j^t)$. Figure 3 presents the process of unlabeled category mixup regularization. First, we should construct convex combinations, denoted as $(\widetilde{\mathbf{x}}^t, \mathcal{M}_\lambda(h(\mathbf{x}_i^t), h(\mathbf{x}_j^t)))$, of pairs of samples $(\mathbf{x}_i^t, \mathbf{x}_j^t)$ and their virtual labels $(h(\mathbf{x}_i^t), h(\mathbf{x}_j^t))$. Then we conduct regularization by enforcing $h(\widetilde{\mathbf{x}}^t)$ to be consistent with $\mathcal{M}_\lambda(h(\mathbf{x}_i^t), h(\mathbf{x}_j^t))$ via a penalty term:

$$\mathcal{L}_t^r(G, C) = \mathbb{E}_{\mathbf{x}_i^t, \mathbf{x}_j^t \sim D^t} Dis(h(\widetilde{\mathbf{x}}^t), \mathcal{M}_\lambda(h(\mathbf{x}_i^t), h(\mathbf{x}_j^t))) \tag{7}$$

where $Dis(\cdot,\cdot)$ denotes the penalty term which can punish the difference between $h(\widetilde{\mathbf{x}}^t)$ and $\mathcal{M}_\lambda(h(\mathbf{x}_i^t), h(\mathbf{x}_j^t))$, encouraging a linear behavior in-between training samples and λ equals to the one used in labeled category mixup regularization. In our experiments, we use $L1$-Norm function as the penalty term. In general, Category mixup regularization can smooth the output distribution by constructing neighboring samples of the training samples and enforce prediction consistency between the neighboring and training samples, which exploits the class-aware information of the target domain in the training process, leading to performance improvement.

Domain Mixup Regularization. In practice, as we use mini-batch SGD in training, samples from the source and target domains alone are usually insufficient for global distribution alignment. Domain mixup has the ability of generating more intermediate samples and exploring more internal structures within each domain. Linear interpolations of each domain can be generated in the input space, and the domain mixup regularization term can be defined as follows:

$$\mathcal{L}_{Adv}^r(G, D) = \mathbb{E}_{\mathbf{x}_i^s,\mathbf{x}_j^s \sim D^s} \log D(G(\widetilde{\mathbf{x}}^s)) + \mathbb{E}_{\mathbf{x}_i^t,\mathbf{x}_j^t \sim D^t} \log(1 - D(G(\widetilde{\mathbf{x}}^t))) \quad (8)$$

where λ is the same as the one used in category mixup regularization. Contrary to prior adversarial-learning-based domain adaptation methods [4,15,24], not only the samples from source and target domains, but also the mixed samples are used to align the global distribution statistics. Finally, in our proposed domain mixup regularization applied to adversarial domain adaptation, the domain-invariance of the latent space is expected to be enhanced, and with category mixup regularization, the learned representations can be more discriminative. Formally, the DMRL method can be formulated as:

$$\min_{C,C}\max_{D} \quad \mathcal{L}_c(G, C) + \lambda_s \mathcal{L}_s^r(G, C) + \lambda_t \mathcal{L}_t^r(G, C) + \lambda_d \mathcal{L}_{Adv}(G, D) + \lambda_r \mathcal{L}_{Adv}^r(G, D)$$
$$(9)$$

where λ_s, λ_t, λ_d and λ_r are hyperparameters for trading off different losses.

3.3 Training Procedure

The training algorithm of DMRL which uses mini-batch SGD is presented in Algorithm 1. In each iteration, we first mix the samples in the input space for the source and target domains, separately. Then the mixed samples and their constitutions are used to guide the adversarial learning and conduct two mixup-based regularizations. The category mixup regularization can enforce consistent prediction constraint, while the domain mixup regularization can enrich the feature patterns in the latent space, so that the learned representation can be both discriminative and domain-invariant. α is a hyperparameter that controls the selection of λ. λ_s, λ_t, λ_d and λ_r are hyperparameters that balance different losses. In our experiments, we set α, λ_s and λ_r as 0.1, 0.0001 and 0.00001. According to our parameter sensitivity analysis, the values of α, λ_s and λ_r do not

---

Algorithm 1. Stochastic gradient descent training algorithm of DMRL

1: **Input:** Source domain: D^s, target domain: D^t and batch size: N.
2: **Output:** Configurations of DMRL
3: **Initialize** α, λ_s, λ_t, λ_d and λ_r
4: **for** number of training iterations **do**
5:     $(\mathbf{x}^s, y^s) \leftarrow$ RANDOMSAMPLE(D^s, N)
6:     $(\mathbf{x}^t) \leftarrow$ RANDOMSAMPLE(D^t, N)
7:     $\lambda \leftarrow$ RANDOMSAMPLE$(Beta(\alpha, \alpha))$
8:     $(\widetilde{x}^s, \widetilde{y}^s) \leftarrow$ Eq. (4, 5)   #get mixed images for the source domain
9:     $(\widetilde{x}^t) \leftarrow$ Eq.(4)   #get mixed images for the target domain
10:     Calculate $l_D = \lambda_d \mathcal{L}_{Adv}(G, D) + \lambda_r \mathcal{L}_{Adv}^r(G, D)$;
        Update D by ascending along gradients ∇l_D.
11:     Calculate $loss = \mathcal{L}_c(G, C) + \lambda_s \mathcal{L}_s^r(G, C) + \lambda_t \mathcal{L}_t^r(G, C) + \lambda_d \mathcal{L}_{Adv}(G, D) + \lambda_r \mathcal{L}_{Adv}^r(G, D)$;
        Update G, C by descending along gradients $\nabla loss$.
12: **end for**

---

influence much the adaptation performance of our approach, and these hyperparameters are kept fixed in all experiments. The value of λ_t has an influence on the adaptation performance, so λ_t is selected via tuning on the unlabeled test data for different tasks. λ_d is adapted according to the strategy from [5].

3.4   Discussion

For the adversarial-learning-based domain adaptation methods, both the domain label and category label play crucial roles in filling the gap between the source and target domains. Specifically, domain labels are used to help make global distribution alignment across different domains, and category labels can enable features to be discriminative [4,15,24]. These two types of information can help reduce the domain discrepancy in different aspects, and complement each other for domain adaptation. However, prior adversarial domain adaptation methods suffer two limitations: (1) The lack of class-aware information of the target domain can make the extracted features less discriminative; (2) As we use mini-batch SGD in training, samples from source and target domains do not allow models to completely explore internal feature patterns in the latent space.

In our method, we conduct two mixup-based regularizations to alleviate the above issues. First, we apply category mixup regularization on source and target domains. Specifically, for unlabeled target data, pseudo-labels are introduced. Since there are obviously false labels as target pseudo-labels, consistent prediction constraint is exploited to suppress the detrimental influence brought by the false target pseudo-labels. Moreover, category mixup regularization can smooth the output distribution of the model by encouraging the model to behave linearly in-between training samples. Then, Mixup [29], as an effective data augmentation technique, is expected to provide extra mixed samples for adversarial learning. Domain mixup regularization is used to explore more internal feature patterns in the latent space, which leads to a more continuous domain-invariant latent distribution. In summary, our proposed approach can learn discriminative and domain-invariant representations and improve the performance of different unsupervised domain adaptation tasks.

4 Experiments

We evaluate our proposed DMRL on unsupervised domain adaptation tasks of four benchmarks, validate the effectiveness of different components in detail and investigate the influences of different hyperparameters.

4.1 Setup

Dataset. We conducted experiments on four domain adaptation benchmarks, **Office-31**, **ImageCLEF-DA**, **VisDA-2017** and **Digits**. Office-31 is a benchmark domain adaptation dataset containing images belonging to 31 classes from three domains: Amazon (A) with 2,817 images, Webcam (W) with 795 images and DSLR (D) with 498 images. We evaluate all methods on six domain adaptation tasks: $\mathbf{A \to W, D \to W, W \to D, A \to D, D \to A}$, and $\mathbf{W \to A}$.

ImageCLEF-DA is a benchmark dataset for ImageCLEF 2014 domain adaptation challenges, which contains 12 categories shared by three domains: Caltech-256 (C), ImageNet ILSVRC 2012 (I), and Pascal VOC 2012 (P). Each domain contains 600 images and 50 images for each category. The three domains in this dataset are of the same size, which is a good complementation of the Office-31 dataset where different domains are of different sizes. We build six domain adaptation tasks: $\mathbf{I \to P, P \to I, I \to C, C \to I, C \to P}$, and $\mathbf{P \to C}$.

VisDA-2017 is a large simulation-to-real dataset, with over 280,000 images of 12 classes in the combined training, validation, and testing domains. The source images were obtained by rendering 3D models of the same object classes as in the real data from different angles and under different lighting conditions. It contains 152,397 synthetic images. The validation and test domains comprise natural images. The validation one has 55,388 images in total. We use the training domain as the source domain and validation domain as the target domain.

For Digits domain adaptation tasks, we explore three digit datasets: **MNIST**, **USPS** and **SVHN**. Each dataset contains digit images of 10 categories (0–9). We adopt the experimental settings of CyCADA [12] with three domain adaptation tasks: MNIST to USPS (**MNIST→USPS**), USPS to MNIST (**USPS→MNIST**), and SVHN to MNIST (**SVHN→MNIST**).

Comparison Methods. We compare our proposed DMRL with state-of-the-art models. For Office-31, we compare with Deep Adaptation Network (DAN) [14], Domain Adversarial Neural Network (DANN) [4], Joint Adaptation Network (JAN) [16], Generate to Adapt (GTA) [21], Multi-Adversarial Domain Adaptation (MADA) [18], Conditional Domain Adaptation Network (CDAN) [15] and Confidence Regularized Self-Training (CRST). For ImageCLEF-DA, we compare with DAN, DANN, JAN, MADA and CDAN. For VisDA-2017, We compare with DANN, DAN, JAN, GTA, Maximum Classifier Discrepancy (MCD)

[20] and CDAN. To further validate our method, we also conduct experiments on digit datasets, including MNIST, USPS and SVHN, we compare with DANN, Adversarial Discriminative Domain Adaptation (ADDA) [24], Cycle-consistent Adversarial Domain Adaptation (CyCADA)[12], MCD, CDAN and Consensus Adversaria Domain Adaptation (CADA) [30].

Implementation Details. We follow standard evaluation protocols for unsupervised domain adaptation [15, 20]. For Office-31 and ImageCLEF-DA datasets, we utilize ResNet50 [10] pre-trained on ImageNet [13] as the backbone. For each domain adaptation task of the Office-31 and ImageCLEF-DA datasets, we report classification results of mean $\pm$ standard error over three random trials. For VisDA-2017 dataset, we follow [20] and use ResNet101 [10] pre-trained on ImageNet [13] as the backbone. For Digits datasets, we use a modified version of Lenet architecture as the base network, and train the models from scratch. For each backbone, we use all its layers up to the second last one as the feature extractor G and replace the last full-connected layer with a task-specific fully-connected layer as the category classifier C. For discriminator, we use the same architecture as DANN [4].

We adopt mini-batch SGD with momentum of 0.9 and the learning rate annealing strategy as [5]: the learning rate is adjusted by $\eta_p = \frac{\eta_0}{(1+\theta p)^\beta}$, where p denotes the process of training epochs that is normalized to be in $[0, 1]$, and we set $\eta_0 = 0.01$, $\theta = 10$, $\beta = 0.75$, which are optimized to promote convergence and low errors on the source domain. λ_d is progressively changed from 0 to 1 by multiplying to $\frac{1-exp(-\delta p)}{1+exp(-\delta p)}$, where $\delta = 10$. For all experiments, we set the hyperparameter α of distribution $Beta(\alpha, \alpha)$ to 0.2 as used in [29]. λ_s and λ_r are fixed as 0.0001 and 0.00001 respectively. λ_t is chosen in the range $\{0.1, 1, 2, 5, 6, 10\}$, we select it on a per-experiment basis relying on unlabeled target data.

4.2   Results

The unsupervised domain adaptation results in terms of classification accuracy in the target domain on Office-31, ImageCLEF-DA, VisDA-2017 and Digits datasets are reported in Table 1, 2 and 3 respectively, with results of comparison methods directly cited from their original papers wherever available.

Results on the Office-31 dataset are presented in Table 1. The results of ResNet-50 trained with only source domain data serve as the lower bound. Our proposed approach can obtain the best performance in four of six tasks: D $\rightarrow$ W, W $\rightarrow$ D, A $\rightarrow$ D, and D $\rightarrow$ A. For task W $\rightarrow$ A, DMRL can produce competitive accuracy comparing with the state-of-the-art. It is noteworthy that DMRL results in improved accuracies in two hard transfer tasks: A $\rightarrow$ D, and D $\rightarrow$ A. For two easier tasks: D $\rightarrow$ W and W $\rightarrow$ D, our approach achieves accuracy no less than 99.0%. Moreover, our method can achieve the best average domain adaptation accuracy on this dataset. Given the fact that the number of samples per category is limited in the Office-31 dataset, these results can demonstrate

Table 1. Accuracy (%) on Office-31.

Method	A → W	D → W	W → D	A → D	D → A	W → A	Avg
ResNet-50 [10]	68.4 ± 0.2	96.7 ± 0.1	99.3 ± 0.1	68.9 ± 0.2	62.5 ± 0.3	60.7 ± 0.3	76.1
DAN [14]	80.5 ± 0.4	97.1 ± 0.2	99.6 ± 0.1	78.6 ± 0.2	63.6 ± 0.3	62.8 ± 0.2	80.4
DANN [4]	82.0 ± 0.4	96.9 ± 0.2	99.1 ± 0.1	79.7 ± 0.4	68.2 ± 0.4	67.4 ± 0.5	82.2
JAN [16]	85.4 ± 0.3	97.4 ± 0.2	99.8 ± 0.2	84.7 ± 0.3	68.6 ± 0.3	70.0 ± 0.4	84.3
GTA [21]	89.5 ± 0.5	97.9 ± 0.3	99.8 ± 0.4	87.7 ± 0.5	72.8 ± 0.3	**71.4 ± 0.4**	86.5
MADA [18]	90.0 ± 0.1	97.4 ± 0.1	99.6 ± 0.1	87.8 ± 0.2	70.3 ± 0.3	66.4 ± 0.3	85.2
CDAN [15]	**93.1 ± 0.2**	98.2 ± 0.2	100.0 ± 0.0	89.8 ± 0.3	70.1 ± 0.4	68.0 ± 0.4	86.6
CRST [31]	89.4 ± 0.7	98.9 ± 0.4	100.0 ± 0.0	88.7 ± 0.8	72.6 ± 0.7	70.9 ± 0.5	86.8
DMRL (Proposed)	90.8 ± 0.3	**99.0 ± 0.2**	100.0 ± 0.0	**93.4 ± 0.5**	**73.0 ± 0.3**	71.2 ± 0.3	**87.9**

that our method manages to improve the generalization ability of adversarial domain adaptation in the target domain.

Table 2. Accuracy (%) on ImageCLEF-DA.

Method	I → P	P → I	I → C	C → I	C → P	P → C	Avg
ResNet-50 [10]	74.8 ± 0.3	83.9 ± 0.1	91.5 ± 0.3	78.0 ± 0.2	65.5 ± 0.3	91.2 ± 0.3	80.7
DAN [14]	74.5 ± 0.4	82.2 ± 0.2	92.8 ± 0.2	86.3 ± 0.4	69.2 ± 0.4	89.8 ± 0.4	82.5
DANN [4]	75.0 ± 0.6	86.0 ± 0.3	96.2 ± 0.4	87.0 ± 0.5	74.3 ± 0.5	91.5 ± 0.6	85.0
JAN [16]	76.8 ± 0.4	88.0 ± 0.2	94.7 ± 0.2	89.5 ± 0.3	74.2 ± 0.3	91.7 + 0.3	85.8
MADA [18]	75.0 ± 0.3	87.9 ± 0.2	96.0 ± 0.3	88.8 ± 0.3	75.2 ± 0.2	92.2 ± 0.3	85.8
CDAN [15]	76.7 ± 0.3	90.6 ± 0.3	97.0 ± 0.4	90.5 ± 0.4	74.5 ± 0.3	93.5 ± 0.4	87.1
DMRL (Proposed)	**77.3 ± 0.4**	**90.7 ± 0.3**	**97.4 ± 0.3**	**91.8 ± 0.3**	**76.0 ± 0.5**	**94.8 ± 0.3**	**88.0**

Results on the ImageCLEF-DA dataset are shown in Table 2, the results reveal several interesting observations: (1) Deep transfer learning methods outperform standard deep learning methods; this validates that domain shifts cannot be captured by deep networks [28]. (2) The three domains in the ImageCLEF-DA dataset are more balanced than those in the Office-31 dataset. We can verify whether the performance of domain adaptation models can be improved when domain size does not change, with these more balanced domains. From Table 2, we can see that our approach can outperform all comparison methods in all transfer tasks but with lower improvements compared to the results of the Office-31 dataset in term of the average accuracy, which validates that the domain size may cause domain shift [16]. (3) our proposed DMRL method can achieve a new state-of-the-art on the ImageCLEF-DA dataset, strongly confirming the effectiveness of our method in aligning the features across domains. Moreover, for three of six tasks: C → I, C → P and P → C, our method can produce results with larger rooms of improvement, which further illustrates the effectiveness of the two mixup-based regularization mechanisms.

Positive results are also obtained on the VisDA-2017 and Digits datasets, as shown in Table 3. For VisDA-2017, our approach achieves the highest accuracy among all compared methods, and exceeds the baseline of ResNet-101 pre-trained

Table 3. Accuracy (%) on digits and VisDA-2017.

Method	MNIST → USPS	USPS → MNIST	SVHN → MNIST	Avg	Method	Synthetic → Real
No Adaptation [12]	82.2	69.6	67.1	73.0	ResNet-101 [10]	52.4
DANN [4]	90.4	94.7	84.2	89.8	DANN [4]	57.4
ADDA [24]	89.4	90.1	86.3	88.6	DAN [14]	61.1
CyCADA [12]	95.6	96.5	90.4	94.2	JAN [16]	65.7
MCD [20]	92.1	90.0	94.2	92.1	GTA [21]	69.5
CDAN [15]	93.9	96.9	88.5	93.1	MCD [20]	71.9
CADA [30]	**96.4**	97.0	90.9	95.6	CDAN [15]	73.7
DMRL (Proposed)	96.1	**99.0**	**96.2**	**97.2**	DMRL (Proposed)	**75.5**

on ImageNet with a great margin. For the Digits datasets, our approach can gain improvements of more than 1.5% on two tasks: USPS → MNIST and SVHN → MNIST, while for the MNIST → USPS task, we can obtain competitive results with the existing approaches.

4.3   Further Analysis

Ablation Study. We conduct ablation study with two domain adaptation tasks on the Digits dataset, MNIST → USPS and USPS → MNIST, to investigate the contributions of different components in DMRL. First, in order to examine the effectiveness of domain mixup (DM) and category mixup (CM), we produce two variants of DMRL: DMRL(w/o DM) and DMRL(w/o CM). Furthermore, since category mixup consists of two components: labeled category mixup (LCM) and unlabeled category mixup (UDM), DMRL(w/o LCM) and DMRL(w/o UCM) are also taken into consideration. In addition, the very basic baseline "No Adaptation", which simply trains the model on the source domain and tests on the target domain, is included in our study as well. The comparison results are presented in Table 4. The results show that both domain mixup and category mixup can contribute to our method, which verify the effectiveness of these two mixup-based regularizations. In particular, category mixup contributes more to the model than domain mixup. When evaluating two components in category mixup, we can see that the model without unlabeled category mixup produces worse classification accuracies for both tasks, demonstrating that class-aware information in the target domain can make a significant contribution in adversarial domain adaptation. All variants produce inferior results, and the full model with two regularizations produces the best results. This validates the contribution of both category and domain mixup regularization terms.

Parameter Sensitivity Analysis. In this section, we discuss the sensitivity of our approach to the values of the hyperparameters α, λ_s, λ_r and λ_t. λ_s, λ_r and

Table 4. Ablation studies.

Method	MNIST → USPS	USPS → MNIST
No Adaptation [12]	82.2	69.6
DMRL(w/o DM)	94.8	97.8
DMRL(w/o CM)	90.3	92.6
DMRL(w/o LCM)	95.0	97.9
DMRL(w/o UCM)	90.7	93.3
DMRL(full)	96.1	99.0

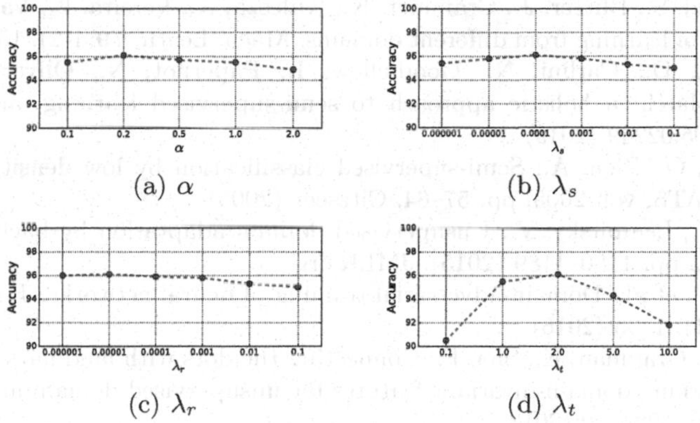

(a) α

(b) λ_s

(c) λ_r

(d) λ_t

Fig. 4. Parameter sensitivity analysis

λ_t are used to trade-off among losses, and α constrains the selection of λ when conducting Mixup. We evaluate these hyperparameters on the Digits dataset, especially, the MNIST → USPS task. When evaluating one hyperparameter, the others are fixed to their default values (e.g., $\alpha = 2$, $\lambda_s = 0.0001$, $\lambda_r = 0.00001$ and $\lambda_t = 2$). α is tested in the range $\{0.1, 0.2, 0.5, 1.0, 2.0\}$, λ_s and λ_r are explored in the range $\{0.000001, 0.00001, 0.0001, 0.001, 0.01, 0.1\}$, and λ_t is evaluated in the range $\{0.1, 1, 2, 5, 6, 10\}$. The experimental results are reported in Fig. 4.

From Fig. 4, it can be observed that the domain adaptation performance is not sensitive to the hyperparameters α, λ_s and λ_r. Consequently, we can set α, λ_s and λ_r as 0.2, 0.0001 and 0.00001 in all experiments. In addition, with the increase of λ_t, the accuracy increases dramatically and reaches the best value at $\lambda_t = 2.0$, then it decreases rapidly. The parameter sensitivity analysis illustrates that a properly selected λ_t can effectively improve the performance.

5   Conclusion

In this paper, we propose a dual mixup regularized learning (DMRL) framework for adversarial domain adaptation. By conducting category and domain mixup on pixel level, the DMRL cannot only guide the classifier in enhancing consistent predictions in-between samples, which can help avoid mismatches and enforce a stronger discriminability of the latent space, but also explore more internal structures in the latent space, which leads to a more continuous latent space. These two mixup-based regularizations can enhance and complement each other to learn discriminative and domain-invariant representations for target task. The experiments demonstrate that the proposed DMRL can effectively gain performance improvements on unsupervised domain adaptation tasks.

References

1. Ben-David, S., Blitzer, J., Crammer, K., Kulesza, A., Pereira, F., Vaughan, J.W.: A theory of learning from different domains. Mach. Learn. **79**(1-2), 151–175 (2010)
2. Berthelot, D., Carlini, N., Goodfellow, I., Papernot, N., Oliver, A., Raffel, C.: MixMatch: a holistic approach to semi-supervised learning. arXiv preprint arXiv:1905.02249 (2019)
3. Chapelle, O., Zien, A.: Semi-supervised classification by low density separation. In: AISTATS, vol. 2005, pp. 57–64. Citeseer (2005)
4. Ganin, Y., Lempitsky, V.: Unsupervised domain adaptation by backpropagation. In: ICML, pp. 1180–1189 (2015). JMLR.org
5. Ganin, Y., et al.: Domain-adversarial training of neural networks. J. Mach. Learn. Res. **17**(1), 1–35 (2016)
6. Gong, B., Grauman, K., Sha, F.: Connecting the dots with landmarks: discriminatively learning domain-invariant features for unsupervised domain adaptation. In: ICML, pp. 222–230 (2013)
7. Gong, B., Shi, Y., Sha, F., Grauman, K.: Geodesic flow kernel for unsupervised domain adaptation. In: CVPR, pp. 2066–2073 (2012)
8. Goodfellow, I., et al.: Generative adversarial nets. In: Advances in Neural Information Processing Systems, pp. 2672–2680 (2014)
9. Gretton, A., Borgwardt, K., Rasch, M., Schölkopf, B., Smola, A.J.: A kernel method for the two-sample-problem. In: Advances in Neural Information Processing Systems, pp. 513–520 (2007)
10. He, K., Zhang, X., Ren, S., Sun, J.: Deep residual learning for image recognition. In: CVPR, pp. 770–778 (2016)
11. Hinton, G., et al.: Deep neural networks for acoustic modeling in speech recognition. IEEE Sig. Process. Mag. **29** (2012)
12. Hoffman, J., et al.: Cycada: cycle-consistent adversarial domain adaptation. In: ICML (2018)
13. Krizhevsky, A., Sutskever, I., Hinton, G.E.: ImageNet classification with deep convolutional neural networks. In: Advances in Neural Information Processing Systems, pp. 1097–1105 (2012)
14. Long, M., Cao, Y., Wang, J., Jordan, M.I.: Learning transferable features with deep adaptation networks. In: ICML, pp. 97–105 (2015). JMLR.org
15. Long, M., Cao, Z., Wang, J., Jordan, M.I.: Conditional adversarial domain adaptation. In: Advances in Neural Information Processing Systems, pp. 1640–1650 (2018)
16. Long, M., Zhu, H., Wang, J., Jordan, M.I.: Deep transfer learning with joint adaptation networks. In: ICML, pp. 2208–2217 (2017). JMLR.org
17. Pan, S.J., Yang, Q.: A survey on transfer learning. IEEE Trans. Knowl. Data Eng. **22**(10), 1345–1359 (2009)

18. Pei, Z., Cao, Z., Long, M., Wang, J.: Multi-adversarial domain adaptation. In: Thirty-Second AAAI Conference on Artificial Intelligence (2018)
19. Saito, K., Ushiku, Y., Harada, T.: Asymmetric tri-training for unsupervised domain adaptation. In: ICML, pp. 2988–2997 (2017). JMLR.org
20. Saito, K., Watanabe, K., Ushiku, Y., Harada, T.: Maximum classifier discrepancy for unsupervised domain adaptation. In: CVPR, pp. 3723–3732 (2018)
21. Sankaranarayanan, S., Balaji, Y., Castillo, C.D., Chellappa, R.: Generate to adapt: aligning domains using generative adversarial networks. In: CVPR, pp. 8503–8512 (2018)
22. Shimodaira, H.: Improving predictive inference under covariate shift by weighting the log-likelihood function. J. Stat. Plann. Inference **90**(2), 227–244 (2000)
23. Silver, D., et al.: Mastering the game of go with deep neural networks and tree search. Nature **529**(7587), 484 (2016)
24. Tzeng, E., Hoffman, J., Saenko, K., Darrell, T.: Adversarial discriminative domain adaptation. In: CVPR, pp. 7167–7176 (2017)
25. Tzeng, E., Hoffman, J., Zhang, N., Saenko, K., Darrell, T.: Deep domain confusion: maximizing for domain invariance. arXiv preprint arXiv:1412.3474 (2014)
26. Verma, V., et al.: Manifold mixup: better representations by interpolating hidden states. arXiv preprint arXiv:1806.05236 (2018)
27. Verma, V., Lamb, A., Kannala, J., Bengio, Y., Lopez-Paz, D.: Interpolation consistency training for semi-supervised learning. arXiv preprint arXiv:1903.03825 (2019)
28. Yosinski, J., Clune, J., Bengio, Y., Lipson, H.: How transferable are features in deep neural networks? In: Advances in Neural Information Processing Systems, pp. 3320–3328 (2014)
29. Zhang, H., Cisse, M., Dauphin, Y.N., Lopez-Paz, D.: mixup: beyond empirical risk minimization. arXiv preprint arXiv:1710.09412 (2017)
30. Zou, H., Zhou, Y., Yang, J., Liu, H., Das, H.P., Spanos, C.J.: Consensus adversarial domain adaptation. In: Proceedings of the AAAI Conference on Artificial Intelligence, vol. 33, pp. 5997–6004 (2019)
31. Zou, Y., Yu, Z., Liu, X., Kumar, B., Wang, J.: Confidence regularized self-training. In: Proceedings of the IEEE International Conference on Computer Vision, pp. 5982–5991 (2019)

Robust and On-the-Fly Dataset Denoising for Image Classification

Jiaming Song[1(✉)], Yann Dauphin[2], Michael Auli[3], and Tengyu Ma[1(✉)]

[1] Stanford University, Stanford, USA
{tsong,tengyuma}@stanford.edu
[2] Google Brain, Mountain View, USA
yann@dauphin.io
[3] Facebook AI Research, New York, USA
michaelauli@fb.com

Abstract. Memorization in over-parameterized neural networks could severely hurt generalization in the presence of mislabeled examples. However, mislabeled examples are hard to avoid in extremely large datasets collected with weak supervision. We address this problem by reasoning counterfactually about the loss distribution of examples with uniform random labels had they were trained with the real examples, and use this information to remove noisy examples from the training set. First, we observe that examples with uniform random labels have higher losses when trained with stochastic gradient descent under large learning rates. Then, we propose to model the loss distribution of the counterfactual examples using only the network parameters, which is able to model such examples with remarkable success. Finally, we propose to remove examples whose loss exceeds a certain quantile of the modeled loss distribution. This leads to On-the-fly Data Denoising (ODD), a simple yet effective algorithm that is robust to mislabeled examples, while introducing almost zero computational overhead compared to standard training. ODD is able to achieve state-of-the-art results on a wide range of datasets including real-world ones such as WebVision and Clothing1M.

1 Introduction

Over-parametrized deep neural networks have remarkable generalization properties while achieving near-zero training error [45]. However, the ability to fit the entire training set is highly undesirable, as a small portion of mislabeled examples in the dataset could severely hurt generalization [1,45]. Meanwhile, an exponential growth in training data size is required to linearly improve generalization in vision [37]; this progress could be hindered if there are mislabeled examples within the dataset.

Work done at Facebook AI research.

Electronic supplementary material The online version of this chapter (https://doi.org/10.1007/978-3-030-58526-6_33) contains supplementary material, which is available to authorized users.

© Springer Nature Switzerland AG 2020
A. Vedaldi et al. (Eds.): ECCV 2020, LNCS 12374, pp. 556–572, 2020.
https://doi.org/10.1007/978-3-030-58526-6_33

Mislabeled examples are to be expected in large datasets that contain millions of examples. Web-based supervision produces noisy labels [21, 26] whereas human labeled datasets sacrifice accuracy for scalability [19]. Therefore, algorithms that are robust to various levels of mislabeled examples are warranted in order to further improve generalization for very large labeled datasets.

In this paper, we are motivated by the observation that crowd-sourcing or web-supervision could have multiple disagreeing sources; in such cases, noisy labels could exhibit higher conditional entropy than the ground truth labels. Since the information about the noisy labels (such as the amount of noise) is often scarce, we pursue *a general approach* by counterfactual reasoning of the behavior of noisy examples with high conditional entropy. Specifically, we reason about the *counterfactual* case of how examples with uniform random noise would behave *had they appeared in the training dataset*, without actually training on such labels. If a real example has higher loss than what most counterfactual examples with uniform random noise would have, then there is reason to believe that this example is likely to contain a noisy label; removing this example would then improve performance on a clean test set.

To reason about the *counterfactual loss distribution* of examples with uniform random noise, we first show that training residual networks with *large learning rates* will create a significant gap between the losses of clean examples and noisy examples. The distribution of training loss over clean examples decrease yet that of the uniformly noisy examples does not change, regardless of the proportion of noisy examples in the dataset. Based on this observation, we propose a distribution that simulates the loss distribution of uniform noisy examples based only on the network parameters. Reasonable thresholds can be derived from percentiles of this distribution, which we can then utilize to denoise the dataset. This is critical in real-world applications, because prior knowledge about the distribution of label noise is often scarce; even if we have such information (such as transition matrices of label noise), algorithms that specifically utilize this information are not scalable when there are thousands of labels.

We proceed to propose *On-the-fly Data Denoising* (ODD, see Fig. 1), a simple and robust method for training with noisy examples based on the implicit regularization effect of stochastic gradient descent. First, we train residual networks with large learning rate schedules and use the resulting losses to separate clean examples from mislabeled ones. This is done by identifying examples whose losses exceed a certain threshold. Finally, we remove these examples from the dataset and continue training until convergence. ODD is a general approach that can be used to train clean dataset as well as noisy datasets with almost no modifications.

Empirically, ODD performs favorably against previous methods in datasets containing *real-world noisy examples*, such as WebVision [21] and Clothing1M [41]. ODD also achieves equal or better accuracy than the state-of-the-art *on clean datasets*, such as CIFAR and ImageNet. We further conduct ablation studies to demonstrate that ODD is robust to different hyperparameters and artificial noise levels. Qualitatively, we demonstrate the effectiveness of ODD by

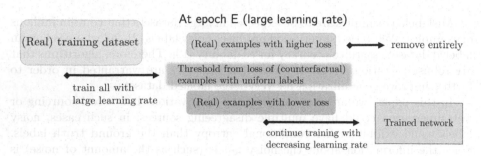

Fig. 1. Pipeline of our method. We utilize the implicit regularization effect of SGD to (counterfactually) reason the loss distribution of examples with uniform label noise. We remove examples that have loss higher than the threshold and train on the remaining examples. There is no assumption that the dataset has to contain uniformly random labels (thus such labels are "counterfactual"); we empirically validate our method on real-world noisy datasets.

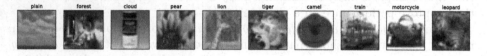

Fig. 2. Mislabeled examples in the CIFAR-100 training set detected by ODD.

detecting mislabeled examples in the "clean" CIFAR-100 dataset without any supervision other than the training labels (Fig. 2). These results suggest that we can use ODD in both clean and noisy datasets with minimum computational overhead to the training algorithm.

2 Problem Setup

The goal of supervised learning is to find a function $f \in \mathcal{F}$ that describes the probability of a random label vector $Y \in \mathcal{Y}$ given a random input vector $X \in \mathcal{X}$, which has underlying joint distribution $P(X, Y)$. Given a loss function $\ell(\mathbf{y}, \hat{\mathbf{y}})$, one could minimize the average of ℓ over P:

$$\mathcal{R}(f) = \int_{\mathcal{X} \times \mathcal{Y}} \ell(\mathbf{y}, f(\mathbf{x})) \, dP(\mathbf{x}, \mathbf{y}), \qquad (1)$$

The joint distribution $P(X, Y)$ is usually unknown, but we could gain access to its samples via a potentially noisy labeling process, such as crowdsourcing [19] or web queries [21].

We denote the training dataset with N examples as $\mathcal{D} = (\mathbf{x}_i, \mathbf{y}_i)_{i \in [N]} = \mathcal{G} \cup \mathcal{B}$. $\mathcal{G}$ represents correctly labeled (clean) examples sampled from $P(X, Y)$. $\mathcal{B}$ represents mislabeled examples that are not sampled from $P(X, Y)$, but from another distribution $Q(X, Y)$. We assume that $\mathcal{G} \cap \mathcal{B} = \varnothing$, as a sample should not be both correctly labeled and mislabeled.

We aim to learn the function f from $\mathcal{D}$ *without knowledge about* $\mathcal{B}, \mathcal{G}$ *or their statistics* (e.g. the amount of mislabeled examples $|\mathcal{B}|$). A typical approach is to pretend that $\mathcal{B} = \varnothing$—i.e., all examples are i.i.d. from $P(X, Y)$—and minimize the empirical risk:

$$\hat{\mathcal{R}}(f) = \frac{1}{N} \sum_{i=1}^{N} \ell(\mathbf{y}, f(\mathbf{x})).$$

If $\mathcal{B} = \varnothing$ is indeed true, then the empirical risk converges to the population risk: $\hat{\mathcal{R}}(f) \to \mathcal{R}(f)$ as $N \to \infty$. However, if $\mathcal{B} \neq \varnothing$, then $\hat{\mathcal{R}}(f)$ is no longer an unbiased estimator of $\mathcal{R}(f)$. Moreover, when the hypothesis class $\mathcal{F}$ contains large neural nets with the number of parameters exceeding N, the empirical risk minimizer could fit the entire training dataset, including the mislabeled examples [45]. Overfitting to wrong labels empirically causes poor generalization. For example, training CIFAR-10 with 20% of uniformly mislabeled examples and a residual network gives a test error of 11.5%, which is significantly higher than the 4.25% error obtained with training on the clean examples.

2.1   Entropy-Based Assumption over Noisy Labels

Therefore, if we were able to identify the clean examples belonging to $\mathcal{G}$, we could vastly improve the generalization on $P(X, Y)$; this requires us to provide valid prior assumptions that could distinguish clean examples from mislabeled ones. We note that these assumptions have to be general enough so as to *not depend on additional assumptions specific to each dataset*. For example, knowledge about noise transition matrices is not allowed.

We assume that for any example $\mathbf{x} \in \mathcal{X}$, the entropy of the clean label distribution is smaller than that of the noisy label distribution:

$$\mathcal{H}(P(Y|X = \mathbf{x})) < \mathcal{H}(Q(Y|X = \mathbf{x})) \quad \forall \mathbf{x} \in \mathcal{X} \tag{2}$$

where the randomness of labeling $Q(Y|X)$ could arise from noisy labelings, such as Mechanical Turk [19]. Let ℓ be the cross entropy loss, then the ERM objective is essentially trying to minimize the KL divergence between the empirical conditional distribution (denoted as $\hat{P}(\mathbf{y}|\mathbf{x})$) and the conditional distribution parametrized by our model (denoted as $p_\theta(\mathbf{y}|\mathbf{x})$):

$$\mathbb{E}_{\hat{P}(\mathbf{y}|\mathbf{x})}[-\log p_\theta(\mathbf{y}|\mathbf{x})] = \mathcal{H}(\hat{P}(\mathbf{y}|\mathbf{x})) + D_{\mathrm{KL}}(\hat{P}(\mathbf{y}|\mathbf{x})\|p_\theta(\mathbf{y}|\mathbf{x})) \tag{3}$$

which is minimized as $D_{\mathrm{KL}} \to 0$; in this case, the cross entropy loss is higher if $\hat{P}$ has higher entropy, which suggests that the mislabeled examples are likely to have higher loss than correct ones.

3   Denoising Datasets On-the-Fly with Counterfactual Thresholds

In the following section, we study the behavior of samples with uniformly random label noise; this allows us to reason about their loss distribution *counterfactually*, and develop suitable thresholds to remove noisy examples that appear in the training set. We denote the conditional distribution of uniformly random label noise as $Q_U(Y|X)$, which is the distribution that maximizes entropy; therefore, any real-world noise distribution $Q(Y|X)$ will have smaller entropy than Q_U.

While it is unreasonable to assume that the label noise is uniformly random in practice, we do not make such assumption over our training set. Instead, we reason about the following counterfactual case:

> *Had the training set contained some examples with uniform random labels, can we characterize the loss distribution of these examples?*

We illustrate how such a counterfactual analysis allows simple and practical algorithms that work even under real-world noisy datasets. The following sections are organized as follows:

1. First, we show that when training ResNets via SGD with large learning rates[1], the training loss of uniform noisy labels and clean labels can be clearly separated. As the learning rate becomes smaller, however, the model would begin to overfit the noisy labels, leading to poor generalization properties.
2. Next, we propose an approach to model the (counterfactual) loss distribution *by only looking at the weights of the network*. We empirically show that this does not depend on the type or the amount of noisy labels in the dataset, making this approach generalize well to various counterfactual scenarios (such as different portions of uniform random labels in the dataset).
3. Finally, we can simply remove all examples that perform worse than a certain percentile of the counterfactual distribution. Since higher entropy examples tend to have higher loss than lower entropy ones, the samples we remove are more likely to be more noisy. In Fig. 2, we empirically demonstrate that the proposed threshold identifies mislabeled samples in CIFAR-100 even without any additional supervision, validating our assumption.

3.1 Separating Mislabeled Examples via SGD

First, we find that training the model with stochastic gradient descent (SGD) with large learning rates (e.g. 0.1) will result in significant discrepancy between the loss statistics of the clean examples and mislabeled examples. To illustrate this, we consider training deep residual networks on CIFAR-100 and ImageNet with different percentages of uniform label noise (20% and 40%), but with large learning rates (close to 0.1), and at specific epochs, we plot the histogram of the loss for each example in Fig. 3.

From the histograms, we may draw two key observations that are useful for cleaning the dataset without knowledge about the exact percentage of noise:

---

[1] Here, "large learning rates" refer to ones that are close to initial learning rates in practice; in standard ResNet training [10], the initial learning rate is 0.1, so we consider it "large"; we use the default momentum parameter of 0.9 for all cases discussed.

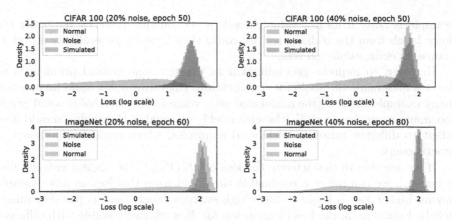

Fig. 3. Histogram of the distributions of losses, where "normal", "noise", and "simulated" denote (real) examples with clean labels, (real) examples with uniform random labels and the counterfactual model $q_n(\ell)$ respectively. $q_n(\ell)$ matches the loss distribution of noisy examples, which have higher loss than clean ones; $q_n(\ell)$ depends only on the network parameters.

- First, the loss distributions of the clean examples and the mislabeled ones have *notable statistical distance.*
- Second, it seems that the loss distribution of the uniform labeled examples are relatively stable, and *does not depend much on the amount of uniform random noise in the training set.*

These observations are consistent with those in [45], as the network starts to fit mislabeled examples when learning rate decrease further; decreasing learning rate is crucial for achieving better generalization on clean datasets.

The working of the implicit regularization of stochastic gradient descent is by and large an open question that attracts much recent attentions [6, 22, 27, 28]. Empirically, it has been observed that large learning rates are beneficial for generalization [18]. Chaudhari and Soatto [4] have argued that SGD iterates converge to limit cycles with entropic regularization proportional to the learning rate and inversely proportional to batch size. Training with large learning rates under fixed batch sizes could then encourage solutions that are more robust to large random perturbations in the parameter space and less likely to overfit to mislabeled examples, which would partially explain our observations in Fig. 3.

Given these empirical and theoretical evidences on the benefits of large learning rates on generalization, we propose to classify correct and mislabeled examples through the loss statistics, and achieve better generalization by removing the examples that are potentially mislabeled.

3.2 Thresholds that Classify Mislabeled Examples

The above observation suggests that it is possible to distinguish clean and noisy examples *via a threshold over the loss value.* In principle, we can claim an example is noisy if its loss value exceeds a certain threshold; by removing the noisy labels from the training set, we could then improve generalization performance on clean validation sets.

However, to improve generalization in practice, one critical problem is to select a reasonable threshold for classification. High thresholds could include too many examples from $\mathcal{B}$ (the mislabeled set), whereas low thresholds could prune too many examples from $\mathcal{G}$ (the clean set); reasonable thresholds should also adapt to different ratios of mislabeled examples, which could be unknown to practitioners.

If we are able to characterize the loss of $Q_U(Y|X)$ (the highest entropy distribution), we can select a reasonable threshold from this loss as any example having higher loss is likely to have high entropy labels (and is possibly mislabeled). From Fig. 3, the loss distribution for $\mathcal{B}$ is relatively stable with different ratios of $|\mathcal{B}|/|\mathcal{D}|$; examples in $\mathcal{B}$ are making little progress when learning rate is large. This suggests a threshold selecting criteria that is *independent of the amount of mislabeled examples in the dataset.*

We propose to characterize the loss distribution of (counterfactual) uniform label noise via the following procedure:

$$
l = -\tilde{\mathbf{y}}_k + \log\left(\sum_{i \in [N]} \exp(\tilde{\mathbf{y}}_i)\right)
$$
$$
\tilde{\mathbf{y}} = \mathrm{fc}(\mathrm{relu}(\tilde{\mathbf{x}})), \tilde{\mathbf{x}} \sim \mathcal{N}(0, I), k \sim \mathrm{Uniform}\{0, \dots, K\}
$$
(4)

We denote this counterfactual distribution model as $q_n(l)$. $q_n(l)$ tries to simulate the behavior of the model (and the loss distribution) with several components.

– k represents a random label from K classes. This simulates the case where $Q(Y|X)$ has the highest entropy, i.e. uniformly random.
– $\mathrm{fc}(\cdot)$ is the final (fully connected) layer of the network and $\mathrm{relu}(\tilde{\mathbf{x}}) = \max(\tilde{\mathbf{x}}, \mathbf{0})$ is the Rectified Linear Unit. This simulates the behavior at the last layer of the network outputs $\tilde{\mathbf{y}}$.
– $\tilde{\mathbf{x}} \sim \mathcal{N}(0, I)$ suggests that the inputs to the last layer has an identity covariance; the scale of the covariance could result from well-conditioned objectives defined via deep residual networks [10], batch normalization [15] and careful initialization [11].

We qualitatively demonstrate the validity of our characterization on CIFAR-100 and ImageNet datasets in Fig. 3, where we plot the histogram of the $q_n(l)$ distribution for CIFAR-100 and ImageNet (our estimates), and compare then with the empirical distribution of the loss of uniform noisy labeled examples (ground truth). The similarities between the noisy loss distribution and simulated loss distribution $q_n(l)$ demonstrate that an accurate characterization of the loss distribution can be made *without prior knowledge of the mislabeled examples.*

To effectively trade-off between precision (correctly identifying noisy examples) and recall (identifying more noisy examples), we define a threshold via the p-th percentile of $q_n(l)$ using the samples generated by Eq. 4; it relates to approximately how much examples in $\mathcal{B}$ we would retain if $Q(Y|X)$ is uniform.

In Sect. 4.5, we show that this method is able to identify different percentages of uniform label noise with high precision.

3.3 A Practical Algorithm for Robust Training

We can utilize this to remove examples that might harm generalization, leading to *On-the-fly Data Denoising* (ODD), a simple algorithm that is robust to mislabeled examples. First, we train all the samples using a relatively large learning rate (e.g. 0.1), which should favor learning clean samples than noisy ones. Next, we use the threshold to cutoff samples that have large loss values, which we deem as having been mislabeled. Finally, we train the remaining samples using the original ERM procedure.

Algorithm 1. On-the-fly Data Denoising

Input: dataset $\mathcal{D}$, model f_θ, percentile p, epoch E, learning rate schedule $\eta(t)$.
for $e = 1 \ldots E$ **do**
    Train on $\mathcal{D}$ with learning rate $\eta(e)$.
end for
$T_p = p$-th percentile of $q_n(\ell)$ in Eq. (4)
Remove "noisy" data according to threshold: $\mathcal{G} = \{(x, y) \mid \ell(y, f_\theta(x)) < T_p\}$
for $e = E + 1 \ldots$ **do**
    Train on $\mathcal{G}$ with learning rate $\eta(e)$.
end for

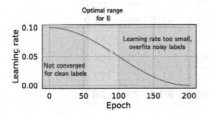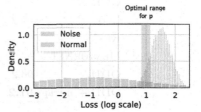

Fig. 4. Hyperparameter selection. (Left) Cosine learning rate schedule across epochs that is typical in ResNet training; we wish to select E before learning rate becomes small, and after training over clean labels have converged. (Right) Histogram of the losses; we wish to select p that does not remove too many clean data, but also removes as many (conterfactually) noisy data as possible.

Hyperparameter Selection. ODD introduces two hyperparameters: E determines the amount of training that separates clean examples from noisy ones; p determines T_p that specifies the trade-off between less noisy examples and more clean examples. We do not explicitly estimate the portion of noise in the dataset, nor do we assume any specific noise model. Moreover, ODD is compatible with existing practices for learning rate schedules, such as stepwise [10] or cosine [24].

In Fig. 4 we demonstrate and discuss how to choose the hyperparameters E and p. For E, we wish to perform ODD operation at a point not too early (to allow enough time for training on clean labels to converge) and not too late (to prevent overfitting noisy labels with small learning rates). For p, we wish to trade-off between keeping as much clean data as possible and removing counterfactually noisy data; selecting $p \in [1, 30]$ typically works for our case.

4   Experiments

We evaluate our method extensively on several clean and noisy datasets including CIFAR-10, CIFAR-100, ImageNet [34], WebVision [21] and Clothing1M [41]. Our experiments consider datasets that are clean, have artificial noise (in CIFAR-10, CIFAR-100 and ImageNet), or have inherent noise from web-supervision (as in the case of WebVision and Clothing1M).

Table 1. Validation accuracy (in percentage) with uniform label noise. $\star$ denotes methods trained with knowledge of 1000 additional clean labels

	CIFAR-10			CIFAR-100		
% mislabeled	0	20	40	0	20	40
ERM	96.3	88.5	84.4	81.6	69.6	55.7
mixup	97.0	93.9	91.7	81.4	71.2	59.4
GCE	–	89.9	87.1	–	66.8	62.7
Luo	96.2	**96.2**	94.9	81.4	80.6	74.2
Ren*	–	–	86.9	–	–	61.4
MentorNet*	–	92.0	89.0	–	73.0	68.0
ODD	96.2	94.7	92.8	81.8	77.2	72.4
ODD + *mixup*	**97.2**	**95.6**	**95.5**	**82.5**	**79.1**	**76.5**

4.1   CIFAR-10 and CIFAR-100

We first evaluate our method on the CIFAR-10 and CIFAR-100 datasets. We train the wide residual network architecture (WRN-28-10) in [44] for 200 epochs with a minibatch size of 128, momentum 0.9 and weight decay 5×10^{-4}. We set $E = 75$ (total number of epochs is 200) and $p = 10$ in our experiments.

Input-Agnostic Label Noise. We first consider label noise that are agnostic to inputs. Following [45], We randomly replace 0%/20%/40% of the training labels to uniformly random ones, and evaluate generalization error on the clean validation set. We compare with the following baselines: Empirical Risk Minimization (ERM, Eq. 1, [8]), MENTORNET [16], REN [32], *mixup* [46], Generalized Cross Entropy (GCE, [47]) and LUO [25], a strong baseline which regularizes the Jacobian of the network. In particular, LUO would spend about two times the computational effort when performing one iteration due to back-propagation through the Jacobian. We also consider using *mixup* training after pruning noisy examples with ODD. From the results in Table 1, ODD + *mixup* outperforms all other algorithms (except for LUO).

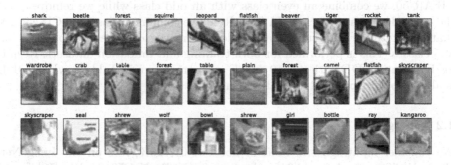

Fig. 5. Examples with label "leopard" that are classified as mislabeled.

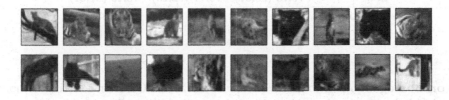

Fig. 6. Random CIFAR-100 examples that are classified as mislabeled.

Mislabeled Examples in CIFAR-100. To demonstrate that ODD can identify mislabeled examples even in practice, we display the examples in CIFAR-100 training set for which our ODD methods identify as noise across 3 random seeds. One of the most common case is the "leopard" label; in fact, 21 "leopard" examples in the training set are perceived as hard, and we show some of them in Fig. 5. It turns out that a lot of the "leopard" examples contains images that clearly contains tigers and black panthers (CIFAR-100 has a label corresponding to "tiger", and only 2 out of 100 "leopard" test images are black panthers). We also demonstrate random examples from the CIFAR-100 that are identified as noise in Fig. 6. The examples identified as noise often contains multiple or ambiguous objects. We include more results in Section A.2.

Table 2. Results on non-homogeneous labels.

% samples removed	CIFAR-50			CIFAR-20		
	0	20	40	0	20	40
ERM	78.5	77.9	77.5	86.4	85.1	84.4
ODD	**79.0**	**78.6**	**78.1**	**86.6**	**85.4**	**84.7**

Table 3. Top-1 (top-5) accuracy on ImageNet.

% mislabeled	0	20	40
ERM	76.61 (93.23)	73.77 (91.49)	71.39 (89.48)
LUO	76.73 (93.31)	75.17 (92.26)	73.19 (91.01)
MENTORNET	–	–	65.1 (85.9)
ODD ($p = 10$)	**76.71** (93.23)	**74.95** (92.11)	**72.49** (90.75)

Non-homogeneous Labels. We evaluate ERM and ODD on a setting without mislabeled examples, but the ratio of classes could vary. To prevent the model from utilizing the number of examples in a class, we combine multiple classes of CIFAR-100 into a single class, creating the CIFAR-20 and CIFAR-50 tasks. In CIFAR-50, we combine an even class with an odd class while we remove $c\%$ of the examples in the odd class. In CIFAR-20, we combine 5 classes in CIFAR-100 that belong to the same super-class while we remove $c\%$ of the examples in 4 out of 5 classes. This is performed for both training and validation datasets. Results for ERM and ODD with $p = 10$ and $E = 75$ are shown in Table 2, where ODD is able to outperform ERM in all settings.

4.2   ImageNet

We consider input-agnostic random noise of 0%/20%/40% on the ImageNet-2012 classification dataset [34], where we train ResNet-50 models [10] for 90 epochs and report top-1 and top-5 validation errors in Table 3. ODD significantly outperforms ERM and MENTORNET in terms of both top-1 and top-5 errors with 20% and 40% label noise, while being comparable to LUO in these scenarios. Our training time is comparable to that of ERM and much more efficient than LUO since we do not need to differentiate through the Jacobian.

4.3   WebVision

We further verify the effectiveness of our method on a real-world noisy dataset. The WebVision-2017 dataset [21] contains 2.4 million of real-world noisy labels, that are crawled from Google and Flickr using the 1,000 labels from the ImageNet-2012 dataset. We consider training ResNet-50 [38] for 50 epochs, and use both WebVision and ImageNet validation sets for 1-crop validation, following

Table 4. Top-1 (top-5) accuracy on WebVision and ImageNet validation sets when trained on WebVision.

Method	WebVision	ImageNet
LASS [1]	66.6 (85.6)	59.0 (80.8)
CleanNet [20]	68.5 (86.5)	60.2 (81.1)
ERM	69.7 (87.0)	62.9 (83.6)
MentorNet [16]	70.8 (88.0)	62.5 (83.0)
CurriculumNet [9]	73.1 (89.2)	64.7 (84.9)
Luo [25]	72.7 (89.5)	64.9 (85.7)
ODD	72.6 (89.3)	64.8 (85.5)

the settings in [16]. We do not use a pretrained model or additional labeled data from ImageNet. In Table 4, we demonstrate superior results than most other competitive methods tailored for learning with noisy labels, which suggests that the approach is empirically valid against noisy labels in the real world.

Our ODD method with $p = 30$ removes 9.3% of the total examples. In comparison, we removed around 1.1% of examples in ImageNet; this suggest that WebVision labels are indeed much noisier than the ImageNet labels since there are more examples removed by the (counterfactual) threshold.

Table 5. Validation accuracy on Clothing1M.

Method	Setting	Accuracy
ERM	Noisy	68.9
GCE	Noisy	69.1
Loss Correction [30]	Noisy	69.2
LCCN [43]	Noisy	71.6
Joint Opt. [39]	Noisy	72.2
DMI [42]	Noisy	72.5
ODD	Noisy	**73.5**
ERM	Clean	75.2
Loss Correction	Noisy + clean	**80.4**
ODD	Noisy + clean	**80.3**

4.4 Clothing1M

Clothing1M [41] contains 1 million examples with noisy labels and 50,000 examples with clean labels of 14 classes. Following procedures from previous work, we use the ResNet-50 architecture pre-trained on ImageNet, with a starting learning rate of 0.001 trained with 10 epochs. We consider three settings, where the dataset contains *clean* labels only, *noisy* labels only, or both types of labels. For ODD, we set $E = 1, p = 1$ for the noisy dataset ($E = 1$ because we fine-tune from ImageNet pre-trained model); we then fine-tune on the clean labels if they are available.

Table 5 suggests our method compares favorably against existing methods such as GCE, Joint Optimization [39], latent class-conditional noise model (LCCN, [43]) and Determinant based Mutual Information (DMI, [42]) on the *noisy* dataset, and is comparable to Loss Correction (LC, [30]) on the *noisy + clean* dataset. We note that the complexity of LC scales quadratically in the number of classes, and it would not be feasible for ImageNet or WebVision.

4.5   Ablation Studies

We include additional ablation studies in Appendix A.1.

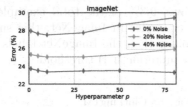 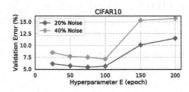

Fig. 7. (Left) ablating p on ImageNet. (Right) ablating E on CIFAR10.

Sensitivity to p. We first evaluate noisy ImageNet classification with varying p. A higher p includes more clean examples at the cost of involving more noisy examples. From Fig. 7 (left), ODD is not very sensitive to p in the range of $[0, 30)$, and empirically $p = 10$ represents the best trade-off.

Sensitivity to E. We evaluate the validation error of ODD on CIFAR with 20% and 40% input-agnoistic label noise where $E \in \{25, 50, 75, 100, 150, 200\}$ ($E = 200$ is equivalent to ERM). The results in Fig. 7 (right) demonstrate that the effect of E on final performance behaves according to our suggestion, which is to perform ODD when the learning rate is high.

5   Related Work

Generalization of SGD Training. The generalization of neural networks trained with SGD depend heavily on learning rate schedules [24]. It has been proposed that wide local minima could result in better generalization [3,13,17]. Several factors could contribute to wider local optima and better generalization, such as smaller minibatch sizes [17], reasonable learning rates [18], longer training time [14], or distance from the initialization point [14]. In the presence of mislabeled examples, changes in optimization landscape [1] could result in bad local minima [45], although it is argued that larger batch sizes could mitigate this effect [33].

Training with Mislabeled Examples. One paradigm involves estimating the noise distribution [23] or confusion matrix [36]. Another line of methods propose to identify and clean the noisy examples [5] through predictions of auxillary networks [30,40] or via binary predictions [29]; the noisy labels are either pruned [2] or replaced with model predictions [31]. Our method is comparable to these approaches, but the key difference is that we leverage the implicit regularization of SGD to identify noisy examples. We note that ODD is different from hard example mining [35] which prunes "easier" examples with lower loss; this does not remove mislabeled examples effectively. The method proposed in [29] is most similar to ours in principle, but is restricted to binary classification settings. Other approaches propose to balance the examples via a pretrained network [16], meta learning [32], or surrogate loss functions [7,39,47]. Some methods require a set of trusted examples [12,41].

ODD has several appealing properties compared to existing methods. First, the thresholds for classifying mislabeled examples from ODD do not rely on estimations of the noise confusion matrix. Next, ODD does not require additional trusted examples. Finally, ODD removes potentially noisy examples on-the-fly; it has little computational overhead compared to standard SGD training.

6 Discussion

We have proposed ODD, a straightforward method for robust training with mislabeled examples. ODD utilizes the implicit regularization effect of stochastic gradient descent, which allows us to reason counterfactually about the loss distribution of examples with uniform label noise. Based on quantiles of this (counterfactual) distribution, we can then prune examples that would potentially harm generalization. Empirical results demonstrate that ODD is able to significantly outperform related methods on a wide range of datasets with artificial and real-world mislabeled examples, maintain competitiveness with ERM on clean datasets, as well as detecting mislabeled examples automatically in CIFAR-100.

The implicit regularization of stochastic gradient descent opens up other research directions for implementing robust algorithms. For example, we could consider removing examples not only once but multiple times or retraining from scratch with the denoised dataset. Moreover, it would be interesting to understand the ODD from additional theoretical viewpoints, such as the effects of large learning rates.

References

1. Arpit, D., et al.: A closer look at memorization in deep networks. arXiv preprint arXiv:1706.05394, June 2017
2. Brodley, C.E., Friedl, M.A., et al.: Identifying and eliminating mislabeled training instances. In: Proceedings of the National Conference on Artificial Intelligence, pp. 799–805 (1996)
3. Chaudhari, P., et al.: Entropy-SGD: biasing gradient descent into wide valleys. arXiv preprint arXiv:1611.01838, November 2016
4. Chaudhari, P., Soatto, S.: Stochastic gradient descent performs variational inference, converges to limit cycles for deep networks. In: 2018 Information Theory and Applications Workshop (ITA), pp. 1–10. IEEE (2018)
5. Cretu, G.F., Stavrou, A., Locasto, M.E., Stolfo, S.J., Keromytis, A.D.: Casting out demons: sanitizing training data for anomaly sensors. In: IEEE Symposium on Security and Privacy, SP 2008, pp. 81–95. IEEE (2008)
6. Du, S.S., Hu, W., Lee, J.D.: Algorithmic regularization in learning deep homogeneous models: layers are automatically balanced. arXiv preprint arXiv:1806.00900, June 2018
7. Ghosh, A., Kumar, H., Sastry, P.S.: Robust loss functions under label noise for deep neural networks. In: AAAI, pp. 1919–1925 (2017)
8. Goyal, P., et al.: Accurate, large minibatch SGD: Training ImageNet in 1 hour. arXiv preprint arXiv:1706.02677, June 2017
9. Guo, S., et al.: CurriculumNet: weakly supervised learning from large-scale web images. In: Ferrari, V., Hebert, M., Sminchisescu, C., Weiss, Y. (eds.) ECCV 2018. LNCS, vol. 11214, pp. 139–154. Springer, Cham (2018). https://doi.org/10.1007/978-3-030-01249-6_9
10. He, K., Zhang, X., Ren, S., Sun, J.: Deep residual learning for image recognition. arXiv preprint arXiv:1512.03385, December 2015
11. He, K., Zhang, X., Ren, S., Sun, J.: Delving deep into rectifiers: Surpassing Human-Level performance on ImageNet classification. arXiv preprint arXiv:1502.01852, February 2015
12. Hendrycks, D., Mazeika, M., Wilson, D., Gimpel, K.: Using trusted data to train deep networks on labels corrupted by severe noise. arXiv preprint arXiv:1802.05300, February 2018
13. Hochreiter, S., Schmidhuber, J.: Simplifying neural nets by discovering flat minima. In: Tesauro, G., Touretzky, D.S., Leen, T.K. (eds.) Advances in Neural Information Processing Systems, vol. 7, pp. 529–536. MIT Press (1995)
14. Hoffer, E., Hubara, I., Soudry, D.: Train longer, generalize better: closing the generalization gap in large batch training of neural networks. In: Guyon, I., et al. (eds.) Advances in Neural Information Processing Systems, vol. 30, pp. 1731–1741. Curran Associates, Inc. (2017)
15. Ioffe, S., Szegedy, C.: Batch normalization: accelerating deep network training by reducing internal covariate shift. arXiv preprint arXiv:1502.03167 (2015)
16. Jiang, L., Zhou, Z., Leung, T., Li, L.J., Fei-Fei, L.: MentorNet: learning data-driven curriculum for very deep neural networks on corrupted labels. arXiv preprint arXiv:1712.05055, December 2017
17. Keskar, N.S., Mudigere, D., Nocedal, J., Smelyanskiy, M., Tang, P.T.P.: On Large-Batch training for deep learning: Generalization gap and sharp minima. arXiv preprint arXiv:1609.04836, September 2016

18. Kleinberg, R., Li, Y., Yuan, Y.: An alternative view: When does SGD escape local minima? arXiv preprint arXiv:1802.06175, February 2018
19. Krishna, R.A., et al.: Embracing error to enable rapid crowdsourcing. In: Proceedings of the 2016 CHI Conference on Human Factors in Computing Systems, CHI 2016, pp. 3167–3179. ACM, New York (2016). https://doi.org/10.1145/2858036.2858115
20. Lee, K.H., He, X., Zhang, L., Yang, L.: CleanNet: transfer learning for scalable image classifier training with label noise. In: Proceedings of the IEEE Conference on Computer Vision and Pattern Recognition, pp. 5447–5456 (2018)
21. Li, W., Wang, L., Li, W., Agustsson, E., Van Gool, L.: WebVision database: visual learning and understanding from web data. arXiv preprint arXiv:1708.02862, August 2017
22. Li, Y., Ma, T., Zhang, H.: Algorithmic regularization in over-parameterized matrix sensing and neural networks with quadratic activations. arXiv preprint arXiv:1712.09203, December 2017
23. Liu, T., Tao, D.: Classification with noisy labels by importance reweighting. arXiv preprint arXiv:1411.7718, November 2014
24. Loshchilov, I., Hutter, F.: SGDR: stochastic gradient descent with warm restarts. arXiv preprint arXiv:1608.03983, August 2016
25. Luo, Y., Zhu, J., Pfister, T.: A simple yet effective baseline for robust deep learning with noisy labels. arXiv preprint arXiv:1909.09338 (2019)
26. Mahajan, D., et al.: Exploring the limits of weakly supervised pretraining. arXiv preprint arXiv:1805.00932, May 2018
27. Mandt, S., Hoffman, M.D., Blei, D.M.: Stochastic gradient descent as approximate Bayesian inference. J. Mach. Learn. Res. 18(1), 4873–4907 (2017)
28. Neyshabur, B.: Implicit regularization in deep learning. arXiv preprint arXiv:1709.01953, September 2017
29. Northcutt, C.G., Wu, T., Chuang, I.L.: Learning with confident examples: rank pruning for robust classification with noisy labels. arXiv preprint arXiv:1705.01936, May 2017
30. Patrini, G., Rozza, A., Menon, A.K., Nock, R., Qu, L.: Making deep neural networks robust to label noise: a loss correction approach. In: 2017 IEEE Conference on Computer Vision and Pattern Recognition (CVPR), pp. 2233–2241. IEEE (2017)
31. Reed, S., Lee, H., Anguelov, D., Szegedy, C., Erhan, D., Rabinovich, A.: Training deep neural networks on noisy labels with bootstrapping. arXiv preprint arXiv:1412.6596, December 2014
32. Ren, M., Zeng, W., Yang, B., Urtasun, R.: Learning to reweight examples for robust deep learning. arXiv preprint arXiv:1803.09050, March 2018
33. Rolnick, D., Veit, A., Belongie, S., Shavit, N.: Deep learning is robust to massive label noise. arXiv preprint arXiv:1705.10694, May 2017
34. Russakovsky, O., et al.: ImageNet large scale visual recognition challenge. Int. J. Comput. Vis. 115(3), 211–252 (2015). https://doi.org/10.1007/s11263-015-0816-y
35. Shrivastava, A., Gupta, A., Girshick, R.: Training region-based object detectors with online hard example mining. In: Proceedings of the IEEE Conference on Computer Vision and Pattern Recognition, pp. 761–769 (2016)
36. Sukhbaatar, S., Bruna, J., Paluri, M., Bourdev, L., Fergus, R.: Training convolutional networks with noisy labels. arXiv preprint arXiv:1406.2080, June 2014
37. Sun, C., Shrivastava, A., Singh, S., Gupta, A.: Revisiting unreasonable effectiveness of data in deep learning era. arXiv preprint arXiv:1707.02968, July 2017

38. Szegedy, C., Ioffe, S., Vanhoucke, V., Alemi, A.: Inception-v4, Inception-ResNet and the impact of residual connections on learning. arXiv preprint arXiv:1602.07261, February 2016
39. Tanaka, D., Ikami, D., Yamasaki, T., Aizawa, K.: Joint optimization framework for learning with noisy labels. arXiv preprint arXiv:1803.11364 (2018)
40. Veit, A., Alldrin, N., Chechik, G., Krasin, I., Gupta, A., Belongie, S.J.: Learning from noisy large-scale datasets with minimal supervision. In: CVPR, pp. 6575–6583 (2017)
41. Xiao, T., Xia, T., Yang, Y., Huang, C., Wang, X.: Learning from massive noisy labeled data for image classification. In: Proceedings of the IEEE Conference on Computer Vision and Pattern Recognition, pp. 2691–2699 (2015)
42. Xu, Y., Cao, P., Kong, Y., Wang, Y.: L_dmi: an information-theoretic noise-robust loss function. arXiv preprint arXiv:1909.03388 (2019)
43. Yao, J., Wu, H., Zhang, Y., Tsang, I.W., Sun, J.: Safeguarded dynamic label regression for noisy supervision. In: AAAI (2019)
44. Zagoruyko, S., Komodakis, N.: Wide residual networks. arXiv preprint arXiv:1605.07146, May 2016
45. Zhang, C., Bengio, S., Hardt, M., Recht, B., Vinyals, O.: Understanding deep learning requires rethinking generalization. arXiv preprint arXiv:1611.03530, November 2016
46. Zhang, H., Cisse, M., Dauphin, Y.N., Lopez-Paz, D.: mixup: beyond empirical risk minimization, October 2017
47. Zhang, Z., Sabuncu, M.: Generalized cross entropy loss for training deep neural networks with noisy labels. In: Advances in Neural Information Processing Systems, pp. 8778–8788 (2018)

Imaging Behind Occluders Using Two-Bounce Light

Connor Henley$^{(\boxtimes)}$, Tomohiro Maeda, Tristan Swedish, and Ramesh Raskar

Massachusetts Institute of Technology, Cambridge, MA 02139, USA
{co24401,tomotomo,tswedish,raskar}@mit.edu

Abstract. We introduce the new non-line-of-sight imaging problem of *imaging behind an occluder*. The behind-an-occluder problem can be solved if the hidden space is flanked by opposing visible surfaces. We illuminate one surface and observe light that scatters off of the opposing surface after traveling through the hidden space. Hidden objects attenuate light that passes through the hidden space, leaving an observable signature that can be used to reconstruct their shape. Our method uses a simple capture setup—we use an eye-safe laser pointer as a light source and off-the-shelf RGB or RGB-D cameras to estimate the geometry of relay surfaces and observe two-bounce light. We analyze the photometric and geometric challenges of this new imaging problem, and develop a robust method that produces high-quality 3D reconstructions in uncontrolled settings where relay surfaces may be non-planar.

Keywords: Non-line-of-sight imaging · Computational photography

1 Introduction

Traditional optical imaging techniques produce images using measurements of light that has propagated directly, along a straight and unoccluded line of sight, from an object or scene of interest to one's imaging sensor. Non-line-of-sight (NLOS) imaging techniques, by contrast, generate images from measurements of light that has traveled from the object or scene of interest via indirect paths that typically include reflections off of intermediate surfaces. NLOS imaging techniques are particularly useful for looking behind things—walls, buildings, vehicles—any opaque surface that blocks one's line of sight to objects at greater depths.

In this paper, we introduce a new NLOS imaging problem, which we will refer to as the problem of *imaging behind an occluder*. The problem can be described as follows: any opaque occluder will block an observer's line of sight to all surfaces that lie behind it, creating a hidden volume that extends to infinite depth behind

Electronic supplementary material The online version of this chapter (https://doi.org/10.1007/978-3-030-58526-6_34) contains supplementary material, which is available to authorized users.

© Springer Nature Switzerland AG 2020
A. Vedaldi et al. (Eds.): ECCV 2020, LNCS 12374, pp. 573–588, 2020.
https://doi.org/10.1007/978-3-030-58526-6_34

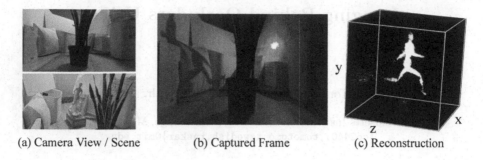

(a) Camera View / Scene   (b) Captured Frame   (c) Reconstruction

Fig. 1. Imaging behind occluders using visible surfaces on opposing sides of a hidden space. (a) Top: A mannequin is hidden from the camera. Bottom: A third person view of the occluding plant. (b) We illuminate points lying to one side of the hidden space. And observe distorted shadows that hidden objects carve out of the two-bounce light signal, which scatters off of the opposing visible surface towards our camera. (c) Euclidean 3D reconstruction of the mannequin.

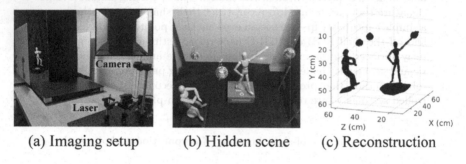

(a) Imaging setup   (b) Hidden scene   (c) Reconstruction

Fig. 2. Our method produces accurate reconstruction of hidden scenes containing multiple objects using a laser pointer and off-the-shelf RGB or RGB-D cameras.

the occluder. In some scenarios, an observer will have an unobstructed view of surfaces that lie on either side of this hidden volume. This scenario is depicted in Fig. 3. Such a geometry immediately suggests an elementary measurement: the observer can illuminate a point lying to one side of the hidden volume, and observe a point lying on the other side. The line segment that connects these two points passes through the hidden space. If no occluding surfaces lie on this segment, then light reflected off of the illuminated point will propagate to the observed point, which will then also appear illuminated. On the other hand, if an occluding surface does lie upon this line segment, then the observed point will lie in that surface's shadow, and appear dim. The problem of imaging behind an occluder is the problem of interpreting these shadows to produce a 3D reconstruction of the hidden scene.

1.1   Contributions

We summarize the contributions of our paper as follows:

1. Formulation of a new non-line-of-sight imaging problem of *imaging behind an occluder*. We discuss challenges and sources of error in *imaging behind an occluder*, particularly in natural settings in which extraction of shadows and 3D estimation of visible surfaces are prone to error
2. Method that solves this problem with a space carving approach that exploits two-bounce light measurements. Ours is the first NLOS method to recover detailed 3D shapes of hidden objects without specialized equipment. We use a laser pointer and off-the-shelf RGB or RGB-D cameras.
3. Reconstruction results that highlight the capability of our method to recover fine structure of hidden objects, capture video reconstructions of moving scenes, and handle non-planar and non-continuous visible surfaces

Behind-an-occluder methods can be applied to a variety of useful tasks, including vision in cluttered environments such as forests, maintaining spatial awareness of the space behind neighboring vehicles on the road, seeing inside buildings during search-and-rescue operations, or determining the 3D shape of the back-facing side of an object. We present additional results that highlight potential applications of our method to autonomous driving and search-and-rescue operations.

2   Related Work

2.1   Non-Line-of-Sight Imaging

Most NLOS imaging methods published to date have attempted to solve the challenging problem of *seeing around the corner*. In the around-the-corner problem, an observer can only learn about the hidden scene from light that is reflected off of intermediate surfaces *after* it has reflected off of or been emitted by the hidden scene itself.

Velten et al. [20,27] demonstrated the first 3D reconstruction of an object hidden around the corner by using time of flight (ToF) measurements to provide constraints on hidden object locations. Since then, many robust and efficient algorithms to image around corners have been proposed [1,9,15,28]. Others exploit speckle patterns [10], spatial coherence [3], and radiometric (intensity) measurements [6,11,24] for NLOS imaging. Around-the-corner methods typically require measurement of three-bounce photons. Our method captures two-bounce photons and does not require ToF measurements to recover the 3D geometry of hidden objects.

Occlusions constrain the set of paths that light can take as it propagates through the hidden scene and back to the observer. *Occlusion-assisted* imaging techniques exploit knowledge of these constraints to produce a forward light transport model that can be inverted to estimate hidden scene properties.

Bouman et al. [5] produce 1D images of hidden scenes by exploiting the occlusions that occur as light propagates past a corner or an edge. Others exploit occlusions that occur within the hidden scene to recover surface albedos [21,26], and 4D light fields [2] around corners. In these techniques, the occluders and hidden scenes are considered separate, and the former is exploited to learn about the latter. In our method, we directly measure the shadows cast by hidden objects—in effect, the hidden object *is* the occluder.

We refer readers to [16] for a comprehensive overview of NLOS imaging.

2.2 Shape from Shadows

In *shadowgram imaging* methods, the shape of an object is estimated from a series of shadows cast onto a planar screen for point sources placed at various locations [22,29]. Shadowgram methods are typically applied in controlled, object scanning setups and require the placement of point sources with a direct line of sight to the scene being imaged. Our method is designed to reconstruct scenes that might be entirely hidden from view, and to utilize shadows projected onto visible surfaces that may be non-planar, have varied surface albedo, and have to be estimated by the observer.

Shadowgram methods exploit the fact that a sharp shadow cast onto a plane by an object that is illuminated by a point source can be interpreted as a 2D *silhouette image* of that object, taken from the perspective of the point source. The problem of determining the three-dimensional shape of an object from a series of two-dimensional silhouettes has been studied extensively by the computer vision community [14,18,25]. With multiple silhouettes, taken from multiple camera points, an object's shape can be confined to an intersection of affine cones, referred to as the *visual hull* [13]. The first shape-from-silhouettes (SfS) algorithm is often credited to Baumgart [4]. Later, Martin and Aggarwal [17] introduced a volumetric space carving approach to SfS in which voxels lying outside the visual cone are "carved" away, forming a visual hull.

We modify our space carving approach so that it is robust to the errors in visible surface geometry and shadow classification. Some researchers have explored robust *probabilistic carving* methods. These methods are typically designed for shape-from-silhouettes applications that require a direct view of an imaged scene [8,12,23], and cannot be applied directly to our problem of space carving from distorted shadows cast onto arbitrary surfaces by objects that are hidden from view.

3 Imaging Behind Occluders

Consider the flatland scene presented in Fig. 3. In this scene, a large fraction of the observer's field of view is blocked by an occluder. We want to know what is behind that occluder—specifically, we want to find out whether any opaque object is occupying the hidden point x.

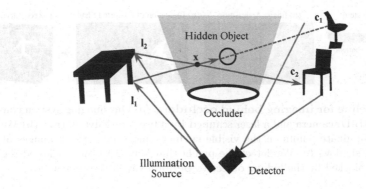

Fig. 3. Visibility Carving Our algorithm reconstructs the shape of hidden objects by determining the set of line segments l_i-c_i that pass through the hidden scene. Here we observe that light can travel from points l_2 to c_2 without occlusion, and thus infer that hidden point x is unoccupied.

3.1 The Elementary Measurement

We illuminate the visible point l_1, and observe point c_1, chosen such that the line segment connecting l_1 and c_1 passes through x. In this case, we note that point c_1 lies in shadow. From this measurement, we can infer that a hidden object must lie *somewhere* along the line segment l_1-c_1, but cannot say for certain that this object is located at point x.

We then take a second measurement, illuminating a different point l_2 and observing point c_2, again chosen such that the line segment l_2-c_2 passes through x. This time the observed point c_2 appears to be illuminated by light that has been diffusely reflected from the visible surface at l_2. This measurement is more informative. If light has traveled from l_2 to c_2 without occlusion, then there cannot be an occluding surface at point x or at any point along the segment l_2-c_2.

This is the elementary measurement that underlies our method for seeing behind occluders. The measurement can be extended to a reconstruction problem by discretizing the hidden space into a grid of points $X = \{x_1, x_2, \ldots, x_n\}$ to be probed. Instead of probing the occupancy of each point x_i with a series of l-c pairs, we can efficiently collect this data by using a camera to observe an *area* on the visible surface rather than observing single points. In this case, every measurement probes a large number of hidden points in the scene simultaneously.

3.2 Method

Our method can be decomposed into three basic steps: data capture, pre-processing, and reconstruction.

In the data capture step, we use a galvo-scanned laser to illuminate points on the visible surface. For each laser scan position, we take two photos: one short exposure photo that allows us to pinpoint the location and (if desired) brightness

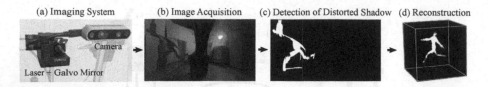

(a) Imaging System   (b) Image Acquisition   (c) Detection of Distorted Shadow   (d) Reconstruction

Fig. 4. A pipeline for imaging behind occluders. (a) Our imaging system consists a RGB (or RGB-D) camera and a laser scanned by a two-axis galvo mirror. (b) We use the laser to illuminate points on the visible surfaces and then capture images of the laser spot and shadow. (c) We detect the pixels that belong to the shadowed regions where light is blocked by the hidden object to perform (d) 3D reconstruction.

of the laser spot, and a second, longer exposure photo that captures the shadows cast onto the opposing visible surface with high contrast. Pictures of our data capture equipment are shown in Fig. 4(a).

In the pre-processing step, we loop through our acquired photo stacks. For each short exposure photo, we pinpoint the image-space position of the laser spot and convert this position to 3D world coordinates. If we are using an RGB camera (as in Sect. 4) we rely on prior knowledge of visible surface geometry to make this estimate. If geometry information is not available, we directly measure the 3D position using an RGB-D camera. For each longer exposure photo, we select a region of interest that we expect will contain cast shadows that are informative about the hidden scene. We classify each pixel in this region as *shadowed* or *lit* or, in some cases, unknown. Our choice of *shadow segmentation* criteria can be made simple (such as a binary threshold) or more complex depending on the complexity of the visible scene. An example of shadow segmentation is shown in Figs. 4(b) and (c).

Finally, in the reconstruction step, we discretize the hidden space into a 3D grid of points or voxels. Given an illumination spot l and its associated shadow image, we project each element of the hidden grid onto the opposing visible surface, using l as the center of projection. We then determine whether the projection of this element lies *inside* the shadowed region of the visible surface or *outside* of the shadows, in a lit region. It is also possible that the projection does not land on the visible surface at all, or lands on a region of ambiguous classification. If the element is a voxel, it may project to a mixture of shadowed and lit pixels. In this case, we are conservative—classifying the voxel as *outside* only if it projects to all lit pixels, and *inside* only if it projects to all shadowed pixels. We perform this test on each element, for each illumination point and shadow image in our stack. We count how many times an element has been classified as inside and as outside. Once all frames have been processed, we use these counts to determine which elements are occupied and which are empty. In the naive version of our algorithm that is used in Sect. 4, an element is assessed as empty if it is classified as outside in at least one frame. In Sects. 5 and 6 we apply a probabilistic thresholding procedure that is robust to errors in the inside/outside test. An example reconstruction result is shown in Fig. 4(d). In this example voxels are colored by probability of occupancy.

4   Reconstructing the Shape of Hidden Objects

In this section, we demonstrate that our method is capable of producing highly detailed reconstructions of stationary hidden scenes, and also demonstrate how our method could be used to capture video reconstructions of hidden objects that are moving. We place hidden objects in a simple testbed consisting of two white relay walls and a black occluding wall. This simple testbed allows us to demonstrate the potential capabilities of our method when errors in pre-processing—that is, visible surface estimation and shadow segmentation errors—are minimal. In later sections, we will address the problem of reconstructing a hidden scene when pre-processing errors are unavoidable.

4.1   Implementation

Our testbed consisted of two white *observation walls*, and a black *occluding wall* placed between the scene-to-be-imaged and all imaging equipment. The observation walls were 61 cm × 76 cm rectangles oriented parallel to one another and 76 cm apart. A photo of the testbed is shown in Fig. 2. We illuminate the two observation walls at a series of points using a green CW laser with a power of ∼5 mW scanned with a two-axis scanning galvo mirror system, and use a Point Grey Blackfly RGB camera to capture images. To segment shadow images into lit and shadowed regions, we binarize the pixels in a region of interest using either a hand-tuned threshold or a threshold set adaptively using Otsu's method [19]. We de-noise this binary image using a bilateral filter and then use a second threshold to separate the filtered images into shadowed and illuminated pixels. We found that this second filtering step resulted in more robust pixel classifications.

In this set of collections, we assume that the geometry of the two observation walls is known to the observer. Since the observation walls are planar, this means that we can measure the position of the laser spot in pixel space and then convert that measurement to 3D world coordinates using a homography. Likewise, to test the inside/outside status of a hidden point $\mathbf{x}$, we find the point at which the ray drawn from l and through $\mathbf{x}$ intersects the opposing observation wall, and then use a homography to convert this point of intersection to pixel space. We then compare the ray's point of intersection with the estimated shadow boundaries in pixel space to determine whether it falls inside or outside of the illuminated region. All points that accrue one or more *outside* classifications are carved out of the reconstructed scene. *Inside* classifications are not used.

4.2   Stationary Hidden Objects

We scanned a variety of stationary objects in this testbed. Photos of three of these objects, along with our estimation of each object's shape, can be seen in

(a) Reconstruction of objects with fine features   (b) Reconstruction of a moving object

Fig. 5. Reconstruction results. (a) Our method recovers the complex geometry of a spring (left) and a plant (right). An additional reconstruction of a disco dance party can be seen in Fig. 2. We use the shadows cast by hidden objects to reconstruct their 3D shape. (b) Snapshots from a 15 FPS video reconstruction of a moving mannequin. Reconstructed frames are colored by time. For clarity, snapshots in this figure are spaced 3 s apart. The full video reconstruction can be found in the supplemental material.

Figs. 2 and 5. Our method reproduces the fine structure of reasonably complex objects. On the left side of Fig. 5(a), we recover the shape of a spring. Our estimate correctly reconstructs the spring as a singular object, with no breaks along the coil. On the right side of Fig. 5(a), we estimate the shape of a plant and are able to resolve individual leaves despite significant self-occlusion. Finally, as shown in Fig. 2(a), we recover the shape of a dance party. We believe that the quality of the estimated scene should be more than sufficient for high level tasks such as pose estimation. We also note that the specular surfaces of the scene's three disco balls had no effect on our ability to recover their shapes.

For each object, we scanned a 9 × 11 grid of illumination points on both the left and right observation walls. Points on these grids were evenly spaced and spanned almost the entire area of each observation wall. The total acquisition time for each object was 27 s. After the acquisition and shadow segmentation, space carving was initiated on a grid of hidden points with 0.25 cm spacing.

4.3   Moving Hidden Object

By reducing the number of scan points per reconstruction, we can capture videos of dynamic hidden scenes. Although these videos do not capture fine detail, we are able to recover the approximate size, shape, and position of a moving object using only four observed shadows per video frame.

Our reconstructed video can be found in the supplemental material. Snapshots of the result are plotted in Fig. 5(b). The hidden object was a mannequin that was slowly moved around the hidden area. We acquire 15 shadows per second, such that the four-point scan pattern can be completed in .27 s. Space carving was executed on a 55 cm × 45 cm × 55 cm grid of hidden points, with a grid spacing of 1 cm.

5 Overcoming Geometric and Photometric Challenges of Imaging Behind Occluders

The physical principle that underlies our method of imaging behind occluders using two-bounce light is simple: if we observe that light can travel directly between two points without occlusion, then we know that no opaque surfaces can be present on the line segment that connects those two points. Unfortunately reality is typically messier than this. Errors in visible surface estimation and shadow segmentation will invariably impact our ability to determine which voxels lie inside and outside of the true hidden scene.

In this section, we discuss how such errors manifest as reconstruction artifacts when a naive carving approach, such as the approach used in Sect. 4, is used. We then apply this insight to develop a probabilistic carving approach that is robust to pre-processing errors. We apply our robust method to reconstruct the shape of a mannequin using shadows cast onto non-planar relay surfaces that must be estimated from noisy depth measurements. We show that our robust algorithm handily outperforms a naive carving approach in this setting.

5.1 Sources of Error in Imaging Behind Occluders

False Carving: In a naive carving approach, a voxel is discarded from the set of points that lie inside a hidden object if just a single ray drawn from a laser spot to an observed "lit" pixel passes through that voxel. The reasoning behind this decision makes physical sense—If a voxel is occupied by an opaque material, then light should not be able to pass through it. In practice, this model is too strict. If the observed pixel is wrongly classified as lit when it in fact lies in shadow, then an entire ray of voxels might be erroneously carved out of the reconstructed scene. Furthermore, if there are errors in our estimates of the laser spot and observed pixel's 3D locations, then the line segment that we carve through the scene will be misaligned with respect to the true propagation path. This will lead to further erroneous carving.

Complex light transport phenomena such as inter-reflections, specular reflections, translucent hidden objects or refraction can also cause false carving because they cannot be explained by a two-bounce scattering model or opaque occlusions. An example of this is shown in Fig. 6(a).

False Positives: In addition to erroneously carving away voxels that are inside of a hidden object, a carving method may also fail to carve away a voxel that lies outside of all hidden objects. In some cases, these *false positives* cannot be prevented—for instance, if an empty voxel happens to lie inside of the hidden

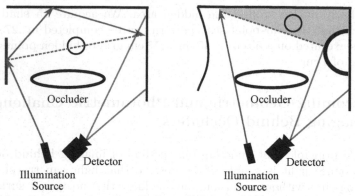

(a) False carving due to specular reflection   (b) False Positives due to unobservable surface

Fig. 6. Limitations on the hidden scene reconstruction. (a) Complex light transport such as specular reflection (blue surface) can result in unintended carving of the space occupied by the hidden object. (b) The shape of visible surfaces determines the recoverable space in the hidden volume. (Color figure online)

scene's visual hull [13]. In other cases an empty voxel may remain uncarved for the same reasons that an occupied voxel may be carved erroneously; that is, due to errors in shadow classification or visible surface estimation. These sources of error tend to produce more false negatives (over-carving) than false positives when most voxels in the scene are probed by a large number of rays. This is because a naive carving method only requires a single *outside* result to carve away a voxel, whereas *all* tests must return an *inside* result for a voxel to be classified as outside of a hidden scene.

When the number of rays that probe a voxel is low, however, the probability of false positive classification becomes more significant. This becomes more of a problem when the visible surface geometry is complex, due to the fact that non-planar surfaces will often occlude themselves from the point of view of the observer. This self-occlusion reduces the set of points on the surface that can be illuminated and observed to probe the hidden scene. In many situations, this self-occlusion renders large swaths of the hidden scene unobservable. A simple example of this is shown in Fig. 6(b). In general, however, any level of self-occlusion will reduce the set of rays that can be probed and this will tend to increase the number of false positives in the reconstructed scene.

5.2   Robust Carving

To improve the robustness of our method to the false *outside* results that cause over-carving, we can relax our acceptance criteria by requiring some number $M > 1$ *outside* results to accrue before a voxel can be carved out of the hidden scene. This relaxed threshold reduces the probability of false carving at the cost of an increased false positive rate. This trade-off is often worth it if a stricter acceptance threshold would lead to excessive overcarving. To reduce the number of false positives, we can add a new acceptance criteria that requires some number $N > 0$ of *inside* results to accept a voxel as lying inside of the hidden scene. This acceptance criteria is motivated by the fact that an occupied voxel that *is* probed should produce *inside* results most of the time, in spite of any errors. This test also provides the added benefit of screening out all voxels in the hidden scene that are unobservable due to visible surface self-occlusion, or for other reasons.

Ideally, we would prefer to set these acceptance criteria using a principled approach that considers the probability of occupancy of each voxel. To this end, we considered a simple probabilistic model inspired by an acceptance test that was employed by Cheung et al. [7] for shape-from-silhouettes. If a voxel is subjected to N inside/outside tests, the result y_j of any single test is treated as a Bernoulli trial that is independent of the other $N - 1$ tests when conditioned on the true state of the probed voxel, which can be either empty (e) or occupied (o). Under this model the probability that a voxel is occupied given m outside results and n inside results is

$$\mathbb{P}(v_i = o | y_1, ... y_N) = \frac{\eta^m (1 - \eta)^n p_o}{(1 - \xi)^m \xi^n p_e + \eta^m (1 - \eta)^n p_o}. \tag{1}$$

Here η is the *miss probability*—that is, the probability that an occupied voxel projects to an illuminated region (is declared *outside*)—and ξ is the *probability of false alarm*—the probability that an empty voxel projects to a shadowed region (is declared *inside*). The values $p_o = \mathbb{P}(v_i = o)$ and $p_e = \mathbb{P}(v_i = e)$ represent the prior probabilities that voxel i is occupied or empty, respectively. A derivation of Eq. (1) is provided in the supplemental material.

To use Eq. (1) as an acceptance criterion, we can assume that all voxels share the same values of η, ξ, p_o, and p_e. Then we can calculate the conditional occupancy probability for all combinations of m and n that we might feasibly encounter in N observations (at most $(N + 1)^2$ values) and store these probabilities in a lookup table. When all inside/outside tests have been completed, we look up the appropriate occupancy probability based on number of inside and outside counts, and accept a voxel as occupied if the probability lies above some threshold.

In theory, the parameters η, ξ, p_o, and p_e could be determined via an empirical analysis of inside/outside test errors. In practice we use them as tuning parameters. A lower value of η signifies a higher level of trust in *outside* results. Reducing this parameter should reduce the rate of false positives and increase the rate of false carving. Setting η equal to zero reproduces the naive carving acceptance criteria. A lower value of ξ strengthens the effect of *inside* results. This means that fewer inside measurements will be required to declare a voxel occupied. The prior probabilities guide the acceptance criterion in the absence of measurement data, and in practice have very little effect on the acceptance result if a voxel has been probed more than a handful of times.

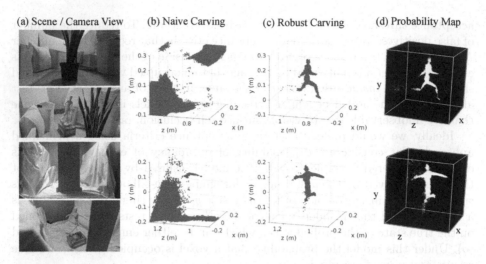

Fig. 7. Imaging behind occluders with complex visible geometries. (a) Scene setup and view from the camera. A mannequin is hidden behind an occluder. (b) A naive carving algorithm removes voxels occupied by the mannequin. (c) A robust carving method demonstrates robustness in reconstructing the hidden object with complex visible surfaces. (d) The probability map of voxel occupancy computed using Eq. (1).

5.3 Implementation

To test our robust carving method, we attempt to reconstruct the shape of a mannequin hidden behind an occluder by observing shadows cast onto non-planar and non-continuous visible surfaces consisting of various white objects (see supplemental material for results obtained when relay surface albedo varies spatially). We acquire the laser spot positions and the shape of the visible surface using a RealSense D435 active stereo RGB-D camera. We acquire shadow images and laser spot positions while the mannequin is present in the hidden scene. We then remove the mannequin to acquire a set of background images to aid in background segmentation. We perform shadow segmentation by applying a per-pixel threshold on the ratio of pixel values in shadow frames and background frames. We note that we were also able to obtain good results without the aid of background measurements. These results as well as the shadow segmentation method that we used can be found in the supplemental material.

We present two reconstructions of the hidden mannequin. Each utilizes the same set of 30 illumination spots and shadow images. The first result, which is shown in Fig. 7(b), was produced by applying a naive carving criterion in which all voxels that are projected to an illuminated region of the visible surface *at least once* are carved away, and inside tests are not used. In this case, the mannequin reconstruction is clearly overcarved and resembles 3D pepper noise. The absence of inside tests also results in a residue of false positives at the periphery of the voxel grid. In Fig. 7(c) we show results obtained when we applied the probabilistic acceptance criteria based on Eq. (1), and in Fig. 7(d) we visualize the voxel occupancy probability map using a maximum intensity projection. In these results we can clearly distinguish the form of the mannequin.

6   Applications

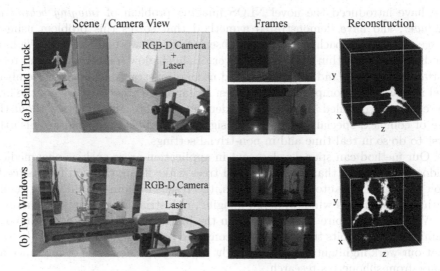

Fig. 8. Demonstration of potential applications of imaging behind occluders. Our technique can be used to (a) detect a person that is behind a truck and (b) reconstruct the a hidden part of rooms that have at least two windows.

We highlight some potential applications of our method by reconstructing hidden scenes embedded in settings that simulate situations that might be encountered in the real world.

Seeing Behind Trucks for Autonomous Vehicles: We want to simulate the ability to detect a child that is playing with a ball behind a parked truck and is about to run out into the street. We observe a truck parked next to a building's wall. A mannequin and ball were placed in front of the truck and out of view of the observer. We illuminate points distributed on the ground plane and on the building's facade, and also observe shadows cast onto these surfaces. We acquire these surfaces with an RGB-D camera and use our robust carving method to reconstruct the shape of the reckless child. Results are shown in Fig. 8(a).

Imaging Between Windows for Search and Rescue: We want to simulate a rescue team's ability to locate humans in rooms. We observe a room with two windows from the outside. The room geometry has three vertical walls that are not parallel. The room also contains clutter seen in the left window. We shine the laser through one window and observe shadows cast onto a wall through the other window. As in the behind-trucks scene, we use an RGB-D camera to acquire the intensity and position of observed and illuminated pixels, and then use our robust carving method to reconstruct a hidden scene consisting of two humans in conflict. Results are shown in Fig. 8(b).

7   Conclusion

We have introduced the novel NLOS imaging problem of *imaging behind an occluder*, and have demonstrated a method that solves this problem using a space carving approach that exploits observations of two-bounce light. We have modified our algorithm to be robust to errors in shadow segmentation and relay surface estimation, and this has enabled us to reconstruct hidden scenes embedded in reasonably complex visible scenes. As far as we are aware, we are the first to recover the detailed 3D shape of hidden, non-line-of-sight objects without the use of complex, specialized devices or significant calibrations, and are also the first to do so in real time and in non-trivial settings.

Our method can spur development in application areas such as new medical endoscopes, drones that can see behind trees while flying through forests, collision avoidance for autonomous vehicles, the rescue of humans hidden in rooms with more than one window, aerial imaging, industrial inspection, and more.

We hope to inspire more research in the imaging and computer vision communities that exploits the information conveyed by cast shadows. We also believe that our work highlights non-line-of-sight imaging as a new area of relevance for shape-from-silhouettes research.

In future work we hope to apply our method outside of the laboratory and in natural, larger-scale scenes. We believe that this is feasible if interference from ambient background light can be reduced by operating at infrared wavelengths or by employing time-modulated illumination. An exploration of how relay surface geometry affects reconstruction quality would also be valuable. Finally, variations on the *behind-occluders* problem could be explored such as reconstruction without depth information, or reconstructing transparent hidden objects.

Acknowledgements:. We thank our reviewers for their helpful comments. This work was supported by DARPA REVEAL (N00014-18-1-2894) and the Media Lab Consortium. TS was supported in part by NSF GRFP (No. 1122374).

References

1. Ahn, B., Dave, A., Veeraraghavan, A., Gkioulekas, I., Sankaranarayanan, A.C.: Convolutional approximations to the general non-line-of-sight imaging operator. In: The IEEE International Conference on Computer Vision (ICCV) (2019)
2. Baradad, M., et al.: Inferring light fields from shadows. In: The IEEE Conference on Computer Vision and Pattern Recognition (CVPR) (2018)
3. Batarseh, M., Sukhov, S., Shen, Z., Gemar, H., Rezvani, R., Dogariu, A.: Passive sensing around the corner using spatial coherence. Nat. Commun. **9**(1), 1–6 (2018)
4. Baumgart, B.: Geometric modeling for computer vision. Ph.D. thesis, Stanford University (1974)

5. Bouman, K.L., et al.: Turning corners into cameras: principles and methods. In: Proceedings of the IEEE International Conference on Computer Vision, vol. 572, pp. 2270–2278 (2017)
6. Chen, W., Daneau, S., Mannan, F., Heide, F.: Steady-state non-line-of-sight imaging. In: The IEEE Conference on Computer Vision and Pattern Recognition (CVPR) (2019)
7. Cheung, G.K., Kanade, T., Bouguet, J.Y., Holler, M.: A real time system for robust 3D voxel reconstruction of human motions. In: Proceedings IEEE Conference on Computer Vision and Pattern Recognition. CVPR, vol. 2, pp. 714–720. IEEE (2000)
8. Franco, J.S., Boyer, E.: Fusion of multiview silhouette cues using a space occupancy grid. In: The IEEE International Conference on Computer Vision (ICCV), vol. 1. IEEE (2005)
9. Heide, F., O'Toole, M., Zang, K., Lindell, D.B., Diamond, S., Wetzstein, G.: Non-line-of-sight imaging with partial occluders and surface normals. ACM Trans. Graph. **38**, 1–10 (2019)
10. Katz, O., Heidmann, P., Fink, M., Gigan, S.: Non-invasive real-time imaging through scattering layers and around corners via speckle correlations. Nat. Photonics **8**, 784–790 (2014)
11. Klein, J., Peters, C., Martín, J., Laurenzis, M., Hullin, M.B.: Tracking objects outside the line of sight using 2D intensity images. Sci. Rep. **6**(1), 1–9 (2016)
12. Landabaso, J.L., Pardas, M., Casas, J.R.: Shape from inconsistent silhouette. Comput. Vis. Image Underst. **112**(2), 210–224 (2008)
13. Laurentini, A.: How far 3D shapes can be understood from 2D silhouettes. IEEE Trans. Pattern Anal. Mach. Intell. **3**(2), 188–195 (1995)
14. Lazebnik, S., Furukawa, Y., Ponce, J.: Projective visual hulls. Int. J. Comput. Vision **74**(2), 137–165 (2007). https://doi.org/10.1007/s11263-006-0008-x
15. Liu, X., et al.: Non-line-of-sight imaging using phasor-field virtual wave optics. Nature, 1–4 (2019)
16. Maeda, T., Satat, G., Swedish, T., Sinha, L., Raskar, R.: Recent advances in imaging around corners. arXiv preprint arXiv:1910.05613 (2019)
17. Martin, W., Aggarwal, J.K.: Volumetric descriptions of objects from multiple views. IEEE Trans. Pattern Anal. Mach. Intell. **5**(2), 150–158 (1983)
18. Matusik, W., Buehler, C., McMillan, L., Gortler, S.: An efficient visual hull computation algorithm (2002)
19. Otsu, N.: A threshold selection method from gray-level histograms. IEEE Trans. Syst. Man Cybern. **9**(1), 62–66 (1979)
20. Raskar, R., Davis, J.: 5D time-light transport matrix: what can we reason about scene properties? (2008)
21. Saunders, C., Murray-Bruce, J., Goyal, V.: Computational periscopy with an ordinary digital camera. Nature **565**, 472 (2019)
22. Savarese, S., Andreetto, M., Rushmeier, H., Bernardini, F., Perona, P.: 3D reconstruction by shadow carving: theory and practical evaluation. Int. J. Comput. Vision **71**(3), 305–336 (2007). https://doi.org/10.1007/s11263-006-8323-9
23. Tabb, A.: Shape from silhouette probability maps: reconstruction of thin objects in the presence of silhouette extraction and calibration error. In: The IEEE International Conference on Computer Vision (ICCV), pp. 161–168 (2013)
24. Tancik, M., Satat, G., Raskar, R.: Flash photography for data-driven hidden scene recovery (2018)

588   C. Henley et al.

25. Tarini, M., Callieri, M., Montani, C., Rocchini, C., Olsson, K., Persson, T.: Marching intersections: an efficient approach to shape-from-silhouette. In: 2002 Proceedings of VMV, pp. 255–262 (2002)
26. Thrampoulidis, C., et al.: Exploiting occlusion in non-line-of-sight active imaging. IEEE Trans. Comput. Imaging **4**, 419–431 (2018)
27. Velten, A., Wilwacher, T., Gupta, O., Veeraraghavan, A., Bawendi, M., Raskar, R.: Recovering three-dimensional shape around a corner using ultrafast time-of-flight imaging. Nat. Commun. **3** (2012). Article Number: 745. https://www.nature.com/articles/ncomms1747?page=2
28. Xin, S., Nousias, S., Kutulakos, K.N., Sankaranarayanan, A.C., Narasimhan, S.G., Gkioulekas, I.: A theory of fermat paths for non-line-of-sight shape reconstruction. In: Proceedings of the IEEE Conference on Computer Vision and Pattern Recognition, pp. 6800–6809 (2019)
29. Yamazaki, S., Narasimhan, S.G., Baker, S., Kanade, T.: The theory and practice of coplanar shadowgram imaging for acquiring visual hulls of intricate objects. Int. J. Comput. Vision **81**(3), 259–280 (2009). https://doi.org/10.1007/s11263-008-0170-4

Improving Object Detection with *Selective* Self-supervised Self-training

Yandong Li[1,2](✉) , Di Huang[2] , Danfeng Qin[2] , Liqiang Wang[1] ,
and Boqing Gong[2]

[1] University of Central Florida, Orlando, USA
lyndon.leeseu@outlook.com, lwang@cs.ucf.edu
[2] Google Inc., Menlo Park, USA
dihuang@google.com, qind@google.com, bgong@google.com

Abstract. We study how to leverage Web images to augment human-curated object detection datasets. Our approach is two-pronged. On the one hand, we retrieve Web images by image-to-image search, which incurs less domain shift from the curated data than other search methods. The Web images are diverse, supplying a wide variety of object poses, appearances, their interactions with the context, etc. On the other hand, we propose a novel learning method motivated by two parallel lines of work that explore unlabeled data for image classification: self-training and self-supervised learning. They fail to improve object detectors in their vanilla forms due to the domain gap between the Web images and curated datasets. To tackle this challenge, we propose a selective net to rectify the supervision signals in Web images. It not only identifies positive bounding boxes but also creates a safe zone for mining hard negative boxes. We report state-of-the-art results on detecting backpacks and chairs from everyday scenes, along with other challenging object classes.

1 Introduction

Object detection is a fundamental task in computer vision. It has achieved unprecedented performance for many objects, partially thanks to the recently developed deep neural detectors [32,47,48,53]. Some detectors have made their way into real-world applications, such as smart mobile phones and self-driving cars.

However, upon a careful investigation into the top three teams' class-wise detection results on 80 common objects in context (COCO) [34], we find that they still fall short of detecting backpacks, handbags, and chairs, among other *functional* objects. As of the paper submission, they report an average precision [34] of less than 0.30 on detecting backpacks and less than 0.40 on chairs.

Y. Li—This work was done while Yandong Li was an intern at Google Inc.

Electronic supplementary material The online version of this chapter (https://doi.org/10.1007/978-3-030-58526-6_35) contains supplementary material, which is available to authorized users.

© Springer Nature Switzerland AG 2020
A. Vedaldi et al. (Eds.): ECCV 2020, LNCS 12374, pp. 589–607, 2020.
https://doi.org/10.1007/978-3-030-58526-6_35

Fig. 1. Chairs (top) and backpacks (bottom) are difficult to detect in daily scenes. (Images and groundtruth boxes (red) are from COCO [34], and the blue boxes are predicted by the best detectors in our experiments. (color figure online)

These man-made objects are defined by their functionalities more than visual appearances, leading to high intra-class variation. Minsky [38] wrote that "there's little we can find in common to all chairs - except for their intended use." Grabner *et al.* [20] quantitatively evaluated the challenges of detecting functional objects like chairs. Another potential reason that contributes to the low performance in detecting backpacks and chairs is that they are too common to draw photographers' attention. Consequently, they often sit out of the camera focus, appear small, and become occluded in context (cf. Fig. 1).

How to improve the detection of backpacks, chairs, and other common, "less eye-catching" functional objects? We believe the answer resides on both the quality of training data and the inductive bias of advanced detectors. In this paper, we focus on the data aspect and mainly study the potential of *unlabeled* Web images for improving the detection results of backpacks and chairs, without heavily taxing human raters.

In other words, we study how to leverage the *unlabeled* Web images to augment human-curated object detection datasets. Web images are diverse and massive, supplying a wide variety of object poses, appearances, their interactions with the context, etc., which may lack in the curated object detection datasets. However, the Web images are *out of the distribution* of the curated datasets. The domain gap between them calls for a careful design of methods to effectively take advantage of the signals in the Web data.

Our approach is two-pronged. On the one hand, we retrieve a big pool of candidate Web images via Google Image (https://images.google.com) by using its image-to-image search. The query set consists of all training images in the original human-curated dataset. Compared with text-based search, which mainly returns iconic photos, the image-based search gives rise to more natural images with diverse scenes, schematically reducing the domain mismatch between the retrieved images and the original datasets.

On the other hand, we propose a novel learning method to utilize the Web images for object detection inspired by self-training [50,66] and self-supervised learning [6,11,18,21,42,64], both of which are popular in semi-supervised learning. Our problem is similar to semi-supervised learning, but there exists a domain gap between the Web images and the curated datasets. We find that the domain gap fails both self-training and self-supervised learning in their vanilla forms because the out-of-domain Web images give rise to many inaccurate candidate boxes and uncalibrated box classification scores. To tackle these challenges, we propose a selective net to identify high-quality positive boxes and a safe zone for mining hard negative boxes [33,48]. It rectifies the supervision signals in Web images, enabling self-training and self-supervised learning to improve neural object detectors by leveraging the Web images.

The main contributions in the paper are as follows. First, we customize self-training for the object detection task by a selective net, which identifies positive bounding boxes and assigns some negative boxes to a safe zone to avoid messing up the hard negative mining in the training of object detectors. Second, we improve the consistency-based [9,28,62] semi-supervised object detection [26] by the selective net under our self-training framework. Third, to the best of our knowledge, this work is the first to explore *unlabeled, out-of-domain* Web images to augment curated object detection datasets. We report state-of-the-art results for detecting backpacks and chairs, along with other challenging objects.

2 Augmenting COCO Detection with Web Images

We augment the training set of COCO detection [34] by retrieving relevant Web images through Google Image (https://images.google.com). We focus on the backpack and chair classes in this paper. They represent non-rigid and rigid man-made objects, respectively, and the existing results of detecting them are still unsatisfactory (less than 0.40 *AP* on COCO as of March 5th, 2020).

COCO-backpack, COCO-chair, Web-backpack, and Web-chair. COCO is a widely used dataset for object detection, which contains 118k training images and 5k validation images [34]. Out of them, there are 8,714 backpacks in 5,528 training images. We name these images the COCO-backpack query set. Similarly, we have a COCO-chair query set that contains 12,774 images with 38,073 chair instances. Using the images in COCO-backpack and COCO-chair to query Google Image, we obtain 70,438 and 186,192 unlabeled Web images named Web-backpack and Web-chair, respectively. We have removed the Web images that are nearly duplicate with any image in the COCO training and validation sets. Figure 2 shows two query images and the retrieved Web images.

Labeling Web-backpack[1]. To facilitate the evaluation of our approach and future research, we label a subset of the Web-backpack images. This subset contains 16,128 images and is selected as follows. We apply a pre-trained R101-FPN object detector [32,63] to all Web-backpack images and then keep the ones that contain at least one backpack box detected with the confidence score higher

---

[1] We use Google Cloud Data Labeling Service for all the labeling work.

Fig. 2. Examples of the top three retrieved images from image-to-image search.

Fig. 3. Labeling errors in COCO-backpack. Two images in one group mean that they contain conflict annotations.

than 0.7. We ask three raters to label each survived images. One rater draws bounding boxes over all backpacks in an image. The other two examine the results sequentially. They modify the boxes if they find any problem with the previous rater's annotation. Please see the supplementary materials for more details of the annotation procedure, including a full annotation instruction we provided to the raters.

Relabeling COCO-backpack (see footnote 1). Using the same annotation procedure above, we also relabel the backpacks in COCO training and validation sets. The main reason for the relabeling is to mitigate the annotation mismatch between Web-backpack and COCO-backpack caused by different annotation protocols. Another reason is that we observe inconsistent bounding boxes in the existing COCO detection dataset. As Fig. 3 shows, some raters label the sling bags, while others do not, and some raters enclose the straps in the bounding boxes and others do not. We still ask three raters to label the COCO-backpack training images. For the validation set, we tighten the quality control and ask five raters to screen each image.

Table 1. Statistics of the Web images in this paper and their counterparts in COCO

Dataset	# Images	# Boxes by COCO	# Boxes by us	# Raters
COCO-backpack	5,528	8,714	7,170	3
COCO-backpack validation	5,000	371	436	5
Web-backpack	70,438	–	–	–
Web-backpack labeled	16,128	–	23,683	3
COCO-chair	12,774	38,073	–	–
COCO-chair validation	5,000	1,771	–	–
Web-chair	186,192	–	–	–

Statistics. Table 1 shows the statistics of the datasets used in this paper. We augment the COCO-backpack (chair) by Web-backpack (chair), whose size is about 15 times as the former. "Web-backpack labeled" is for evaluation only.

3   *Selective* Self-supervised Self-training

In this section, we describe our learning method named Selective Self-Supervised Self-training for Object Detection (S^4OD). Without loss of generality, we consider detecting only one class of objects from an input image. Denote by $\mathcal{D} = \{(I_1, \mathcal{T}_1), (I_2, \mathcal{T}_2), ..., (I_n, \mathcal{T}_n)\}$ the labeled image set and $\mathcal{U} = \{\tilde{I}_1, \tilde{I}_2, ..., \tilde{I}_m\}$ the crawled Web images, where $t_i^j = \{x_i^j, y_i^j, w_i^j, h_i^j\} \in \mathcal{T}_i$ contains the top-left coordinate, width, and height of the j-th ground-truth bounding box in the i-th image I_i. The labeled image set $\mathcal{D}$ is significantly smaller than the set of Web images $\mathcal{U}$. Besides, there exists a domain shift between the two sets, although we have tried to mitigate the mismatch by using the image-to-image search. Finally, some Web images could contain zero objects of the class being considered. In the following, we first customize vanilla self-training for object detection, discuss its limitations and fixations by a selective net, and then arrive at the full S^4OD algorithm. Figure 4 illustrates a diagram of our approach.

3.1   Self-training for Object Detection (SOD)

Given the labeled set $\mathcal{D}$ and unlabeled set $\mathcal{U}$, it is natural to test how self-training performs, especially given that it has recently achieved remarkable results [66] on ImageNet [8]. Following the procedure in [66], we first train a teacher object detector $f(I, \theta_t^*)$ from the labeled images, where θ_t^* stands for the network weights. We then produce pseudo boxes for each unlabeled Web image $\tilde{I}_i \in \mathcal{U}$:

$$\tilde{\mathcal{T}}_i, \tilde{\mathcal{S}}_i \leftarrow f(\tilde{I}_i, \theta_t^*), \quad i = 1, ..., m \tag{1}$$

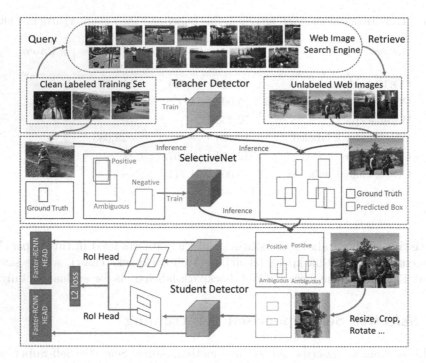

Fig. 4. Overview of the proposed approach. Top: using a small curated dataset to retrieve relevant Web images and to train a teacher detector. Middle: training a selective net to group pseudo boxes predicted by the teacher into positive, negative, and ambiguous groups. Bottom: learning a student detector from the Web-augmented training set with a self-supervised loss.

where each pseudo box $\widetilde{t}_i^j \in \widetilde{\mathcal{T}}_i$ is also associated with a confidence score $\widetilde{s}_i^j \in \widetilde{\mathcal{S}}_i$. We obtain the confidence score from the detector's classification head. Finally, we train a student detector in a pre-training-and-fine-tuning manner. The idea is to pre-train the student detector using the Web images $\mathcal{U}$ along with the pseudo bounding boxes, followed by fine-tuning it on the curated set $\mathcal{D}$.

Modern object detectors generate hundreds of object candidates per image to ensure high recall even after non-maximum suppression [48], implying that many of the predicted pseudo boxes are incorrect. Traditional self-training used in image classification [50,66] disregards low-confidence labels when they train the student model. In the same spirit, we only keep the pseudo boxes whose confidence scores are higher than 0.7.

3.2 Selective Self-training for Object Detection (S²OD)

In SOD described above, a crucial step is the selection of pseudo boxes by thresholding the confidence scores. The effectiveness of SOD largely depends on the quality of the selected boxes, which, unfortunately, poorly correlates with the confidence score. Figure 5 shows some examples where the pseudo boxes tightly

Fig. 5. Top: images in the COCO training set with groundtruth boxes. Bottom: boxes, their IoUs with the groundtruth, and confidence scores predicted by the teacher detector. The correlation between IoU and confidence is low.

bound the backpacks, but the teacher detector assigns them very low confidence scores. As a result, those boxes would be removed before SOD trains the student detector, under-utilizing the Web images $\mathcal{U}$. What's worse is that the mistakenly removed boxes could be discovered as false hard negatives during training.

To tackle the challenges, we propose a selective net g to calibrate the confidence scores of the pseudo boxes in the Web images $\mathcal{U}$. The main idea is to automatically group the boxes into three categories: positive, negative, and ambiguity. Denote by $g \circ \widetilde{T}_i$ the grouping results of the pseudo boxes in $\widetilde{T}_i$ for the i-th Web image $\widetilde{\mathbf{I}}_i$. We pre-train the student detector $f(\mathbf{I}, \theta_s^*)$ by the following (and then fine-tune it on the curated training set $\mathcal{D}$),

$$\theta_s^* \leftarrow \arg\min_{\theta_s} \frac{1}{m} \sum_{i=1}^{m} \ell(g \circ \widetilde{T}_i, f(\widetilde{\mathbf{I}}_i, \theta_s)), \tag{2}$$

where $\ell(\cdot, \cdot)$ is the conventional loss for training object detectors. For Faster-RCNN [48], the loss consists of regression, classification, objectiveness, etc. All the positive boxes predicted by our selective net g are used to activate those loss terms. In contrast, the ambiguous boxes, which could be correct but missed by the selective net, create a safe zone and do not contribute to any of the loss terms. This safe zone is especially useful when the learning algorithm has a hard negative mining scheme built in because it excludes potentially false "hard negatives" that fall in this ambiguity group.

Preparing Training Data for the Selective Net g. How do we learn the selective net g without knowing any groundtruth labels of the Web images? We seek answers by revisiting the labeled training set $\mathcal{D}$ instead. After we learn the teacher detector, we apply it to the training images in $\mathcal{D}$ and obtain a large pool of pseudo boxes. We assign each pseudo box $\widetilde{t}_i^j$ in the i-th image to one of the three groups by comparing it to the groundtruth boxes T_i,

$$g \circ \widetilde{t}_i^j = \begin{cases} \text{Negative,} & \max_{t \in T_i} IoU(\widetilde{t}_i^j, t)) \leq \gamma_l, \\ \text{Positive,} & \max_{t \in T_i} IoU(\widetilde{t}_i^j, t)) \geq \gamma_h, \\ \text{Ambiguity,} & \text{otherwise,} \end{cases} \quad (3)$$

where IoU is the intersection-over-union, a common evaluation metric in object detection and semantic segmentation [12,34], and $\gamma_h = 0.6$ and $\gamma_l = 0.05$ are two thresholds for IoU as opposed for the confidence scores. Interestingly, we can choose γ_h by using the COCO evaluation protocol [34] as follows. Considering all the boxes in the positive group as the teacher detector's final output, we can compute their mean average precision (mAP) over the labeled images $\mathcal{D}$. We choose the threshold of $\gamma_h = 0.6$ that maximizes the mAP.

Preparing Features for the Selective Net g. We accumulate all potentially useful features to represent a pseudo box so that the selective net can have enough information to group the boxes. The feature vector for a box $\widetilde{t}_i^j$ is

$$\left(f_{RoI}(\widetilde{t}_i^j, \mathbf{I}_i, \theta_t^*), \widetilde{s}_i^j, \widetilde{x}_i^j/W_i, \widetilde{y}_i^j/H_i, \widetilde{w}_i^j/W_i, \widetilde{h}_i^j/H_i, W_i, H_i \right) \quad (4)$$

where $f_{RoI}(\widetilde{t}_i^j, \mathbf{I}_i, \theta_t^*)$ is the RoI-pooled features [48] from the teacher detector, $\widetilde{s}_i^j$ is the confidence score, W_i and H_i are respectively the width and height of the i-th image, and the others are normalized box coordinate and size.

Training the Selective Net g. With the training data (Eq. (3)) and features (Eq. (4)) of the pseudo boxes, we learn the selective net by a three-way cross-entropy loss. We employ a straightforward architecture for the selective net. It comprises two towers. One is to process the normalized RoI-pooled features, and the other is to encode the remaining box features. They are both one-layer perceptrons with 512 and 128 hidden units, respectively. We then concatenate and feed their outputs into a three-way classifier.

One may concern that applying the teacher detector to the original training set $\mathcal{D}$ may not give rise to informative training data for the selective net because the detector could have "overfitted" the training set. Somehow surprisingly, we find that it is extremely difficult to overfit a detector to the training set, probably due to inconsistent human annotations of the bounding boxes. At best, the detector plays the role of an "average rater" who still cannot reach 100% mAP on the training set whose bounding boxes are provided by different users.

3.3   Selective Self-supervised Self-training Object Detection (S^{4}OD)

Finally, we boost S^{2}OD by a self-supervised loss based on two considerations. One is that Xie et al. demonstrate that it is beneficial to enforce the student network to learn more knowledge than what the teacher provides (e.g., robustness to artificial noise) [66]. The other is that Jeong et al. show the effectiveness of adding a consistency regularization to semi-supervised object detection [26].

More concretely, we add the following loss to Eq. (2) for each Web image $\widetilde{\mathbf{I}}_i$,

$$\ell_i := \sum_{\widetilde{t}_i^j \in \widetilde{\mathcal{T}}_i(\text{Positive})} \left\| f_{RoI}(\widetilde{t}_i^j, \widetilde{\mathbf{I}}_i, \theta_s) - f_{RoI}(t_i^j, \mathbf{I}_i, \theta_s) \right\|_2 + \ell(g \circ \mathcal{T}_i, f(\mathbf{I}_i, \theta_s)), \quad (5)$$

which consists of a consistency term borrowed from [26] and the same detection loss as Eq. (2) yet over a transformed Web image $\mathbf{I}_i$ — we explain the details below. They are additional cues to the pseudo boxes provided by the teacher. By learning harder than the teacher from all the supervision using the extra Web data, we expect the student detector to outperform the teacher.

Given a noisy Web image $\widetilde{\mathbf{I}} \in \mathcal{U}$, we use the selective net g to pick up positive boxes $\widetilde{\mathcal{T}}(\text{Positive})$ and limit the consistency loss over them. This small change from [26], which is feasible due to the selective net, turns out vital to the final performance. We transform a Web image $\widetilde{\mathbf{I}}$ to $\mathbf{I}$ by randomly choosing an operation from {rotation by 90, 180, or 270°} × {horizontal flip or not} × {random crop}. We use the crop ratio 0.9 and always avoid cutting through positive boxes. Accordingly, we can also obtain the transformed pseudo boxes $t^j \in \mathcal{T}$. The bottom panel of Fig. 4 exemplifies this transformation procedure.

4 Related Work

Weakly Supervised Object Detection. To reduce the dependence on large-scale human-curated data of state-of-the-art object detectors [19,22,32,35,47, 48,75], weakly supervised object detection (WSOD) [2,5,54,55,70,73] resort to image-level annotations to help localize objects. Besides, motion cues in videos are also explored by [27,45,52,71]. In addition, Tao et al. [57] incorporate Web images to learn a good feature representation for WSOD. Fine-grained segmentation [17,31] can also be used to guide WSOD. Although great progress has been made in WSOD, there is still quite a gap for them to catch up its supervised counterpart. The performance of fully supervised methods [48,63] are about 25 points better in terms of mean average precision compared to the weakly supervised ones [73].

Semi-supervised Classification. Semi-supervised learning [3,4,28,36,39,46, 58,65] jointly explore labeled and unlabeled data in training machine learning models. The consistency of model predictions or latent representations is an effective tool in semi-supervised learning [4,36,39,65]. The consistency regularization [28,39,58] reach state-of-the-art performance [41]. Besides, pseudo-label based approaches are also popular in semi-supervised learning [1,24,29,51]. In this vein, Xie et al. [66] use self-training [49,50,67,72] and aggressively inject noise in the learning procedure. They report state-of-the-art results on ImageNet with 300M unlabeled images.

Semi-supervised Object Detection. In the literature, the semi-supervised object detection could refer to two categories of work: **weakly semi-supervised detectors** [16,56,68,69] and **complete semi-supervised detectors** [26,60]. The former uses fully annotated data with box annotations and weakly labeled data with only image-level annotations. The latter learns by using unlabeled data in combination with the box-level labeled data. There are only a few works in the complete semi-supervised detection. Wang *et al.* [60] present a self-supervised sample mining process in active learning to get reliable region proposals for enhancing the object detector. Jeong *et al.* [26] introduce a consistency loss over unlabeled images.

Unsupervised/Self-supervised Representation Learning. Unsupervised/ self-supervised representation learning [10,13–15,21,74] from unlabeled data has attracted great attention recently. Some of them define a wide range of pretext tasks, such as recovering the input under some corruption [44,59], predicting rotation [18] or patch orderings [10,40] of an exemplar image, and tracking [61] or segmenting objects [43] in videos. Others utilize contrastive learning [6,11,21,42,64] by maximizing agreement between differently augmented views of the same data example. High-quality visual representations can help object detection [18,37], and self-supervised learning has been applied to replace the supervised ImageNet pretraining [25,43] for object detection. In addition, Lee *et al.* [30] propose a set of auxiliary tasks to make better use of scarce labels.

We study how to improve object detection with unlabeled, out-of-distribution Web images. While this problem is similar to semi-supervised object detection, neither self-training nor the self-supervised consistency loss, which are popular in semi-supervised learning, can cope with the Web images (cf. Table 2 where SOD and CSD can barely improve over the baselines (BD)). A major reason is that the out-of-distribution Web images give rise to many inaccurate candidate boxes and uncalibrated box classification scores. To tackle these challenges, we propose a selective net to identify high-quality (both positive and negative) boxes.

5   Experiments

We augment the COCO-backpack and COCO-chair training images with unlabeled Web images and run extensive experiments to test our approach on them. We note that the Web images are out of the distribution of COCO [34]. Some of them may contain no backpack or chair at all, and we rely on our selective net to identify the useful pseudo boxes in them produced by a teacher detector.

Besides, since S^4OD is readily applicable to vanilla semi-supervised object detection, whose labeled and unlabeled sets follow the same distribution, we also test it following the experiment protocol of [26].

Table 2. Comparison results for detecting backpacks and chairs.

Dataset	Method	Web	$AP@[.5,.95]$	$AP@.5$	$AP@.75$	AP_S	AP_M	AP_L
COCO-backpack	R101-FPN [32]	✗	16.45	33.63	15.26	18.29	18.23	17.61
	BD [32]	✗	17.34	34.71	15.13	19.03	19.46	21.31
	SOD [50]	✓	17.62	33.83	16.71	18.98	21.88	15.93
	CSD [26]	✓	17.87	32.97	16.55	20.42	22.05	20.45
	CSD-Selective (ours)	✓	18.32	33.35	17.05	**20.84**	21.08	21.11
	S²OD (ours)	✓	18.28	35.35	18.46	19.23	21.76	**23.94**
	S⁴OD (ours)	✓	**19.48**	**36.25**	**19.01**	20.31	**23.47**	18.53
COCO-chair	R101-FPN [32]	✗	28.28	48.93	28.57	19.14	33.21	42.06
	BD [32]	✗	29.56	49.15	30.70	20.16	36.39	41.99
	SOD [50]	✓	30.01	49.05	31.53	20.32	37.10	43.12
	CSD [26]	✓	30.19	48.58	31.92	20.77	37.19	43.74
	CSD-Selective (ours)	✓	30.95	49.75	32.11	21.03	38.02	44.96
	S²OD (ours)	✓	30.95	50.25	31.92	**21.19**	37.82	45.48
	S⁴OD (ours)	✓	**31.74**	**51.15**	**33.29**	20.57	**38.66**	**46.93**

5.1 Augmenting COCO-backpack(chair) with Web-backpack(chair)

We compare S^4OD to the following competing methods.

R101-FPN [32]: We use R101-FPN implemented in Detectron2 [63] as our object detector. Its AP on COCO-validation is 42.03, which is on par with the state of the arts. However, this detector still performs unsatisfactorily on the backpack and chair classes. As shown in Table 2, it only achieves 16.45 AP on detecting backpacks and 28.28 AP on chairs.

BD [32]: We finetune R101-FPN on COCO-backpack and COCO-chair, respectively, by changing the 80-class detector to a binary backpack or chair detector. We observe that the binary detector outperforms its 80-class counterpart by about 1% (0.9% AP for backpacks and 1.3% AP for chairs). We denote by BD the binary backpack (chair) detector and use it as the teacher detector for the remaining experiments.

CSD [26]: We include the recently published consistency-based semi-supervised object detection (CSD) in the experiments. We carefully re-implement it by using the same backbone detector (R101-FPN) as ours. CSD imposes the consistency loss over all candidate boxes.

CSD-Selective: We also report the results of applying our selective net to CSD. We remove the loss terms in CSD over the boxes of negative and ambiguity groups predicted by the selective net. Table 2 shows that it increases the performance of the original CSD by 0.45 AP on detecting backpacks and 0.76 AP on chairs.

SOD [50]: We presented a vanilla self-training procedure for object detection (SOD) in Sect. 3.1. Unlike self-training in image classification [66], we find that SOD is sensitive to the threshold of confidence scores probably because the out-of-distribution Web images make the scores highly uncalibrated. We

Positive boxes with low confidences | Negative boxes with high confidences | Ambiguous boxes

Fig. 6. The two panels on the left: Positive (negative) boxes chosen by the selective net yet with low (high) confidence scores predicted by the teacher detector. Rightmost: Ambiguous boxes according to the selective net.

Text-to-image search | COCO-backpack training images | Image-to-image search

Fig. 7. Example Web images for augmenting COCO-backpack.

test all thresholds between 0.5 and 0.9 (with an interval of 0.2) and find that SOD can only beat BD at the threshold of 0.7 (reported in Table 2).

S²OD (ours): This is an improved method over SOD by employing the selective net (cf. Sect. 3.2). It is also an ablated version of our S⁴OD by removing the self-supervised loss.

Table 2 presents the comparison results on the validation sets of both COCO-backpack and COCO-chair. We can see that S^2OD performs consistently better than its teacher detector (BD), the vanilla self-training (SOD), and the consistency-based self-supervised learning (CSD). Our full approach (S^4OD) brings additional gains and gives rise to 19.48 AP on detecting COCO backpacks and 31.74 AP on detecting chairs — about 2% better than its original teacher (BD). Besides, the improvements of S^2OD over SOD and CSD-selective over CSD both attribute to the selective net. Finally, Fig. 6 shows some cases where the selective net correctly groups low(high)-confidence boxes into the positive (negative) group.

In addition to the overall comparison results in Table 2, we next ablate our approach and examine some key components. We also report the "upper-bound" results for augmenting COCO-backpack with Web-backpack by using the bounding box labels we collected for a subset of Web-backpack.

Web-backpack vs. Text-to-Image Search. The image-to-image search for Web images is the very first step of our approach, and it is superior over the text-to-image search in various aspects. As the leftmost panel in Fig. 7 shows, most top-ranked Web images are iconic with salient objects sitting in clean backgrounds if we search using class names. For COCO [34], the images are

Table 3. Results of S^4OD with various Web data augmentations to COCO-backpack.

Method	$AP@[.5, .95]$	$AP@.5$	$AP@.75$	AP_S	AP_M	AP_L
BD [32]	17.34	34.71	15.13	19.03	19.46	21.31
S^4OD-text	17.37	34.26	14.46	18.77	20.10	17.66
S^4OD w/ 1/3 Web-backpack	18.40	34.12	17.91	19.25	21.20	21.56
S^4OD w/ 2/3 Web-backpack	19.08	35.35	17.96	21.04	20.15	21.27
S^4OD w/ full Web-backpack	19.48	36.25	19.01	20.31	**23.47**	18.53
S^4OD-2_{nd} Iteration	**19.52**	**36.44**	**19.07**	**21.70**	20.90	**24.90**

Table 4. Comparison results on the **relabeled** COCO-backpack.

Method	$AP@[.5, .95]$	$AP@.5$	$AP@.75$	AP_S	AP_M	AP_L
BD	18.75	36.03	16.98	11.90	21.62	31.58
Upper bound	22.63	40.89	21.15	15.97	26.53	32.11
S^4OD	20.27	37.55	20.49	13.51	24.04	31.17

collected from Flickr by complex object-object and object-scene queries. Using the same technique, we can retrieve more natural images. However, they are mostly recent and come from diverse sources, exhibiting a clear domain shift from the COCO dataset which is about six years old. In contrast, the image-to-image search well balances between the number of the retrieved Web images and their domain similarity to the query images (cf. the right panel in Fig. 7).

Table 3 compares image-to-image search with text-to-image search by their effects on the final results. The S^4OD-text row is the results obtained using 230k Web images crawled by text-to-image. While they are slightly better than BD's results, they are significantly worse than what S^4OD achieves with 70k Web images retrieved by image-to-image search (cf. Table 1).

Web-backpack: The Size Matters. We study how the number of the unlabeled Web images influences the proposed S^4OD by training with 1/3 and 2/3 of the crawled Web-backpack. As shown in Table 3, S^4OD-1/3 can improve over BD by 1% AP, and S^4OD-2/3 is better than BD by 1.7% AP. In contrast, S^4OD with the full Web-backpack leads to about 2.1% AP improvement. Overall, we see that the more Web images, the larger boost in performance, implying that it is worth studying the data pipeline and expanding its coverage in future work.

Iterating S^4OD. What happens if we use the detector trained by S^4OD as the teacher and train another student detector using S^4OD again? S^4OD-2_{nd} Iteration in Table 3 outperforms the single-iteration version only by a very small

Table 5. Comparison results on VOC2007 (* the numbers reported in [26])

Method	Labeled data	Unlabeled data	AP	Gain
Supervised [7,48]	VOC07	–	*73.9/74.1	–
Supervised [7,48]	VOC07&12	–	*79.4/80.3	–
CSD [26]	VOC07	VOC12	*74.7	*0.8
SOD	VOC07	VOC12	74.8	0.7
S^2OD	VOC07	VOC12	75.8	1.7
S^4OD	VOC07	VOC12	**76.4**	**2.3**

margin. It is probably because, in the second iteration, the student detector does not receive more supervision than what the teacher BD provides. We will explore some stochastic self-supervised losses in future work to enforce the student to learn more than what the teacher provides at every iteration.

An "Upper Bound" of S^4OD. We further investigate the effectiveness of S^4OD by comparing it to an "upper bound". We run the experiments using the labeled subset of Web-backpack and the relabeled COCO-backpack to have consistent annotations across the two datasets. Recall that we have fixed some inconsistent bounding boxes in the original COCO-backpack in the relabeling process. As a result, if we train BD [48] on the relabeled COCO-backpack and evaluate on the relabeled validation set, the AP is 18.75 (cf. Table 4), in contrast to 17.34 AP (cf. Table 2) by BD trained and evaluated using the original COCO-backpack. Using this BD as the teacher, we train a student detector using S^4OD. By the "upper bound", we pool the labeled subset of Web-backpack and the relabeled COCO-backpack together and then train BD. The results in Table 4 indicate that S^4OD is almost right in the middle of the lower-bound (BD) and the upper bound. The gap between S^4OD and the upper bound is small. We will study better learning methods to close the gap and how the Web data volume could impact the performance in future work.

5.2 Semi-supervised Object Detection on PASCAL VOC

It is straightforward to extend S^4OD to multiple object detection tasks. Following the experiment protocol in [26], we further validate it in the setting of semi-supervised object detection, whose labeled and unlabeled sets are drawn from the same distribution. We use PASCAL VOC2007 trainval (5,011 images) as the labeled set and PASCAL VOC2012 trainval (11,540 images) as the unlabeled set. There are 20 classes of objects to detect. We test all detectors on the test set of PASCAL VOC2007 (4,952 images). In order to make our results comparable to what are reported in [26], we switch to Faster-RCNN [48] with ResNet-50 [23] as the base detector.

Table 5 shows the comparison results. We include both our results and those reported in [25] and mark the latter by *. Considering the object detector trained on VOC2007 as a baseline and the one trained on both datasets with labels as an upper bound, our S^4OD is right in the middle, outperforming CSD by 1.5 AP. Although we propose the selective net mainly to handle noisy Web images, it is delightful that the resulting method also works well with the clean, in-domain VOC2012 images. It is probably because the consistency loss in CSD drives the detector toward high-entropy, inaccurate predictions (cf. more discussions in [66]). Our selective net avoids this caveat by supplying high-quality boxes to the consistency loss.

6 Conclusion

In this paper, we propose a novel approach to improving object detection with massive unlabeled Web images. We collect the Web images with image-to-image search, leading to smaller domain mismatch between the retrieved Web images and the curated dataset than text-to-image search does. Besides, we incorporate a principled selective net into self-training rather than threshold confidence scores as a simple heuristic for selecting bounding boxes. Moreover, we impose a self-supervised training loss over the high-quality boxes chosen by the selective net to make better use of the Web images. The improvement in detecting challenging objects is significant over the competing methods, Our approach works consistently well on not only the Web-augmented object detection but also the traditional semi-supervised object detection.

References

1. Arazo, E., Ortego, D., Albert, P., O'Connor, N.E., McGuinness, K.: Pseudo-labeling and confirmation bias in deep semi-supervised learning. arXiv preprint arXiv:1908.02983 (2019)
2. Arun, A., Jawahar, C., Kumar, M.P.: Dissimilarity coefficient based weakly supervised object detection. In: Proceedings of the IEEE Conference on Computer Vision and Pattern Recognition, pp. 9432–9441 (2019)
3. Bachman, P., Alsharif, O., Precup, D.: Learning with pseudo-ensembles. In: Advances in Neural Information Processing Systems, pp. 3365–3373 (2014)
4. Berthelot, D., Carlini, N., Goodfellow, I., Papernot, N., Oliver, A., Raffel, C.A.: Mixmatch: a holistic approach to semi-supervised learning. In: Advances in Neural Information Processing Systems, pp. 5050–5060 (2019)
5. Bilen, H., Vedaldi, A.: Weakly supervised deep detection networks. In: Proceedings of the IEEE Conference on Computer Vision and Pattern Recognition, pp. 2846–2854 (2016)
6. Chen, T., Kornblith, S., Norouzi, M., Hinton, G.: A simple framework for contrastive learning of visual representations. arXiv preprint arXiv:2002.05709 (2020)
7. Dai, J., Li, Y., He, K., Sun, J.: R-FCN: object detection via region-based fully convolutional networks. In: Advances in Neural Information Processing Systems, pp. 379–387 (2016)
8. Deng, J., Dong, W., Socher, R., Li, L.J., Li, K., Fei-Fei, L.: ImageNet: a large-scale hierarchical image database. In: 2009 IEEE Conference on Computer Vision and Pattern Recognition, pp. 248–255. IEEE (2009)

9. Ding, Y., Wang, L., Fan, D., Gong, B.: A semi-supervised two-stage approach to learning from noisy labels. In: 2018 IEEE Winter Conference on Applications of Computer Vision (WACV), pp. 1215–1224. IEEE (2018)
10. Doersch, C., Gupta, A., Efros, A.A.: Unsupervised visual representation learning by context prediction. In: Proceedings of the IEEE International Conference on Computer Vision, pp. 1422–1430 (2015)
11. Donahue, J., Simonyan, K.: Large scale adversarial representation learning. In: Advances in Neural Information Processing Systems, pp. 10541–10551 (2019)
12. Everingham, M., Van Gool, L., Williams, C.K., Winn, J., Zisserman, A.: The pascal visual object classes (VOC) challenge. Int. J. Comput. Vis. **88**(2), 303–338 (2010). https://doi.org/10.1007/s11263-009-0275-4
13. Gan, C., Sun, C., Duan, L., Gong, B.: Webly-supervised video recognition by mutually voting for relevant web images and web video frames. In: Leibe, B., Matas, J., Sebe, N., Welling, M. (eds.) ECCV 2016. LNCS, vol. 9907, pp. 849–866. Springer, Cham (2016). https://doi.org/10.1007/978-3-319-46487-9_52
14. Gan, C., Yao, T., Yang, K., Yang, Y., Mei, T.: You lead, we exceed: labor-free video concept learning by jointly exploiting web videos and images. In: CVPR, pp. 923–932 (2016)
15. Gan, C., Zhao, H., Chen, P., Cox, D., Torralba, A.: Self-supervised moving vehicle tracking with stereo sound. In: Proceedings of the IEEE International Conference on Computer Vision, pp. 7053–7062 (2019)
16. Gao, J., Wang, J., Dai, S., Li, L.J., Nevatia, R.: Note-RCNN: noise tolerant ensemble RCNN for semi-supervised object detection. In: Proceedings of the IEEE International Conference on Computer Vision, pp. 9508–9517 (2019)
17. Gao, Y., et al.: C-MIDN: coupled multiple instance detection network with segmentation guidance for weakly supervised object detection. In: Proceedings of the IEEE International Conference on Computer Vision, pp. 9834–9843 (2019)
18. Gidaris, S., Singh, P., Komodakis, N.: Unsupervised representation learning by predicting image rotations. arXiv preprint arXiv:1803.07728 (2018)
19. Girshick, R.: Fast R-CNN. In: Proceedings of the IEEE International Conference on Computer Vision, pp. 1440–1448 (2015)
20. Grabner, H., Gall, J., Van Gool, L.: What makes a chair a chair? In: CVPR 2011, pp. 1529–1536. IEEE (2011)
21. He, K., Fan, H., Wu, Y., Xie, S., Girshick, R.: Momentum contrast for unsupervised visual representation learning. arXiv preprint arXiv:1911.05722 (2019)
22. He, K., Gkioxari, G., Dollár, P., Girshick, R.: Mask R-CNN. In: Proceedings of the IEEE International Conference on Computer Vision, pp. 2961–2969 (2017)
23. He, K., Zhang, X., Ren, S., Sun, J.: Deep residual learning for image recognition. In: Proceedings of the IEEE Conference on Computer Vision and Pattern Recognition, pp. 770–778 (2016)
24. Iscen, A., Tolias, G., Avrithis, Y., Chum, O.: Label propagation for deep semi-supervised learning. In: Proceedings of the IEEE Conference on Computer Vision and Pattern Recognition, pp. 5070–5079 (2019)
25. Jenni, S., Favaro, P.: Self-supervised feature learning by learning to spot artifacts. In: Proceedings of the IEEE Conference on Computer Vision and Pattern Recognition, pp. 2733–2742 (2018)
26. Jeong, J., Lee, S., Kim, J., Kwak, N.: Consistency-based semi-supervised learning for object detection. In: Advances in Neural Information Processing Systems, pp. 10758–10767 (2019)

27. Kumar Singh, K., Xiao, F., Jae Lee, Y.: Track and transfer: watching videos to simulate strong human supervision for weakly-supervised object detection. In: Proceedings of the IEEE Conference on Computer Vision and Pattern Recognition, pp. 3548–3556 (2016)
28. Laine, S., Aila, T.: Temporal ensembling for semi-supervised learning. arXiv preprint arXiv:1610.02242 (2016)
29. Lee, D.H.: Pseudo-label: the simple and efficient semi-supervised learning method for deep neural networks. In: Workshop on Challenges in Representation Learning, ICML, vol. 3, p. 2 (2013)
30. Lee, W., Na, J., Kim, G.: Multi-task self-supervised object detection via recycling of bounding box annotations. In: Proceedings of the IEEE Conference on Computer Vision and Pattern Recognition, pp. 4984–4993 (2019)
31. Li, X., Kan, M., Shan, S., Chen, X.: Weakly supervised object detection with segmentation collaboration. In: Proceedings of the IEEE International Conference on Computer Vision, pp. 9735–9744 (2019)
32. Lin, T.Y., Dollár, P., Girshick, R., He, K., Hariharan, B., Belongie, S.: Feature pyramid networks for object detection. In: Proceedings of the IEEE Conference on Computer Vision and Pattern Recognition, pp. 2117–2125 (2017)
33. Lin, T.Y., Goyal, P., Girshick, R., He, K., Dollár, P.: Focal loss for dense object detection. In: Proceedings of the IEEE International Conference on Computer Vision, pp. 2980–2988 (2017)
34. Lin, T.-Y., et al.: Microsoft COCO: common objects in context. In: Fleet, D., Pajdla, T., Schiele, B., Tuytelaars, T. (eds.) ECCV 2014. LNCS, vol. 8693, pp. 740–755. Springer, Cham (2014). https://doi.org/10.1007/978-3-319-10602-1_48
35. Liu, W., et al.: SSD: single shot MultiBox detector. In: Leibe, B., Matas, J., Sebe, N., Welling, M. (eds.) ECCV 2016. LNCS, vol. 9905, pp. 21–37. Springer, Cham (2016). https://doi.org/10.1007/978-3-319-46448-0_2
36. Luo, Y., Zhu, J., Li, M., Ren, Y., Zhang, B.: Smooth neighbors on teacher graphs for semi-supervised learning. In: Proceedings of the IEEE Conference on Computer Vision and Pattern Recognition, pp. 8896–8905 (2018)
37. Mahajan, D., et al.: Exploring the limits of weakly supervised pretraining. In: Ferrari, V., Hebert, M., Sminchisescu, C., Weiss, Y. (eds.) ECCV 2018. LNCS, vol. 11206, pp. 185–201. Springer, Cham (2018). https://doi.org/10.1007/978-3-030-01216-8_12
38. Minsky, M.: Society of Mind. Simon and Schuster, New York (1988)
39. Miyato, T., Maeda, S.I., Koyama, M., Ishii, S.: Virtual adversarial training: a regularization method for supervised and semi-supervised learning. IEEE Trans. Pattern Anal. Mach. Intell. 41(8), 1979–1993 (2018)
40. Noroozi, M., Favaro, P.: Unsupervised learning of visual representations by solving jigsaw puzzles. In: Leibe, B., Matas, J., Sebe, N., Welling, M. (eds.) ECCV 2016. LNCS, vol. 9910, pp. 69–84. Springer, Cham (2016). https://doi.org/10.1007/978-3-319-46466-4_5
41. Oliver, A., Odena, A., Raffel, C.A., Cubuk, E.D., Goodfellow, I.: Realistic evaluation of deep semi-supervised learning algorithms. In: Advances in Neural Information Processing Systems, pp. 3235–3246 (2018)
42. van den Oord, A., Li, Y., Vinyals, O.: Representation learning with contrastive predictive coding. arXiv preprint arXiv:1807.03748 (2018)
43. Pathak, D., Girshick, R., Dollár, P., Darrell, T., Hariharan, B.: Learning features by watching objects move. In: Proceedings of the IEEE Conference on Computer Vision and Pattern Recognition, pp. 2701–2710 (2017)

44. Pathak, D., Krahenbuhl, P., Donahue, J., Darrell, T., Efros, A.A.: Context encoders: feature learning by inpainting. In: Proceedings of the IEEE Conference on Computer Vision and Pattern Recognition, pp. 2536–2544 (2016)
45. Prest, A., Leistner, C., Civera, J., Schmid, C., Ferrari, V.: Learning object class detectors from weakly annotated video. In: 2012 IEEE Conference on Computer Vision and Pattern Recognition, pp. 3282–3289. IEEE (2012)
46. Rasmus, A., Berglund, M., Honkala, M., Valpola, H., Raiko, T.: Semi-supervised learning with ladder networks. In: Advances in Neural Information Processing Systems, pp. 3546–3554 (2015)
47. Redmon, J., Farhadi, A.: YOLO9000: better, faster, stronger. In: Proceedings of the IEEE Conference on Computer Vision and Pattern Recognition, pp. 7263–7271 (2017)
48. Ren, S., He, K., Girshick, R., Sun, J.: Faster R-CNN: towards real-time object detection with region proposal networks. In: Advances in Neural Information Processing Systems, pp. 91–99 (2015)
49. Riloff, E., Wiebe, J.: Learning extraction patterns for subjective expressions. In: Proceedings of the 2003 Conference on Empirical Methods in Natural Language Processing, pp. 105–112 (2003)
50. Scudder, H.: Probability of error of some adaptive pattern-recognition machines. IEEE Trans. Inf. Theory 11(3), 363–371 (1965)
51. Shi, W., Gong, Y., Ding, C., Ma, Z., Tao, X., Zheng, N.: Transductive semi-supervised deep learning using min-max features. In: Ferrari, V., Hebert, M., Sminchisescu, C., Weiss, Y. (eds.) ECCV 2018. LNCS, vol. 11209, pp. 311–327. Springer, Cham (2018). https://doi.org/10.1007/978-3-030-01228-1_19
52. Singh, K.K., Lee, Y.J.: You reap what you sow: using videos to generate high precision object proposals for weakly-supervised object detection. In: Proceedings of the IEEE Conference on Computer Vision and Pattern Recognition, pp. 9414–9422 (2019)
53. Tan, M., Pang, R., Le, Q.V.: EfficientDet: scalable and efficient object detection. In: Proceedings of the IEEE/CVF Conference on Computer Vision and Pattern Recognition, pp. 10781–10790 (2020)
54. Tang, P., Wang, X., Bai, X., Liu, W.: Multiple instance detection network with online instance classifier refinement. In: Proceedings of the IEEE Conference on Computer Vision and Pattern Recognition, pp. 2843–2851 (2017)
55. Tang, P., et al.: Weakly supervised region proposal network and object detection. In: Ferrari, V., Hebert, M., Sminchisescu, C., Weiss, Y. (eds.) ECCV 2018. LNCS, vol. 11215, pp. 370–386. Springer, Cham (2018). https://doi.org/10.1007/978-3-030-01252-6_22
56. Tang, Y., Wang, J., Gao, B., Dellandréa, E., Gaizauskas, R., Chen, L.: Large scale semi-supervised object detection using visual and semantic knowledge transfer. In: Proceedings of the IEEE Conference on Computer Vision and Pattern Recognition, pp. 2119–2128 (2016)
57. Tao, Q., Yang, H., Cai, J.: Zero-annotation object detection with web knowledge transfer. In: Ferrari, V., Hebert, M., Sminchisescu, C., Weiss, Y. (eds.) ECCV 2018. LNCS, vol. 11215, pp. 387–403. Springer, Cham (2018). https://doi.org/10.1007/978-3-030-01252-6_23
58. Tarvainen, A., Valpola, H.: Mean teachers are better role models: weight-averaged consistency targets improve semi-supervised deep learning results. In: Advances in Neural Information Processing Systems, pp. 1195–1204 (2017)

59. Vincent, P., Larochelle, H., Bengio, Y., Manzagol, P.A.: Extracting and composing robust features with denoising autoencoders. In: Proceedings of the 25th International Conference on Machine Learning, pp. 1096–1103 (2008)
60. Wang, K., Yan, X., Zhang, D., Zhang, L., Lin, L.: Towards human-machine cooperation: self-supervised sample mining for object detection. In: Proceedings of the IEEE Conference on Computer Vision and Pattern Recognition, pp. 1605–1613 (2018)
61. Wang, X., Gupta, A.: Unsupervised learning of visual representations using videos. In: Proceedings of the IEEE International Conference on Computer Vision, pp. 2794–2802 (2015)
62. Wei, X., Gong, B., Liu, Z., Lu, W., Wang, L.: Improving the improved training of Wasserstein GANs: a consistency term and its dual effect. arXiv preprint arXiv:1803.01541 (2018)
63. Wu, Y., Kirillov, A., Massa, F., Lo, W.Y., Girshick, R.: Detectron2 (2019). https://github.com/facebookresearch/detectron2
64. Wu, Z., Xiong, Y., Yu, S., Lin, D.: Unsupervised feature learning via non-parametric instance-level discrimination. arXiv preprint arXiv:1805.01978 (2018)
65. Xie, Q., Dai, Z., Hovy, E., Luong, M.T., Le, Q.V.: Unsupervised data augmentation. arXiv preprint arXiv:1904.12848 (2019)
66. Xie, Q., Hovy, E., Luong, M.T., Le, Q.V.: Self-training with noisy student improves ImageNet classification. arXiv preprint arXiv:1911.04252 (2019)
67. Yalniz, I.Z., Jégou, H., Chen, K., Paluri, M., Mahajan, D.: Billion-scale semi-supervised learning for image classification. arXiv preprint arXiv:1905.00546 (2019)
68. Yan, Z., Liang, J., Pan, W., Li, J., Zhang, C.: Weakly-and semi-supervised object detection with expectation-maximization algorithm. arXiv preprint arXiv:1702.08740 (2017)
69. Yang, H., Wu, H., Chen, H.: Detecting 11K classes: large scale object detection without fine-grained bounding boxes. In: Proceedings of the IEEE International Conference on Computer Vision, pp. 9805–9813 (2019)
70. Yang, K., Li, D., Dou, Y.: Towards precise end-to-end weakly supervised object detection network. In: Proceedings of the IEEE International Conference on Computer Vision, pp. 8372–8381 (2019)
71. Yang, Z., Mahajan, D., Ghadiyaram, D., Nevatia, R., Ramanathan, V.: Activity driven weakly supervised object detection. In: Proceedings of the IEEE Conference on Computer Vision and Pattern Recognition, pp. 2917–2926 (2019)
72. Yarowsky, D.: Unsupervised word sense disambiguation rivaling supervised methods. In: 33rd Annual Meeting of the Association for Computational Linguistics, pp. 189–196 (1995)
73. Zeng, Z., Liu, B., Fu, J., Chao, H., Zhang, L.: WSOD2: learning bottom-up and top-down objectness distillation for weakly-supervised object detection. In: Proceedings of the IEEE International Conference on Computer Vision, pp. 8292–8300 (2019)
74. Zhao, H., Gan, C., Rouditchenko, A., Vondrick, C., McDermott, J., Torralba, A.: The sound of pixels. In: Ferrari, V., Hebert, M., Sminchisescu, C., Weiss, Y. (eds.) ECCV 2018. LNCS, vol. 11205, pp. 587–604. Springer, Cham (2018). https://doi.org/10.1007/978-3-030-01246-5_35
75. Zhou, X., Wang, D., Krähenbühl, P.: Objects as points. arXiv preprint arXiv:1904.07850 (2019)

Deep Local Shapes: Learning Local SDF Priors for Detailed 3D Reconstruction

Rohan Chabra[1,3]($\boxtimes$), Jan E. Lenssen[2,3], Eddy Ilg[3], Tanner Schmidt[3], Julian Straub[3], Steven Lovegrove[3], and Richard Newcombe[3]

[1] University of North Carolina at Chapel Hill, Chapel Hill, USA
rohanc@cs.unc.edu
[2] TU Dortmund University, Dortmund, Germany
[3] Facebook Reality Labs, Redmond, WA, USA

Abstract. Efficiently reconstructing complex and intricate surfaces at scale is a long-standing goal in machine perception. To address this problem we introduce Deep Local Shapes (DeepLS), a deep shape representation that enables encoding and reconstruction of high-quality 3D shapes without prohibitive memory requirements. DeepLS replaces the dense volumetric signed distance function (SDF) representation used in traditional surface reconstruction systems with a set of locally learned continuous SDFs defined by a neural network, inspired by recent work such as DeepSDF. Unlike DeepSDF, which represents an object-level SDF with a neural network and a single latent code, we store a grid of independent latent codes, each responsible for storing information about surfaces in a small local neighborhood. This decomposition of scenes into local shapes simplifies the prior distribution that the network must learn, and also enables efficient inference. We demonstrate the effectiveness and generalization power of DeepLS by showing object shape encoding and reconstructions of full scenes, where DeepLS delivers high compression, accuracy, and local shape completion.

1 Introduction

A signed distance function (SDF) represents three-dimensional surfaces as the zero-level set of a continuous scalar field. This representation has been used by many classical methods to represent and optimize geometry based on raw sensor observations [12,31,36,37,46]. In a typical use case, an SDF is approximated by storing values on a regularly-spaced voxel grid and computing intermediate values using linear interpolation. Depth observations can then be used to infer these values and a series of such observations are combined to infer the most likely SDF using a process called fusion.

R. Chabra and J. E. Lenssen—Work performed during an internship at Facebook Reality Labs.

Electronic supplementary material The online version of this chapter (https://doi.org/10.1007/978-3-030-58526-6_36) contains supplementary material, which is available to authorized users.

© Springer Nature Switzerland AG 2020
A. Vedaldi et al. (Eds.): ECCV 2020, LNCS 12374, pp. 608–625, 2020.
https://doi.org/10.1007/978-3-030-58526-6_36

Fig. 1. Reconstruction performed by our Deep Local Shapes (DeepLS) of the Burghers of Calais scene [57]. DeepLS represents surface geometry as a sparse set of local latent codes in a voxel grid, as shown on the right. Each code compresses a local volumetric SDF function, which is reconstructed by an implicit neural network decoder.

Voxelized SDFs have been widely adopted and used successfully in a number of applications, but they have some fundamental limitations. First, the dense voxel representation requires significant amounts of memory (typically on a resource-constrained parallel computing device), which imposes constraints on resolution and the spatial extent that can be represented. These limits on resolution, as well as sensor limitations, typically lead to surface estimates that are missing thin structures and fine surface details. Second, as a non-parametric representation, SDF fusion can only infer surfaces that have been directly observed. Some surfaces are difficult or impossible for a typical range sensor to capture, and observing every surface in a typical environment is a challenging task. As a result, reconstructions produced by SDF fusion are often incomplete.

Recently, deep neural networks have been explored as an alternative representation for signed distance functions. According to the universal approximation theorem [25], a neural network can be used to approximate any continuous function, including signed distance functions [10,34,35,39]. With such models, the level of detail that can be represented is limited only by the capacity and architecture of the network. In addition, a neural network can be made to represent not a single surface but a family of surfaces by conditioning the function on a latent code. Such a network can then be used as a parametric model for estimating the most likely surface given only partial noisy observations. Incorporating shape priors in this way allows us to move from the maximum likelihood (ML) estimation of classical reconstruction techniques to potentially more robust reconstruction via maximum a posteriori (MAP) inference.

These neural network representations have their own limitations, however. Most of the prior work on learning SDFs is object-centric and does not trivially scale to the detail required for scene-level representations. This is likely due to the global co-dependence of the SDF values at any two locations in space, which are computed using a shared network and a shared parameterization. Furthermore, while the ability of these networks to learn distributions over classes of shapes allows for robust completion of novel instances from known classes, it does not easily generalize to novel classes or objects, which would be necessary for applications in scene reconstruction. In scanned real-world scenes, the diversity of objects and object setups is usually too high to be covered by an object-centric training data distribution.

Contribution. In this work, we introduce Deep Local Shapes (DeepLS) to combine the benefits of both worlds, exposing a trade-off between the prior-based MAP inference of memory efficient deep global representations (e.g., DeepSDF), and the detail preservation of computationally efficient, explicit volumetric SDFs. We divide space into local volumes, each with a small latent code representing signed distance functions in local coordinate frames (Fig. 1). These voxels can be larger than is typical in fusion systems without sacrificing on the level of surface detail that can be represented (c.f. Sect. 5.2), increasing memory efficiency. The proposed representation has several favorable properties, which are verified in our evaluations on several types of input data:

1. It relies on readily available local shape patches as training data and generalizes to a large variety of shapes,
2. provides significantly finer reconstruction and orders of magnitude faster inference than global, object-centric methods like DeepSDF, and
3. outperforms existing approaches in dense 3D reconstruction from partial observations, showing thin details with significantly better surface completion and high compression.

2   Related Work

The key contribution of this paper is the application of learned local shape priors for reconstruction of 3D surfaces. This section will therefore discuss related work on traditional representations for surface reconstruction, learned shape representations, and local shape priors.

2.1   Traditional Shape Representations

Traditionally, scene representation methods can broadly be categorized into two categories, namely local and global approaches.

Local Approaches. Most implicit surface representations from unorganized point sets are based on Blinn's idea of blending local implicit primitives [6]. Hope et al. [24] explicitly defined implicit surfaces by the tangent of the normals of the input points. Ohtake et al. [38] established more control over the local shape functions using quadratic surface fitting and blended these in a multi-scale partition of unity scheme. Curless and Levoy [12] introduced volumetric integration of scalar approximations of implicit SDFs in regular grids. This technique was further extended into real-time systems [31,36,37,46]. Surfaces are also shown to be represented by surfels, i.e. oriented planar surface patches [30,40,51].

Global Approaches. Global implicit function approximation methods aim to approximate single continuous signed distance functions using, for example, kernel-based techniques [8,16,28,49]. Visibility or free space methods estimate which subset of 3D space is occupied, often by subdividing space into distinct tetrahedra [4,26,32]. These methods aim to solve for a globally view consistent surface representation.

Our work falls into the local surface representation category. It is related to the partition of unity approach [38], however, instead of using quadratic functions as local shapes, we use data-driven local priors to approximate implicit SDFs, which are robust to noise and can locally complete supported surfaces. While we also experimented with partition of unity blending of neighboring local shapes, we found it to be not required in practice, since our training formulation already includes border consistency (c.f. Sect. 4.1), thus saving function evaluations during decoding. In comparison to volumetric SDF integration methods, such as SDF Fusion [37], our approach provides better shape completion and denoising, while at the same time uses less memory to store the representation. Unlike point- or surfel-based methods, our method leads to smooth and connected surfaces.

2.2 Learned Shape Representations

Recently there has been lot of work on 3D shape learning using deep neural networks. This class of work can also be classified into four categories: point-based methods, mesh-based methods, voxel-based methods and continuous implicit function-based methods.

Points. The methods use generative point cloud models for scene representation [3,55,56]. Typically, a neural network is trained to directly regress 3D coordinates of points in the point cloud.

Voxels. These methods provide non-parametric shape representation using 3D voxel grids which store either occupancy [11,53] or SDF information [14,33,47], similarly to the traditional techniques discussed above. These methods thus inherit the limitations of traditional voxel representations with respect to high memory requirements. Octree-based methods [23,42,48] relax the compute and memory limitations of dense voxel methods to some degree and have been shown on voxel resolutions of up to 512^3.

Meshes. These methods use existing [44] or learned [5,21] parameterization techniques to describe 3D surfaces by morphing 2D planes. When using mesh representations, there is a tradeoff between the ability to support arbitrary topology and the ability to reconstruct smooth and connected surfaces. Works such as [5,44] are variations on deforming a sphere into more complex 3D shape, which produces smooth and connected shapes but limits the topology to shapes that are homeomorphic to the sphere. AtlasNet, on the other hand, warps multiple 2D planes into 3D which together form a shape of arbitrary topology, but this results in disconnected surfaces. Other works, such as Scan2Mesh [13] and Mesh-RCNN [20], use deep networks to predict meshes corresponding to range scans or RGB images, respectively.

Implicit Functions. Very recently, there has been significant work on learning continuous implicit functions for shape representations. Occupancy Networks [34] and PiFU [43] represent shapes using continuous indicator functions which specify which subset of 3D space the shapes occupy. Similarly, DeepSDF [39] approximates shapes using Signed Distance Fields. We adopt the DeepSDF model as the backbone architecture for our local shape network.

Much of the work in this area has focused on learning object-level representations. This is especially useful when given partial observations of a known class, as the learned priors can often complete the shape with surprising accuracy. However, this also introduces two key difficulties. First, the object-level context means that generalization will be limited by the extent of the training set – objects outside of the training distribution may not be well reconstructed. Second, object-level methods do not trivially scale to full scenes composed of many objects as well as surfaces (e.g. walls and floors). In contrast, DeepLS maintains separate representations for small, distinct regions of space, which allows it to scale easily. Furthermore, the local representation makes it easier to compile a representative training set; at a small scale most surfaces have similar structure.

2.3   Local Shape Priors

In early work on using local shape priors, Gal et al. [17] used a database of local surface patches to match partial shape observations. However, the ability to match general observations was limited by the size of the database as the patches could not be interpolated. Ricao et al. [41] used both PCA and a learned autoencoder to map SDF subvolumes to lower-dimensional representations, approaching local shape priors from the perspective of compression. With this approach the SDF must be computed by fusion first, which serves as an information bottleneck limiting the ability to develop priors over fine-grained structures. In another work, Xu et al. [54] developed an object-level learned shape representation using a network that maps from images to SDFs . This representation is conditioned on and therefore not independent of the observed image. Williams et al. [52] showed recently that a deep network can be used to fit a representation of a surface by training and evaluating on the same point cloud, using a local chart for each point which is then combined to form a surface atlas. Their results are on complete point clouds in which the task is simply to densify and denoise, whereas we also show that our priors can locally complete surfaces that were not observed. Other work on object-level shape representation has explored representations in which shapes are composed of smaller parts. Structured implicit functions used anisotropic Gaussian kernels to compose global implicit shape representations [19]. Similarly, CvxNets compose shapes using a collection of convex subshapes [15]. Like ours, both of these methods show the promise of compositional shape modelling, but surface detail was limited by the models used. The work of Genova et al. [18] combines a set of irregularly positioned implicit functions to improve details in full object reconstruction. Similar to our work, concurrent work of Jiang et al. [27] proposes to use local implicit functions in a grid for detailed 3D reconstruction.

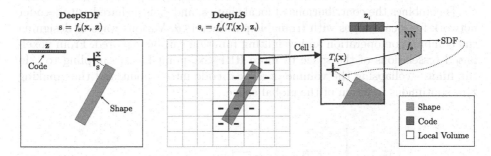

Fig. 2. 2D example of DeepSDF [39] and DeepLS (ours). DeepSDF provides global shape codes (left). We use the DeepSDF idea for local shape codes (center). Our approach requires a matrix of low-dimensional code vectors which in total require less storage than the global version. The gray codes are an indicator for empty space. The SDF to the surface is predicted using a fully-connected network that receives the local code and coordinates as input.

3   Review of DeepSDF

We will briefly review DeepSDF [39]. Let $f_\theta(\mathbf{x}, \mathbf{z})$ be a signed surface distance function modeled as a fully-connected neural network with trainable parameters θ and shape code $\mathbf{z}$. Then a shape $\mathcal{S}$ is defined as the zero level set of $f_\theta(\mathbf{x}, \mathbf{z})$:

$$\mathcal{S} = \{\mathbf{x} \in \mathbb{R}^3 \mid f_\theta(\mathbf{x}, \mathbf{z}) = 0\}. \tag{1}$$

In order to simultaneously train for a variety of shapes, a $\mathbf{z}$ is optimized for each shape while network parameters θ are shared for the whole set of shapes.

4   Deep Local Shapes

The key idea of DeepLS is to compose complex general shapes and scenes from a collection of simpler local shapes as depicted in Fig. 2. Scenes and shapes of arbitrary complexity cannot be described with a compact fixed length shape code such as used by DeepSDF. Instead it is more efficient and flexible to encode the space of smaller local shapes and to compose the global shape from an adaptable amount of local codes.

To describe a surface S in $\mathbb{R}^3$ using DeepLS, we first define a partition of the space into local volumes $V_i \subseteq \mathbb{R}^3$ with associated local coordinate systems. Like in DeepSDF, but at a local level, we describe the surface in each local volume using a code $\mathbf{z}_i$. With the transformation $T_i(\mathbf{x})$ of the global location $\mathbf{x}$ into the local coordinate system, the global surface S is described as the zero level set

$$S = \left\{ \mathbf{x} \in \mathbb{R}^3 \mid \bigoplus_i w(\mathbf{x}, V_i) f_\theta \left(T_i(\mathbf{x}), \mathbf{z}_i \right) = 0 \right\}, \tag{2}$$

where $w(\mathbf{x}, V_i)$ weighs the contribution of the ith local shape to the global shape S, $\bigoplus$ combines the contributions of local shapes, and f_θ is a shared autodecoder network for local shapes with trainable parameters θ. Various ways of designing the combination operation and weighting function can be explored. From voxel-based tesselations of the space to more RBF-like point-based sampling to – in the limit – collapsing the volume of a local code into a point and thus making $\mathbf{z}_i$ a continuous function of the global space.

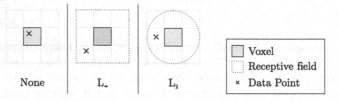

None L_∞ L_2

☐ Voxel
⋯ Receptive field
× Data Point

Fig. 3. Square (L_∞ norm) and spherical (L_2 norm) for the extended receptive fields for training local codes.

Here we focus on exploring the straight forward way of defining local shape codes over a sparsely allocated voxels V_i of the 3D space as illustrated in Fig. 2. We define $T_i(\mathbf{x}) := \mathbf{x} - \mathbf{x}_i$, transforming a global point $\mathbf{x}$ into the local coordinate system of voxel V_i by subtracting its center $\mathbf{x}_i$. The weighting function becomes the indicator function over the volume of voxel V_i. Thus, DeepLS describes the global surface S as:

$$S = \left\{ \mathbf{x} \in \mathbb{R}^3 \mid \sum_i \mathbb{1}_{\mathbf{x} \in V_i} f_\theta \left(T_i(\mathbf{x}), \mathbf{z}_i \right) = 0 \right\}. \tag{3}$$

4.1   Shape Border Consistency

We found that with the proposed division of space (i.e. disjoint voxels for local shapes) leads to inconsistent surface estimates at the voxel boundaries. One possible solution is to choose w as partition of unity [38] basis functions with local support to combine the decoded SDF values. We experimented with trilinear interpolation as an instance of this. However, this method increases the number of required decoder evaluations to query an SDF value by a factor of eight.

Instead, we keep the indicator function and train decoder weights and codes such that a local shape is correct beyond the bounds of one voxel, by using training pairs from neighboring voxels. Then, the SDF values on the voxel boundaries are accurately computable from any of the abutting local shapes. We experimented with spheres (i.e. L_2 norm) and voxels (i.e. L_∞ norm) (c.f. Fig. 3) for the definition range of extended local shapes and found that using an L_∞ norm with a radius of 1.5 times the voxel side-length provides a good trade-off between accuracy (fighting border artifacts) and efficiency (c.f. Sect. 5).

4.2 Deep Local Shapes Training and Inference

Given a set of SDF pairs $\{(\mathbf{x}_j, s_j)\}_{j=1}^{N}$, sampled from a set of training shapes, we aim to optimize both the parameters θ of the shared shape decoder $f_\theta(\cdot)$ and all local shape codes $\{\mathbf{z}_i\}$ during training and only the codes during inference.

Let $\mathcal{X}_i = \{\mathbf{x}_j \mid L(T_i(\mathbf{x}_j)) < r\}$ denote the set of all training samples $\mathbf{x}_j$, falling within a radius r of voxel i with local code $\mathbf{z}_i$ under the distance metric L. We train DeepLS by minimizing the negative log posterior over the training data $\mathcal{X}_i$:

$$\arg\min_{\theta, \{\mathbf{z}_i\}} \sum_i \sum_{\mathbf{x}_j \in \mathcal{X}_i} \|f_\theta(T_i(\mathbf{x}_j), \mathbf{z}_i) - s_j\|_1 + \frac{1}{\sigma^2}\|\mathbf{z}_i\|_2^2 .$$

In order to encode a new scene or shape into a set of local codes, we fix decoder weights θ and find the maximum a-posteriori codes $\mathbf{z}_i$ as

$$\arg\min_{\mathbf{z}_i} \sum_{\mathbf{x}_j \in \mathcal{X}_i} \|f_\theta(T_i(\mathbf{x}_j), \mathbf{z}_i) - s_j\|_1 + \frac{1}{\sigma^2}\|\mathbf{z}_i\|_2^2 , \tag{4}$$

given partial observation samples $\{(\mathbf{x}_j, s_j)\}_{j=1}^{M}$ with $\mathcal{X}_i$ defined as above.

4.3 Point Sampling

For sampling data pairs $(\mathbf{x}_j, s_j)$, we distinguish between sampling from meshes and depth observations. For meshes, the method proposed by Park et al. [39] is used. For depth observations, we estimate normals from the depth map and sample points in 3D that are displaced slightly along the normal direction, where the SDF value is assumed to be the magnitude of displacement. In addition to those samples, we obtain free space samples along the observation rays. The process is described formally in the supplemental materials.

5   Experiments

The experiment section is structured as follows. First, we compare DeepLS against recent deep learning methods (e.g. DeepSDF, AtlasNet) in Sect. 5.1. Then, we present results for scene reconstruction and compare them against related approaches on both synthetic and real scenes in Sect. 5.2.

Experiment Setup. The models used in the following experiments were trained on a set of local shape patches, obtained from 200 primitive shapes (e.g. cuboids and ellipsoids) and a total of 1000 shapes from the Warehouse [1] dataset (200 each for the airplane, table, chair, lamp, and sofa classes). Our decoder is a four layer MLP, mapping from latent codes of size 128 to the SDF value. We present examples from the training set, several additional results, comparisons and further details about the experimental setup in the supplemental materials.

5.1 Object Reconstruction

3D Warehouse [1]. We quantitatively evaluate surface reconstruction accuracy of DeepLS and other shape learning methods on various classes from the 3D Warehouse [1]. Quantitative results for the chamfer distance error are shown in Table 1. As can be seen DeepLS improves over related approaches by approximately one order of magnitude. It should be noted that this is not a comparison between equal methods since the other methods infer a global, object-level representation that comes with other advantages. Also, the parameter distribution varies significantly (c.f. Table 1). Nonetheless, it proves that local shapes lead to superior reconstruction quality and that implicit functions modeled by a deep neural network are capable of representing fine details. Qualitatively, DeepLS encodes and reconstructs much finer surface details as can be seen in Fig. 4.

Fig. 4. Comparison between our DeepLS and DeepSDF on 3D Warehouse [1] dataset.

Table 1. Comparison for reconstructing shapes from the 3D Warehouse [1] test set, using the Chamfer distance. Results with additional metrics are similar as detailed in the supplemental materials. Note that due to the much smaller decoder, DeepLS is also more than one order of magnitude faster in decoding (querying SDF values).

Method	Shape category					Decoder	Represent.	Decoding
	Chair	Plane	Table	Lamp	Sofa	Params	Params	Time (s)
Atl.Net-Sph. [21]	0.752	0.188	0.725	2.381	0.445	3.6 M	1.0 K	**0.01**
Atl.Net-25 [21]	0.368	0.216	0.328	1.182	0.411	43.5 M	1.0 K	0.32
DeepSDF [39]	0.204	0.143	0.553	0.832	0.132	1.8 M	**0.3 K**	16.65
DeepLS	**0.030**	**0.018**	**0.032**	**0.078**	**0.044**	**0.05 M**	312 K	1.25

Efficiency Evaluation on Stanford Bunny [2]. Further, we show the superior inference efficiency of DeepLS with a simple experiment, illustrated in Fig. 5. A DeepLS model was trained on a dataset composed only of randomly oriented primitive shapes. It is used to infer local codes that pose an implicit representation of the Stanford Bunny. Training and inference together took just one minute on a single GPU. The result is an RMSE of only 0.03% relative to the length of the diagonal of the minimal ground truth bounding box, highlighting the ability of DeepLS to generalize to novel shapes. To achieve the same surface error with a DeepSDF model (jointly training latent code and decoder on the bunny) requires over 8 days of GPU time, showing that the high compression rates and object-level completion capabilities of DeepSDF and related techniques come at the cost of long training and inference times. This is likely caused at least in part by gradient computation amongst all training samples, which we avoid by subdividing physical space and optimizing local representations in parallel.

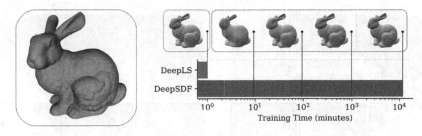

Fig. 5. A comparison of the efficiency of DeepLS and DeepSDF. With DeepLS, a model trained for one minute is capable of reconstructing the Stanford Bunny [2] in full detail. We then trained a DeepSDF model to represent the same signed distance function corresponding to the Stanford Bunny until it reaches the same accuracy. This took over 8 days of GPU time (note the log scale of the plot).

Table 2. Surface reconstruction accuracy of DeepLS and TSDF Fusion [37] on the synthetic ICL-NUIM dataset [22] benchmark.

Method	Mean	kt0	kt1	kt2	kt3
TSDF fusion	5.42 mm	5.35 mm	5.88 mm	5.17 mm	5.27 mm
DeepLS	**4.92 mm**	**5.15 mm**	**5.48 mm**	**4.32 mm**	**4.71 mm**

5.2   Scene Reconstruction

We evaluate the ability of DeepLS to reconstruct at scene scale using synthetic (in order to provide quantitative comparisons) and real depth scans. For synthetic scans, we use the ICL-NUIM RGBD benchmark dataset [22]. The evaluation on real scans is done using the 3D Scene Dataset [57]. For quantitative evaluation, the asymmetric Chamfer distance metric provided by the benchmark [22] is used.

Synthetic ICL-NUIM Dataset Evaluation. We provide quantitative measurements of surface reconstruction quality on all four ICL-NUIM sequences [22] (CC BY 3.0, Handa, A., Whelan, T., McDonald, J., Davison) in Table 2, where each system has been tuned for lowest surface error. Please note that for efficiency reasons, we compared DeepLS with TSDF Fusion implementation provided by Newcombe et al. [37], the original work can be found in [12]. We also show results qualitatively in Fig. 6 and show additional results, e.g. on data with artificial noise, in the supplemental materials. Most surface reconstruction techniques involve a tradeoff between surface accuracy and completeness. For TSDF Fusion [37], this tradeoff is driven by choosing a truncation distance and the minimum confidence at which surfaces are extracted by marching cubes. With DeepLS, we only extract surfaces up to some fixed distance from the nearest observed depth point, and this threshold is what trades off accuracy and completion of our system. For a full and fair comparison, we derived a pareto-optimal

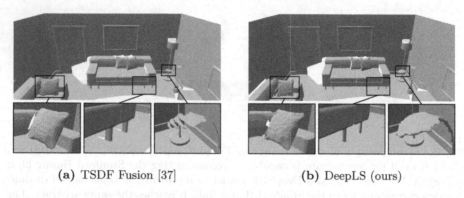

(a) TSDF Fusion [37]   (b) DeepLS (ours)

Fig. 6. Qualitative results of TSDF Fusion [37] and DeepLS for scene reconstruction on a synthetic ICL-NUIM [22] scene. The highlighted areas indicate the ability of DeepLS to handle oblique viewing angles, partial observation, and thin structures.

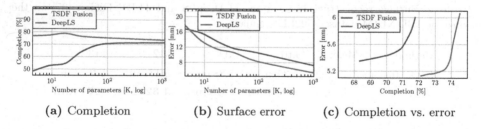

(a) Completion   (b) Surface error   (c) Completion vs. error

Fig. 7. Comparison of completion (a) and surface error (b) as a function of representation parameters on a synthetic scene from the ICL-NUIM [22] dataset. In contrast to TSDF Fusion, DeepLS maintains reconstruction completeness almost independent of compression rate. On the reconstructed surfaces (which is 50% less for TSDF Fusion) the surface error decreases for both methods (c.f. Fig. 8). Plot (c) shows the trend of surface error vs. mesh completion. DeepLS consistently shows higher completion at the same surface error.

curve by varying these parameters for the two methods on the 'kt0' sequence of the ICL-NUIM benchmark and plot the results in Fig. 7. We measure completion by computing the fraction of ground truth points for which there is a reconstructed point within 7 mm. Generally, DeepLS can reconstruct more complete surfaces at the same level of accuracy as SDF Fusion.

The number of representation parameters used by DeepLS is theoretically independent of the rendering resolution and only depends on the resolution of the local shapes. In contrast, traditional volumetric scene reconstruction methods such as TSDF Fusion have a tight coupling between number of parameters and the desired rendering resolution. We investigate the relationship between representation size per unit volume of DeepLS and TSDF Fusion by evaluating the surface error and completeness as a function of the number of parameters. As a starting point we choose a representation that uses 8^3 parameters per 5.6cm × 5.6cm × 5.6cm volume (7 mm voxel resolution). To increase compression we increase the voxel size for TSDF Fusion and the local shape code volume size for DeepLS. We provide the quantitative and qualitative analysis of the scene reconstruction results with varying representation size in Fig. 7 (a and b) and

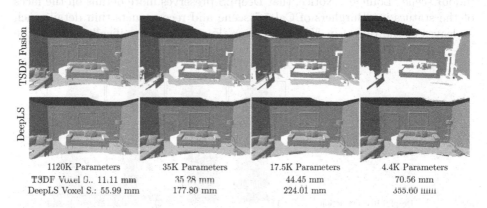

	1120K Parameters	35K Parameters	17.5K Parameters	4.4K Parameters
TSDF Voxel S.:	11.11 mm	35.28 mm	44.45 mm	70.56 mm
DeepLS Voxel S.:	55.99 mm	177.80 mm	224.01 mm	355.00 mm

Fig. 8. Qualitative analysis of representation size with DeepLS and TSDF Fusion [37] on a synthetic scene in the ICL-NUIM [22] dataset. DeepLS is able to retain details at higher compression rates (lower number of parameters). It achieves these compression rates by using bigger local shape voxels, leading to a stronger influence of the priors.

Table 3. Quantitative evaluation of DeepLS with TSDF Fusion on 3D Scene Dataset [57]. The error is measured in mm and *Comp* (completion) corresponds to the percentage of ground truth surfaces that have reconstructed surfaces within 7 mm. Results suggest that DeepLS produces more accurate and complete 3D reconstruction in comparison to volumetric fusion methods on real depth acquisition datasets.

Method	Burghers		Lounge		CopyRoom		StoneWall		TotemPole	
	Error	Comp	Error	Comp	Error	Comp	Error	Comp	Error	Comp
TSDF F. [12]	10.11	85.46	11.71	85.17	12.35	83.99	14.23	91.02	13.03	83.73
DeepLS	**5.74**	**95.78**	**7.38**	**96.00**	**10.09**	**99.70**	**6.45**	**91.37**	**8.97**	**87.23**

Fig. 8 respectively. The plots in Fig. 7 show conclusively that TSDF Fusion drops to about 50% less complete reconstructions while DeepLS maintains completeness even at the highest compression rate, using only 4.4K parameters for the full scene. Quantitatively, TSDF Fusion also achieves low surface error for high compression. However, this can be contributed to the used ICL-NUIM benchmark metric, which does not strongly punish missing surfaces.

Evaluation on Real Scans. We evaluate DeepLS on the 3D Scene Dataset [57], which contains several scenes captured by commodity structured light sensors, and a challenging scan of thin objects. In order to also provide quantitative errors we assume the reconstruction performed by volumetric fusion [12] of all depth frames to be the *ground truth*. We then apply DeepLS and TSDF fusion on a small subset of depth frames, taking every 10th frame in the capture sequence. The quantitative results of this comparison are detailed in Table 3 for various scenes. It is shown that DeepLS produces both more accurate and more complete 3D reconstructions. Furthermore, we provide qualitative examples of this experiment in Fig. 9 for the outdoor scene "Burghers of Calais" and for the indoor scene "Lounge". Notice, that DeepLS preserves more details on the faces of the statues in "Burghers of Calais" scene and reconstructs thin details such as leaves of the plants in "Lounge" scene. Further, we specifically analyse the

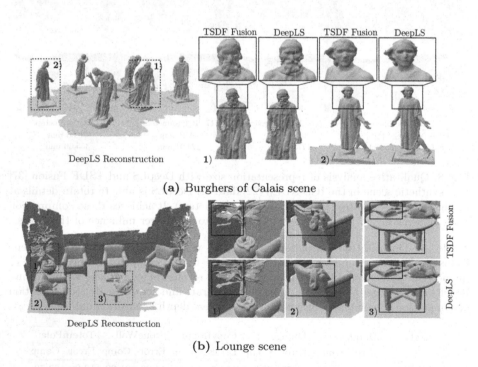

(a) Burghers of Calais scene

(b) Lounge scene

Fig. 9. Qualitative results for DeepLS and TSDF Fusion [37] on two scenes of the 3D Scene Dataset [57]. Best viewed with zoom in the digital version of the paper.

strength of DeepLS in representing and completing thin local geometry. We collected a scan from an object consisting of two thin circles kept on a stool with long but thin cylindrical legs (see Fig. 10). The 3D points were generated by a structured light sensor [9,45,50]. It was scanned from limited directions leading to very sparse set of points on the stool's surface and legs. We compared our results on this dataset to several 3D methods including TSDF Fusion [37], Multi-level Partition of Unity (MPU) [38], Smooth Signed Distance Function [7], Poisson Surface Reconstruction [29], PFS [49] and TSR [4]. We found that due to lack of points and thin surfaces most of the methods failed to either represent details or complete the model. MPU [38], which fits quadratic functions in local grids and is very related to our work, fails in this experiment (see Fig. 10b). This indicates that our learned shape priors are more robust than fixed parameterized functions. Methods such as PSR [29], SSD [7] and PFS [49] fit global implicit function to represent shapes. These methods made the thin shapes thicker than they should be. Moreover, they also had issues completely reconstructing the thin rings on top of the stool. TSR [4] was able to fit to the available points

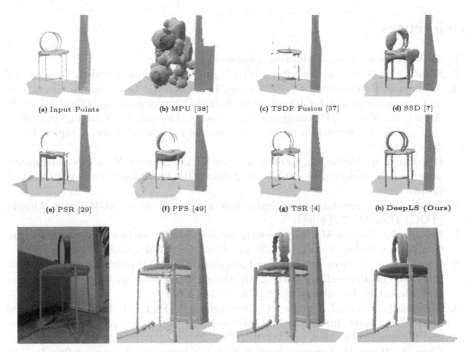

(a) Input Points   (b) MPU [38]   (c) TSDF Fusion [37]   (d) SSD [7]

(e) PSR [29]   (f) PFS [49]   (g) TSR [4]   (h) DeepLS (Ours)

(i) Close-up view. From left to right: RGB, Input Points, TSR and DeepLS reconstructions.

Fig. 10. Qualitative comparison of DeepLS against other 3D Reconstruction techniques on a very challenging thin and incomplete scan. Most of the methods fail to build thin surfaces in this dataset. TSR fits to the thin parts but is unable to complete structures such as the bottom and cylindrical legs of the stool. In contrast, DeepLS reconstructs thin structures and also completes them.

but is unable to complete structures such as bottom surface of the stool and it's cylindrical legs, where no observations exist. This shows how our method utilizes local shape priors to complete partially scanned shapes.

6   Conclusion

In this work we presented DeepLS, a method to combine the benefits of volumetric fusion and deep shape priors for 3D surface reconstruction from depth observations. A key to the success of this approach is the decomposition of large surfaces into local shapes. This decomposition allowed us to reconstruct surfaces with higher accuracy and finer detail than traditional SDF fusion techniques, while simultaneously completing unobserved surfaces, all using less memory than storing the full SDF volume would require. Compared to recent object-centric shape learning approaches, our local shape decomposition leads to greater efficiency for both training and inference while improving surface reconstruction accuracy by an order of magnitude.

References

1. 3D Warehouse. https://3dwarehouse.sketchup.com/
2. Stanford 3D Scanning Repository. http://graphics.stanford.edu/data/3Dscanrep/
3. Achlioptas, P., Diamanti, O., Mitliagkas, I., Guibas, L.: Learning representations and generative models for 3D point clouds (2017). arXiv preprint arXiv:1707.02392
4. Aroudj, S., Seemann, P., Langguth, F., Guthe, S., Goesele, M.: Visibility-consistent thin surface reconstruction using multi-scale kernels. ACM Trans. Graph. (TOG) **36**(6), 187 (2017)
5. Ben-Hamu, H., Maron, H., Kezurer, I., Avineri, G., Lipman, Y.: Multi-chart generative surface modeling. In: SIGGRAPH Asia 2018 Technical Papers, p. 215. ACM (2018)
6. Blinn, J.F.: A generalization of algebraic surface drawing. ACM Trans. Graph. (TOG) **1**(3), 235–256 (1982)
7. Calakli, F., Taubin, G.: Ssd: smooth signed distance surface reconstruction. In: Computer Graphics Forum. vol. 30, pp. 1993–2002. Wiley Online Library (2011)
8. Carr, J.C., et al.: Reconstruction and representation of 3D objects with radial basis functions. In: Proceedings of the 28th Annual Conference on Computer Graphics and Interactive Techniques, pp. 67–76. ACM (2001)
9. Chabra, R., Straub, J., Sweeney, C., Newcombe, R., Fuchs, H.: Stereodrnet: dilated residual stereonet. In: Proceedings of the IEEE Conference on Computer Vision and Pattern Recognition, pp. 11786–11795 (2019)
10. Chen, Z., Zhang, H.: Learning implicit fields for generative shape modeling. In: The IEEE Conference on Computer Vision and Pattern Recognition (CVPR) (2019)
11. Choy, C.B., Xu, D., Gwak, J.Y., Chen, K., Savarese, S.: 3D-R2N2: a unified approach for single and multi-view 3D object reconstruction. In: Leibe, B., Matas, J., Sebe, N., Welling, M. (eds.) ECCV 2016. LNCS, vol. 9912, pp. 628–644. Springer, Cham (2016). https://doi.org/10.1007/978-3-319-46484-8_38
12. Curless, B., Levoy, M.: A volumetric method for building complex models from range images (1996)

13. Dai, A., Nießner, M.: Scan2mesh: from unstructured range scans to 3D meshes. In: Proceedings of the IEEE Conference on Computer Vision and Pattern Recognition. pp. 5574–5583 (2019)
14. Dai, A., Ruizhongtai Qi, C., Nießner, M.: Shape completion using 3D-encoder-predictor CNNs and shape synthesis. In: Proceedings of the IEEE Conference on Computer Vision and Pattern Recognition, pp. 5868–5877 (2017)
15. Deng, B., Genova, K., Yazdani, S., Bouaziz, S., Hinton, G., Tagliasacchi, A.: Cvxnets: learnable convex decomposition (2019)
16. Fuhrmann, S., Goesele, M.: Floating scale surface reconstruction. ACM Trans. Graph. (ToG) **33**(4), 46 (2014)
17. Gal, R., Shamir, A., Hassner, T., Pauly, M., Cohen-Or, D.: Surface reconstruction using local shape priors. In: Symposium on Geometry Processing, No. CONF, pp. 253–262 (2007)
18. Genova, K., Cole, F., Sud, A., Sarna, A., Funkhouser, T.: Deep structured implicit functions (2019). arXiv preprint arXiv:1912.06126
19. Genova, K., Cole, F., Vlasic, D., Sarna, A., Freeman, W.T., Funkhouser, T.: Learning shape templates with structured implicit functions (2019). arXiv preprint arXiv:1904.06447
20. Gkioxari, G., Malik, J., Johnson, J.: Mesh r-cnn (2019). arXiv preprint arXiv:1906.02739
21. Groueix, T., Fisher, M., Kim, V.G., Russell, B.C., Aubry, M.: Atlasnet: apapier-m\ ach\ approach to learning 3D surfacegeneration (2018). arXiv preprint arXiv:1802.05384
22. Handa, A., Whelan, T., McDonald, J., Davison, A.: A benchmark for RGB-D visual odometry, 3D reconstruction and SLAM. In: IEEE International Conference on Robotics and Automation, ICRA. Hong Kong, China (2014)
23. Häne, C., Tulsiani, S., Malik, J.: Hierarchical surface prediction for 3D object reconstruction. In: 2017 International Conference on 3D Vision (3DV), pp. 412–420. IEEE (2017)
24. Hoppe, H., DeRose, T., Duchamp, T., McDonald, J., Stuetzle, W.: Surface reconstruction from unorganized points. In: Proceedings of the 19th Annual Conference on Computer Graphics and Interactive Techniques, pp. 71–78 (1992)
25. Hornik, K., Stinchcombe, M., White, H.: Multilayer feedforward networks are universal approximators. Neural Netw. **2**(5), 359–366 (1989)
26. Jancosek, M., Pajdla, T.: Multi-view reconstruction preserving weakly-supported surfaces. In: CVPR 2011, pp. 3121–3128. IEEE (2011)
27. Jiang, C., Sud, A., Makadia, A., Huang, J., Nießner, M., Funkhouser, T.: Local implicit grid representations for 3D scenes. In: Proceedings of the IEEE Conference on Computer Vision and Pattern Recognition (2020)
28. Kazhdan, M., Bolitho, M., Hoppe, H.: Poisson surface reconstruction. In: Proceedings of the Fourth Eurographics Symposium on Geometry Processing, vol. 7 (2006)
29. Kazhdan, M., Hoppe, H.: Screened poisson surface reconstruction. ACM Trans. Graph. (ToG) **32**(3), 1–13 (2013)
30. Keller, M., Lefloch, D., Lambers, M., Izadi, S., Weyrich, T., Kolb, A.: Real-time 3D reconstruction in dynamic scenes using point-based fusion. In: 2013 International Conference on 3D Vision-3DV 2013, pp. 1–8. IEEE (2013)
31. Klein, G., Murray, D.: Parallel tracking and mapping for small ar workspaces. In: Proceedings of the 2007 6th IEEE and ACM International Symposium on Mixed and Augmented Reality, pp. 1–10. IEEE Computer Society (2007)

32. Labatut, P., Pons, J.P., Keriven, R.: Robust and efficient surface reconstruction from range data. In: Computer Graphics Forum, vol. 28, pp. 2275–2290. Wiley Online Library (2009)
33. Liao, Y., Donne, S., Geiger, A.: Deep marching cubes: learning explicit surface representations. In: Proceedings of the IEEE Conference on Computer Vision and Pattern Recognition, pp. 2916–2925 (2018)
34. Mescheder, L., Oechsle, M., Niemeyer, M., Nowozin, S., Geiger, A.: Occupancy networks: learning 3D reconstruction in function space. In: Proceedings of the IEEE Conference on Computer Vision and Pattern Recognition, pp. 4460–4470 (2019)
35. Michalkiewicz, M., Pontes, J.K., Jack, D., Baktashmotlagh, M., Eriksson, A.: Deep level sets: implicit surface representations for 3D shape inference (2019). arXiv preprint arXiv:1901.06802
36. Newcombe, R.A., Davison, A.J.: Live dense reconstruction with a single moving camera. In: 2010 IEEE Computer Society Conference on Computer Vision and Pattern Recognition, pp. 1498–1505. IEEE (2010)
37. Newcombe, R.A., et al.: Kinectfusion: real-time dense surface mapping and tracking. In: 2011 10th IEEE International Symposium on Mixed and Augmented Reality (ISMAR), pp. 127–136. IEEE (2011)
38. Ohtake, Y., Belyaev, A., Alexa, M., Turk, G., Seidel, H.P.: Multi-level partition of unity implicits. In: Acm Siggraph 2005 Courses, pp. 173-es (2005)
39. Park, J.J., Florence, P., Straub, J., Newcombe, R., Lovegrove, S.: Deepsdf: learning continuous signed distance functions for shape representation (2019). arXiv preprint arXiv:1901.05103
40. Pfister, H., Zwicker, M., Van Baar, J., Gross, M.: Surfels: surface elements as rendering primitives. In: Proceedings of the 27th Annual Conference on Computer Graphics and Interactive Techniques, pp. 335–342. ACM Press/Addison-Wesley Publishing Co. (2000)
41. Ricao Canelhas, D., Schaffernicht, E., Stoyanov, T., Lilienthal, A., Davison, A.: Compressed voxel-based mapping using unsupervised learning. Robotics 6(3), 15 (2017)
42. Riegler, G., Osman Ulusoy, A., Geiger, A.: Octnet: learning deep 3D representations at high resolutions. In: Proceedings of the IEEE Conference on Computer Vision and Pattern Recognition, pp. 3577–3586 (2017)
43. Saito, S., Huang, Z., Natsume, R., Morishima, S., Kanazawa, A., Li, H.: Pifu: pixel-aligned implicit function for high-resolution clothed human digitization (2019). arXiv preprint arXiv:1905.05172
44. Sinha, A., Bai, J., Ramani, K.: Deep learning 3D shape surfaces using geometry images. In: Leibe, B., Matas, J., Sebe, N., Welling, M. (eds.) ECCV 2016. LNCS, vol. 9910, pp. 223–240. Springer, Cham (2016). https://doi.org/10.1007/978-3-319-46466-4_14
45. Straub, J., et al.: The Replica dataset: a digital replica of indoor spaces (2019). arXiv preprint arXiv:1906.05797
46. Stühmer, J., Gumhold, S., Cremers, D.: Real-time dense geometry from a handheld camera. In: Goesele, M., Roth, S., Kuijper, A., Schiele, B., Schindler, K. (eds.) DAGM 2010. LNCS, vol. 6376, pp. 11–20. Springer, Heidelberg (2010). https://doi.org/10.1007/978-3-642-15986-2_2
47. Stutz, D., Geiger, A.: Learning 3D shape completion from laser scan data with weak supervision. In: Proceedings of the IEEE Conference on Computer Vision and Pattern Recognition, pp. 1955–1964 (2018)

48. Tatarchenko, M., Dosovitskiy, A., Brox, T.: Octree generating networks: efficient convolutional architectures for high-resolution 3D outputs. In: Proceedings of the IEEE International Conference on Computer Vision, pp. 2088–2096 (2017)
49. Ummenhofer, B., Brox, T.: Global, dense multiscale reconstruction for a billion points. In: Proceedings of the IEEE International Conference on Computer Vision, pp. 1341–1349 (2015)
50. Whelan, T., et al.: Reconstructing scenes with mirror and glass surfaces. ACM Trans. Graph. (TOG) **37**(4), 102 (2018)
51. Whelan, T., Leutenegger, S., Salas-Moreno, R., Glocker, B., Davison, A.: Elastic-fusion: Dense slam without a pose graph. Robotics: Science and Systems (2015)
52. Williams, F., Schneider, T., Silva, C., Zorin, D., Bruna, J., Panozzo, D.: Deep geometric prior for surface reconstruction. In: Proceedings of the IEEE Conference on Computer Vision and Pattern Recognition, pp. 10130–10139 (2019)
53. Wu, Z., et al.: 3d shapenets: a deep representation for volumetric shapes. In: Proceedings of the IEEE Conference on Computer Vision and Pattern Recognition, pp. 1912–1920 (2015)
54. Xu, Q., Wang, W., Ceylan, D., Mech, R., Neumann, U.: Disn: deep implicit surface network for high-quality single-view 3D reconstruction (2019). arXiv preprint arXiv:1905.10711
55. Yang, Y., Feng, C., Shen, Y., Tian, D.: Foldingnet: interpretable unsupervised learning on 3D point clouds (2017). arXiv preprint arXiv:1712.07262
56. Yuan, W., Khot, T., Held, D., Mertz, C., Hebert, M.: Pcn: point completion network. In: 2018 International Conference on 3D Vision (3DV), pp. 728–737. IEEE (2018)
57. Zhou, Q.Y., Koltun, V.: Dense scene reconstruction with points of interest. ACM Trans. Graph. (ToG) **32**(4), 1–8 (2013)

Info3D: Representation Learning on 3D Objects Using Mutual Information Maximization and Contrastive Learning

Aditya Sanghi[✉]

Autodesk AI Lab, Toronto, Canada
aditya.sanghi@autodesk.com

Abstract. A major endeavor of computer vision is to represent, understand and extract structure from 3D data. Towards this goal, unsupervised learning is a powerful and necessary tool. Most current unsupervised methods for 3D shape analysis use datasets that are aligned, require objects to be reconstructed and suffer from deteriorated performance on downstream tasks. To solve these issues, we propose to extend the InfoMax and contrastive learning principles on 3D shapes. We show that we can maximize the mutual information between 3D objects and their "chunks" to improve the representations in aligned datasets. Furthermore, we can achieve rotation invariance in SO(3) group by maximizing the mutual information between the 3D objects and their geometric transformed versions. Finally, we conduct several experiments such as clustering, transfer learning, shape retrieval, and achieve state of art results.

Keywords: 3D shape analysis · Unsupervised learning · Rotation invariance · InfoMax · Contrastive learning

1 Introduction

Recently, several unsupervised methods have managed to extract powerful features for 3D objects such as in [1,16,30,49] and [26]. However, these methods assume all 3D objects are aligned and have the same pose in the given category. In real world scenarios, this is not the case. For example, when a robot is identifying and picking up an object, the object is in an unknown pose. Even in online repositories of 3D shapes, most of the data is randomly oriented as users create objects in different poses. To use these methods effectively we require to align all objects for a given category which is a very expensive and time consuming process.

Furthermore, these unsupervised methods require reconstruction of 3D shapes which is not ideal for many reasons. Firstly, it is not always feasible to

Electronic supplementary material The online version of this chapter (https://doi.org/10.1007/978-3-030-58526-6_37) contains supplementary material, which is available to authorized users.

© Springer Nature Switzerland AG 2020
A. Vedaldi et al. (Eds.): ECCV 2020, LNCS 12374, pp. 626–642, 2020.
https://doi.org/10.1007/978-3-030-58526-6_37

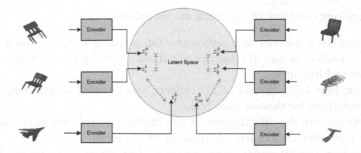

Fig. 1. General idea of the method. We try to bring the 3D shape and the different view of the 3D shape closer in latent space while pushing away the representation of other objects in the dataset further away.

reconstruct the 3D representation of a shape. For example, due to the discrete nature of meshes, reconstructing their representation may not be attainable. Moreover, in cases where we need invariant representation, it's hard to reconstruct back the shape from this invariant representation. For example, if you wanted to create rotation invariant embeddings, you need to lose pose information from the embeddings of 3D objects. If you lose the pose information, it will not be possible to reconstruct the shape back as you need to reconstruct it with that given pose.

To overcome the challenges mentioned above we propose a decoder-free, unsupervised representation learning mechanism which is rotation insensitive. The method we introduce takes inspiration from the Contrastive Predictive Coding [29] and Deep InfoMax [43] approach. These methods usually require a different "view" of the object, which is used to maximize the mutual information with the object. This other view can be different modalities, data augmentation of the object, local substructure of the object, etc.

We consider two different views of a given object in this work. First, we consider maximizing the mutual information between a local chunk of a 3D object and the whole 3D object. The intuition for using this view is that the 3D shape is forced to learn about its local region as it has to distinguish it from other parts of different objects. This greatly enhances the representation learnt in aligned objects. Second, we consider maximizing the mutual information between 3D shape and a geometric transformed version of the 3D shape. The advantage of maximizing the mutual information in this scenario is that it can create global geometry invariant representations. This is very useful in the case of achieving rotation insensitive representation in SO(3). Figure 1 illustrates the rough intuition behind the method. Note, despite using objects from different category in the figure, we push away every other shape in the dataset. This might even include object from the same category. This method can be thought as instance discrimination [48].

The key contributions of our work are as follows:

- We introduce a decoder-free representation learning method which can easily be extended to any 3D descriptor without needing to construct complex decoder architectures.
- We show how local chunks of 3D objects can be used to get very effective representations for downstream tasks.
- We demonstrate the effectiveness of the method on rotated inputs and show how it is insensitive to such rotations in SO(3) group.
- We conduct several experiments to show the efficacy of our method and achieve state of art results in transfer learning, clustering and semi-supervised learning.

2 Background

Representation Learning on 3D objects. Much progress has been made to learn good representations of 3D objects in an unsupervised manner which can then be used in several downstream tasks. In point clouds, works such as [1,16,18,36,49] and [52] have been proposed. For voxels and implicit representations, work such as [10,26,27,30,46] create powerful representation features which are used for several downstream tasks such as shape completion, shape generation and classification. Recently, there has been a lot of progress in auto-encoding meshes such as in [11,40,40]. One disadvantage of the above approaches are that they require you to reconstruct or generate the 3D shapes. As stated earlier, it might be expensive or not possible to reconstruct the shape. A recent approach [51] does not require to reconstruct the shape and instead does two stages of training. It first uses parts of a shape to learn features using contrastive learning and then uses pseudo-clusters to cluster all the data. Our method does not require two stages of training, and furthermore allows us to create rotation invariant embeddings.

Maximizing Mutual Information and Contrastive Learning. A lot of methods have used mutual information to do unsupervised learning. Historically, works such as [4–6,24] have explored the InfoMax principle and mutual information maximization. More recently [29] proposed the Contrastive Predicting Coding framework which uses the embeddings to capture maximal information about future samples. The Deep InfoMax (DIM) [43] approach is similar but has the advantage of doing orderless autoregression. In concurrent to these works, works such as [48] and [53], have extended these ideas from the metric learning point of view. These methods were extended in [3,19,42] by considering multiple views of the data and achieved state of the art representation learning on images. The InfoMax principle has been extended to graphs [39,43] and for state representation in reinforcement learning [2]. Our method is inspired by these methods and we extend them to 3D representations. For the multiple views, we use geometric transformations and chunks of a 3D object. Finally, in concurrent to our work several news works such as [9,28] have been proposed.

Rotation Invariance on 3D objects. Traditionally, several methods have focused on hand engineered features to get local rotation invariant descriptors, such as [22,34,37]. More recently, methods such as [13] and [14], first, get local invariant features by encoding local geometry into patches and then use autoencoder to reconstruct the local features. These approaches require creating hand-crafted local features and need normal information. Methods such as MVCNN [38], Rot-SO-Net [23] and [35] explicitly force invariance by taking multiple poses of the object and aggregating them over the poses. However, such methods only work on discrete rotations or can only reconstruct objects rotated along one axis. A lot of deep learning methods have also attempted to use equivariance based architectures to achieve local and global rotation equivariance in 2D and 3D data. Methods such as [12,15,41,45] either use constrained filters that achieve rotation equivariance or use filter orbit which are themselves equivariant. It is usually difficult to create such architectures for different 3D representations. Furthermore, to generate invariant representation to rotation from equivariant representation usually requires a post-processing step. Our method uses the InfoMax loss with rotation transformation to enforce rotation invariance. This method can be easily extended to voxels, implicit representation and meshes without needing to create complex architectures.

3   Methodology

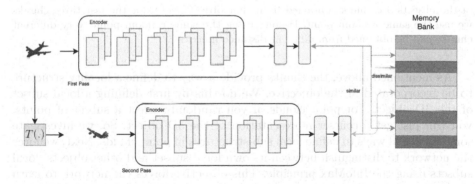

Fig. 2. Overview of the method. A 3D object and different view of that object is encoded using the same encoder. The features across these views are made similar while a memory bank is used for negative examples. We also store the features obtained in the memory bank.

Our goal is to extract good features from 3D shapes in an unsupervised manner. To achieve this, we maximize the mutual information on the features extracted from a 3D shape and a different view of the 3D shape. We consider two such views in this work for achieving different purposes. First, we consider using a "chunk" of a 3D object to improve the representation of aligned shapes. The goal of using chunks of a 3D object is to incorporate some form of locality and structure in input into the objective. We investigate different ways to extract a chunk of a 3D object and show how this can improve the representations learned. Next, we consider geometric transformation as the second view. Intuitively, the goal is to make the shape and its geometric transformed version closer in the latent space while distancing itself from other objects. We show how this can create transformation invariant embeddings to a large extent. We discuss this method in the context of point clouds but it should be straightforward to extend it to other 3D representations. The method in a pictorial form is shown in Fig. 2.

Let x_i and $T(x_i)$ represent a 3D shape and the object obtained after applying the transformation as mentioned above. We use an encoder, $f(.)$, to transform x_i and $T(x_i)$ into the latent representation z_i^a and z_i^b. Note that this encoder also includes a final normalizing layer which makes the embeddings unit vectors. In the next sections, we detail the motivation and mechanism for using chunks of a 3D object and geometric transformations as second view.

3.1   Local Chunks

| Axis chopped chunk | Cosine distance chunk | Euclidean distance chunk | Chebyshev distance chunk |

Fig. 3. Types of chunks. The first object represents a sample 3D object. The rest of the objects are chunks obtained from that object. Note, for the last three chunks we use the same random point. Despite using the same random point, very different chunks can be obtained from different distance metrics.

As mentioned above, the chunks provide a way to define a locality structure to be incorporated in the objective. We do this by first defining a local subset of the 3D object. For point clouds, if you randomly select a subset of points, you will just get a coarser representation of the 3D object. So, we investigate some potential ways to define local sub-structures in point clouds. Next, we force the network to distinguish between its own local subset and other objects' local subsets using the InfoMax principle. This objective forces the network to learn about its own local sub-structures and creates more informative embeddings.

We define chunks by considering two mechanisms. In the first approach, we randomly select a point from the point cloud. Then, we use a distance measure in Euclidean space to select a subset of points. The distance measure we consider are Euclidean distance, cosine distance and Chebyshev distance. Once we select a subset of points, we normalize them using a bounding sphere. In the second approach, we take a chunk of a 3D object based on chopping the object randomly along the cartesian axes. The chunk is again normalized using a bounding sphere. The different chunks obtained from a sample 3D object are shown in Fig. 3.

3.2   Geometric Transformation

In this work, we consider several different geometric transformed versions of a 3D object. This can be thought of as a form of data augmentation. However, our method differs from traditional use of data augmentation by explicitly influencing the latent space to create better embeddings rather than implicitly hoping that it would create meaningful representation. Furthermore, we are doing data augmentation in an unsupervised setting, so the increased cost of augmentation can be shared across several tasks instead of just one task such as in a supervised setting.

Though several other data augmentation methods can be used with the Info-Max principle, we consider geometric transformations for two major reasons. First, it is very trivial to apply an affine transformation to a 3D object. We simply need to multiply a transformation matrix. Second, we can use the rotation affine transformation to create embeddings which are less sensitive to alignment.

In this paper, we only consider translation, rotation along z axis, rotation in $SO(3)$ rotation group, uniform scale and non-uniform scale as our geometric augmentations. We also combine different transformations together to create more complex transformations. When we learn representations from unaligned datasets, we always rotate the object before applying a transformation to ensure it is not sensitive to rotation.

3.3   InfoNCE Objective

To estimate mutual information we use the InfoNCE [29] objective. Let us consider N samples from some unknown joint distribution $p(x,y)$. For this objective, we need to construct positive samples and negative samples. The positive examples, are sampled from the joint distribution of $p(x,y)$ and negative samples from the product of marginals $p(x)\ p(y)$. The objective is to learn a critic function, $h(.)$, by increasing the probability of positive examples and decreasing the probability for negative examples. The bound is given by

$$I_{NCE} = \sum_{i=1}^{N} \log \frac{h(x_i, y_i)}{\sum_{j=1}^{N} h(x_i, y_j)} \tag{1}$$

In our case, the positive samples are constructed by using the shape, x_i, and a different view, $T(x_i)$, of the shape. We construct negative examples by uniformly sampling, k pairs, over the whole transformed version of the dataset. Note, this procedure can lead to objects from the same category being part of negative examples. We consider a batch size of N. The critic function is defined as exponential of bi-linear function of $f(x_i)$ and $f(T(x_i))$. We parameterize this function using W. Note that the critic can be defined on the global features from $f(.)$ or the intermediate features of $f(.)$. We can also modulate the distribution using the parameter τ. The critic function is shown below

$$h(x_i, T(x_i)) = \exp(f(x_i) W f(T(x_i))/\tau) \tag{2}$$

We now consider the objective where we maximize the mutual information between the global representations of x and $T(x)$. That is, we maximize the mutual information between features from the latter layers of the encoder. The loss is shown below

$$L = \sum_{i=1}^{N} - \log \frac{h(x_i, T(x_i))}{h(x_i, T(x_i)) + \sum_{j=1}^{k} h(x_i, T(x_j))} \tag{3}$$

In theory, more negative examples, k, should lead to a tighter mutual information bound. One way of achieving this involves using large batch size, which might not be ideal. To avoid this, we take inspiration from [42,48,53] and use a memory bank to store data from previous batches. This allows us to use large number of negative examples. Increasing the number of negative examples leads to prohibitive cost in computing the softmax. Hence, we use the Noise-Contrastive estimation [17] to approximate the above loss as in [42]. The loss is as shown below

$$L_{NCE} = \frac{1}{N} \sum_{i=1}^{N} \left(- \log \left[h(x_i, T(x_i)) \right] - \sum_{j=1}^{k} \log \left[1 - h(x_i, T(x_j)) \right] \right) \tag{4}$$

4   Experiments

We conduct several experiments to test the efficacy of our method and the representations learned by the encoder on both aligned and unaligned shape datasets. We divided the experiment section into three parts. In the first section, we conduct experiments on aligned datasets and show the effectiveness of our method. In the second section, we discuss representation learning on rotated 3D shapes. Finally, in the last section, we look at different hyperparameters and factors affecting our method.

Training Details. For most of the experiments we use a batch size of 32, sample 2048 points on the shapes and use the ADAM optimizer [20]. We use a learning rate of 0.0001. For ModelNet40, we run the experiment for 250 epochs in the case of aligned datasets whereas we run it for 750 epochs for unaligned datasets. As ModelNet10 is a smaller dataset, we run 750 epochs for aligned dataset and 1000 epochs for unaligned dataset. We set the number of negative examples to 512 and use 0.07 as temperature parameter. For ShapeNet v1/v2 dataset, we run it for 200 epochs and use 2000 negative examples. For aligned datasets we use the features from the 6th layer in our model whereas for unaligned datasets we use features from the 7th layer. Furthermore, for aligned datasets we use the cosine distance based chunks for clustering task and axis chopped chunks for all other experiments as the second view whereas for rotated datasets we use rotation in any SO(3) rotation group plus translation as the second view. The choice of these parameters are further discussed in the ablation study section.

We also do early stopping if the network has converged. More details about the encoder structures and training details are in the appendix section. Finally, for ABC dataset [21] we use a batch size of 64 and use 1024 sample points on the surface.

Baseline Setup. For the tasks of clustering, rotation invariance and shape retrieval, we compare our method with three important works on representation learning on point clouds: FoldingNet [49], Latent-GAN Autoencoder [1] and AtlasNet [16]. For all three baselines, we use similar training conditions as mentioned above, except we train the three models on ShapeNet [8] for 750 epochs and ModelNet40 [47] for 1500 epochs for unaligned dataset. For aligned dataset we train ModelNet40 for 500 epochs. We use the PointNet encoder as the encoders for these baselines. Note, this is different from the respective paper implementations. More details regarding the architecture used for the baselines can be found in appendix.

Data Preparation. All our experiments are conducted on ModelNet10 [47], ModelNet40 [47], ShapeNet v1/v2 [8] and ABC dataset [21]. In some of the above datasets objects are aligned according to their categories, so to unalign them we randomly generate a quanterion and rotate them in SO(3) space. During unsupervised training, for each epoch, we rotate the shape differently so that the network can see different poses of the same shape. However, when we test this on the downstream tasks, we only rotate the dataset once. This is to ensure that we only test the effectiveness of unsupervised learning part rather than the effectiveness of the downstream part.

4.1 Representation Learning on Aligned Shapes

In this section, we demonstrate how our method performs when we do representation learning on aligned shapes. Here we compare with well established baselines and show the advantages of our method. Moreover, we also show how the embeddings obtained from our method are more clusterable then autoencoder methods. Note for clustering experiment we use the cosine distance based chunks whereas for all other experiments we use the axis chopped chunks as the second view.

Transfer Learning, Semi-supervised Learning and Pre-training. A well established benchmark used for unsupervised learning is transfer learning. We follow the same procedure as [1] and [49]. We first use unsupervised learning to train from the ShapeNet v1 dataset. We then train a Linear SVM using the training dataset of ModelNet40. In Table 1 we report the accuracy score on the test set of ModelNet40. Furthermore, we use the pre-trained weights from training ShapeNet dataset and initialize the pointnet classifier with those weights. We compare the results with randomly initialized weights by reporting

Table 1. Results on aligned datasets. Left table represents the transfer learning results on ModelNet40 whereas the right table represents the supervised learning results.

Unsup. method	Acc	Sup. method	Acc
3D-GAN [46]	83.3	PointNet [32]	89.2
Latent-GAN [1]	85.7	DeepSets [50]	90.3
ClusterNet [51]	86.8	PointNet++ [33]	90.7
FoldingNet [49]	88.4	DGCNN [44]	93.5
Multi-Task PC [18]	89.1	Relational PC [25]	**93.6**
Recon. PC(PointNet) [36]	87.3	PointNet (Pretrained)	*90.20*
Recon. PC(DGCNN) [36]	90.6	DGCNN (Pretrained)	*93.03*
Ours (PointNet)	*89.8*		
Ours (DGCNN)	**91.6**		

Table 2. Semi-supervised results on ModelNet40.

Method	1%	2%	5%	20%	100%
FoldingNet [49]	56.15	67.05	75.97	84.06	88.41
3D Cap. Net [52]	59.24	67.67	76.49	84.48	88.91
ours (best)	**59.66**	**71.06**	**80.48**	**87.66**	**91.64**
ours (mean)	54.42 ± 3.77	66.34 ± 2.99	77.12 ± 1.51	86.851 ± 0.816	91.57 ± 0.06

the classification accuracy in Table 1. Finally, we test our method in limited data scenarios. We compare our method to [49] and [52] as mentioned in the appendix of [52]. It is not clear on how they select the subset of the data. This especially matters when we take very limited data, as you can have as less as 0 to 3 shapes per category. So we report both the best and mean accuracy over 10 runs of choosing a random subset. The results are reported in Table 2.

It can be seen from the Table 1 that we achieve state of the art results on transfer learning benchmark. We beat current state of the art by 1% when we use DGCNN encoder with our method. Furthermore, using a simple encoder like PointNet beats many previous unsupervised methods with complex architectures and surprisingly beats the original PointNet supervised learning benchmark. Initializing our model with pre-trained weights also helps in achieving high accuracy in very less epochs. We can achieve 91% accuracy within 3 epochs whereas it takes about 32 epochs for random initialize weights. This is shown in more detail in test accuracy training curve in the appendix. Finally, we perform very well in limited data scenarios. We can achieve about 87% accuracy with 20% of labelled data.

Clusterable Representation. A good unsupervised method would create a seperatable manifold associated with object classes [7]. A good way to test this is to see how easily the data can be naturally clustered. We train the network by first using our unsupervised method and then use K-means algorithm on embeddings obtained. We use the implementation present in sklearn [31]. We set the number of clusters equal to 40 and use rest of the default parameters. To test the associations of the embeddings with the object classes we use adjusted mutual information metric (AMI). The results are shown in the aligned column of Table 3. We use the training set of ModelNet40 for the embeddings.

Table 3. Clustering on aligned and unaligned dataset.

Method	Aligned (AMI)	Unaligned (AMI)
Latent-GAN [1]	0.646 ± 0.001	0.197 ± 0.001
AtlasNet [16]	0.654 ± 0.001	0.197 ± 0.002
FoldingNet [49]	0.666 ± 0.001	0.141 ± 0.001
Ours (PointNet)	$\mathbf{0.677 \pm 0.002}$	$\mathbf{0.496 \pm 0.004}$

As seen in Table 3, our method produces more clusterable embeddings than autoencoder methods. This can be surprising as we are doing instance discrimination and trying to push away every other object in dataset. Our intuition is that, as the neural network is compressing the 3D objects into a lower dimension, the network has to arrange the embeddings strategically which we believe leads to semantic categories being closer in space compared to other objects from different categories.

4.2 Representation Learning on Rotated Shapes

In this part, the goal is to test how our method would compare to autoencoder baselines on datasets which are randomly rotated in SO(3) space. As mentioned earlier, most data in real-world scenarios are unaligned. Note that we use rotation in SO(3) rotation group plus translation as the second view for this section.

Simple Rotation Invariance Experiment. We create a simple experiment to test the sensitivity of the embedding of the shape with respect to the pose of the shape. We conduct the experiment by randomly selecting 10 shapes from ShapeNet v1 dataset and then randomly rotating them 50 times in SO(3) space generating 50 separate objects with different poses per shape. We then generate embeddings for all these objects and then apply clustering. We use t-SNE to give a visual representation of the clustering in $\mathbb{R}^2$ as shown in Fig. 4. We compare with the Latent-GAN [1] model.

It can be seen from Fig. 4 that our model manages to cluster objects and their different poses together. In contrast, Latent-GAN model fails to create meaningful clusters. To quantify this we use the k-means algorithm on the embeddings. In terms of AMI metric, our model achieves 1.0 whereas the baseline achieves

0.555 score. This implies that our method successfully learns to space objects and their different poses in close proximity in latent space, leading to less sensitive embeddings for downstream task.

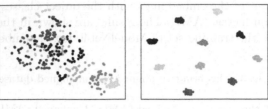

Baseline Autoencoder   Our Model

Fig. 4. t-SNE visualization of rotation invariance check. Figure illustrating how 3D shapes and their random poses are clustered in our method but fail to cluster in the baseline method.

Clusterable Representations. We also test how well our method does on clustering of embeddings when the object is rotated in SO(3) space. The experimental setup is similar to above section and the results are shown in the last column of Table 3. It can be seen that our method significantly out performs the autoencoder baselines. This illustrates that autoencoder baseline are very sensitive to rotation and incorporating some form of rotation invariance into the objective can lead to significant improvement in the embeddings obtained. We also show how this can affect transfer learning results, which are present in the appendix.

Shape Retrieval. In many applications, retrieving an object similar to a query object is very useful irrespective of their poses. For such applications, we take embedding of a query object from ShapeNet v1 dataset and ABC dataset, and retrieve the 5 nearest neighbours for a given shape by using the euclidean distance. The results are shown in Fig. 5. We again compare with Latent-GAN baseline model.

It can be seen from the first row of Fig. 5 that the objects retrieved by the autoencoder baselines are affected by the pose of the object. That is the object retrieved is similar in pose and also sometimes from a different category. Whereas our method manages to bring more semantically similar objects with different poses as shown in second row. The last two rows show sample objects retrieved on ABC dataset. More examples are shown in the appendix.

4.3   Ablation Studies

We detail the affect of using different architectures, transformations and chunks on the performance of our method. We run most of the experiments on Model-Net40 dataset and use the task of clustering. We run the clustering algorithm

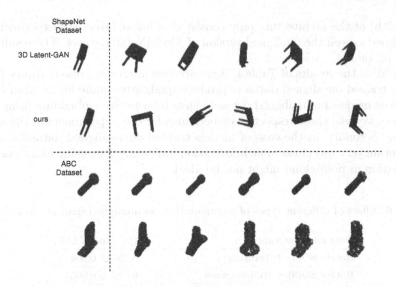

Fig. 5. Shape retrieval. First object in the row represents the query object and next five object are the retrieved objects.

Table 4. Layer wise embeddings clustering accuracy.

Data type	Layer 1	Layer 2	Layer 3	Layer 4	Layer 5	Layer 6	Layer 7
Aligned	0.504 ± 0.001	0.510 ± 0.002	0.517 + 0.002	0.535 ± 0.002	0.563 ± 0.004	**0.633 ± 0.002**	0.401 ± 0.005
Unaligned	0.107 ± 0.003	0.106 ± 0.002	0.112 ± 0.004	0.144 ± 0.002	0.179 ± 0.002	0.404 ± 0.002	**0.496 ± 0.004**

3 times because of the stochastic nature of k-means. We report the mean and standard deviation on the set of experiments. All the results are measured in AMI.

Table 5. Effect of different types of augmentation on aligned data of ModelNet40.

Data transformation	Aligned (AMI)
Translate (T)	0.633 ± 0.002
Axis chopped chunk	0.660 ± 0.002
Euclidean distance chunk	0.670 ± 0.003
Cosine distance chunk	**0.677 ± 0.002**
Chebyshev distance chunk	0.671 ± 0.001

Choice of Layer. We see the effect of choosing different layers to obtain the embeddings on the task of clustering. The goal of this experiment is to empirically test which layer contains the most information about the shape. The last layer (7th) of the architecture only consist of a linear layer. We experiment on the aligned as well the unaligned version of ModelNet40 dataset. The results are shown in Table 4.

Based on the results of Table 4, there are two interesting observations. First, models trained on aligned datasets produce qualitative embeddings from Layer 6 whereas models on unaligned datasets have informative embedding from layer 7. Hence, we take those respective embeddings for our experiments in the above sections. Secondly, in the case of models trained on unaligned datasets, Layer 1–5 contain very less clusterable information indicating that using exact position information in pointcloud might not be ideal.

Table 6. Effect of different types of augmentation on unaligned data of ModelNet40.

Data augmentation	Unaligned (AMI)
Rotate SO(3) + Translate	0.496 ± 0.004
Rotate SO(3) + Uniform scale	0.313 ± 0.005
Rotate SO(3) + Random scale	0.407 ± 0.005
Rotate SO(3) + Uniform scale + translate	$\mathbf{0.500 \pm 0.001}$
Rotate SO(3) + Random scale + translate	0.485 ± 0.004

Different Types of Geometric Transformation and Chunk Selection. In this section, we investigate effectiveness of using different ways of obtaining the chunk from a 3D object and geometric transformation of a 3D object for mutual information maximization. We do separate experiments for aligned and unaligned dataset on ModelNet40 dataset. The transformations for aligned dataset are shown in Table 5 whereas for unaligned dataset it is shown in Table 6. In the case of uniform scaling, we scale the object uniformly across the three axis in the range of 0.5 to 1.5 units of the original object. For the translation data augmentation we randomly translation between -0.2 to $+0.2$ along each axis. We also do comparison on the task of transfer learning for aligned dataset as shown in Table 7. Based on the results from the mentioned tables, we choose our transformations for a given task.

Effect of Chunk Size. The experiment illustrates the trade off between local and global information. If the chunk size is big more global information will be incorporate and the network might fail to capture locality. If the chunk size is small we will capture finer details of the object but will affect the accuracy due to the contrasting nature of the algorithm. The results are shown in Table 8.

Table 7. Effect of different types of augmentation on transfer learning of ModelNet40.

Encoder (Data augmentation)	Transfer learning (Acc. (%))
PointNet (Translate)	87.8
PointNet (Axis chopped chunk)	89.8
DGCNN (Axis chopped chunk)	**91.6**
DGCNN (Euclidean distance based chunk)	90.9
DGCNN (Cosine distance based chunk)	90.8
DGCNN (Chebyshev distance based chunk)	91.3

Table 8. Effect of different chunk size.

Chunk size	Aligned (AMI)
128	0.653 ± 0.002
256	0.665 ± 0.003
512	$\mathbf{0.677 \pm 0.002}$
768	0.658 ± 0.008
1024	0.665 ± 0.001

5   Conclusion

In this paper, we investigated using different views of 3D objects to create effective embeddings which generalizes well to different downstream tasks. We showed how considering local substructure in the objective is very effective while considering rotation as a different view can create rotation invariant embeddings. In terms of future work, we would like to explore how our method generalizes to other tasks such as segmentation and part detection. Secondly, we would like to investigate other views of 3D object such as surface normals. Finally, we would like to extend this method to other 3D representations such as meshes and voxels.

References

1. Achlioptas, P., Diamanti, O., Mitliagkas, I., Guibas, L.: Learning representations and generative models for 3D point clouds (2017). arXiv preprint arXiv:1707.02392
2. Anand, A., Racah, E., Ozair, S., Bengio, Y., Côté, M.A., Hjelm, R.D.: Unsupervised state representation learning in atari (2019). arXiv preprint arXiv:1906.08226
3. Bachman, P., Hjelm, R.D., Buchwalter, W.: Learning representations by maximizing mutual information across views (2019). arXiv preprint arXiv:1906.00910
4. Becker, S.: An information-theoretic unsupervised learning algorithm for neural networks. University of Toronto (1992)
5. Becker, S.: Mutual information maximization: models of cortical self-organization. Netw. Comput. Neural Syst. **7**(1), 7–31 (1996)
6. Bell, A.J., Sejnowski, T.J.: An information-maximization approach to blind separation and blind deconvolution. Neural Comput. **7**(6), 1129–1159 (1995)

7. Bengio, Y., Courville, A., Vincent, P.: Representation learning: a review and new perspectives. IEEE Trans. Pattern Anal. Mach. Intell. **35**(8), 1798–1828 (2013)
8. Chang, A.X., et al.: Shapenet: an information-rich 3D model repository (2015). arXiv preprint arXiv:1512.03012
9. Chen, T., Kornblith, S., Norouzi, M., Hinton, G.: A simple framework for contrastive learning of visual representations (2020). arXiv preprint arXiv:2002.05709
10. Chen, Z., Zhang, H.: Learning implicit fields for generative shape modeling. In: Proceedings of the IEEE Conference on Computer Vision and Pattern Recognition, pp. 5939–5948 (2019)
11. Cheng, S., Bronstein, M., Zhou, Y., Kotsia, I., Pantic, M., Zafeiriou, S.: Meshgan: non-linear 3D morphable models of faces (2019). arXiv preprint arXiv:1903.10384
12. Cohen, T., Welling, M.: Group equivariant convolutional networks. In: International Conference on Machine Learning, pp. 2990–2999 (2016)
13. Deng, H., Birdal, T., Ilic, S.: Ppf-foldnet: unsupervised learning of rotation invariant 3D local descriptors. In: Proceedings of the European Conference on Computer Vision (ECCV), pp. 602–618 (2018)
14. Deng, H., Birdal, T., Ilic, S.: 3D local features for direct pairwise registration (2019). arXiv preprint arXiv:1904.04281
15. Esteves, C., Xu, Y., Allen-Blanchette, C., Daniilidis, K.: Equivariant multi-view networks (2019). arXiv preprint arXiv:1904.00993
16. Groueix, T., Fisher, M., Kim, V.G., Russell, B.C., Aubry, M.: Atlasnet: a papier-m\ˆ ach\'e approach to learning 3D surface generation (2018). arXiv preprint arXiv:1802.05384
17. Gutmann, M., Hyvärinen, A.: Noise-contrastive estimation: A new estimation principle for unnormalized statistical models. In: Proceedings of the Thirteenth International Conference on Artificial Intelligence and Statistics, pp. 297–304 (2010)
18. Hassani, K., Haley, M.: Unsupervised multi-task feature learning on point clouds. In: Proceedings of the IEEE International Conference on Computer Vision, pp. 8160–8171 (2019)
19. Hénaff, O.J., Razavi, A., Doersch, C., Eslami, S., Oord, A.V.d.: Data-efficient image recognition with contrastive predictive coding (2019). arXiv preprint arXiv:1905.09272
20. Kingma, D.P., Ba, J.: Adam: A method for stochastic optimization (2014). arXiv preprint arXiv:1412.6980
21. Koch, S., et al.: Abc: a big cad model dataset for geometric deep learning. In: Proceedings of the IEEE Conference on Computer Vision and Pattern Recognition, pp. 9601–9611 (2019)
22. Lazebnik, S., Schmid, C., Ponce, J.: Semi-local affine parts for object recognition (2004)
23. Li, J., Bi, Y., Lee, G.H.: Discrete rotation equivariance for point cloud recognition (2019). arXiv preprint arXiv:1904.00319
24. Linsker, R.: An application of the principle of maximum information preservation to linear systems. In: Advances in Neural Information Processing Systems, pp. 186–194 (1989)
25. Liu, Y., Fan, B., Xiang, S., Pan, C.: Relation-shape convolutional neural network for point cloud analysis. In: Proceedings of the IEEE Conference on Computer Vision and Pattern Recognition, pp. 8895–8904 (2019)
26. Mescheder, L., Oechsle, M., Niemeyer, M., Nowozin, S., Geiger, A.: Occupancy networks: Learning 3D struction in function space. In: Proceedings of the IEEE Conference on Computer Vision and Pattern Recognition, pp. 4460–4470 (2019)

27. Michalkiewicz, M., Pontes, J.K., Jack, D., Baktashmotlagh, M., Eriksson, A.: Deep level sets: Implicit surface representations for 3D shape inference (2019). arXiv preprint arXiv:1901.06802

28. Misra, I., van der Maaten, L.: Self-supervised learning of pretext-invariant representations (2019). arXiv preprint arXiv:1912.01991

29. Oord, A.V.d., Li, Y., Vinyals, O.: Representation learning with contrastive predictive coding (2018). arXiv preprint arXiv:1807.03748

30. Park, J.J., Florence, P., Straub, J., Newcombe, R., Lovegrove, S.: Deepsdf: learning continuous signed distance functions for shape representation (2019). arXiv preprint arXiv:1901.05103

31. Pedregosa, F., et al.: Scikit-learn: machine learning in python. J. Mach. Learn. Res. **12**(Oct), 2825–2830 (2011)

32. Qi, C.R., Su, H., Mo, K., Guibas, L.J.: Pointnet: deep learning on point sets for 3d classification and segmentation. In: Proceedings of the IEEE Conference on Computer Vision and Pattern Recognition, pp. 652–660 (2017)

33. Qi, C.R., Yi, L., Su, H., Guibas, L.J.: Pointnet++: deep hierarchical feature learning on point sets in a metric space. In: Advances in Neural Information Processing Systems, pp. 5099–5108 (2017)

34. Rublee, E., Rabaud, V., Konolige, K., Bradski, G.R.: Orb: an efficient alternative to sift or surf. In: ICCV, vol. 11, p. 2. Citeseer (2011)

35. Sanghi, A., Danielyan, A.: Towards 3d rotation invariant embeddings

36. Sauder, J., Sievers, B.: Context prediction for unsupervised deep learning on point clouds (2019). arXiv preprint arXiv:1901.08396

37. Steder, B., Rusu, R.B., Konolige, K., Burgard, W.: Narf: 3D range image features for object recognition. In: Workshop on Defining and Solving Realistic Perception Problems in Personal Robotics at the IEEE/RSJ International Conference on Intelligent Robots and Systems (IROS), vol. 44 (2010)

38. Su, H., Maji, S., Kalogerakis, E., Learned-Miller, E.: Multi-view convolutional neural networks for 3d shape recognition. In: Proceedings of the IEEE International Conference on Computer Vision, pp. 945–953 (2015)

39. Sun, F.Y., Hoffmann, J., Tang, J.: Infograph: unsupervised and semi-supervised graph-level representation learning via mutual information maximization (2019). arXiv preprint arXiv:1908.01000

40. Tan, Q., Gao, L., Lai, Y.K., Xia, S.: Variational autoencoders for deforming 3D mesh models. In: Proceedings of the IEEE Conference on Computer Vision and Pattern Recognition, pp. 5841–5850 (2018)

41. Thomas, N., et al.: Tensor field networks: rotation-and translation-equivariant neural networks for 3D point clouds (2018). arXiv preprint arXiv:1802.08219

42. Tian, Y., Krishnan, D., Isola, P.: Contrastive multiview coding (2019). arXiv preprint arXiv:1906.05849

43. Veličković, P., Fedus, W., Hamilton, W.L., Liò, P., Bengio, Y., Hjelm, R.D.: Deep graph infomax (2018). arXiv preprint arXiv:1809.10341

44. Wang, Y., Sun, Y., Liu, Z., Sarma, S.E., Bronstein, M.M., Solomon, J.M.: Dynamic graph CNN for learning on point clouds. ACM Trans. Graph. (TOG) **38**(5), 146 (2019)

45. Worrall, D.E., Garbin, S.J., Turmukhambetov, D., Brostow, G.J.: Harmonic networks: deep translation and rotation equivariance. In: Proceedings of the IEEE Conference on Computer Vision and Pattern Recognition, pp. 5028–5037 (2017)

46. Wu, J., Zhang, C., Xue, T., Freeman, B., Tenenbaum, J.: Learning a probabilistic latent space of object shapes via 3D generative-adversarial modeling. In: Advances in Neural Information Processing Systems, pp. 82–90 (2016)

47. Wu, Z., et al.: 3D shapenets: A deep representation for volumetric shapes. In: Proceedings of the IEEE Conference on Computer Vision and Pattern Recognition, pp. 1912–1920 (2015)
48. Wu, Z., Xiong, Y., Yu, S., Lin, D.: Unsupervised feature learning via non-parametric instance-level discrimination (2018). arXiv preprint arXiv:1805.01978
49. Yang, Y., Feng, C., Shen, Y., Tian, D.: Foldingnet: point cloud auto-encoder via deep grid deformation. In: Proceedings of the IEEE Conference on Computer Vision and Pattern Recognition, pp. 206–215 (2018)
50. Zaheer, M., Kottur, S., Ravanbakhsh, S., Poczos, B., Salakhutdinov, R.R., Smola, A.J.: Deep sets. In: Advances in Neural Information Processing Systems, pp. 3391–3401 (2017)
51. Zhang, L., Zhu, Z.: Unsupervised feature learning for point cloud understanding by contrasting and clustering using graph convolutional neural networks. In: 2019 International Conference on 3D Vision (3DV), pp. 395–404. IEEE (2019)
52. Zhao, Y., Birdal, T., Deng, H., Tombari, F.: 3D point capsule networks. In: Proceedings of the IEEE Conference on Computer Vision and Pattern Recognition, pp. 1009–1018 (2019)
53. Zhuang, C., Zhai, A.L., Yamins, D.: Local aggregation for unsupervised learning of visual embeddings. In: Proceedings of the IEEE International Conference on Computer Vision, pp. 6002–6012 (2019)

Adversarial Data Augmentation via Deformation Statistics

Sahin Olut[1]([⊠]), Zhengyang Shen[1], Zhenlin Xu[1], Samuel Gerber[2], and Marc Niethammer[1]

[1] UNC Chapel Hill, Chapel Hill, USA
{olut,zyshen,zhenlinx,mn}@cs.unc.edu
[2] Kitware Inc., New York, USA
samuel.gerber@kitware.com

Abstract. Deep learning models have been successful in computer vision and medical image analysis. However, training these models frequently requires large labeled image sets whose creation is often very time and labor intensive, for example, in the context of 3D segmentations. Approaches capable of training deep segmentation networks with a limited number of labeled samples are therefore highly desirable. Data augmentation or semi-supervised approaches are commonly used to cope with limited labeled training data. However, the augmentation strategies for many existing approaches are either hand-engineered or require computationally demanding searches. To that end, we explore an augmentation strategy which builds statistical deformation models from unlabeled data via principal component analysis and uses the resulting statistical deformation space to augment the labeled training samples. Specifically, we obtain transformations via deep registration models. This allows for an intuitive control over plausible deformation magnitudes via the statistical model and, if combined with an appropriate deformation model, yields spatially regular transformations. To optimally augment a dataset we use an adversarial strategy integrated into our statistical deformation model. We demonstrate the effectiveness of our approach for the segmentation of knee cartilage from 3D magnetic resonance images. We show favorable performance to state-of-the-art augmentation approaches.

Keywords: Data augmentation · Image registration · Segmentation

Image segmentation is an important task in computer vision and medical image analysis, for example, to localize objects of interest or to plan treatments and surgeries. Deep neural networks (DNNs) achieve state-of-the art segmentation performance in these domains [19,20,44]. However, training DNNs typically relies on large datasets with labeled structures of interest. In many cases, and for medical segmentation problems in particular, labeled training data is scarce, as obtaining manual segmentations is costly and requires expertise [35]. To allow for training of well-performing DNNs from a limited number of segmentations, various data augmentation strategies have been proposed [24]. Augmentation strate-

© Springer Nature Switzerland AG 2020
A. Vedaldi et al. (Eds.): ECCV 2020, LNCS 12374, pp. 643–659, 2020.
https://doi.org/10.1007/978-3-030-58526-6_38

gies range from pre-defined random transformations [1,23,25,27] to learning-based approaches [14,15,42]. Random transformations usually include intensity changes such as contrast enhancement, brightness adjustments, as well as random deformations, e.g., affine transformations. These methods are often difficult to tune as they do not directly estimate image variations observable in real data [10,24,43]. Generative Adversarial Networks (GANs) [12] are also widely used to generate data augmentations [5,18,33]. However, the proposed methods do not explicitly model deformation spaces and hence only have indirect control over augmentation realism.

In fact, existing learning-based data augmentation techniques are generally not based on statistical deformation models as correspondences between random natural image pairs might not be meaningful. However, if such deformations can be established, as is frequently the case for medical images of the same anatomy within or across patients, they may be used to create plausible deformations for data augmentation [18,43]. While statistical deformation models have a long history in computer vision [7,8] and medical image analysis [28,34] they have not been well explored in the context of data augmentation.

Our proposed data augmentation method (*AdvEigAug*) uses learned deformation statistics as a sensible constraint within an adversarial data augmentation setting. Specifically, we make the following contributions:

1) We explicitly model the deformation distribution via principal component analysis (PCA) to guide data augmentation. This allows us to estimate reasonable deformation ranges for augmentation.
2) We propose to efficiently estimate this PCA deformation space via deep image registration models. We explore PCA models on displacement and momentum fields, where the momentum fields assure spatial regularity via the integration of a partial differential equation model.
3) We integrate our PCA model into an adversarial formulation to select deformations for augmentation which are challenging for the segmentation.
4) We extensively evaluate our augmentation approach and show favorable performance with respect to state-of-the-art augmentation approaches.

The manuscript is organized as follows: Sect. 1 gives an overview of related work; Sect. 2 describes our *AdvEigAug* technique; Sect. 3 describes our experimental setup; Sect. 4 presents and discusses the results of our method.

1   Related Work

We focus our discussion here on related data augmentation and semi-supervised learning approaches that use adversarial training or image registrations.

Adversarial Training. It is well known that adversarial examples created by locally perturbing an input with imperceptible changes may drastically affect image classification results [30]. But it has also been shown that training DNNs with adversarial examples can improve DNN robustness and performance [13], which is our goal. Here, we focus on adversarial training via spatial transformations.

Engstrom *et al.* [11] use rigid transformations to generate adversarial examples for data augmentations, whereas Kanbak *et al.* [21] develop *ManiFool*, an adversarial training technique which is based on affine transformations. A more general adversarial deformation strategy is pursued by Xiao *et al.* [37] where a displacement field is optimized to cause misclassifications. Smoothness of the displacement field is encouraged via regularization.

All these approaches focus on classification instead of segmentation. Furthermore, transformation models are prescribed rather than inferred from observed transformations in the data and selecting a deformation magnitude requires an iterative approach. Our *AdvEigAug* approach instead explores using statistical deformation models obtained from image pairs by fluid- or displacement based registrations. The statistical model also results in a clear guidance for the perturbation magnitudes, thereby eliminating the need for iterations.

Data Augmentation and Semi-supervised Learning via Registration.
Image registration is widely used for atlas based segmentation [17]. As it allows estimating deformations between unlabeled image pairs it has also been used to create plausible data deformations for data augmentation. Via *semi-supervised learning* this allows training deep segmentation networks with very limited labeled training data by exploiting large unlabeled image sets. Such approaches have successfully been used in medical image segmentation [5,38,43].

For example, Zhao *et al.* [43] train spatial and appearance transformation networks to synthesize labeled images which can then be used to train a segmentation network. Chaitanya *et al.* [5] model appearance variations and spatial transformations via GANs for few-shot image segmentation. Xu *et al.* [38] use a semi-supervised approach to jointly train a segmentation and a registration network. This allows the segmentation network to benefit from the transformations generated via the registration network while simultaneously improving registration results by including the obtained segmentations in the image similarity loss.

However, the approaches above do not employ adversarial samples for training segmentation networks and do not use statistical deformation models. Instead, our *AdvEigAug* approach captures deformation statistics via a PCA model which is efficiently estimated, *separately for each sample*, via deep registration networks and integrated into an adversarial training strategy.

2   Method

Our approach is an the adversarial training scheme. In particular, it builds upon, extends, and combines two different base strategies: (1) the estimation of a statistical deformation model and (2) the creation of adversarial examples that can fool a segmentation network and improve its training. Our proposed *AdvEigAug* approach builds upon the *ManiFool* augmentation idea [21]. *ManiFool* is limited to adversarial deformations for classification which are subject to an affine transformations model. The adversarial directions are based on the gradient with respect to the classification loss, but how far to move in this gradient direction requires tuning. Our approach on the other hand estimates a statistical deformation space which can be much richer than an affine transformation. In fact, we propose two efficient ways of computing these spaces, both based on non-parametric registration approaches: an approach based on a deep registration network directly predicting displacement fields [3] and another deep network predicting the initial momentum of the large deformation diffeomorphic metric mapping model (LDDMM) [4,31]. The benefit of the LDDMM model is that it can assure spatially regular (diffeomorphic) spatial transformations. Furthermore, as we estimate a statistical model we can sample from it and we also have a direct measure of the range of deformations which are consistent with deformations observed in the data. Note that these deformation networks can use labeled data [2,38] to improve registration accuracies, but can also be trained directly on intensity images [3,31,40], which is the strategy we follow here.

As our approach is related to *ManiFool* and the augmentation approach in [11] we introduce our reformulation of their concepts for image segmentation and affine transformations in Sect. 2.1. Section 2.2 introduces our *AdvEigAug* approach.

2.1   Baseline Method: AdvAffine

Our baseline method is related to [11,21] where rigid and affine transformations (in 2D) are generated with the goal of fooling a classifier. We extend the approach to 3D affine transformations and apply it to image segmentation instead of classification. While this is a minor change in terms of the loss function (see details below) it precludes simple approaches for step-size selection, i.e., to determine how strong of an affine transformation to apply. For example, while the *ManiFool* approach takes adversarial steps until the classification label changes such an approach is no longer appropriate when dealing with image segmentations as one is now dealing with a set of labels instead of one.

Specifically, we assume we have a deep neural segmentation network, NN, which, given an input image, I, results in a segmentation $\hat{y}$. We also assume we have a segmentation loss, $loss(y, \hat{y})$ (typically a binary cross-entropy loss averaged over the image volume; or a classification loss in [11]), where y is the target segmentation. Our goal is to determine how to spatially transform an input image I so that it is maximally detrimental (i.e., adversarial) for the loss to be minimized. We parameterize the affine transformation as

$$\Phi^{-1}(x; \theta) = (E + A)x + b, \ A, E \in \mathbb{R}^{d \times d}, \ b, x \in \mathbb{R}^d, \tag{1}$$

where d denotes the spatial dimension ($d = 3$ in our experiments), x denotes the spatial coordinate, $\theta = \{A, b\}$ are the affine transformation parameters, and E is the identity matrix, which allows us to easily start from the identity transform

$(A = 0, b = 0)$. The transformed image is then $I \circ \Phi^{-1}$. To compute an adversarial direction we perform gradient ascent with respect to the loss

$$\mathcal{L}(\theta) = loss(y, \hat{y}), \text{ s.t. } \hat{y} = NN(I \circ \Phi^{-1}(\cdot; \theta)). \tag{2}$$

It is unclear how far to step into the ascent direction for segmentation problems: we simply take t steps with a chosen learning rate, Δt, i.e., $\theta_{t+1} = \theta_t + \Delta t \frac{\partial \mathcal{L}(\theta_t)}{\partial \theta}$.

This process is repeated for each sample in the training data set. The resulting affine transformations are then applied to the images and their segmentations (using nearest neighbor interpolation) to augment the training dataset.

2.2 Proposed Method: AdvEigAug

Figure 1 gives an overview of our proposed approach. We use a statistical deformation model to capture plausible anatomical variations and efficiently estimate them via two different deep registration approaches [9,31]. These statistical deformation models are integrated into an adversarial strategy which selects a deformation direction to challenge the segmentation loss.

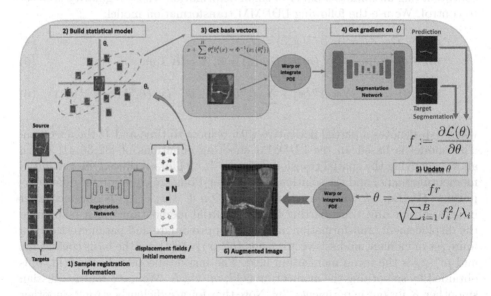

Fig. 1. Overview: We first build the deformation model (1–3) and then obtain an adversarial deformation from the segmentation network (4–6). Our method generates samples (warped images and warped segmentations) that are difficult to segment, but that help training the segmentation network.

Deformation Models. Following the approach in Sect. 2.1 we want to obtain adversarial samples subject to a given transformation model. However, instead of specifying this transformation model (e.g., an affine model as in *AdvAffine*) we want to learn plausible models from data.

Displacement Deformation Model. We parameterize the transformation via a small number of learned displacement field basis elements:

$$\Phi^{-1}(x; \{\theta_i^d\}) = x + u(x) \approx x + \mu^d(x) + \sum_{i=1}^{B} \theta_i^d b_i^d(x), \qquad (3)$$

where $u(x)$ is the displacement field which we approximate by B basis elements. Here, $\mu^d(x)$ is the mean displacement field, $\{b_i^d(x)\}$ is a set of basis displacement fields, and $\{\theta_i^d\}$ are the basis coefficients with respect to which we will compute the adversarial direction. These basis fields could be specified, but we estimate them from sample image pairs (see *Statistical Model* paragraph below).

Fluid Deformation Model. Specifying the transformation via basis displacement fields, $\{b_i^d(x)\}$ as in Eq. 3 might result in transformations that are not spatially smooth or invertible. For example, for large values of the coefficients $\{\theta_i\}$ folds might occur. This can be avoided by transformation models based on fluid mechanics [16]. Foldings are avoided in such models by indirectly representing the transformation via integration of a velocity field. Essentially, this amounts to concatenating an infinite number of small deformation whose regularity is easier to control. We use the following LDDMM transformation model:

$$\phi_t^{-1} + D\phi^{-1}v = 0, \ \phi^{-1}(x,0) = x, \qquad (4)$$

$$m_t + div(v)m + Dv^T(m) + Dm(v) = 0, \ v = K \star m, \qquad (5)$$

$$m(x,0) = m_0(x) \approx \mu^m(x) + \sum_{i=1}^{B} \theta_i^m b_i^m(x), \ \Phi^{-1}(x; \{\theta_i^m\}) = \phi^{-1}(x,1), \ (6)$$

where $(\cdot)_t$ denotes a partial derivative with respect to time and D the Jacobian. This model is based on the LDDMM shooting equations of [32,36,41] which allow specifying the spatial transformation $\phi^{-1}(x,t)$ for all times t via the initial momentum $m_0(x)$ and the solution of the Euler-Poincaré equation for diffeomorphisms [41] (Eq. 5). Our desired transformation $\Phi^{-1}(x; \{\theta_i^m\})$ is obtained after integration for unit time starting from the initial momentum which (similar to the displacement transformation model) we approximate and parameterize via a finite set of momentum basis vector fields $\{b_i^m(x)\}$; $\{\theta_i^m\}$ are the basis coefficients and $\mu^m(x)$ is the mean momentum field. The instantaneous velocity field, v, is obtained by smoothing the momentum field, m, via K. We use a multi-Gaussian smoother K for our experiments [26]. Note that for a sufficiently strong smoother K the resulting spatial transformations are assured to be diffeomorphic. We will see that this is a convenient property for our augmentation strategy.

Statistical Model. The introduced deformation models require the specification of their mean displacement and momentum vector fields $(\mu^d(x), \mu^m(x))$ as well as of their basis vector fields $(\{b_i^d(x)\}, \{b_i^m(x)\})$. We learn them from data. Specifically, given a source image I we register it to N target images $\{I_i\}$ based on the displacement or the fluid deformation model respectively (without the finite basis approximation). This results in a set of N displacement fields $\{u_i(x)\}$ or initial momentum fields $\{m_i(x)\}$ from which we can build the statistical model. Specifically, we obtain the set of basis vectors $(\{b_i^d(x)\}, \{b_i^m(x)\})$ via principal component analysis (PCA) based on the covariance matrix, C, of the displacement or initial momentum fields. As we estimate these spaces relative to a source image I (which defines the tangent space for the transformations) we typically set $\mu^d(x) = 0$ and $\mu^m(x) = 0$. In the following we denote the set of (eigenvalue,eigenvector) pairs of the covariance matrix as $\{(\lambda_i, e_i(x))\}$, where eigenvalues are ordered in descending order, i.e., $\lambda_i \geq \lambda_{i+1} \geq 0$. The same analysis will hold for the displacement-based and the fluid transformation models. The only difference will be how to obtain the transformations from the basis representations. As computing registrations by numerical optimization is costly, we approximate their solution via deep-learning registration models [3,31]. These can rapidly predict the displacement or initial momentum fields.

Algorithm 1. *AdvEigAug*

Inputs: I_0, a set of (zero-centered) displacement fields for X registering I_0 to a set of target images, y segmentation mask

Outputs: augmented image I_{aug}, y_{aug} mask for augmented image

\# *Compute the Gramian matrix*

$\mathbf{G} = \mathbf{X}\mathbf{X}^{\mathbf{T}}$

$\{\lambda_i\}, \{e_i\} = \text{eigendecompose}(\mathbf{G})$

\# *Compute deformation & warp image*

$\{\theta_i\}_1^B \leftarrow 0$

$\phi^{-1}(x) \leftarrow x + \mu^d(x) + \sum_{i=1}^{B} \theta_i e_i(x)$

$\hat{\mathbf{I}}_0 = \mathbf{I}_0 \circ \phi^{-1}(\mathbf{x})$

\# *Get the gradient wrt. θ*

$\hat{y} = \text{predict\_segmentation}(\hat{\mathbf{I}}_0)$

$f = \nabla_\theta(\text{segmentation\_loss}(y, \hat{y}))$

\# *Update θ*

$\theta \leftarrow \dfrac{f|r|}{\sqrt{\sum_{i=1}^{B} f_i^2/\lambda_i}}$.

$\phi^{-1}(x) \leftarrow x + \mu^d(x) + \sum_{i=1}^{B} \theta_i e_i(x)$

\# *Warp image and mask*

$\mathbf{I}_{aug} = \mathbf{I}_0 \circ \phi^{-1}(\mathbf{x})$

$\mathbf{y}_{aug} = \mathbf{y} \circ \phi^{-1}(\mathbf{x})$

Low-Rank Approximation. Given a set of N centered displacement or initial momentum vector fields (linearized into column vectors) we can write the covariance matrix, C, and its low-rank approximation C_l as

$$C = \sum_{i=1}^{N} \lambda_i e_i e_i^T \approx \sum_{i=1}^{B} \lambda_i e_i e_i^T = C_B. \quad (7)$$

We use the first B eigenvectors to define the basis vectors for our deformation models. As for the *AdvAffine* approach of Sect. 2.1 we can then compute the gradient of the segmentation loss with respect to the transformation parameters

$$f := \frac{\partial \mathcal{L}(\theta)}{\partial \theta}. \quad (8)$$

The only change is in the deformation model where θ is now either $\{\theta_i^d\}$ or $\{\theta_i^m\}$ to parameterize the displacement or the initial momentum field respectively. Everything else stays unchanged. A key benefit of our approach, however, is that the low-rank covariance matrix, C_B induces a B-dimensional Gaussian distribution on the basis coefficient $\{\theta_i\}$, which are by construction decorrelated and have variances $\{\lambda_i\}$, i.e., they are normally distributed according to $N(0, C_m)$, where $C_m = diag(\{\lambda_i\})$. As we will see next, this is beneficial to define step-sizes in the adversarial direction which are consistent with the observed data.

Adversarial Direction. So far our transformation models allow more flexible, data-driven transformations than the affine transformation model of Sect. 2.1, but it is still unclear how far one should move in the adversarial gradient direction, f (Eq. 8). However, now that we have a statistical deformation model we can use it to obtain deformation parameters θ which are consistent with the range of values that should be expected based on the statistical model. Specifically, we want to specify how far we move away from the mean in terms of standard deviations of the distribution, while moving in the adversarial direction. To move $r \geq 0$ standard deviations away we scale the adversarial gradient as follows

$$\theta = \frac{fr}{\sqrt{\sum_{i=1}^{B} f_i^2 / \lambda_i}}. \tag{9}$$

It is easy to see why this is the case.

Proof. We first transform the gradient direction (with respect to the basis coefficients, θ), f, to Mahalanobis space via $C_m^{-\frac{1}{2}}$ in which the *coefficients*, θ, are distributed according to $N(0, I)$. We are only interested in the direction and not the magnitude of the adversarial gradient, f. Hence, we normalize it to obtain

$$\overline{f} = \frac{C_m^{-\frac{1}{2}} f}{\|C_m^{-\frac{1}{2}} f\|_2}. \tag{10}$$

In this space, we scale the transformed coefficients by r to move r standard deviations out and subsequently transform back to the original space by multiplying with the inverse transform $C_m^{\frac{1}{2}}$ resulting in

$$\theta = C_m^{\frac{1}{2}} \overline{f} r = C_m^{\frac{1}{2}} \frac{C_m^{-\frac{1}{2}} f r}{\|C_m^{-\frac{1}{2}} f\|_2} = \frac{fr}{\|C_m^{-\frac{1}{2}} f\|_2} = \frac{fr}{\sqrt{\sum_{i=1}^{B} f_i^2 / \lambda_i}}. \tag{11}$$

Given an adversarial direction, f, this allows us to sample a set of transformation coefficients, θ, which are consistent with this adversarial direction *and* which have a desired magnitude, r, as estimated via the statistical deformation model. Hence, in contrast to the *AdvAffine* approach of Sect. 2.1 there is now an intuitive way for specifying how large the adversarial deformations should be so that they remain consistent with the observed deformations between image pairs. Figure 1 shows an example illustration of determining such a scaled adversarial direction. Pseudo-code for the overall augmentation algorithm for the displacement-based deformation model is given in Algorithm 2.2. The fluid-flow-based approach follows similarly, but requires integration and backpropagation through Eqs. 4–6. The resulting deformations are then applied to the training image and its segmentation (using nearest neighbor interpolation) to augment the training dataset.

3   Experiments

We investigate the performance of different augmentation strategies for the segmentation of knee cartilage from the 3D magnetic resonance images (MRIs) of the Osteoarthritis Initiative[1]. All images are affinely registered to a knee atlas. The original images are of size $384 \times 384 \times 160$ with a voxel size of $0.36 \times 0.36 \times 0.7$ mm^3. We normalize the intensities of each image such that the 0.1 percentile and the 99.9 percentile are mapped to $\{0, 1\}$ and clamp values that are smaller to 0 and larger to 1 to avoid outliers. We do not bias-field correct these images as it has been shown in [39] that it is not required for this dataset. This also justifies our choice to test the *Brainstorm* approach without its appearance-normalization part. To be able to store the 3D data in the 11 GB of our NVIDIA GTX 2080Ti GPU we re-sample the input images and their segmentations to $190 \times 190 \times 152$. Note that this resampling might introduce slight evaluation bias with respect to the manual segmentations drawn on the original full-resolution images. However, our objective is to compare augmentation approaches which would all be equally affected, hence, the relative comparisons between methods remain fair. To assure that we can compare between methods we use the same hyperparameters that we use in the *NoAug* experiments across all experiments. All segmentation networks are trained with a learning rate of 0.001 using Adam [22] where $\beta_1 = 0.5$ and $\beta_2 = 0.99$. For the displacement field network, we use a learning rate of 0.0002. For the momentum generation network, we use the same settings as in [32].

3.1   Registration and Segmentation Networks

We use a 3D-UNet [6] with 5 encoder and decoder layers. The number of feature channels in the encoder layers are $[16, 32, 32, 32, 32]$ and the number of feature channels in the decoder layers are $[32, 32, 32, 8, 8]$. Each convolutional block is followed by batch normalization and ReLU activation except for the last one.

Segmentation Network. We use binary cross-entropy loss for simplicity and jointly segment the femoral and tibial cartilage. However, our approach generalizes to any loss function, e.g., to multiple class labels.

Registration Networks. We train two registration networks: to predict (1) displacement fields and (2) the initial momentum of EPDiff (Eq. 5).

Displacement Field Network. For the displacement field network we follow the VoxelMorph architecture [3], but use a modified loss of the form

$$\mathcal{L}_{disp}(u(x)) = \mathcal{L}_{reg}^d(u(x)) + \lambda^d \mathcal{L}_{sim}(I_T, I \circ (x + u(x))). \tag{12}$$

---

[1] https://nda.nih.gov/oai/.

where u is the displacement field predicted by the network for a given source image, I, and target image I_T; $\mathcal{L}_{sim}$ is the image similarity measure (we use normalized cross correlation (NCC)) and $\mathcal{L}_{reg}$ is the regularizer for which we choose the bending energy [29] to keep transformations smooth:

$$\mathcal{L}_{reg}^d(u(x)) = \frac{1}{N} \sum_x \sum_{i=1}^d \|H(u_i(x))\|_F^2, \tag{13}$$

where $\|\cdot\|_F$ is the Frobenius norm, $H(u_i(x))$ is the Hessian of the i-th component of $u(x)$, N the number of voxels, and d is the spatial dimension ($d = 3$). We set $\lambda^d = 200$.

Initial Momentum Network. We use the LDDMM network of [32], which predicts the initial momentum as an intermediate output. The network loss is

$$\mathcal{L}_m(m_0(x)) = \mathcal{L}_{reg}^m(m_0(x)) + \lambda^m \mathcal{L}_{sim}(I_T, I \circ \Phi^{-1}), \text{ s.t. Eqs. 4–6 hold}, \tag{14}$$

where $m_0(x)$ is the initial momentum (not yet approximated via a finite basis for the network loss), $\mathcal{L}_{reg} = \langle m_0, K * m_0 \rangle$, Φ^{-1} is the spatial transform induced by $m_0(x)$ and we use a Localized NCC similarity loss [31]. See [32] for details.

3.2   Experimental Design

Our goal is to demonstrate that (1) building a statistical deformation model is useful, that (2) using adversarial samples based on the statistical deformation model improves segmentation results, and that (3) *AdvEigAug* outperforms other augmentation approaches, in particular, approaches that do not attempt to learn deformation models from data. We focus on a few-shot learning setting.

Methods We Compare to and Rationale for Comparisons

NoAug uses no augmentation. It only uses the segmented training samples. All augmentation approaches are expected to outperform this method.

RandDeform is our representative of an augmentation strategy using various spatial transformations which are *not* inferred from data and hence is expected to show inferior performance to approaches that do. Specifically, we randomly rotate images between $\pm 15°$ in all axes and apply random translations by up to 15 voxels in each space direction. We simultaneously apply random displacement fields. Translations and rotations are drawn from uniform distributions. The displacement fields are obtained by Gaussian smoothing of random displacement fields drawn from a unit normal distribution.

AdvAffine is described in detail in Sect. 2.1. It is our baseline adversarial augmentation approach. It lacks our ability to determine a desirable deformation magnitude, uses a simple affine transformation, and does not build a statistical deformation model. Our various *AdvEigAug* approaches directly compete with *AdvAffine* and favorable performance indicates that using statistical deformation models can be beneficial for augmentation. We choose $t = 20$ and $\Delta t = 0.25$ to select the deformation magnitude for *AdvAffine*.

Brainstorm [43] relates to our *AdvEigAug* approach in the sense that it uses unlabeled images via registrations (we use the displacement field network of Sect. 3.1 for its implementation) to create realistic deformations for augmentation. In contrast to *AdvEigAug* it is not an adversarial approach nor does it build a statistical deformation model and hence it relies on the set of deformations that can be observed between images. We will show that these design choices matter by comparing to *AdvEigAug* directly and to a non-adversarial variant *EigAug* (explained next). *Brainstorm* also uses an intensity transfer network. However, as we are working on one specific dataset without strong appearance differences we only use the deformation component of *Brainstorm* for a more direct comparison.

EigAug is our *AdvEigAug* approach when replacing the adversarial gradient direction, f by a random direction. We compare against this approach to (1) show that modeling the deformation space is beneficial (e.g., with respect to *Brainstorm*) and (2) that using the adversarial direction of *AdvEigAug* yields improvements.

AdvEigAug is our proposed approach and described in detail in Sect. 2.2.

Upper Bound. We use the segmentations for the entire training set ($n = 200$) to train the segmentation network. We regard the resulting segmentation accuracy as the quasi upper-bound for achievable segmentation accuracy.

Randomization and Augmentation Strategies

Randomization. *AdvEigAug* and *EigAug* require the selection of, r, which specifies how many standard deviations away from the mean the parameterized deformation is. In our experiments we explore fixing $r \in \{1, 2, 3\}$ as well as randomly selecting it: $r \sim U(1.5, 3)$. We hypothesize that randomly selecting r results in a richer set of deformations and hence in better segmentation performance.

Covariance and Basis Vector Computations. *AdvEigAug* and *EigAug* also require the computation of the covariance matrix based on displacement or momentum fields. We recompute this covariance matrix for *every* augmented sample we create to assure sufficient variability of deformation directions. Specifically, for a given training sample we randomly select N images (we do not use or need their segmentations) and register the training sample to these images using one of our registration networks to obtain the displacement or momentum fields respectively. We choose the top two eigendirections ($B = 2$) as our basis vectors. Clearly, using more samples results in more stable results, but it also limits the space being explored. In the extreme case one would use all available image pairs to estimate the covariance matrix which would then limit the variability. We therefore use only $N = 10$ samples to estimate these directions.

Offline Augmentation. For almost all our experiments we employ an *offline* augmentation strategy. We train the segmentation network with the original training data for 700 epochs. We then create a set of adversarial examples. Specifically, we create 5 adversarial samples for each training sample. We then continue training with the augmented training set for another 1,300 epochs. To assure that training is not biased too strongly to the adversarial samples [24] we down-weight the adversarial examples by 1/5 in the loss.

Online Augmentation. We also explore an online augmentation approach for *AdvEigAug* only to determine if the successive introduction of adversarial samples is beneficial. As for the offline approach, we first train the segmentation network for 700 epochs. We then add 1 adversarial example per original training sample and continue training for 150 epochs. We repeat this augmentation step 4 more times and finally train for another 550 epochs. As for the offline variant adversarial samples are down-weighted. We down-weight by $\frac{1}{k}$, where k is the kth addition of adversarial samples. This assures that at all times the adversarial samples are balanced with the original training data.

Data. We split our dataset into a test set with segmentations ($n = 100$), a validation set with segmentations ($n = 50$), as well as a training set ($n = 200$) of which we use $n = 4$ or $n = 8$ as the original training set including their segmentations. This training set (excluding their segmentations) is also used to obtain deformations for *Brainstorm* and the deformation spaces for *AdvEigAug* and *EigAug*.

4   Results and Discussion

Table 1 shows Dice scores and surface distance measures between the obtained segmentations and the manual knee cartilage segmentations on the test set ($n = 100$). As we focus on training models with small numbers of manual segmentations we show results for training with 4 and 8 segmentations respectively. Images without segmentations were also used during training of the *Brainstorm*, *EigAug*, and *AdvEigAug* models. All results are averages over 5 random runs; standard deviations are reported in parentheses.

Overall Performance. Our proposed L-$AdvEigAug_{\text{rand\_r}}^{\text{online}}$ approach performs best in terms of Dice scores and surface distances demonstrating the effectiveness of using statistical deformation models to capture plausible anatomical variations in low sample settings in combination with a randomized deformation approach and an LDDMM-based deformation model which can assure spatially well behaved deformations. In particular, our *AdvEigAug* approaches outperform all competing baseline approaches we tested: no augmentation (*NoAug*); augmentations with random deformations (*RandDeform*); *Brainstorm* which also uses unsegmented images to obtain deformations via registrations; and a competing adversarial approach using affine transformations only (*AdvAffine*). Note that our images are all already affinely aligned to an atlas, hence an improvement by *AdvAffine* was not expected in this setting. However, it illustrates that the more flexible deformation models of *AdvEigAug* are indeed beneficial.

Table 1. Reported mean Dice scores and surface distance measures in *mm* (average surface distance (ASD), 50th, and 95th-percentile) over 5 training runs. Surface distances are measured based on the binarized predictions. Standard deviations are in parentheses. Prefix L denotes training with the LDDMM model, prefix Adv indicates using adversarial samples. Dice scores for the approaches were tested for statistical difference with respect to the L-AdvEigAug$_{rand\_r}^{online}$ approach (bold) using a paired t-test. At a significance level $\alpha = 0.05$ with 14 different tests per measure results in statistical significance based on Bonferroni correction for $p < \alpha/14 \approx 0.0035$; † shows statistical significance where $p < 1e^{-3}$ and * where $p < 1e^{-6}$.

Experiment	4 training samples				8 training samples			
	Dice in %	ASD	50%	95%	Dice in %	ASD	50%	95%
NoAug	75.2(0.31)*	1.79	0.134	20.17	77.1(0.28)*	1.17	0.093	9.14
RandDeform	76.3(0.34)*	1.80	0.171	14.72	78.1(0.25)*	0.87	0.072	4.12
Brainstorm [43]	77.7(0.20)*	1.17	0.087	8.80	79.4(0.17)*	0.71	0.031	**3.17**
AdvAffine	75.1(0.49)*	1.62	0.102	13.78	77.8(0.37)*	1.19	0.042	5.13
EigAug$_{rand\_r}$	73.3(0.81)*	1.76	0.352	9.37	76.2(0.13)*	0.86	0.139	5.15
EigAug$_{fix\_r=1}$	73.4(0.45)*	1.45	0.241	12.04	77.8(0.12)*	0.92	0.086	5.36
EigAug$_{fix\_r=2}$	73.6(0.40)*	1.46	0.217	10.32	77.7(0.10)*	0.85	**0.011**	5.17
EigAug$_{fix\_r=3}$	73.0(0.40)*	1.44	0.232	11.05	77.3(0.30)*	0.98	0.147	5.24
AdvEigAug$_{fix\_r=1}$	75.6(0.31)*	1.77	0.231	10.12	77.8(0.23)*	0.97	0.074	6.78
AdvEigAug$_{fix\_r=2}$	76.1(0.20)*	1.24	0.128	9.28	78.7(0.21)*	0.88	0.031	4.76
AdvEigAug$_{fix\_r=3}$	74.9(0.15)*	1.75	0.287	11.18	76.9(0.27)*	1.22	0.113	9.99
AdvEigAug$_{rand\_r}$	78.4(0.25)†	0.85	0.091	7.14	81.0(0.15)†	0.65	0.031	3.77
AdvEigAug$_{rand\_r}^{online}$	78.2(0.23)†	0.92	0.071	6.44	81.1(0.16)†	0.58	0.025	3.51
L-AdvEigAug$_{rand\_r}$	78.5(0.17)	0.79	0.083	4.68	81.1(0.11)	0.55	0.021	3.85
L-AdvEigAug$_{rand\_r}^{online}$	**79.1(0.14)**	**0.72**	**0.020**	**3.72**	**81.2(0.12)**	**0.51**	0.023	3.46
Upper bound	82.5(0.2)	0.47	0.019	2.42				

Adversarial Sample Effect. Comparing the adversarial *AdvEigAug* to the corresponding non-adversarial *EigAug* results directly shows the impact of the adversarial samples. In general, the adversarial examples result in improved Dice scores for the 4 training sample cases for $r \in \{1, 2, 3\}$ and in similar Dice scores for the 8 training sample case. Surface distances appear comparable. The exception is the randomized *AdvEigAug$_{rand\_r}$* strategy which performs uniformly better on all measures than $EigAug_{rand\_r}$ and all its fixed r value variants.

Randomization of Deformation Magnitude, r. Table 1 shows that randomizing $r \sim U(1.5, 3)$ improves performance over using a fixed r in terms of Dice scores and surface distances. As we draw 5 augmented samples per training sample, randomization occurs for fixed r only via the computation of the covariance matrices based on the randomly drawn images. The random r strategy, however, introduces additional randomness through the random deformation magnitude and can therefore explore the deformation space more fully.

Online vs Offline. The online and offline variants of our approach show comparable performance, though we observe a slight improvement for the 95-th percentile surface distance for the online approach. This might be because the online approach allows improving upon earlier adversarial examples during training.

LDDMM vs Displacement Field Networks. The LDDMM parameterization assures a well-behaved diffeomorphic transformation regardless of the chosen deformation magnitude, r. In contrast the displacement field approach might create very strong deformations even to the point of creating foldings. Figure reffig:augmentedspssamples shows that augmented samples created via the LDDMM model tend to indeed be better behaved than the ones created via the displacement field model. While slight, this benefit appears to be born out by the validation results in Table tab:results which generally show higher Dice scores and lower surface distance measures for the LDDMM models, in particular for the lowest sample size ($n = 4$).

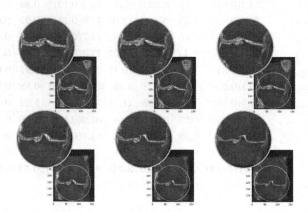

Fig. 2. Augmentation with LDDMM (top row) and displacement field network (bottom row). Left to right: augmentations with $r = 1$, $r = 2$, $r = 3$. LDDMM produces well behaved transformations even for large values of r.

5   Conclusion

We introduced an end-to-end data augmentation framework guided by a statistical deformation model. To the best of our knowledge, this is the first work that studies statistical deformation modelling in conjunction with data augmentation. We proposed two variants of our method, one using a fluid- and the other a displacement-based deformation model. We showed that such statistical deformation models allow an intuitive control over deformation magnitudes and that a combination of randomizing the deformation magnitude, online training, and a fluid-based deformation model performs best for our segmentation task.

Acknowledgements. Research reported in this publication was supported by the National Institutes of Health (NIH) under award numbers NIH 1R41MH118845 and NIH 1R01AR072013. The content is solely the responsibility of the authors and does not necessarily represent the official views of the NIH.

References

1. Akkus, Z., Galimzianova, A., Hoogi, A., Rubin, D.L., Erickson, B.J.: Deep learning for brain MRI segmentation: state of the art and future directions. J. Dig. Imaging **30**(4), 449–459 (2017)
2. Balakrishnan, G., Zhao, A., Sabuncu, M., Guttag, J., Dalca, A.V.: VoxelMorph: a learning framework for deformable medical image registration. IEEE TMI Trans. Med. Imaging **38**, 1788–1800 (2019)
3. Balakrishnan, G., Zhao, A., Sabuncu, M.R., Guttag, J., Dalca, A.V.: VoxelMorph: a learning framework for deformable medical image registration. IEEE Trans. Med. Imaging **38**(8), 1788–1800 (2019)
4. Beg, M.F., Miller, M.I., Trouvé, A., Younes, L.: Computing large deformation metric mappings via geodesic flows of diffeomorphisms. Int. J. Comput. Vision **61**(2), 139–157 (2005)
5. Chaitanya, K., Karani, N., Baumgartner, C.F., Becker, A., Donati, O., Konukoglu, E.: Semi-supervised and task-driven data augmentation. In: Chung, A.C.S., Gee, J.C., Yushkevich, P.A., Bao, S. (eds.) IPMI 2019. LNCS, vol. 11492, pp. 29–41. Springer, Cham (2019). https://doi.org/10.1007/978-3-030-20351-1_3
6. Çiçek, Ö., Abdulkadir, A., Lienkamp, S.S., Brox, T., Ronneberger, O.: 3D U-Net: learning dense volumetric segmentation from sparse annotation. In: Ourselin, S., Joskowicz, L., Sabuncu, M.R., Unal, G., Wells, W. (eds.) MICCAI 2016. LNCS, vol. 9901, pp. 424–432. Springer, Cham (2016). https://doi.org/10.1007/978-3-319-46723-8_49
7. Cootes, T.F., Taylor, C.J.: Statistical models of appearance for medical image analysis and computer vision. In: Medical Imaging 2001: Image Processing, vol. 4322, pp. 236–248. International Society for Optics and Photonics (2001)
8. Cootes, T.F., Taylor, C.J., Cooper, D.H., Graham, J.: Active shape models-their training and application. Comput. Vis. Image Underst. **61**(1), 38–59 (1995)
9. Dalca, A.V., Balakrishnan, G., Guttag, J., Sabuncu, M.R.: Unsupervised Learning for fast probabilistic diffeomorphic registration. In: Frangi, A.F., Schnabel, J.A., Davatzikos, C., Alberola-López, C., Fichtinger, G. (eds.) MICCAI 2018. LNCS, vol. 11070, pp. 729–738. Springer, Cham (2018). https://doi.org/10.1007/978-3-030-00928-1_82

10. Eaton-Rosen, Z., Bragman, F., Ourselin, S., Cardoso, M.J.: Improving data augmentation for medical image segmentation (2018)
11. Engstrom, L., Tran, B., Tsipras, D., Schmidt, L., Madry, A.: A rotation and a translation suffice: Fooling CNNs with simple transformations (2017). arXiv preprint arXiv:1712.02779
12. Goodfellow, I., et al.: Generative adversarial nets. In: Advances in Neural Information Processing Systems, pp. 2672–2680 (2014)
13. Goodfellow, I.J., Shlens, J., Szegedy, C.: Explaining and harnessing adversarial examples (2014). arXiv preprint arXiv:1412.6572
14. Hataya, R., Zdenek, J., Yoshizoe, K., Nakayama, H.: Faster AutoAugment: learning augmentation strategies using backpropagation (2019). arXiv preprint arXiv:1911.06987
15. Ho, D., Liang, E., Stoica, I., Abbeel, P., Chen, X.: Population based augmentation: Efficient learning of augmentation policy schedules (2019). arXiv preprint arXiv:1905.05393
16. Holden, M.: A review of geometric transformations for nonrigid body registration. IEEE Trans. Med. Imaging 27(1), 111–128 (2007)
17. Iglesias, J.E., Sabuncu, M.R.: Multi-atlas segmentation of biomedical images: a survey. Med. Image Anal. 24(1), 205–219 (2015)
18. Jendele, L., Skopek, O., Becker, A.S., Konukoglu, E.: Adversarial augmentation for enhancing classification of mammography images (2019). arXiv preprint arXiv:1902.07762
19. Kamnitsas, K., et al.: DeepMedic for brain tumor segmentation. In: Crimi, A., Menze, B., Maier, O., Reyes, M., Winzeck, S., Handels, H. (eds.) Brainlesion: Glioma, Multiple Sclerosis, Stroke and Traumatic Brain Injuries, BrainLes 2016. Lecture Notes in Computer Science, vol. 10154, pp. 138–149. Springer, Cham (2016). https://doi.org/10.1007/978-3-319-55524-9_14
20. Kamnitsas, K., Ledig, C., Newcombe, V.F., Simpson, J.P., Kane, A.D., Menon, D.K., Rueckert, D., Glocker, B.: Efficient multi-scale 3D CNN with fully connected CRF for accurate brain lesion segmentation. Med. Image Anal. 36, 61–78 (2017)
21. Kanbak, C., Moosavi-Dezfooli, S.M., Frossard, P.: Geometric robustness of deep networks: analysis and improvement. In: Proceedings of the IEEE Conference on Computer Vision and Pattern Recognition, pp. 4441–4449 (2018)
22. Kingma, D.P., Ba, J.: Adam: A method for stochastic optimization (2014). arXiv preprint arXiv:1412.6980
23. Moeskops, P., Viergever, M.A., Mendrik, A.M., De Vries, L.S., Benders, M.J., Išgum, I.: Automatic segmentation of MR brain images with a convolutional neural network. IEEE Trans. Med. Imaging 35(5), 1252–1261 (2016)
24. Paschali, M., et al.: Data augmentation with manifold exploring geometric transformations for increased performance and robustness (2019). arXiv preprint arXiv:1901.04420
25. Pereira, S., Pinto, A., Alves, V., Silva, C.A.: Brain tumor segmentation using convolutional neural networks in MRI images. IEEE Trans. Med. Imaging 35(5), 1240–1251 (2016)
26. Risser, L., Vialard, F.X., Wolz, R., Murgasova, M., Holm, D.D., Rueckert, D.: Simultaneous multi-scale registration using large deformation diffeomorphic metric mapping. IEEE Trans. Med. Imaging 30(10), 1746–1759 (2011)
27. Roth, H.R., et al.: DeepOrgan: multi-level deep convolutional networks for automated pancreas segmentation. In: Navab, N., Hornegger, J., Wells, W.M., Frangi, A.F. (eds.) MICCAI 2015. LNCS, vol. 9349, pp. 556–564. Springer, Cham (2015). https://doi.org/10.1007/978-3-319-24553-9_68

28. Rueckert, D., Frangi, A.F., Schnabel, J.A.: Automatic construction of 3-D statistical deformation models of the brain using nonrigid registration. IEEE Trans. Med. Imaging **22**(8), 1014–1025 (2003)
29. Rueckert, D., Sonoda, L.I., Hayes, C., Hill, D.L., Leach, M.O., Hawkes, D.J.: Nonrigid registration using free-form deformations: application to breast MR images. IEEE Trans. Med. Imaging **18**(8), 712–721 (1999)
30. Shaham, U., Yamada, Y., Negahban, S.: Understanding adversarial training: increasing local stability of supervised models through robust optimization. Neurocomputing **307**, 195–204 (2018)
31. Shen, Z., Han, X., Xu, Z., Niethammer, M.: Networks for joint affine and nonparametric image registration. In: Proceedings of the IEEE Conference on Computer Vision and Pattern Recognition, pp. 4224–4233 (2019)
32. Shen, Z., Vialard, F.X., Niethammer, M.: Region-specific diffeomorphic metric mapping (2019). arXiv preprint arXiv:1906.00139
33. Shin, H.C., et al.: Medical image synthesis for data augmentation and anonymization using generative adversarial networks. In: Gooya, A., Goksel, O., Oguz, I., Burgos, N. (eds.) SASHIMI 2018. LNCS, vol. 11037, pp. 1–11. Springer, Cham (2018). https://doi.org/10.1007/978-3-030-00536-8_1
34. Sotiras, A., Davatzikos, C., Paragios, N.: Deformable medical image registration: a survey. IEEE Trans. Med. Imaging **32**(7), 1153–1190 (2013)
35. Tajbakhsh, N., Jeyaseelan, L., Li, Q., Chiang, J., Wu, Z., Ding, X.: Embracing imperfect datasets: a review of deep learning solutions for medical image segmentation (2019). arXiv preprint arXiv:1908.10454
36. Vialard, F.X., Risser, L., Rueckert, D., Cotter, C.J.: Diffeomorphic 3D image registration via geodesic shooting using an efficient adjoint calculation. Int. J. Comput. Vis. **97**(2), 229–241 (2012)
37. Xiao, C., Zhu, J.Y., Li, B., He, W., Liu, M., Song, D.: Spatially transformed adversarial examples (2018). arXiv preprint arXiv:1801.02612
38. Xu, Z., Niethammer, M.: Deepatlas: joint semi-supervised learning of image registration and segmentation (2019). arXiv preprint arXiv:1904.08465
39. Xu, Z., Shen, Z., Niethammer, M.: Contextual additive networks to efficiently boost 3D image segmentations. In: Stoyanov, D., et al. (eds.) DLMIA/ML-CDS -2018. LNCS, vol. 11045, pp. 92–100. Springer, Cham (2018). https://doi.org/10.1007/978-3-030-00889-5_11
40. Yang, X., Kwitt, R., Styner, M., Niethammer, M.: Quicksilver: fast predictive image registration-a deep learning approach. NeuroImage **158**, 378–396 (2017)
41. Younes, L., Arrate, F., Miller, M.I.: Evolutions equations in computational anatomy. NeuroImage **45**(1), S40–S50 (2009)
42. Zhang, X., Wang, Q., Zhang, J., Zhong, Z.: Adversarial AutoAugment (2019). arXiv preprint arXiv:1912.11188
43. Zhao, A., Balakrishnan, G., Durand, F., Guttag, J.V., Dalca, A.V.: Data augmentation using learned transformations for one-shot medical image segmentation. In: Proceedings of the IEEE Conference on Computer Vision and Pattern Recognition, pp. 8543–8553 (2019)
44. Zhou, X.Y., Guo, Y., Shen, M., Yang, G.Z.: Artificial intelligence in surgery (2019). arXiv preprint arXiv:2001.00627

Neural Predictor for Neural Architecture Search

Wei Wen[1,2]([✉]), Hanxiao Liu[1], Yiran Chen[2], Hai Li[2], Gabriel Bender[1], and Pieter-Jan Kindermans[1]

[1] Google Brain, Mountain View, USA
[2] Duke University, Durham, USA
wei.wen@alumini.duke.edu

Abstract. Neural Architecture Search methods are effective but often use complex algorithms to come up with the best architecture. We propose an approach with three basic steps that is conceptually much simpler. First we train N random architectures to generate N (architecture, validation accuracy) pairs and use them to train a regression model that predicts accuracies for architectures. Next, we use this regression model to predict the validation accuracies of a large number of random architectures. Finally, we train the top-K predicted architectures and deploy the model with the best validation result. While this approach seems simple, it is more than 20× as sample efficient as Regularized Evolution on the NASBench-101 benchmark. On ImageNet, it approaches the efficiency of more complex and restrictive approaches based on weight sharing such as ProxylessNAS while being fully (embarrassingly) parallelizable and friendly to hyper-parameter tuning.

Keywords: Neural architecture search · Automated machine learning · Graph neural networks · NASBench-101 · Mobile models · ImageNet

1 Introduction

Early Neural Architecture Search (NAS) methods showed impressive results, allowing researchers to automatically find high-quality neural networks within human-defined search spaces [17,18,22,23]. However, these early methods required thousands of models to be trained from scratch to run a single search, making the methods prohibitively expensive for most practitioners. Thus, algorithms which can improve the *sample efficiency* are of high value. A second consideration when designing a NAS algorithm is *friendliness to hyper-parameter*

W. Wen—Work done as a Research Intern and Student Researcher in Google Brain.

Electronic supplementary material The online version of this chapter (https://doi.org/10.1007/978-3-030-58526-6_39) contains supplementary material, which is available to authorized users.

© Springer Nature Switzerland AG 2020
A. Vedaldi et al. (Eds.): ECCV 2020, LNCS 12374, pp. 660–676, 2020.
https://doi.org/10.1007/978-3-030-58526-6_39

tuning. In many existing approaches – such as those based on Reinforcement Learning (RL) [22] or Evolutionary Algorithms (EA) [17], trying out a new set of hyper-parameters for the search algorithm requires us to train and evaluate a new set of neural network from scratch. Ideally, we would be able to train a single set of neural networks, evaluate them once, and then use the results to try out many different hyper-parameter configurations for the search algorithm. A third design consideration of NAS is *full parallelizability* (or *embarassing paralleliz- ability)*: existing methods based on RL [22], EA [17] and Bayesian Optimization (BO) [6] are complex to implement, requiring complex coordination between tens or hundreds of workers when collecting reward/fitness/acquisitiothe n dur- ing a search. An fully parallelizable algorithm can avoid this coordination and accelerate the search when idle computing resources are available. Ideally, we pursue an algorithms with the above merits – *sample efficiency, friendliness to hyper-parameter tuning,* and *full parallelizability.*

To design a NAS algorithm possessing the three merits described above, we investigate how well it is possible to do using a combination of two tech- niques which are ubiquitous in the ML community: supervised learning and random sampling. We show that an algorithm that intelligently combines these two approaches can be surprisingly effective in practice. With an infinite com- pute budget, a very simple but naïve approach to architecture search would be to sample tons of random architectures, train and evaluate each one, and then select the architectures with the best validation set accuracies for deployment; this is a straightforward application of the ubiquitous random search heuris- tic. It is *friendly to hyper-parameter tuning* and is *fully parallelizable.* However, the *efficiency* (computational requirements) of this approach makes it infeasible in practice. For example, to exhaustively train and evaluate all of the 423, 624 architectures in the NASBench-101 [21], it would take roughly 25 years of TPU training time. Only a small number of companies and corporate research labs can afford this much compute, and it is far out of reach for most ML practitioners.

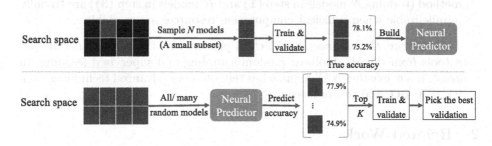

Fig. 1. Building (top) and applying (bottom) the Neural Predictor.

One way to alleviate this is to identify a small subset of promising models. If this is done with reasonably high precision (i.e., most models selected are indeed of high quality) then we can train and validate just this limited set of models to reliably select a good one for deployment. To achieve this, the proposed Neural Predictor uses the following steps to perform an architecture search:

(1) Build a predictor by training N random architectures to obtain N (architecture, validation accuracy) pairs. Use this data to train a regressor.

(2) Quality prediction using the regression model over a large set of random architectures. Select the K most promising architectures for final validation.

(3) Final validation of the top K architectures by training them. Then we select the architecture with the highest validation accuracy to deploy.

The workflow is illustrated in Fig. 1. In this setup, the first step is a traditional regression problem where we first generate a dataset of N samples to train on. The second step can be carried out efficiently because evaluating a model using the predictor is cheap and parallelizable. The third step is nothing more than traditional validation where we only evaluate a well curated set of K models. While the method outlined above might seem straightforward, this solution is surprisingly effective and satisfies the three goals discussed above:

- *Efficiency*: The Neural Predictor strongly outperforms random search on NASBench-101. It is also about 22.83 times as sample-efficient as Regularized Evolution – the best performing method in the NASBench-101 paper. The Neural Predictor can easily handle different search spaces. In addition to NASBench-101, we evaluated it on the ProxylessNAS [5] search space and found that the predicted architecture is as accurate as ProxylessNAS and clearly better than random search.
- *Friendliness to hyper-parameter tuning*: All hyper-parameters of the regression model are cross validated by the dataset collected just once in step **(1)**. The cost of tuning those hyper-parameters is small because the predictor model is small.
- *Full parallelizability*: The most computationally intensive components of the method (training N models in step **(1)** and K models in step **(3)**) are trivially parallelizable when sufficient computation resources are available.

Furthermore, the architecture selection process uses two of the most ubiquitous tools from the ML toolbox: random sampling and supervised learning. In contrast, many existing NAS approaches rely on more advanced techniques such as RL, EA, BO, and weight sharing.

2   Related Work

Neural Architecture Search was proposed to automate the design of neural networks, by searching models in a design space using techniques such as RL [22], EA [18] or BO [6,9]. A clear limitation of the early approaches is their computation efficiency. Thus, recent methods often focus on efficient NAS by using weight sharing [2,4,5,14,16]. As aforementioned, when comparing with previous works, Neural Predictor is friendly to hyper-parameter tuning, conceptually simple and fully parallelizable. Moreover, our Neural Predictor can potentially work as a surrogate model to accelerate accuracy acquisition of architectures during the search in RL, EA and BO, or during the candidate architecture evaluation in weight sharing approaches [2].

The idea of predictive models of the accuracy, which we use, has been explored in prior works. In [7] an LSTM was used to generate a feature representation of an architecture, which was subsequently used to predict the quality. The one-shot approach by Bender et al. [2] used a weight sharing model to predict the accuracy of an individually trained architecture. Baker et al. [1] used predictive models to perform early stopping to speed up architecture and hyperparameter optimization. NAO [15] used both a learned representation and a predictor to search for high quality architecture. PNAS [13] progressively trained a predictor to accelerate the search. A key difference between PNAS and Neural Predictor is that in PNAS the predictor is only a small component in a large traditional NAS system. In PNAS, the predictor and the models are trained over time as the architectures become more complex. Because of this, the PNAS approach cannot be completely parallelized. Another popular approach combined a predictor with Bayesian Optimization [6]. Unlike above methods, which used ν-SVR/random forests [1], multi-layer perceptrons [13], LSTMs [7,15] or Gaussian processes [6], we use a Graph Convolutional Network (GCN) [10] for our regression model. GCNs are naturally permutation-invariant, capturing the intuition that an architecture under different node permutations should have the same predicted accuracy. Furthermore, we show strong results can be achieved without the use of advanced techniques such as RL, EA, BO or NAO [15], enabling our merits of *friendliness to hyper-parameter tuning* and *full parallelizability*. Finally, because the effectiveness of NAS has been questioned [12], we include random baseline to show that the search spaces used in this work are meaningful and that while the proposed approach is simple, it is clearly better than a random approach.

While preparing the final version of this paper, we found a concurrent work [19] which also used a GCN to predict accuracy. However, a *node* in their graph is a model in the search space; as the search space is usually huge, their method is hard to scale up. In our design, a *graph* represents a model and the graph size is approximately proportional to the number of layers in a model. Therefore, our Neural Predictor is able to scale to huge search spaces, such as the ProxylessNAS ImageNet search space with the size of 6.64×10^{17}.

3   Neural Predictor

The core idea behind the Neural Predictor is that carrying out the actual training and validation process is the most reliable way to find the best model. The goal of the Neural Predictor is to provide us with a curated list of promising models for final validation prior to deployment. The entire Neural Predictor process is outlined below.

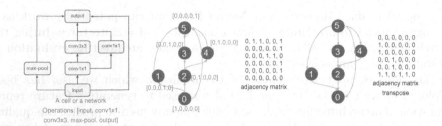

Fig. 2. An illustration of graph and node representations. Left: A neural network architecture with 5 candidate operations per node. Each node is represented by a one-hot code of its operation. The one-hot codes are inputs of a bidirectional GCN, which takes into account both the original adjacency matrix (middle) and its transpose (right).

Step 1: Build the predictor using N samples. We train N models to obtain a small dataset of (architecture, validation accuracy) pairs. The dataset is then used to train a regression model that maps an architecture to a predicted validation accuracy.

Step 2: Quality prediction. Because architecture evaluation using the learned predictor is efficient and trivially parallelizable, we use it to rapidly predict the accuracies of a large number of random architectures. We then select the top K predicted architectures for final validation.

Step 3: Final validation on K samples. We train and validate the top K models in the traditional way. This allows us to select the best model based on the actual validation accuracy. Even if our predictor is somewhat noisy, this step allows us to use a more reliable measurement to select our final architecture for deployment.

Training $N+K$ models is by far the most computationally expensive part of the Neural Predictor. If we assume a constant compute budget, N and K are key hyper-parameters which need to be set; we will discuss this next. Note that the two most expensive steps (Step 1 and Step 3) are both fully parallelizable.

3.1 Hyper-parameters in the Workflow

Hyper-parameters for model training are always needed if we train a single model in the search space. In this respect the Neural Predictor is no different from other methods. We found that using the same hyper-parameters for all models we train is an effective strategy, and was also used in NASBench-101.

Trade-off Between N and K for a Fixed Budget: For a given compute budget, the total number of architectures we train and evaluate, $N + K$, must remain fixed. However, we can trade off between N (the number of samples used to train the predictor) and K (the number of samples used for final evaluation). If N is too small, the predictor's outputs will be very noisy and will not provide a reliable signal for the search. As we increase N, the predictor will become more accurate; however, increasing N requires us to decrease K. If K is small, the predictor must very reliably identify high-quality models from the search space. As K increases, we will be able to tolerate larger noise in the predictor, and the predictor's ability to precisely rank architectures in the search space will become less important. Because it is difficult to theoretically predict the optimal trade-off between N and K, we will investigate this in the experimental setting.

To find a lower bound on N we can start with a small number of samples, and iteratively increase N until we observe a good cross-validation accuracy. This means that contrast to some other methods such as Regularized Evolution [17], ENAS [16], NASNet [23], ProxylessNAS [5], there is no need to repeat the entire search experiment in order to tune this hyper-parameter. The same applies to the hyper-parameters and the architecture of the Neural Predictor itself.

The hyper-parameters of the Neural Predictor can be optimized by cross-validation using the N training samples. The cost of training a neural predictor including the hyper-parameter tuning is negligible compared to the cost of training image models. It takes 25 seconds to train a neural predictor on $N = 172$ samples from NASBench-101. The mean (resp. median) training time of a CIFAR-10 model in NASBench-101 is 32 (resp. 26) minutes. At the cost of training two CIFAR-10 models (about one hour), we could try 144 hyper-parameter configurations for the neural predictor. In contrast, RL or EA require us to train more models in order to try out a new hyper-parameter configuration.

3.2   Modeling by Graph Convolutional Networks

We tried many options for the architecture of the predictor. We find Graph Convolutional Networks (GCNs) work best. Due to space constraints we will limit our discussion to GCNs. A comparison against other regression models is in the supplementary material. Graph Convolutional Networks (GCNs) are good at learning representations for graph-structured data [10,20] such as a neural network architecture. The graph convolutional model we use is based on [10], which assumes undirected graphs. We will modify their approach to handle neural architectures represented as directed graphs.

We start with a D_0-dimensional representation for each of the I nodes in the graph, giving us an initial feature vector $V_0 \in \mathbb{R}^{I \times D_0}$. For each node we use a one-hot vector representing the selected operation. An example for NASBench-101 is shown in Fig. 2. The node representation is iteratively updated using Graph Convolutional Layers. Each layer uses an adjacency matrix $A \in \mathbb{R}^{I \times I}$ based on the node connectivity and a trainable weight matrix $W_l \in \mathbb{R}^{D_l \times D_{l+1}}$:

$$V_{l+1} = \text{ReLU}\left(A V_l W_l\right). \tag{1}$$

Following previous work [10], we add an identity matrix to A (corresponding to self cycles) and normalize it using the node degree.

The original GCNs [10] assume undirected graphs. When applied to a directed acyclic graph, the directed adjacency matrix allows information to flow only in a single direction. To make information flow both ways, we always use the average of two GCN layers: one where we use $\mathbf{A}$ to propagate information in the forward directions and another where we use $\mathbf{A}^T$ to reverse the direction:

$$V_{l+1} = \frac{1}{2}\text{ReLU}\left(\mathbf{A}V_l W_l^+\right) + \frac{1}{2}\text{ReLU}\left(\mathbf{A}^T V_l W_l^-\right).$$

Figure 2 shows an example of how the adjacency matrices are constructed (without normalization or self-cycles).

GCNs are able to learn high quality node representations by stacking multiple of these layers together. Since we are more interested in the accuracy of the overall network (a global property), we take the average over node representations from the final graph convolutional layer and attach one or more fully connected layers to obtain the desired output. Details are provided in the supplementary.

4 Experiments

In this section we will discuss two studies. First we will analyze the Neural Predictor's behavior in the controlled environment from NASBench-101 [21]. Afterwards we will use our approach to search for high quality mobile models in the ProxylessNAS search space [5].

4.1 NASBench-101

NASBench-101 [21] is a dataset used to benchmark NAS algorithms. The goal is to come up with a high quality architecture as efficiently as possible. The dataset has the following properties: **(1)** train time, validation and test accuracy are provided for all 423,624 models in the search space; **(2)** each model was trained and evaluated three times. This allows us to look at the variance across runs; **(3)** all models were trained in a consistent manner, preventing biases from the implementation from skewing results; **(4)** NASBench-101 recommends using only validation accuracies during a search, and reserving test accuracies for the final report; this is important to avoid overfitting.

NASBench-101 uses a cell-based NAS [23] on CIFAR-10 [11]. Each cell is a Directed Acyclic Graph (DAG) with up to 7 nodes. There is an input node, an output node and up to 5 interior nodes. Each interior node can be a 1×1 convolution (`conv1x1`), 3×3 convolution (`conv3x3`) or max-pooling op (`max-pool`). One example is shown in Fig. 2 (left). In each experiment, we use the validation accuracy from a *single run*[1] as a search signal. The single run is uniformly sampled from these three records. This simulates training the architecture once. Test accuracy is only used for reporting the accuracy on the model that was selected at the end of a search. For that model we use the *mean* test accuracy over three runs as the "ground truth" measure of accuracy.

---

[1] In the training dataset of our Neural Predictor, this means that each model's accuracy label is sampled once and fixed across all epochs.

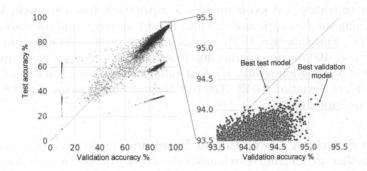

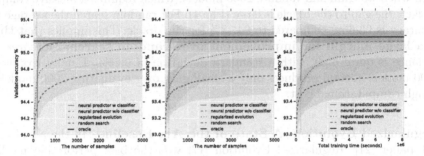

Fig. 3. (Left) Validation vs. test accuracy in NASBench-101. (Right) Zoomed in on the highly accurate region. Each model (point) is the validation accuracy from a single training run. Test accuracies are averaged over three runs. This plot demonstrates that even knowing the validation accuracy of every possible model is not sufficient to predict which model will perform best on the *test* set.

Fig. 4. Comparison of search efficiency among oracle, random search, Regularized Evolution and our Neural Predictor (with and without a two stage regressor). All experiments are averaged over 600 runs. The x-axis represents the total compute budget $N + K$. The vertical dotted line is at $N = 172$ and represents the number of samples (or total training time) used to build our Neural Predictor. From this line on we start from $K = 1$ and increase it as we use more architectures for final validation. The shaded region indicates standard deviation of each search method.

Oracle: An Upper Bound Baseline. Under the assumption of infinite compute, a traditional machine learning approach would be to train and validate all possible architectures to select the best one. We refer to this baseline as the "oracle" method. Figure 3 plots the validation versus the test accuracy for all models. The model that the oracle method would select based on the validation accuracy of 95.15% has a test set accuracy of 94.08%. This means that **the oracle does not select the model with the highest test set accuracy.** The global optimum on the test set is 94.32%. However, since this model cannot be found using extensive validation, one should not expect this model to be found using any NAS algorithm. A more reasonable goal is to reliably select a model that has similar quality to the one selected by the oracle. Furthermore, it is important to realize that **even an oracle approach has variance.** We have three training runs for each model, which allows us to run multiple variations of the "oracle". This simulates the impact of random variations on the final result. Averaged over 100 oracle experiments, where in each experiment we randomly select one of 3 validation results, the best validation accuracy has a mean 95.13% and a standard deviation 0.03%. The test accuracy has a mean of 94.18% and a standard deviation 0.07%.

Random Search: A Lower Bound Baseline. Recently, Li *et al.* [12] questioned whether architecture search methods actually outperform random search. Because this depends heavily on the search space and Li *et al.* [12] did not investigate the NASBench-101 search space, we need to check this ourselves. Therefore we replicate the random baseline from NASBench-101 by sampling architectures without replacement. After training, we pick the architecture with the highest validation accuracy and report its result on the test set. Here we observe that even when we train and validate 2000 models, which requires a massive compute budget, the gap to the oracle is large (Fig. 4). For random search the average test accuracy is 93.66% compared to 94.18% for the oracle. This implies that **there is a large margin for improvement over random search.** Moreover, the variance is quite high, with a standard deviation of 0.25%. Finally, evaluating 5000 models in total produces only a small gain over evaluating 2000 models at a high computational cost.

Regularized Evolution: A State of the Art Baseline. In the NASBench-101 paper [21], Regularized Evolution [17] was the best performing method. We replicated those experiments using the open source code and their hyperparameter settings (available in the supplementary material). **Regularized evolution is significantly better than random** as shown in Fig. 4. However even after 2000 models are trained, it is still clearly worse than the oracle (on average) with an accuracy of 93.97% and a standard deviation of 0.26%.

Neural Predictor. Having set our baselines, we now describe the precise Neural Predictor setup and evaluation. The graph representation of a model is a DAG with upto 7 nodes. Each node is represented by an one-hot code of "[input, conv1x1, conv3x3, max-pool, output]". The GCN has three Graph Convolutional layers with the constant node representation size D and one hidden fully-connected layer with output size 128. Finally, the accuracy we need to predict is limited to a finite range. While it is not that common for regression, we can force the network to make predictions in this finite range by using a sigmoid at the output layer. Specifically, we use a sigmoid function that is scaled and shifted such that its output accuracy is always between 10% and 100%.

All hyper-parameters for the predictor are first optimized using cross-validation where $\frac{1}{3}N$ samples were used for validation. After setting the hyper-parameters, we use all N samples to train the final predictor. At this point we heuristically increase the node representation size D of the predictor such that the number of parameters in the Neural Predictor is also 1.5× as large. Specific N and D values and other training details are in the supplementary material. The models selected for final evaluation are always trained in the same way, regardless of the method (baseline or predictor) that selected the model, to ensure fairness.

A two Stage Predictor. Looking at a small dataset of $N = 172$ models in Fig. 6 (left) during cross-validation,[2] we realized that for NASBench-101 a two stage predictor is needed. The NasBench-101 dataset contains many models that are not stable during training or perform very poorly (e.g. a model with only pooling operations). The two stage predictor, shown in Fig. 5, filters obviously bad models first by predicting whether each model will achieve an accuracy above 91%. This allows the the second stage to focus on a narrower accuracy range. It makes training the regression model easier, which in turn makes it more reliable. Both stages share the same GCN architecture but have different output layers. A classifier trained on these $N = 172$ models has a low False Negative Rate as shown in Fig. 6 (right). This implies that the classifier will filter out very few actually good models.

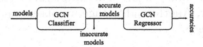

Fig. 5. Neural Predictor on NASBench-101. It is a cascade of a classifier and a regressor. The classifier filters out inaccurate models and the regressor predicts accuracies of accurate models.

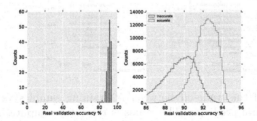

Fig. 6. The classifier filtering out inaccurate models in NASBench. 172 models (left) are sampled to build the classifier, which is tested by unseen data (right).

---

[2] In our implementation, we split the NASBench-101 dataset to 10,000 shards and each shard has 43 samples. The $N = 172$ comes from a random 4 shards.

The Two Stage Approach Improves the Results but is not Required. If we only use a single stage, the MSE for the validation accuracy is 1.95 (averaged over 10 random splits). By introducing the filtering stage this reduces to 0.66. In Fig. 4 we observe that even without the filtering stage the predictor clearly outperforms the random search baseline and regularized evolution. Therefore the two stage approach should be seen as a non-essential fine-tuning of the proposed method. Using only a single stage is more elegant, but adding the second stage gives additional performance benefits.

Results Using N=172 (or 0.04% of the search space) for training are shown in Figure 4. We used $N = 172$ models to train the predictor. Then we vary K, the number of architectures with the highest predicted accuracies to be trained and validated to select the best one. Therefore, "the number of samples" in the figure equals $N + K$ for Neural Predictor. In Fig. 4 (left), our Neural Predictor significantly outperforms Regularized Evolution in terms of sample efficiency. The mean validation accuracy is comparable to that of the oracle after about 1000 samples. The sample efficiency in validation accuracy transfers well to test accuracy in terms of both the total number of trained models in Fig. 4 (middle) and wall-clock time in Fig. 4 (right). After 5000 samples, Regularized Evolution reaches validation and test accuracies of 95.06% and 94.04% respectively; our predictor can reach the same validation accuracy 12.40× faster and the same test accuracy 22.83× faster. Another advantage we observe is that Neural Predictor has small search variance.

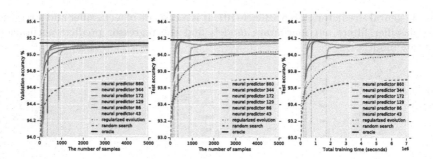

Fig. 7. Analysis of the trade-off between N training samples vs K final validation samples in the neural predictor. The x-axis is the total compute budget $N + K$. The vertical lines indicate different choices for N – the number of training samples and the point where we start validating K models. All experiments are averaged over 600 runs.

N vs K and Ablation Study. We next consider the problem of choosing an optimal value of N when the total number of models we're permitted to train, $N + K$, is fixed. Figure 7 summarizes our study on N. A Neural Predictor underperforms with a very small N (43 or 86), as it cannot predict accurately enough which models are interesting to evaluate. Finally, we consider the case where N is large (e.g., $N = 860$) but K is small. In this case we clearly see that the increase in quality of the GCN cannot compensate for the decrease in evaluation budget. Note that in Fig. 3 we have shown that some models are higher ranked according to validation accuracies than test accuracies. This can cause the test accuracy to degrade as we increase K in Fig. 7.

4.2 ImageNet Experiments

While the NASBench-101 dataset allows us to look at the behavior in a well controlled environment, it does not allow us to evaluate whether the approach generalizes to larger scale problems. It also does not address the issue of finding high-quality inference time constrained models. Therefore, to demonstrate that our approach is more widely applicable we look at this use case in our second set of experiments on ImageNet [8] with the ProxylessNAS search space [5]. We will compare our results to a random baseline and our own reproduction of the ProxylessNAS search. In this search space, the goal is to find a good model that has an inference time between 75 ms and 85 ms on a Pixel-1 phone.

Search Space. The ProxylessNAS search space does not have the cell-based structure from NASBench-101; it instead requires independent choices for the individual layers. There is a visualization of the search space in the supplementary material. The layers are divided up into blocks, each of which has its own fixed resolution and fixed number of output filters. We search over which layers to skip and what operations to use in each layer. (The first layer of a block is always present.) There are approximately 6.64×10^{17} models in the search space.

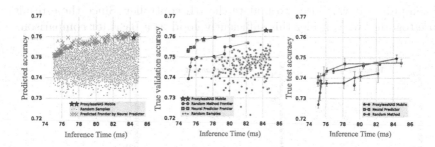

Fig. 8. Comparison between random search, ProxylessNAS and Neural Predictor. **Left:** Frontier models predicted by the Neural Predictor. **Middle:** Validation accuracy of each found frontier models. **Right:** Test accuracy of each found frontier model. (Each test accuracy is averaged over 5 training runs under different initial weight values. Error bars with 95% confidence interval are also plotted in the figure.)

672   W. Wen et al.

Baselines. Because this search space is so large, we cannot generate the oracle baseline; we must instead rely on the random search and ProxylessNAS reimplementations we discuss next.

The random search baseline samples 256 models with inference times between 75ms and 85ms. All these models are trained for 90 epochs. We then look at which models are Pareto optimal (i.e. have good trade-offs between inference time and validation quality). All Pareto optimal models were then trained for 360 epochs to be evaluated on the test set. The results are shown in Fig. 8. Implementation details are in the supplementary.

ProxylessNAS [5] is an efficient NAS algorithm based on weight sharing and RL. It trains a large neural network where different paths can be switched on or off to mimic specific architectures in the search space. Our baselines were obtained using reproductions of their search algorithm and RL reward function, as well as one of their best searched models. These reproductions come from the TuNAS [3] codebase. Our reproduction of their model (with slightly improved training hyper-parameters) achieves 74.9% accuracy, compared with 74.6% in the original paper. Across 5 independent runs, the *search algorithm* finds models with an average accuracy of 75.0% with a variance of 0.1%; latencies are comparable to that of the ProxylessNAS model. These results are near state-of-the-art for mobile CPUs. The RL controllers for ProxylessNAS and TuNAS assume that for a single network all decisions can be made independently (i.e. the probability distribution over architectures is factorized). To train the shared weights of the large model, TuNAS repeatedly (i) samples an architecture from the RL controller and (ii) trains it for a single step. To update the RL controller, another batch is sampled. This time the batch is evaluated on the validation set, and this result is used in combination with additional information (i.e. the latency) to compute a reward used to update the RL controller. Since these results are close together we consider this sufficiently good as a basis for comparison.

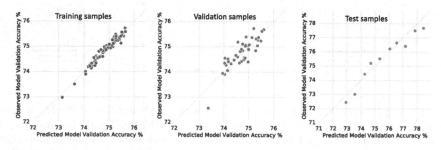

Fig. 9. The performance of Neural Predictor on training, validation and test samples.

Neural Predictor. Overall, we use the same basic pipeline as in the NASBench-101 experiments. However, because the models in the ProxylessNAS search space are much more stable than those in NASBench-101, we only need a single stage predictor. To transfer from NASBench-101 to the ProxylessNAS search space, all we have to do is to modify the graph and its node representations. The ProxylessNAS search space is just a linear graph, and the node representation at the input is nothing more than a one hot vector with length 7. This allows us to describe all architectures.

Training and Validating the Neural Predictor. To build the neural predictor we randomly sample 119 models; 79 samples are used for training and 40 samples are for validation. To find the GCN's hyperparameters we average validation MSE scores over 10 random training and validation splits. Based on this we select a GCN with 18 Graph Convolutional layers with node representation size 96, and with two fully-connected layers with hidden sizes 512 and 128 on top of the mean node representations in the last Graph Convolutional layer. After all hyper-parameters are finalized, we train our GCN with all 119 samples.

Our validation also showed that for ImageNet experiments, no classifier is needed to filter inaccurate models. This is because model accuracies lie within a relatively small range as shown in Fig. 9 (left and middle). Our final settings for the Neural Predictor achieved on MSE 0.109 ± 0.028 averaged over 10 validation runs. Figure 9 shows an example of the correlation between true accuracy and predicted accuracy for training samples (left) and validation samples (middle). For validation samples, the Kendall rank correlation coefficient is 0.649 and the R^2 score is 0.648895.

Looking at the Predictive Performance of the Predictor. In Fig. 9 (right), we test the generalization of our Neural Predictor to unseen test architectures. We first randomly sample 100,000 models from the ProxylessNAS search space without inference time constraint and predict their accuracies. We then pick the model with minimum predicted accuracy (72.94%), the model with maximum predicted accuracy (78.45%), and 8 additional models which are evenly spaced between those two endpoints. We train those 10 models to obtain their true accuracies. In Fig. 9 (right), the Kendall rank correlation coefficient is 0.956 and the R^2 score is 0.929. More interesting, although our training dataset never observed models with accuracies higher than 76%, our Neural Predictor can still successfully predict the 5 models with accuracies higher than 76%. This demonstrates the generalization of our Neural Predictor to unseen data.

Finding High Quality Mobile Sized Models. We now use the predictor to select frontier models with good trade-offs between accuracies and inference times. We randomly sample $N = 112,000$ models with inference times between 75ms and 85ms, and predict their accuracies as shown in Fig. 8 (left). As a sanity check, we also predict the quality of the ProxylessNAS model. The predicted validation accuracy is 76.0% and close to its true accuracy 76.3%.

The next step is selecting K Pareto optimal models. However, because the predictor can make mistakes, we need a soft version of Pareto optimality. To do so, we sort the models based on increasing inference time. In the regular definition, a model is Pareto optimal if no faster model has higher quality. In our setup, we define a model as"soft-Pareto optimal" when the predicted accuracy is higher than the minimum of the previous J models. In our experiments we set $J = 6$. This leaves us with 137 promising models in green in Fig. 8 (left). All $K = 137$ models are then trained and validated. This allows us to obtain a traditional Pareto frontier as shown in Fig. 8 (middle). The architectures of those true frontier models are included in the supplementary material.

The Neural Predictor Outperforms the Random Baseline and is Comparable to ProxylessNAS. Recall that the random baseline trained 256 models. This is the same number of models we trained in total for our method ($N = 119$ models for training the predictor and $K = 137$ models selected for final validation). For test accuracy comparisons, we train the frontier models in Fig. 8 (middle) for 360 epochs. The results are shown in Fig. 8 (right). Now we observe that the gap between the Pareto frontier of the neural predictor and the random baseline is stable on unseen data. Note that, unlike Neural Predictor which obtained frontier models in a *single* search, to obtain a frontier for ProxylessNAS, one should run *multiple* searches to reduce/increase the inference time. We opt to reduce filters in each layer to 0.92× for a faster model. The results show that the Neural Predictor and ProxylessNAS perform comparably.

The Resource Cost of the Neural Predictor vs. ProxylessNAS. Directly comparing the resource cost of the Neural Predictor versus ProxylessNAS is difficult. Training all $N + K = 256$ models for the entire Neural Predictor experiment took 47.5 times as much compute as a single ProxylessNAS search. However, in practice the gap is actually much smaller because optimizing the hyper-parameters of the Neural Predictor has negligible cost. Trying out a new hyper-parameter configuration for ProxylessNAS requires a full search. In our experiments we needed to run 7 searches to fine-tune the hyper-parameters for ProxylessNAS when we made a modification to the search space. This makes the Neural Predictor at most 7 times as expensive as ProxylessNAS. On top of that, the Neural Predictor is more effective at targeting different latency targets than ProxylessNAS, which needs a search per target. This could reduce the gap even more. Finally, in the ideal case we can parallelize model training for the Neural Predictor ($N = 119$ models for training in parallel followed by $K = 136$ for validation in parallel). This would finish in half the time of a ProxylessNAS run. Based on the analysis above, we believe that the Neural Predictor and ProxylessNAS are complementary. The method of choice will depend on the effort on tuning hyper-parameters, the complexity of implementing the search space (with weight sharing) and the available resources. The biggest advantage of our approach, and the most remarkable result to us, is how effective the Neural Predictor is given the simplicity of the method.

Acknowledgements. We would like to thank Chris Ying, Ken Caluwaerts, Esteban Real, Jon Shlens and Quoc Le for valuable input and discussions.

References

1. Baker, B., Gupta, O., Raskar, R., Naik, N.: Accelerating neural architecture search using performance prediction (2017). arXiv preprint arXiv:1705.10823
2. Bender, G., Kindermans, P.J., Zoph, B., Vasudevan, V., Le, Q.: Understanding and simplifying one-shot architecture search. In: International Conference on Machine Learning, pp. 549–558 (2018)
3. Bender, G., et al.: Can weight sharing outperform random architecture search? an investigation with tunas. In: Proceedings of the IEEE/CVF Conference on Computer Vision and Pattern Recognition, pp. 14323–14332 (2020)
4. Brock, A., Lim, T., Ritchie, J.M., Weston, N.: Smash: one-shot model architecture search through hypernetworks (2017). arXiv preprint arXiv:1708.05344
5. Cai, H., Zhu, L., Han, S.: Proxylessnas: direct neural architecture search on target task and hardware (2018). arXiv preprint arXiv:1812.00332
6. Dai, X., et al.: Chamnet: towards efficient network design through platform-aware model adaptation. In: Proceedings of the IEEE Conference on Computer Vision and Pattern Recognition, pp. 11398–11407 (2019)
7. Deng, B., Yan, J., Lin, D.: Peephole: predicting network performance before training (2017). arXiv preprint arXiv:1712.03351
8. Deng, J., Dong, W., Socher, R., Li, L.J., Li, K., Fei-Fei, L.: Imagenet: a large-scale hierarchical image database. In: 2009 IEEE Conference on Computer Vision and Pattern Recognition, pp. 248–255. IEEE (2009)
9. Kandasamy, K., Neiswanger, W., Schneider, J., Poczos, B., Xing, E.P.: Neural architecture search with bayesian optimisation and optimal transport. In: Advances in Neural Information Processing Systems, pp. 2016–2025 (2018)
10. Kipf, T.N., Welling, M.: Semi-supervised classification with graph convolutional networks (2016). arXiv preprint arXiv:1609.02907
11. Krizhevsky, A., Hinton, G., et al.: Learning multiple layers of features from tiny images. Technical report, Citeseer (2009)
12. Li, L., Talwalkar, A.: Random search and reproducibility for neural architecture search (2019). arXiv preprint arXiv:1902.07638
13. Liu, C., et al.: Progressive neural architecture search. In: Proceedings of the European Conference on Computer Vision (ECCV), pp. 19–34 (2018)
14. Liu, H., Simonyan, K., Yang, Y.: Darts: differentiable architecture search (2018). arXiv preprint arXiv:1806.09055
15. Luo, R., Tian, F., Qin, T., Chen, E., Liu, T.Y.: Neural architecture optimization. In: Advances in Neural Information Processing Systems, pp. 7816–7827 (2018)
16. Pham, H., Guan, M.Y., Zoph, B., Le, Q.V., Dean, J.: Efficient neural architecture search via parameter sharing (2018). arXiv preprint arXiv:1802.03268
17. Real, E., Aggarwal, A., Huang, Y., Le, Q.V.: Regularized evolution for image classifier architecture search. Proceedings of the AAAI Conference on Artificial Intelligence, vol. 33, pp. 4780–4789 (2019)
18. Real, E., et al.: Large-scale evolution of image classifiers. In: Proceedings of the 34th International Conference on Machine Learning, vol. 70, pp. 2902–2911. JMLR. org (2017)
19. Tang, Y., et al.: A semi-supervised assessor of neural architectures. In: IEEE/CVF Conference on Computer Vision and Pattern Recognition (CVPR) (2020)

20. Veličković, P., Cucurull, G., Casanova, A., Romero, A., Lio, P., Bengio, Y.: Graph attention networks (2017). arXiv preprint arXiv:1710.10903
21. Ying, C., Klein, A., Real, E., Christiansen, E., Murphy, K., Hutter, F.: Nas-bench-101: towards reproducible neural architecture search (2019). arXiv preprint arXiv:1902.09635
22. Zoph, B., Le, Q.V.: Neural architecture search with reinforcement learning (2016). arXiv preprint arXiv:1611.01578
23. Zoph, B., Vasudevan, V., Shlens, J., Le, Q.V.: Learning transferable architectures for scalable image recognition. In: Proceedings of the IEEE Conference on Computer Vision and Pattern Recognition, pp. 8697–8710 (2018)

Learning Permutation Invariant Representations Using Memory Networks

Shivam Kalra[1], Mohammed Adnan[1,2], Graham Taylor[2,3],
and H. R. Tizhoosh[1,2(✉)]

[1] Kimia Lab, University of Waterloo, Waterloo, Canada
{shivam.kalra,m7adnan,tizhoosh}@uwaterloo.ca
[2] Vector Institute for Artificial Intelligence, MaRS Centre, Toronto, Canada
[3] School of Engineering, University of Guelph, Guelph, Canada
gwtaylor@uoquelph.ca

Abstract. Many real-world tasks such as classification of digital histopathology images and 3D object detection involve learning from a set of instances. In these cases, only a group of instances or a set, collectively, contains meaningful information and therefore only the sets have labels, and not individual data instances. In this work, we present a permutation invariant neural network called *Memory-based Exchangeable Model (MEM)* for learning universal set functions. The MEM model consists of memory units that embed an input sequence to high-level features enabling it to learn inter-dependencies among instances through a self-attention mechanism. We evaluated the learning ability of MEM on various toy datasets, point cloud classification, and classification of whole slide images (WSIs) into two subtypes of the lung cancer—Lung Adenocarcinoma, and Lung Squamous Cell Carcinoma. We systematically extracted patches from WSIs of the lung, downloaded from The Cancer Genome Atlas (TCGA) dataset, the largest public repository of WSIs, achieving a competitive accuracy of 84.84% for classification of two subtypes of lung cancer. The results on other datasets are promising as well, and demonstrate the efficacy of our model.

Keywords: Permutation invariant models · Multi Instance Learning · Whole slide image classification · Medical images

1 Introduction

Deep artificial neural networks have achieved impressive performance for representation learning tasks. The majority of these deep architectures take a single instance as an input. Recurrent Neural Networks (RNNs) are a popular approach to learn representations from sequential ordered instances. However, the lack of

---

S. Kalra and M. Adnan—Authors have equal contribution.

© Springer Nature Switzerland AG 2020
A. Vedaldi et al. (Eds.): ECCV 2020, LNCS 12374, pp. 677–693, 2020.
https://doi.org/10.1007/978-3-030-58526-6_40

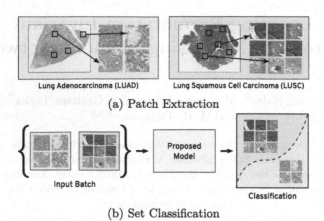

(a) Patch Extraction

(b) Set Classification

Fig. 1. An exemplar application of learning permutation invariant representation for disease classification of Whole-Slide Images (WSIs). (a) A set of patches are extracted from each WSI of patients with lung cancer. (b) The sets of patches are fed to the proposed model for classification of the sub-type of lung cancer—LUAD versus LUSC. The model classifies on a per set basis. This form of learning is known as Multi Instance Learning (MIL).

permutation invariance renders RNNs ineffective for exchangeable or unordered sequences. We often need to learn representations of unordered sequential data, or exchangeable sequences in many practical scenarios such as Multiple Instance Learning (MIL). In the MIL scenario, a label is associated with a set, instead of a single data instance. One of the application of MIL is classification of high resolution histopathology images, called whole slide images (WSIs). Each WSI is a gigapixel image with size $\approx$ 50,000 $\times$ 50,000 pixels. The labels are generally associated with the entire WSI instead of patch, region, or pixel level. MIL algorithms can be used to learn representations of these WSIs by disassembling them into multiple representative patches [1,14,15].

In this paper, we propose a novel architecture for exchangeable sequences incorporating attention over the instances to learn inter-dependencies. We use the results from Deep Sets [39] to construct a permutation invariant model for learning set representations. Our main contribution is a sequence-to-sequence permutation invariant layer called **Memory Block**. The proposed model uses a series of connected memory block layers, to model complex dependencies within an input set using a self attention mechanism. We validate our model using a toy datasets and two real-world applications. The real world applications include, i) point cloud classification, and ii) classification of WSI into two sub-type of lung cancers—Lung Adenocarcinoma (LUAD)/Lung Squamous Cell Carcinoma (LUSC) (see Fig. 1).

The paper is structured as follows: Sect. 2 discusses related and recent works. We cover the mathematical concepts for exchangeable models in Sect. 3. We explain our approach and experimental results in Sect. 4 and Sect. 5.

2   Related Work

In statistics, exchangeability has been long studied. de Finetti studied exchangeable random variables and showed that sequence of infinite exchangeable random variables can be factorised to independent and identically distributed mixtures conditioned on some parameter θ. Bayesian sets [8] introduced a method to model exchangeable sequences of binary random variables by analytically computing the integrals in de Finetti's theorem. Orbanz et al. [22] used de Finetti's theorem for Bayesian modelling of graphs, matrices, and other data that can be modeled by random structures. Considerable work has also been done on partially exchangeable random variables [2].

Symmetry in neural networks was first proposed by Shawe et al. [26] under the name *Symmetry Network*. They proposed that invariance can be achieved by weight-preserving automorphisms of a neural network. Ravanbaksh et al. proposed a similar method for equivariance network through parameter sharing [24]. Bloem Reddy et al. [4] studied the concept of symmetry and exchangeability for neural networks in detail and established similarity between functional and probabilistic symmetry, and obtained generative functional representations of joint and conditional probability distributions that are invariant or equivariant under the action of a compact group. Zhou et al. [41] proposed treating instances in a set as non identical and independent samples for multi instance problem.

Most of the work published in recent years have focused on ordered sets. Vinyals et al. introduced Order Matter: Sequence to Sequence for Sets in 2016 to learn a sequence to sequence mapping. Many related models and key contributions have been proposed that uses the idea of external memories like RNNSearch [3], Memory Networks [32,34] and Neural Turing Machines [10]. Recent interest in exchangeable models was developed due to their application in MIL. Deep Symmetry Networks [7] used kernel-based interpolation to tractably tie parameters and pool over symmetry spaces of any dimension. Deep Sets [39] by Zaheer et al. proposed a permutation invariant model. They proved that any pooling operation (mean, sum, max or similar) on individual features is a universal approximator for any set function. They also showed that any permutation invariant model follows de Finetti's theorem. Work has also been done on learning point cloud classification which is an example of MIL problem. Deep Learning with Sets and Point Cloud [23] used parameter sharing to get a equivariant layer. Another important paper on exchangeable model is Set Transformer. Set Transformer [19] by Lee et al. used results from Zaheer et al. [39] and proposed a Transformer [32] inspired permutation invariant neural network. The Set Transformer uses attention mechanisms to attend to inputs in order to invoke activation. Instead of using averaging over instances like in Deep Sets, the Set Transformer uses a parametric aggregating function pool which can adapt to the problem at hand. Another way to handle exchangeable data is to modify RNNs to operate on exchangeable data. BRUNO [18] is a model for exchangeable data and makes use of deep features learned from observations so as to model complex data types such as images. To achieve this, they constructed a bijective mapping between random variables $x_i \in X$ in the observation space and features $z_i \in Z$, and explicitly define an exchangeable model for the sequences $z_1, z_2, z_3, \ldots, z_n$. Deep Amortized Clustering [20] proposed using Set Transformers to cluster sets of points with only few forward passes. Deep Set Prediction Networks [40] introduced an interesting approach to predict sets from a feature vector which is in contrast to predicting an output using sets.

MIL for Histopathological Image Analysis. Exchangeable models are useful for histopathological images analysis as ground-truth labeling is expensive and labels are available at WSI instead of at the pixel level. A small pathology lab may process $\approx$10,000 WSIs/year, producing a vast amount of data, presenting a unique opportunity for MIL methods. Dismantling a WSI into smaller patches is a common practice; these patches can be used for MIL. The authors in [12] used attention-based pooling to infer important patches for cancer classification. A large amount of partially labeled data in histopathology can be used to discover hidden patterns of clinical importance [17]. Authors in [28] used MIL for breast cancer classification. A permutation invariant operator introduced by [30,31] was applied to pathology images. Recently, graph CNNs have been successfully used for representation learning of WSIs [1]. These compact and robust representations of WSIs can be further used for various clinical applications such as image-based search to make well-informed diagnostic decisions [14,15].

3   Background

This section explains the general concepts of exchangeability, its relation to de Finetti's theorem, and briefly discusses Memory Networks.

Exchangeable Sequence. A sequence of random variables $x_1, \ldots, x_n$ is exchangeable if the joint probability distribution does not change on permutation of the elements in a set. Mathematically, if $P(x_1, \ldots, x_n) = P(x_{\pi(1)}, \ldots, x_{\pi(n)})$ for a permutation function π, then the sequence $x_1, \ldots, x_n$ is exchangeable.

Exchangeable Models. A model is said to be exchangeable if the output of the model is invariant to the permutation of its inputs. Exchangeability implies that the information provided by each instance x_i is independent of the order in which they are presented. Exchangeable models can be of two types depending on the application: i) permutation invariant, and ii) permutation equivariant.

A model represented by a function $f : X \rightarrow Y$ where X is a set, is said to be permutation equivariant if permutation of input instances permutes the output labels with the same permutation π. Mathematically, a permutation-equivariant model is represented as,

$$f(x_{\pi(1)}, x_{\pi(2)}, \ldots, x_{\pi(n)}) = [y_{\pi(1)}, y_{\pi(2)}, \ldots, y_{\pi(n)}]. \tag{1}$$

Similarly, a function is permutation invariant if permutation of input instances does not change the output of the model. Mathematically,

$$f(x_1, x_2, \ldots, x_n) = f(x_{\pi(1)}, x_{\pi(2)}, \ldots, x_{\pi(n)}). \tag{2}$$

Deep Sets [39] incorporate a permutation-invariant model to learn arbitrary set functions by pooling in a latent space. The authors further showed that any pooling operation such as averaging and max on individual instances of a set can be used as a universal approximator for any arbitrary set function. The authors proved the following two results about permutation invariant models.

Theorem 1. A function $f(x)$ operating on a set $X = \{x_1, \ldots, x_n\}$ having elements from a countable universe, is a valid set function, i.e., invariant to the permutation of instances in X, if it can be decomposed to $\rho\left(\sum \phi(x)\right)$, for any function ϕ and ρ.

Theorem 2. Assume the elements are from a compact set in $\mathbb{R}^d$, i.e., possibly uncountable, and the set size is fixed to M. Then any continuous function operating on a set X, i.e., $f : \mathbb{R}^{d \times M} \to \mathbb{R}$ which is permutation invariant to the elements in X can be approximated arbitrarily close in the form of $\rho \sum(\phi(x))$.

The Theorem 1 is linked to de Finetti's theorem, which states that a random infinitely exchangeable sequence can be factorised into mixture densities conditioned on some parameter θ which captures the underlying generative process:

$$P(x_1, \ldots, x_n) = \int p(\theta) \prod_{i=1}^{n} p(x_i|\theta) \ d(\theta). \tag{3}$$

Memory Networks. The idea of using an external memory for relational learning tasks was introduced by Weston et al. [34]. Later, an end-to-end trainable model was proposed by Sukhbaatar et al. [29]. Memory networks enable learning of dependencies among instances of a set by providing an explicit memory representation for each instance in the sequence. The idea of self attention is popularized by [32], these models are known as *transformers*, widely used in NLP applications. The proposed MEM model uses the self-attention (similar to transformers) within memory vectors, aggregated using a pooling operation (weighted averaging) to form a permutation-invariant representation (based on Theorems 1 and 2). The next section explains it in details.

4 Proposed Approach

This section discusses the motivations, components, and offers an analysis of the proposed Memory-based Exchangeable Model (MEM) capable of learning permutation invariant representation of sets and unordered sequences.

4.1 Motivation

In order to learn an efficient representation for a set of instances, it is important to focus on instances which are "important" for a given task at hand, i.e., we need to attend to specific instances more than other instances. We therefore use

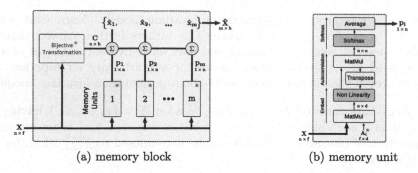

(a) memory block   (b) memory unit

Fig. 2. X is an input sequence containing n number of f-dimensional vectors. (a) The **memory block** is a sequence-to-sequence model that takes X and returns another sequence $\hat{X}$. The output $\hat{X}$ is a permutation-invariant representation of X. A bijective transformation model (an autoencoder) converts the input X to a permutation-equivariant sequence C. The weighted sum of C is computed over different probability distributions p_i from memory units. The hyper-parameters of a memory block are i) dimensions of the bijective transformation h, and ii) number of memory units m. (b) The **memory unit** has A_i, an embedding matrix (trainable parameters) that transforms elements of X to a d-dimensional space (memories). The output p_i is a probability distribution over the input X, also known as attention. The memory unit has a single hyper-parameter d, i.e. the dimension of the embedding space. (* represents learnable parameters.)

the memory network to learn an attention mapping for each instance. Memory networks are conventionally used for NLP for mapping questions posted in natural language to an answer [29,34]. We exploit the idea of having *memories* which can learn *key* features shared by one or more instances. Through these *key* features, the model can learn inter-dependencies using transformer style self-attention mechanism. As inter-dependencies are learnt, a set can be condensed into a compact vector such that a MLP can be used for a classification or regression learning.

4.2   Model Components

MEM is composed of four sequentially connected units: i) a feature extraction model, ii) memory units, iii) memory blocks, and iv) fully connected layers to predict the output.

A *memory block* is the main component of MEM and learns a permutation invariant representation of a given input sequence. Multiple memory blocks can be stacked together for modeling complex relationships and dependencies in exchangeable data. The memory block is made of memory units and a bijective transformation unit shown in Fig. 2.

Memory Unit. A memory unit transforms a given input sequence to an attention vector. The higher attention value represents the higher "importance" of the corresponding element of the input sequence. Essentially, it captures the relationships among different elements of the input. Multiple memory units enable the memory block to capture many complex dependencies and relationships among the elements. Each memory unit consists of an embedding matrix A_i that transforms a f-dimensional input vector x_j to a d-dimensional memory vector u_{ij}, as follows:

$$u_{ij} = \rho(x_j A_i),$$

where ρ is some non-linearity. The memory vectors are stacked to form a matrix $U_i = [u_{i0}, \ldots, u_{in}]$ of the shape $(n \times d)$. The relative degree of correlations among the memory vectors are computed using cross-correlation followed by a column-wise softmax and then taking a row-wise average, as follows:

$$S_i = \text{column-wise-softmax}(U_i U_i^T),$$
$$p_i = \text{row-wise-average}(S_i), \tag{4}$$

The p_i is the final output vector $(1 \times n)$ from the ith memory unit U_i, as shown in Fig. 2. The purpose of memory unit is to embed feature vectors into another space that could correspond to a distinct "attribute" or "characteristic" of instances. The cross correlation or the calculated attention vector represents the instances which are highly suggestive of those "attributes" or "characteristic". We do not normalize memory vectors as magnitude of these vectors may play an important role during the cross correlation.

Memory Block. A memory block is a sequence-to-sequence model, i.e., it transforms a given input sequence $X = x_1, \ldots, x_n$ to another representative sequence $\hat{X} = \hat{x}_1, \ldots, \hat{x}_m$. The output sequence is invariant to the element-wise permutations of the input sequence. A memory block contains m number of memory units. In a memory block, each memory unit takes a sequential data as an input and generates an attention vector. These attention vectors are subsequently used to compute the final output sequence. The schematic diagram of a memory block is shown in Fig. 2a.

The final output sequence $\hat{X}$ of a memory block is computed as a weighted sum of C with the probability distributions $p_1, \ldots, p_m$ from all the m memory units where C is a bijective transformation of X learned using an autoencoder. Each memory block has its own autoencoder model to learn the bijective mapping. The ith element $\hat{x}_i$ of the output sequence $\hat{X}$ is computed as matrix multiplication of p_i and C, as follows:

$$\hat{x}_i = p_i C,$$

where, p_i is the output of ith memory unit given by (4).

The bijective transformation from $X \mapsto C$ enables equivariant correspondence between the elements of the two sequences X & $\hat{X}$, and maps two different elements in the input sequence to different elements in the output sequence. It must be noted that bijective transformation is permutation equivariant not invariant. The reconstruction maintains one-to-one mapping between X and C. The final output sequence from a memory block is permutation invariant as it uses matrix multiplication between p_i (attention) and C (Fig. 3).

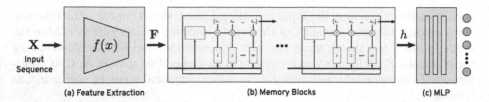

(a) Feature Extraction   (b) Memory Blocks   (c) MLP

Fig. 3. The overall architecture of the proposed Memory-based Exchangeable Model (MEM). The input to the model is a sequence, for e.g., a sequence of images or vectors. Each element of the input sequence X is passed through **(a)** feature extractor (CNN or MLP) to extract a sequence of feature vectors F, which is passed to **(c)** sequentially connected memory blocks. A memory block outputs another sequence which is a permutation-invariant representation of the input sequence. The output from the last memory block is vectorized and given to **(c)** MLP layers for classification/regression.

4.3   Model Architecture

1. Each element of a given input sequence $X = x_1, \ldots, x_n$ is passed through a feature extraction model to produce a sequence of feature vectors $F = f_1, \ldots, f_n$.
2. The feature sequence F is then passed through a memory block to obtain another sequence $\hat{X}$ which is a permutation-invariant representation of the input sequence. The number of elements in the sequence $\hat{X}$ depends on the number of memory unit in the memory block layer.
3. Multiple memory blocks can be stacked in series. The output from the last memory block is either vectorized or pooled, which is subsequently passed to a MLP layer for classification or regression.

4.4   Analysis

This section discusses the mathematical properties of our model. We use theorems from Deep Sets [39] to prove that our model is permutation invariant and universal approximator for arbitrary set functions.

Property 1. Memory units are permutation equivariant.
Consider an input sequence $X = x_1 \ldots x_n$. Since, for each memory unit,

$$\mathbf{U_i} = [\rho(x_o \mathbf{A_i}), \rho(x_1 \mathbf{A_i}), \ldots, \rho(x_n \mathbf{A_i})]$$

By Eq. (1), $\mathbf{U_i}$ is permutation equivariant and thus S_i in (4) is permutation equivariant. Finally, the attention vector p_i is calculated by averaging all rows, therefore the final output of memory unit p_i is permutation equivariant.

Property 2. Memory Blocks are permutation invariant.
A memory block layer consisting of m memory units generates a sequence $\hat{X} = \hat{x}_1, \ldots, \hat{x}_m$ where $\hat{x}_i$ can be written as

$$\hat{x}_i = p_i \mathbf{C}$$

Since both $\mathbf{C}$ and p_i are permutation equivariant, therefore, $\hat{x}_i$, which is calculated by matrix multiplication of p_i and $\mathbf{C}$, is permutation invariant.

5   Experiments

We performed two series of experiments comparing MEM against the simple pooling operations proposed by Deep Sets [39]. In the first series of experiments, we established the learning ability of the proposed model using toy datasets. For the second series, we used two real-world dataset, i) classification of subtypes of lung cancer against the largest public dataset of histopathology whole slide images (WSIs) [33], and ii) 3-D object classification using Point Cloud Dataset [36].

Model Comparison. We compared the performance of MEM against Deep Sets [39]. We use same the feature extractor for both Deep Sets and MEM, and experimented with different choices of pooling operations—max, mean, dot product, and sum. MEM also has a special pooling "$mb1$", which is a memory block with a single memory unit in the last hidden layer. Therefore, we tested 9 different models for each experiment—five configurations of our model, and four configurations of Deep Sets. We tried to achieve the best performance by varying the hyper-parameters for each of the configuration of both MEM and Deep Sets. We found that MEM had higher learning capacity, therefore higher number of parameters resulted in better accuracy for MEM but not necessarily for Deep Set. We denote the common feature extractor as **FF** and Deep Sets as **DS** in the discussion below. The other approaches that are compared have been appropriately cited.

5.1   Toy Datasets

To demonstrate the advantage of MEM over simple pooling operations, we consider four toy problems, involving regression and classification over sets. We constructed these toy datasets using the MNIST dataset.

Sum of Even Digits. Sum of even digits is a regression problem over the set of images containing handwritten digits from MNIST. For a given set of images $X = \{x_1, \ldots, x_n\}$, the goal is to find the sum of all even digits. We used the Mean Absolute Error (MAE). We split the MNIST dataset into 70–30% training, and testing data-sets, respectively. We sampled 100,000 sets of 2 to 10 images from the training data. For testing, we sampled 10,000 sets of images containing m number of images per set where $m \in [2, 10]$. Figur 4 shows the performance of MEM against simple pooling operations with respect to the number of images in the set.

Prime Sum. Prime Sum is a classification problem over a set of MNIST images. A set is labeled positive if it contains any two digits such that their sum is a prime number. We constructed the dataset by randomly sampling five images from the MNIST dataset. We constructed the training data with 20,000 sets

Table 1. Results on the toy datasets for different configurations of MEM and feature pooling. It must be noted that for maximum of set, the configuration FF + Max (DS) achieves the best accuracy but it may predict the output perfectly by learning the identity function therefore we highlighted second best configuration FF + Dotprod (DS) as well.

Methods	Sum of even digits		Prime sum	Counting unique images		Maximum of set		Gaussian clustering
	Accuracy	MAE	Accuracy	Accuracy	MAE	Accuracy	MAE	NLL
FF + MEM + MB1 (ours)	0.9367 ± 0.0016	0.2516 ± 0.0105	0.9438 ± 0.0043	0.7108 ± 0.0084	0.3931 ± 0.0080	0.9326 ± 0.0036	0.1449 ± 0.0068	**1.348**
FF + MEM + Mean (ours)	0.9355 ± 0.0015	0.2437 ± 0.0087	0.7208 ± 0.0217	0.4264 ± 0.0062	0.9525 ± 0.0109	0.9445 ± 0.0035	0.1073 ± 0.0067	1.523
FF + MEM + Max (ours)	**0.9431 ± 0.0020**	**0.2295 ± 0.0098**	0.9361 ± 0.0060	0.6888 ± 0.0066	0.4140 ± 0.0079	0.9498 ± 0.0022	0.1086 ± 0.0060	1.388
FF + MEM + Dotprod (ours)	0.8411 ± 0.0045	0.3932 ± 0.0065	**0.9450 ± 0.0086**	**0.7284 ± 0.0055**	**0.3664 ± 0.0037**	**0.9517 ± 0.0041**	**0.0999 ± 0.0097**	1.363
FF + MEM + Sum (ours)	0.9353 ± 0.0022	0.2739 ± 0.0081	0.6652 ± 0.0389	0.3138 ± 0.0094	1.3696 ± 0.0151	0.9430 ± 0.0031	0.1318 ± 0.0058	1.611
FF + Mean (DS)	0.9159 ± 0.0019	0.2958 ± 0.0049	0.5280 ± 0.0078	0.3140 ± 0.0071	1.2169 ± 0.0136	0.3223 ± 0.0075	1.0029 ± 0.0155	2.182
FF + Max (DS)	0.6291 ± 0.0047	1.3292 ± 0.0211	0.9257 ± 0.0033	0.7088 ± 0.0060	0.3933 ± 0.0059	**0.9585 ± 0.0012**	**0.0742 ± 0.0032**	1.608
FF + Dotprod (DS)	0.1503 ± 0.0015	1.8015 ± 0.0016	0.9224 ± 0.0028	**0.7254 ± 0.0063**	**0.3726 ± 0.0054**	**0.9548 ± 0.0017**	0.1355 ± 0.0027	8.538
FF + Sum (DS)	0.6333 ± 0.0043	0.5763 ± 0.0069	0.5264 ± 0.0050	0.2982 ± 0.0042	1.3415 ± 0.0169	0.3344 ± 0.0038	0.9645 ± 0.0111	12.05

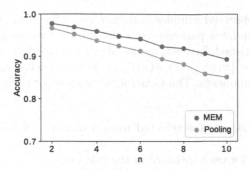

Fig. 4. Comparison of MEM and feature pooling on a regression problem involving finding the sum of even digits within a set of MNIST images. Each point corresponds to the best configurations for the two models.

randomly sampled from the training data of MNIST. For testing, we randomly sampled 5,000 sets from the testing data of MNIST. The results are reported in the second column of Table 1 that shows the robustness of memory block.

Maximum of a Set. Maximum of a set is a regression problem to predict the highest digit present in a set of images from MNIST. We constructed a set of five images by randomly selecting samples from MNIST dataset. The label for each set is the largest number present in the set. For example, images of $\{2, 5, 3, 3, 6\}$ is labeled as 6. We constructed 20,000 training sets and for testing we randomly sampled 5,000. The detailed comparison of accuracy and MAE between different models is given in the second last column of Table 1. We found that FF+Max learns the identity mapping and thus results in a very high accuracy. In all the training sessions, we consistently obtained the training accuracy of 100% for the FF+Max configuration, whereas MEM generalizes better than the Deep Sets.

Counting Unique Images. Counting unique images is a regression problem over a set. This task involves counting unique objects in a set of images from fashion MNIST dataset [37]. We constructed the training data by selecting a set, as follows:

1. Let n be the number of total images and u be the number of unique images.
2. Randomly select an integer n between 2 and 10.
3. Randomly select another integer u between 1 and n.
4. Select u number of unique objects from fashion-MNIST training data.
5. Then add n-u number of randomly selected objects from the previous step.

The task is to count unique objects u in a given set. The results are shown in the third column of Table 1.

Amortized Gaussian Clustering. Amortized Gaussian clustering is a regression problem that involves estimating the parameters of a population of Mixture of Gaussian (MoG). Similar to Set Transformer [19], we test our model's ability to learn parameters of a Gaussian Mixture with k components such that the likelihood of the observed samples is maximum. This is in contrast to the EM algorithm which updates parameters of the mixture recursively until the stopping criterion is satisfied. Instead, we use MEM to directly predict parameters of a MoG i.e. $f(x; \theta) = \{\pi(x), (\mu(x), \sigma(x))_{j=1}^k\}$. For simplicity we sample from MoG with only four components. The Generative process for each training dataset is as follows

1. Mean of each Gaussian is selected from a uniform distribution i.e. $\mu_{j=1}^k \sim$ Unif$(0, 8)$.
2. Select a cluster for each instance in the set, i.e.,

$$\pi \sim \text{Dir}([1, 1]^T); z_i \sim \text{Categorical}(\pi)$$

3. Generate data from an univariate Gaussian $\sim \mathcal{N}(\mu_{z_i}, 0.3)$.

We created a dataset of 20,000 sets each consisting of 500 points sampled from different MoGs. Results in Table 1 show that MEM is significantly better than Deep Sets.

5.2 Real World Datasets

To show the robustness and scalability of the model for the real-world problems, we have validated MEM on two larger datasets. Firstly, we tested our model on a point cloud dataset for predicting the object type from the set of 3D coordinates. Secondly, we used the largest public repository of histopathology images (TCGA) [33] to differentiate between two main sub-types of lung cancer. Without any significant effort in extracting histologically relevant features and fine-tuning, we achieved a remarkable accuracy of 84.84% on 5-fold validation.

Point Cloud Classification. We evaluated MEM on a more complex classification task using ModelNet40 [36] point cloud dataset. The dataset consists of 40 different objects or classes embedded in a three dimensional space as points. We produce point-clouds with 100 points (x, y, z-coordinates) each from the mesh representation of objects using the point-cloud library's sampling routine [25][1]. We compare the performance against various other models reported in Table 2. We experimented with different configurations of our model and found that FF+MB1 works best for 100 points cloud classification. We achieves the classification accuracy of 85.21% using 100 points. Our model performs better than Deep Sets and Set Transformer for the same number of instances, showing the effectiveness of having attention from memories.

Lung Cancer Subtype Classification. Lung Adenocarcinoma (LUAD) and Lung Squamous Cell Carcinoma (LUSC) are two main types of non-small cell lung cancer (NSCLC) that account for 65–70% of all lung cancers [9]. Classifying patients accurately is important for prognosis and therapy decisions. Automated classification of these two main subtypes of NSCLC is a crucial step to build

---

[1] We obtained the training and test datasets from Zaheer et al. [39].

Table 2. Test accuracy for the point cloud classification on different instance sizes using various methods. MEM with configuration FF + MEM + MB1 achieves 85.21% accuracy for the instance size of 100 which is best compared to others.

Configuration	Instance size	Accuracy
3DShapeNet [36]	30^3	0.77
Deep set [39]	100	0.8200
VoxNet [21]	32^2	0.8310
3D GAN [35]	64^3	0.833
Set Transformer [19]	100	0.8454
Set Transformer [19]	1000	0.8915
Deep set [39]	5000	0.9
MVCNN [27]	$164 \times 164 \times 12$	0.901
Set Transformer [19]	5000	0.9040
VRN Ensemble [5]	32^3	0.9554
FF + MEM + MB1 (Ours)	**100**	**0.8521**

computerized decision support and triaging systems. We present a two-staged method to differentiate LUAD and LUSC for whole slide images, short WSIs, that are very large images. Firstly, we implement a method to systematically sample patches/tiles from WSIs. Next, we extract image features from these patches using DenseNet [11]. We then use MEM to learn the representation of a set of patches for each WSI.

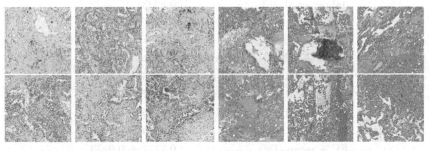

(a) Lung Adenocarcinoma   (b) Lung Squamous Cell Carcinoma

Fig. 5. The patches extracted from two WSIs of patients with (a) LUAD and (b) LUSC. Each slide roughly contains 500 patches.

To the best of our knowledge, this is the first ever study conducted on all the lung cancer slides in TCGA dataset (comprising of 2 TB of data consisting of 2.5 million patches of size 1000 × 1000 pixels). All research works in literature use a subset of the WSIs with their own test-train split instead of cross validation, making it difficult to compare against them. However, we have achieved greater than or similar to all existing research works without utilizing any expert's opinions (pathologists) or domain-specific techniques. We used 2,580 WSIs from TCGA public repository [33] with 1,249, and 1,331 slides for LUAD and LUSC, respectively. We process each WSI as follows.

1. **Tissue Extraction.** Every WSI contains a bright background that generally contains irrelevant (non-tissue) pixel information. We removed non-tissue regions using color thresholds.
2. **Selecting Representative Patches.** Segmented tissue is then divided into patches. All the patches are then grouped into a pre-set number of categories (classes) via a clustering method. A 10% of all clustered patches are uniformly randomly selected distributed within each class to assemble *representative patches*. Six of these representative patches for each class (LUAD and LUSC) is shown in Fig. 5.
3. **Feature Set.** A set of features for each WSI is created by converting its representative patches into image features. We use DenseNet [11] as the feature extraction model. There are a different number of feature vectors for each WSI.

Table 3. Accuracy for LUAD vs LUSC classification for various methods. For our experiments, we conducted comprehensive 5-fold cross validation accuracy whereas other methods have used non-standardized test set.

Methods	Accuracy
Coudray et al. [6]	**0.85**
Jabber et al. [13]	0.8333
Khosravi et al. [16]	0.83
Yu et al. [38]	0.75
FF + MEM + Sum (ours)	**0.8484 ± 0.0210**
FF + MEM + Mean (ours)	0.8465 ± 0.0225
FF + MEM + MB1 (ours)	0.8457 ± 0.0219
FF + MEM + Dotprod (ours)	0.6345 ± 0.0739
FF + sum (DS)	0.5159 ± 0.0120
FF + mean (DS)	0.7777 ± 0.0273
FF + dotprod (DS)	0.4112 ± 0.0121

The results are shown in Table 3. We achieved the maximum accuracy of 84.84% with FF + MEM + Sum configuration. It is difficult to compare our approach against other approaches in literature due to non-standardization of the dataset. Coudray et al. [6] used the TCGA dataset with around 1,634 slides to classify LUAD and LUSC. They achieved AUC of 0.947 using patches at 20×. We achieved a similar AUC of 0.94 for one of the folds and average AUC of **0.91**. In fact, without any training they achieved the similar accuracy as our model (around 85%). It is important to note that we did not do any fine-tuning or utilize any form of input from an expert/pathologist. Instead, we extracted diverse patches and let the model learn to differentiate between two sub-types by "attending" relevant ones. Another study by Jaber et al. [13] uses cell density maps, achieving an accuracy of 83.33% and AUC of 0.9068. However, they used much smaller portion of the TCGA, i.e., 338 TCGA diagnostic WSIs (164 LUAD and 174 LUSC) were used to train, and 150 (71 LUAD and 79 LUSC).

6   Conclusions

In this paper, we introduced a Memory-based Exchangeable Model (MEM) for learning permutation invariant representations. The proposed method uses attention mechanisms over "memories" (higher order features) for modelling complicated interactions among elements of a set. Typically for MIL, instances are treated as independently and identically distributed. However, instances are rarely independent in real tasks, and we overcome this limitation using an "attention" mechanism in memory units, that exploits relations among instances. We also prove that the MEM is permutation invariant. We achieved good performance on all problems that require exploiting instance relationships. Our model scales well on real world problems as well, achieving an accuracy score of 84.84% on classifying lung cancer sub-types on the largest public repository of histopathology images.

References

1. Adnan, M., Kalra, S., Tizhoosh, H.R.: Representation learning of histopathology images using graph neural networks, p. 8
2. Aldous, D.J.: Representations for partially exchangeable arrays of random variables. J. Multivar. Anal. **11**(4), 581–598 (1981)
3. Bahdanau, D., Cho, K., Bengio, Y.: Neural machine translation by jointly learning to align and translate. arXiv preprint arXiv:1409.0473 (2014)
4. Bloem-Reddy, B., Teh, Y.W.: Probabilistic symmetry and invariant neural networks. arXiv preprint arXiv:1901.06082 (2019)
5. Brock, A., Lim, T., Ritchie, J.M., Weston, N.: Generative and Discriminative Voxel Modeling with Convolutional Neural Networks. http://arxiv.org/abs/1608.04236 (2016)
6. Coudray, N., et al.: Classification and mutation prediction from non-small cell lung cancer histopathology images using deep learning. Nat. Med. **24**(10), 1559 (2018)
7. Gens, R., Domingos, P.M.: Deep symmetry networks. In: Advances in Neural Information Processing Systems, pp. 2537–2545 (2014)
8. Ghahramani, Z., Heller, K.A.: Bayesian sets. In: Advances in Neural Information Processing Systems, pp. 435–442 (2006)
9. Graham, S., Shaban, M., Qaiser, T., Koohbanani, N.A., Khurram, S.A., Rajpoot, N.: Classification of lung cancer histology images using patch-level summary statistics. In: Medical Imaging 2018: Digital Pathology, vol. 10581, p. 1058119. International Society for Optics and Photonics (2018)

10. Graves, A., Wayne, G., Danihelka, I.: Neural turing machines. arXiv preprint arXiv:1410.5401 (2014)
11. Huang, G., Liu, Z., Van Der Maaten, L., Weinberger, K.Q.: Densely connected convolutional networks. In: Proceedings of the IEEE Conference on Computer Vision and Pattern Recognition, pp. 4700–4708 (2017)
12. Ilse, M., Tomczak, J.M., Welling, M.: Chapter 22 - Deep multiple instance learning for digital histopathology. In: Zhou, S.K., Rueckert, D., Fichtinger, G. (eds.) Handbook of Medical Image Computing and Computer Assisted Intervention, pp. 521–546. Academic Press. https://doi.org/10.1016/B978-0-12-816176-0.00027-2
13. Jaber, M.I., Beziaeva, L., Szeto, C.W., Elshimali, J., Rabizadeh, S., Song, B.: Automated adeno/squamous-cell NSCLC classification from diagnostic slide images: a deep-learning framework utilizing cell-density maps (2019)
14. Kalra, S., et al.: Yottixel – an image search engine for large archives of histopathology whole slide images **65**, 101757. https://doi.org/10.1016/j.media.2020.101757
15. Kalra, S., et al.: Pan-cancer diagnostic consensus through searching archival histopathology images using artificial intelligence **3**(1), 1–15. https://doi.org/10.1038/s41746-020-0238-2
16. Khosravi, P., Kazemi, E., Imielinski, M., Elemento, O., Hajirasouliha, I.: Deep convolutional neural networks enable discrimination of heterogeneous digital pathology images. EBioMedicine **27**, 317–328 (2018)
17. Komura, D., Ishikawa, S.: Machine learning methods for histopathological image analysis **16**, 34–42. https://doi.org/10.1016/j.csbj.2018.01.001
18. Korshunova, I., Degrave, J., Huszár, F., Gal, Y., Gretton, A., Dambre, J.: Bruno: a deep recurrent model for exchangeable data. In: Advances in Neural Information Processing Systems, pp. 7190–7198 (2018)
19. Lee, J., Lee, Y., Kim, J., Kosiorek, A.R., Choi, S., Teh, Y.W.: Set transformer. CoRR abs/1810.00825 (2018). http://arxiv.org/abs/1810.00825
20. Lee, J., Lee, Y., Teh, Y.W.: Deep amortized clustering. arXiv preprint arXiv:1909.13433 (2019)
21. Maturana, D., Scherer, S.: VoxNet: a 3D convolutional neural network for real-time object recognition. In: 2015 IEEE/RSJ International Conference on Intelligent Robots and Systems (IROS), pp. 922–928. https://doi.org/10.1109/IROS.2015.7353481
22. Orbanz, P., Roy, D.M.: Bayesian models of graphs, arrays and other exchangeable random structures. IEEE Trans. Pattern Anal. Mach. Intell. **37**(2), 437–461 (2014)
23. Ravanbakhsh, S., Schneider, J., Poczos, B.: Deep learning with sets and point clouds. arXiv preprint arXiv:1611.04500 (2016)
24. Ravanbakhsh, S., Schneider, J., Poczos, B.: Equivariance through parameter-sharing. In: Proceedings of the 34th International Conference on Machine Learning-Volume 70, pp. 2892–2901. JMLR. org (2017)
25. Rusu, R., Cousins, S.: 3D is here: point cloud library (PCL). In: 2011 IEEE International Conference on Robotics and Automation (ICRA), pp. 1–4, May 2011. https://doi.org/10.1109/ICRA.2011.5980567, http://ieeexplore.ieee.org/xpl/login.jsp?tp=&arnumber=5980567&url=http%3A%2F%2Fieeexplore.ieee.org%2Fxpls%2Fabs_all.jsp%3Farnumber%3D5980567
26. Shawe-Taylor, J.: Symmetries and discriminability in feedforward network architectures. IEEE Trans. Neural Netw. **4**(5), 816–826 (1993)
27. Su, H., Maji, S., Kalogerakis, E., Learned-Miller, E.G.: Multi-view convolutional neural networks for 3D shape recognition. In: Proceedings of the ICCV (2015)

28. Sudharshan, P.J., Petitjean, C., Spanhol, F., Oliveira, L.E., Heutte, L., Honeine, P.: Multiple instance learning for histopathological breast cancer image classification **117**, 103–111. https://doi.org/10.1016/j.eswa.2018.09.049

29. Sukhbaatar, S., Weston, J., Fergus, R., et al.: End-to-end memory networks. In: Advances in Neural Information Processing Systems, pp. 2440–2448 (2015)

30. Tomczak, J.M., Ilse, M., Welling, M.: Deep Learning with Permutation-invariant Operator for Multi-instance Histopathology Classification. http://arxiv.org/abs/1712.00310

31. Tomczak, J.M., et al.: Histopathological classification of precursor lesions of esophageal adenocarcinoma: a deep multiple instance learning approach (2018)

32. Vaswani, A., et al.: Attention is all you need. In: Advances in Neural Information Processing Systems, pp. 5998–6008 (2017)

33. Weinstein, J.N., et al.: The cancer genome atlas pan-cancer analysis project. Nat. Genet. **45**(10), 1113 (2013)

34. Weston, J., Chopra, S., Bordes, A.: Memory networks. arXiv preprint arXiv:1410.3916 (2014)

35. Wu, J., Zhang, C., Xue, T., Freeman, W.T., Tenenbaum, J.B.: Learning a Probabilistic Latent Space of Object Shapes via 3D Generative-Adversarial Modeling. http://arxiv.org/abs/1610.07584

36. Wu, Z., et al.: 3D ShapeNets: a deep representation for volumetric shapes. In: Proceedings of the IEEE Conference on Computer Vision and Pattern Recognition, pp. 1912–1920 (2015)

37. Xiao, H., Rasul, K., Vollgraf, R.: Fashion-mnist: a novel image dataset for benchmarking machine learning algorithms. arXiv preprint arXiv:1708.07747 (2017)

38. Yu, K.H., et al.: Predicting non-small cell lung cancer prognosis by fully automated microscopic pathology image features. Nat. Commun. **7**, 12474 (2016)

39. Zaheer, M., Kottur, S., Ravanbakhsh, S., Poczos, B., Salakhutdinov, R.R., Smola, A.J.: Deep sets. In: Advances in Neural Information Processing Systems, pp. 3391–3401 (2017)

40. Zhang, Y., Hare, J., Prügel-Bennett, A.: Deep set prediction networks. arXiv preprint arXiv:1906.06565 (2019)

41. Zhou, Z.H., Sun, Y.Y., Li, Y.F.: Multi-instance learning by treating instances as non-IID samples. In: Proceedings of the 26th annual international conference on machine learning, pp. 1249–1256. ACM (2009)

Feature Space Augmentation
for Long-Tailed Data

Peng Chu[1]([⊠]), Xiao Bian[2], Shaopeng Liu[3], and Haibin Ling[1,4]

[1] Temple University, Philadelphia, USA
pchu@temple.edu
[2] Google Inc., Mountain View, USA
xbian@google.com
[3] GE Research, Niskayuna, USA
sliu@ge.com
[4] Stony Brook University, Stony Brook, USA
hling@cs.stonybrook.edu

Abstract. Real-world data often follow a long-tailed distribution as the frequency of each class is typically different. For example, a dataset can have a large number of under-represented classes and a few classes with more than sufficient data. However, a model to represent the dataset is usually expected to have reasonably homogeneous performances across classes. Introducing class-balanced loss and advanced methods on data re-sampling and augmentation are among the best practices to alleviate the data imbalance problem. However, the other part of the problem about the under-represented classes will have to rely on additional knowledge to recover the missing information.

In this work, we present a novel approach to address the long-tailed problem by augmenting the under-represented classes in the feature space with the features learned from the classes with ample samples. In particular, we decompose the features of each class into a class-generic component and a class-specific component using class activation maps. Novel samples of under-represented classes are then generated on the fly during training stages by fusing the class-specific features from the under-represented classes with the class-generic features from confusing classes. Our results on different datasets such as iNaturalist, ImageNet-LT, Places-LT and a long-tailed version of CIFAR have shown the state of the art performances.

1 Introduction

Deep neural networks have shown considerable success in a wide variety of visual recognition tasks. Its effectiveness and generalizability have been well proved by

P. Chu and X. Bian—Work was done at GE Research.

Electronic supplementary material The online version of this chapter (https:// doi.org/10.1007/978-3-030-58526-6_41) contains supplementary material, which is available to authorized users.

© Springer Nature Switzerland AG 2020
A. Vedaldi et al. (Eds.): ECCV 2020, LNCS 12374, pp. 694–710, 2020.
https://doi.org/10.1007/978-3-030-58526-6_41

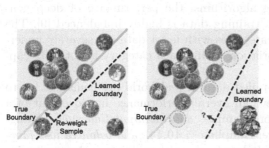

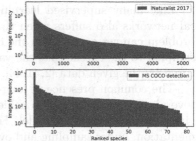

Fig. 1. Left: With limited but well-spread data, the optimal decision boundary search can be recovered by sample re-weighting/loss balancing. Right: Without sufficient sample coverage, the "optimal direction" to move the decision boundary becomes unclear. In this paper, augmented samples are generated to recover the underlying distribution.

Fig. 2. The sorted sample size of each class from different dataset follows similar long-tailed distribution.

many state-of-the-art work [16,19,25,30] and a wide variety of real-world applications in different industries [3,6,12,49]. However, there is often one underlying condition that each category of interest needs to be well represented.

To quantify the "representativeness" of data can be a challenging problem itself. In practice, it is usually scrutinized using different heuristics, and one common criterion could be the balance of a dataset. Indeed, many public datasets are intentionally organized to have the same number of samples from each class [24,33]. For problems such as segmentation and detection which are hard to ensure exact balanced data, it is always preferable to ensure good data coverage in a way that the rare classes still have sufficient data and are hence well represented [13,40].

However, real-world visual understanding problems are often fine-grained and long-tailed. To achieve human-level visual understanding, it almost implies the ability to distinguish the subtle differences between fine-grained categories and to robustly handle the presence of rare categories [1]. In fact, these two properties of real-world data usually accompany each other as a large number of fine-grained categories often leads to a highly imbalanced dataset, as illustrated in Fig. 2. For example, in iNaturalist dataset 2017 [40] for species classification, there are a total of 5089 classes with the largest classes more than 1000 samples and the smallest classes fewer than 10. In iNaturalist competition 2019, even with an effort to filter out species that have insufficient observations and to further cap the maximum class size to be 500, there is still serious imbalance in the dataset as the smallest classes around 10 samples. Similar data distribution can be observed in other applications, such as a UAV-based object detection dataset [49] and COCO [28].

Like many supervised learning algorithms, the performance of deep neural networks also suffers when the training data is highly imbalanced [9]. The problem can get worse when the categories with fewer data are severely under-sampled to the extent that the variation within each category is not fully captured by the given data [2,41,44].

The common presence of long-tailed data in real-world problems has led to several effective practices to achieve an overall performance improvement of a given machine learning model. For example, *data manipulation* such as augmentation, under-sampling and over-sampling [4,10,14], and *balanced loss function design* (*e.g.*, focal-loss [27] and class-balanced loss [9]), are the two mainstream approaches. These practices often improve the performance reasonably yet the improvement deteriorates when certain categories are severely underrepresented, as shown in Fig. 1. Specifically, these methods are often designed to move the class decision boundary to reduce the bias introduced by imbalanced classes. However, when a class is severely under-represented such that it is hard to draw its complete data distribution, finding the right direction to adjust the decision boundary becomes challenging. We therefore focus on exploring the information learned from the head classes (the ones with ample samples) to help the tail classes (the under-represented ones) in a long-tailed dataset.

In this work, we present a novel method to address the long-tailed data classification problem by augmenting the tail classes in the feature space using the information from the head classes. In particular, we insert an attention unit with the help of the class activation map (CAM) [46] to filter out class-specific features and class-generic features from each class. For the samples of each tail class, we augment the high-level features (from high-level deep network layers) by mixing its class-specific features with the class-generic features from the head classes. This method is based on two underlying assumptions: 1) information from the head classes, represented as class-generic features, can help to recover the distribution of tail classes; and 2) the class-generic and class-specific features can be extracted and re-mixed to generate novel samples in the high-level feature space due to a more "linear" representation at that level. We have designed an end-to-end training pipeline to efficiently perform such feature space augmentation, and evaluated our method on artificially created long-tailed CIFAR-10 and CIFAR-100 datasets [24], ImageNet-LT, Places-LT [29] and naturally long-tailed datasets such as iNaturalist 2017 & 2018 [40]. Our approach has shown the state of the art performance on these long-tailed datasets compared to other mainstream deep learning models on data imbalance problems.

2   Related Work

In this section, we first discuss the two directly related approaches, *learning with balanced loss* and *data augmentation*, and then discuss the difference and relation of our approach to few-shot learning and transfer learning.

2.1   Learning with Balanced Loss

One of the most common and often most effective practices is *learning with balanced loss*. The key to such approaches is to counter the effect of a skewed data distribution by adjusting the weights of the samples from the small classes in the loss function. It is typically accomplished by: 1) over-sampling or under-sampling [4,14,22,37,50] to achieve an even data distribution across various classes, and/or 2) assigning proper weights to the loss terms corresponding to the tail classes [9,11,20,23,27,35,39,45,48].

Specifically, these approaches treat the issue of long-tailed data as an optimization problem such that an "optimal" classification boundary can be recovered by carefully adjusting the weight/frequency of each data point in the training set. They typically have the advantage of a relatively clean implementation by either adjusting the loss function [9,27] or manipulating the input batch [4,14,37,50], and hence were widely adopted in practice [27,37,50]. However, when the samples of the tail classes are far from sufficient to recover the true distribution, the performance of such methods deteriorates [9].

Two works along this direction, class-balanced loss [9] and focal loss [27] draw our attention in particular for their generic applications on deep learning models. Specifically, focal loss weights the loss term of each sample based on the probability generated from the last soft-max layer [27]. It implicitly gives higher weights to samples from the tail classes to counter the bias introduced by the sample size. In [9], the concept of the effective number is introduced to calculate the weight of each class in the loss term.

Note that our approach can be used jointly with approaches such as focal loss [27] and potentially gain the benefits from both. For example, we can use the feature space augmentation approach to facilitate the performance of the tail classes, and at the same time, balanced loss methods such as focal loss can give higher weights to the hard examples regardless of the class label during training.

2.2   Data Synthesis and Augmentation

Generating and synthesizing new samples to compensate the small sample size of a tail class is a natural way to improve the performance of deep learning models on long-tailed data. These samples can be either generated from similar samples [5] or synthesized based on the given information of a dataset [15,18,50]. The general application of data augmentation in different deep learning models also boosts the interest of developing more sophisticated data augmentation methods. For example, in [7], an input image is partitioned into local regions which are shuffled during training. In [8], local regions of an image are replaced by unlabeled data to generate synthetic images to help training. In [41], a parametric generative model takes noise and existed samples to hallucinate new samples to support training.

As directly manipulating the raw input images may as well introduce unexpected noise, feature vectors are instead generated by training a function to learn the relation between a pair of samples from one class and applies it to the samples in another [17]. Furthermore, recent progress in generative adversarial networks (GAN) have inspired advanced methods using generative models to address the data insufficiency problem [43].

In contrast to the existing approaches on augmenting feature vectors, we focus on modeling the feature space itself rather than training a heavily parametric model that applies to all different classes. The decomposition of feature space is then used to formulate novel training samples in the feature space on the fly.

2.3 Transfer Learning

Past works in the domain of transfer learning and few-shot learning [2,31,32,42, 44,47] have been conducted to solve the long-tailed problem. Our work shares a similar assumption with these works that the information from the head classes can be used to help the tail classes. However, we explicitly distinguish the generic features and specific features from each class instead of making strong assumptions on the general transferability of knowledge from the head classes to the tail classes. Specifically, in [42], a meta-network is trained to predict the many-shot parameters from few-shot model parameters using data from the head classes with the assumption that the model parameters from different classes share a similar dynamic behavior even if the size of the training set varies. In [2], the representation is shared in general with different embedding approaches across different classes. In [44], the variance of the head classes is learned and transferred to the tail classes with the underlying assumption that each class has its own mean but a shared variance. In [32], visually similar classes are clustered together in order to reduce the level of data imbalance. Knowledge can then be transferred from each cluster to its sub-classes during the fine tuning stage of deep networks for object detection specifically.

In comparison, we intentionally separate the features of each class into class-specific features and class-generic features. Only class-generic features from head classes are seen as transferable knowledge and are hence used for feature space augmentation on the tail classes.

3   The Problem of Long Tail

In this section, we first analyze the underlying issues of long-tailed data that affect model performance (Sect. 3.1), and then explore deeper into the feature space of DNNs and illustrate a novel way to alleviate the problem (Sect. 3.2).

3.1   Two Reasons of Model Performance Drop

Long-tailed data hurt the performance of learning-based classification models mainly due to the following two issues: (1) data imbalance which is relatively easy to solve, and (2) missing coverage of the data distribution caused by limited data, which is harder to deal with.

   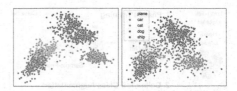

Fig. 3. The difference between the two "optimal" decision boundaries.

Fig. 4. Left: class-specific features, Right: class-generic features.

The data imbalance issue has been discussed in several recent works [4,9, 14,27] with good solutions proposed to minimize its impact. This problem is essentially about the bias introduced by the different number of samples in the dataset. With carefully designed sampling schemes and/or loss weights, we can compensate this negative impact and move the classification decision boundary in the right direction. For example, a common practice of training on an imbalanced dataset is to over-sample the small classes or under-sample the large classes [10]. This is built upon the assumption that the underlying decision boundary is indeed well-defined with the given data, and hence with careful adjustment we can find its optimal location.

However, when there is simply no sufficient data for the tail classes to recover their underlying distribution, the problem of finding an optimal decision boundary becomes ill-defined. In this scenario, it becomes extremely difficult to guess the location of the decision boundary without recovering the distribution first. We hypothesize that the knowledge obtained from the head classes can help with solving the issue.

We further elaborate the issue in Fig. 3 by plotting the feature distribution of 4 classes in CIFAR-10. The features are from the last fully-connected (FC) layer of ResNet-18 and then embedded in 2-D space. When the ship class is under-represented, as shown in the left graph, simply moving the decision boundary will not provide the optimal decision boundary (as shown in the right graph) as if there were sufficient samples.

3.2 Class Activation Map and Feature Decomposition

With limited data in the tail classes and ample data in the head classes, it seems natural to use the knowledge learned from the head classes to help recovering the missing information in the tail classes. However, we have to be careful to differentiate the class-generic information that can be used to recover the distribution of the tail classes from the class-specific information that may mislead the recovery of the distribution of the tail classes.

Inspired by the recent works on attention and visual explanation [36,46], we find that deep neural network features can be decomposed into two such components in a similar fashion. In particular, let us define class activation map M_c of class c as in [46],

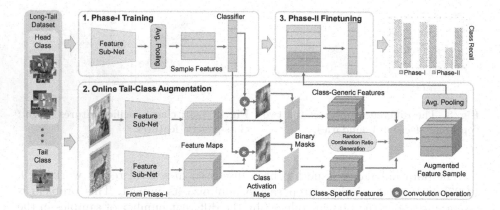

Fig. 5. Overview of the proposed two-phase learning scheme.

$$M_c(x,y) = \sum_k w_k^c f_k(x,y), \tag{1}$$

where $f_k(x,y)$ is the feature vector in location (x,y) of channel k, and w_k^c the weights of the last layer of classifier corresponding to class c. The larger value of $M_c(x,y)$, the more important of feature vector at (x,y) is to class c, and vice versa.

We further normalize the value of $M_c(x,y)$ to the range of 0 and 1. Therefore, given a pair of thresholds $0 < \tau_s, \tau_g < 1$, we can decompose the class activation map M_c into two parts, M_c^s and M_c^g, to separate the feature vectors into class-specific features and class-generic features as follows,

$$M_c^s = \mathrm{sgn}(M_c - \tau_s) \odot M_c, \tag{2}$$
$$M_c^g = \mathrm{sgn}(\tau_g - M_c) \odot M_c, \tag{3}$$

where $\odot$ is the Hadamard product between two tensors, $\mathrm{sgn}(x) = 1$ for $x \geq 0$ and $\mathrm{sgn}(x) = 0$ for $x < 0$.

Figure 4 shows the scatter plot of class-generic features and class-specific features of different classes from CIFAR-10 (More results can be seen in the supplemental material). We can see that after decomposition, even when embedded in a 2-D space, the class-specific features are clearly more separated than class-generic features. In general, we have observed a much stronger correlation between class-generic features than class-specific features across different classes and different datasets. These results further substantiate our approach on using class-generic features to augment the tail classes during training.

4   Method

We propose a two-phase training scheme to leverage the class-generic information to recover the distribution of tail classes, as shown in Fig. 5. In Phase-I, samples from all classes are used to learn the feature representation and a base classifier. In Phase-II, online feature space augmentation is applied to generate novel samples for tail classes.

4.1 Initial Feature Learning

In the Phase-I training, we use all images in the dataset to learn the feature sub-network and the base classifier. In order to calculate the class activation maps in the following steps, we choose a network architecture that contains a single FC layer as the final classifier, which takes input from a global average pooling layer as illustrated in Fig. 5. A number of the modern deep convolutional neural network architectures fit into this category, e.g., ResNet [19], DenseNet [21], MobileNet [34], and EfficientNet [38].

4.2 Feature Space Augmentation

With the pre-trained feature sub-network and the classifier, augmented samples can be generated in the feature space on the fly by mixing the class-specific features from the given tail class and the class-generic features.

One question we need to address is that, given a tail class, how to choose the classes from which the class-generic features will be extracted. A naive solution would be to randomly select the classes from the training dataset. However, the class-generic features from different classes may vary with each other, and such features of a randomly selected class cannot always guarantee a good recovery of the classification decision boundary. From the perspective of the optimal classification decision boundary, we observe that the "nearby" classes in the feature space, i.e. the most "confusing" classes with respect to the given tail class, have the biggest impact on recovering the previously ill-defined decision boundary, as seen in Fig. 3. Specifically, we calculate the classification scores for all other classes for each training sample in a given tail class, and then find its top N_f confusing classes by ranking the average classification scores of other classes over all samples within the tail class.

As described in Sect. 3.2, we use class activation maps to separate the class-generic and class-specific information from a given image. As shown in Fig. 5, the feature sub-network trained in Phase-I is used to extract feature maps for each input image. The weights of the linear classifier trained in Phase-I are adapted to form the 1×1 convolutional filter for each class. For each input image, the filter associated with the ground truth class is applied on its feature maps to generate the class activation map which is further normalized to the range of $[0, 1]$ for consistency. Two independent thresholds τ_g and τ_s are used to extract the corresponding binarized masks for class-generic features and class-specific features following Eq. 2.

The class-generic information in the confusing classes is then leveraged to generate the augmented samples of each tail class in order to recover its intrinsic data distribution. Directly blending information at the pixel level often introduces artificial edges and hence imposes bias to the augmented samples. We,

---

Algorithm 1. Online Feature Augmentation

1: **Input:** All training images features $\mathbb{F}$ and their CAM $\mathbb{M}$.
2: **Output:** Training batch with augmented feature samples b_{out}.
3: Initialize output batch b_{out}.
4: **for** $i = 1, \ldots, N_t$ **do**
5:      $\mathbf{F}_c \leftarrow$ Draw one sample from tail classes
6:      Append $\mathbf{F}_c$ to b_{out}
7:      $M_c^s \leftarrow M_c > \tau_s$
8:      $\mathbf{F}_c^s \leftarrow M_c^s \odot \mathbf{F}_c$
9:      $\{u\} \leftarrow$ Find confusing classes for class c
10:     **for** u in $\{u\}$ **do**
11:         $\mathbf{F}_u \leftarrow$ Draw one sample from class u
12:         $M_u^g \leftarrow M_u < \tau_g$
13:         $\mathbf{F}_u^g \leftarrow M_u^g \odot \mathbf{F}_u$
14:         Generate combination ratio $\gamma \in (0, 1)$
                ▷ Total L spatial locations in $\mathbf{F}_i$
                ▷ Draw with repeat and excluding all zeros feature
15:         $\{\mathbf{f}_c^s\} \leftarrow$ Draw γL feature vectors from $\mathbf{F}_c^s$
16:         $\{\mathbf{f}_u^g\} \leftarrow$ Draw $(1 - \gamma)L$ feature vectors from $\mathbf{F}_u^g$
17:         $\mathbf{F}_c^{aug} \leftarrow$ Merge $\{\mathbf{f}_c^s\}$ and $\{\mathbf{f}_u^g\}$
18:         Append $\mathbf{F}_c^{aug}$ to b_{out}
19:     **end for**
20:     $\{\mathbf{F}_k\} \leftarrow$ Draw $N_t(1 + N_a)$ samples from head classes
21:     Append $\{\mathbf{F}_k\}$ to b_{out}
22: **end for**

---

therefore, conduct the fusion in the feature space to suppress the noise and potential bias. In particular, for each real sample in the tail class, we sample N_a images from its N_f confusing classes. The class-specific features from the sample are then combined with the class-generic features from the N_a samples in a linear way. A random combination ratio is generated to guide the fusion by randomly drawing class-generic and class-specific feature vectors to form an augmented sample for the tail class. By randomly modulating the combination ratio between the class-generic and class-specific features, the sample variance is built into this augmentation procedure. In the end, a total of N_a augmented samples are generated for each real sample from the tail class.

4.3 Fine Tuning with Online Augmented Samples

The augmented samples are generated online to fine tune the network trained in Phase-I to improve the performance of the tail classes. In each batch, we sample N_t images from the tail classes, and generate N_a augmented samples online for each of the real samples, which creates a batch including $N_t(1 + N_a)$ samples from the tail classes. The same number of images are also randomly drawn from the head classes to balance the distribution. Thus, a batch of size $2N_t(1 + N_a)$ is generated online for each fine tuning iteration. We summarize this process in Algorithm 1.

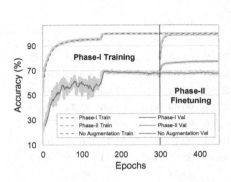

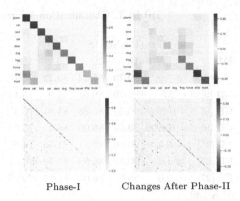

Fig. 6. Learning Curve for long-tailed CIFAR-10 with an imbalance factor of 200 using ResNet-18. The overall accuracy of the validation dataset is illustrated.

Fig. 7. Confusion matrix for CIFAR-10 (upper) and CIFAR-100 (bottom) at IM 200 using ResNet-18.

Fine tuning is performed on the layers after the features being extracted. Since the augmentation is conducted in the feature space, augmented samples can be generated at any stage of the network. However, the deeper features, compared to its shallow counterparts, are more linearly separable, which greatly help the fusion of features from the tail classes and their confusing classes. Moreover, richer spatial information in the lower-level feature maps may introduce artifacts to bias the model training. We analyze the detailed effect of augmenting samples at different depths in Sect. 5.4. We choose the features right before the last average pooling layer to help with the classification performance and at the same time to realize a simple design. Since the average pooling layer accumulates features in all spatial locations, the spatial distribution of class-generic and class-specific features become irrelevant in the augmented samples. Therefore, when combining, only the ratio between the two types of features needs to be given. Finally, we use the augmented batches to fine tune the FC classifier layer as shown in Fig. 5.

5   Experiments

We conduct experiments on the artificially created long-tailed CIFAR dataset [9] with various simulated imbalance factors, ImageNet-LT [29], Places-LT [29] and the real world long-tailed iNaturalist 2017 and 2018 [40] datasets to validate the proposed method. Deep residual network (ResNet) with various depth are employed in our experiments.

Table 1. Classification accuracy on long-tailed CIFAR-10.

IM	ResNet-18					ResNet-34				
	10	20	50	100	200	10	20	50	100	200
Baseline	90.73	87.24	82.32	75.16	70.22	91.03	87.32	82.74	78.58	71.42
CB [9] $\beta = 0.9$	90.79	86.61	81.9	75.16	69.16	91.03	87.18	82.48	75.99	70.0
CB [9] $\beta = 0.999$	90.54	86.83	81.81	76.4	69.83	90.74	87.24	81.66	74.85	70.08
CB [9] $\beta = 0.9999$	89.61	86.05	80.4	75.04	69.21	90.69	86.9	81.06	75.74	68.79
FL [27] $\gamma = 0.5$	90.66	86.61	81.55	74.99	69.06	90.76	87.18	81.91	76.5	69.87
FL [27] $\gamma = 1.0$	90.59	86.83	81.79	74.07	68.23	90.7	87.24	81.34	76.44	70.02
FL [27] $\gamma = 2.0$	90.5	86.05	81.25	75.13	68.27	90.08	86.9	82.44	75.58	69.87
SLA [26]	–	–	–	–	–	89.58	–	–	80.24	–
Ours	**91.75**	**88.54**	**84.51**	**80.57**	**77.06**	**91.2**	**89.26**	**84.49**	**82.06**	**75.52**

Table 2. Classification accuracy on long-tailed CIFAR-100.

IM	ResNet-18					ResNet-34				
	10	20	50	100	200	10	20	50	100	200
Baseline	62.59	57.09	48.55	43.65	38.87	63.87	57.55	48.07	43.55	37.5
CB [9] $\beta = 0.9$	63.1	57.02	48.15	43.51	38.58	64.14	58.03	48.44	42.94	38.84
CB [9] $\beta = 0.999$	61.76	55.3	44.28	32.19	26.61	63.05	54.13	40.89	32.65	26.2
CB [9] $\beta = 0.9999$	60.71	53.93	42.02	31.32	25.91	62.28	53.64	40.03	29.82	26.63
FL [27] $\gamma = 0.5$	62.64	57.02	47.9	42.82	38.73	64.36	58.45	48.31	42.72	36.18
FL [27] $\gamma = 1.0$	62.85	57.22	47.76	42.81	40.47	64.83	58.78	48.24	42.64	37.29
FL [27] $\gamma = 2.0$	63.37	57.15	47.0	42.18	40.31	64.48	58.55	47.47	43.33	38.11
SLA [26]	–	–	–	–	–	59.89	–	–	45.53	–
Ours	**65.08**	**58.69**	**51.9**	**46.57**	**42.84**	**65.29**	**59.75**	**52.17**	**48.51**	**41.46**

5.1 Long-Tailed CIFAR

To demonstrate the effectiveness of the proposed method, long-tailed versions of CIFAR dataset are generated following the protocol mentioned in [9] as $IM = \max(\{N_i\})/\min(\{N_i\})$. Five datasets of different imbalance factors, $\{10, 20, 50, 100, 200\}$, are created for both CIFAR-10 and CIFAR-100, where an imbalance factor is defined as where N_i is the number of training samples of the i-th class. ResNet with depth 18 and 34 are adapted for this experiment. We use the original validation set of the CIFAR-10 and CIFAR-100 to evaluate the performance.

The baseline network and the proposed method are implemented in PyTorch and run on a Xeon CPU 2.1 GHz and Tesla V100 GPU. The initial learning rate for Phase-I is 0.1 and decreases by 1/10 every 150 epochs. The feature sub-network and base classifier are trained for 300 epochs. In Phase-II, the learning rate is fixed at 0.001. The classifier is fine tuned for 6,400 iterations with a batch size of 128. For each real sample in tail classes, we choose $N_a = N_f = 3$.

Table 3. Top-1 classification accuracy on ImageNet-LT and Places-LT.

	ImageNet-LT				Places-LT			
	>100 Many	⩽100 & >20 Medium	<20 Few	Overall	>100 Many	⩽100 & >20 Medium	<20 Few	Overall
Plain Model [29]	40.9	10.7	0.4	20.9	**45.9**	22.4	0.36	27.2
Lifted Loss [31]	35.8	30.4	17.9	30.8	41.1	35.4	24	35.2
FL [27]	36.4	29.9	16	30.5	41.1	34.8	22.4	34.6
Range Loss [45]	35.8	30.3	17.6	30.7	41.1	35.4	23.2	35.1
FSLwF [15]	40.9	22.1	15	28.4	43.9	29.9	**29.5**	34.9
OLTR [29]	43.2	**35.1**	**18.5**	**35.6**	44.7	37	25.3	35.9
Ours	**47.3**	31.6	14.7	35.2	42.8	**37.5**	22.7	**36.4**
Ours + FL	47.0	31.3	16.8	35.3	42.2	36.4	24.0	36.0

A sample learning curve for long-tailed CIFAR-10 with an imbalance factor of 200 using ResNet-18 is shown in Fig. 6. The performance of the proposed method is compared with a baseline setting of the same learning rate but without the feature space augmentation. After the Phase-II feature space augmentation, the accuracy of the proposed method on the validation set increases about 7% during the fine tuning stage, while no noticeable change in accuracy for the baseline setting is observed.

To further illustrate the improvement, the confusion matrix before Phase-II and its changes are shown in Fig. 7. After Phase-I training, tail classes show poor accuracy on the validation set due to insufficient training samples in those classes. Most mis-classified samples fall into the first several head classes, where most training samples belong to, as indicated in the left bottom corner of the confusion matrix. After Phase-II fine tuning, significant improvement is observed for the diagonal elements of the tail classes. The off-diagonal elements decrease accordingly. Although the accuracy of the head classes decrease slightly, due to dramatic improvement in the tail classes, the overall accuracy still increases.

The complete classification performance on different imbalance factors of the two dataset are shown in Table 1 and 2. The method using the same ResNet with cross-entropy loss and traditional data augmentation on input images is referred as Baseline in Table 1 and 2. In our experiments, we compare our method with the state of the arts on addressing the long-tailed problem, including Class-balanced (CB) loss [9] based method and Focal Loss (FL) from [27] with various choices of hyper-parameters and augmentation based method [26]. Our method outperforms all other methods in both datasets.

5.2   ImageNet-LT and Places-LT Dataset

We also evaluate the proposed method on two constructed large-scale long-tailed datasets ImageNet-LT and Places-LT [29]. ImageNet-LT is a subset of the ILSVRC2012 dataset. Its training set is drawn from the original training set following the Pareto distribution with $\alpha = 6$, which results 115.8K images from

Table 4. Top-1 classification accuracy on iNaturalist (*: results from literature).

		iNaturalist 2017	iNaturalist 2018
Baseline	ResNet-50	60.50	62.27
	ResNet-101	61.81	65.19
	ResNet-152	65.12	66.17
CB [9]	ResNet-50*	58.08	61.12
	ResNet-101*	60.94	63.88
	ResNet-152	64.75	66.97
Ours	ResNet-50	**61.96**	**65.91**
	ResNet-101	**64.16**	**68.39**
	ResNet-152	**66.58**	**69.08**

1000 categories with a maximum of 1280 images per category and a minimum of 5 images per category ($IM = 256$). The original validation set with 50K images is used as test set in our experiments. Places-LT dataset is constructed similarly with ImageNet-LT from Places-2 dataset. Finally, 184.5K images from 365 categories are collected, where the largest class contains 4980 images while the smallest ones with 5 images ($IM = 996$). The test set contains 36.5K images.

For fair comparison, we use the same scratch ResNet-10 for ImageNet-LT and pre-trained ResNet-152 for Places-LT as in [29]. The numerical results and comparison with other peer methods are reported in Table 3. We also evaluate the combination of other balanced loss methods with the proposed method in these experiments. The different losses are applied in the Phase-I training. We use "Ours+FL" to refer the experiments using Focal Loss and "Ours" for ordinary cross-entropy loss. Both of our methods achieve comparable performance with the state-of-the-art method.

Note that, "Ours+FL" shows better performance than "Ours" in the ImageNet-LT dataset while "Ours" is better in Places-LT. Feature maps generated in the shallow network as ResNet-10 is not as sparse as in ResNet-152. Therefore, as explained in Sect. 4.3, the feature space augmentation delivers more performance boost to ResNet-152. On the other hand, our class balanced training batch generation achieves a similar effect as other balanced loss methods in the Phase-II fine tuning. But applying those losses in the Phase-I may still improve the performance when poor Phase-I performance affects CAM quality.

5.3 iNaturalist

iNaturalist is a real-world fine-grained species classification dataset. Its 2017 version contains 579,184 training images of 5,089 categories, and its 2018 version [40] has 437,513 training samples in 8,142 classes. The imbalance factor for iNaturalist 2017 is 435 and 500 for iNaturalist 2018. For both versions, there are three validation samples for each class. We adapt ResNet-50, ResNet-101 and

ResNet-152 in our experiments, all with 224×224 input image size. The similar training strategy with CIFAR datasets is adapted for iNaturalist. In Phase-I, the starting learning rate is 0.1 and reduced every 30 epochs for total of 100 epochs. In Phase-II, fine tuning is performed with a fixed learning rate of 0.001 for 200 iterations. The top-1 classification accuracy for the validation set of the two datasets are reported in Table 4. We compare the proposed method with class-balanced cross-entropy loss on ResNet-152 and class-balanced focal loss on ResNet-101/50. Our method has shown the best performance in all the settings.

5.4   Ablation Analysis

One major hyper-parameter in the proposed method is how to separate the head classes from the tail classes. Specifically, the classes are first sorted in the descent order by the number of training samples in each class as illustrated in Fig. 2. The first h classes are chosen as head classes. In order to unify the choice between different imbalance factors of datasets, we introduce $h_r \in (0, 1)$ which is the ratio between the number of samples in the head classes and the total number of samples. Different h_r choices against the Phase-II classification accuracy are evaluated in Fig. 8. The curves among different datasets show peaks around $h_r = 0.95$. On the left of peaks, fewer samples or classes are used as the head class, and thus class-generic features cannot be drawn sufficiently for feature augmentation. On the right side, fewer classes are selected as the tail classes, and therefore some classes with insufficient training samples will not be fine tuned with augmented samples. For consistency, we choose the minimum of h that satisfies $h_r \geq 0.9$ in all the CIFAR experiments.

We also investigate the classification performance when applying the feature space augmentation at different depths of the network. The ResNet architecture we adapted usually consists of four convolutional blocks. We plug our feature space augmentation after each of the last three convolutional blocks of ResNet-18 on CIFAR dataset. When augmenting features after Block2 and Block3, class-specific features in $\mathbf{F}_u^g$ are replaced with the class-specific features in $\mathbf{F}_c^g$ with random ratio to generate augmented samples, where spatial information of $\mathbf{F}_u^g$ is preserved. The corresponding classification accuracy after Phase-II fine tuning is shown in Fig. 9. From Fig. 9, one can observe that feature augmentation after Block4 gains the best performance among different datasets and imbalance factors. The feature maps closer to the input side contain more spatial information, which also introduces additional artifacts into the augmented samples. Features generated by Block4 are directly passed into the global pooling layer where the noise in the spatial dimension introduced by augmentation can be eliminated. Moreover, the linearity of the high-level feature space helps the final linear operation of the fusion. We, therefore, apply the feature space augmentation after Block4. We also compare the performance of only sampling balanced finetuning batch without augmentation applied, which is refered as "No Aug" in Fig. 9.

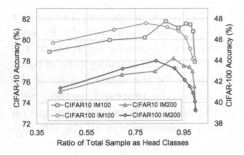

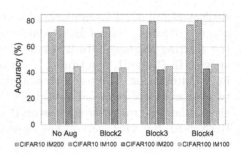

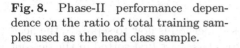

Fig. 8. Phase-II performance dependence on the ratio of total training samples used as the head class sample.

Fig. 9. Classification accuracy by applying feature space augmentation at different depth of ResNet-18.

6   Conclusion

In this paper, we propose a novel learning scheme to address the problem of training the deep convolutional neural network based classifier with long-tailed datasets. In detail, by combining the class-generic features in head classes with class-specific features in tail classes, augmented samples are online generated to enhance the performance of the tail classes. Results on long-tailed version CIFAR-10/100, ImageNet-LT, Places-LT and real-world long-tailed iNaturalist 2017/2018 datasets have shown the effectiveness of proposed method.

Acknowledgment. Ling was supported in part by US NSF Grants 1814745, 1618398, and 2002434.

References

1. Akata, Z., Reed, S., Walter, D., Lee, H., Schiele, B.: Evaluation of output embeddings for fine-grained image classification. In: CVPR (2015)
2. Bengio, S.: Sharing representations for long tail computer vision problems. In: ICMI (2015)
3. Bian, X., Lim, S.N., Zhou, N.: Multiscale fully convolutional network with application to industrial inspection. In: WACV (2016)
4. Buda, M., Maki, A., Mazurowski, M.A.: A systematic study of the class imbalance problem in convolutional neural networks. Neural Netw. **106**, 249–259 (2018)
5. Chawla, N.V., Bowyer, K.W., Hall, L.O., Kegelmeyer, W.P.: SMOTE: synthetic minority over-sampling technique. JAIR **16**, 321–357 (2002)
6. Chen, C., Seff, A., Kornhauser, A., Xiao, J.: Deepdriving: learning affordance for direct perception in autonomous driving. In: ICCV (2015)
7. Chen, Y., Bai, Y., Zhang, W., Mei, T.: Destruction and construction learning for fine-grained image recognition. In: CVPR (2019)
8. Chen, Z., Fu, Y., Chen, K., Jiang, Y.G.: Image block augmentation for one-shot learning. In: AAAI (2019)
9. Cui, Y., Jia, M., Lin, T.Y., Song, Y., Belongie, S.: Class-balanced loss based on effective number of samples. In: CVPR (2019)

10. Drummond, C., Holte, R.C., et al.: C4. 5, class imbalance, and cost sensitivity: why under-sampling beats over-sampling. In: Workshop on Learning from Imbalanced Datasets II (2003)
11. Elkan, C.: The foundations of cost-sensitive learning (2001)
12. Esteva, A., et al.: A guide to deep learning in healthcare. Nat. Med. **25**, 24–29 (2019)
13. Everingham, M., Van Gool, L., Williams, C.K., Winn, J., Zisserman, A.: The pascal visual object classes (VOC) challenge. IJCV **88**, 303–338 (2010)
14. Geifman, Y., El-Yaniv, R.: Deep active learning over the long tail. arXiv:1711.00941 (2017)
15. Gidaris, S., Komodakis, N.: Dynamic few-shot visual learning without forgetting. In: CVPR (2018)
16. Girshick, R.: Fast R-CNN. In: ICCV (2015)
17. Hariharan, B., Girshick, R.: Low-shot visual recognition by shrinking and hallucinating features. In: CVPR (2017)
18. He, H., Bai, Y., Garcia, E.A., Li, S.: ADASYN: adaptive synthetic sampling approach for imbalanced learning. In: 2008 IEEE International Joint Conference on Neural Networks (2008)
19. He, K., Zhang, X., Ren, S., Sun, J.: Deep residual learning for image recognition. In: CVPR (2016)
20. Huang, C., Li, Y., Change Loy, C., Tang, X.: Learning deep representation for imbalanced classification. In: CVPR (2016)
21. Huang, G., Liu, Z., Van Der Maaten, L., Weinberger, K.Q.: Densely connected convolutional networks. In: CVPR (2017)
22. Kang, B., et al.: Decoupling representation and classifier for long-tailed recognition. In: ICLR (2020)
23. Khan, S.H., Hayat, M., Bennamoun, M., Sohel, F.A., Togneri, R.: Cost-sensitive learning of deep feature representations from imbalanced data. TNNLS **29**, 3573–3587 (2017)
24. Krizhevsky, A., Hinton, G., et al.: Learning multiple layers of features from tiny images. Technical report, Citeseer (2009)
25. Krizhevsky, A., Sutskever, I., Hinton, G.E.: Imagenet classification with deep convolutional neural networks. In: NIPS (2012)
26. Lee, H., Hwang, S.J., Shin, J.: Rethinking data augmentation: Self-supervision and self-distillation. arXiv preprint arXiv:1910.05872 (2019)
27. Lin, T.Y., Goyal, P., Girshick, R., He, K., Dollár, P.: Focal loss for dense object detection. In: ICCV (2017)
28. Lin, T.-Y., et al.: Microsoft COCO: common objects in context. In: Fleet, D., Pajdla, T., Schiele, B., Tuytelaars, T. (eds.) ECCV 2014. LNCS, vol. 8693, pp. 740–755. Springer, Cham (2014). https://doi.org/10.1007/978-3-319-10602-1_48
29. Liu, Z., Miao, Z., Zhan, X., Wang, J., Gong, B., Yu, S.X.: Large-scale long-tailed recognition in an open world. In: CVPR (2019)
30. Long, J., Shelhamer, E., Darrell, T.: Fully convolutional networks for semantic segmentation. In: CVPR (2015)
31. Oh Song, H., Xiang, Y., Jegelka, S., Savarese, S.: Deep metric learning via lifted structured feature embedding. In: CVPR (2016)
32. Ouyang, W., Wang, X., Zhang, C., Yang, X.: Factors in finetuning deep model for object detection with long-tail distribution. In: CVPR (2016)
33. Russakovsky, O., et al.: ImageNet large scale visual recognition challenge. Int. J. Comput. Vis. **115**(3), 211–252 (2015). https://doi.org/10.1007/s11263-015-0816-y

34. Sandler, M., Howard, A., Zhu, M., Zhmoginov, A., Chen, L.C.: Mobilenetv 2: inverted residuals and linear bottlenecks. In: CVPR (2018)
35. Sarafianos, N., Xu, X., Kakadiaris, I.A.: Deep imbalanced attribute classification using visual attention aggregation. In: ECCV (2018)
36. Selvaraju, R.R., Cogswell, M., Das, A., Vedantam, R., Parikh, D., Batra, D.: Grad-cam: Visual explanations from deep networks via gradient-based localization. In: ICCV (2017)
37. Shen, L., Lin, Z., Huang, Q.: Relay backpropagation for effective learning of deep convolutional neural networks. In: ECCV (2016)
38. Tan, M., Le, Q.: EfficientNet: rethinking model scaling for convolutional neural networks. In: ICML (2019)
39. Ting, K.M.: A comparative study of cost-sensitive boosting algorithms. In: ICML (2000)
40. Van Horn, G., et al.: The inaturalist species classification and detection dataset. In: CVPR (2018)
41. Wang, Y.X., Girshick, R., Hebert, M., Hariharan, B.: Low-shot learning from imaginary data. In: CVPR (2018)
42. Wang, Y.X., Ramanan, D., Hebert, M.: Learning to model the tail. In: NIPS (2017)
43. Xian, Y., Lorenz, T., Schiele, B., Akata, Z.: Feature generating networks for zero-shot learning. In: CVPR (2018)
44. Yin, X., Yu, X., Sohn, K., Liu, X., Chandraker, M.: Feature transfer learning for deep face recognition with long-tail data. arXiv:1803.09014 (2018)
45. Zhang, X., Fang, Z., Wen, Y., Li, Z., Qiao, Y.: Range loss for deep face recognition with long-tailed training data. In: ICCV (2017)
46. Zhou, B., Khosla, A., Lapedriza, A., Oliva, A., Torralba, A.: Learning deep features for discriminative localization. In: CVPR (2016)
47. Zhou, B., Cui, Q., Wei, X.S., Chen, Z.M.: BBN: bilateral-branch network with cumulative learning for long-tailed visual recognition. In: CVPR (2020)
48. Zhou, Z.H., Liu, X.Y.: Training cost-sensitive neural networks with methods addressing the class imbalance problem. TKDE **18**, 63–77 (2005)
49. Zhu, P., et al.: VisDrone-VDT2018: The vision meets drone video detection and tracking challenge results. In: ECCV (2018)
50. Zou, Y., Yu, Z., Vijaya Kumar, B., Wang, J.: Unsupervised domain adaptation for semantic segmentation via class-balanced self-training. In: ECCV (2018)

Laying the Foundations of Deep Long-Term Crowd Flow Prediction

Samuel S. Sohn[1]([⊠]), Honglu Zhou[1], Seonghyeon Moon[1], Sejong Yoon[2], Vladimir Pavlovic[1], and Mubbasir Kapadia[1]

[1] Rutgers University, Piscataway, USA
{sss286,hz289,sm2062,vladimir,mk1353}@cs.rutgers.edu
[2] The College of New Jersey, Ewing, USA
yoons@tcnj.edu

Abstract. Predicting the crowd behavior in complex environments is a key requirement for crowd and disaster management, architectural design, and urban planning. Given a crowd's immediate state, current approaches must be successively repeated over multiple time-steps for long-term predictions, leading to compute expensive and error-prone results. However, most applications require the ability to accurately predict hundreds of possible simulation outcomes (e.g., under different environment and crowd situations) at real-time rates, for which these approaches are prohibitively expensive. We propose the first deep framework to instantly predict the *long-term flow* of crowds in arbitrarily large, realistic environments. Central to our approach are a novel representation CAGE, which efficiently encodes crowd scenarios into compact, fixed-size representations that losslessly represent the environment, and a modified SegNet architecture for instant long-term crowd flow prediction. We conduct comprehensive experiments on novel synthetic and real datasets. Our results indicate that our approach is able to capture the essence of real crowd movement over very long time periods, while generalizing to never-before-seen environments and crowd contexts. The associated Supplementary Material, models, and datasets are available at github.com/SSSohn/LTCF.

Keywords: Vision applications and systems · Datasets and evaluation

1 Introduction

Predicting the behavior of human crowds in complex environments is a key requirement in a multitude of application areas, including crowd and disaster management, architectural design, and urban planning [38]. In recent years, crowds research has been receiving increasing attention from the computer vision community [3, 6, 13, 30, 33, 40, 46, 48, 52, 54]. These works have largely focused on the esti-

Electronic supplementary material The online version of this chapter (https://doi.org/10.1007/978-3-030-58526-6_42) contains supplementary material, which is available to authorized users.

© Springer Nature Switzerland AG 2020
A. Vedaldi et al. (Eds.): ECCV 2020, LNCS 12374, pp. 711–728, 2020.
https://doi.org/10.1007/978-3-030-58526-6_42

mation of a crowd's size, while some others [1, 2, 14, 43, 55] have tackled the prediction of "crowd flow" at a fixed future time-step. Crowd flow is the cumulative distribution of crowd dynamics over an environment for a certain time interval. This notion has been well-studied because it is critical for understanding and managing the movement of large crowds in confined spaces [8]. In contrast to these tasks, we formulate the novel task of *long-term crowd flow prediction*. In this context, the term "long-term" indicates that crowd flow is predicted over the full simulation duration (i.e., until all agents have reached their goals), instead of over a small, fixed duration. The long-term task requires repetitions of the comparatively short-term task, which will naturally lead to error propagation over long prediction durations and become prohibitively expensive for real-time applications. Some current approaches are unable to predict long-term crowd flow because only one snapshot of the initial crowd configuration is given [1, 2]. Others present a tradeoff between computational efficiency, prediction accuracy, and scenario complexity for predictions on arbitrary large, complex environments with high-density crowds (in terms of the environment and crowd size) [43, 55], each of which is critical for real-time decision-making in real-world applications, such as evacuation planning. In order to escape this continuum, we tackle this new problem as a whole (not as a repeating subproblem like others have) and present the first accurate and instantaneous prediction of long-term crowd movement characteristics in complex environmental and crowd scenarios [17].

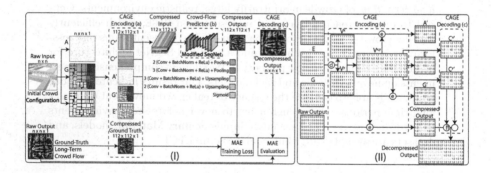

Fig. 1. Proposed approach. (I) Deep long-term crowd flow prediction framework. (II) Process for CAGE encoding and decoding.

We propose a unified deep framework (Fig. 1) to perform long-term crowd flow prediction in realistic scenarios where environments can be arbitrarily large and complex, and crowd densities can be both low and high. This is made possible through the following main contributions: (1) We propose a novel CAGE representation consisting of Capacity, Agent, Goal, and Environment-oriented information, which can compress environments of arbitrary size and high complexity, into a fixed-size image-like representation while preserving the key environmental characteristics, which are critical for predicting the crowd movement.

(2) In order to overcome the challenge of obtaining real-world data, we develop a synthetic dataset of procedurally generated environments, dynamically simulated crowd flows, and statically derived "proxy" crowd flows (which have more error but are more efficient to compute), for model training and evaluation. (3) While the framework is model-agnostic, we have specifically chosen the efficient SegNet architecture [4], modified to accept as input the CAGE representation (a 3D tensor) and predict the crowd flow. We demonstrate the efficacy of CAGE in encoding environments of any size and the evaluate prediction accuracy across environmental conditions, amounts of compression, crowd densities (from sparse to extremely dense crowds), and contexts (from crowds moving toward a single goal to multiple goals) using models trained on proxy and simulated crowd flow data. We showcase the model's ability to capture the essence of real human crowd behavior while permitting extrapolation to new and unseen situations. Our research lays a strong foundation towards the application of deep long-term crowd-flow prediction in applications including disaster and security management, computer-aided architectural design, and urban planning.

2 Related Work

Crowds research has been gaining attention from within the computer vision community in tackling problems such as pedestrian detection [3,48,52], crowd counting [6,30,33,40,54], crowd density estimation [13], and crowd flow prediction. Crowd flow prediction, our focus, is a spatio-temporal problem, which bears a similarity to crowd density estimation. The estimation of a crowd's density is normally performed on a camera image or video of a crowd, and the task is to determine the crowd's distribution over a space [21,25,42,51,53]. Meanwhile, the prediction of crowd flow attempts to determine the aggregation of crowd densities across the temporal axis, which encodes the crowd's movement [1,2,14,16,43,49,55]. The problem can take two forms: predictions in the short-term or long-term. Regardless of whether by rule based simulators or deep-learning-based models, tackling long-term crowd flow prediction with short-term approaches relies on successive, expensive predictions of the crowd's future state over short time intervals (i.e, predicting the crowd motion at the next time step, then using the predicted crowd motion to continue predicting until termination). Computational time and resources are wasted in using short-term predictors for the cases where only the aggregated long-term analysis is necessary (e.g., identifying locations in the environment that will cause congestion). In contrast to the previous works, we predict crowd behaviors in the long term.

Convolutional Neural Networks (CNN) have been prevalently deployed in the domain of crowds [19,24,36,39,47], with applications in crowd counting [12,23,27, 41,44,45] and density estimation [15,20,22,50]. CNNs have also shown promise for predicting crowd dynamics [18,49]. Behavior-CNN [46] models pedestrian behaviors in crowded scenes with applications to walking path prediction, destination prediction, and tracking. ST-ResNet [49] and STRCNs [16] forecast the inflow and outflow of crowds in every region of a city. CSRNet [19] can understand and recognize highly congested scenes. Variants of CNNs (with recurrence) are also used to predict short-term crowd dynamics [14,16,55]. Given the complexity of the scenarios that we propose to tackle, we cannot rely on the same datasets as prior works, since these are not representative of concerted crowd flows at different densities. This compels us to generate our own dataset. We employ a deep convolutional encoder-decoder architecture for the instant prediction of long-term crowd flow. Our task is similar to crowd density estimation in terms of the visual format of the output. However, we do not generate a density map for a static crowd configuration as aforementioned works have. Instead, we predict the aggregated change in a crowd's configuration over a long, future time interval.

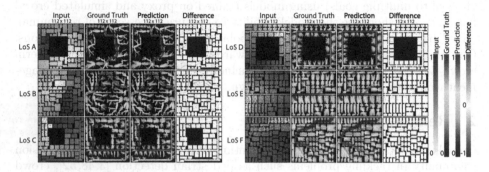

Fig. 2. Model performance on different density levels in small environments (no compression involved). See Sect. 4.3 for color-coding.

3   Method

We formulate the problem of *long-term crowd flow prediction* as follows:

Definition 1. *(Long-Term Crowd Flow Prediction): Given an initial crowd scenario, consisting of environment, agent, and goal information, at time t_0 and time t_l at which the last agent will reach its goal,* **Long-Term Crowd Flow Prediction** *aims to predict the aggregated crowd flow, i.e., the log-frequency with which agents appear in each location of the environment from t_0 to t_l.*

3.1   Preliminaries

In predicting long-term crowd flow, we do not limit ourselves with a fixed-size environment or simplistic crowd configurations. We aim specifically to make these predictions for an arbitrary crowd configuration in a built environment of any size that moves toward arbitrary goal locations. As raw input, we are given a matrix $\mathbf{E} \in \{0,1\}^{n \times n}$, which stores a 2D discretization of a built environment where each cell's width and height represent the size of one person. The diameter of the agents is 0.6×0.6 m^2 in the simulated data, which makes the environment size between $0.6n \times 0.6n$ m^2. In matrix $\mathbf{E} \in \{0,1\}^{n \times n}$, navigable cells have value 0 and non-navigable cells have value 1. We add paddings with value 1 for environments that are not in a square shape. Another matrix $\mathbf{A} \in \{0, 1\}^{n \times n}$ is used to represent the unique initial agent locations, where the cells occupied by an agent have value 1, and all others are 0. The matrix $\mathbb{G} \in \{0, 1\}^{n \times n}$ stores in each cell whether the cell contains one of the agents' goals. We use index pair (i, j) to denote a location in the matrices, where $i = 1..n$ and $j = 1..n$. The crowd flow prediction output $\mathbf{Y} \in [0, 1]^{n \times n}$ is calculated for each navigable cell in environment $\mathbf{E}$ as the normalized logarithm of the number of times any agent appears in the cell (i, j). As defined earlier, long-term crowd flow aggregates the crowd states across all time steps in the interval. In this work's datasets, we have configured the time steps to be 0.02 s.

We further represent $\mathbb{G} \in \{0, 1\}^{n \times n}$ with the matrix $\mathbf{G} \in [0, 1]^{n \times n}$, which stores in each cell the normalized path distance to its closest goal location. Agents can have multiple goals, which are encoded into $\mathbf{G}$ by the following procedure. We transform each goal location $g := (i, j)$ into a distance matrix $\mathbf{D}^{(g)} \in [0, \infty)^{n \times n}$, where each cell is the non-normalized shortest path length from that cell to the closest goal using Theta* [29]. The resulting distance information $\mathbf{D}^{(g)}$ for each goal is then aggregated into $\mathbf{G}$ by taking the minimum value across the distance matrices for each cell, $\mathbf{G} = \min_g \mathbf{D}^{(g)}$, which is subsequently normalized. The resulting $\mathbf{G}$ encodes not only where goals are located (i.e., wherever $\mathbf{G}_{i,j} = 0$), but also how far an agent is from its goal. Figure 4 visualizes multiple goals in its second row. The implication of $\mathbf{G}$ is that agents can perform pathfinding to find the shortest path to its goal, which is ubiquitous among crowd simulators.

The resulting input representation is $\mathbf{X} = [\mathbf{A}, \mathbf{G}, \mathbf{E}]$, for which the scale of the environment is dependent on the diameter of agents (which is equal to the width of a cell in $\mathbf{E}$). This means that a cell can represent an exact capacity of one agent per cell. However, there is a certain inflexibility to this representation for predicting crowd flow $\mathbf{Y}$ from $\mathbf{X}$. With a fixed number n, $\mathbf{E}$ cannot capture the size of an arbitrary input environment. We resolve this problem by converting the variable-size agent-centric representation $\mathbf{X}$ into a fixed-size environment-centric one $\mathbf{X}'$ through our novel CAGE encoding and decoding techniques. The proposed framework bridges these two representations (Fig. 1). The pipeline consists of three major stages: (a) the CAGE-encoding of raw data, (b) the prediction of crowd flow, and (c) the CAGE-decoding of the prediction. The predictor takes as input the CAGE representation $\mathbf{X}'$, which is computed by compressing the original input $\mathbf{X}$ to a fixed-size representation compatible with the predictive model's input (Sect. 3.3). If compression has occurred, then the decompression stage is needed in order to view the predicted crowd flow $\hat{\mathbf{Y}}'$ on the original environment $\hat{\mathbf{Y}}$. Next, we describe CAGE and propose a variant to the CNN architecture for instant long-term crowd flow prediction.

3.2 CAGE Representation

Figure 1 (II) illustrates the CAGE encoding and decoding process, described below.

Encoding. Given an axis-aligned real environment in its discretized form $\mathbf{E}^{n \times n}$, we apply a compression method that preserves environmental information and reduces the dimensions of $\mathbf{E}$ to $n' \times n'$, where $n' <= n^1$. This is achieved by maintaining the local navigability between cells, while warping their capacities (i.e., the maximum number of agents that can occupy them). For this, we require two additional "channels" of information $\mathbf{C}^x \in [0,1]^{n' \times n'}$ and $\mathbf{C}^y \in [0,1]^{n' \times n'}$ for storing capacities: one for capacity along the x-axis and the other for capacity along the y-axis. Before compression, both $\mathbf{C}^x$ and $\mathbf{C}^y$ consists of ones, meaning that each cell has a capacity of 1 agent per cell. $\mathbf{C}^x$ and $\mathbf{C}^y$ for compressed environments are calculated as follows. First, based on $\mathbf{E}$, we introduce another two $n \times n$ matrices, $\mathbf{V}^x$ and $\mathbf{V}^y$, which encode the visibility of $\mathbf{E}$, i.e., in each navigable cell (i,j), $\mathbf{V}^x{}_{i,j}$ stores the number of visible cells along the x-axis (the number of successive navigable cells without being blocked by any non-navigable cells along the x-axis). $\mathbf{V}^y$ is computed in a similar way but along the y-axis (Fig. 1 II.d). Combining $\mathbf{V}^x$ and $\mathbf{V}^y$, we form matrix $\mathbf{V}^{x,y}$ of dimension $n \times n$, in which cell (i,j) stores values from $\mathbf{V}^x$ and $\mathbf{V}^y$ in the form of $(\mathbf{V}^x{}_{i,j}, \mathbf{V}^y{}_{i,j})$. The $p \times q$ successive cells (forming a region of size $p \times q$) in $\mathbf{V}^{x,y}$ that share the same value are the cells that can be merged into a single cell in the compressed environment. A proof of the above procedure's optimality has been provided in the Supplementary Material. As the capacity of that single cell (newly compressed) along the x-axis corresponds to q agents per cell and along the y-axis corresponds to p agents per cell, the value of that single cell becomes $1/q$ in $\mathbf{C}^x$ and $1/p$ in $\mathbf{C}^y$. The total capacity of that single cell (newly compressed) is $p \times q$. By manipulating $\mathbf{C}^x$ and $\mathbf{C}^y$, regions in $\mathbf{E}$ are able to be compressed. After the above procedure, navigable regions in the original environment are compressed and become smaller, but the compressed environment remains the same size as the original environment with dimensions $n \times n$. However, there is an increase in the amount of non-navigable cells in the borders. The outer rows and columns that are entirely non-navigable can be removed, thus decreasing the dimensions of the compressed environment by $(n - n')$ along both the x- and y-axes. In this way, $\mathbf{E}$ is now transformed into $\mathbf{E}'$. For our framework, we have used $n' = 112$ and an agent radius of 0.3, which allows us to encode environments from size 67×67 m^2 to at least 336×336 m^2 in our dataset, but this size can exceed beyond $17,136 \times 17,136$ m^2. By explicitly encoding the amount of compression, our predictive model does not need to infer the scale of the input. Other works such as Switch-CNN [35] rely on giving different scales to different regressors.

The CAGE encoding of $\mathbf{E}$ affects channels $\mathbf{A}$ and $\mathbf{G}$ as well. For the compressed environment, $\mathbf{A}' \in [0,1]^{n \times n}$ represents agent density, instead of binary agent presence. The compressed cell's value in $\mathbf{A}'$ is the total number of agents among the corresponding original cells in the uncompressed environment divided by the total capacity of that compressed cell. For the $\mathbf{G}' \in [0,1]^{n \times n}$ of a compressed environment, each of its cells takes on the minimum $\mathbf{G}$-value of the original cells that are being compressed into this single cell, as this is consistent with the aforementioned use of $\mathbf{G}$, i.e., in finding the shortest path. Although some information from $\mathbf{A}$ and $\mathbf{G}$ is lost during compression, erroneous compression of $\mathbf{E}$ can more adversely affect

---

[1] If $n < n'$, $\mathbf{A}$, $\mathbf{E}$, and $\mathbf{G}$ are padded with row and columns to become $n' \times n'$. In this case, $\mathbf{A}'$, $\mathbf{E}'$, and $\mathbf{G}'$ are equal to $\mathbf{A}$, $\mathbf{E}$, and $\mathbf{G}$. $\mathbf{C}^x$ and $\mathbf{C}^y$ are matrix of dimension $n' \times n'$ with all entry values as 1.

the crowd flow prediction (e.g., a thin wall disappearing), so we primarily seek to encode $\mathbf{E}$ losslessly, while maintaining a fixed bound of the imprecision in $\mathbf{A}'$ and $\mathbf{G}'$, limited to the decompressed region. The resulting CAGE representation consultutes of 5 channels, $\mathbf{X}' = [\mathbf{C}^x, \mathbf{C}^y, \mathbf{A}', \mathbf{G}', \mathbf{E}']$, where the prime symbol indicates that a matrix has been compressed to $n' \times n'$.

When training our predictive model, CAGE is also used to encode the ground truth crowd flow $\mathbf{Y}$ into compressed crowd flow $\mathbf{Y}'$ (Fig. 1 II.e), for the cases of model input being compressed. As a region of cells in $\mathbf{Y}$ is compressed into a single cell in $\mathbf{Y}'$, we sum up the crowd flow values in the original cells and divide by the total capacity of the compressed cell. This CAGE encoding technique is now equipped to represent a nearly infinite number of compressed environments and crowd scenarios within $n' \times n'$. Other image resampling techniques cannot do the same without losing some environment information. We have evaluated CAGE with other techniques such as nearest-neighbor, area, bilinear, bicubic, and Lanczos resampling for environments from our dataset, and we find that all other techniques consistently produce artifacts such as blurring [31], while CAGE is pixel-perfect (provided in the Supplementary Material). While CAGE is not guaranteed to compress a *random* environment, typical real-world environments are *structured* and, therefore, more likely to be compressed to $n' \times n'$. Furthermore, we provide a variant of the compression procedure for non-axis-aligned environments (Supplementary Material Fig. 3) that yields the same representation with some loss of environmental information.

Decoding. After the crowd flow is predicted on the compressed representation $\mathbf{X}'$ (denoted by $\hat{\mathbf{Y}}'$), decompression can be achieved by using $\mathbf{C}^x$ and $\mathbf{C}^y$ (Fig. 1 II.f). The value of a compressed cell in $\hat{\mathbf{Y}}'$ is duplicated back to the original cells that formed this compressed cell. In this way, we can turn $\hat{\mathbf{Y}}'$ with dimension $n' \times n'$ into $\hat{\mathbf{Y}}$ of dimension $n \times n$. CAGE considers each cell value in $\hat{\mathbf{Y}}'$ as the *average* crowd flow in a cell, and when CAGE decodes $\hat{\mathbf{Y}}'$, it recovers the original amount of crowd flow that was in the same compressed region. If prediction is perfectly accurate, the sum of values in $\hat{\mathbf{Y}}$ is equal to the sum of values in $\mathbf{Y}$. This can produce some aliasing artifacts, but since crowd flow is smooth, it is less affected by this.

3.3 Crowd-Flow Prediction

CAGE-encoded representations are effectively 5-channel images, which motivates our use of Convolutional Neural Networks. For the predictor (b) in Fig. 1, we modified SegNet [4] to facilitate the deep prediction of the complex crowd flows. We favor ourselves in modifying SegNet out of the following reasons: (1) SegNet is a symmetrical architecture with the ability to encode image-like representations to low dimensions, and map the low-resolution representation back to its full input resolution, perfectly serving our purposes. The mapping also has superiority in delineating objects of small sizes and in retaining boundary information for structured prediction without having as many trainable parameters as models such as U-Net [34]. (2) SegNet has the capability to extract, preserve, and understand appearance, shape, and the spatial-relationship information from the input image representation, which is beneficial for us to differentiate agents, obstacles, and navigable areas, and to capture the spatial-temporal correlation behind the scene. (3) SegNet is smaller and easier to train in an end-to-end fashion than many other CNN architectures such as [26,28]. It is designed to be efficient in memory and computational time especially in inference, which is crucial for our application scenarios (e.g. predicting long-term crowd flow to aid timely plan-making for emergency evacuations).

Network Architecture. The modified SegNet still has a symmetrical convolutional encoder-decoder architecture like the original SegNet. We made the following 2 changes. First, we removed the 3 outermost layers from both the encoder and decoder to allow inputs and outputs with dimensions of 112×112, instead of the original 224×224. This is because 224×224 is unnecessarily large for our dataset. The architecture of the modified SegNet can be seen in Fig. 1 (I), in which all convolutional layers utilize 3×3 convolution kernels and 1 padding. Second, the original SegNet is designed for image segmentation. Therefore, its last layer is a *softmax* for pixel-wise labeling. The prediction of long-term crowd flow is a regression task, so we have substituted the last layer with a *sigmoid* layer to predict the compressed output.

Training Details and Loss Function. For the modified SegNet, the encoder and decoder weights are all initialized following [10]. Stochastic gradient descent (SGD) with a fixed learning rate of 0.01 and momentum of 0.9 is used as the optimizer [5]. The batch size is set to 64, and the training set is shuffled before each epoch. We train the model with 200 epochs and use the same set of hyper-parameters for each of the following model variants. We select model hyper-parameters according to the validation performance. The batch size, learning rate and the momentum value are chosen empirically to obtain the best validation result from sets $\{32, 64, 128\}$, $\{0.1, 0.01, 0.001\}$, and $\{0.8, 0.9\}$ respectively. Adaptive learning rates have been considered but we find that the fixed one remains to be the most effective. We use *mean absolute error* as the objective loss function. The size of the training set (10% for validation) and the size of the testing set vary for each model variant, which we will describe in detail in Sect. 5.

4   Experimental Preliminaries

4.1   Synthetic Datasets

We have compiled a large dataset of synthetic environments and crowd movement for model training and evaluation.

Synthetic Environments. The work in [7] establishes a comprehensive taxonomy of floorplans based on observations of the exterior shapes and interior organizations of modern architecture. We have re-purposed this taxonomy in order to generate floorplans instead of classifying them. For this study, a dataset of over 84,000 floorplans was procedurally generated. We describe the particulars of our generator in the Supplementary Material.

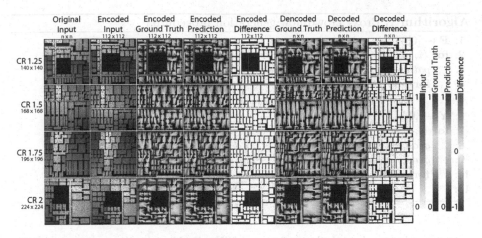

Fig. 3. Model performance on different compression rates (CR). (Color Figure online)

Simulated Crowd Flow. The crowd flow data that we use as ground truth must either be taken from the real world or simulated. It is quite difficult to acquire real-world crowd flows in varied environments and scenarios, so we have instead opted to simulate crowd flow using the Social Force model (SF) [11]. Our choice of the SF model is based on the model's computational efficiency and widespread use to this day in fields such as architecture, urban planning, and transportation engineering.

The simulated crowd flow has been generated under two crowd scenarios: scenarios with a single goal where agents are uniformly distributed across the environment, and multi-goal and non-uniform scenarios (MGNU), where agents share multiple goals and are distributed in clusters. Using an open-source simulator [37], it takes over 167 h to produce 7,800 crowd flow images in a 112×112 environment, making it prohibitively expensive to produce larger dataset with larger environments. Therefore, it is desirable to have a proxy that has key characteristics of simulated crowd flow, but is less expensive to generate.

Proxy Crowd Flow. Using proxy crowd flow, we aim to capture two major features of SF crowd flow: that (1) agents rely on the shortest path, and (2) as agents move closer to their goal, they become clustered with other agents, which causes congestion at bottlenecks and increases crowd flow closer to goals. The average mean absolute error between proxy and SF crowd flow across 600 varied goal, environment, and agent scenarios is 0.073 ± 0.025. In the context of the visualization, each primary/secondary color occupies $60°$ of the hue spectrum, and this error translates to an expected hue error of $26.28° \pm 9°$ for each pixel, which is very small. A visual comparison between proxy and SF crowd flows can be found in the Supplementary Material.

Given raw input $\mathbf{X} = [\mathbf{A}, \mathbf{G}, \mathbf{E}]$, the proxy crowd flow is generated using Algorithm 1. Each agent in $\mathbf{A}$ has its shortest path of cells planned to its goal (line 3), and has its path overlaid onto a frequency matrix $\mathbf{F}$ (line 5). The frequency matrix

Algorithm 1. Proxy crowd flow generation procedure.

1: $\mathbf{F} \leftarrow [0]^{n \times n}$
2: **for** $(i,j) \in \mathbf{A} \mid \mathbf{A}_{i,j} = 1$ **do**
3:     $path \leftarrow ShortestPathToClosestGoal(i,j)$
4:     **for** $(a,b) \in path$ **do**
5:         $\mathbf{F}_{a,b} \leftarrow \mathbf{F}_{a,b} + 1$
6: $\mathbf{Y} \leftarrow [0]^{n \times n}$
7: **for** $(i,j) \in \mathbf{F}$ **do**
8:     $groupSize \leftarrow \log(1 + \mathbf{F}_{i,j})$
9:     $group \leftarrow GetClosestCells(i,j,groupSize)$
10:     **for** $(a,b) \in group$ **do**
11:         $\mathbf{Y}_{a,b} \leftarrow \mathbf{Y}_{a,b} + \mathbf{F}_{i,j}$

---

is then used to produce the latter feature of SF crowd flow by clustering agents (line 9) and increasing crowd flow closer to goals (line 11). Function $GetClosestCells$ returns a set of $groupSize$-many cells with closest path distance to cell (i,j). The proxy crowd flow is generated under the crowd scenario where there is a single goal in the environment and agents are uniformly distributed. It takes little over 2 h to generate the same 7,800 scenarios as the SF crowd flow (167 h).

4.2 Real-World Datasets

NARF Dataset. We evaluate our model on real-world environments with simulated crowd flows. These environments include King's Cross, Prentice Women's Hospital, University of Economics Vienna's Library and Learning Centre, and Tate Britain. The environments are non-axis-aligned real-world floorplans (NARF), and serve to evaluate the ability for our models, which were trained only on axis-aligned environments, to generalize.

Stanford Crowd Flow. The Stanford crowd trajectory dataset [1] provides a large set of real pedestrian trajectories collected at a train station of size $25\,\mathrm{m} \times 100\,\mathrm{m}$ for $12 \times 2\,\mathrm{h}$ (totally 12 days, for each day, 7:00 a.m. to 8:00 a.m. and 5:00 p.m. to 6:00 p.m. are recorded) by a set of distributed cameras. We follow the same data processing strategy introduced in [32] to retrieve the pedestrians' trajectories and the environment layout. We select the data of the first day and choose 5 min as the time interval to bin the two-hour data by.

4.3 Evaluation Protocols

Qualitative Evaluation. We use a visualization scheme for comparing two types of crowd flows (typically ground truth and predictions). For example, in Fig. 3, the first two columns show the CAGE input, where the white-blue gradient corresponds to $\mathbf{G}$, the transparent-to-red gradient corresponds to $\mathbf{A}$, and the black pixels show where $\mathbf{E}$ is non-navigable. The next two columns show crowd flow using the jet gradient, and the following column computes the signed difference by subtracting

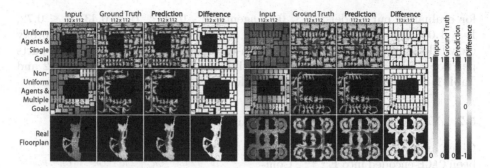

Fig. 4. Model performance on SF crowd flows for synthetic environment (first two rows) and real-world floorplan (last row).

Table 1. Average mean absolute error of models trained without compression.

	$PanL^{PX}$			$PanL^{SF}$					$PanL^{SF+}$			
CR-Train	1	1	1	1	1	1	1	1	1	1	1	1
Flow-Train	Proxy	Proxy	Proxy	SF	SF	SF	SF	SF	SF + MGNU	SF + MGNU	SF + MGNU	SF + MGNU
Flow-Test	Proxy	SF	Stanford	Proxy	SF	MGNU	NARF	Stanford	SF	MGNU	NARF	Stanford
LoS A	**0.0186**	0.0707	0.0160	0.0665	0.0250	0.0336	0.0672	**0.0141**	**0.0200**	**0.0084**	**0.0514**	0.0142
LoS B	**0.0179**	0.0777	0.0151	0.0720	0.0261	0.0341	0.0649	0.0149	**0.0216**	**0.0088**	**0.0537**	0.0144
LoS C	**0.0167**	0.0752	0.0162	0.0700	0.0256	0.0346	0.0635	0.0152	**0.0215**	**0.0099**	**0.0544**	0.0145
LoS D	**0.0161**	0.0738	0.0172	0.0700	0.0254	0.0345	0.0607	0.0158	**0.0222**	**0.0095**	**0.0546**	0.0148
LoS E	**0.0147**	0.0732	N/A	0.0711	0.0250	0.0326	0.0594	N/A	**0.0221**	**0.0121**	**0.0546**	N/A
LoS F	**0.0143**	0.0722	N/A	0.0732	0.0241	0.0346	0.0570	N/A	**0.0220**	**0.0095**	**0.0542**	N/A

the third column from the fourth column, where negative values are false negatives (FN) that underpredict crowd flow, and positive values are false positives (FP) that produce phantom crowd flow where there should be either less or none. The signed difference is visualized using a blue-white-red gradient, where FN is blue, perfect accuracy is white, and FP is red.

Quantitative Evaluation. We use the widely-adopted Mean Absolute Error (MAE) and Mean Squared Error (MSE) for quantitative evaluation [24,27,44,50] [15,23,42,45]. For each test scenario, we calculate the MAE and MSE of each test data point and then average over all test data points in that test set. We choose MAE to report instead of MSE because MSE is often being criticized for throwing the scale off and penalizing more for outliers, it is also biased towards higher deviations. In general MAE is more robust compared to MSE.

5   Experiments and Evaluation

5.1   Varied Density and No Compression

Level of Service (LoS) [9] describes flow relationships with respect to volume or density in a pedestrian environment. Six Levels of Service are defined (A to F) from sparse to dense crowds. We use the following LoS density conditions for evaluating our models: A ≤ 0.27 agents/m^2, $0.31 < B \leq 0.43$, C ≤ 0.72, D ≤ 1.08, E ≤ 2.17, F > 2.17. We have trained a model $PanL^{PX}$ using proxy crowd flow data from LoS A to F (train-test split of $12k : 3k$ with equal parts per LoS). The qualitative and quantitative results are presented in Fig. 2 and Table 1 respectively. Model $PanL^{PX}$ performs well across all density levels.

5.2   Varied Compression Rate with Varied Density

Given an environment $\mathbf{E}^{n \times n}$ that has been CAGE-encoded into $\mathbf{E}^{n' \times n'}$, Compression Rate (CR) is defined as n/n'. We evaluate CAGE and the effect of compression rate on proxy crowd flow prediction for crowds of varied density levels (LoS A to F). We first train independent models on data with each of the following compression rates $\{1.0, 1.25, 1.5, 1.75, 2.0\}$. The models are $PanL^{PX}$, $CR^{1.25}$, $CR^{1.5}$, $CR^{1.75}$, and CR^2 respectively (train-test split is $12k : 3k$ with equal parts per LoS). We test the 5 models on all compression rates. The results shown in Table 2 indicate that models trained with samples from CR 1.25 and CR

	$PanL^{PX}$	$CR^{1.25}$	$CR^{1.5}$	$CR^{1.75}$	CR^2	$PanCR$
CR-Train	1	1.25	1.5	1.75	2	1~2
Flow-Train	Proxy	Proxy	Proxy	Proxy	Proxy	Proxy
Flow-Test	Proxy	Proxy	Proxy	Proxy	Proxy	Proxy
LoS A	0.0728	0.0346	0.0358	0.0357	0.1270	**0.0322**
LoS B	0.0582	0.0312	0.0320	0.0326	0.1147	**0.0289**
LoS C	0.0509	0.0296	0.0300	0.0308	0.1248	**0.0269**
LoS D	0.0475	0.0274	0.0279	0.0285	0.1345	**0.0248**
LoS E	0.0402	0.0232	0.0237	0.0241	0.0997	**0.0210**
LoS F	0.0389	0.0212	0.0224	0.0211	0.0795	**0.0211**
CR 1.0	**0.0164**	0.0190	0.0201	0.0261	0.1011	0.0167
CR 1.25	0.0433	0.0247	0.0255	0.0280	0.1057	**0.0236**
CR 1.5	0.0577	0.0288	0.0293	0.0291	0.1097	**0.0273**
CR 1.75	0.0667	0.0321	0.0326	0.0301	0.1128	**0.0298**
CR 2.0	0.0728	0.0348	0.0355	**0.0313**	0.1415	0.0317

Table 2. Average Mean Absolute Error of models trained with CAGE under different compression rate (CR).

1.75 perform better across compression rates, alluding to the benefits of a model trained across different compression rates.

Following these observations, we train a model on a dataset of all compression rates and all density levels (train-test split is $18k : 3k$ with equal parts per CR and LoS). The qualitative results of this multi-density and multi-compression-rate model ($PanCR$) can be found in Fig. 3. We also compare $PanCR$ with the models trained on a single compression rate quantitatively in Table 2. Model $PanCR$ is able to successfully predict proxy crowd flow across different environment sizes and crowd densities.

5.3   Results on Simulated Crowd Flow

Substitute Proxy with Simulated. Using model $PanL^{SF}$, we evaluate the performance of our approach on Social Force simulated crowd flows for all density levels on environments with a compression rate of 1 (train-test split is $7.2k : 0.6k$ with equal parts per LoS). Qualitative results are shown in the first row of Fig. 4, and a quantitative comparison between $PanL^{PX}$ and $PanL^{SF}$ can be seen in Table 1. Our model achieves high performance on the SF crowd flow prediction task, showing results that we theorized to be more difficult to achieve than for proxy crowd flow. This finding indicates that our approach is applicable to realistic and complex simulated crowd flows.

Multi-Goal and Non-Uniform Scenario (MGNU). We further explore the model's capability for more complex crowd scenarios, where agents or people in the crowd are targeting different goals simultaneously and are initially clustered instead of being uniformly distributed (MGNU scenarios). We test $PanL^{SF}$, which was trained with only SF flows for single goals and uniform agent distributions, on 600 cases of SF simulated MGNU scenarios. We also trained a model $PanL^{SF+}$ with the same amount of total data but 50% being SF flows for single goals and uniform agents, and the other 50% being SF flows for multiples goals and clustered agents. The quantitative comparison between $PanL^{SF}$ and $PanL^{SF+}$ can be seen in Table 1. $PanL^{SF+}$ performs better than $PanL^{SF}$ on the more challenging MGNU scenarios, while still performing well on the single-goal and uniform distribution scenarios. The qualitative results are shown in the second row of Fig. 4.

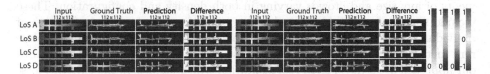

Fig. 5. Model $PanL^{SF+}$ on real-world crowd flows (Stanford Crowd Trajectory dataset).

5.4   Results on Real-World Data

Table 1 and Fig. 4 illustrate the prediction performance of our models on the NARF dataset. Though trained with only axis-aligned crowd configurations, our model generalizes well to non-axis-aligned unseen real environments. We test $PanL^{PX}$, $PanL^{SF}$ and $PanL^{SF+}$ on Stanford Crowd Trajectory dataset.

The quantitative results can be found in Table 1 and the qualitative results of $PanL^{SF+}$ can be seen in Fig. 5. $PanL^{SF+}$ reproduces the crowd movement patterns exhibited in real crowds at different density levels. $PanL^{PX}$ (trained with proxy crowd flows) also achieved comparable performance, indicating that proxy crowd flows are adequate for gaining a general understanding of real crowd flows. More results on Stanford Crowd Trajectory dataset can be found in the Supplementary Material.

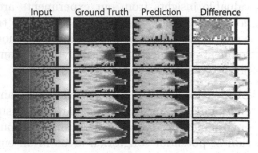

Fig. 6. Crowd flow prediction for different bottleneck widths.

5.5 Case Study

We demonstrate the potential of our prediction platform to be used as part of computer-aided design pipelines. In this simple study, we use $PanL^{SF+}$ to investigate the change in predicted crowd flow in a simple egress scenario through a bottleneck that gradually widens, where the goal is centered on the right side of the bottleneck (Fig. 6). The results are consistent with our expectation that widening the bottleneck would decrease congestion, and the result for an inaccessible goal (top) is quite surprising, since no training data has been provided for this scenario.

6 Conclusion

We have formulated the novel task of long-term crowd flow prediction and proposed the first deep framework for instantly and accurately predicting long-term crowd movement patterns (crowd flow) in scenarios involving complex environments of any size and crowds varying in both density and configuration. These include sparse crowds which require the prediction of microscopic features of individuals, extremely dense crowds which are prohibitively challenging for current methods, and heterogeneous crowds with different destinations. Central to our approach are a novel CAGE representation which facilitates the efficient, lossless encoding of environments, and a modified SegNet architecture for crowd flow prediction. We have developed a novel dataset of synthetic environments and two forms of crowd flows (dynamically simulated crowd flows and statically derived proxy crowd flows) for model training and evaluation, offsetting the challenges of obtaining sufficient data from real human crowds. This dataset acts as a gateway into a new and exciting crowds research area within computer vision and establishes the baseline for others to compare their advancements against. Our models are able to accurately and efficiently predict proxy and simulated crowd flow across a large variety of environments and crowd contexts, while also generalizing to new situations (i.e., real crowd flows and real environments). Two valuable insights from our experiments are that (1) training on proxy crowd flows yields similar predictive accuracy to training on simulated crowd flows for a real-world dataset and (2) training our model on axis-aligned, rectangular environments provides a surprisingly good transfer to testing on never-before-seen, non-axis-aligned environments (Table 1).

Limitations and Future Work. CAGE can handle arbitrarily large axis-aligned environments, but makes no guarantee to successfully compress any axis-aligned environment into a pre-defined size. Although it facilitates the accurate prediction of potential congestion areas in an environment and is thus adequate for applications (e.g. crowd and disaster management), CAGE-decoding may cause aliasing artifacts on the crowd flow.

Acknowledgements. The project was funded in part by NSF IIS-1703883 and NSF S&AS-1723869.

References

1. Alahi, A., Goel, K., Ramanathan, V., Robicquet, A., Fei-Fei, L., Savarese, S.: Social LSTM: human trajectory prediction in crowded spaces. In: Proceedings of the IEEE Conference on Computer Vision and Pattern Recognition, pp. 961–971 (2016)
2. Amirian, J., Hayet, J.B., Pettre, J.: Social ways: learning multi-modal distributions of pedestrian trajectories with gans. In: The IEEE Conference on Computer Vision and Pattern Recognition (CVPR) Workshops, June 2019
3. Modiri Assari, S., Idrees, H., Shah, M.: Human re-identification in crowd videos using personal, social and environmental constraints. In: Leibe, B., Matas, J., Sebe, N., Welling, M. (eds.) ECCV 2016. LNCS, vol. 9906, pp. 119–136. Springer, Cham (2016). https://doi.org/10.1007/978-3-319-46475-6_8
4. Badrinarayanan, V., Kendall, A., Cipolla, R.: Segnet: A deep convolutional encoder-decoder architecture for image segmentation. CoRR abs/1511.00561 (2015). http://arxiv.org/abs/1511.00561
5. Bottou, L.: Large-scale machine learning with stochastic gradient descent. In: Lechevallier Y., Saporta G. (eds.) Proceedings of COMPSTAT'2010, pp. 177–186. Springer, Heidelberg (2010). https://doi.org/10.1007/978-3-7908-2604-3_16
6. Cao, X., Wang, Z., Zhao, Y., Su, F.: Scale aggregation network for accurate and efficient crowd counting. In: The European Conference on Computer Vision (ECCV), September 2018
7. Dogan, T., Saratsis, E., Reinhart, C.: The optimization potential of floor-plan typologies in early design energy modeling, December 2015
8. Fang, Z., Lo, S., Lu, J.: On the relationship between crowd density and movement velocity. Fire Saf. J. **38**(3), 271–283 (2003)
9. Fruin, J.J.: Pedestrian planning and design. Technical report (1971)
10. He, K., Zhang, X., Ren, S., Sun, J.: Delving deep into rectifiers: surpassing human-level performance on imagenet classification. In: Proceedings of the IEEE International Conference on , pp. 1026–1034 (2015) 8
11. Helbing, D., Molnár, P.: Social force model for pedestrian dynamics. Phys. Rev. E **51**, 4282–4286 (1995). https://doi.org/10.1103/PhysRevE.51.4282
12. Huang, S., et al.: Body structure aware deep crowd counting. IEEE Trans. Image Process. **27**(3), 1049–1059 (2017)
13. Idrees, H., et al.: Composition loss for counting, density map estimation and localization in dense crowds. In: Ferrari, V., Hebert, M., Sminchisescu, C., Weiss, Y. (eds.) ECCV 2018. LNCS, vol. 11206. Springer, Cham (2018). https://doi.org/10.1007/978-3-030-01216-8
14. Jiang, R., et al.: Deepurbanevent: a system for predicting citywide crowd dynamics at big events. In: Proceedings of the 25th ACM SIGKDD International Conference on Knowledge Discovery & Data Mining, pp. 2114–2122. ACM (2019)
15. Jiang, X., et al.: Crowd counting and density estimation by trellis encoder-decoder networks. In: Proceedings of the IEEE Conference on Computer Vision and Pattern Recognition, pp. 6133–6142 (2019)
16. Jin, W., Lin, Y., Wu, Z., Wan, H.: Spatio-temporal recurrent convolutional networks for citywide short-term crowd flows prediction. In: Proceedings of the 2nd International Conference on Compute and Data Analysis, pp. 28–35. ACM (2018)
17. Kapadia, M., Pelechano, N., Allbeck, J., Badler, N.: Virtual crowds: steps toward behavioral realism. Synthesis lectures on visual computing: computer graphics, animation, computational photography, and imaging **7**(4), 1–270 (2015)

18. Li, C., Zhang, Z., Sun Lee, W., Hee Lee, G.: Convolutional sequence to sequence model for human dynamics. In: Proceedings of the IEEE Conference on Computer Vision and Pattern Recognition, pp. 5226–5234 (2018)
19. Li, Y., Zhang, X., Chen, D.: CSRNet: dilated convolutional neural networks for understanding the highly congested scenes. In: Proceedings of the IEEE Conference on Computer Vision and Pattern Recognition, pp. 1091–1100 (2018)
20. Lian, D., Li, J., Zheng, J., Luo, W., Gao, S.: Density map regression guided detection network for RGB-D crowd counting and localization. In: Proceedings of the IEEE Conference on Computer Vision and Pattern Recognition, pp. 1821–1830 (2019)
21. Liu, C., Weng, X., Mu, Y.: Recurrent attentive zooming for joint crowd counting and precise localization. In: Proceedings of the IEEE Conference on Computer Vision and Pattern Recognition, pp. 1217–1226 (2019)
22. Liu, J., Gao, C., Meng, D., Hauptmann, A.G.: Decidenet: counting varying density crowds through attention guided detection and density estimation. In: Proceedings of the IEEE Conference on Computer Vision and Pattern Recognition, pp. 5197–5206 (2018)
23. Liu, L., Qiu, Z., Li, G., Liu, S., Ouyang, W., Lin, L.: Crowd counting with deep structured scale integration network. In: Proceedings of the IEEE International Conference on Computer Vision, pp. 1774–1783 (2019)
24. Liu, N., Long, Y., Zou, C., Niu, Q., Pan, L., Wu, H.: ADCrowdNet: an attention-injective deformable convolutional network for crowd understanding. In: Proceedings of the IEEE Conference on Computer Vision and Pattern Recognition, pp. 3225–3234 (2019)
25. Liu, S., Huang, D., Wang, Y.: Adaptive NMS: refining pedestrian detection in a crowd. In: Proceedings of the IEEE Conference on Computer Vision and Pattern Recognition, pp. 6459–6468 (2019)
26. Liu, W., Rabinovich, A., Berg, A.C.: Parsenet: Looking wider to see better. arXiv preprint arXiv:1506.04579 (2015)
27. Liu, W., Salzmann, M., Fua, P.: Context-aware crowd counting. In: Proceedings of the IEEE Conference on Computer Vision and Pattern Recognition, pp. 5099–5108 (2019)
28. Long, J., Shelhamer, E., Darrell, T.: Fully convolutional networks for semantic segmentation. In: Proceedings of the IEEE conference on computer vision and pattern recognition, pp. 3431–3440 (2015)
29. Nash, A., Daniel, K., Koenig, S., Felner, A.: Theta$^{\wedge *}$: any-angle path planning on grids. AAAI 7, 1177–1183 (2007)
30. Oñoro-Rubio, D., López-Sastre, R.J.: Towards perspective-free object counting with deep learning. In: Leibe, B., Matas, J., Sebe, N., Welling, M. (eds.) ECCV 2016. LNCS, vol. 9911, pp. 615–629. Springer, Cham (2016). https://doi.org/10.1007/978-3-319-46478-7_38
31. Parker, J.A., Kenyon, R.V., Troxel, D.E.: Comparison of interpolating methods for image resampling. IEEE Trans. Med. Imaging 2(1), 31–39 (1983)
32. Qiao, G., Zhou, H., Kapadia, M., Yoon, S., Pavlovic, V.: Scenario generalization of data-driven imitation models in crowd simulation. In: Motion, Interaction and Games, p. 36. ACM (2019)
33. Ranjan, V., Le, H., Hoai, M.: Iterative crowd counting. In: Ferrari, V., Hebert, M., Sminchisescu, C., Weiss, Y. (eds.) ECCV 2018. LNCS, vol. 11211, pp. 278–293. Springer, Cham (2018). https://doi.org/10.1007/978-3-030-01234-2_17

34. Ronneberger, O., Fischer, P., Brox, T.: U-Net: convolutional networks for biomedical image segmentation. In: Navab, N., Hornegger, J., Wells, W.M., Frangi, A.F. (eds.) MICCAI 2015. LNCS, vol. 9351, pp. 234–241. Springer, Cham (2015). https://doi.org/10.1007/978-3-319-24574-4_28

35. Sam, D.B., Surya, S., Babu, R.V.: Switching convolutional neural network for crowd counting. In: 2017 IEEE Conference on Computer Vision and Pattern Recognition (CVPR), pp. 4031–4039. IEEE (2017)

36. Sindagi, V., Patel, V.M.: A survey of recent advances in CNN-based single image crowd counting and density estimation. CoRR abs/1707.01202 (2017). http://arxiv.org/abs/1707.01202

37. Singh, S., Kapadia, M., Faloutsos, P., Reinman, G.: An open framework for developing, evaluating, and sharing steering algorithms. In: Egges, A., Geraerts, R., Overmars, M. (eds.) MIG 2009. LNCS, vol. 5884, pp. 158–169. Springer, Heidelberg (2009). https://doi.org/10.1007/978-3-642-10347-6_15

38. Thalmann, D., Musse, S.R.: Crowd Simulation, 2nd edn. Springer, London (2013). https://doi.org/10.1007/978-1-4471-4450-2

39. Tripathi, G., Singh, K., Vishwakarma, D.K.: Convolutional neural networks for crowd behaviour analysis: a survey. Vis. Comput. 35(5), 753–776 (2019)

40. Walach, E., Wolf, L.: Learning to count with CNN boosting. In: Leibe, B., Matas, J., Sebe, N., Welling, M. (eds.) ECCV 2016. LNCS, vol. 9906, pp. 660–676. Springer, Cham (2016). https://doi.org/10.1007/978-3-319-46475-6_41

41. Wan, J., Luo, W., Wu, B., Chan, A.B., Liu, W.: Residual regression with semantic prior for crowd counting. In: Proceedings of the IEEE Conference on Computer Vision and Pattern Recognition, pp. 4036–4045 (2019)

42. Wang, Q., Gao, J., Lin, W., Yuan, Y.: Learning from synthetic data for crowd counting in the wild. In: Proceedings of the IEEE Conference on Computer Vision and Pattern Recognition, pp. 8198–8207 (2019)

43. Wolinski, D., Lin, M.C., Pettré, J.: Warpdriver: Context-aware probabilistic motion prediction for crowd simulation. ACM Trans. Graph. 35(6), 164:1–164:11 (2016). https://doi.org/10.1145/2980179.2982442

44. Xu, C., Qiu, K., Fu, J., Bai, S., Xu, Y., Bai, X.: Learn to scale: generating multipolar normalized density maps for crowd counting. In: Proceedings of the IEEE International Conference on Computer Vision, pp. 8382–8390 (2019)

45. Yan, Z., et al.: Perspective-guided convolution networks for crowd counting. In: Proceedings of the IEEE International Conference on Computer Vision, pp. 952–961 (2019)

46. Yi, S., Li, H., Wang, X.: Pedestrian behavior understanding and prediction with deep neural networks. In: Leibe, B., Matas, J., Sebe, N., Welling, M. (eds.) ECCV 2016. LNCS, vol. 9905, pp. 263–279. Springer, Cham (2016). https://doi.org/10.1007/978-3-319-46448-0_16

47. Yogameena, B., Nagananthini, C.: Computer vision based crowd disaster avoidance system: a survey. Int. J. Disaster Risk Reduction 22, 95–129 (2017)

48. Yun, S., Yun, K., Choi, J., Choi, J.Y.: Density-aware pedestrian proposal networks for robust people detection in crowded scenes. In: Hua, G., Jégou, H. (eds.) ECCV 2016. LNCS, vol. 9914, pp. 643–654. Springer, Cham (2016). https://doi.org/10.1007/978-3-319-48881-3_45

49. Zhang, J., Zheng, Y., Qi, D.: Deep spatio-temporal residual networks for citywide crowd flows prediction. In: Thirty-First AAAI Conference on Artificial Intelligence (2017)

50. Zhang, Q., Chan, A.B.: Wide-area crowd counting via ground-plane density maps and multi-view fusion CNNS. In: Proceedings of the IEEE Conference on Computer Vision and Pattern Recognition, pp. 8297–8306 (2019)
51. Zhang, S., Wu, G., Costeira, J.P., Moura, J.M.: Understanding traffic density from large-scale web camera data. In: Proceedings of the IEEE Conference on Computer Vision and Pattern Recognition, pp. 5898–5907 (2017)
52. Zhang, S., Wen, L., Bian, X., Lei, Z., Li, S.Z.: Occlusion-aware R-CNN: detecting pedestrians in a crowd. In: The European Conference on Computer Vision (ECCV), September 2018
53. Zhao, M., Zhang, J., Zhang, C., Zhang, W.: Leveraging heterogeneous auxiliary tasks to assist crowd counting. In: Proceedings of the IEEE Conference on Computer Vision and Pattern Recognition, pp. 12736–12745 (2019)
54. Zhao, Z., Li, H., Zhao, R., Wang, X.: Crossing-line crowd counting with two-phase deep neural networks. In: Leibe, B., Matas, J., Sebe, N., Welling, M. (eds.) ECCV 2016. LNCS, vol. 9912, pp. 712–726. Springer, Cham (2016). https://doi.org/10.1007/978-3-319-46484-8_43
55. Zonoozi, A., jae Kim, J., li Li, X., Cong, G.: Periodic-CRN: a convolutional recurrent model for crowd density prediction with recurring periodic patterns. In: IJCAI (2018)

Weakly-Supervised Action Localization with Expectation-Maximization Multi-Instance Learning

Zhekun Luo[1], Devin Guillory[1], Baifeng Shi[2], Wei Ke[3], Fang Wan[4],
Trevor Darrell[1], and Huijuan Xu[1(✉)]

[1] University of California, Berkeley, USA
{zhekun_luo,dguillory,trevordarrell,huijuan}@eecs.berkeley.edu
[2] Peking University, Beijing, China
bfshi@pku.edu.cn
[3] Carnegie Mellon University, Pittsburgh, USA
weik@andrew.cmu.edu
[4] Chinese Academy of Sciences, Beijing, China
wanfang@ucas.ac.cn

Abstract. Weakly-supervised action localization requires training a model to localize the action segments in the video given only video level action label. It can be solved under the Multiple Instance Learning (MIL) framework, where a bag (video) contains multiple instances (action segments). Since only the bag's label is known, the main challenge is assigning which key instances within the bag to trigger the bag's label. Most previous models use attention-based approaches applying attentions to generate the bag's representation from instances, and then train it via the bag's classification. These models, however, implicitly violate the MIL assumption that instances in negative bags should be uniformly negative. In this work, we explicitly model the key instances assignment as a hidden variable and adopt an Expectation-Maximization (EM) framework. We derive two pseudo-label generation schemes to model the E and M process and iteratively optimize the likelihood lower bound. We show that our EM-MIL approach more accurately models both the learning objective and the MIL assumptions. It achieves state-of-the-art performance on two standard benchmarks, THUMOS14 and ActivityNet1.2.

Keywords: Weakly-supervised learning · Action localization · Multiple instance learning

1 Introduction

As the growth of video content accelerates, it becomes increasingly necessary to improve video understanding ability with less annotation effort. Since videos can contain a large number of frames, the cost of identifying the exact start and end frames of each action is high (frame-level) in comparison to just labeling what

© Springer Nature Switzerland AG 2020
A. Vedaldi et al. (Eds.): ECCV 2020, LNCS 12374, pp. 729–745, 2020.
https://doi.org/10.1007/978-3-030-58526-6_43

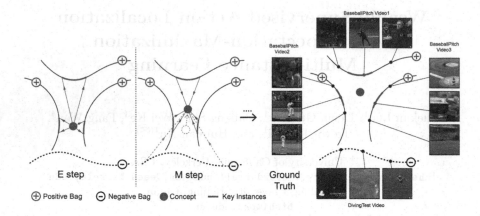

Fig. 1. Each curve represents a bag and points on the curve represent instances in the bag. We aim to find a concept point such that each positive bag contains some key instances close to it while all instances in the negative bags are far from it. In E step we use the current concept to pick key instances for each positive bag. In M step we use key instances and negative bags to update the concept.

actions the video contains (video-level). Researchers are motivated to explore approaches that do not require per-frame annotations. In this work, we focus on weakly-supervised action localization paradigm, using only video-level action labels to learn activity recognition and localization. This problem can be framed as a special case of the **Multiple Instance Learning (MIL)** problem [4]: a bag contains multiple instances; Instances' labels collectively generate the bag's label, and only the bag's label is available during training. In our task, each video represents bag, and the clips of the video represent the instances inside the bag. The key challenge here is to handle **key instance assignment** during training – to identify which instances within the bag trigger the bag's label.

Most previous works used **attention-based** approaches to model the key instance assignment process. They used attention weights to combine instance-level classification to produce the bag's classification. Models of this form are then trained via standard classification procedures. The learned attention weights imply the contribution of each instance to the bag's label, and thus can be used to localize the positive instances (action clips) [17,26]. While promising results have been observed, models of this variety tend to produce incomplete action proposals [13,31], that only part of the action is detected. This is also a common problem in attention-based weakly-supervised object detection [11,25]. We argue that this problem is due to a misspecification of the MIL-objective. Attention weights, which indicate key instances' assignment, should be our optimization target. But in an attention-MIL framework, attention is learned as a by-product when conducting classification for bags. As a result, the attention module tends to only pick the most discriminative parts of the action or object to correctly classify a bag, due to the fact that the loss and training signal come from the bag's classification.

Inspired by traditional MIL literature, we adopt a different method to tackle weakly-supervised action localization using the Expectation–Maximization framework. Historically, Expectation–Maximization (EM) or similar iterative estimation processes have been used to solve the MIL problems [4,5,35] before the deep learning era. Motivated by these works, we explicitly model key instance assignment as a hidden variable and optimize this as our target. Shown in Fig. 1, we adopt the EM algorithm to solve the interlocking steps of key instance assignment and action concept classification. To formulate our learning objective, we derive two pseudo-label generating schemes to model the E and M process respectively. We show that our alternating update process optimizes a lower bound of the MIL-objective. We also find that previous attention-MIL models implicitly violate the MIL assumptions. They apply attention to negative bags, while the MIL assumption states that instances in negative bags are uniformly negative. We show that our method can better model the data generating procedure of both positive and negative bags. It achieves state-of-the-art performance with a simple architecture, suggesting its potential to be extended to many practical settings. The main contributions of this paper are:

- We propose to adapt the Expectation–Maximization MIL framework to weakly supervised action localization task. We derive two novel pseudo-label generating schemes to model the E and M process respectively.[1]
- We show that previous attention-MIL models implicitly violate the MIL assumptions, and our method better model the background information.
- Our model is evaluated on two standard benchmarks, THUMOS14 and ActivityNet1.2, and achieves state of the art results.

2   Related Work

Weakly-Supervised Action Localization. Weakly supervised action localization learns to localize activities inside videos when only action class labels are available. UntrimmedNet [26] first used attention to model the contribution of each clip to a video-level action label. It performs classification separately at each clip, and predicts video's label through a weighted combination of clips' scores. Later the STPN model [17] proposed that instead of combining clips' scores, it uses attention to combine clips' features into a video-level feature vector and conducts classification from there. [8] generalizes a framework for these attention-based approaches and formalizes such combination as a permutation-invariant aggregation function. W-TALC [19] proposed a regularization to enforce action periods of the same class must share similar features. It is also noticed that attention-MIL methods tend to produce incomplete localization results. To tackle that, a series of papers [22,23,33,38] took the adversarial erasing idea to improve the detection completeness by hiding the most discriminative parts. [31] conducted sub-samplings based on activation to suppress the

---

[1] Code: https://github.com/airmachine/EM-MIL-WeaklyActionDetection.

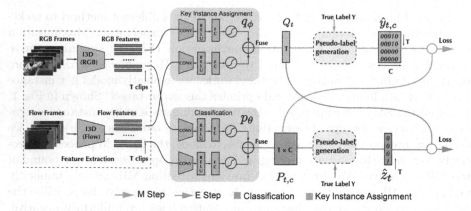

Fig. 2. Our EM-MIL model architecture builds on fixed two-stream I3D features, and alternates between updating the key-instance assignment branch q_ϕ (E Step) and the classification branch p_θ (M Step). We use the classification score and key instance assignment result to generate pseudo-labels for each other (detailed in Sects. 3.1 and refM), and alternate freezing one branch to train the other.

dominant response of the discriminative action parts. To model complete actions, [13] proposed to use a multi-branch network with each branch handling distinctive action parts. To generate action proposals, they combine per-clip attention and classification scores to form the Temporal Class Activation Sequence (T-CAS [17]) and group the high activation clips. Another type of models [14,21] train a boundary predictor based on pre-trained T-CAS scores to output the action start and end point without grouping (Fig. 2).

Some previous methods in weakly-supervised object or action localization involve iterative refinement, but their training processes and objectives are different from our Expectation–Maximization method. RefineLoc [1]'s training contains several passes. It uses the result of the ith pass as supervision for the $(i+1)^{th}$ pass and trains a new model from scratch iteratively. [24] uses a similar approach in image objection detection but stacks all passes together. Our approach differs from these in the following ways: Their self-supervision and iterative refinement happen between **different passes**. In each pass all modules are trained jointly till converge. In comparison, we adopts an EM framework which explicitly models key instance assignment as hidden variables. Our pseudo-labeling and alternating training happen between **different modules** of the same model. Thus our model requires only one pass. In addition, as discussed in Sect. 3.4, they handle the attention in negative bags different to us.

Traditional Multi-Instance Learning Methods. The Multiple Instance Learning problem was first defined by Dietterich et al. [4], who proposed the iterated discrimination algorithm. It starts from a point in the feature space and iteratively searches for the smallest box covering at least one point (instance) per positive bag and avoiding all points in negative bags. [15] sets up the Diverse

Density framework. They defined a point in the feature space to be the positive concept. Every positive bag ("diverse") contains at least one instance close to the concept while all instances in the negative bags are far from it (in terms of some distance metric). They modeled the likelihood of a concept using Gaussian Mixture models along with a Noisy-OR probability estimation. [34] then applied AdaBoost to this Noisy-OR model and [10]'s ISR model, and derived two MIL loss functions. [5] adapted the K-nearest neighbors method to the Diverse Density framework. Later [35] proposed the EM-DD algorithm, combing Expectation Maximization process and the Diverse Density metric. These early works did not involve neural networks and were not applied over the high-dimensional task of action localization. Many of them involve modeling key instances assignment as hidden variable and use iterative optimization. They also differ from the predominant attention-MIL paradigm in how they treat negative instances. We view these distinctions as motivation to explore our approach.

3   Method

Multiple Instance Learning (MIL) is a supervised learning problem where instead of one instance X being matched to one label y, a bag or set of multiple instances $[X_1, X_2, X_3, ...]$ are matched to single label y. In the binary MIL setting, a bag's label is positive if at least one instance in the bag is positive. Therefore a bag is negative only if all instances in the bag are negative.

In our task, following the best practice of previous works [17,19,26], we divide a long video into multiple 15-frame clips. Then a video corresponds to a bag (bag-level video label is given), and the clips of the video represent the instances inside the bag (instance-level clip labels are missing). Each video (bag) contains T video clips (instances), denoted by $\mathbf{X} = \{\mathbf{x}_t\}_{t=1}^T$, where $\mathbf{x}_t \in \mathbb{R}^d$ is the feature of clip t. We represent the video's action label in one hot way, where $y_c = 1$ if the video contains clips of action c, otherwise $y_c = 0$, $c \in \{1, 2, \cdots, C\}$ (each video can contain multiple action classes). In the MIL setting, label of each video is determined by the labels of clips it contains. To be specific, we assign a binary variable $z_t \in \{0, 1\}$ to each clip t, denoting whether clip t is responsible for the generation of video-level label. $\mathbf{z} = \{z_t\}_{t=1}^T$ models the assignment of **key instances' scope**. Video-level label is generated with probability:

$$p_\theta(y_c = 1|\mathbf{X}, \mathbf{z}) = \sigma_{t \in \{1, \cdots, T\}}\{ p_\theta(y_{c,t} = 1|\mathbf{x}_t) \cdot [z_t = 1] \} \tag{1}$$

where $[z_t = 1]$ is the indicator function for assignment. $p_\theta(y_{c,t} = 1|\mathbf{x}_t)$ is the probability (parameterized by θ) that clip t belongs to class c. The closer clip t is to the concept, the higher $p_\theta(y_{c,t} = 1|\mathbf{x}_t)$ is. σ is a permutation-invariant operator, e.g. maximum [36] or mean operator [8].

In our temporal action localization problem, we propose to first estimate the probability of $z_t = 1$ with an estimator $q_\phi(z_t = 1|\mathbf{x}_t)$ parameterized by ϕ, and then choose the clips with high estimated likelihood as our action segments. Since $\{z_t\}$ are latent variables with no ground truth, we optimize q_ϕ through maximization of the variational lower bound:

$$\log p_\theta(y_c|\mathbf{X}) = KL(q_\phi(\boldsymbol{z}|\mathbf{X}) \ || \ p_\theta(\boldsymbol{z}|\mathbf{X}, y_c)) + \int q_\phi(\boldsymbol{z}|\mathbf{X}) \log \frac{p_\theta(\boldsymbol{z}, y_c|\mathbf{X})}{q_\phi(\boldsymbol{z}|\mathbf{X})} d\boldsymbol{z}$$

$$\geq \int q_\phi(\boldsymbol{z}|\mathbf{X}) \log p_\theta(\boldsymbol{z}, y_c|\mathbf{X}) d\boldsymbol{z} + H(q_\phi(\boldsymbol{z}|\mathbf{X})), \tag{2}$$

where $H(q_\phi(\boldsymbol{z}|\mathbf{X}))$ is entropy of q_ϕ. By maximizing the lower bound, we are actually optimizing the likelihood of y_c given $\mathbf{X}$. In this work, we adopt the Expectation-Maximization (EM) algorithm, and optimize the lower bound by updating θ and ϕ alternately. To be specific, we first update ϕ by minimizing $KL(q_\phi(\boldsymbol{z}|\mathbf{X}) \ || \ p_\theta(\boldsymbol{z}|\mathbf{X}, y_c))$ and tighten the lower bound in E step, and update θ through maximization of the lower bound in M step. In the following subsections, we will first get into details of updating θ and ϕ in E step and M step separately, and then sum up the whole algorithm.

3.1 E Step

In E step, we update ϕ by minimizing $KL(q_\phi(\boldsymbol{z}|\mathbf{X}) \ || \ p_\theta(\boldsymbol{z}|\mathbf{X}, y_c))$ and tighten the lower bound in Eq. 2. As in previous works [17,18], we approximate $q_\phi(\boldsymbol{z}|\mathbf{X})$ with $\prod_t q_\phi(z_t|\mathbf{x}_t)$ assuming the independence between different clips, where $q_\phi(z_t|\mathbf{x}_t)$ is estimated by neural network with parameter ϕ on each clip. Thus we only have to minimize $KL(q_\phi(z_t|\mathbf{x}_t) \ || \ p_\theta(z_t|\mathbf{x}_t, y_c))$ for each clip t. Following the literature, we assume that the posterior $p_\theta(z_t|\mathbf{x}_t, y_c)$ is proportional to the classification score $p_\theta(y_c|\mathbf{x}_t)$. Then we propose to update q_ϕ with pseudo label generated from classification score. Specifically, dynamic thresholds are calculated based on the instance classification scores to generate pseudo-labels for q_ϕ. If an instance has a classification score over the threshold for any ground truth class within the video, the instance is treated as a positive example; otherwise, it is treated as a negative example. The pseudo label is formulated as follows:

$$\hat{z}_t = \begin{cases} 1, \text{ if } \sum_{c=1}^{C} \mathbb{1}(P_{t,c} > \overline{P}_{1:T,c} \ \wedge \ y_c = 1) > 0 \\ 0, \text{ otherwise} \end{cases} \tag{3}$$

where $P_{t,c} = p_\theta(y_c|\mathbf{x}_t)$ and $\overline{P}_{1:T,c}$ is the mean of $P_{t,c}$ over temporal axis. Then we update q_ϕ using binary cross entropy (BCE) loss and the updating process is illustrated in Fig. 3.

$$\mathcal{L}(q_\phi) = -\hat{z}_t \log q_\phi(z_t|\mathbf{x}_t) - (1 - \hat{z}_t) \log(1 - q_\phi(z_t|\mathbf{x}_t)). \tag{4}$$

3.2 M Step

In M step, we update p_θ through optimization of the lower bound in Eq. 2. Since $H(q_\phi(\boldsymbol{z}|\mathbf{X}))$ is constant wrt θ, we only optimize $\int q_\phi(\boldsymbol{z}|\mathbf{X}) \log p_\theta(\boldsymbol{z}, y_c|\mathbf{X}) d\boldsymbol{z}$, which is equivalent to optimize the classification performance given key instance assignment $q_\phi(\boldsymbol{z}|\mathbf{X})$. To this end, we use the class-agnostic key-instance assigning module q_ϕ and the ground truth video-level labels to generate a $T \times C$ pseudo-label map which discriminates between foreground and background clips within the same video. Similarly, our pseudo-label generation procedure calculates a dynamic threshold based on the distribution of instance-assignment scores for each video clip. It assigns positive classifications for all instances whose scores are higher than the threshold, and negative classifications for all instances whose scores are below or instances in negative bags. The pseudo label is given by:

$$\hat{y}_{t,c} = \begin{cases} 1, \text{ if } y_c = 1 \text{ and } Q_t > \overline{Q}_{1:T} + \gamma \cdot (\max(Q_t) - \min(Q_t)) \\ 0, \text{ otherwise} \end{cases}, \qquad (5)$$

where $Q_t = q_\phi(z_t|\mathbf{x}_t)$ and $\overline{Q}_{1:T}$ is the mean of Q_t over temporal axis. The threshold hyper-parameter γ implies a distribution priori on how similar the same action exhibits across several videos. Then we update p_θ with BCE loss and the updating process is illustrated in Fig. 4.

$$\mathcal{L}(p_\theta) = -\hat{y}_{t,c} \log p_\theta(y_c|x_t) - (1 - \hat{y}_{t,c}) \log(1 - p_\theta(y_c|x_t)). \qquad (6)$$

3.3 Overall Algorithm

We summarize our EM-style algorithm in Algorithm 1. We update the key-instance assigning module q_ϕ and classification module p_θ alternately. In E step we freeze the classification p_θ and update q_ϕ using pseudo labels from p_θ. In M step we optimize classification based on q_ϕ. Two steps are processed alternately to maximize the likelihood $\log p_\theta(y_c|\mathbf{X})$, and meanwhile optimize the localization results.

Algorithm 1: EM-MIL Weakly-Supervised Activity Localization

Initialization: learning rate β, classification threshold γ
classifier parameters θ, attention parameters ϕ
while θ, ϕ *has not converged* **do**
    #*E step*
    for $(\mathbf{X}, y_c)$ *in train set* **do**
        $P_{t,c} \leftarrow p_\theta(y_c|\mathbf{x}_t)$;
        $\phi \leftarrow \phi - \beta \cdot \nabla_\phi \mathcal{L}(q_\phi)$;
    end
    #*M step*
    for $(\mathbf{X}, y_c)$ *in train set* **do**
        $Q_t \leftarrow q_\phi(z_t|\mathbf{x}_t)$;
        $\theta \leftarrow \theta - \beta \cdot \nabla_\theta \mathcal{L}(p_\theta)$;
    end
end

3.4 Comparison with Previous Methods

After careful examination of Eqs. 3 and 5, we find that our pseudo-labeling process Q_t and $\hat{y}_{t,c}$ can also be interpreted as a special kind of attention. Denote loss function by $\mathcal{L}$, then in Eq. 5, the loss is calculated as

$$\mathcal{L}[\, p_\theta(\mathbf{y}|\mathbf{x}), \, \mathcal{F}(\mathbf{Q}, \mathbf{y}) \,] \qquad (7)$$

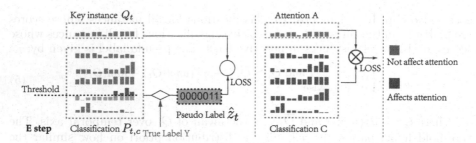

Fig. 3. In our EM-MIL model only the foreground classification score $P_{t,c}$ affects the key instance pseudo label $\hat{z}_t$ (left), while in previous models all-class classification scores contribute to the attention weights (right).

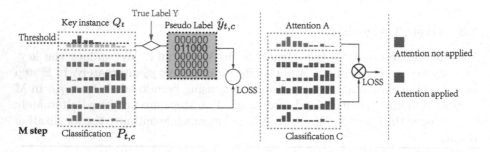

Fig. 4. Our EM-MIL model (left) uses key instance assignment Q_t to generate pseudo classification labels $\hat{y}_{t,c}$ only for the foreground classes, while in previous models such as UntrimmedNet (right) attentions are applied to all classes.

$\mathcal{F}$ is the pseudo label generation function in Eq. 5, $\mathbf{Q}, \mathbf{y}, \mathbf{x}$ is the compact expression of Q_t, y_c, x_t. On the other hand, if we denote attention and classification score as $\mathbf{a}, \mathbf{c}$, the loss for a typical attention-based model like [26] is:

$$\mathcal{L}[\ \sigma(\mathbf{c} \odot \mathbf{a}),\ \mathbf{y}\ ] \tag{8}$$

Here σ is the aggregation operator [8], such as *reduce_sum* or *reduce_max*. Comparing Eq. 7 to Eq. 8, it is easy to see that they can be matched. $p_\theta(\mathbf{y}|\mathbf{x})$ is classification score ($\mathbf{c}$), and $\mathbf{Q}$ can be seen as special attention (corresponds to $\mathbf{a}$). In M step it attends to the key instance it estimates. But compared to previous attention-MIL methods, Eq. 3 shows that this "attention" only happens in positive bags. We believe it better aligns with the MIL assumption, which says that all instances in negative bags are uniformly negative. Previous methods that applies attention to negative bags implicitly assumes that some instances are more negative than others. This violates the MIL assumption. The differences between our attention and theirs are illustrated in Figs. 3 and 4. In addition, in Eq. 5, this "attention" is a threshold-based hard attention. Clips below the threshold are classified as background with high confidence, while clips above the threshold are weighted equally and re-scored in the next iteration. The use of hard pseudo labels allows for the distinct treatment of positive and negative instances that would be more complex to enforce with soft-boundaries. We initialize our training procedure by labeling every clip in a positive bag to be 1 and gradually narrow down the search scope. Such training process maintains high recalls for action clips in each E-M iteration. It prevents attention from focusing on the discriminative parts too quickly, thus increases the proposal completeness.

Another way to compare our methods with previous ones is through the lens of the MIL framework. As discussed in [2], MIL problem has two setting: instance-level vs bag-level. The instance level setting prioritizes classification precision of instance over bag's, and vice versa. Our task aligns with the instance setting as the primary goal is action localization (equivalent to clips' classification). Previous attention-MIL models like [17,19,26] treat instance-localization as the by-product of an accurate bag-level classification system, which align with the bag-level MIL setting. By modeling the problem through an instance-level MIL framework our approach more accurately models the target objective. This change in objective function and optimization procedure allows substantial improvement in performance.

3.5 Inference

At test time, we use another branch for video-level classification and use our model for localization as in previous work [21]. For classification branch, we used a plain UntrimmedNet [26] with soft attention for the THUMOS14 dataset and the W-TALC [19] for the ActivityNet1.2 dataset. We run a forward pass with our model to get the localization score L by fusing instance assignment score Q_t and classification score $P_{t,c}$.

$$L_t = \lambda * Q_t + (1 - \lambda) * P_{t,c}, \tag{9}$$

where λ is set to be 0.8 through grid search in THUMOS14 dataset and 0.3 in the ActivityNet1.2 dataset. In the Experiment Sec. 4.2 we analyze the impact of different of λ. We threshold the L_t score to get prediction y'_i for each clip using the same scheme as in Eq. 5. Then we group the clips above the threshold to get the temporal start and end point of the action proposal.

4   Experiments

In this section, we evaluate our EM-MIL model on two large-scale temporal activity detection datasets: THUMOS14 [9] and ActivityNet1.2 [7]. Section 4.1 introduces experimental setup of these datasets, the evaluation metrics and the implementation details. Section 4.2 compares weakly localization results between our proposed model and the state-of-the-art models on both THUMOS14 and ActivityNet1.2 datasets, and visualizes some localization results. Section 4.3 shows the ablation studies for each component of our model on THUMOS14 dataset.

4.1   Experimental Setup

Datasets: The THUMOS14 [9] activity detection dataset contains over 24 h of videos from 20 different athletic activities. The train set contains 2765 trimmed videos, while the validation set and the test set contains 200 and 213 untrimmed videos respectively. We use the validation set as train data and report weakly-supervised temporal activity localization results on the test set. This dataset is particularly challenging as it consists of very long videos with multiple activity instances of very small duration. Most videos contain multiple activity instances of the same activity class. In addition, some videos contain activity instances from different classes.

The ActivityNet [7] dataset consists three versions. We use the ActivityNet1.2 version which contains a total of around 10000 videos including 4819 train videos, 2383 validation videos, and 2480 withheld test videos for challenge purpose. We report the weakly-supervised temporal activity localization results on the validation videos. In ActivityNet1.2, around 99% videos contain activity instances of a single class. Many of the videos have activity instances covering more than half of the duration. Compared to THUMOS14, this is a large-scale dataset, both in terms of the number of activities involved and the amount of videos.

Evaluation Metric: The weakly-supervised temporal activity localization results are evaluated in terms of mean Average Precision (mAP) with different temporal Intersection over Union (tIoU) thresholds, which is denoted as mAP@α where α is the threshold. Average mAP at 10 evenly distributed tIoU thresholds between 0.5 and 0.95 is also commonly used in the literature.

Implementation Details: Video frames are sampled at 12 fps (for THUMOS14) or 25 fps (for ActivityNet1.2). For each frame, we perform the center crop of size 224×224 after re-scaling the shorter dimension to 256 and construct video clips for every 15 frames. We extract the features of the clips using the publicly released, two-stream I3D model pretrained on Kinetics dataset [3]. We use the feature map from $Mixed\_5c$ layer as feature representation. For optical flow stream, TV-L1 flow [27,32] is used as the input.

Our model is implemented in pyTorch and trained using Adam optimizer with initial learning rate 0.0001 for both datasets. For the THUMOS14 dataset, we train the model by alternating E/M step every 10 epochs in the first 30 epochs. Then we raise the learning rate to 4 times larger and decrease the alternating cycle to 1 epoch for another 35 epochs. For ActivityNet1.2 dataset, we use a similar training approach but the alternating cycle is 5 epochs and the learning rate is constant. We use our model to generate instance assignment Q_t and classification score $P_{t,c}$ separately for RGB and Flow branch. Then, we fuse the RGB/Flow score by weighted averaging. The threshold hyper-parameter γ in Eq. 5 is set to 0.15 for THUMOS14 dataset and 0 for ActivityNet1.2 dataset. Intuitively, the value of γ reflects how similar the same action exhibits across several videos, and should be negatively correlated with the variance of the action's feature distribution. We also explore different γ in the range of $[0.05, 0.2]$,

Table 1. Our EM-MIL detection results on THUMOS14 in percentage. mAP at different tIoU thresholds α are reported. The top half shows fully-supervised methods while the bottom half shows weakly-supervised ones including ours. EM-MIL-UNT represents the result using UntrimmedNet's [26] features.

Supervision	Models	α						
		0.1	0.2	0.3	0.4	0.5	0.6	0.7
Fully-supervised	CDC [20]	-	-	40.1	29.4	23.3	13.1	7.9
	R-C3D [28]	54.5	51.5	44.8	35.6	28.9	-	-
	Gao et al. [6]	-	-	50.1	41.3	31.0	19.1	9.9
	SSN [37]	66.0	59.4	51.9	41.0	29.8	19.6	10.7
	Xu et al. [29]	56.9	54.7	51.2	43.0	36.1	-	-
	BSN [12]	-	-	53.5	45.0	36.9	28.4	20.0
Weakly-supervised	Hide [22]	36.4	27.8	19.5	12.7	6.8	-	-
	UntrimmedNet [26]	44.4	37.7	28.2	21.1	13.7	-	-
	STPN [17]	52.0	44.7	35.5	25.8	16.9	9.9	4.3
	Autoloc [21]	-	-	35.8	29.0	21.2	13.4	5.8
	W-TALC [19]	55.2	49.6	40.1	31.1	22.8	-	7.6
	RefineLoc-I3D [1]	-	-	40.8	-	23.1	-	5.3
	Liu et al. [14]	-	-	37.0	30.9	23.9	13.9	7.1
	Yu et al. [30]	-	-	39.5	-	24.5	-	7.1
	3C-Net [16]	59.1	53.5	44.2	34.1	26.6	-	8.1
	Nguyen et al. [18]	**64.2**	**59.5**	**49.1**	**38.4**	27.5	17.3	8.6
	EM-MIL (ours)	59.1	52.7	45.5	36.8	**30.5**	**22.7**	**16.4**
	EM-MIL-UNT (ours)	59.0	50.4	42.7	34.5	27.2	18.9	10.2

mAP@tIoU=0.5 varies between 29.0% and 30.5% in THUMOS14 dataset, compared to the previous SOTA 26.8% [18] using the same training data.

4.2 Comparison with State-of-the-Art Approaches

Results on THUMOS14 Dataset: We compare our model's results on the THUMOS14 dataset with state-of-the-art results in Table 1. Our model outperforms all the previous published models and achieves a new state-of-the-art result at mAP@0.5, **30.5%**. This result is achieved by our simple EM training policy and the pseudo-labeling scheme, without auxiliary losses to regularize the learning process. Compared to the best result among the six recent models [1,16–19,30] using the same two-stream I3D feature extraction backbone as our model, we get 3% significant improvement at mAP@0.5. We also tried using UntrimmedNet's feature on our model (denoted as EM-MIL-UNT in Table 1), and got a mAP@0.5 of 27.2% which still improves significantly over previous models (e.g.. [14,21,26]) using the same feature backbone. Our model also shows

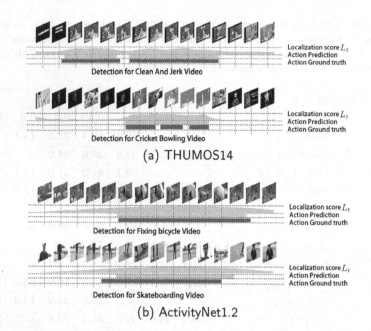

Fig. 5. Qualitative visualization. (a) and (b) show results for two videos each on THU-MOS14 and ActivityNet1.2, a good prediction example (top) and a bad one (bottom). Ground truth activity segments are marked in red. Localization score distribution L_t and predicted activity segments are in blue. (Color figure online)

more significant improvement at high threshold metrics tIoU=0.6 and tIoU=0.7, which implies that our action proposals are more complete. On the other hand, our performance is slightly worse in the low IoU metrics.

Several examples' qualitative results are shown in Fig. 5(a). For each example, we show the video, intermediate score map L_t from our model, final activity detection result and ground truth temporal segment annotation. In the first example of *Clean and Jerk*, we localize the activity correctly with almost 100% overlap. We also show one bad prediction from our model in the second example, where our model overestimates the *Cricket Bowling* activity duration by 20%, as an effect of the interactive shrinkage training process which first labels every instance positive. Our model greatly resolves the incompleteness problem for activity detection in videos containing multiple action segments, while in some cases it might also bring in additional false positives. In addition, our model is also highly time efficient: in THUMOS14 our model trains for 65 epochs, taking 64.7 s on two TITAN RTX GPUs. We have run the released code for AutoLoc [21] and W-TALC [19] on the same machine with their recommended training procedures. Their training times are 44.5 s and 6051.2 s, respectively. All experiments used pre-computed features and [21]'s training required additional pretrained CAS scores.

Table 2. Detection results on ActivityNet1.2 in terms of mAP@{0.5, 0.7, 0.9} and average mAP at tIoU thresholds $\alpha \in (0.5, 0.95)$ with step 0.05 (in percentage). It shows both fully-supervised method and weakly-supervised ones.

Supervision	Models	α			avg. mAP
		0.5	0.7	0.9	
Fully-supervised	SSN [37]	41.3	30.4	13.2	26.6
Weakly-supervised	UntrimmedNet [26]	7.4	3.9	1.2	3.6
	Autoloc [21]	27.3	17.5	6.8	16.0
	W-TALC [19]	37.0	14.6	4.2	18.0
	3C-Net [16]	37.2	**23.7**	**9.2**	**21.7**
	Liu et al. [14]	37.1	23.4	9.2	21.6
	TSM [30]	28.3	18.9	7.5	17.1
	EM-MIL (ours)	**37.4**	23.1	2.0	20.3

Results on ActivityNet1.2 Dataset: We compare our model's results on the ActivityNet1.2 dataset with previous results in Table 2. Our model outperforms previously published models in mAP@0.5 and gets the value of **37.4%**. Despite the state-of-the-art result in mAP@0.5, our model performs worse in high tIoU metrics, which is the opposite to what we observed on THUMOS14 dataset. We further investigate the reason for different result trends on both datasets. Videos in the THUMOS14 dataset contains multiple action segments, each segment with relatively short duration. It has high localization requirement where our model outperforms pervious ones at high tIoU. Unlike THUMOS14, most videos (> 99%) in the ActivityNet1.2 dataset have only one action class, and most of these videos have only a few activity segments which compose a big portion of the whole video duration. Thus videos in ActivityNet1.2 dataset can be regarded as trimmed actions in certain extent. We speculate that the action localization performance in the ActivityNet1.2 dataset depends more on the classification module, which might be the bottleneck for our model. This speculation also correlates with the different λ values in Eq. 9 when calculating localization score on THUMOS14 and ActivityNet1.2 datasets. According to our model's assumption, key instance assignment score Q_t implies the action clips and higher weight for this part facilitates the localization. On THUMOS14, the weight λ for the key instance assignment score Q_t is set to be a high value 0.8. But for ActivityNet1.2, the classification score $P_{t,c}$ has a higher weight (0.7), implying that the model mostly relies on classification to succeed on this dataset. For further illustration, we also visualize some good and bad detection results from ActivityNet1.2 dataset in Fig. 5(b).

4.3 Ablation Studies

We ablate our pseudo label generation scheme and Expectation-Maximization alternating training method on THUMOS14 dataset with mAP@0.5 in Table 3.

Table 3. Ablation results for the pseudo labeling and EM alternating training on THUMOS14 dataset in terms of mAP@0.5 (%).

Ablation models	Pseudo label	Alternating training	mAP@0.5
Alternating model		✓	24.5
Pseudo labeling model	✓		26.8
Full Model	✓	✓	30.5

Ablation on the Pseudo Labeling: We first ablate on the pseudo labeling scheme for $\hat{z}_t$ and $\hat{y}_{t,c}$, and include the results in Table 3. We switch our learning to be supervised by an attention-MIL loss based on softmax function, similar to [17,26]. In the E step, classification scores of all classes contribute collectively to the attention weights. In the M step, attention weights are applied equally to both positive and negative videos without paying special attention to the bag's label. Compared to the "Alternating model" doing alternating training but with a plain attention, "Full Model" improves mAP@0.5 from 24.5% to 30.5%. This indicates the usefulness of the proposed pseudo labeling strategy. It models the key instance assignment explicitly and aligns with the MIL assumption better.

Ablation on the EM Alternating Training Technique: We also evaluate the effectiveness of Expectation-Maximization alternating training compared to joint optimization. The EM training method iteratively estimates the key instance assignment, then maximizes the video classification accuracy, and achieves better activity detection performance. "Full Model" improves mAP@0.5 from 26.8% to 30.5% compared to "Pseudo labeling" model with joint optimization. The same training process can be potentially applied on other MIL based models for weakly-supervised object detection task to improve accuracy as well.

5 Conclusion

We propose a EM-MIL framework with pseudo labeling and alternating training for weakly-supervised action detection in video. Our EM-MIL framework is motivated by traditional MIL literature which is under-explored in deep learning settings. By allowing us to explicitly model latent variables, this framework improves our control over the learning objective of the instance-level MIL, which leads to state of the art performance. While this work uses a relatively simple pseudo-labeling scheme to implement the EM method, more sophisticated EM methods can be designed, e.g.. explicitly parameterize the latent distribution for instances and directly optimize the instance likelihood in E and M steps. Incorporating the video's temporal structure is also a promising direction for further performance improvement.

Acknowledgement. Prof. Darrell's group was supported in part by DoD, BAIR and BDD.

References

1. Alwassel, H., Heilbron, F.C., Thabet, A., Ghanem, B.: Refineloc: iterative refinement for weakly-supervised action localization. arXiv preprint arXiv:1904.00227 (2019)
2. Carbonneau, M.A., Cheplygina, V., Granger, E., Gagnon, G.: Multiple instance learning: a survey of problem characteristics and applications. Pattern Recogn. **77**, 329–353 (2018)
3. Carreira, J., Zisserman, A.: Quo vadis, action recognition? a new model and the kinetics dataset. In: 2017 IEEE Conference on Computer Vision and Pattern Recognition (CVPR), pp. 4724–4733 (2017)
4. Dietterich, T., Lathrop, R., Lozano-Perez, T.: Solving the multiple instance problem with axis-parallel rectangles. Artif. Intell. **89**, 31–71 (1997)
5. Dooly, D.R., Zhang, Q., Goldman, S.A., Amar, R.A., Brodley, E., Danyluk, A.: Multiple-instance learning of real-valued data. J. Mach. Learn. Res. **3**, 3–10 (2001)
6. Gao, J., Yang, Z., Nevatia, R.: Cascaded boundary regression for temporal action detection. arXiv preprint arXiv:1705.01180 (2017)
7. Heilbron, F.C., Escorcia, V., Ghanem, B., Niebles, J.C.: ActivityNet: a large-scale video benchmark for human activity understanding. In: IEEE Conference on Computer Vision and Pattern Recognition, pp. 961–970 (2015)
8. Ilse, M., Tomczak, J.M., Welling, M.: Attention-based deep multiple instance learning. arXiv preprint arXiv:1802.04712 (2018)
9. Jiang, Y.G., et al.: THUMOS challenge: action recognition with a large number of classes. http://crcv.ucf.edu/THUMOS14/ (2014)
10. Keeler, J.D., Rumelhart, D.E., Leow, W.K.: Integrated segmentation and recognition of hand-printed numerals. In: Lippmann, R.P., Moody, J.E., Touretzky, D.S. (eds.) Advances in Neural Information Processing Systems 3, pp. 557–563. Morgan-Kaufmann (1991)
11. Li, X., Kan, M., Shan, S., Chen, X.: Weakly supervised object detection with segmentation collaboration. In: The IEEE International Conference on Computer Vision (ICCV), October 2019
12. Lin, T., Zhao, X., Su, H., Wang, C., Yang, M.: BSN: boundary sensitive network for temporal action proposal generation. In: Ferrari, V., Hebert, M., Sminchisescu, C., Weiss, Y. (eds.) ECCV 2018. LNCS, vol. 11208, pp. 3–21. Springer, Cham (2018). https://doi.org/10.1007/978-3-030-01225-0_1
13. Liu, D., Jiang, T., Wang, Y.: Completeness modeling and context separation for weakly supervised temporal action localization. In: Proceedings of the IEEE Conference on Computer Vision and Pattern Recognition, pp. 1298–1307 (2019)
14. Liu, Z., et al.: Weakly supervised temporal action localization through contrast based evaluation networks. In: Proceedings of the IEEE International Conference on Computer Vision, pp. 3899–3908 (2019)
15. Maron, O., Lozano-Pérez, T.: A framework for multiple-instance learning. In: Jordan, M.I., Kearns, M.J., Solla, S.A. (eds.) Advances in Neural Information Processing Systems 10, pp. 570–576. MIT Press (1998)
16. Narayan, S., Cholakkal, H., Khan, F.S., Shao, L.: 3c-net: category count and center loss for weakly-supervised action localization. In: Proceedings of the IEEE International Conference on Computer Vision, pp. 8679–8687 (2019)
17. Nguyen, P., Liu, T., Prasad, G., Han, B.: Weakly supervised action localization by sparse temporal pooling network. In: Proceedings of the IEEE Conference on Computer Vision and Pattern Recognition, pp. 6752–6761 (2018)

18. Nguyen, P.X., Ramanan, D., Fowlkes, C.C.: Weakly-supervised action localization with background modeling. In: Proceedings of the IEEE International Conference on Computer Vision, pp. 5502–5511 (2019)
19. Paul, S., Roy, S., Roy-Chowdhury, A.K.: W-TALC: weakly-supervised temporal activity localization and classification. In: Ferrari, V., Hebert, M., Sminchisescu, C., Weiss, Y. (eds.) ECCV 2018. LNCS, vol. 11208, pp. 588–607. Springer, Cham (2018). https://doi.org/10.1007/978-3-030-01225-0_35
20. Shou, Z., Chan, J., Zareian, A., Miyazawa, K., Chang, S.F.: CDC: convolutional-de-convolutional networks for precise temporal action localization in untrimmed videos. In: Proceedings of the IEEE Conference on Computer Vision and Pattern Recognition, pp. 5734–5743 (2017)
21. Shou, Z., Gao, H., Zhang, L., Miyazawa, K., Chang, S.-F.: AutoLoc: weakly-supervised temporal action localization in untrimmed videos. In: Ferrari, V., Hebert, M., Sminchisescu, C., Weiss, Y. (eds.) ECCV 2018. LNCS, vol. 11220, pp. 162–179. Springer, Cham (2018). https://doi.org/10.1007/978-3-030-01270-0_10
22. Singh, K.K., Lee, Y.J.: Hide-and-seek: forcing a network to be meticulous for weakly-supervised object and action localization. In: 2017 IEEE International Conference on Computer Vision (ICCV), pp. 3544–3553. IEEE (2017)
23. Su, H., Zhao, X., Lin, T.: Cascaded pyramid mining network for weakly supervised temporal action localization. In: Jawahar, C.V., Li, H., Mori, G., Schindler, K. (eds.) ACCV 2018. LNCS, vol. 11362, pp. 558–574. Springer, Cham (2019). https://doi.org/10.1007/978-3-030-20890-5_36
24. Tang, P., Wang, X., Bai, X., Liu, W.: Multiple instance detection network with online instance classifier refinement. In: Proceedings of the IEEE Conference on Computer Vision and Pattern Recognition, pp. 2843–2851 (2017)
25. Wan, F., Liu, C., Ke, W., Ji, X., Jiao, J., Ye, Q.: C-MIL: continuation multiple instance learning for weakly supervised object detection. In: CVPR, pp. 2199–2208 (2019)
26. Wang, L., Xiong, Y., Lin, D., Van Gool, L.: Untrimmednets for weakly supervised action recognition and detection. In: Proceedings of the IEEE conference on Computer Vision and Pattern Recognition, pp. 4325–4334 (2017)
27. Wang, L., et al.: Temporal segment networks: towards good practices for deep action recognition. In: Leibe, B., Matas, J., Sebe, N., Welling, M. (eds.) ECCV 2016. LNCS, vol. 9912, pp. 20–36. Springer, Cham (2016). https://doi.org/10.1007/978-3-319-46484-8_2
28. Xu, H., Das, A., Saenko, K.: R-C3D: region convolutional 3D network for temporal activity detection. In: Proceedings of the IEEE International Conference on Computer Vision, pp. 5783–5792 (2017)
29. Xu, H., Das, A., Saenko, K.: Two-stream region convolutional 3D network for temporal activity detection. IEEE Trans. Pattern Anal. Mach. Intell. 41(10), 2319–2332 (2019)
30. Yu, T., Ren, Z., Li, Y., Yan, E., Xu, N., Yuan, J.: Temporal structure mining for weakly supervised action detection. In: Proceedings of the IEEE International Conference on Computer Vision, pp. 5522–5531 (2019)
31. Yuan, Y., Lyu, Y., Shen, X., Tsang, I.W., Yeung, D.Y.: Marginalized average attentional network for weakly-supervised learning. arXiv preprint arXiv:1905.08586 (2019)
32. Zach, C., Pock, T., Bischof, H.: A duality based approach for realtime TV-L^1 optical flow. In: Hamprecht, F.A., Schnörr, C., Jähne, B. (eds.) DAGM 2007. LNCS, vol. 4713, pp. 214–223. Springer, Heidelberg (2007). https://doi.org/10.1007/978-3-540-74936-3_22

33. Zeng, R., Gan, C., Chen, P., Huang, W., Wu, Q., Tan, M.: Breaking winner-takes-all: iterative-winners-out networks for weakly supervised temporal action localization. IEEE Trans. Image Process. **28**(12), 5797–5808 (2019)
34. Zhang, C., Platt, J.C., Viola, P.A.: Multiple instance boosting for object detection. In: Weiss, Y., Schölkopf, B., Platt, J.C. (eds.) Advances in Neural Information Processing Systems 18, pp. 1417–1424. MIT Press (2006)
35. Zhang, Q., Goldman, S.A.: EM-DD: an improved multiple-instance learning technique. In: Dietterich, T.G., Becker, S., Ghahramani, Z. (eds.) Advances in Neural Information Processing Systems 14, pp. 1073–1080. MIT Press (2002)
36. Zhang, Q., Goldman, S.A.: EM-DD: an improved multiple-instance learning technique. In: Advances in Neural Information Processing Systems, pp. 1073–1080 (2002)
37. Zhao, Y., Xiong, Y., Wang, L., Wu, Z., Tang, X., Lin, D.: Temporal action detection with structured segment networks. In: Proceedings of the IEEE International Conference on Computer Vision, pp. 2914–2923 (2017)
38. Zhong, J.X., Li, N., Kong, W., Zhang, T., Li, T.H., Li, G.: Step-by-step erasion, one-by-one collection: a weakly supervised temporal action detector. arXiv preprint arXiv:1807.02929 (2018)

Fairness by Learning Orthogonal Disentangled Representations

Mhd Hasan Sarhan[1,2]([✉]) [iD], Nassir Navab[1,3], Abouzar Eslami[1] [iD],
and Shadi Albarqouni[1,4] [iD]

[1] Computer Aided Medical Procedures, Technical University of Munich,
Munich, Germany
hasan.sarhan@tum.de
[2] Carl Zeiss Meditec AG, Munich, Germany
[3] Computer Aided Medical Procedures, Johns Hopkins University, Baltimore, USA
[4] Computer Vision Lab, ETH Zurich, Zurich, Switzerland

Abstract. Learning discriminative powerful representations is a crucial step for machine learning systems. Introducing invariance against arbitrary nuisance or sensitive attributes while performing well on specific tasks is an important problem in representation learning. This is mostly approached by purging the sensitive information from learned representations. In this paper, we propose a novel disentanglement approach to invariant representation problem. We disentangle the meaningful and sensitive representations by enforcing orthogonality constraints as a proxy for independence. We explicitly enforce the meaningful representation to be agnostic to sensitive information by entropy maximization. The proposed approach is evaluated on five publicly available datasets and compared with state of the art methods for learning fairness and invariance achieving the state of the art performance on three datasets and comparable performance on the rest. Further, we perform an ablative study to evaluate the effect of each component.

Keywords: Representation learning · Disentangled representation · Fairness in machine learning

1 Introduction

Learning representations that are useful for downstream tasks yet robust against arbitrary nuisance factors is a challenging problem. Automated systems powered by machine learning techniques are corner stones for decision support systems such as granting loans, advertising, and medical diagnostics. Deep neural networks learn powerful representations that encapsulate the extracted variations in the data. Since these networks learn from historical data, they are prone to represent the past biases and the learnt representations might contain information

Electronic supplementary material The online version of this chapter (https://doi.org/10.1007/978-3-030-58526-6_44) contains supplementary material, which is available to authorized users.

© Springer Nature Switzerland AG 2020
A. Vedaldi et al. (Eds.): ECCV 2020, LNCS 12374, pp. 746–761, 2020.
https://doi.org/10.1007/978-3-030-58526-6_44

that were not intended to be released. This has raised various concerns regarding fairness, bias and discrimination in statistical inference algorithms [17]. The European union has recently released their "Ethics guidelines for trustworthy AI" report[1] where it is stated that unfairness and biases must be avoided.

Since a few years, the community has been investigating to learn a latent representation z that well describes a target observed variable y (e.g. Annual salary) while being robust against a sensitive attribute s (e.g. Gender or race). This nuisance could be independent from the target task which is termed as a domain adaptation problem. One example is the identification of faces y regardless of the illumination conditions s. In the other case termed fair representation learning s and y are not independent. This could be the case with y being the credit risk of a person while s is age or gender. Such relation between these variables could be due to past biases that are inherently in the data. This independence is assumed to hold when building fair classification models. Although this assumption is over-optimistic as these factors are probably not independent, we wish to find a representation z that is independent from s which justifies the usage of such a prior belief [18]. This is mostly approached by approximations of mutual information scores between z and s and force the two variables to minimize this score either in an adversarial [16,25] or non-adversarial [14,18] manner. These methods while performing well on various datasets, are still limited by either convergence instability problems in case of adversarial solutions or hindered performance compared to the adversarial counterpart. Learning disentangled representations has been proven to be beneficial to learning fairer representations compared to general purpose representations [13]. We use this concept to disentangle the components of the learned representations. Moreover, we treat the s and y as separate independent generative factors and decompose the learned representation in such a way that each representation holds information related to the respective generative factor. This is achieved by enforcing orthogonality between the representations as a relaxation for the independence constraint. We hypothesize that decomposing the latent code into target code z_T and residual sensitive z_S code would be beneficial for limiting the leakage of sensitive information into z_T by redirecting it to z_S while keeping it informative about some target task that we are interested in.

We propose a framework for learning invariant fair representations by decomposing learned representations into target and residual/sensitive representations. We impose disentanglement on the components of each code and impose orthogonality constraint on the two learned representations as a proxy for independence. The learned target representation is explicitly enforced to be agnostic to sensitive information by maximizing the entropy of sensitive information in z_T.

Our contributions are three-folds:

- Decomposition of target and sensitive data into two orthogonal representations to promote better mitigation of sensitive information leakage.

---

[1] Ethics guidelines for trustworthy AI, https://ec.europa.eu/digital-single-market/en/news/ethics-guidelines-trustworthy-ai.

- Promote disentanglement property to split hidden generative factors of each learned code.
- Enforce the target representation to be agnostic of sensitive information by maximizing the entropy.

2 Related Work

Learning fair and invariant representations has a long history. Earlier strategies involved changing the examples to ensure fair representation of the all groups. This relies on the assumption that equalized opportunities in the training set would generalize to the test set. Such techniques are referred to as data massaging techniques [9,19]. These approaches may suffer of under-utilization of data or complications on the logistics of data collection. Later, Zemel *et al.* [26] proposed a semi-supervised fair clustering technique to learn a representation space where data points are clustered such that each cluster contains similar proportions of the protected groups. One drawback is that the clustering constraint limits the power of a distributed representation. To solve this, Louizos *et al.* [14] presented the Variational Fair Autoencoder (VFAE) where a model is trained to learn a representation that is informative enough yet invariant to some nuisance variables. This invariance is approached through Maximum Mean Discrepancy (MMD) penalty. The learned sensitive-information-free representation could be later used for any subsequent processing such as classification of a target task. After the success of Generative Adversarial Networks (GANs) [7], multiple approaches leveraged this learning paradigm to produce robust invariant representations [5,16,25,27]. The problem setup in these approaches is a minimax game between an encoder that learns a representation for a target task and an adversary that extracts sensitive information from the learned representation. In this case, the encoder minimizes the negative log-likelihood of the adversary while the adversary is forced to extract sensitive information alternatively. While methods relying on adversarial zeros-sum game of negative log-likelihood minimization and maximization perform well in the literature, they sometimes suffer from convergence problems and require additional regularization terms to stabilize the training. To overcome these problems, Roy *et al.* [21] posed the problem as an adversarial non-zero sum game where the encoder and discriminator have competing objectives that optimize for different metrics. This is achieved by adding an entropy loss that forces the discriminator to be un-informed about sensitive information. It is worth noting that it is argued by [18] that adversarial training for fairness and invariance is unnecessary and sometimes leads to counter productive results. Hence, they approximated the mutual information between the latent representation and sensitive information using a variational upper bound. Lastly, Creager *et al.* [3] proposed a fair representation learning model by disentanglement, their model has the advantage of flexibly changing sensitive information at test time and combine multiple sensitive attributes to achieve subgroup fairness. In their work, independence is enforced adversarially by utilizing a discriminator to distinguish simulated independent representations (fake) from learned representations (real).

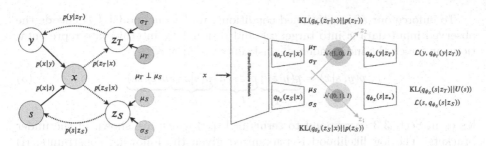

Fig. 1. Left: The graphical model of our proposed method. Right: Our framework encode the input data to intermediate target and residual (sensitive) representations, parameterized by μ and σ. Samples from the estimated posteriors are fed to the discriminators to predict the target and sensitive labels.

3 Methodology

let $\mathcal{X}$ be the dataset of individuals from all groups and $\boldsymbol{x} \in \mathbb{R}^D$ be an input sample. Each input is associated with a target attribute $\boldsymbol{y} = \{y_1, \ldots, y_n\} \in \mathbb{R}^n$ with n classes, and a sensitive attribute $\boldsymbol{s} = \{s_1, \ldots, s_m\} \in \mathbb{R}^m$ with m classes. Our goal is to learn an encoder that maps input $\boldsymbol{x}$ to two low-dimensional representations $\boldsymbol{z_T} \in \mathbb{R}^{d_T}$, $\boldsymbol{z_S} \in \mathbb{R}^{d_S}$. Ideally $\boldsymbol{z_T}$ must contain information regarding target attribute while mitigating leakage about the sensitive attribute and $\boldsymbol{z_S}$ contains residual information that is related to the sensitive attribute.

3.1 Fairness Definition

One of the common definition of fairness that has been proposed in the literature [1, 20, 21, 25] is simply requiring the sensitive information to be statistically independent from the target. Mathematically, the prediction of a classifier $p(\boldsymbol{y}|\boldsymbol{x})$ must be independent from the sensitive information, which is expressed as follows

$$p(\boldsymbol{y}|\boldsymbol{x}) = p(\boldsymbol{y}|\boldsymbol{x}, \boldsymbol{s}) \tag{1}$$

For example, in the German credit dataset, we need to predict the credit behaviour of the bank account holder regardless the sensitive information, such as gender, age ...etc. In other words, $p(\boldsymbol{y} = \text{good credit risk}|\boldsymbol{x}, \boldsymbol{s} = \text{male})$ should be equal to $p(\boldsymbol{y} = \text{good credit risk}|\boldsymbol{x}, \boldsymbol{s} = \text{female})$. The main objective is to learn fair data representations that are i) informative enough for the downstream task, and ii) independent from the sensitive information.

3.2 Problem Formulation

To promote the independence of the generative factors, *i.e.* target and sensitive information, we aim to maximize the log likelihood of the conditional distribution $\log p(\boldsymbol{y}, \boldsymbol{s}|\boldsymbol{x})$, given the fairness assumption in Eq. 1

$$p(\boldsymbol{y}, \boldsymbol{s}|\boldsymbol{x}) = \frac{p(\boldsymbol{y}|\boldsymbol{x}, \boldsymbol{s})p(\boldsymbol{x}|\boldsymbol{s})p(\boldsymbol{s})}{p(\boldsymbol{x})} = p(\boldsymbol{s}|\boldsymbol{x})p(\boldsymbol{y}|\boldsymbol{x}) \tag{2}$$

To enforce our aforementioned conditions, we let our model $f(\cdot)$ encode the observed input data x into target z_T and residual z_S intermediate representations on which the independence constraints are applied,

$$p(y, s|x) = \underbrace{p(y|z_T)}_{L_T} \underbrace{p(z_T|x)}_{L_{z_T}} \underbrace{p(s|z_S)}_{L_S} \underbrace{p(z_S|x)}_{L_{z_S}} \tag{3}$$

losses in Sect. 3.3 correspond to terms in Eq. 3 which are shown in the under brackets. The log likelihood is maximized given the following constraints; (i) $p(z_S|x)$ is statistically independent from $p(z_T|x)$, and (ii) z_T is agnostic to sensitive information s. Our objective function J can be written as

$$J = -\log p(y, s|x) \text{ s.t. } \mathrm{MI}(z_T, z_S) = 0 \text{ and } \mathrm{KL}(p(s|z_T), \mathcal{U}) = 0 \tag{4}$$

where $\mathcal{U}(s)$ is the uniform distribution.

3.3 Fairness by Learning Orthogonal and Disentangled Representations

As depicted in Fig. 1, our observed data x is fed to a shared encoder $f(x; \theta)$, then projected into two subspaces producing our target, and residual (sensitive) representations using the encoders; $q_{\theta_T}(z_T|x)$, and $q_{\theta_S}(z_S|x)$, respectively, where θ is shared parameter, *i.e.* $\theta = \theta_S \cap \theta_T$. Each representation is fed to the corresponding discriminator; target discriminator, $q_{\phi_T}(y|z_T)$, and sensitive discriminator $q_{\phi_S}(s|z_S)$. Both discriminators and encoders are trained in supervised fashion to minimize the following loss,

$$\mathcal{L}_T(\theta_T, \phi_T) = \mathrm{KL}(p(y|x) \parallel q_{\phi_T}(y|z_T)), \tag{5}$$

$$\mathcal{L}_S(\theta_S^*, \phi_S) = \mathrm{KL}(p(s|x) \parallel q_{\phi_S}(s|z_S)), \tag{6}$$

where $\theta_S^* = \theta_S \backslash \theta$.

To ensure that our target representation does not encode any leakage of the sensitive information, we follow Roy *et al.* [21] in maximizing the entropy of the sensitive discriminator given the target representation $q_{\phi_S}(s|z_T)$ as

$$\mathcal{L}_E(\phi_S, \theta_T) = \mathrm{KL}(q_{\phi_S}(s|z_T) \parallel \mathcal{U}(s)). \tag{7}$$

We relax the independence assumption by enforcing i) disentanglement property, and ii) the orthogonality of the corresponding representations.

To promote the (i) disentanglement property on the target representation, we first need to estimate the distribution $p(z_T|x)$ and enforce some sort of independence among the latent factors,

$$p(z_T|x) = \frac{p(x|z_T)p(z_T)}{p(x)}, \text{ s.t. } p(z_T) = \prod_{i=1}^{N_T} p(z_T^i). \tag{8}$$

Since $p(z_T|x)$ is intractable, we employ the Variational Inference, thanks to the re-paramterization trick [11], and let our model output the distribution parameters; $\boldsymbol{\mu}_T$, and $\boldsymbol{\sigma}_T$, and minimize the KL-divergence between posterior $q_{\theta_T}(\boldsymbol{z}_T|\boldsymbol{x})$ and prior $p(\boldsymbol{z}_T)$ distributions as

$$\mathcal{L}_{z_T}(\theta_T) = \mathrm{KL}(q_{\theta_T}(\boldsymbol{z}_T|\boldsymbol{x}) \parallel p(\boldsymbol{z}_T)), \tag{9}$$

where $p(\boldsymbol{z}_T) = \prod_{i=1}^{N_T} p(z_T^i) = \mathcal{N}(\boldsymbol{0}, I)$, and $q_{\theta_T}(\boldsymbol{z}_T|\boldsymbol{x}) = \mathcal{N}(\boldsymbol{z}_T; \boldsymbol{\mu}_T, diag(\boldsymbol{\sigma_T}^2))$. Similarly, we enforce the same constraints on the residual (sensitive) representation $\boldsymbol{z}_S$ and minimize the KL-divergence as $\mathcal{L}_{z_S}(\theta_S) = \mathrm{KL}(q_{\theta_S}(\boldsymbol{z}_S|\boldsymbol{x}) \parallel p(\boldsymbol{z}_S))$.

To enforce the (ii) orthogonality between the target and residual (sensitive) representations,$i.e.$ $\boldsymbol{\mu}_S \perp \boldsymbol{\mu}_T$, we hard code the means of the prior distributions to orthogonal means. In this way, we implicitly enforce the weight parameters to project the representations into orthogonal subspaces. To illustrate this in 2-dimensional space, we set the prior distributions to $p(\boldsymbol{z}_S) = \mathcal{N}([0,1]^T, I)$, and $p(\boldsymbol{z}_T) = \mathcal{N}([1,0]^T, I)$ ($cf.$ Fig. 1).

To summarize, an additional loss term is introduced to the objective function promoting both Orthogonality and Disentanglement properties, denoted *Orthogonal-Disentangled* loss,

$$\mathcal{L}_{OD}(\theta_T, \theta_S) = \mathcal{L}_{z_T}(\theta_T) + \mathcal{L}_{z_S}(\theta_S). \tag{10}$$

A variant of this loss without the property of orthogonality, denoted *Disentangled* loss, is also introduced for the purpose of ablative study (See Sect. 4.3).

3.4   Overall Objective Function

To summarize, our overall objective function is

$$\arg\min_{\theta_T, \theta_S, \phi_T, \phi_S} \mathcal{L}_T(\theta_T, \phi_T) + \mathcal{L}_S(\theta_S^*, \phi_S) + \lambda_E \mathcal{L}_E(\phi_S, \theta_T) + \lambda_{OD} \mathcal{L}_{OD}(\theta_T, \theta_S)$$
$$\tag{11}$$

where λ_E, and λ_{OD} are hyper-parameters to weigh the *Entropy* loss and the *Orthogonal-Disentangled* loss, respectively. A sensitivity analysis on the hyper-parameters is presented in Sec. 4.5.

4   Experiments

In this section, the performance of the learned representations using our method will be evaluated and compared against various state of the art methods in the domain. First, we present the experimental setup by describing the five datasets used for validation, the model implementation details for each dataset, and design of the experiments. We then compare the model performance with state of the art fair representation models on the datasets. We perform an ablative study to monitor the effect of each added component on the overall performance. We then evaluate the models qualitatively by showing t-SNE projections of the learned representations. Lastly, we perform a sensitivity analysis to study the effect of hyper-parameters on the training.

Algorithm 1. Learning Orthogonal Disentangled Fair Representations

---

Require: Maximum Epochs E_{max}, Step size t_s, $\lambda_{OD}, \lambda_E, \gamma_{OD}, \gamma_E, p(z_T), p(z_S)$
Ensure: $z_S \perp z_T$
   Initialize: $\theta_T, \theta_S, \phi_T, \phi_S \leftarrow \theta_T^{(0)}, \theta_S^{(0)}, \phi_T^{(0)}, \phi_S^{(0)}$
   for $t = 1, 2, \ldots, E_{max}$ **do**
     $[\mu_T, \sigma_T] = q_{\theta_T}(z_T | x)$
     $[\mu_S, \sigma_S] = q_{\theta_S}(z_S | x)$
     sample $z_T \sim \mathcal{N}(\mu_T, diag(\sigma_T{}^2))$
     sample $z_S \sim \mathcal{N}(\mu_S, diag(\sigma_S{}^2))$
     compute $\mathcal{L}_{z_T}(\theta_T) = \mathrm{KL}(q_{\theta_T}(z_T | x) \| p(z_T))$
     compute $\mathcal{L}_{z_S}(\theta_S) = \mathrm{KL}(q_{\theta_S}(z_S | x) \| p(z_S))$
     compute $\mathcal{L}_T(\theta_T, \phi_T) = -\sum p(y|x) \log[q_{\phi_T}(y|z_T)]$
     compute $\mathcal{L}_S(\theta_S^*, \phi_S) = -\sum p(s|x) \log[q_{\phi_S}(s|z_S)]$
     compute $\mathcal{L}_E(\phi_S, \theta_T) = \sum q_{\phi_S}(s|z_T) \log[q_{\phi_S}(s|z_T)]$
     update $\lambda_{OD} \leftarrow \lambda_{OD}\gamma_{OD}^{t/t_s}$
     update $\lambda_E \leftarrow \lambda_E\gamma_E^{t/t_s}$
     $\mathcal{L}_{OD}(\theta_T, \theta_S) = \mathcal{L}_{z_T}(\theta_T) + \mathcal{L}_{z_S}(\theta_S)$
     $J(\theta_T, \theta_S, \phi_T, \phi_S) = \mathcal{L}_T(\theta_T, \phi_T) + \mathcal{L}_S(\theta_S^*, \phi_S) + \lambda_E\mathcal{L}_E(\phi_S, \theta_T) + \lambda_{OD}\mathcal{L}_{OD}(\theta_T, \theta_S)$
     update $\theta_T, \theta_S, \phi_T, \phi_S \leftarrow arg \min J(\theta_T, \theta_S, \phi_T, \phi_S)$
   end for
   return $\theta_T, \theta_S, \phi_T, \phi_S$

4.1 Experimental Setup

Tabular Data: For evaluating fair classification, we use two datasets from the UCI repository [4], namely, the German and the Adult datasets. The German credit dataset consists of 1000 samples each with 20 attributes, and the target task is to classify a bank account holder having good or bad credit risk. The sensitive attribute is the gender of the bank account holder. The adult dataset contains 45,222 samples each with 14 attributes. The target task is a binary classification of annual income being more or less than $50,000$ and again gender is the sensitive attribute.

Visual Data: To examine the model learned invariance on visual data, we have used the application of illumination invariant face classification. Ideally, we want the representation to contain information about the subject's identity without holding information regarding illumination direction. For this purpose, the extended YaleB dataset is used [6]. The dataset contains the face images of 38 subjects under five different light source direction conditions (upper right, lower right, lower left, upper left, and front). The target task is the identification of the subject while the light source condition is considered the sensitive attribute.

CIFAR Data: Following Roy et al. [21], we have created a binary target task from CIFAR-10 dataset [12]. The original dataset contains 10 classes we refer to as fine classes, we divide the 10 classes into two categories living and non-living classes and refer to this split as coarse classes. It is expected that living objects have common visual proprieties that differ from non-living ones. The target task is the classification of the coarse classes while not revealing information about the fine classes. With a similar concept, we divide the 100 fine classes of CIFAR-100 dataset into 20 coarse classes that cluster similar concepts into one category. For example, the coarse class 'aquatic mammals' contains the fine classes 'beaver', 'dolphin', 'otter', 'seal', and 'whale'. For the full details of the split, the reader is referred to [21] or the supplementary materials of this manuscript. The target task for CIFAR-100 is the classification of the coarse classes while mitigating information leakage regarding the sensitive fine classes.

Implementation Details: For the Adult and German datasets, we follow the setup appeared in [21] by having a 1-hidden-layer neural network as encoder, the discriminator has two hidden layer and the target predictor is a logistic regression layer. Each hidden layer contains 64 units. The size of the representation is 2. The learning rate for all components is 10^{-3} and weight decay is 5×10^{-4}.

For the Extended YaleB dataset, we use an experimental setup similar to Xie *et al.* [25] and Louizos *et al.* [14] by using the same train/test split strategy. We used $38 \times 5 = 190$ samples for training and 1096 for testing. The model setup is similar to [21,25], the encoder consisted of one layer, target predictor is one linear layer and the discriminator is neural network with two hidden layers each contains 100 units. The parameters are trained using Adam optimizer with a learning rate of 10^{-4} and weight decay of 5×10^{-2}.

Similar to [21], we employed ResNet-18 [8] architecture for training the encoder on the two CIFAR datasets. For the discriminator and target classifiers, we employed a neural network with two hidden layers (256 and 128 neurons). For the encoder, we set the learning rate to 10^{-4} and weight decay to 10^{-2}. For the target and discriminator networks, the learning rate and weight decay were set to 10^{-2} and 10^{-3}, respectively. Adam optimizer [10] is used in all experiments.

Experiments Design: We address two questions in the experiments First, is *how much information about the sensitive attributes is retained in the learned representation z_T?*. Ideally, z_T would not contain any sensitive attribute information. This is evaluated by training a classifier with the same architecture as the discriminator network on sensitive attributes classification task. The closer the accuracy to a naive majority label predictor, the better the model is. This classifier is trained with z_T as input after the encoder, target, and discriminator had been trained and frozen. Second, is *how well the learned representation z_T performs in identifying target attributes?*. To this end, we train a classifier similar to the target on the learned representation z_T to detect the target attributes. We also visualize the representations z_T and z_S by using their t-SNE projections to show how the learned representations describe target attributes while being agnostic to the sensitive information.

Table 1. Results on CIFAR-10 and CIFAR-100 datasets.

	CIFAR-10		CIFAR-100	
	Target Acc. ↑	Sensitive Acc. ↓	Target Acc. ↑	Sensitive Acc. ↓
Baseline	0.9775	0.2344	**0.7199**	0.3069
Xie et al. [25] (trade-off #1)	0.9752	0.2083	0.7132	0.1543
Roy et al. [21] (trade-off #1)	**0.9778**	0.2344	0.7117	0.1688
Xie et al. [25] (trade-off #2)	0.9735	0.2064	0.7040	0.1484
Roy et al. [21] (trade-off #2)	0.9679	0.2114	0.7050	0.1643
Ours	0.9725	**0.1907**	0.7074	**0.1447**

4.2 Comparison with State of the Art

We compare the proposed approach against various state of the art methods on the five presented datasets. We first train the model with Algorithm 1 while changing hyper-parameters between runs. We choose the best performing model in terms of the trade-off between target and sensitive classification accuracy based on z_T. We then compare it with various state of the art methods for sensitive information leakage and retaining target information.

CIFAR Datasets: We compare the proposed approach with two other state of the art methods on the CIFAR-10 and CIFAR-100 datasets, namely Xie *et al.* [25] and Roy *et al.* [21]. We examine two different trade-off points of both approaches. The first trade-off point is the one with best target accuracy reported by the model while the second trade-off point is the one with the target accuracy closest to ours for a more fair comparison. The lower the target accuracy in the trade-off the better (lower) the sensitive accuracy is. We can see when the target accuracies are comparable, our model performs better in preventing sensitive information leakage to the representation z_T. Hence, the proposed method has a better trade-off on the target and sensitive accuracy for both CIFAR-10 and CIFAR-100 datasets. However, the peak target performance is comparable but lower than the peak target performance of the studied methods (Table 1).

Extended YaleB Dataset: For the illumination invariant classification task on the extended YaleB dataset, the proposed method is compared with the logistic regression baseline (LR), variational fair autoencoder VFAE [14], Xie *et al.* [25] and Roy *et al.* [21]. The results are shown in Fig. 2 on the right hand side. The proposed model performs best on the target attribute classification while having the closest performance to the majority classification line (dashed line in Fig. 2). The majority line is the trivial baseline of predicting the majority label. The closer the sensitive accuracy to the majority line the better the model is in hiding sensitive information from z_T. This means the learned representation is powerful at identifying subject in the images regardless of illumination conditions. To assess this visually, refer to Sect. 4.4 for qualitative analysis.

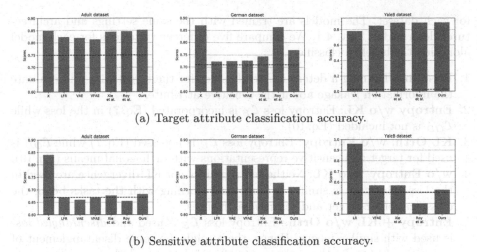

(a) Target attribute classification accuracy.

(b) Sensitive attribute classification accuracy.

Fig. 2. Results on Adult, German, and extended YaleB datasets. The dashed black line represent a naive majority classifier that predicts the majority label.

Tabular Datasets: On the Adult and German datasets, we compare with LFR [26], vanilla VAE [11], variational fair autoencoder [14], Xie *et al.* [25] and Roy *et al.* [21]. The results of these comparisons are shown in Fig. 2. On the German dataset, we observe a very good performance in hiding sensitive information with 71% accuracy compared to 72.7% in [21]. On the target task, the model performs well compared to other models except for [21] which does marginally better than the rest. On the Adult dataset, our proposed model performs better than the aforementioned models on the target task while leaking slightly more information compared to other methods and the majority line at 67%. Our method has 68.26% sensitive accuracy while LFR, VAE, vFAE, Xie *et al.*, and Roy *et al.* have 67%, 66%, 67%, 67.7%, and 65.5% sensitive accuracy, respectively.

Generally, we observe that the proposed model performs well on all datasets with state of the art performance on visual datasets (CIFAR-10, CIFAR-100, YaleB). This suggests that such a model could lead to more fair/invariant representation without large sacrifices on downstream tasks.

4.3   Ablative Study

In this section, we evaluate the contributions provided in the paper by eliminating parts of the loss function and study how each part affects the training in terms of target and sensitive accuracy. To this end, we used the best performing models after hyper-parameter search when training for all contributions for each dataset. The models are trained with the same settings and architectures described in Sect. 4.1. We compare five different variations for each model alongside the baseline classifier:

1. **Baseline**: Training a deterministic classifier for the target task and evaluate the information leakage about the sensitive attribute.
2. **Entropy w/o KL**: Entropy loss $\mathcal{L}_E$ is incorporated (Eq. 7) in the loss while $\mathcal{L}_{OD}$ is not included (Eq. 10).
3. **KL Orth. w/o Entropy**: Entropy loss $\mathcal{L}_E$ is not used (Eq. 7) while $\mathcal{L}_{OD}$ is used for target and sensitive representations with orthogonal means (Eq. 10).
4. **w/o Entropy w/o KL**: Neither entropy loss nor KL divergence are used in the loss. This case is similar ti multi-task learning with the tasks being the classification of target and sensitive attributes.
5. **Entropy + KL w/o Orth.**: Entropy loss $\mathcal{L}_E$ is used and *disentangled loss* is used with similar means. Hence, there might be some disentanglement of generative factors in the components of each latent code but no constraints are applied to force disentanglement of the two representations.
6. **Entropy + KL Orth.**: All contributions are included.

The results of the ablative study are shown in Fig. 3.

– For the *sensitive class accuracy*, it is desirable to have a lower accuracy in distinguishing sensitive attributes. Compared to the baseline, we observe that adding entropy loss and orthogonality constraints on the representations lowers the discriminative power of the learned representation regarding sensitive information. This is valid on all studied datasets except for CIFAR-10 where orthogonality constraint without entropy produced better representations for hiding sensitive information with a small drop (0.26%) on the target task performance. In the rest of the cases, having either entropy loss or KL loss only does not bring noticeable performance gains compared to a multi-task learning paradigm. This could be attributed to the fact that orthogonality on its own does not enforce independence of random variables and another constraint is needed to encourage independent latent variables (*i.e.* entropy loss).
– Comparing baseline with **w/o Entropy w/o KL** case answers the important question *"Does multi-task learning with no constraints on representations bring any added value in mitigating sensitive information leakage?"*. In three out of the five studied datasets, it is the case. We can see lower accuracy in identifying sensitive information by using the learned target representation as input to a classifier while having no constraints on the relationship between the sensitive and target representations during the training process of the encoder. Simply, adding an auxiliary classifier to the target classifier and force it to learn information about sensitive attributes hides some sensitive data from the target classifier.

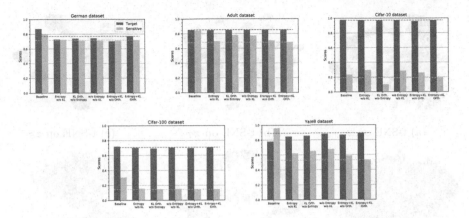

Fig. 3. Ablative study. Dark gray and light gray dashed lines represent the accuracy results on the target and sensitive task respectively for the "Entropy + KL Orth." model.

- Regarding **target accuracy**, the proposed model does not suffer from large drops in target performance when disentangling target from sensitive information. This could be seen by comparing target accuracy between the baseline and **Entropy+KL Orth.** columns. The largest drop in target performance compared to no privacy baseline is seen on the German dataset. This could be because of the very high dependence between gender and granting good or bad credit to a subject in the dataset and the small amount of subjects in the dataset.

4.4   Qualitative Analysis

We visualize the learned embeddings using t-SNE [15] projections for the extended YaleB and CIFAR-10 datasets (*cf.* Fig. 4). We use the image space, z_T, z_S as inputs to the projection to visualize what type of information is held within each representation. We also show the label of each image with regards to the target task to make it easier to investigate the clusters. For the extended YaleB, we see that, using the image space x, the images are clustered mostly depending on their illumination conditions. However, when using z_T, the images are not clustered according lighting conditions but rather, mostly based on the subject identity. Moreover, the visualization of representation z_S shows that the representation contains information about the sensitive class. For the CIFAR-10 dataset, using the image space basically clusters the images on the dominant color. When using z_T, it is clear that the target information is separated where the right side represent the non-living objects, and the left to inside part represents the living objects. What should be observed in z_T, is that within each target class, the fine classes are mixed and indistinguishable as we see cars, boats and trucks mixed in the right hand side of the figure, for example. The representation z_S has some information about the target class and also has the residual

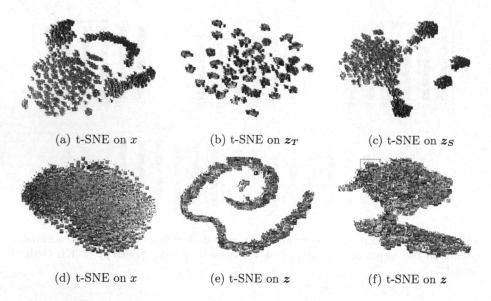

(a) t-SNE on x (b) t-SNE on z_T (c) t-SNE on z_S

(d) t-SNE on x (e) t-SNE on z (f) t-SNE on z

Fig. 4. t-SNE visualization of the extended YaleB faces (top) and CIFAR-10 (bottom) images. Figure is better seen in color and high resolution. (Color figure online)

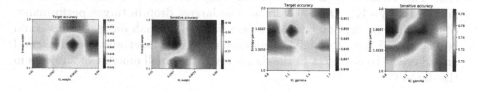

Fig. 5. Sensitivity analysis on the Adult dataset

information about the fine classes as we see in the annotated red rectangle. A group of horses images are clustered together, then few dogs' images are clustered under it, then followed by birds. This shows that z_S has captured some sensitive information while z_T is more agnostic to the sensitive fine classes.

4.5 Sensitivity Analysis

To analyze the effect of hyper-parameters choices on the sensitive and target accuracy, we show heatmaps of how the performance changes when the studied hyper-parameters are changed. The investigated hyper-parameters are KL weight (λ_{OD}), Entropy Weight (λ_E), KL gamma (γ_{OD}), and Entropy gamma (γ_E). We show the results on the Adult dataset. We can see that the sensitive accuracy is sensitive to λ_{OD} more than λ_E as changes in λ_E do not induce much change on the sensitive accuracy. A similar trend is not visible on the target accuracy. Regarding the choice of γ_{OD} and γ_E, we can see that the sensitive leakage is highly affected by these hyper-parameters and the results vary when changed. However, a more robust performance is observed on the target classification task (Fig. 5).

5 Conclusions and Future Work

In this work, we have proposed a novel model for learning invariant representations by decomposing the learned codes into sensitive and target representation. We imposed orthogonality and disentanglement constraints on the representations and forced the target representation to be uninformative of the sensitive information by maximizing sensitive entropy. The proposed approach is evaluated on five datasets and compared with the state of the art models. The results show that our proposed model performs better than state of the art models on three datasets and performed comparably on the other two. We observe better hiding of sensitive information while affecting the target accuracy minimally. This goes in line with our hypothesis that decomposing the two representations and enforcing orthogonality could solve the information leakage problem by redirecting the information into the sensitive representation. One current limitation of this work is that it requires a target task to learn the disentanglement which could be avoided by learning the reconstruction as an auxiliary task similar to other privacy-preserving applications [24]. A direction worth investigating is replacing the pre-definition of the orthogonal sub-spaces priory by learning orthogonality intrinsically with low-rank constraints on the learned representations [22]. Another direction for future work could be focusing on the disentanglement part of the framework. The current disentanglement of factors of generation in the learned representations could be improved by using other disentanglement frameworks [2,23] that are capable of better disentanglement.

Acknowledgments. S.A. is supported by the PRIME programme of the German Academic Exchange Service (DAAD) with funds from the German Federal Ministry of Education and Research (BMBF).

References

1. Barocas, S., Hardt, M., Narayanan, A.: Fairness and Machine Learning. fairmlbook.org (2019). http://www.fairmlbook.org
2. Chen, R.T., Li, X., Grosse, R.B., Duvenaud, D.K.: Isolating sources of disentanglement in variational autoencoders. In: Advances in Neural Information Processing Systems, pp. 2610–2620 (2018)
3. Creager, E., et al.: Flexibly fair representation learning by disentanglement. arXiv preprint arXiv:1906.02589 (2019)
4. Dua, D., Graff, C.: UCI machine learning repository (2017)
5. Edwards, H., Storkey, A.: Censoring representations with an adversary. arXiv preprint arXiv:1511.05897 (2015)

6. Georghiades, A.S., Belhumeur, P.N., Kriegman, D.J.: From few to many: illumination cone models for face recognition under variable lighting and pose. IEEE Trans. Pattern Anal. Mach. Intell. **23**(6), 643–660 (2001)

7. Goodfellow, I., et al.: Generative adversarial nets. In: Advances in Neural Information Processing Systems, pp. 2672–2680 (2014)

8. He, K., Zhang, X., Ren, S., Sun, J.: Identity mappings in deep residual networks. In: Leibe, B., Matas, J., Sebe, N., Welling, M. (eds.) ECCV 2016. LNCS, vol. 9908, pp. 630–645. Springer, Cham (2016). https://doi.org/10.1007/978-3-319-46493-0_38

9. Kamiran, F., Calders, T.: Classifying without discriminating. In: 2009 2nd International Conference on Computer, Control and Communication, pp. 1–6. IEEE (2009)

10. Kingma, D.P., Ba, J.: Adam: a method for stochastic optimization. arXiv preprint arXiv:1412.6980 (2014)

11. Kingma, D.P., Welling, M.: Auto-encoding variational Bayes. arXiv preprint arXiv:1312.6114 (2013)

12. Krizhevsky, A., Hinton, G., et al.: Learning multiple layers of features from tiny images (2009)

13. Locatello, F., Abbati, G., Rainforth, T., Bauer, S., Schölkopf, B., Bachem, O.: On the fairness of disentangled representations. In: Advances in Neural Information Processing Systems, pp. 14584–14597 (2019)

14. Louizos, C., Swersky, K., Li, Y., Welling, M., Zemel, R.: The variational fair autoencoder. arXiv preprint arXiv:1511.00830 (2015)

15. Maaten, L.v.d., Hinton, G.: Visualizing data using t-SNE. J. Mach. Learn. Res. **9**, 2579–2605 (2008)

16. Madras, D., Creager, E., Pitassi, T., Zemel, R.: Learning adversarially fair and transferable representations. arXiv preprint arXiv:1802.06309 (2018)

17. Mehrabi, N., Morstatter, F., Saxena, N., Lerman, K., Galstyan, A.: A survey on bias and fairness in machine learning. arXiv preprint arXiv:1908.09635 (2019)

18. Moyer, D., Gao, S., Brekelmans, R., Galstyan, A., Ver Steeg, G.: Invariant representations without adversarial training. In: Advances in Neural Information Processing Systems, pp. 9084–9093 (2018)

19. Pedreshi, D., Ruggieri, S., Turini, F.: Discrimination-aware data mining. In: Proceedings of the 14th ACM SIGKDD International Conference on Knowledge Discovery and Data Mining, pp. 560–568 (2008)

20. Quadrianto, N., Sharmanska, V., Thomas, O.: Discovering fair representations in the data domain. In: Proceedings of the IEEE Conference on Computer Vision and Pattern Recognition, pp. 8227–8236 (2019)

21. Roy, P.C., Boddeti, V.N.: Mitigating information leakage in image representations: a maximum entropy approach. In: Proceedings of the IEEE Conference on Computer Vision and Pattern Recognition, pp. 2586–2594 (2019)

22. Sanyal, A., Kanade, V., Torr, P.H., Dokania, P.K.: Robustness via deep low-rank representations. arXiv preprint arXiv:1804.07090 (2018)

23. Sarhan, M.H., Eslami, A., Navab, N., Albarqouni, S.: Learning interpretable disentangled representations using adversarial VAEs. In: Wang, Q., et al. (eds.) DART/MIL3ID -2019. LNCS, vol. 11795, pp. 37–44. Springer, Cham (2019). https://doi.org/10.1007/978-3-030-33391-1_5

24. Xiao, T., Tsai, Y.H., Sohn, K., Chandraker, M., Yang, M.H.: Adversarial learning of privacy-preserving and task-oriented representations. arXiv preprint arXiv:1911.10143 (2019)

25. Xie, Q., Dai, Z., Du, Y., Hovy, E., Neubig, G.: Controllable invariance through adversarial feature learning. In: Advances in Neural Information Processing Systems, pp. 585–596 (2017)
26. Zemel, R., Wu, Y., Swersky, K., Pitassi, T., Dwork, C.: Learning fair representations. In: International Conference on Machine Learning, pp. 325–333 (2013)
27. Zhang, B.H., Lemoine, B., Mitchell, M.: Mitigating unwanted biases with adversarial learning. In: Proceedings of the 2018 AAAI/ACM Conference on AI, Ethics, and Society, pp. 335–340 (2018)

Self-supervision with Superpixels: Training Few-Shot Medical Image Segmentation Without Annotation

Cheng Ouyang$^{(\boxtimes)}$, Carlo Biffi, Chen Chen, Turkay Kart, Huaqi Qiu,
and Daniel Rueckert

BioMedIA Group, Department of Computing, Imperial College London, London, UK
c.ouyang@imperial.ac.uk

Abstract. Few-shot semantic segmentation (FSS) has great potential for medical imaging applications. Most of the existing FSS techniques require abundant annotated semantic classes for training. However, these methods may not be applicable for medical images due to the lack of annotations. To address this problem we make several contributions: (1) A novel self-supervised FSS framework for medical images in order to eliminate the requirement for annotations during training. Additionally, superpixel-based pseudo-labels are generated to provide supervision; (2) An adaptive local prototype pooling module plugged into prototypical networks, to solve the common challenging foreground-background imbalance problem in medical image segmentation; (3) We demonstrate the general applicability of the proposed approach for medical images using three different tasks: abdominal organ segmentation for CT and MRI, as well as cardiac segmentation for MRI. Our results show that, for medical image segmentation, the proposed method outperforms conventional FSS methods which require manual annotations for training.

1 Introduction

Automated medical image segmentation is a key step for a vast number of clinical procedures and medical imaging studies, including disease diagnosis and follow-up [1–3], treatment planning [4,5] and population studies [6,7]. Fully supervised deep learning based segmentation models can achieve good results when trained on abundant labeled data. However, the training of these networks in medical imaging is often impractical due to the following two reasons: there is often a lack of sufficiently large amount of expert-annotated data for training due the considerable clinical expertise, cost and time associated with annotation; This problem is further exacerbated by differences in image acquisition procedures

---

C. Biffi, C. Chen and T. Kart—Equal contribution.

Electronic supplementary material The online version of this chapter (https://doi.org/10.1007/978-3-030-58526-6_45) contains supplementary material, which is available to authorized users.

A. Vedaldi et al. (Eds.): ECCV 2020, LNCS 12374, pp. 762–780, 2020.
https://doi.org/10.1007/978-3-030-58526-6_45

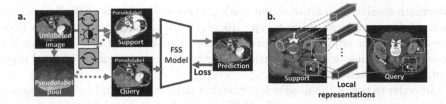

Fig. 1. (a). Proposed superpixel-based self-supervised learning. For each unlabeled image, pseudolabels are generated on superpixels. In each iteration during training, a randomly selected pseudolabel and the original image serve as the candidate for both support and query. Then, random transforms (marked in blue boxes) are applied between the support and the query. The self-supervision task is designed as segmenting the pseudolabel on the query with reference to the support, despite the transforms applied in between. (b). The proposed ALPNet solves the class-imbalance-induced ambiguity problem by adaptively extracting multiple local representations of the large background class (in blue). Each of them only represents a local region of background. (Color figure online)

across medical devices and hospitals, often resulting in datasets containing few manually labeled images; Moreover, the number of possible segmentation targets (different anatomical structures, different types of lesions, etc.) are countless. It is impractical to cover every single unseen class by training a new, specific model.

As a potential solution to these two challenges, few-shot learning has been proposed [8–13]. During *inference*, a few-shot learning model distills a discriminative representation of an unseen class from only a few labeled examples (usually denoted as *support*) to make predictions for unlabeled examples (usually denoted as *query*) without the need for re-training the model. If applying few-shot learning to medical images, segmenting a rare or novel lesion can be potentially efficiently achieved using only a few labeled examples.

However, *training* an existing few-shot semantic segmentation (FSS) model for medical imaging has not had much success in the past, as most of FSS methods rely on a large training dataset with many annotated training classes to avoid overfitting [14–19, 21–25]. In order to bypass this unmet need of annotation, we propose to train an FSS model on unlabeled images instead via self-supervised learning, an unsupervised technique that learns generalizable image representations by solving a carefully designed task [26–33]. Another challenge for a lot of state-of-the-art FSS network architectures is the loss of local information within a spatially variant class in their learned representations. This problem is in particular magnified in medical images since extreme foreground-background imbalance commonly exists in medical images. As shown in Fig. 1(b), the background class is large and spatially inhomogeneous whereas the foreground class (in purple) is small and homogeneous. Under this scenario, an ambiguity in prediction on foreground-background boundary might happen if the distinct appearance information of different local regions (or saying, parts) in the background is unreasonably averaged out. Unfortunately, this loss of intra-class local information exists in a lot of recent works, where each class is spatially averaged into a 1-D representation prototype [16,18,19,34] or weight vectors of a linear classifier [17]. In adjust to this problem, we instead encourage the network to preserve intra-class local information, by extracting an ensemble of local representations for each class.

In order to break the deadlock of training data scarcity and to boost segmentation accuracy, we propose SSL-ALPNet, a self-supervised few-shot semantic segmentation framework for medical imaging. The proposed framework exploits *superpixel-based self-supervised learning* (SSL), using superpixels for eliminating the need for manual annotations, and an *adaptive local prototype pooling* enpowered prototypical network (ALPNet), improving segmentation accuracy by preserving local information in learned representations. As shown in Fig. 1(a), to ensure image representations learned through self-supervision are well-generalizable to real semantic classes, we generate pseudo-semantic labels using superpixels, which are compact building blocks for semantic objects [35–37]. In addition, to improve the discriminative ability of learned image representations, we formulate the self-supervision task as one-superpixel-against-the-rest segmentation. Moreover, to enforce invariance in representations between support and query, which is crucial for few-shot segmentation in real-world, we synthesis variants in shape and intensity by applying random geometric and intensity transforms between support and query. In our experiments, we observed that by purely training with SSL, our network outperforms those trained with manual annotated classes by considerable margins. Besides, as shown in Fig. 1, to boosts segmentation accuracy, we designed adaptive local prototype module (ALP) for preserving local information of each class in their prototypical representations. This is achieved by extracting an ensemble of local representation prototypes, each focuses on a different region. Of note, the number of prototypes are allocated adaptively by the network based on the spatial size of each class. By this mean, ALP alleviates ambiguity in segmentation caused by insufficient local information.

Overall, the proposed SSL-ALPNet framework has the following major advantages: Firstly, compared with current state-of-the-art few-shot segmentation methods which in general rely on a large number of annotated classes for training, the proposed method eliminates the need for annotated training data instead. By completely detaching representation extraction from manual labeling, the proposed method potentially expands the application of FSS in annotation-scarce medical images. In addition, unlike most of self-supervised learning methods for segmentation where fine-tuning on labeled data is still required before testing [27–30,32,38], the proposed method requires no fine-tuning after SSL. Moreover, compared to some of novel modules [39–41] used in FSS where slight performance gain are at the cost of heavy computations, the proposed ALP is simple and efficient in contrast to its significant performance boosting. No trainable parameters is contained in ALP.

Our contributions are summarized as follows:

- We propose SSL-ALPNet, the first work that explores self-supervised learning for few-shot medical image segmentation, to the best of our knowledge. It outperforms peer FSS methods, which usually require training with manual annotations, by merely training on unlabeled images.
- We propose adaptive local prototype pooling, a local representation computation module that significantly boosts performance of the state-of-the-art prototypical networks on medical images.
- We for the first time evaluated FSS on different imaging modalities, segmentation classes and with the presence of patient pathologies. The established evaluation strategy not only highlights wide applicability of our work, but also facilitates future works that seek to evaluate FSS in a more realistic scenario.

2   Related Work

2.1   Few-Shot Semantic Segmentation

Recent work by [31] firstly introduces self-supervised learning into few-shot image classification. However, few-shot segmentation is often more challenging: dense prediction needs to be performed at a pixel level. To fully exploit information in limited support data, most of popular FSS methods directly inject support to the network as guiding signals [20,21,42,43], or construct discriminative representations from support as reference to segment query [15–19]. The pioneering work [15] learns to generate classifier weights from support; [17] extends weights generation to multi-scale. [14] instead directly use support to condition segmentation on query by fusing their feature maps. Exploiting network components such as attention modules [39,44] and graph networks [40,41], recent works boost segmentation accuracy [21] and enable FSS with coarse-level supervisions [22,24,42]. Exploiting learning-based optimization, [23,25] combine meta-learning with FSS. However, almost all of these methods assume abundant annotated (including weakly annotated) training data to be available, making them difficult to translate to segmentation scenarios in medical imaging.

One main stream of FSS called *prototypical networks* focuses on exploiting *representation prototypes* of semantic classes extracted from the support. These prototypes are utilized to make similarity-based prediction [8,18,34] on query, or to tune representations of query [16]. Recently, prototypical alignment network (PANet) [18] has achieved state-of-the-art performance on natural images. This is achieved simply with a generic convolutional network and an alignment regularization. However, these works aim to improve performance on training-classes-abundant natural images. Their methodologies focus on network design. Our work, by contrast, focuses on utilizing unlabeled medical image for training by exploiting innovative training strategies and pesudolabels. Nevertheless, since PANet is one of state-of-the-art and is conceptually simple, we take this method as our baseline to highlight our self-supervised learning as a generic training strategy.

In medical imaging, most of recent works on few-shot segmentation only focus on training with less data [45–49]. These methods usually still require retraining before applying to unseen classes, and therefore they are out-of-scope in our discussion. Without retraining on unseen classes, the SE-Net [43] introduces squeeze and excite blocks [50] to [14]. To the best of our knowledge, it is the first FSS model specially designed for medical images, with which we compared our method in experiments.

2.2 Self-supervised Learning in Semantic Segmentation

A series of self-supervision tasks have been proposed for semantic segmentation. Most of these works focus on intuitive handcrafted supervision tasks including spatial transform prediction [51], image impainting [32], patch reordering [27], image colorization [33], difference detection [52], motion interpolation [53] and so on. Similar methods have been applied to medical images [38,54–56]. However, most of these works still require a second-stage fine-tuning after initializing with weights learned from self-supervision. In addition, features learned from handcrafted tasks may not be sufficiently generalizable to semantic segmentation, as two tasks might not be strongly related [57]. In contrast, in our work, segmenting superpixel-based pseudolabels is directly related to segmenting real objects. This is because superpixels are compact building blocks for semantic masks for real objects. Recent works [48,58,59] on medical imaging rely on second-order optimization [60]. These works differ from our work in key method and task.

Our proposed SSL technique shares a similar spirit as [61] (or arguably, as some recent works on contrastive learning [62–65]) in methodology. Both methods encourage invariance in image representation by intentionally creating variants. While [61] focuses on visual information clustering, we focus on the practical but challenging few-shot medical image segmentation problem.

2.3 Superpixel Segmentation

Superpixels are small, compact image segments which are usually piece-wise smooth [35,66]. Superpixels are generated by clustering local pixels using statistical models with respect to low-level image features. These models include Gaussian mixture [37] and graph cut [67]. In this work, we employed off-the-shelf, efficient and unsupervised graph-cut-based algorithm by [68]. Compared with the popular SLIC method [37], superpixels generated by [68] are more diverse in shape. Training with these superpixels intuitively improves generalizability of the network to unseen classes in various shapes.

3 Method

We first introduce problem formulation for few-shot semantic segentation (FSS). Then, the ALPNet architecture is introduced with a focus on adaptive local

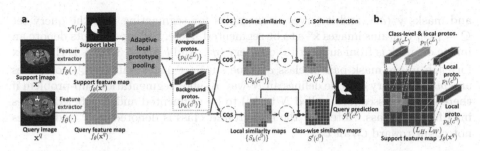

Fig. 2. (a). Workflow of the proposed network: The feature extractor $f_\theta(\cdot)$ takes the support image and query image as input to generate feature maps $f_\theta(\mathbf{x}^s)$ for support and $f_\theta(\mathbf{x}^q)$ for query. The proposed adaptive local prototype pooling module then takes support feature map and support label as input to obtain an ensemble of representation prototypes $p_k(c^j)$'s. These prototypes are used as references for comparing with query feature map $f_\theta(\mathbf{x}^q)$. Similarity maps generated by these comparisons are fused together to form the final segmentation. This figure illustrates a 1-way segmentation setting, where c^L is the foreground class, c^0 is the background. (b). Illustration of the adaptive local prototype pooling module: Local prototypes are calculated by spatially averaging support feature maps within pooling windows (orange boxes); class-level prototypes are averaged under the entire support label (purple region). (Color figure online)

prototype pooling and the corresponding inference process. We highlight our superpixel-based self-supervised learning (SSL) with details in pseudolabel generation process and episode formation in Sect. 3.3. Finally, we introduce the overall end-to-end training objective under the proposed SSL technique. Of note, after the proposed self-supervised learning, ALPNet can be directly applied to unseen classes with its weights fixed, and with reference to a few human-labeled support slices. There is no fine-tuning required in this testing phase.

3.1   Problem Formulation

The aim of few-shot segmentation is to obtain a model that can segment an unseen semantic class, by just learning from a few labeled images of this unseen class during inference without retraining the model. In few-shot segmentation, a training set $\mathcal{D}_{tr}$ containing images with training semantic classes $\mathcal{C}_{tr}$ (e.g., $\mathcal{C}_{tr} = \{liver, spleen, spine\}$), and a testing set $\mathcal{D}_{te}$ of images containing testing unseen classes $\mathcal{C}_{te}$ (e.g., $\mathcal{C}_{te} = \{heart, kidney\}$), are given, where $\mathcal{C}_{tr} \cap \mathcal{C}_{te} = \emptyset$. The task is to train a segmentation model on $\mathcal{D}_{tr}$ (e.g. labeled images of livers, spleens and spines) that can segment semantic classes $\mathcal{C}_{te}$ in images in $\mathcal{D}_{te}$, given a few annotated examples of $\mathcal{C}_{te}$ (e.g. to segment $kidney$ with reference to a few labeled images of kidney), without re-training. $\mathcal{D}_{tr} = \{(\mathbf{x}, \mathbf{y}(c^{\hat{j}}))\}$ is composed of images $\mathbf{x} \in \mathcal{X}$ and corresponding binary masks $\mathbf{y}(c^{\hat{j}})$'s $\in \mathcal{Y}$ of classes $c^{\hat{j}} \in \mathcal{C}_{tr}$, where $\hat{j} = 1, 2, 3, ..., N$ is the class index. $\mathcal{D}_{te}$ is defined in the same way but for testing images and masks with $\mathcal{C}_{te}$. In each inference pass, a $support$ set $\mathcal{S}$ and a $query$ set $\mathcal{Q}$ are given. The support $\mathcal{S} = \{(\mathbf{x}_l^s, \mathbf{y}_l^s(c^{\hat{j}}))\}$ contains images $\mathbf{x}_l^s$

and masks $\mathbf{y}_l^s(c^{\hat{j}})$, and it serves as examples for segmenting $c^{\hat{j}}$'s; the query set $\mathcal{Q} = \{\mathbf{x}^q\}$ contains images $\mathbf{x}^q$'s to be segmented. Here, the superscripts denote an image or mask is from support (s) or query (q). And $l = 1, 2, 3, ..., K$ is the index for each image-mask pair of class $c^{\hat{j}}$. One support-query pair $(\mathcal{S}, \mathcal{Q})$ comprises an *episode*. Every episode defines an N-way K-shot segmentation sub-problem if there are N classes (also called N tasks) to be segmented and K labeled images in $\mathcal{S}$ for each class. Note that the background class is denoted as c^0 and it does not count toward $\mathcal{C}_{tr}$ or $\mathcal{C}_{te}$.

3.2   Network Architecture

Overview: Our network is composed of: (a) a generic *feature extractor* network $f_\theta(\cdot) : \mathcal{X} \rightarrow \mathcal{E}$ parameterized by θ, where $\mathcal{E}$ is the representation space (i.e. feature space) on which segmentation operates; (b) the proposed *adaptive local prototype pooling module* (ALP) $g(\cdot, \cdot) : \mathcal{E} \times \mathcal{Y} \rightarrow \mathcal{E}$ for extracting representation *prototypes* from support features and labels; (c) and a *similarity based classifier* $sim(\cdot, \cdot) : \mathcal{E} \times \mathcal{E} \rightarrow \mathcal{Y}$ for segmentation by comparing prototypes and query features.

As shown in Fig. 2, in inference, the feature extractor network $f_\theta(\cdot)$ provides ALP with feature maps by mapping both $\mathbf{x}_l^s$'s and $\mathbf{x}^q$'s to feature space $\mathcal{E}$, producing feature maps $\{(f_\theta(\mathbf{x}^q), f_\theta(\mathbf{x}_l^s))\} \in \mathcal{E}$. ALP takes each $(f_\theta(\mathbf{x}_l^s), \mathbf{y}_l^s(c^{\hat{j}}))$ pair as input to compute both *local prototypes* and *class-level prototypes* of semantic class $c^{\hat{j}}$ and background c^0. These prototypes will later be used as references of each class for segmenting query images. Prototypes of all c^j's forms a prototype ensemble $\mathcal{P} = \{p_k(c^j)\}, j = 0, 1, 2, ..., N$ where k is prototype index and $k \geq 1$ for each c^j. This prototype ensemble is used by the classifier $sim(\cdot, \cdot)$ to predict the segmentation for the query image, saying $\hat{\mathbf{y}}^q = sim(\mathcal{P}, f_\theta(\mathbf{x}^q))$. This is achieved by first measuring similarities between each $p_k(c^j)$'s and query feature map $f_\theta(\mathbf{x}^q)$, and then fusing these similarities together.

Adaptive Local Prototype Pooling: In contrast to previous works [17,18,34], where intra-class local information is unreasonably spatially averaged out underneath the semantic mask, we propose to preserve local information in prototypes by introducing adaptive local prototype pooling module (ALP). In ALP, each *local prototype* is only computed within a *local pooling window* overlaid on the support and only represents one part of object-of-interest.

Specifically, we perform average pooling with a pooling window size (L_H, L_W) on each $f_\theta(\mathbf{x}_l^s) \in \mathbb{R}^{D \times H \times W}$ where (H, W) is the spatial size and D is the channel depth. Of note, (L_H, L_W) determines the spatial extent under which each local prototype is calculated in the representation space $\mathcal{E}$. The obtained local prototype $p_{l,mn}(c)$ with undecided class c at spatial location (m, n) of the average-pooled feature map is given by

$$p_{l,mn}(c) = \text{avgpool}(f_\theta(\mathbf{x}_l^s))(m, n) = \frac{1}{L_H L_W} \sum_h \sum_w f_\theta(\mathbf{x}_l^s)(h, w), \quad (1)$$

where $mL_H \leq h < (m+1)L_H, \ nL_W \leq w < (n+1)L_W$.

To decide the class c of each $p_{l,mn}(c)$, we average-pool the binary mask $\mathbf{y}_l^s(c^{\hat{j}})$ of the foreground class $c^{\hat{j}}$ to the same size $(\frac{H}{L_H}, \frac{W}{L_W})$. Let $y_{l,mn}^a$ be the value of $\mathbf{y}_l^s(c^{\hat{j}})$ after average pooling at location (m, n), c is assigned as:

$$c = \begin{cases} c^0 & y_{l,mn}^a < T \\ c^{\hat{j}} & y_{l,mn}^a \geq T \end{cases} \quad \text{where} \quad y_{l,mn}^a = \text{avgpool}(\mathbf{y}_l^s(c^{\hat{j}}))(m, n). \tag{2}$$

T is the lower-bound threshold for foreground which is empirically set to 0.95.

To ensure at least one prototype is generated for objects smaller than the pooling window (L_H, L_W), we also compute a *class-level prototype* $p_l^g(c^{\hat{j}})$ using masked average pooling [18,19]:

$$p_l^g(c^{\hat{j}}) = \frac{\sum\limits_{h,w} \mathbf{y}_l^s(c^{\hat{j}})(h, w) f_\theta(\mathbf{x}_l^s)(h, w)}{\sum\limits_{h,w} \mathbf{y}_l^s(c^{\hat{j}})(h, w)}. \tag{3}$$

In the end, $p_{l,mn}(c^{\hat{j}})$'s and $p_l^g(c^{\hat{j}})$'s are re-indexed with subscript k's for convenience, and hence comprise the representation prototype ensemble $\mathcal{P} = \{p_k(c^{\hat{j}})\}$. This ensemble therefore preserves more intra-class local distinctions by explicitly representing different local regions into separate prototypes.

Similarity-Based Segmentation: The similarity-based classifier $sim(\cdot, \cdot)$ is designed to make dense prediction on query by exploiting local image information in $\mathcal{P}$. This is achieved by firstly matching each prototype to a corresponding local region in query, and then fusing the local similarities together.

As a loose interpretation, to segment a large *liver* in query, in the first stage, a local prototype $p_k(c^L)$ with class $c^L = liver$, whose pooling window falls over the *right lobe* of the *liver* particularly finds a similar region which looks like a *right lobe* in query (instead of matching the entire *liver*). Then, to get an entire liver, results from *right lobe*, and *left lobe* are fused together to form a *liver*.

Specifically, $sim(\cdot, \cdot)$ first takes query feature map $f_\theta(\mathbf{x}^q)$ and prototype ensemble $\mathcal{P} = \{p_k(c^{\hat{j}})\}$ as input to compute *local similarity maps* $S_k(c^{\hat{j}})$'s between $f_\theta(\mathbf{x}^q)$ and all $p_k(c^{\hat{j}})$'s respectively. Each entry $S_k(c^{\hat{j}})(h, w)$ at spatial location (h, w) corresponding to $f_\theta(\mathbf{x}^q)$ is given by

$$S_k(c^{\hat{j}})(h, w) = \alpha p_k(c^{\hat{j}}) \odot f_\theta(\mathbf{x}^q)(h, w), \tag{4}$$

where $\odot$ denotes cosine similarly, which is bounded, same as in [18]: $a \odot b = \frac{\langle a,b \rangle}{\|a\|_2 \|b\|_2}$, $a, b \in \mathbb{R}^{D \times 1 \times 1}$, α is a multiplier, which helps gradients to backpropagate in training [69]. In our experiments, α is set to 20, same as in [18].

Then, to obtain similarity maps (unnormalized) with respect to each class $c^{\hat{j}}$ as a whole, local similarity maps $S_k(c^{\hat{j}})$'s are fused for each class separately into *class-wise similarities* $S'(c^{\hat{j}})$, this is done through a softmax function:

$$S'(c^{\hat{j}})(h, w) = \sum_k S_k(c^{\hat{j}})(h, w) \, \text{softmax}[S_k(c^{\hat{j}})(h, w)]. \tag{5}$$

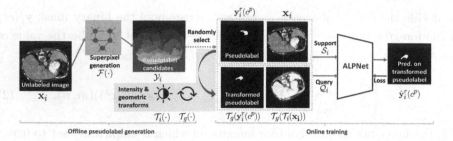

Fig. 3. Workflow of the proposed superpixel-based self-supervised learning technique.

$\text{softmax}_k[S_k(c^j)(h,w)]$ refers to the operation of first stacking all $S_k(c^j)(h,w)$'s along channel dimension and then computing softmax function along channels.

To obtain the final dense prediction, in the end, class-wise similarities are normalized into probabilities:

$$\hat{\mathbf{y}}^q(h,w) = \text{softmax}_j[S'(c^j)(h,w)]. \tag{6}$$

3.3 Superpixel-Based Self-supervised Learning

To obtain accurate and robust results, two properties are highly desirable for similarity-based classifiers. For each class, the representations should be *clustered* in order to be discriminative under a similarity metric; meanwhile, these representations should be *invariant* across images (in our case any combinations of support and query) to ensure robustness in prediction [61].

These two properties are encouraged by the proposed superpixel-based self-supervised learning (SSL). As annotations for real semantic classes are unavailable, SSL exploits pseudolabels to enforce *clustering* at a superpixel-level. This is naturally achieved by back-propagating segmentation loss via cosine-similarity-based classifier. Here, the superpixel-level clustering property can be transferred to real semantic classes, since one semantic mask is usually composed of several superpixels [35,36]. Additionally, to encourage representations to be invariant against shape and intensity differences between images, we perform geometric and intensity transforms between support and query. This is because shape and intensity are the largest sources of variations in medical images [70].

The proposed SSL framework consists of two phases: offline pseudolabel generation and online training. The entire workflow can be seen in Fig. 3.

Unsupervised Pseudolabel Generation: To obtain candidates for pseudolabels, a collection of superpixels $\mathcal{Y}_i = \mathcal{F}(\mathbf{x}_i)$ are generated for every image $\mathbf{x}_i$. This is efficiently done with the unsupervised algorithm [68] denoted by $\mathcal{F}(\cdot)$.

Online Episode Composition: For each episode i, an image $\mathbf{x}_i$ and a randomly chosen superpixel $\mathbf{y}_i^r(c^p) \in \mathcal{Y}_i$ form the support $\mathcal{S}_i = \{(\mathbf{x}_i, \mathbf{y}_i^r(c^p))\}$. Here $\mathbf{y}_i^r(c^p)$ is a binary mask with index $r = 1, 2, 3, ..., |\mathcal{Y}_i|$ and c^p denotes the *pseudolabel* class (corresponding background mask $\mathbf{y}_i^r(c^0)$ is given by $1 - \mathbf{y}_i^r(c^p)$). Meanwhile, the query set $\mathcal{Q}_i = \{(\mathcal{T}_g(\mathcal{T}_i(\mathbf{x}_i)))\}, \mathcal{T}_g(\mathbf{y}_i^r(c^p)))$ is constructed by applying random geometric and intensity transforms: $\mathcal{T}_g(\cdot)$ and $\mathcal{T}_i(\cdot)$ to the support. By this mean, each $(\mathcal{S}_i, \mathcal{Q}_i)$ forms a 1-way 1-shot segmentation problem. In practice, $\mathcal{T}_g(\cdot)$ includes affine and elastic transforms, $\mathcal{T}_i(\cdot)$ is gamma transform.

End-to-End Training: The network is trained end-to-end, where each iteration i takes an episode $(\mathcal{S}_i, \mathcal{Q}_i)$ as input. Cross entropy loss is employed where the segmentation loss $\mathcal{L}_{seg}^i$ for each iteration is written as:

$$\mathcal{L}_{seg}^i(\theta; \mathcal{S}_i, \mathcal{Q}_i) = -\frac{1}{HW} \sum_h^H \sum_w^W \sum_{j \in \{0,p\}} \mathcal{T}_g(\mathbf{y}_i^r(c^j))(h, w) \log(\hat{\mathbf{y}}_i^r(c^j)(h, w)), \quad (7)$$

where $\hat{\mathbf{y}}_i^r(c^p)$ is the prediction of query pseudolabel $\mathcal{T}_g(\mathbf{y}_i^r(c^p))$ and is obtained as described in Sect. 3.2. In practice, weightings of 0.05 and 1.0 are given to c^0 and c^p separately for mitigating class imbalance. We also inherited the *prototypical alignment regularization* in [18]: taking prediction as support, i.e. $\mathcal{S}' = (\mathcal{T}_g(\mathcal{T}_i(\mathbf{x}_i)), \hat{\mathbf{y}}_i^r(c^p))$, it should correctly segment the original support image $\mathbf{x}_i$. This is presented as

$$\mathcal{L}_{reg}^i(\theta; \mathcal{S}_i', \mathcal{S}_i) = -\frac{1}{HW} \sum_h^H \sum_w^W \sum_{j \in \{0,p\}} \mathbf{y}_i^r(c^j)(h, w) \log(\bar{\mathbf{y}}_i^r(c^j)(h, w)), \quad (8)$$

where $\bar{\mathbf{y}}_i^r(c^p)$ is the prediction of $\mathbf{y}_i^r(c^p)$ taking $\mathbf{x}_i$ as query.

Overall, the loss function for each epioside is:

$$\mathcal{L}^i(\theta; \mathcal{S}_i, \mathcal{Q}_i) = \mathcal{L}_{seg}^i + \lambda \mathcal{L}_{reg}^i, \quad (9)$$

where λ controls strength of regularization as in [18].

After self-supervised learning, the network can be directly used for inference on unseen classes.

4   Experiments

Datasets: To demonstrate the general applicability of our proposed method under different imaging modalities, segmentation classes and health conditions of the subject, we performed evaluations under three scenarios: abdominal organs segmentation for CT and MRI (Abd-CT and Abd-MRI) and cardiac segmentation for MRI (Card-MRI). All three datasets contain rich information outside their regions-of-interests, which benefits SSL by providing sources of superpixels. Specifically,

– **Abd-CT** is from MICCAI 2015 Multi-Atlas Abdomen Labeling challenge [71]. It contains 30 3D abdominal CT scans. Of note, this is a clinical dataset containing patients with various pathologies and variations in intensity distributions between scans.

- **Abd-MRI** is from ISBI 2019 Combined Healthy Abdominal Organ Segmentation Challenge (Task 5) [72]. It contains 20 3D T2-SPIR MRI scans.
- **Card-MRI** is from MICCAI 2019 Multi-sequence Cardiac MRI Segmentation Challenge (bSSFP fold) [73], with 35 clinical 3D cardiac MRI scans.

To unify experiment settings, all images are re-formated as 2D axial (Abd-CT and Abd-MRI) or 2D short-axis (Card-MRI) slices, and resized to 256×256 pixels. Prepossessings are applied following common practices. Each 2D slice is repeated for three times in channel dimension to fit into the network.

To comparatively evaluate the results on classes with various shapes, locations and textures between partically-pathologic, imhomogeneous Abd-CT and all-healthy, homogeneous Abd-MRI, we construct a shared label set containing left kidney, right kidney, spleen and liver; For Card-MRI, the label set contains left-ventricle blood pool (LV-BP), left-ventricle myocardium (LV-MYO) and right-ventricle (RV). In all experiments, we perform five-fold cross-validation.

Evaluation: To measure the overlapping between prediction and ground truth, we employ Dice score (0–100, 0: mismatch; 100: perfectly match), which is commonly used in medical image segmentation researches. To evaluate 2D segmentation on 3D volumetric images, we follow the evaluation protocol established by [43]. In a 3D image, for each class c^j, images between the top slice and the bottom slice containing c^j are divided into C equally-spaced chunks. The middle slice in each chunk from the support scan is used as reference for segmenting all the slices in corresponding chunk in query. In our experiments C is set to be 3. Of note, the support and query scans are from different patients.

To evaluate generalization ability to unseen testing classes, beyond the standard few-shot segmentation experiment setting for medical images established by [43] (**setting 1**), where testing class might appear as background in training data, we introduce a **setting 2**. In setting 2, we force testing classes (even unlabeled) to be completely unseen by removing any image that contains a testing class, from the training dataset.

Labels are therefore partitioned differently according to the settings and types of supervision. In setting 1, when training with SSL, no label partitioning is required for training. When training with annotated images, each time we take one class for testing and the rest for training. To observe if the learned representations encode spatial concepts like left and right, we deliberately group ⟨left/ right kidney⟩ to appear together in training or testing. In setting 2, as ⟨spleen, liver⟩, or ⟨left/ right kidney⟩ usually appear together in a 2D slice respectively, we group them into *upper* abdomen and *lower* abdomen groups separately. In each experiment all slices containing the testing group will be removed from training data. For Card-MRI, only setting 1 is examined as all the labels usually appear together in 2D slices, making label exclusion impossible.

To simulate the scarcity of labeled data in clinical practice, all our experiments in this section are performed under 1-way 1-shot setting.

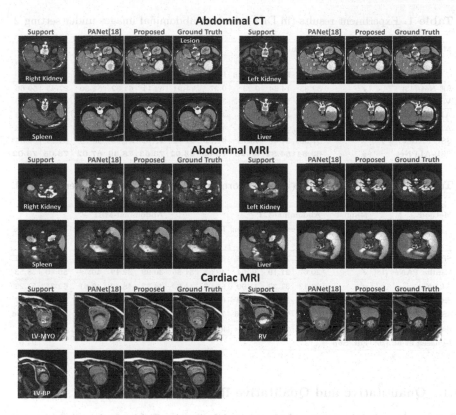

Fig. 4. Qualitative results of our method on all three combinations of imaging modalities and segmentation tasks. The proposed method achieves desirable segmentation results which are close to ground truth. To highlight the strong generalization ability, examples of results from the proposed method on Abd-CT and Abd-MRI are from setting 2, where images containing testing classes are strictly excluded in training set even though they are unlabeled. See supplemental materials for more examples.

Implementation Details: The network is implemented with PyTorch based on official PANet implementation[1] [18]. To obtain high spatial resolutions in feature maps, $f_\theta(\cdot)$ is configured as an off-the-shelf fully-convolutional ResNet101, which is pre-trained on part of MS-COCO for higher segmentation performance [18,74] (same for vanilla PANet in our experiments). It takes a $3 \times 256 \times 256$ image as input and produces a $256 \times 32 \times 32$ feature map. Local pooling window (L_H, L_W) for prototypes is set to 4×4 for training and 2×2 for inference on feature maps. The loss in Eq. 9 is minimized for 100k iterations using stochastic gradient descent with a batch size of 1. The learning rate is 0.001 with a stepping decay rate of 0.98 per 1000 iterations. The self-supervised training takes ~3h on a single Nvidia RTX 2080Ti GPU, consuming 2.8 GBs of memory.

---

[1] https://github.com/kaixin96/PANet.

Table 1. Experiment results (in Dice score) on abdominal images under setting 2.

Method	Manual Anno.?	Abdominal-CT					Abdominal-MRI				
		Lower		Upper		Mean	Lower		Upper		Mean
		LK	RK	Spleen	Liver		LK	RK	Spleen	Liver	
SE-Net [43]	✓	32.83	14.34	0.23	0.27	11.91	62.11	61.32	51.80	27.43	50.66
Vanilla PANet [18]	✓	32.34	17.37	29.59	38.42	29.43	53.45	38.64	50.90	42.26	46.33
ALPNet-init	-	13.90	11.61	16.39	41.71	20.90	19.28	14.93	23.76	37.73	23.93
ALPNet	✓	34.96	30.40	27.73	47.37	35.11	53.21	58.99	52.18	37.32	50.43
SSL-PANet	✗	37.58	34.69	43.73	61.71	44.42	47.71	47.95	58.73	64.99	54.85
SSL-ALPNet	✗	**63.34**	**54.82**	**60.25**	**73.65**	**63.02**	**73.63**	**78.39**	**67.02**	**73.05**	**73.02**

Table 2. Experiment results (in Dice score) on abdominal images under setting 1.

Method	Manual Anno.?	Abdominal-CT					Abdominal-MRI				
		Kidneys		Spleen	Liver	Mean	Kidneys		Spleen	Liver	Mean
		LK	RK				LK	RK			
SE-Net [43]	✓	24.42	12.51	43.66	35.42	29.00	45.78	47.96	47.30	29.02	42.51
Vanilla PANet [18]	✓	20.67	21.19	36.04	49.55	31.86	30.99	32.19	40.58	50.40	38.53
ALPNet	✓	29.12	31.32	41.00	65.07	41.63	44.73	48.42	49.61	62.35	51.28
SSL-PANet	✗	56.52	50.42	55.72	60.86	57.88	58.83	60.81	61.32	71.73	63.17
SSL-ALPNet	✗	**72.36**	**71.81**	**70.96**	**78.29**	**73.35**	**81.92**	**85.18**	**72.18**	**76.10**	**78.84**
Zhou et al. [75]	Ful. Sup	95.3	92.0	96.8	97.4	95.4			-		
Isenseen et al. [76]	Ful. Sup			-					-		94.6

4.1   Quantitative and Qualitative Results

Comparison with State-of-the-Art Methods: Tables 1, 2 and 3 show the comparisons of our method with vanilla PANet, one of state-of-the-art methods on natural images and SE-Net[2] [43], the lastest FSS method for medical images. Without using any manual annotation, our proposed SSL-ALPNet consistently outperforms them by an average Dice score of >25. As shown in Fig. 4, the proposed framework yields satisfying results on organs with various shapes, sizes and intensities. Of note, for all evaluated methods, results on Abd-MRI are in general higher than those on Abd-CT. This not surprising as Abd-MRI is more homogeneous, and most of organs in Abd-MRI have distinct contrast to surrounding tissues, which helps to reduce ambiguity at boundaries.

Importantly, Table 1 demonstrates the strong generalization ability of our method to unseen classes. This implies that the proposed superpixel-based self-supervised learning has successfully trained the network to learn more diverse and generalizable image representations from unlabeled images.

The upperbounds obtained by fully-supervised learning on all labeled images are shown in Table 2 for reference.

Performance Boosts by ALP and SSL: The separate performance gains obtained by introducing adaptive local prototype pooling or self-supervised learning can be observed in rows *ALPNet* and *SSL-PANet* in Tables 1, 2 and

---

[2] https://github.com/abhi4ssj/few-shot-segmentation.

Table 3. Experiment results (in Dice score) on cardiac images under setting 1.

Method	Manual anno.?	LV-BP	LV-MYO	RV	Mean
SE-Net [43]	✓	58.04	25.18	12.86	32.03
Vanilla PANet [18]	✓	53.64	35.72	39.52	42.96
ALPNet	✓	73.08	49.53	58.50	60.34
SSL-PANet	✗	70.43	46.79	69.52	62.25
SSL-ALPNet	✗	**83.99**	**66.74**	**79.96**	**76.90**

Table 4. Ablation study on types transformations.

Int	Geo	LK	RK	Spleen	Liver	Mean
✗	✗	45.49	48.40	53.05	73.60	55.13
✓	✗	55.56	49.12	59.20	73.39	59.31
✗	✓	59.32	51.45	57.74	**78.93**	61.86
✓	✓	**63.34**	**54.82**	**60.25**	73.65	**63.02**

Table 5. Ablation study on minimum pseudolabel sizes.

Min. size (px)	LK	RK	Spleen	Liver	Mean
100	52.92	47.45	53.16	68.40	55.49
400	**63.34**	**54.82**	**60.25**	73.65	**63.02**
1600	51.74	44.83	56.99	**74.73**	57.08
Avg. Size in 2D (px)	798	799	1602	5061	

3. These results suggests both SSL and ALP contribute greatly. The performance gains of SSL highlight the benefit of a well-designed training strategy that encourages learning generalizable features, which is usually overlooked in recent few-shot segmentation methods. More importantly, the synergy between them (*SSL-ALPNet*) leads to significant performance gains by learning richer image representations and by constructing more effective inductive bias. To be assured that MS-COCO initialization alone cannot do FSS, we also include the results when the ALPNet is directly tested after initialization, shown in *ALPNet-init* in Table 1.

Robustness Under Patient Pathology: As shown in Fig. 4, despite the large dark lesion on right kidney in Abd-CT, the proposed method stably produces satisfying results.

4.2 Ablation Studies

Ablation studies are performed on Abd-CT under setting 2. This scenario is challenging but close to clinical scenario in practice.

Importance of Transforms Between the Support and Query: To demonstrate the importance of geometric and intensity transformations in our method, we performed ablation studies as shown in Table 4. Unsurprisingly, the highest and lowest overall results are obtained by applying both or no transforms, proving the effectiveness of introducing random transforms. Interestingly, applying intensity transform even hurts performance on liver. This implies that the configuration of intensity transforms in our experiments may deviate from the actual intensity distribution of livers in the dataset.

Effect of Pseudolabel Sizes: To investigate the effect of pseudolabel sizes on performance, we experimented with pseudolabel sets with different minimum superpixel sizes. Table 5 shows that the granularity of superpixels should be reasonably smaller than sizes of actual semantic labels. This implies that too-coarse or too-fine-grained pseudolabels might divert the granularity of clusters in the learned representation space from that of real semantic classes.

5   Conclusion

In this work, we propose a self-supervised few-shot segmentation framework for medical imaging. The proposed method successfully outperforms state-of-the-art methods without requiring any manual labeling for training. In addition, it demonstrates strong generalization to unseen semantic classes in our experiments. Moreover, the proposed superpixel-based self-supervision technique provides an effective way for image representation learning, opening up new possibilities for future works in semi-supervised and unsupervised image segmentation.

Acknowledgements. This work is supported by the EPSRC Programme Grant EP/P001009/1. This work is also supported by the UK Research and Innovation London Medical Imaging and Artificial Intelligence Centre for Value Based Healthcare. The authors would like to thank Konstantinos Kamnitsas and Zeju Li for insightful comments.

References

1. Pham, D.L., Xu, C., Prince, J.L.: Current methods in medical image segmentation. Annu. Rev. Biomed. Eng. **2**(1), 315–337 (2000)
2. Sharma, N., Aggarwal, L.M.: Automated medical image segmentation techniques. J. Med. Phys. Assoc. Med. Phys. India **35**(1), 3 (2010)
3. Zhang, D., Wang, Y., Zhou, L., Yuan, H., Shen, D., Alzheimer's Disease Neuroimaging, et al.: Multimodal classification of Alzheimer's disease and mild cognitive impairment. Neuroimage **55**(3), 856–867 (2011)
4. Ei Naqa, I., et al.: Concurrent multimodality image segmentation by active contours for radiotherapy treatment planning a. Med. Phys. **34**(12), 4738–4749 (2007)
5. Zaidi, H., El Naqa, I.: Pet-guided delineation of radiation therapy treatment volumes: a survey of image segmentation techniques. Eur. J. Nucl. Med. Mol. Imaging **37**(11), 2165–2187 (2010)
6. De Leeuw, F., et al.: Prevalence of cerebral white matter lesions in elderly people: a population based magnetic resonance imaging study. the rotterdam scan study. J. Neurol. Neurosurg. Psychiatry **70**(1), 9–14 (2001)
7. Petersen, S.E., et al.: Imaging in population science: cardiovascular magnetic resonance in 100,000 participants of UK biobank-rationale, challenges and approaches. J. Cardiovasc. Magn. Reson. **15**(1), 46 (2013)
8. Snell, J., Swersky, K., Zemel, R.: Prototypical networks for few-shot learning. In: Advances in Neural Information Processing Systems, pp. 4077–4087 (2017)
9. Sung, F., Yang, Y., Zhang, L., Xiang, T., Torr, P.H., Hospedales, T.M.: Learning to compare: relation network for few-shot learning. In: Proceedings of the IEEE Conference on Computer Vision and Pattern Recognition, pp. 1199–1208 (2018)

10. Garcia, V., Bruna, J.: Few-shot learning with graph neural networks. arXiv preprint arXiv:1711.04043 (2017)
11. Vinyals, O., Blundell, C., Lillicrap, T., Wierstra, D., et al.: Matching networks for one shot learning. In: Advances in Neural Information Processing Systems, pp. 3630–3638 (2016)
12. Fei-Fei, L., Fergus, R., Perona, P.: One-shot learning of object categories. IEEE Trans. Pattern Anal. Mach. Intell. **28**(4), 594–611 (2006)
13. Lake, B., Salakhutdinov, R., Gross, J., Tenenbaum, J.: One shot learning of simple visual concepts. In: Proceedings of the Annual Meeting of the Cognitive Science Society, vol. 33 (2011)
14. Rakelly, K., Shelhamer, E., Darrell, T., Efros, A., Levine, S.: Conditional networks for few-shot semantic segmentation. In: 6th International Conference on Learning Representations, ICLR 2018, Vancouver, BC, Canada, 30 April–3 May, 2018, Workshop Track Proceedings (2018)
15. Shaban, A., Bansal, S., Liu, Z., Essa, I., Boots, B.: One-shot learning for semantic segmentation. arXiv preprint arXiv:1709.03410 (2017)
16. Dong, N., Xing, E.: Few-shot semantic segmentation with prototype learning. In: British Machine Vision Conference 2018, BMVC 2018, Northumbria University, Newcastle, UK, 3–6 September 2018, vol. 3, p. 79(2018)
17. Siam, M., Oreshkin, B.N., Jagersand, M.: AMP: adaptive masked proxies for few-shot segmentation. In: Proceedings of the IEEE International Conference on Computer Vision, pp. 5249–5258 (2019)
18. Wang, K., Liew, J.H., Zou, Y., Zhou, D., Feng, J.: Panet: few-shot image semantic segmentation with prototype alignment. In: Proceedings of the IEEE International Conference on Computer Vision, pp. 9197–9206 (2019)
19. Zhang, X., Wei, Y., Yang, Y., Huang, T.: Sg-one: Similarity guidance network for one-shot semantic segmentation. arXiv preprint arXiv:1810.09091 (2018)
20. Rakelly, K., Shelhamer, E., Darrell, T., Efros, A.A., Levine, S.: Few-shot segmentation propagation with guided networks. arXiv preprint arXiv:1806.07373 (2018)
21. Zhang, C., Lin, G., Liu, F., Guo, J., Wu, Q., Yao, R.: Pyramid graph networks with connection attentions for region-based one-shot semantic segmentation. In: Proceedings of the IEEE International Conference on Computer Vision, pp. 9587–9595 (2019)
22. Siam, M., Doraiswamy, N., Oreshkin, B.N., Yao, H., Jagersand, M.: Weakly supervised few-shot object segmentation using co-attention with visual and semantic inputs. arXiv preprint arXiv:2001.09540 (2020)
23. Tian, P., Wu, Z., Qi, L., Wang, L., Shi, Y., Gao, Y.: Differentiable meta-learning model for few-shot semantic segmentation. arXiv preprint arXiv:1911.10371 (2019)
24. Hu, T., Mettes, P., Huang, J.H., Snoek, C.G.: Silco: show a few images, localize the common object. In: Proceedings of the IEEE International Conference on Computer Vision, pp. 5067–5076 (2019)
25. Hendryx, S.M., Leach, A.B., Hein, P.D., Morrison, C.T.: Meta-learning initializations for image segmentation. arXiv preprint arXiv:1912.06290 (2019)
26. Lieb, D., Lookingbill, A., Thrun, S.: Adaptive road following using self-supervised learning and reverse optical flow. In: Robotics: Science and Systems, pp. 273–280 (2005)
27. Doersch, C., Gupta, A., Efros, A.A.: Unsupervised visual representation learning by context prediction. In: Proceedings of the IEEE International Conference on Computer Vision, pp. 1422–1430 (2015)

28. Dosovitskiy, A., Springenberg, J.T., Riedmiller, M., Brox, T.: Discriminative unsupervised feature learning with convolutional neural networks. In: Advances in Neural Information Processing Systems, pp. 766–774 (2014)
29. Larsson, G., Maire, M., Shakhnarovich, G.: Learning representations for automatic colorization. In: Leibe, B., Matas, J., Sebe, N., Welling, M. (eds.) ECCV 2016. LNCS, vol. 9908, pp. 577–593. Springer, Cham (2016). https://doi.org/10.1007/978-3-319-46493-0_35
30. Noroozi, M., Favaro, P.: Unsupervised learning of visual representations by solving jigsaw puzzles. In: Leibe, B., Matas, J., Sebe, N., Welling, M. (eds.) ECCV 2016. LNCS, vol. 9910, pp. 69–84. Springer, Cham (2016). https://doi.org/10.1007/978-3-319-46466-4_5
31. Gidaris, S., Bursuc, A., Komodakis, N., Pérez, P., Cord, M.: Boosting few-shot visual learning with self-supervision. In: Proceedings of the IEEE International Conference on Computer Vision, pp. 8059–8068 (2019)
32. Pathak, D., Krahenbuhl, P., Donahue, J., Darrell, T., Efros, A.A.: Context encoders: feature learning by inpainting. In: Proceedings of the IEEE Conference on Computer Vision and Pattern Recognition, pp. 2536–2544 (2016)
33. Zhang, R., Isola, P., Efros, A.A.: Colorful image colorization. In: Leibe, B., Matas, J., Sebe, N., Welling, M. (eds.) ECCV 2016. LNCS, vol. 9907, pp. 649–666. Springer, Cham (2016). https://doi.org/10.1007/978-3-319-46487-9_40
34. Liu, J., Qin, Y.: Prototype refinement network for few-shot segmentation. arXiv preprint arXiv:2002.03579 (2020)
35. Ren, X., Malik, J.: Learning a classification model for segmentation. In: null, p. 10. IEEE (2003)
36. Stutz, D., Hermans, A., Leibe, B.: Superpixels: an evaluation of the state-of-the-art. Comput. Vis. Image Underst. **166**, 1–27 (2018)
37. Achanta, R., Shaji, A., Smith, K., Lucchi, A., Fua, P., Süsstrunk, S.: Slic superpixels. Technical report, EPFL (2010)
38. Zhou, Z., Sodha, V., Rahman Siddiquee, M.M., Feng, R., Tajbakhsh, N., Gotway, M.B., Liang, J.: Models genesis: generic autodidactic models for 3D medical image analysis. In: Shen, D., Liu, T., Peters, T.M., Staib, L.H., Essert, C., Zhou, S., Yap, P.-T., Khan, A. (eds.) MICCAI 2019. LNCS, vol. 11767, pp. 384–393. Springer, Cham (2019). https://doi.org/10.1007/978-3-030-32251-9_42
39. Vaswani, A., et al.: Attention is all you need. In: Advances in Neural Information Processing Systems, pp. 5998–6008 (2017)
40. Schlichtkrull, M., et al.: Modeling relational data with graph convolutional networks. In: Gangemi, A., et al. (eds.) ESWC 2018. LNCS, vol. 10843, pp. 593–607. Springer, Cham (2018). https://doi.org/10.1007/978-3-319-93417-4_38
41. Kipf, T.N., Welling, M.: Semi-supervised classification with graph convolutional networks. arXiv preprint arXiv:1609.02907 (2016)
42. Zhang, C., Lin, G., Liu, F., Yao, R., Shen, C.: Canet: class-agnostic segmentation networks with iterative refinement and attentive few-shot learning. In: Proceedings of the IEEE Conference on Computer Vision and Pattern Recognition, pp. 5217–5226 (2019)
43. Roy, A.G., Siddiqui, S., Pölsterl, S., Navab, N., Wachinger, C.: 'squeeze & excite' guided few-shot segmentation of volumetric images. Med. Image Anal. **59**, 101587 (2020)
44. Hu, T., Yang, P., Zhang, C., Yu, G., Mu, Y., Snoek, C.G.: Attention-based multicontext guiding for few-shot semantic segmentation. In: Proceedings of the AAAI Conference on Artificial Intelligence, vol. 33, pp. 8441–8448 (2019)

45. Zhao, A., Balakrishnan, G., Durand, F., Guttag, J.V., Dalca, A.V.: Data augmentation using learned transforms for one-shot medical image segmentation. In: CVPR (2019)
46. Mondal, A.K., Dolz, J., Desrosiers, C.: Few-shot 3D multi-modal medical image segmentation using generative adversarial learning. arXiv preprint arXiv:1810.12241 (2018)
47. Ouyang, C., Kamnitsas, K., Biffi, C., Duan, J., Rueckert, D.: Data efficient unsupervised domain adaptation for cross-modality image segmentation. In: Shen, D., et al. (eds.) MICCAI 2019. LNCS, vol. 11765, pp. 669–677. Springer, Cham (2019). https://doi.org/10.1007/978-3-030-32245-8_74
48. Yu, H., et al.: Foal: fast online adaptive learning for cardiac motion estimation. In: Proceedings of the IEEE/CVF Conference on Computer Vision and Pattern Recognition, pp. 4313–4323(2020)
49. Chen, C., et al.: Realistic adversarial data augmentation for MR image segmentation. arXiv preprint arXiv:2006.13322 (2020)
50. Hu, J., Shen, L., Sun, G.: Squeeze-and-excitation networks. In: Proceedings of the IEEE Conference on Computer Vision and Pattern Recognition, pp. 7132–7141 (2018)
51. Doersch, C., Zisserman, A.: Multi-task self-supervised visual learning. In: Proceedings of the IEEE International Conference on Computer Vision, pp. 2051–2060 (2017)
52. Shimoda, W., Yanai, K.: Self-supervised difference detection for weakly-supervised semantic segmentation. In: Proceedings of the IEEE International Conference on Computer Vision, pp. 5208–5217 (2019)
53. Zhan, X., Pan, X., Liu, Z., Lin, D., Loy, C.C.: Self-supervised learning via conditional motion propagation. In: Proceedings of the IEEE Conference on Computer Vision and Pattern Recognition, pp. 1881–1889 (2019)
54. Jamaludin, A., Kadir, T., Zisserman, A.: Self-supervised learning for Spinal MRIs. DLMIA/ML-CDS -2017. LNCS, vol. 10553, pp. 294–302. Springer, Cham (2017). https://doi.org/10.1007/978-3-319-67558-9_34
55. Bai, W., et al.: Self-supervised learning for cardiac MR image segmentation by anatomical position prediction. In: Shen, D., et al. (eds.) MICCAI 2019. LNCS, vol. 11765, pp. 541–549. Springer, Cham (2019). https://doi.org/10.1007/978-3-030-32245-8_60
56. Chen, L., et al.: Self-supervised learning for medical image analysis using image context restoration. Med. Image Anal. 58, 101539 (2019)
57. Zamir, A.R., Sax, A., Shen, W., Guibas, L.J., Malik, J., Savarese, S.: Taskonomy: disentangling task transfer learning. In: Proceedings of the IEEE Conference on Computer Vision and Pattern Recognition, pp. 3712–3722 (2018)
58. Dou, Q., de Castro, D.C., Kamnitsas, K., Glocker, B.: Domain generalization via model-agnostic learning of semantic features. In: Advances in Neural Information Processing Systems, pp. 6447–6458 (2019)
59. Wu, Y., Rosca, M., Lillicrap, T.: Deep compressed sensing. arXiv preprint arXiv:1905.06723 (2019)
60. Finn, C., Abbeel, P., Levine, S.: Model-agnostic meta-learning for fast adaptation of deep networks. arXiv preprint arXiv:1703.03400 (2017)
61. Ji, X., Henriques, J.F., Vedaldi, A.: Invariant information clustering for unsupervised image classification and segmentation. In: Proceedings of the IEEE International Conference on Computer Vision, pp. 9865–9874 (2019)

62. Wu, Z., Xiong, Y., Yu, S.X., Lin, D.: Unsupervised feature learning via nonparametric instance discrimination. In: Proceedings of the IEEE Conference on Computer Vision and Pattern Recognition, pp. 3733–3742 (2018)
63. Bachman, P., Hjelm, R.D., Buchwalter, W.: Learning representations by maximizing mutual information across views. In: Advances in Neural Information Processing Systems, pp. 15535–15545 (2019)
64. He, K., Fan, H., Wu, Y., Xie, S., Girshick, R.: Momentum contrast for unsupervised visual representation learning. In: Proceedings of the IEEE/CVF Conference on Computer Vision and Pattern Recognition, pp. 9729–9738 (2020)
65. Chen, X., Fan, H., Girshick, R., He, K.: Improved baselines with momentum contrastive learning. arXiv preprint arXiv:2003.04297 (2020)
66. Mumford, D., Shah, J.: Optimal approximations by piecewise smooth functions and associated variational problems. Commun. Pure Appl. Math. **42**(5), 577–685 (1989)
67. Liu, M.Y., Tuzel, O., Ramalingam, S., Chellappa, R.: Entropy rate superpixel segmentation. In: CVPR 2011, pp. 2097–2104. IEEE (2011)
68. Felzenszwalb, P.F., Huttenlocher, D.P.: Efficient graph-based image segmentation. Int. J. Comput. Vision **59**(2), 167–181 (2004)
69. Oreshkin, B., López, P.R., Lacoste, A.: Tadam: task dependent adaptive metric for improved few-shot learning. In: Advances in Neural Information Processing Systems, pp. 721–731 (2018)
70. Heimann, T., Meinzer, H.P.: Statistical shape models for 3D medical image segmentation: a review. Med. Image Anal. **13**(4), 543–563 (2009)
71. Landman, B., Xu, Z., Igelsias, J., Styner, M., Langerak, T., Klein, A.: MICCAI multi-Atlas labeling beyond the cranial vault-workshop and challenge (2015)
72. Kavur, A.E., et al.: Chaos challenge-combined (CT-MR) healthy abdominal organ segmentation. arXiv preprint arXiv:2001.06535 (2020)
73. Zhuang, X.: Multivariate mixture model for myocardial segmentation combining multi-source images. IEEE Trans. Pattern Anal. Mach. Intell. **41**(12), 2933–2946 (2018)
74. Shin, H.C., et al.: Deep convolutional neural networks for computer-aided detection: CNN architectures, dataset characteristics and transfer learning. IEEE Trans. Med. Imaging **35**(5), 1285–1298 (2016)
75. Zhou, Y., et al.: Prior-aware neural network for partially-supervised multi-organ segmentation. In: Proceedings of the IEEE International Conference on Computer Vision, pp. 10672–10681 (2019)
76. Isensee, F., et al.: nnU-Net: self-adapting framework for U-Net-based medical image segmentation. arXiv preprint arXiv:1809.10486 (2018)

On Diverse Asynchronous Activity Anticipation

He Zhao[✉] and Richard P. Wildes

York University, Toronto, ON, Canada
{zhufl,wildes}@cse.yorku.ca

Abstract. We investigate the joint anticipation of long-term activity labels and their corresponding times with the aim of improving both the naturalness and diversity of predictions. We address these matters using Conditional Adversarial Generative Networks for Discrete Sequences. Central to our approach is a reexamination of the unavoidable sample quality vs. diversity tradeoff of the recently emerged Gumbel-Softmax relaxation based GAN on discrete data. In particular, we ameliorate this trade-off with a simple but effective sample distance regularizer. Moreover, we provide a unified approach to inference of activity labels and their times so that a single integrated optimization succeeds for both. With this novel approach in hand, we demonstrate the effectiveness of the resulting discrete sequential GAN on multimodal activity anticipation. We evaluate the approach on three standard datasets and show that it outperforms previous approaches in terms of both accuracy and diversity, thereby yielding a new state-of-the-art in activity anticipation.

1 Introduction

Activity anticipation has drawn considerable recent attention and has evolved from relatively simple next frame prediction to more challenging asynchronous sequences of activities and times that occur further into the future [34,35]. One particular characteristic of activities, uncertainty, is also gaining consideration [1,37]. For example, given certain starting procedures of preparing a meal in a kitchen, initial actions *take_egg, butter_pan* can be followed by *crack_egg, fry_egg* for a fried egg breakfast, or alternatively *spoon_flour, pour_milk, etc.* for a pancake breakfast, as well as many other possibilities as in Fig. 1. Indeed, not only the activities, but also the durations can vary (*e.g.* cooking time differs between *soft* and *hard* boiled eggs). This phenomenon, referred to as diversity in prediction, is one of the core abilities of the human prediction system, while it is much less well captured in contemporary computer vision.

Recent efforts on increasing diversity have concentrated on generative models. GANs [13] are especially notable for generating near realistic (*i.e.* high quality)

Electronic supplementary material The online version of this chapter (https://doi.org/10.1007/978-3-030-58526-6_46) contains supplementary material, which is available to authorized users.

© Springer Nature Switzerland AG 2020
A. Vedaldi et al. (Eds.): ECCV 2020, LNCS 12374, pp. 781–799, 2020.
https://doi.org/10.1007/978-3-030-58526-6_46

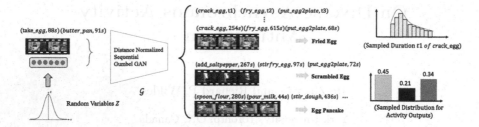

Fig. 1. Proposed activity anticipation process overview. An initial observation of an activity, as a discrete sequence of actions and times (*e.g.* take_egg, 88s, *etc.*) and a random variable, z, input to a sequence generator, G. The generator combines adversarial learning with Gumbel sampling and normalized distance regularization so that sampled outputs are both diverse and realistic.

samples in the continuous domain [3,26,27,57], To date, however, activity anticipation has barely benefited from these developments. The main issue lies in the discrepancy between generating in continuous and discrete spaces, with action labels most naturally being discrete (*e.g.* one-hot representations) while times typically are taken as continuous (*e.g.* real valued numbers). This innate disparity has led to different learning and inference strategies as well as difficulties in encompassing them within adversarial learning.

Moreover, the general task of discrete data generation confronts challenges: Its typical procedure involves data sampling, which is not differentiable and thus standard gradient-based learning techniques cannot be applied directly in these settings [39]. To bypass this limitation, recent efforts have introduced reinforcement learning (RL) [55] to estimate a policy gradient and *Gumbel-Softmax relaxation* [16] that performs reparameterization to allow gradient flow. However, both of these approaches introduce their own limitations: It is unclear how RL based approaches affect diversity; reparameterizing incurs a trade-off between quality and diversity, which prevents generating both realistic and diverse (*i.e.* multimodal) outputs. As noted by others [31,53], insufficient diversity mainly derives from typical mode-collapse issues in GANs [13].

In response to the above challenges, we focus on developing a practical and effective approach that applies a GAN on sequences of action-time pairs, while alleviating mode-collapse in adversarial discrete sequence generation. To achieve this result, we begin by revisiting the representation of time to cast it as a discrete variable and relate this approach to alternatives in the recent literature. Notably, standard activity anticipation datasets (*e.g.* [6,22,47]) provide times in integer units; so, our discrete temporal representation is more consistent with that format than a continuous representation and also allows us to unify inference of activities and their times. Next, we merge the *Normalized Distance Regularizer* [31,53], which serves to combat mode-collapse, with *Gumbel-Softmax* based discrete sequential GANs, which provides realistic instance generation. We then formulate activity anticipation as a conditional adversarial generation problem in the joint action-time discrete domain. Specifically, partial video information in the format of discrete sequences of categorical action labels and their temporal durations are taken as input to generate the remainder of the activity in the same format, *i.e.* sequential action labels and their temporal durations. Empirical evaluation shows that our approach yields new state-of-the-art performance on three popular activity anticipation datasets. Figure 1 provides an overview. Our implementation is available at our project page.

2 Related Research

With the previous success of action recognitions, *e.g.* [4,9,44,52], some research has refocused on two natural extensions, action prediction and activity anticipation. The former refers to recognizing actions as early as possible, while the latter to predicting subsequent future actions. In either case, much work relied on visual feature representation learning from raw videos [11,20,21,43,45,46,51,56]. Alternatively, to better capture future state-space dynamics in complex scenes (*e.g.* pedestrian trajectories or sports player motions), higher-level abstractions (*e.g.* point tracks) have shown to be more practical than raw inputs [29,30,38,48,54].

Indeed, recent approaches that depend on high-level abstraction from videos (*e.g.* semantic labels and times) have also shown promising results for both near and long term activity as well as caption anticipation [2,17,34,35]. Along similar lines, but with a focus on uncertainty modeling, work has been developed that learned a parameterized variational temporal point process to capture the distributions of activity categories and starting times of single immediately following actions [37]. Other work that focused on uncertaintly modeling used beam search to make final selection from a pool of action candidates [42] as well as multi-label learning [10]. Yet other work predicted all subsequent activities and corresponding durations in a stochastic manner [1]. Our research is most similar to the final example in predicting all subsequent actions and durations with a special focus on predictions that are both realistic (high quality) and diverse.

During learning, action labels and times are usually separated [1,2,17,34,35], *e.g.* action labels are taken as classifications and optimized under cross-entropy (CE), while times as real-valued variables and optimized with mean squre error (MSE). CE suffers from overly high resemblance to dominant groundtruth, while suppressing other reasonable possibilities [2,5] and MSE is known for producing blurred outcomes [36]. Recently, superior results have been demonstrated by taking time as a discrete (integer) input to learn exponential-family distributions that are optimized with negative log-likelihood, but still detached from label optimizations [37]. Other work also has shown that framing time as a discrete variable can yield strong results[28,32]. Our approach extends such work by taking input time as a discrete variable, but outputs a softmax discrete random variable and further integrates action and time learning in an adversarial framework to yield naturally unified predictions.

Arguably, the major obstacle to further progress in exploiting categorical models of activity anticipation is the inadequacy of existing models for discrete sequence generation. Still, some progress has been made. Two groups ([16,33])

simultaneously proposed softmax relaxation for discrete random variables; however, they only evaluated on simple datasets under variational inference. Recent efforts in text generation use adversarial generative networks by estimating gradients via a reinforcement policy; however, such approaches are known for training difficulties and high gradient variance [55]. Other work managed to obtain success with the adversarial Gumbel-trick [39], which, however, incurrs a tradeoff between quality and diversity; see Supplemental Material for a detailed example.

Different from previous work, we successfully model activity anticipation in an adversarial generative fashion in the discrete domain. We give a solution for generating outputs having both quality and diversity under Gumbel relaxation via incorporation of a distance regularizer. The upgraded relaxation framework is applied to conditional discrete sequence generation. To the best of our knowledge, ours is the first study that explores a treatment to mode collapse in the discrete sequence generation domain, including an integrated discrete representation of action labels and times and use of conditional GANs for activity anticipation.

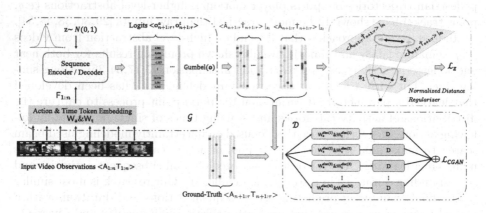

Fig. 2. Depiction of our diversity improved discrete sequential GAN architecture. For both training and inference, input action and time duration token sequences $<A_{1:n}T_{1:n}>$ are embedded via learnable matrices W_a and W_t, (1), to generate a compact feature representation, $F_{1:n}$. $F_{1:n}$ is processed iteratively along with latent noise signals, z, by an encoder (2) and decoder (3) (RNN or LSTM) along with a linear mapping, (4), to produce logits $<o^a_{n+1:\tau}, o^t_{n+1:\tau}>$. At inference time (not shown), the logits are simply sampled with argmax, (5), to produce predicted activity sequences. At training time (shown), Gumbel sampling, (10), is performed to facilitate gradient-based learning. Pairs of samples, $<\hat{A}_{n+1:\tau}\hat{T}_{n+1:\tau}>|_{z_1}$ and $<\hat{A}_{n+1:\tau}\hat{T}_{n+1:\tau}>|_{z_2}$, produced with different latents, z_1 and z_2, are compared under a normalized distance regularizer, (12), to define a loss, $\mathcal{L}_z$, that encourges a generator capable of diverse ouputs. In complement, a sampled output is adversarially compared, (13), to groundtruth, $<A_{n+1:\tau}T_{n+1:\tau}>$, by a discriminator, $\mathcal{D}$, under multiple embeddings, $W_a^{Disc(i)}, W_t^{Disc(i)}$, to define an additional loss, $\mathcal{L}_{CGAN}$, that encourages a generator capable of realistic outputs.

3   Technical Approach

3.1   Overview

We seek to anticipate what will happen next from an observed initial part of a video, *c.f.* [2,17,37]. We are interested in not just anticipating a single action that immediately follows the observed video, but potentially an entire sequence of subsequent actions. Specifically, the input contains a sequence of n observed action tokens $\mathsf{A}_{1:n} = (\mathbf{a}_1, \ldots, \mathbf{a}_n)$, with $\mathbf{a}_i$ the ith token in the sequence. Actions are taken to belong to an action vocabulary of size V, *i.e.* $\mathbf{a}_i \in \mathbb{R}^V$, and each $\mathbf{a}_i$ is a one-hot vector action encoding. Corresponding to each action sequence, $\mathsf{A}_{1:n}$, is a time sequence, $\mathsf{T}_{1:n} = (\mathbf{t}_1, \ldots, \mathbf{t}_n)$ with $\mathbf{t}_i$ the temporal length (*e.g.* in units of seconds) of the ith action belonging to a duration vocabulary of size H, $\mathbf{t}_i \in \mathbb{R}^H$. Each $\mathbf{t}_i$ is a one-hot vector encoding of the duration of action at position i in the sequence, taken from the vocabulary of temporal durations.

Our temporal representation is motivated by a desire to treat encoding, decoding and adversarial learning in a consistent fashion across input, output and groundtruth and will be shown to yield state-of-the-art results. In all three of the considered standard datasets, the time data format (*c.g.* seconds) is given as discrete. Previous work [1,17,37], using MLPs for time decoding of future activities causes difficulty in learning, given discrete groundtruth. Also, recursive predictions that rely on preceding results suffer in such settings, as only discrete values are in the training data. We match our representation to the data by enforcing one-hots for prediction. It has been shown beneficial to treat such formatting as categorical under adversarial learing in allied domains [5,24,39].

We formulate the task as: Given such a partial sequence as an ordered pair, $< \mathsf{A}_{1:n}, \mathsf{T}_{1:n} >$, we produce a distribution over subsequent action-temporal duration pairings, $< \mathsf{A}_{n+1:\tau}, \mathsf{T}_{n+1:\tau} >$, up to and including time τ. Our activity anticipation pipeline has three parts: 1) an input sequential token embedding module, 2) an output sequential generator, $\mathcal{G}$, and 3) a ConvNet based discriminator, $\mathcal{D}$, for adversarial learning. We describe each separately in the following.

3.2   Dual Token Embedding

As widely discussed in the NLP and image/video captioning literatures, the expressive power of raw discrete tokens can be greatly enhanced through dictionary embeddings in higher dimensional continuous feature spaces, *c.f.* [2,17,37]. To this end, each entry of the initial pairing of action and temporal tokens $\mathbf{a}_i$ and $\mathbf{t}_i$ are projected to higher dimension continuous spaces, $\Re^a$ and $\Re^t$, resp., via dictionary mapping matrices W_a and W_t. Here, W_a, of dimension $\Re^a \times V$, is the mapping for actions, while W_t, of dimension $\Re^t \times H$, is the mapping for times. The embedding matrices are constructed such that their j^{th} columns are the mappings for actions and times indicated by the one-hot vector encodings of actions and times. (Note that for a given observation pair, $< \mathbf{a}_i, \mathbf{t}_i >$, j generally is different for the action and time.) Our overall embeddings are then given as

$$\mathbf{f}_i = Cat\left[\mathsf{W}_a\mathbf{a}_i, \mathsf{W}_t\mathbf{t}_i\right] \tag{1}$$

with $\mathbf{f}_i$ associating action and temporal features at the ith index via concatenation of their respective embedding vectors. Analogous to the input sequences, $\mathsf{A}_{1:n}$ and $\mathsf{T}_{1:n}$, we define embedded sequences, $\mathsf{F}_{1:n} = (\mathbf{f}_1, \ldots, \mathbf{f}_n)$.

3.3 Sequence Generation

For computing outputs, $<\mathsf{A}_{n+1:\tau}, \mathsf{T}_{n+1:\tau}>$, given a sequence of input feature embeddings, $\mathsf{F}_{1:n}$, we use the *seq2seq* generator [49]. To lend insight into the generality of our approach, we consider two different backbones, RNN and LSTM, as instantiating mechanisms for the generator and results for both are reported in Sect. 4. While this design choice might be replaced by additional alternatives (*e.g.* a temporal convolutional network (TCN) [7,40] or relational memory [39,41]), RNN and LSTM are general sequence learning structures also adopted by our main comparison approaches [1,2]. In the following, we make use of a general notation such that *Sequence* stands for either RNN or LSTM.

Sequence generation proceeds by sequentially encoding and decoding the embedded features, followed by mapping to logits and sampling. More specifically, the components of the input feature sequence, $(\mathbf{f}_1, \ldots, \mathbf{f}_n)$, are iteratively injected into an encoder, *Sequence^{enc}*, to obtain a fixed dimensional representation $\mathbf{v}$, given by the last hidden state of *Sequence^{enc}*, according to

$$\mathbf{h}_k^{enc} = Sequence^{enc}(\mathbf{f}_k, \mathbf{h}_{k-1}^{enc}), 1 < k < n \quad \text{and} \quad \mathbf{v} = \mathbf{h}_n^{enc}, \qquad (2)$$

$\mathbf{h}_k^{enc}$ the kth hidden encoder state and the initial hidden state, $\mathbf{h}_0$, randomly set.

Next, a decoding sequence model, *Sequence^{dec}*, whose initial hidden state is set to $\mathbf{v}$ and first input variable to the last observation $\mathbf{f}_n$, is used for producing output hidden state sequence $(\mathbf{h}_{n+1}^{dec}, \ldots, \mathbf{h}_\tau^{dec})$ according to

$$\mathbf{h}_{k+1}^{dec} = Sequence^{dec}(\mathbf{f}_k, \mathbf{h}_k^{dec}), n < k < \tau - 1. \qquad (3)$$

Then, a pair of linear transformations defined by matrices W_o^ϕ and vectors $\mathbf{b}_o^\phi$, $\phi \in \{a, t\}$, maps hidden states, $\mathbf{h}_k^{dec}$, to logits according to

$$\left. \begin{array}{l} \mathbf{o}_k^a = \mathsf{W}_o^a \mathbf{h}_k^{dec} + \mathbf{b}_o^a \\ \mathbf{o}_k^t = \mathsf{W}_o^t \mathbf{h}_k^{dec} + \mathbf{b}_o^t \end{array} \right\} \quad n + 1 \leq k \leq \tau, \qquad (4)$$

where $\mathbf{o}_k^a$ and $\mathbf{o}_k^t$ are vectors representing the logits for each entry of action or time vocabulary, *i.e.* they are of dimensions $\Re^a \times V$ and $\Re^t \times H$, resp.

Finally, the logits, (4), are sampled in one of two ways, depending on whether the system is operating in inference or training mode. At inference time, traditional discrete data sampling is used to select the most likely probability index, which is then reformatted as one-hot vectors according to

$$\left. \begin{array}{l} \mathbf{a}_k = one\_hot \left(\underset{i}{\mathrm{argmax}}(\mathbf{o}_k^a[i]) \right) \\ \\ \mathbf{t}_k = one\_hot \left(\underset{i}{\mathrm{argmax}}(\mathbf{o}_k^t[i]) \right) \end{array} \right\} \quad n + 1 \leq k \leq \tau, \qquad (5)$$

with $\mathbf{o}_k^a[i]$ and $\mathbf{o}_k^t[i]$ selecting the ith element of $\mathbf{o}_k^a$ and $\mathbf{o}_k^t$, resp.

Unfortunately, the argmax operator incurs zero derivative with respect to parameters of operations coming before it, *i.e.* in the present context those of the embedding matrices and generator, which interferes with gradient-based training. Therefore, to obtain differentiable one-hot vectors from logits $\mathbf{o}_k$ during training, we adopt the Gumbel-Softmax relaxation technique [16,33] that mimics one-hot vectors from categorical distributions. In particular, we replace (5) with Gumbel sampling, (10), which yields the output discrete tokens according to

$$\left.\begin{array}{l} \mathbf{a}_k[i] = \mathsf{Gumbel}(\mathbf{o}_k^a[i]) \\ \mathbf{t}_k[i] = \mathsf{Gumbel}(\mathbf{o}_k^t[i]) \end{array}\right\} \quad n+1 \leq k \leq \tau. \tag{6}$$

For both sampling techniques (5) and (6), for each iteration on k, we use

$$\mathbf{f}_k = Cat[\mathsf{W_a a}_k, \mathsf{W_t t}_k], n+1 \leq k \leq \tau \tag{7}$$

in the encoding, (2), to reinitialize subsequent decoding (3), logits mapping (4), and sampling (5) or (6), to give the next generation of action/time tokens. Having produced $\mathbf{a}_k$ and $\mathbf{t}_k$ for $n+1 \leq k \leq \tau$ via (6), we define $\mathsf{A}_{n+1:\tau} = (\mathbf{a}_{n+1},\ldots,\mathbf{a}_\tau)$, $\mathsf{T}_{n+1:\tau} = (\mathbf{t}_{n+1},\ldots,\mathbf{t}_\tau)$ and ultimately $<\mathsf{A}_{n+1:\tau},\mathsf{T}_{n+1:\tau}>$ as output.

Overall, the generating process, $\mathcal{G}$, of our approach entails sequential application of the embedding (1), encoding (2), decoding (3), logits mapping (4) and sampling, (5) or (6), to each element of the input, $<\mathsf{A}_{1:n},\mathsf{T}_{1:n}>$, formalized as

$$<\mathsf{A}_{n+1:\tau},\mathsf{T}_{n+1:\tau}> = \mathcal{G}(<\mathsf{A}_{1:n},\mathsf{T}_{1:n}>, z; \theta_g) \tag{8}$$

where θ_g includes learnable parameters of $Sequence^{enc}$, $Sequence^{dec}$, $\mathsf{W}_\phi, \mathsf{W}_o^\phi, \mathbf{b}_o^\phi$, $\phi \in \{a,t\}$, and z is a randomized input that encourages diversity during sampling; see Sect. 3.4. From a probabilistic view, the overall sequence generation procedure can be seen to operate as a Markov process,

$$p(\mathsf{A}_{n+1:\tau},\mathsf{T}_{n+1:\tau}|\mathsf{A}_{1:n},\mathsf{T}_{1:n}) = \prod_{k=n+1}^{\tau} p(\mathbf{a}_k,\mathbf{t}_k|\mathbf{v},\mathbf{f}_n,h_{n+1},...,h_{k-1}). \tag{9}$$

3.4   Adversarial Learning and the Discriminator

Gumbel Sampling. During training, sampling of the logits is formulated as Gumbel-noise reparameterization, referred to as the *Gumbel-Max trick* [14,33], to allow gradients to flow for end-to-end optimization according to

$$\mathsf{Gumbel}(\mathbf{o}_k[i]) \equiv \frac{\exp((\log(\mathbf{o}_k[i] + \mathbf{g}_k[i])\alpha)}{\sum_{j=1}^m \exp((\log(\mathbf{o}_k[j]) + \mathbf{g}_k[j])\alpha)}, \tag{10}$$

with m the dimension of $\mathbf{o}_k$ and each element of $\mathbf{g}_k$ drawn from the *i.i.d.* standard Gumbel distribution [14], $-\log(-\log U_k)$, with U_k drawn from $unif(0,1)$.

In the definition of Gumbel, (10), α is a tunable parameter, called the *temperature*, that controls the similarity of the approximated one-hot vector, (6) to the actual one-hot vector, (5). Intuitively, smaller α provides more accurate outputs for the discriminator, thereby leading to better sample quality. However, the associated drawback is high gradient variance causing mode-collapse [39]. To combat this situation, one can adopt a temperature annealing strategy to encourage diversity [16,33], which is heuristic in terms of how much and when the temperature should be annealed. Instead, we make use of the recently proposed *Normalized Distance Regularizer* [31,53] to augment (10), as detailed next.

Normalized Distance Regularizer. To get diversity in generated sequences, we use a *Normalized Distance Regularizer* [31,53]. This regularizer operates via pairwise distances between various outputs from the same input according to

$$\max_{\mathcal{G}} \mathcal{L}_z(\mathcal{G}) = \mathbb{E}_{z_1, z_2} \left[\min \left(\frac{\|\mathcal{G}(Z, z_1) - \mathcal{G}(Z, z_2)\|}{\|z_1 - z_2\|} \right) \right], \quad (11)$$

where Z is input (*e.g.* in our case, an observed action-time sequence pair, $<A_{1:n}, T_{1:n}>$), $\mathbb{E}[\cdot]$ is the expectation operator, $\|\cdot\|$ is a norm and $z_i \sim N(0, 1)$ is a random latent variable that converts the deterministic model into a stochastic one. The normalized distance regularizer, (11), operates such that the denominator, $\|z_1 - z_2\|$, encourages the two random codes to be close in latent space, while the numerator, $\|\mathcal{G}(Z, z_1) - \mathcal{G}(Z, z_2)\|$, encourages the outputs to be distant in the output space. The intuition is to encourage $\mathcal{G}$ to visit more important modes given an input variable, but with minimum traveling effort in the sampling space. In this way, distinctive outcomes are circumscribed within a tighter latent space, so that the model is less likely to ignore some modes [31,53].

In application to our case, the overall generated sequence $<A_{n+1:\tau}, T_{n+1:\tau}>$ $= \mathcal{G}(<A_{1:n}, T_{1:n}>, z)$, is used to calculate the l_1 distance between samples,

$$\mathcal{L}_z(\mathcal{G}) = \quad \mathbb{E}_{z_1, z_2} \left[\frac{\|\mathcal{G}(<A_{1:n}, T_{1:n}>, z_1) - \mathcal{G}(<A_{1:n}, T_{1:n}>, z_2)\|_{l_1}}{\|z_1 - z_2\|} \right], \quad (12)$$

where $\|\cdot\|_{l_1}$ denotes use of the l_1 norm; explicit dependence of $\mathcal{G}$ on θ is suppressed for conciseness. Details of how (12) is realized are in the Supplement.

It appears that our solution to preserving both quality and diversity in discrete sequence generation is novel. Beyond evidence for the effectiveness of the approach from Sect. 4 experiments, the Supplement has a detailed toy example.

Multi-embedding Discriminator. To train our generator, $\mathcal{G}$, adversarially, we refer its output to a discriminator, $\mathcal{D}$. Successful examples of deep discriminators for sequence generation include DNNs, ConvNets and RNNs [18,24,50], amongst which incorporation of a ConvNet classifier with an ensemble of multiple embeddings has achieved top performance [39]. Notably that approach has shown an ability to dispose of discriminator pretraining. We adopt that approach.

Given a generated output sequence, $<\hat{A}_{n+1:\tau}, \hat{T}_{n+1:\tau}>$, and real (*i.e.* ground truth) sequence, $<A_{n+1:\tau}, T_{n+1:\tau}>$, we embed the elements of the sequences analogous to the embeddings of Sect. 3.2, and execute our discriminator, $\mathcal{D}$.

Let W_a^{Disc} and W_t^{Disc}, be discriminator embedding matrices, distinct from, but analogous to the embedding matrices used earlier, (1). Then, the process unfolds in three steps. First, for each pair $<a_i, t_i>, i \in \{n+1, \tau\}$, and $<\hat{a}_i, \hat{t}_i>, i \in \{n+1, \tau\}$ we produce embedded vectors f_i^{Disc} and $\hat{f}_i^{\text{Disc}}$ by replacing W_a and W_t with W_a^{Disc} and W_t^{Disc}, resp. in (1). Second, the complete sets of produced embeddings f_i^{Disc} and $\hat{f}_i^{\text{Disc}}$ are converted into matrices, $F_{n+1:\tau}^{\text{Disc}}$ and $\hat{F}_{n+1:\tau}^{\text{Disc}}$, whose i^{th} columns are f_i^{Disc} and $\hat{f}_i^{\text{Disc}}$, resp. Third, the resulting $F_{n+1,\tau}$ and $\hat{F}_{n+1:\tau}$ are processed akin to 2D images with a discriminator ConvNet, $\mathcal{D}$, consisting of multiple convolutional, nonlinear activation and max-pooling layers with a fully connected layer at the top to produce a label, $i.e.$ 1 $vs.$ 0 for real $vs.$ generated.

Finally, to account for the desired multiple embeddings, the entire three step process is repeated M times with distinct embeddings, $W_a^{\text{Disc}(i)}$ and $W_t^{\text{Disc}(i)}$, $i \in \{1, \ldots, M\}$, and the final adversarial loss from $\mathcal{D}$ comes by averaging across the results of those embeddings according to

$$l_{\mathcal{D}} = \frac{1}{M} \sum_{i=1}^{M} \log(1 - \mathcal{D}(\hat{F}_{n+1,\tau}^{\text{Disc}(i)}, 0; \theta_d)) + \log(\mathcal{D}(F_{n+1,\tau}^{\text{Disc}(i)}, 1; \theta_d)), \quad (13)$$

with θ_d the learnable parameters of the M matrix embeddings and $\mathcal{D}$.

Combined Objective. Our learning goal is to generate samples that are both realistic (high quality) and diverse. To this end, we merge the standard adversarial learning formula (for quality)

$$\mathcal{L}_{CGAN}(\mathcal{G}, \mathcal{D}; \theta_g, \theta_d) = \mathbb{E}_y[\log \mathcal{D}(y, 1; \theta_d)] + \mathbb{E}_{x,z}[\log(1 - \mathcal{D}(\mathcal{G}(x, z; \theta_g), 0; \theta_d))], \quad (14)$$

which in our case specializes to $l_{\mathcal{D}}$, (13), with the normalized distance regularizer, (12), for diversity, to consider

$$\min_{\mathcal{G}} \max_{\mathcal{D}} \lambda_1 \mathcal{L}_{CGAN}(\mathcal{G}, \mathcal{D}; \theta_g, \theta_d) - \lambda_2 \mathcal{L}_z(\mathcal{G}, \theta_g), \quad (15)$$

which we optimize over θ_g and θ_d according to methods documented under implementation details. The overall learning process is summarized in Fig. 2.

4   Empirical Evaluation

4.1   Datasets

We evaluate on three standard datasets.

Breakfast Dataset contains 1,712 videos of 52 actors making breakfast using 48 fine-grained actions. We follow the data processing strategy used elsewhere for ConvNet modeling [2], which generates 4 sets of training examples by using the first 10%, 20%, 30% and 50% of the video, resp., as observation and the rest of the video as ground-truth for prediction evaluation. This procedure yields 7305 training samples and 253 testing samples. We report average accuracy across four splits, as elsewhere [23].

50Salads Dataset has 50 videos of 25 subjects, each making 2 salads using 17 fine grained actions. The same processing used for the Breakfast dataset is adopted here, which leads to 160 training samples and 40 testing samples. We perform 5-fold cross-validation for evaluation, using previous splits [25].

Epic-Kitchens Dataset contains 1st-person videos of 32 actors in 32 kitchens performing unscripted daily activities. There are 272 training videos recorded from 28 subjects, with the rest used for testing. In total there are 128 actions. We follow the preprocessing strategy used previously that randomly split the training video into 7 subsets, each containing 4 subjects [17]. The final result is averaged across all subsets. For each train and test input, an arbitrary clip is sampled by applying temporal windows of 30 s and the following 60 s is taken as groundtruth prediction.

4.2   Implementation Details.

Groundtruth Time Format. We conform to previous temporal formats [2,37]. For Breakfast and 50Salads, we use temporal intervals in units of seconds as observed in the original videos. For example, in the minutes long *egg_cake* video from Breakfast, the 3rd action in the vocabulary is *cracking_egg* and occupies 15 s, which we correspondingly represent as (3, 15). Accordingly, the column entry size of the time embedding matrix, W_t, is set to the maximum action length across the entire dataset, which is 5791 s for Breakfast and 4149 s for 50Salads, by our counting. For Epic-Kitchens, time units are also in seconds. Random temporal crops of 90 s are extracted from the videos, with the first 30 s taken as observation and the remaining 60 for prediction. So, the column size of W_t is 90, *c.f.* [17]. We set the output length to 25 and 20 for Breakfast and 50Salads, resp., based on the maximum sample length in each dataset and pad zeros for shorter samples. The generated length is 60 for Epic-Kitchen by default, *c.f.* [17].

Network Configurations. We set the embedding dimension for input actions and times to be 32 while the hidden states of $Sequence^{enc}$ and $Sequence^{dec}$ to have dimension 128. The number of embeddings for the discriminator is $M = 64$, each of dimension 32. For noise, z, we sample a 32-dimensional random latent variable from the $N(0, 1)$ distribution, *c.f.* [16] and concatenate it to the input at each time step of the generator, $\mathcal{G}$, *c.f.* [53]. For the ConvNet based discriminator, $\mathcal{D}$, four 2D convolutional layers with 2×2 kernels are followed by ReLU [12] and max-pooling with factor of 2 subsampling. A fully connected layer is appended on top for 1 *vs.* 0 classification for groundtruth/real *vs.* generated, resp. These configurations are applied identically for both RNN and LSTM backbones.

Training. Sequential GANs suffer from training difficulties that usually necessitates pretraining with maximum likelihood estimation (MLE) for both generator and discriminator [8,15,55]. Recently, RelGAN [39] successfully avoided discriminator pretraining, while still relying on it for the generator under the Gumbel GAN structure. We also found pretraining unnecessary for our discriminator, but

Table 1. Breakfast dataset protocol 1 results with dense anticipation mean over classes (MoC) accuracy. Avg stands for averaged results across 16 samplings, while Max stands for taking the best result among 16 samples. Best Avg is highlighted.

Observation	20%				30%			
Prediction	10%	20%	30%	50%	10%	20%	30%	50%
RNN [2]	0.6035	0.5044	0.4528	0.4042	0.6145	0.5025	0.4490	0.4175
CNN [2]	0.5797	0.4912	0.4403	0.3926	0.6032	0.5014	0.4518	0.4051
TOS-Dense [17]	0.6446	0.5627	0.5015	0.4399	0.6595	0.5594	0.4914	0.4423
RNN-HMM (**Avg**) [1]	0.5039	0.4171	0.3779	0.3278	0.5125	0.4294	0.3833	0.3307
RNN-HMM (**Max**) [1]	0.7884	0.7284	0.6629	0.6345	0.8200	0.7283	0.6913	0.6239
Ours-RNN (**Avg**)	0.7101	0.6200	0.5421	0.4383	0.7304	0.7053	0.6257	0.5119
Ours-RNN (**Max**)	0.8076	0.7010	0.6649	0.6283	0.8210	0.7539	0.7134	0.6251
Ours-LSTM (**Avg**)	**0.7222**	**0.6240**	**0.5622**	**0.4595**	**0.7414**	**0.7132**	**0.6530**	**0.5238**
Ours-LSTM (**Max**)	0.8208	0.7059	0.6851	0.6406	0.8336	0.7685	0.7213	0.6406

still begin with generator pretraining using MLE. Both pretraining and adversarial training employ the Adam optimizer [19]. Mini-batch-Stochastic Gradient Descent is used with a learning rate of $1e^{-2}$ for pre-training and $1e^{-4}$ for adversarial training, both with exponential decay of $1e^{-5}$. Empirically, we set λ_1, λ_2 to a ratio of 1:1. For both training and testing we input groundtruth tokens (action and time labels), with subsequent groundtruth beyond the input (*i.e.* prediction) used for training only, as with the compared results [2,17,37].

4.3 Anticipation Results Across Datasets

Breakfast Dataset Protocol 1. When evaluating on this dataset, we first conduct experiments under the setting where partial observation video is obtained by arbitrarily cutting from the entire sequence [1,2,17]. Results are reported for combinations of observation (20% and 30%) *vs.* prediction (10%, 20%, 30% and 50%). Table 1 shows comparisons. As our model captures uncertainty via use of randomized latents, we report two evaluation metrics, average and maximum (*i.e.* mean or max accuracy across multiple samples, 16 for results shown).

It is seen that our approach using RNN or LSTM outperforms the alternatives in almost all settings by an average of 5−8% accuracy under the average metric, with the exception being the most extreme case, *i.e.* 50% prediction from 20% observation, where TOS-Dense [17] performs slightly better (by merely 0.16%) than our RNN backbone; however, our LSTM backbone still wins by 0.96%. Given more information to start with, as in 30% observation, both of our models uniformly surpass the others. The biggest difference between our instantiations reside in remote predictions, *i.e.* 30% and 50%, affirming that LSTM generally is better than RNN in long-term performance [17]

Compared with another stochastic approach, RNN-HMM [1], our RNN model achieves similar performance on the **Max** metric while notably excelling on the **Avg** metric. Under both metrics our LSTM model is the top performer.

Table 2. Breakfast dataset protocol 2 results. Note that larger accuracy is better, while smaller MAE is better. For LL, smaller absolute value is better. APP-VAE+ and APP-VAE− refer to APP-VAE with and without a learned prior, resp.

Model	Loss	Stoch. var.	Accuracy	MAE	LL
TD-LSTM [37]	CE + MSE	−	53.64	173.76	−
APP-LSTM [37]	CE + NLL	−	61.39	152.17	−6.668
RNN-HMM [1]	CE + MSE	✓	57.80	−	−
APP-VAE− [37]	NLL+KL	✓	27.09	270.75	≥−9.427
APP-VAE+ [37]	NLL+KL	✓	62.20	142.65	≥−5.944
Ours-RNN	Adv+NormDist	✓	**68.46**	**87.58**	**≥−4.851**
Ours-LSTM	Adv+NormDist	✓	**68.96**	**87.07**	**≥−4.836**

Arguably, our model owns a similar base architecture as [2] and is less complicated than TOS-Dense [17], which introduced skip-connection and attention, yet achieves better results. We attribute this pattern to two reasons: 1) Our approach avoids error propagation coming with iterative long-term generation [1,2], where anticipation errors in early stages are propagated to the end. 2) Our adversarial learning focuses on the quality of overall generated sequences, rather than any single prediction. The discriminator pushes the generator to produce outputs that are more realistic as a whole.

Breakfast Dataset Protocol 2. To further demonstrate our strength in capturing uncertainty, we present an experiment using the protocol from Action Point Process (APP-VAE) [37], which aims to capture stochastics of single next action anticipation, but with a variational inference structure. In this scenario, the input sequence is obtained by clipping the whole video at the exact end point of random actions. Log-Likelihood (LL) is used as a metric for measuring the approximated posterior via standard importance sampling $c.f.$ [37]. Our method is accommodated to this evaluation by selecting the immediate next action from the generated sequence. To report a lower bound LL, we chose the worst score (largest absolute value) from multiple runs. Details of the LL evaluation are presented in the Supplemental Material. Results are shown in Table 2.

Both our models achieves better LL compared to the previous best of −5.944, with our LSTM at −4.836 slightly outperforming our RNN, after averaging inference results from test sets. Since this protocol focuses on a single next action, results from our two backbones are close. We also calculate the Mean Classification Accuracy over the entire anticipated sequence for action category performance and find our results outperform APP-VAE by ≈7%. For temporal duration estimation, we use Mean Absolute Error (MAE) as the evaluation metric and find that our approach reduces the time length estimation error by ≈40%.

50 Salads Dataset. We adopt the same experimental design as for Breakfast Protocol 1; results are in Table 3. As in the previous experiment, both our models achieve better average performance than the alternatives for most cases,

Table 3. Results for the 50Salads dataset with dense anticipation mean over classes (MoC) accuracy.

Observation	20%				30%			
Prediction	10%	20%	30%	50%	10%	20%	30%	50%
RNN [2]	0.4230	0.3119	0.2522	0.1682	0.4419	0.2951	0.1996	0.1038
CNN [2]	0.3608	0.2762	0.2143	0.1548	0.3736	0.2478	0.2078	0.1405
TOS-Dense [17]	0.4512	0.3323	0.2759	0.1727	**0.4640**	0.3480	0.2524	0.1384
RNN-HMM (**Avg**) [1]	0.3495	0.2805	0.2408	0.1541	0.3315	0.2465	0.1884	0.1434
RNN-HMM (**Max**) [1]	0.7489	0.5875	0.4607	0.3571	0.6739	0.5237	0.4673	0.3664
Ours-RNN (**Avg**)	0.4571	0.3517	0.3182	0.2095	0.4597	0.3591	0.3073	0.1746
Ours-RNN (**Max**)	0.4903	0.3998	0.3596	0.2537	0.4926	0.4582	0.3572	0.2646
Ours-LSTM (**Avg**)	**0.4663**	**0.3562**	**0.3191**	**0.2137**	0.4613	**0.3637**	**0.3310**	**0.1945**
Ours-LSTM (**Max**)	0.5150	0.3845	0.3606	0.2762	0.5079	0.4754	0.3783	0.2908

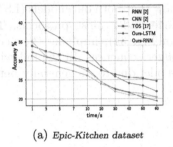

(a) *Epic-Kitchen dataset*

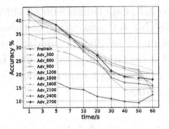

(b) *Ablation on Epic-Kitchen dataset*

Fig. 3. Comparison table for the Epic-Kitchen dataset (a). Ablation study on Ours-LSTM for the Epic-Kitchen dataset. Adv_* indicates * adversarial training epochs (b).

but slightly worse at the 30% observation *vs.* 10% prediction ratio (LSTM worse by merely 0.27%). When it comes to 30% and 50% prediction scales, our models hold a larger lead of >5%. Comparably, the CNN approach [2] also anticipates the whole sequence in one-shot but does so less effectively. A critical difference from ours is that they optimize category and temporal length with the MSE loss, which is known to produce blurred outcomes [13]. Instead, we directly work in the discrete domain with adversarial learning for sharper outcomes, which provides further support for our discrete representation of time.

If we choose the best results across multiple samples, our accuracy is further boosted by ≈6.5%, which while an improvement over mean accuracy, is not as large as for Breakfast and not as good as the best max results [1]. The difference likely owes to 50Salads being much less multimodal compared to Breakfast and the 50Salads dataset being too small (merely 160 training samples) for learning GANs well. Although our RNN model has similar structure to [1], the discriminator incurs extra parameters and therefore requires more training data.

Epic-Kitchens Dataset. Figure 3a shows results for Epic-Kitchens as prediction accuracy at specific times in seconds, *i.e.* (1, 3, 5, 7, 10, 20, 30, 40, 50, 60) with both backbones. Our LSTM results are the top performer when prediction time is up to 20 s. Afterward, we are on par with TOS [17] at 30s, but then trail by an average of ≈1.7% out to 60s. The largest difference happens at 60s, where the gap is 3%. Resembling observations by comparable work in related areas [8,15,39,55], we find the quality of generated samples decreases as the required prediction length increases. Still, our approach outperforms classic RNN and CNN approaches at all prediction times. It is notable that TOS uses a joint learning of long-term prediction with recognition of current observations [17], whereas RNN/CNN [2] and ours solely focus on sequential modelling. These results might indicate that the auxiliary term helps stablize long-term performance. Our RNN results outperform alternatives (except our LSTM) only at one temporal position, *i.e.* 1s, and otherwise are analogous to the alternatives [2].

Overall, experiments on all three datasets show that both our models (LSTM and RNN) set new state-of-the-art performance under most protocols. The exception is at distant predictions (*e.g.* 60 s in Fig. 3a), implying the necessity of a better temporal model (LSTM) for longer sequences. In the following, we restrict further experiments to the LSTM model due to its generally better performance.

4.4   Analysis of Adversarial Learning

We now show the benefits of adversarial learning by examining the accuracy curve at various training stages using Epic-Kitchens, reporting averages across 16 samples. We use Epic-Kitchens as our above experiment with this dataset showed the largest performance change with prediction time; Fig. 3b has results. Just after MLE pretraining, accuracy is lower than our comparison approaches, *c.f.* Fig. 3a. As we start adversarial learning, accuracy immediately rises and does so essentially monotonically with increased training epochs for lower prediction times, and begins to converge at 1200 epochs. For epochs beyond 1500, there is evidence of overfitting at longer prediction times, as performance decreases.

4.5   Analysis of Number of Samples

We now examine the influence of the number of samples on accuracy using the Breakfast dataset as it is larger than the others; see Table 4. We see the largest accuracy increases occur as samples go 2 to 8 and (less so) 8 to 16. Afterwards, improvements grow much slower and performance is stable. In accord with Sect. 4.4 results, adjacent predictions (*e.g.* 10%) benefit more than remote ones (*e.g.* 20–50%). Empirically, we find 16 samples work for both accuracy and efficiency.

Table 4. Influence of various number of samples for Breakfast dataset, Protocol 1, with dense anticipation mean over classes (MoC) accuracy under Avg metric and LSTM backbone.

Observation	20%				30%			
Prediction	10%	20%	30%	50%	10%	20%	30%	50%
2 Samples	0.6157	0.4525	0.3541	0.2382	0.6408	0.4839	0.3775	0.2570
8 Samples	0.7060	0.5950	0.4395	0.3927	0.7032	0.6564	0.5716	0.4320
16 Samples	0.7222	0.6240	0.5622	0.4595	0.7414	0.7132	0.6530	0.5238
32 Samples	0.7464	0.6319	0.5475	0.4643	0.7611	0.7155	0.6566	0.5219
64 Samples	0.7572	0.6325	0.5547	0.4674	0.7758	0.7150	0.6650	0.5291

4.6 Analysis of Normalized Distance Regularization and Diversity

We now provide experimental analysis of the influence our model has on the quality-diversity trade-off. Here, we use the Breakfast dataset because of its large number of videos and relatively high diversity. To evaluate quality, we adopt the accuracy metric, as in Table 2. To evaluate diversity, we calculate the averaged pairwise cosine distance amongst 10 samples, $c.f.$ [53] as well as LL, as in Table 2. Both metrics are presented for pure adversarial learning (Adv) and with the aide of normalized distance based adversarial learning (Adv + NormDist); Table 5 has results. With pure adversarial learning (*i.e.* no normalized distance regularizer), too small or too large of a temperature, *e.g.* $\alpha \in \{0.1, 10, 100, 1000\}$ leads to unsatisfactory results. Best results are obtained when $\alpha=1$; however, the diversity score is not as good as with higher α and the quality score merely equals that of deterministic APP-LSTM in Table 2, row 2.

Table 5. Comparison across various temperatures as well as impact of the normalized distance regularizer. $+$ indicates improvement of adding the normalized distance regularizer (NormDist) to pure adversarial learning (Adv). LSTM backbone is adopted.

Temperature	Loss	LL	Diversity	Accuracy
$\alpha = 0.1$	Adv	-10.83	0.033	23.01
$\alpha = 1$	Adv	$\mathbf{-6.77}$	0.124	**60.98**
$\alpha = 10$	Adv	-8.42	0.345	35.19
$\alpha = 100$	Adv	-8.52	0.423	33.78
$\alpha = 1000$	Adv	-11.46	0.582	19.97
$\alpha = 0.1$	Adv + NormDist	-9.04	0.062 ($+$0.029)	27.61 ($+$4.60)
$\alpha = 1$	Adv + NormDist	$\mathbf{-4.83}$	0.262 ($+$0.138)	**68.96** ($+$7.89)
$\alpha = 10$	Adv + NormDist	-8.19	0.358 ($+$0.013)	36.64 ($+$1.45)

After joint training with the distance regularizer, quality and diversity scores grow together. Especially, when $\alpha = 1$ the model receives the most benefit in terms of quality and diversity (+2% for LL, +0.138 for cosine distance and +7.89% for accuracy), while with $\alpha = 10$ the model yields smaller improvements. The reason for this pattern is that large α values already yield adequate diversity; so, there is little for the distance regularizer to offer. In the case of $\alpha \geq 10$, however, LL suffers noticably. Theoretically, small temperature yields accurate one-hot approximation; however, in our experiments $\alpha = 0.1$ does not perform better, possibly because even with our regularizer, gradient variance is too high.

4.7   Visualization of Diversity and Quality

Figure 4 shows sampled outcomes for our full approach *vs.* our approach lacking normalized distance regularization. Without regularization the small temperature ($\alpha = 1$) samples lack diversity (*i.e. Take_butter* is always generated after *Cut_bun*), whereas large temperature ($\alpha = 100$) samples are diverse to the point of being unrealistic (*e.g. Add_teabag* generated after *Cut_bun*), *i.e.* they lack quality. In contrast, our full approach gives both plausible and diverse outputs.

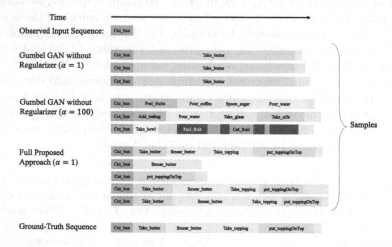

Fig. 4. Visualization of multimodal samples from our full model *vs.* our model lacking normalized distance regularization. See Supplemental Material for additional examples.

5   Summary

In activity anticipation, it is desirable to produce results that realistically capture the potentially multimodal distribution of future actions given initial observations. We have responded with a novel model combining a sequential GAN (for realistic predictions) with a sample distance regularizer (for diverse predictions). By unifying action and time representations, the proposed model can produce outputs that are both realistic and diverse, thereby yielding new state-of-the-art performance on standard activity anticipation datasets. We demonstrated our approach with two different sequence generation backbones (RNN and LSTM).

References

1. Abu Farha, Y., Gall, J.: Uncertainty-aware anticipation of activities. In: ICCVW (2019)
2. Abu Farha, Y., Richard, A., Gall, J.: When will you do what?-Anticipating temporal occurrences of activities. In: CVPR (2018)
3. Babaeizadeh, M., Finn, C., Erhan, D., Campbell, R.H., Levine, S.: Stochastic variational video prediction. arXiv preprint arXiv:1710.11252 (2017)
4. Carreira, J., Zisserman, A.: Quo vadis, action recognition? A new model and the kinetics dataset. In: CVPR (2017)
5. Dai, B., Fidler, S., Urtasun, R., Lin, D.: Towards diverse and natural image descriptions via a conditional GAN. In: ICCV (2017)
6. Damen, D., et al.: Scaling egocentric vision: the EPIC-KITCHENS dataset. In: ECCV (2018)
7. Dauphin, Y.N., Fan, A., Auli, M., Grangier, D.: Language modeling with gated convolutional networks. In: ICML (2017)
8. Fedus, W., Goodfellow, I., Dai, A.M.: Maskgan: better text generation via filling in the_. arXiv preprint arXiv:1801.07736 (2018)
9. Feichtenhofer, C., Pinz, A., Wildes, R.: Spatiotemporal residual networks for video action recognition. In: NIPS (2016)
10. Furnari, A., Battiato, S., Maria Farinella, G.: Leveraging uncertainty to rethink loss functions and evaluation measures for egocentric action anticipation. In: ECCVW (2018)
11. Gao, J., Yang, Z., Nevatia, R.: Red: reinforced encoder-decoder networks for action anticipation. In: BMVC (2017)
12. Glorot, X., Bordes, A., Bengio, Y.: Deep sparse rectifier neural networks. In: AISTAT (2011)
13. Goodfellow, I., et al.: Generative adversarial nets. In: NIPS (2014)
14. Gumbel, E.J.: Statistical Theory of Extreme Values and Some Practical Applications: A Series of Lectures, vol. 33. US Government Printing Office, Washington (1948)
15. Guo, J., Lu, S., Cai, H., Zhang, W., Yu, Y., Wang, J.: Long text generation via adversarial training with leaked information. In: AAAI (2018)
16. Jang, E., Gu, S., Poole, B.: Categorical reparameterization with gumbel-softmax. In: ICLR (2017)
17. Ke, Q., Fritz, M., Schiele, B.: Time-conditioned action anticipation in one shot. In: CVPR (2019)
18. Kim, Y.: Convolutional neural networks for sentence classification. arXiv preprint arXiv:1408.5882 (2014)
19. Kingma, D.P., Ba, J.: Adam: A method for stochastic optimization. arXiv preprint arXiv:1412.6980 (2014)
20. Kong, Y., Tao, Z., Fu, Y.: Deep sequential context networks for action prediction. In: CVPR (2017)
21. Kong, Y., Tao, Z., Fu, Y.: Adversarial action prediction networks. IEEE TPAMI (2019)
22. Koppula, H.S., Saxena, A.: Anticipating human activities using object affordances for reactive robotic response. IEEE TPAMI **38**(1), 14–29 (2015)
23. Kuehne, H., Arslan, A., Serre, T.: The language of actions: recovering the syntax and semantics of goal-directed human activities. In: CVPR (2014)

24. Lai, S., Xu, L., Liu, K., Zhao, J.: Recurrent convolutional neural networks for text classification. In: AAAI (2015)
25. Lea, C., Flynn, M.D., Vidal, R., Reiter, A., Hager, G.D.: Temporal convolutional networks for action segmentation and detection. In: CVPR (2017)
26. Ledig, C., et al.: Photo-realistic single image super-resolution using a generative adversarial network. In: CVPR (2017)
27. Lee, A.X., Zhang, R., Ebert, F., Abbeel, P., Finn, C., Levine, S.: Stochastic adversarial video prediction. arXiv preprint arXiv:1804.01523 (2018)
28. Li, Y., Du, N., Bengio, S.: Time-dependent representation for neural event sequence prediction. In: ICLRW (2018)
29. Liang, J., Jiang, L., Murphy, K., Yu, T., Hauptmann, A.: The garden of forking paths: towards multi-future trajectory prediction. In: CVPR, pp. 10508–10518 (2020)
30. Liang, J., Jiang, L., Niebles, J.C., Hauptmann, A.G., Fei-Fei, L.: Peeking into the future: Predicting future person activities and locations in videos. In: CVPRW, pp. 5725–5734 (2019)
31. Liu, S., Zhang, X., Wangni, J., Shi, J.: Normalized diversification. In: CVPR (2019)
32. Lukas, N., Andrew, Z., Vedaldi, A.: Future event prediction: if and when. In: CVPRW, pp. 14424–14432 (2019)
33. Maddison, C.J., Mnih, A., Teh, Y.W.: The concrete distribution: a continuous relaxation of discrete random variables. In: ICLR (2017)
34. Mahmud, T., Billah, M., Hasan, M., Roy-Chowdhury, A.K.: Captioning near-future activity sequences. arXiv preprint arXiv:1908.00943 (2019)
35. Mahmud, T., Hasan, M., Roy-Chowdhury, A.K.: Joint prediction of activity labels and starting times in untrimmed videos. In: ICCV (2017)
36. Mathieu, M., Couprie, C., LeCun, Y.: Deep multi-scale video prediction beyond mean square error. arXiv preprint arXiv:1511.05440 (2015)
37. Mehrasa, N., Jyothi, A.A., Durand, T., He, J., Sigal, L., Mori, G.: A variational auto-encoder model for stochastic point processes. In: CVPR (2019)
38. Mohamed, A., Qian, K., Elhoseiny, M., Claudel, C.: Social-STGCNN: a social spatio-temporal graph convolutional neural network for human trajectory prediction. In: CVPR, pp. 14424–14432 (2020)
39. Nie, W., Narodytska, N., Patel, A.: ReLGAN: relational generative adversarial networks for text generation. In: ICLR (2019)
40. Oord, A.V.D., et al.: Wavenet: A generative model for raw audio. arXiv preprint arXiv:1609.03499 (2016)
41. Santoro, A., et al.: Relational recurrent neural networks. In: NIPS (2018)
42. Schydlo, P., Rakovic, M., Jamone, L., Santos-Victor, J.: Anticipation in human-robot cooperation: a recurrent neural network approach for multiple action sequences prediction. In: ICRA (2018)
43. Shi, Y., Fernando, B., Hartley, R.: Action anticipation with RBF kernelized feature mapping RNN. In: ECCV (2018)
44. Simonyan, K., Zisserman, A.: Two-stream convolutional networks for action recognition in videos. In: NIPS (2014)
45. Singh, G., Saha, S., Cuzzolin, F.: Predicting action tubes. In: ECCVW (2018)
46. Singh, G., Saha, S., Sapienza, M., Torr, P.H., Cuzzolin, F.: Online real-time multiple spatiotemporal action localisation and prediction. In: ICCV (2017)
47. Stein, S., McKenna, S.J.: Combining embedded accelerometers with computer vision for recognizing food preparation activities. In: UbiComp (2013)
48. Sun, C., Shrivastava, A., Vondrick, C., Sukthankar, R., Murphy, K., Schmid, C.: Relational action forecasting. In: CVPR, pp. 273–283 (2019)

49. Sutskever, I., Vinyals, O., Le, Q.V.: Sequence to sequence learning with neural networks. In: NIPS (2014)
50. Veselỳ, K., Ghoshal, A., Burget, L., Povey, D.: Sequence-discriminative training of deep neural networks. In: Interspeech (2013)
51. Vondrick, C., Pirsiavash, H., Torralba, A.: Anticipating visual representations from unlabeled video. In: CVPR (2016)
52. Wang, L., et al.: Temporal segment networks: towards good practices for deep action recognition. In: ECCV (2016)
53. Yang, D., Hong, S., Jang, Y., Zhao, T., Lee, H.: Diversity-sensitive conditional generative adversarial networks. In: ICLR (2019)
54. Yeh, R.A., Schwing, A.G., Huang, J., Murphy, K.: Diverse generation for multi-agent sports games. In: CVPR, pp. 4610–4619 (2019)
55. Yu, L., Zhang, W., Wang, J., Yu, Y.: SeqGAN: sequence generative adversarial nets with policy gradient. In: AAAI (2017)
56. Zhao, H., Wildes, R.P.: Spatiotemporal feature residual propagation for action prediction. In: ICCV (2019)
57. Zhu, J.Y., Park, T., Isola, P., Efros, A.A.: Unpaired image-to-image translation using cycle-consistent adversarial networks. In: ICCV (2017)

Author Index

Printed in the United States
By Bookmasters